D1061475

Digital Pictures

Representation, Compression, and Standards

Second Edition

Arun N. Netravali and
Barry G. Haskell
AT&T Bell Laboratories
Holmdel, New Jersey

PLENUM PRESS • NEW YORK AND LONDON

Library of Congress Cataloging-in-Publication Data

Netravali, Arun N.
Digital pictures : representation, compression, and standards / Arun N. Netravali and Barry G. Haskell. -- 2nd ed.
p. cm. -- (Applications of communications theory)
Includes bibliographical references and index.
ISBN 0-306-44917-X
1. Digital communications. 2. Image processing--Digital techniques. I. Haskell, Barry G. II. Title. III. Series.
TK5103.7.N47 1994
621.36'7'01154--dc20 94-42988
CIP

ISBN 0-306-44917-X

Plenum Press is a division of Plenum Publishing Corporation
233 Spring Street, New York, N.Y. 10013

10 9 8 7 6 5 4 3 2 1

Printed in the United States of America

to CHITRA and ANN

Preface

For thousands of years mankind has been creating pictures that attempt to portray real or imagined scenes as perceived by human vision. Cave drawings, paintings and photographs are able to stimulate the visual system and conjure up thoughts of faraway places, imagined situations or pleasant sensations. The art of motion picture creation has advanced to the point where viewers often undergo intense emotional experiences. On-the-spot news coverage gives the impression of actually witnessing events as they unfold.

Relatively recently, other forms of visual information have been invented that do not, in themselves, stimulate the eye. For example, voltage variations in an electrical signal, as in television, can represent in analogous fashion the brightness variations in a picture. In this form the visual information can be stored on magnetic tape or transmitted over long distances, and, at least for engineering purposes, it is often much more useful than other forms that do stimulate human vision.

With the evolution of digital techniques for information processing, storage, and transmission, the need arises for digital representation of visual information, that is, the representation of images by a sequence of integer numbers (usually binary). In this form, computer processing and digital circuit techniques can be utilized that were undreamed of only a short time ago. Machine manipulation and interpretation of visual information becomes possible. Sophisticated techniques can be employed for efficient storage of images. And processing methods can be used to significantly reduce the costs of picture transmission.

Many methods have evolved for representing (or *coding*) visual information into a stream of numbers suitable for digital processing. Some have the virtue of easy implementation, while others have special features that make them well suited for specific applications. Most of the coding techniques are based on underlying theoretical notions; however, many of them also utilize ad hoc intuitive techniques appropriate for specific situations.

Corresponding to each *coding* algorithm is a method of *decoding* or reconstructing the desired visual information from the coded data. In many cases, this means producing a *replica* image that, upon viewing, resembles the *original* image. In other situations, the reconstructed information may only be locational or diagrammatical, as for example an infrared heat map of the human body.

The performance of the coding and decoding is, in large part, determined by how well the reconstructed visual information achieves the goals of the system. Unfortunately, these goals are often only vaguely defined or may not be in a form useful for designing the system. They may involve such considerations as viewer satisfaction or the ability to perform some complicated task based on the information presented. Thus, designers of visual information processing systems frequently find themselves involved not only in new digitization techniques, but also in new methods of subjective or objective evaluation.

Another aspect of coder/decoder performance is its efficiency. This term means different things in different contexts, but it usually relates to the data reduction capability. In general, the aim of most coding algorithms is to produce as little digital data as possible while, at the same time, accomplishing the given overall goals of the system. This (somewhat simplified) is the *Image Coding Problem.*

In the following chapters we present the theoretical, intuitive and practical bases for coding images into digital form. Discussion of techniques specific to certain applications will follow and be intermingled with many examples. Some variations on these themes will also be mentioned.

Acknowledgements

During the fifteen years or so it took to produce this book and its revision we naturally received advice, encouragement and criticism from many quarters. We cannot mention you all, but please be assured that we are eternally grateful for your help.

Over the years, successive debuggings were carried out by the very cooperative students in classes we gave at Rutgers University, City College of New York and Columbia University. The revision was debugged by Professor Yao Wang's class at Polytechnic University.

More knowledgeable criticism was given by a number of experts in the field who managed to find the time to read early versions of the manuscript. They include Edward H. Adelson, C. Chapin Cutler, Rui J. de Figueiredo, Robert M. Gray, N. S. Jayant, Thomas Kailath, Murat Kunt, Theodore F. Leibfried, Jr., Jae S. Lim, Donald E. Pearson, Peter Pirsch, William F. Schreiber and David H. Staelin.

First drafts of Chapters 7-9 were prepared by Don Duttweiler (JBIG), Glen Cash (JPEG), Hsueh-Ming Hang (P*64), R. Aravind (MPEG1) and Atul Puri (MPEG2).

Many thanks for help in preparing the manuscript to Helen Baker, Jay Boris, Diane Bush, Donna Carroll, Elsie Edelman, Suzanne Fernandez, Linda Jensen, Virginia King, Suzanne Smith and Arlene Surek.

Finally, a special thanks to our colleagues C. Chapin Cutler, John Limb and Robert Lucky for their support and inspiration throughout this project.

Contents

1

Numerical Representation of Visual Information

1.1 Visual Information

The ability to see is one of the truly remarkable characteristics of living beings. It enables them to perceive and assimilate in a very short time an incredible amount of knowledge about the world around them. The scope and variety of that which can pass through the eye and be interpreted by the brain is nothing short of astounding. Mankind has increased this basic capability by inventing devices that can detect electromagnetic radiation at wavelengths far outside the range of normal vision and at energy levels orders of magnitude below what the eye is able to perceive by itself. By the use of X-rays or sound waves it is possible to "see" inside objects and into places that have been invisible to living beings since the dawn of creation. Ultra-fast photography can stop a speeding bullet or freeze a flying humming bird's wing.

It is, thus, with some degree of trepidation that we introduce the concept of *visual information,* because in the broadest sense, the overall significance of the term is simply overwhelming; and if we had to take into account all of the ramifications of visual information, the situation would be hopeless from the outset. We must, therefore, make some qualifications if we are to deal with visual information within the capability of today's technology and that of the foreseeable future.

The first restriction we shall impose is that of *finite image size.* That is, we assume that the viewer receives his or her visual information as if looking through a rectangular window of finite dimensions as shown in Fig. 1.1.1. Cameras, microscopes and telescopes, for example, all have finite fields of view, and this assumption is usually necessary in dealing with real-world systems that can handle only finite amounts of information.

We further assume that the viewer is incapable of depth perception on his own. That is, in the scene being viewed he cannot tell how far away objects are

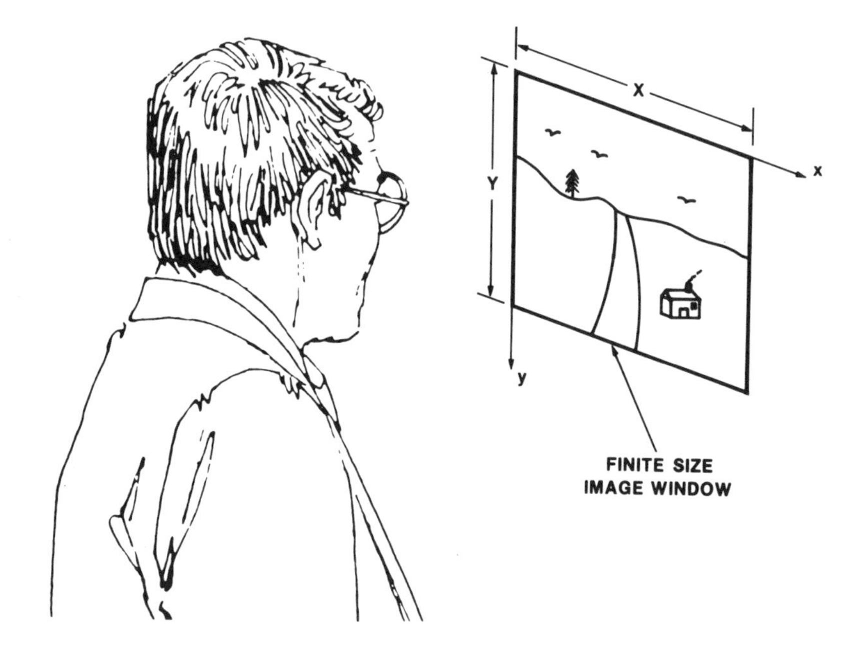

Fig. 1.1.1 The imaging window is of finite size, with no depth. That is, at any instant of time the visual information is completely determined by the wavelengths and amplitudes of the light passing through each point (x,y) of the window.

by the normal use of binocular vision or by changing the focus of his eyes. However, one can *infer* depth in a number of ways. For example, if objects in the scene move, then nearby moving objects will pass in front of those objects further away. If the viewer is in a moving automobile then the viewing angle will change with time while looking through the car window. And finally, if an optical lens is placed between the window and the scene, then depth can be inferred by changing the lens focus. Thus, the world outside the window may change, and the viewer's house or car may move about, but the viewer must sit still and passively accept whatever visual information enters through the window.

This scenario may sound a bit dismal. But, in fact, this model describes an overwhelming proportion of systems that handle visual information, including television, photographs, x-rays, etc. In this setup, the visual information is determined completely by the wavelengths and amplitudes of light that passes through each point of the window and reach the viewer's eye. If the world outside were to be removed and a projector installed that reproduced exactly the light distribution on the window, the viewer inside would not be able to tell the difference.

Thus, the problem of numerically representing visual information is reduced to that of representing the distribution of light energy and wavelengths on the finite area of the window. Let us defer until later the topic of color information and assume that the image perceived by the viewer is *monochromatic*. It is determined completely by the perceived light energy (weighted sum of energy at perceivable wavelengths) passing through each point on the window and reaching the viewer's eye. If we impose cartesian coordinates on the window, we can represent perceived light energy or *intensity** at point x,y by $B(x,y)$. For the moment we will consider only one instance of time, adding later the parameter t when dealing with images that vary with time. Thus, $B(x,y)$ represents the monochromatic visual information or *image* at the instant of time under consideration.

1.2 Representation by a Finite Amount of Data

For certain synthetic images, $B(x,y)$ can be specified exactly. For example, a light rectangle on a dark background might be described by

---

* Intensity is defined rather loosely here. Various imaging devices have different light sensitivities, which can result in different values for $B(x,y)$ for the same image.

$$B(x,y) = \begin{cases} 1 & a \le x \le b,\ c \le y \le d \\ 0 & \text{elsewhere} \end{cases} \tag{1.2.1}$$

However, for most images that occur in real-life situations, no such exact specification is possible with a finite amount of numerical data. An approximation of $B(x,y)$ must be made if it is to be dealt with by practical systems.

Our task is to study methods of representing $B(x,y)$ by a finite amount of data. Since number bases can be changed without loss of information, we may assume binary digital data. In this form the data is eminently well suited for such applications as transmission via digital communication facilities, storage within digital memory media or processing by digital circuitry or computers. Again, such systems can only handle finite amounts of data at a time, so the problem is to represent $B(x,y)$ by a finite number of binary digits, commonly called bits. If $B(x,y)$ changes with time, then a finite number of bits per unit time, i.e., a finite bit-rate, is required.

Thus, in moving from the idealized scenario to a practical situation, some visual information is invariably lost. In the first place, $B(x,y)$ can only be measured to finite accuracy since physical devices have finite optical resolution and light intensities can only be measured to within a certain percentage error. And even if $B(x,y)$ could be obtained with extremely high accuracy, one might choose not to preserve all of the available visual information if it is not needed. For each application it is necessary to decide how accurate a representation of $B(x,y)$ is needed in order to accomplish the specified goals of the system. Generally, a higher representational accuracy requires a larger number of data bits and more costly apparatus for image measurement and processing. Thus, it is economically advantageous to estimate in advance the minimum representational fidelity needed for a particular job and to design the processing system accordingly.

Deciding what fidelity to use in a given situation is not a small task. The problem can sometimes be described mathematically as follows: Starting with $B(x,y)$ and some proposed digitization scheme that conforms to given economic constraints, construct the corresponding representational data bits. Then, from these data, *reconstruct* an approximate replica $\tilde{B}(x,y)$ of the original visual information. Next we assume the existence of a *distortion measure* $D(B,\tilde{B}) \ge 0$ that indicates how badly $\tilde{B}(x,y)$ serves as an approximation of $B(x,y)$ for the particular goals at hand. For example, if television is to be transmitted through a digital channel, $D(B,\tilde{B})$ might be the number of viewers who call in and complain about picture quality. Or if satellite pictures are to be used to measure wheat acreage, $D(B,\tilde{B})$ might be the percentage of ground area falsely classified. Thus, with the aid of this distortion function, the performance of various digitization (coding) schemes can be compared on the basis of their distortions.

While this description of the coding problem is concise, there are several difficulties associated with it. First, visual information processing systems are almost always designed to handle a variety of images. And different coding schemes produce different distortions with different images. In many situations there may exist some images that have too large a $D(B,\tilde{B})$ for *all* coding schemes that are economically feasible for a particular application.

A second difficulty is that it may not be possible to find a distortion function $D(B,\tilde{B})$ that is agreeable to all interested parties while, at the same time, being simple enough to be useful in designing a system. For example, in the aforementioned television situation there is no way to tell for a new proposed coding scheme how many viewers will be dissatisfied with the picture quality unless a system is built and actually tried out on the public.

A third difficulty is that in some applications it does not make sense to speak of a replica image $\tilde{B}$, reconstructed from the coded data. In the wheat acreage example, it may be that satellite on-board processors only transmit sizes and spectral measurements of suspected agricultural fields. In this situation, where the techniques of pattern recognition overlap with those of image coding, a reconstructed replica image $\tilde{B}$ makes less sense.

In addition to bit-rate and fidelity considerations, there are other aspects entering into the evaluation of a coding system, which further serves to cloud the issue. For example, system complexity, reliability and maintainability are almost always major factors in cost and performance evaluation. Also in many cases, compatibility with existing systems is a requirement. Thus, deciding on the overall performance required to meet given goals is far from an exact science, and system design often involves the time honored "try-it-and-see-how-it-works" approach.

1.3 Spatial Sampling Representation

Since it is not feasible to represent $B(x,y)$ with absolute fidelity, we naturally ask ourselves what might be the next best thing to do? At a given point (x,y) we know it is possible to represent the intensity B only with finite accuracy, i.e., with a finite number of bits. Thus, if a finite set of points (x,y) is chosen for representation, and the brightness of each point is represented by a finite number of bits, i.e., *quantized*, then we have, in fact, a representation (coding) of $B(x,y)$ by a finite amount of binary data. If time is a parameter, and this whole process is periodically repeated, then data are produced at a finite *bit-rate*. This procedure of sampling and quantization[1.3.1] is often called *Pulse Code Modulation* or *PCM coding* even though, strictly speaking, PCM implies modulation and transmission of the resulting bits from one place to another.

The set of sampling points might be a rectangular array consisting of rows and columns of samples, as shown in Fig. 1.3.1a, or every other row might be shifted horizontally by one-half sampling period to achieve a somewhat more uniform spatial sampling as shown in Fig. 1.3.1b.

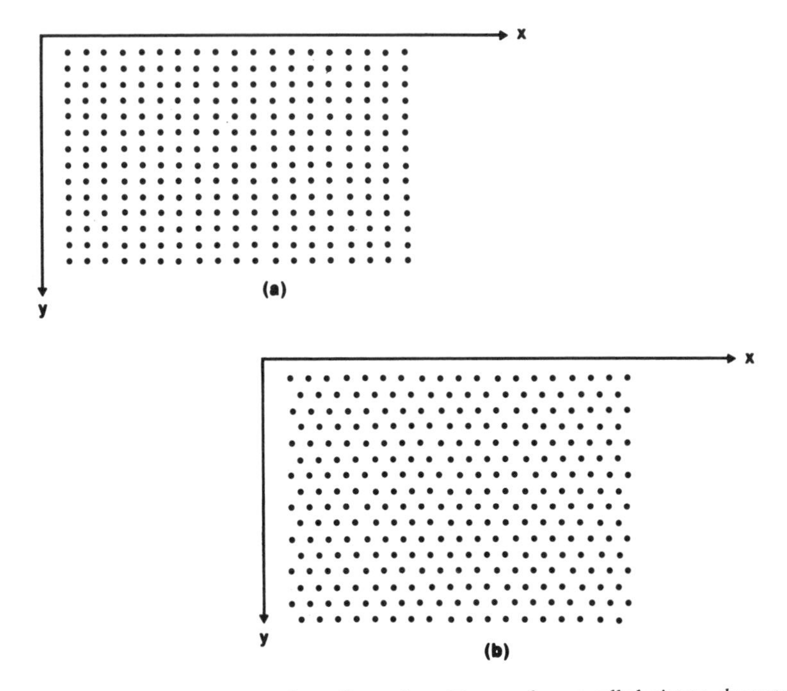

Fig. 1.3.1 (a) A rectangular array of sampling points. The samples are called *picture elements*, or *pels* for short. (b) An offset array of sampling points.

In some applications the sample locations are chosen at random, or they might be chosen to lie in certain "regions of interest" according to some given criterion.

We call the samples *picture elements*, or *pels* for short. [Some authors prefer the slightly longer term *pixels*.] The term may be used to refer to fixed samples in the image plane, e.g., "the 2×2 block of pels in the lower left-hand corner of the image," or it may refer to samples of an object in the scene, e.g., "the pels of a baseball moving from left to right." The terms "pel x,y" and "value of $B(x,y)$" will be used interchangeably, as will other apparent violations of strict mathematical notation.

Intuitively, the pels should be close enough together to portray accurately the intensity variations that are of interest and are perceivable to the viewer at the chosen viewing distance from the window. If the pels are chosen too close together, however, an unnecessarily large number of bits may be produced. Thus, the required sampling density depends on the intensity variations that the system must accommodate. This is intimately related to the concept of *spatial resolution* that arises so often in the optical sciences. Resolution can be specified in a variety of ways, many of which are application dependent and seemingly unrelated. One metric of resolution is the minimum resolvable distance between two lines or two dots in the image. Another, which we shall deal with later, is the maximum frequency content of the image when dealing with Fourier spectral representations. Sometimes the maximum number of resolvable lines in the entire image is used to specify resolution. In any case, no matter which measure is used, the required resolution of the system is the determining factor in choosing a sampling density.

Having decided upon a sampling density by resolution considerations or economic constraints, there may exist in $B(x,y)$ some spatial intensity variations that are too rapid to be accurately portrayed by the chosen sampling. This may occur, for example, if optical components have much more resolution capability than is actually needed. These "too rapid" transitions can actually be harmful to the accurate representation of the desired visual information. Consider a satellite picture of wheat fields, some of which contain a few stones. If this scene is sampled with only a few samples per field, and some of the samples fall on the stones, then a very misleading representation is obtained for those fields. This phenomenon is called *pre-aliasing*, a term which we will define later when discussing the Fourier frequency domain treatment of sampled data theory. Pre-aliasing can also be caused by noise in imaging devices.

In order to avoid pre-aliasing, it is better not to sample the image intensity at precise locations (x,y). Instead, we measure the *average* intensity of a small region surrounding each point (x,y). With this approach, intensity transitions that are too small or too rapid to be of interest do not affect very much the final representation of the visual information. For example, an image area might be partitioned into very small contiguous rectangular areas, and the average brightness measured within each one to define the pel value.

A method that is often easier to implement is to form a *weighted* average over a small neighborhood of each sampling point. For this purpose a camera weighting function $h_c(u,v)$ is needed that peaks at $(u=0,\ v=0)$ and falls off in all directions away from the origin. For example, the Gaussian function

$$K\ exp(-au^2-bv^2) \qquad |u|,|v| < \delta \tag{1.3.1}$$

where K is a normalizing constant, is quite suitable for many applications. Gaussian weighting is also characteristic of many imaging devices. Using this function we first form an averaged or *filtered* image

$$\overline{B}(x,y) = \int_{y-\delta}^{y+\delta}\int_{x-\delta}^{x+\delta} B(w,z)\,h_c(x-w,y-z)\,dwdz \tag{1.3.2}$$

where the intensity at each point is replaced by a weighted average over a small surrounding neighborhood. This process, illustrated along the x direction in Fig. 1.3.2, is called *pre-filtering,* since the image is filtered prior to sampling. $\overline{B}(x,y)$ is the filtered image, and $h_c(u,v)$ is the camera pre-filtering function.* The filtered image is then sampled at a set of suitably chosen sampling points $\{(x_k,y_k),\ k=1...N_S\}$, and each resulting intensity is approximated† to a finite number of bits, i.e., *quantized*, to give the finite-bit representation $\{\overline{B}(x_k,y_k)\}$. Fig. 1.3.3 shows an example of a noisy image that is pre-filtered to avoid pre-aliasing in the sampled pels.

At first, it may seem that filtering the entire image prior to sampling is more complicated than performing the area-averaging only at the points to be sampled. In fact, this is not usually the case. For example, if the imaging system contains optical components, then filtering is easily carried out by proper design of the imaging lens. If a photograph is sampled by a thin light beam, then filtering can be accomplished by slightly enlarging the cross-sectional area of the scanning beam.

Thus, the amount of visual information is reduced in a quantized spatial sampling representation in three ways. First, only a finite number of intensity samples are represented. Second, rapid intensity variations are removed by pre-filtering and are, therefore, absent in the final representation. And third, intensity values are quantized to finite accuracy.

---

* $h_c(u,v)$ is also known as the impulse response of the filter, since setting $B(w,z) = \delta(w)\delta(z)$ produces $h_c(x,y)$ at the output of the filter. $\delta(\cdot)$ is the Dirac delta function.

† For the time being we assume that quantized values are very close to the original unquantized values, i.e., the error due to quantization is negligible.

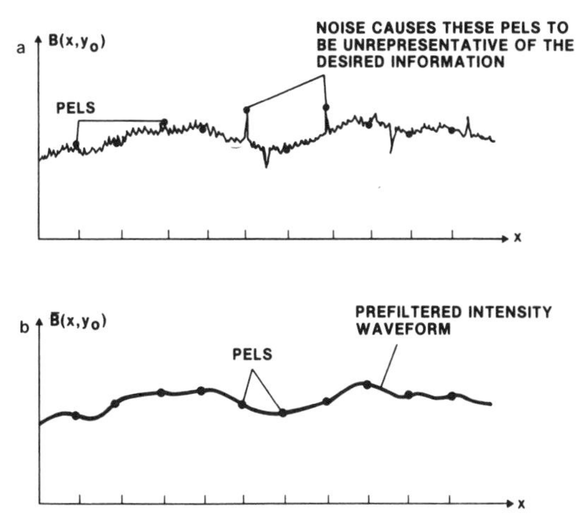

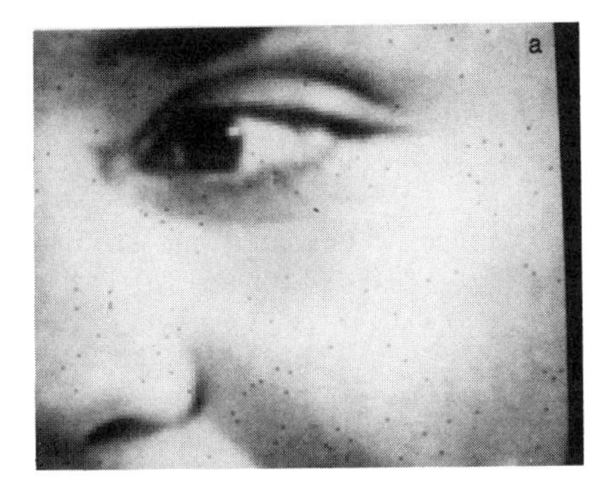

Fig. 1.3.2 Pre-filtering is necessary to avoid pre-aliasing. Here we show intensity variation along the x direction. a) Noise and excess picture detail cause some pels to be unrepresentative of the desired information. b) Pre-filtering has removed unneeded information, and all pels accurately represent the desired data. $\bar{B}$ is obtained using Eq. (1.3.2).

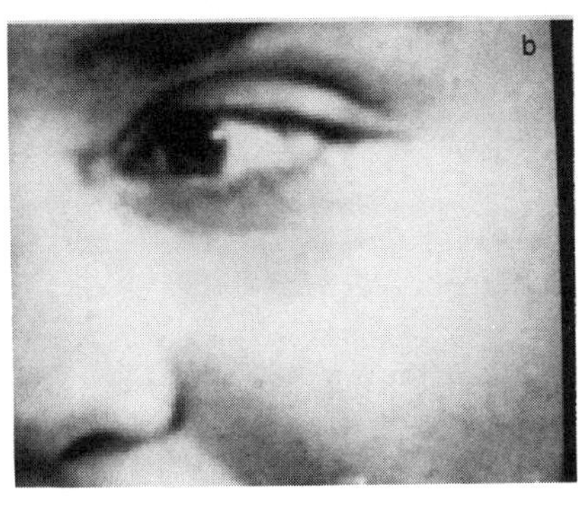

Fig. 1.3.3 a) Noisy image. Samples are liable to be corrupted significantly by the noise. b) Pre-filtered image. The effect of the noise on the pel values is markedly reduced.

Given the quantized pels $\{\bar{B}(x_k,y_k)\}$ of the filtered image, we may wish to generate a reconstructed replica image $\tilde{B}(x,y)$. The main difficulty in this is deciding what values of intensity to assign at points (x,y) that are not sampling points. We might, for example, choose zero, or we might assign the intensity value of the nearest sampling point. In both of these cases, as illustrated in Fig. 1.3.4, the image $\tilde{B}(x,y)$ will contain artificial patterns that may be quite unlike anything that appears in the original image $B(x,y)$. In the former case, $\tilde{B}(x,y)$ is an assemblage of pin pricks as is found in newspaper photographs, and in the latter case shown in Fig. 1.3.5(a), discontinuities appear along lines that are equidistant from two adjacent sampling points. The presence of such artificial patterns in the reconstructed image is called *post-aliasing*. It is quite different than the pre-aliasing that arises from sampling images containing intensity variations that are too rapid.

Post-aliasing in the reconstructed image $\tilde{B}(x,y)$ can be ameliorated by interpolating (or averaging) between pels to achieve relatively smooth intensity transitions as shown in Fig. 1.3.5(b). This interpolation may be carried out at the display using an interpolation function $h_d(u,v)$ that peaks at the origin and falls off away from the origin. The reconstructed replica image is then given by

$$\tilde{B}(x,y) = \sum_k \bar{B}(x_k,y_k)\, h_d(x-x_k,y-y_k) \tag{1.3.3}$$

This is usually a continuous function of x and y, and artificial patterns are absent in the reconstructed image as we see in Fig. 1.3.5(b).

For a rectangular array of pels, as in Fig. 1.3.1a, a commonly used reconstruction is linear interpolation. If we define the linear sawtooth function

$$S(z,\Delta) = 1 - \left|\frac{z}{\Delta}\right| \quad \text{for} \quad \left|\frac{z}{\Delta}\right| \leq 1 \tag{1.3.4}$$

and the horizontal and vertical pel spacings are Δ_x and Δ_y, respectively, then two-dimensional bilinear interpolation is accomplished by using

$$h_d(u,v) = S(u,\Delta_x)\cdot S(v,\Delta_y) \tag{1.3.5}$$

Higher order polynomial interpolations are somewhat smoother. For example, a quadratic interpolation results from

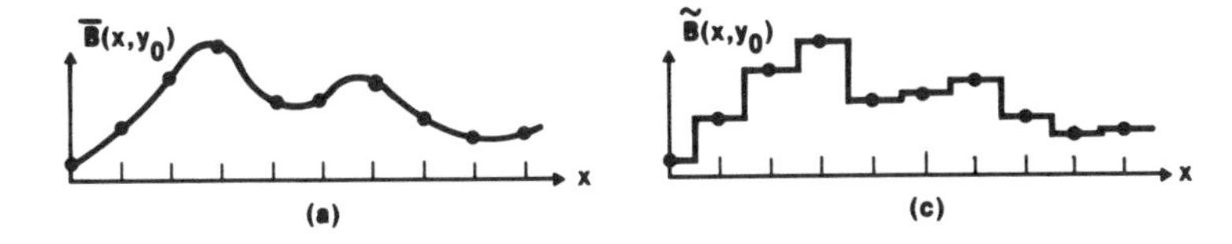

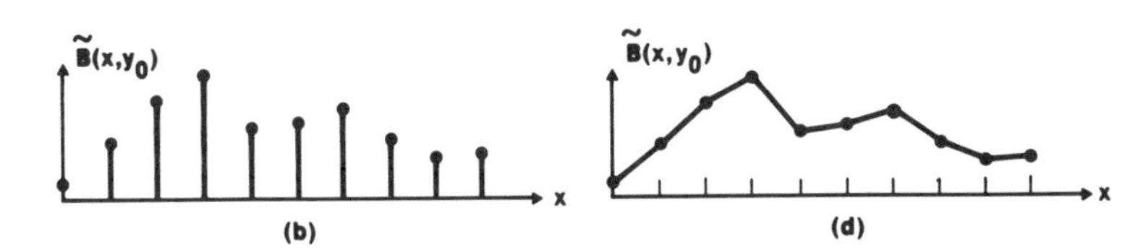

Fig. 1.3.4 Image reconstruction from the pel values without post-filtering. Again we show intensity variation along the x direction. a) Original waveform $\overline{B}$ from which pels are obtained. b) Reconstruction with zero between the pels. c) Sample and hold reconstruction using the value of the nearest pel. d) Linear interpolation.

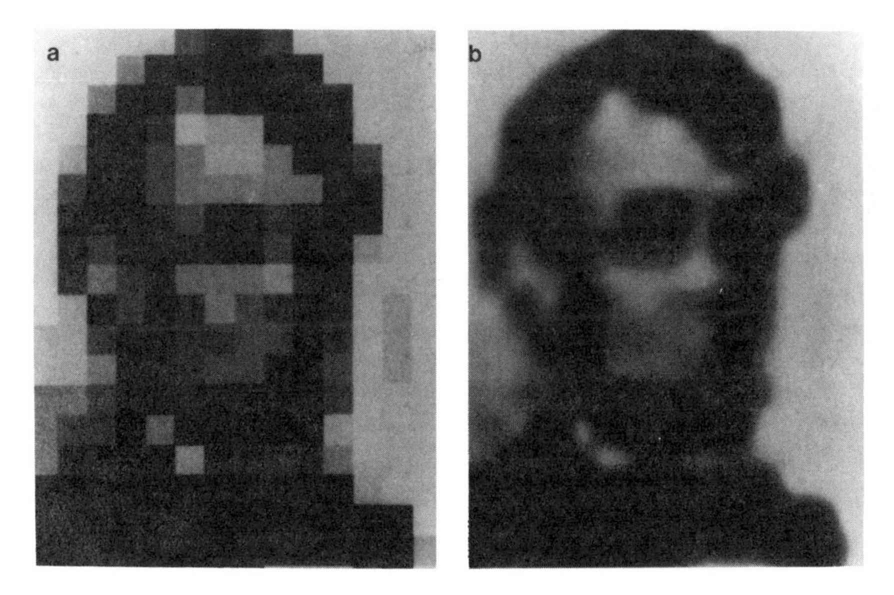

Fig. 1.3.5 Post-filtering is necessary to avoid post-aliasing. a) Artificial post-aliasing patterns caused by nearest-neighbor reconstruction obliterate meaningful information. b) Post-filtering has removed the patterns revealing the true picture (from Harmon and Julesz [1.3.2]).

$$S(z,\Delta) = \begin{cases} \frac{3}{4} - \left|\frac{z}{\Delta}\right|^2 & \text{for } \left|\frac{z}{\Delta}\right| \le \frac{1}{2} \\ \frac{9}{8} - \frac{3}{2}\left|\frac{z}{\Delta}\right| + \frac{1}{2}\left|\frac{z}{\Delta}\right|^2 & \text{for } \frac{1}{2} \le \left|\frac{z}{\Delta}\right| \le \frac{3}{2} \end{cases} \tag{1.3.6}$$

A cubic interpolation that has continuous first derivative and passes through all of the original data points results from

$$S(z,\Delta) = \begin{cases} 1 - \frac{5}{2}\left|\frac{z}{\Delta}\right|^2 + \frac{3}{2}\left|\frac{z}{\Delta}\right|^3 & \text{for } \left|\frac{z}{\Delta}\right| \le 1 \\ 2 - 4\left|\frac{z}{\Delta}\right| + \frac{5}{2}\left|\frac{z}{\Delta}\right|^2 - \frac{1}{2}\left|\frac{z}{\Delta}\right|^3 & \text{for } 1 \le \left|\frac{z}{\Delta}\right| \le 2 \end{cases} \tag{1.3.7}$$

A somewhat smoother interpolation that has continuous first *and* second derivatives results from the so-called cubic B-spline[1.5.5]

$$S(z,\Delta) = \begin{cases} \frac{2}{3} - \left|\frac{z}{\Delta}\right|^2 + \frac{1}{2}\left|\frac{z}{\Delta}\right|^3 & \text{for } \left|\frac{z}{\Delta}\right| \le 1 \\ \frac{4}{3} - 2\left|\frac{z}{\Delta}\right| + \left|\frac{z}{\Delta}\right|^2 - \frac{1}{6}\left|\frac{z}{\Delta}\right|^3 & \text{for } 1 \le \left|\frac{z}{\Delta}\right| \le 2 \end{cases} \tag{1.3.8}$$

For positive pel values, the B-spline interpolation is always positive. Fig. 1.3.6 shows the two-dimensional interpolation functions $h_d(u,v)$ obtained from the above one-dimensional interpolations.

This process of interpolation is called *post-filtering* since it is essentially a smoothing of the spatially discrete pels of an image. In fact, if a *sampled image* is defined using Dirac delta functions

$$B_S(x,y) = \sum_{k=1}^{N_s} \bar{B}(x_k,y_k)\,\delta(x-x_k)\,\delta(y-y_k) \tag{1.3.9}$$

then the reconstructed image of Eq. (1.3.3) is given exactly by

$$\tilde{B}(x,y) = \int\int B_S(w,z)\,h_d(x-w,y-z)\,dwdz \tag{1.3.10}$$

which looks much more like the filtering operation previously described in Eq.

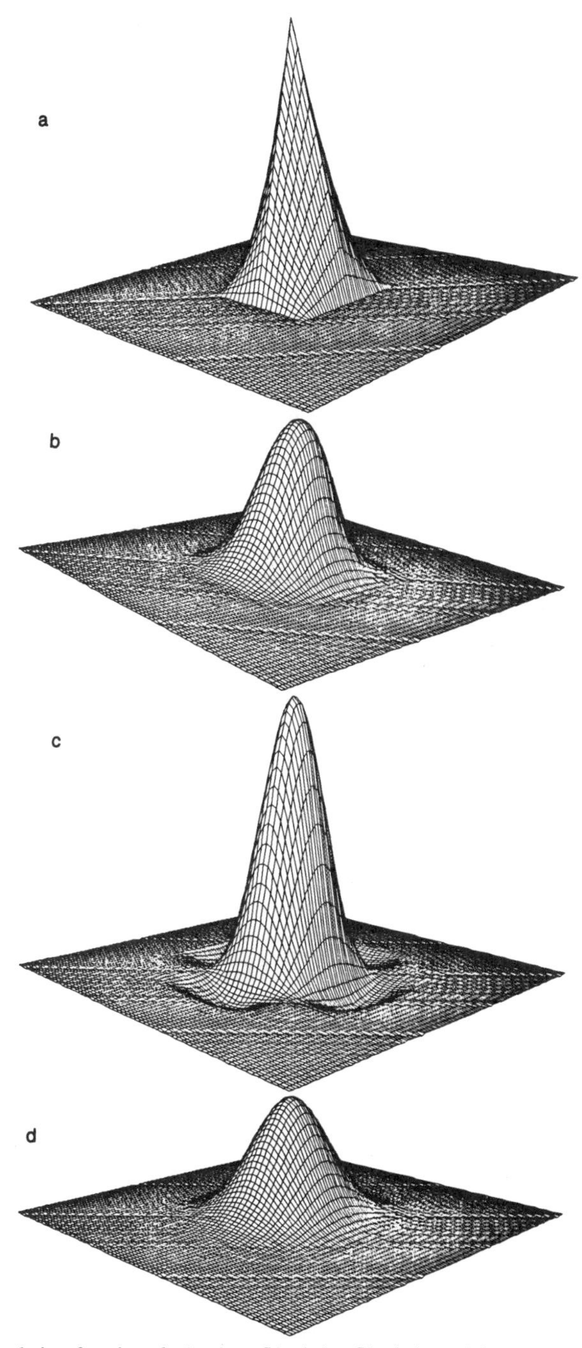

Fig. 1.3.6 Interpolation functions $h_d(u,v) = S(u,\Delta_x) \cdot S(v,\Delta_y)$ used for reconstructing images from their pel values. a) Bilinear interpolation, Eq. (1.3.4). b) Biquadratic polynomial interpolation, Eq. (1.3.6). c) Bicubic polynomial interpolation, Eq. (1.3.7). d) Bicubic B-spline interpolation, Eq. (1.3.8).

(1.3.2). Thus, not only is filtering generally required at the camera before sampling, it is also needed at the display after sampling if a reconstructed replica image is to be generated from the quantized intensity samples. Fig. 1.3.7 shows the entire sequence of sampling and reconstruction.

1.4 Raster Scan

Raster scanning, shown in Fig. 1.4.1, is the most often used form of spatial sampling representation. Raster scanning converts the two-dimensional image intensity into a one-dimensional waveform. Basically, the filtered image (having dimensions X by Y) is segmented into N_y adjacent horizontal strips (called *lines*), and starting at the top of the picture, the intensity along the center of each strip is measured to yield an intensity waveform that is a one-dimensional function of the time parameter t. That is, the image is *scanned* one line at a time, sequentially, left-to-right, top-to-bottom. For example, in a flying spot scanner a small spot of light scans across a photograph, and the reflected energy at any given position is a measure of the intensity at that point. In a television camera, an electron beam scans across an electrically photosensitive target upon which light from the image is focused.

If we require T_L seconds to scan one line,* then scanning the entire image requires $N_y T_L$ seconds. Functionally, we may write the one-dimensional intensity waveform as

$$B(t) = \bar{B}(x_t, y_t) \tag{1.4.1}$$

where during the ith scan line, i.e.,

$$(i-1)T_L < t < i\,T_L \tag{1.4.2}$$

we have†

$$y_t = \frac{(i-0.5)Y}{N_y}$$

$$x_t = [t-(i-1)T_L]\frac{X}{T_L} = \frac{tX}{T_L} \text{ modulo } X \tag{1.4.3}$$

* Each line may contain some blanked, undisplayed region on the right side to allow time for the scanning spot to retrace to the beginning of the next line. The bottom of the image may also be blanked to allow time for vertical retrace, as shown in Fig. 1.4.1.

† In many imaging devices $y_t = tY/N_y T_L$, i.e., vertical scanning is also a linear function of time.

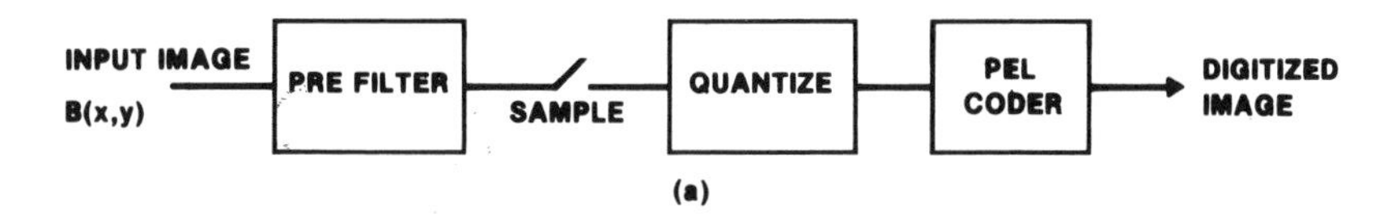

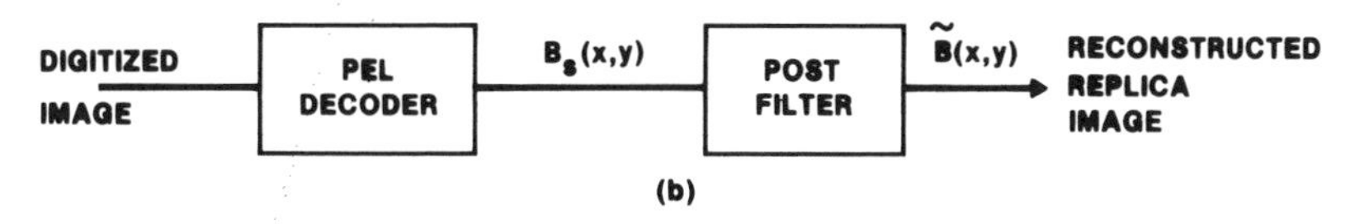

Fig. 1.3.7 Quantized spatial sampling of images generally requires pre-filtering prior to sampling, as shown in a, and post-filtering following reconstruction, as shown in *b*. Inadequate pre-filtering leads to pre-aliasing, whereas inadequate post-filtering causes post-aliasing.

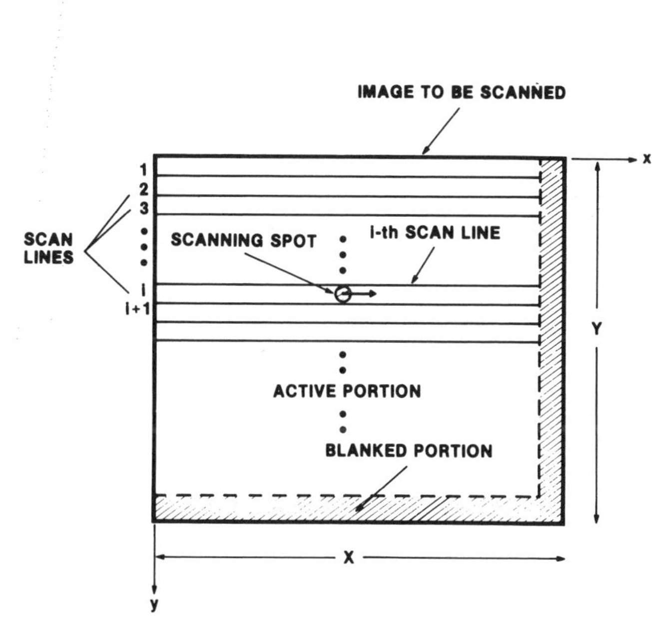

Fig. 1.4.1 Raster scan of an image. The scanning spot moves across each scan line, left-to-right, and at any given position performs a local spatial average to give the image intensity at that point. The blanked region on the right and bottom allows retrace time for the scanning spot.

The scanning spot itself is partly responsible for local spatial averaging of the image, as in Eq. (1.3.2), since the measured intensity at time t is actually a weighted average over the area of the spot centered at (x_t,y_t). However, with some camera devices, e.g., TV cameras that use electron beam scanning, it is very difficult to control the scanning spot pre-filtering function $h_c(u,v)$, with the result that the image $\bar{B}(x,y)$ may be insufficiently pre-filtered. Optical scanning devices that use a beam of light are much better in this regard. Of course, spatial pre-filtering may also be accomplished in most camera devices by slightly defocusing the optical lens.

Filtering may also be carried out on the one-dimensional signal $B(t)$ by means of the relation

$$\bar{B}(t) = \int B(\tau)h_c(t-\tau)\,d\tau \qquad (1.4.4)$$

where $h_c(\tau)$ is a suitable pre-filtering function that is nonzero only near $\tau = 0$. This process is easily accomplished using electrical signal filters, and can be very useful in reducing noise caused by camera devices. However, for spatial filtering it has the drawback of only filtering the image in the horizontal direction unless a much more complicated $h_c(\cdot)$ is used. Pre-aliasing in the vertical direction may still exist.

The one-dimensional signal $\bar{B}(t)$ is then sampled at suitably chosen sampling points $\{t_k\}$, and the resulting intensities are quantized to give the finite-bit representation $\{\bar{B}(t_k)\}$. If $\bar{B}(t)$ is sampled at a rate that is an integer multiple of the line scanning rate, e.g., N_x/T_L, then a rectangular N_y by N_x array of sampling points is produced. Otherwise, the samples in successive lines are spatially displaced with respect to each other.

An image may be reconstructed from the $N_S = N_y \cdot N_x$ samples $\{\bar{B}(t_k)\}$, as before, by defining the *sampled signal*

$$B_S(t) = \sum_{k=1}^{N_S} \bar{B}(t_k)\,\delta(t-t_k) \qquad (1.4.5)$$

and using the one-dimensional display post-filtering

$$\tilde{B}(t) = \int B_S(\tau)h_d(t-\tau)\,d\tau \triangleq \tilde{B}(x_t,y_t) \qquad (1.4.6)$$

This signal is then passed to a raster scanned display device that has a scanning spot characteristic $h_d(u,v)$, perhaps similar to that of the camera. The final post-filtered image $\tilde{B}(x,y)$ is then produced by the spatial filtering operation

$$\tilde{B}(x,y) = \int \tilde{B}(t)h_d(x-x_t,y-y_t)\,dt \qquad (1.4.7)$$

Fig. 1.4.2 Post-aliasing of a raster scanned display. The raster scan-lines of the display are visible due to insufficient post-filtering in the vertical direction.

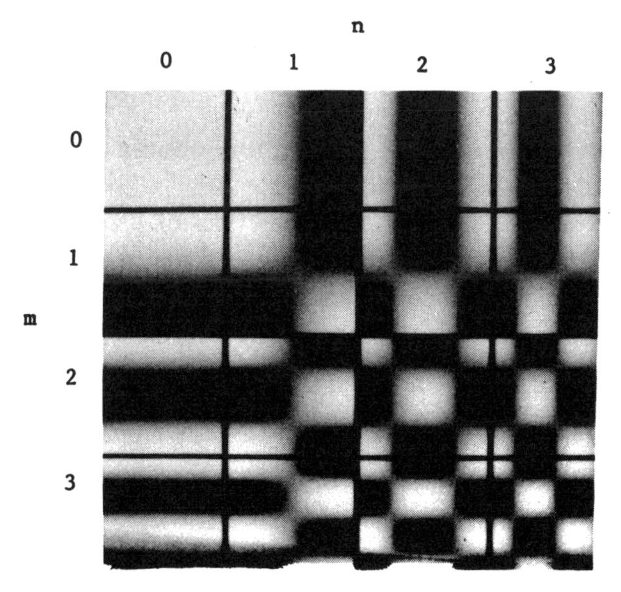

Fig. 1.5.1 Low order basis images $\psi_{mn}(x,y)$ for the cosine series representation. Real images may be represented as linear combinations of these basis images.

With raster scanned display devices, filtering in the vertical direction must be done either by the viewer's eye or by proper control of the scanning beam and, if present, optical components. Such an effort is often not completely successful, resulting in visible post-aliasing in the form of a visible scan-line structure in the displayed image. For example, the electron beam in a cathode ray tube (CRT) has a variable shape that depends on the picture intensity at each point, as shown in Fig. 1.4.2. This precludes the use of beam defocusing to obtain vertical smoothing, with the result that individual scan lines are quite visible on most monochrome and many color CRT displays. This visibility can be reduced by wobbling the scanning beam up and down slightly as it traverses the screen.

1.5 Fourier-Frequency Domain

Spatial sampling and quantization of the image intensity is not the only method of representing visual information by a finite amount of data. If the original image $B(x,y)$ is transformed to another domain of representation prior to sampling and quantization, then in some cases a more convenient representation or coding can be obtained.

1.5.1 Fourier Series

For example, suppose the image boundary is rectangular, i.e.,

$$\begin{aligned} 0 \le x \le X \\ 0 \le y \le Y \end{aligned} \tag{1.5.1}$$

and for x,y in this range we define the complex exponential function

$$\phi_{mn}(x,y) = \frac{1}{\sqrt{XY}} \exp 2\pi\sqrt{-1}\left[\frac{nx}{X} + \frac{my}{Y}\right] \tag{1.5.2}$$

Then for all reasonable images we are likely to encounter in practice, it is possible to write the well-known Fourier series representation[1.5.1]

$$B(x,y) = \sum_{m=-\infty}^{\infty} \sum_{n=-\infty}^{\infty} c_{mn}\phi_{mn}(x,y) \tag{1.5.3}$$

$$c_{mn} = \int_0^X \int_0^Y B(x,y)\phi'_{mn}(x,y)\,dydx \tag{1.5.4}$$

where $'$ denotes the complex conjugate.

Here, $\{\phi_{mn}(x,y)\}$ is a set of orthonormal basis images. It is called orthonormal because*

$$\int_0^X \int_0^Y \phi_{mn}(x,y)\,\phi'_{k\ell}(x,y)\,dydx = \delta_{mk}\,\delta_{n\ell} \tag{1.5.5}$$

The word basis is used because $\{\phi_{mn}(x,y)\}$ can be thought of as a set of canonical images which, when combined properly as in Eq. (1.5.3), yields any image $B(x,y)$ of practical interest.

Other sets of orthonormal basis functions are also useful in certain situations. For example, a cosine series representation results from the basis images

$$\psi_{mn}(x,y) = \frac{k_m k_n}{\sqrt{XY}} \cos\pi n\frac{x}{X} \cos\pi m\frac{y}{Y} \quad m,n \ge 0 \tag{1.5.6}$$

where

$$k_\ell = \begin{cases} 1 & \ell=0 \\ \sqrt{2} & \ell>0 \end{cases}$$

The low order cosine series basis images are shown in Fig. 1.5.1. More will be said in later chapters about other transform domains.

Representing $B(x,y)$ in terms of Fourier exponential basis images, as in Eq. (1.5.2), amounts to a decomposition of the image into *spatial frequency* components, that is, a spectral representation. For each basis image $\phi_{mn}(x,y)$, the x-direction and y-direction spatial frequencies are, respectively,

$$f_x = \frac{n}{X} \quad \text{cycles per unit length}$$

$$f_y = \frac{m}{Y} \quad \text{cycles per unit length} \tag{1.5.7}$$

The overall spatial frequency of a two-dimensional periodic pattern is defined as

$$f_s = \sqrt{f_x^2 + f_y^2} \tag{1.5.8}$$

In general, the components at high spatial frequencies represent rapid intensity changes in the image, such as occur near edges of objects in the scene.

---

* The Kronecker delta function is defined as $\delta_{ij} = 1$ for $i=j$, $\delta_{ij} = 0$ otherwise.

Components at low spatial frequencies represent slower and smoother changes in intensity. Negative frequencies have the same physical significance as their positive counterparts.

If a finite-bit representation of $B(x,y)$ is desired in the frequency domain, then only a finite number of frequencies can be accommodated. If the viewing distance is such that the viewer cannot perceive intensity transitions that are faster than a certain amount, then the frequency range chosen for representation might look like

$$|f_x| < F_x$$

$$|f_y| < F_y \tag{1.5.9}$$

By choosing F_x and F_y sufficiently large, any desired representational accuracy can be achieved.

If, on the other hand, absolute intensity is of little importance, but the positions of rapid intensity transitions must be maintained, such as at edges of objects in the scene, then the appropriate frequency range might be

$$F_x < |f_x| < G_x$$

$$F_y < |f_y| < G_y \tag{1.5.10}$$

This case might arise in a pattern recognition system.

After the appropriate range of frequencies is chosen, one method of coding consists simply of quantizing and representing in binary form the coefficients c_{mn} in Eq. (1.5.3) that correspond to that frequency range.

1.5.2 Bandlimited Images — Sampling Theorem

If an image possesses frequency components only within a finite range, it is said to be *bandlimited*. For example,

$$\bar{B}(x,y) = \sum_{m=-M}^{M} \sum_{n=-N}^{N} c_{mn} \phi_{mn}(x,y) \tag{1.5.11}$$

is bandlimited to the frequency range*

* Frequency bounds are chosen to be between harmonics for mathematical convenience.

$$|f_x| < \frac{N+0.5}{X} = F_x \text{ cycles/length}$$

$$|f_y| < \frac{M+0.5}{Y} = F_y \text{ cycles/length} \qquad (1.5.12)$$

For this image, the number of nonzero coefficients c_{mn} is given by $N_S = (2M+1)(2N+1)$. In order to evaluate these coefficients, the integral Eq. (1.5.4) could be used. However, in most situations this approach is much too difficult to implement.

Practical methods for evaluating the Fourier coefficients c_{mn} are based on spatial samples of the bandlimited image $\bar{B}(x,y)$ at N_S pels $\{(x_k,y_k), k = 1...N_S\}$. For each of the pels (x_k, y_k), we have from Eq. (1.5.11)

$$\bar{B}(x_k,y_k) = \sum_{|m|\le M} \sum_{|n|\le N} c_{mn}\phi_{mn}(x_k,y_k) \qquad (1.5.13)$$

Now, if the pels are chosen such that the N_S Eqs. (1.5.13) are all linearly independent, then it is possible to solve for the N_S unknown coefficients c_{mn} without having to evaluate any integrals.

One such set of sampling points consists of a rectangular array of $N_x = 2N+1$ columns and $N_y = 2M+1$ rows. In this case $N_S = N_x \times N_y$ and it can be easily shown for $|m|,|k| \le M$ and $|n|,|\ell| \le N$ that

$$\sum_{i=0}^{N_y-1} \sum_{j=0}^{N_x-1} \phi_{mn}(x_j,y_i)\phi'_{k\ell}(x_j,y_i) = \frac{N_S}{XY}\delta_{mk}\,\delta_{n\ell} \qquad (1.5.14)$$

where $x_j = \frac{X}{N_x}j$ and $y_i = \frac{Y}{N_y}i$. Applying this result in Eq. (1.5.13) yields

$$c_{mn} = \frac{XY}{N_s} \sum_{i=0}^{N_y-1} \sum_{j=0}^{N_x-1} \bar{B}(x_j,y_i)\phi'_{mn}(x_j,y_i) \qquad (1.5.15)$$

which, as we shall see later, is related to the so-called *Discrete Fourier Transform* (DFT) for sampled data sequences.

Thus, we see that sampling $\bar{B}(x,y)$ using a rectangular array of pels is sufficient to completely specify the bandlimited image. $N_x = 2N+1$ samples should be taken horizontally and $N_y = 2M+1$ vertically. The respective sampling densities are, then

$$D_x = \frac{2N+1}{X} = 2F_x \text{ samples/length}$$

$$D_y = \frac{2M+1}{Y} = 2F_y \text{ samples/length} \tag{1.5.16}$$

Comparing Eqs. (1.5.12) and (1.5.16) we see that each sampling density is slightly larger than *twice* the corresponding highest spatial frequency contained in the image. This is the essence of the well-known *Nyquist sampling theorem*[1.5.1] that applies to an infinitely long function of time and states:

> If $g(t)$ is bandlimited to a frequency range W, that is, its Integral Fourier Transform* $G(f)$ is zero for frequencies $f \geq W$, then $g(t)$ can be completely specified by samples taken at the *Nyquist* sampling rate of $2W$ samples per unit time. Specifically,**
>
> $$g(t) = \sum_{k=-\infty}^{\infty} g(\frac{k}{2W}) \text{ sinc } (2Wt-k) \tag{1.5.17}$$

We are only considering images of finite size. Combining Eqs. (1.5.11) and (1.5.15) results in

$$\bar{B}(x,y) = \sum_{\substack{|m| \leq M \\ |n| \leq N}} \phi_{mn}(x,y) \frac{XY}{N_S} \sum_{i,j} \bar{B}(x_j,y_i)\ \phi'_{mn}(x_j,y_i)$$

$$= \sum_{i,j} \bar{B}(x_j,y_i) \frac{XY}{N_S} \sum_{\substack{|m| \leq M \\ |n| \leq N}} \phi_{mn}(x,y)\phi'_{mn}(x_j,y_i) \tag{1.5.18}$$

Thus, the two-dimensional, finite-size version of the sampling theorem becomes

$$\bar{B}(x,y) = \sum_{i=0}^{2M} \sum_{j=0}^{2N} \bar{B}(x_j,y_i)\ \chi(x-x_j,y-y_i) \tag{1.5.19}$$

where

---

* The Integral Fourier Transform is an extension of the Fourier Series for functions $g(t)$ of infinite duration: $G(f) \triangleq \int_{-\infty}^{\infty} g(t)\exp(-2\pi\sqrt{-1}\ ft)\,dt$, $0 \leq f < \infty$. Images of infinite duration can be formed by periodic replication of the finite size image.

** $\text{sinc}(u) = \frac{\sin \pi u}{\pi u}$, u in radians.

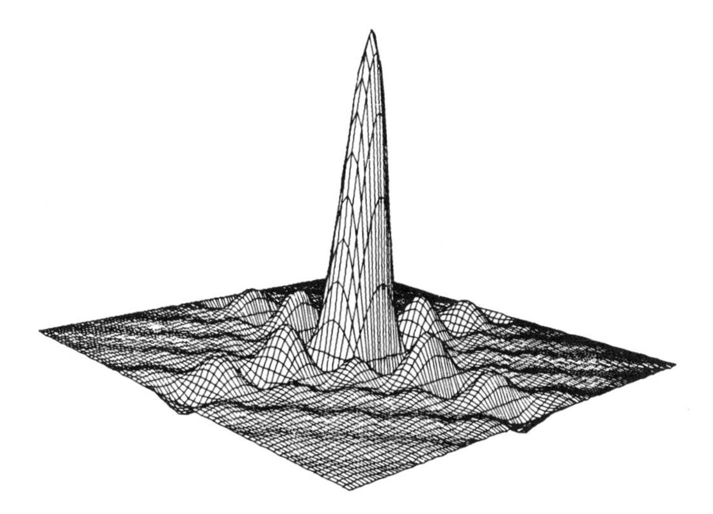

Fig. 1.5.2 Two-dimensional function $\chi(u,v)$ used in the sampling theorem reconstruction of bandlimited, finite sized images. It is similar to those of Fig. 1.3.6 except here, the reconstruction is exact.

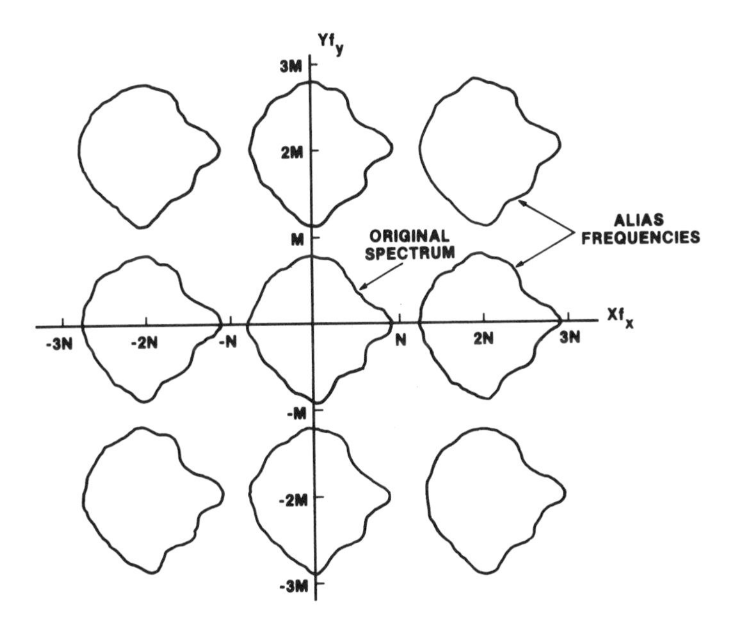

Fig. 1.5.3 Fourier frequency spectrum (top view) of the sampled image $B_S(x,y)$. Xf_x and Yf_y are horizontal and vertical cycles, respectively. It is periodic and of infinite extent. The low-frequency portion is the spectrum of the original bandlimited image $\bar{B}(x,y)$. The remaining portions are *alias* frequencies caused by the sampling process.

$$\chi(u,v) \triangleq \frac{1}{N_S} \sum_{\substack{|m|\leq M \\ |n|\leq N}} \exp 2\pi\sqrt{-1}\left\{\frac{nu}{X} + \frac{mv}{Y}\right\} \tag{1.5.20}$$

Using formulas for sums of geometric series, it is shown easily that

$$\chi(u,v) = \frac{\sin[\pi(2N+1)\frac{u}{X}]\ \sin[\pi(2M+1)\frac{v}{Y}]}{N_S\ \sin(\pi\frac{u}{X})\ \sin(\pi\frac{v}{Y})} \tag{1.5.21}$$

The function $\chi(u,v)$ falls off fairly rapidly from its peak at the origin as shown in Fig. 1.5.2. This means that in Eq. (1.5.19) the most significant terms are those for which (x_j,y_i) are in close proximity to x,y. Thus, Eq. (1.5.19) evaluates $\bar{B}(x,y)$ as a weighted sum, or average, of nearby samples. This is exactly the same as the interpolative reconstruction operation of Eq. (1.3.3) used for spatial sampling, except that here, $\bar{B}$ is strictly bandlimited, and the reconstruction is exact. The oscillatory nature of the $\chi(u,v)$ weighting function may seem a bit odd, but it turns out that this behavior is necessary in order to portray adequately the higher frequency components. Other weighting functions can also be used with reasonably good results. In particular, functions of more limited duration, such as shown in Fig. 1.3.6, are more practical than the $\chi(u,v)$ function. More information can be found regarding this aspect by consulting the theory of filtering of sampled data waveforms.[1.5.2]

1.5.3 Frequency Domain Filtering

We next consider the question of what to do if $B(x,y)$ is not bandlimited, i.e., if it contains spatial frequencies that are higher than needed for a particular application. In the discussion of spatial sampling in the last section, we saw that intensity transitions that are too rapid could be removed by pre-filtering. The same approach can be applied to the problem of removing spatial frequency components that are too high to be of interest. Specifically, given

$$B(x,y) = \sum_{m,n=-\infty}^{\infty} c_{mn}\phi_{mn}(x,y) \tag{1.5.22}$$

we wish to remove terms for which $|m| > M$ or $|n| > N$ in order that the frequency range be bandlimited. Pre-filtering with a weighting function $h_c(\cdot\ ,\ \cdot)$ as in Eq. (1.3.2) yields

$$\bar{B}(x,y) = \int_o^X \int_o^Y B(w,z)h_c(x-w,y-z)\,dzdw \tag{1.5.23}$$

It is shown easily that the desired results are obtained if we choose

$$h_c(u,v) = \frac{N_S}{XY}\chi(u,v) = \frac{1}{\sqrt{XY}} \sum_{-M}^{M} \sum_{-N}^{N} \phi_{mn}(u,v) \tag{1.5.24}$$

For then, from Eq. (1.5.2)

$$h_c(x-w,y-z) = \frac{1}{\sqrt{XY}} \sum_{-M}^{M} \sum_{-N}^{N} \phi_{mn}(x-w,y-z)$$

$$= \sum_{-M}^{M} \sum_{-N}^{N} \phi_{mn}(x,y)\phi'_{mn}(w,z) \tag{1.5.25}$$

Substituting this into Eq. (1.5.23) and using the orthonormality condition of Eq. (1.5.5) we get the desired bandlimited image

$$\overline{B}(x,y) = \sum_{-M}^{M} \sum_{-N}^{N} c_{mn}\phi_{mn}(x,y) \tag{1.5.26}$$

Thus, Eq. (1.5.24) defines a low-pass pre-filter whose shape is shown in Fig. 1.5.2. Note that $\chi(u,v)$ is very much like a two dimensional sinc function, except that it is periodic in u and v with periods X and Y, respectively.

Image filtering has one important difference compared to filtering infinitely long, one-dimensional functions of time. With images, sharp discontinuities typically appear at the boundaries. These discontinuities often give rise to high frequency components that have little or nothing to do with the information content of the image itself.* If an attempt is made to filter out these frequencies, then undesirable perturbations may occur in $\overline{B}(x,y)$ near the boundary. This is called the *Gibbs phenomenon*. However, if the image and allowable bandwidth are large enough, then these "edge effects" are usually ignored. More will be said on this topic in Chapter 5.

1.5.4 Frequency Spectrum of the Sampled Image

Suppose we have a bandlimited image $\overline{B}(x,y)$ that has frequency components c_{mn} only for $|m| \le M$, $|n| \le N$, and we use a rectangular sampling array to construct the *sampled* image

* The cosine transform basis images of Eq. (1.5.6) are somewhat better in this regard, and produce fewer high frequency components at edges of images.

$$B_S(x,y) = \sum_{i=0}^{2M} \sum_{j=0}^{2N} \bar{B}(x_j,y_i)\ \delta(x-x_j)\delta(y-y_i) \tag{1.5.27}$$

where $x_j = \dfrac{X}{2N+1}\ j$ and $y_i = \dfrac{Y}{2M+1}\ i$. The Fourier frequency components of $B_S(x,y)$ are from Eq. (1.5.4)

$$c_{mn}^S = \int_0^X \int_0^Y B_S(x,y)\,\phi'_{mn}(x,y)\,dydx \tag{1.5.28}$$

It is easily shown that the frequency spectrum of $B_S(x,y)$ is periodic and of infinite extent. For example,

$$c_{m+2M+1,n}^S = \int_0^X \int_0^Y B_S(x,y)\,\phi'_{mn}(x,y)$$

$$\exp[-2\pi\sqrt{-1}\,(2M+1)\frac{y}{Y}]\,dydx$$

$$= \int_0^X \int_0^Y \sum_{i,j} \bar{B}(x_j,y_i)\delta(x-x_j)\delta(y-y_i)\phi'_{mn}(x,y)$$

$$\exp[-2\pi\sqrt{-1}\,(2M+1)\frac{y_i}{Y}]\,dydx$$

$$= \int_0^X \int_0^Y \sum_{i,j} \bar{B}(x_j,y_i)\delta(x-x_j)\delta(y-y_i)\phi'_{mn}(x,y)$$

$$\exp[-2\pi\sqrt{-1}\,i]\ dydx$$

$$\triangleq c_{mn}^S \tag{1.5.29}$$

The frequency spectrum of $B_S(x,y)$ is shown in Fig. 1.5.3.

From Eqs. (1.5.19) and (1.5.27) we see that the bandlimited image can be written in terms of the sampled image as

$$\bar{B}(x,y) = \int_0^X \int_0^Y B_S(w,z)\,\chi(x-w,y-z)\,dzdw \tag{1.5.30}$$

But from Eq. (1.5.24) we saw that $\chi(u,v)$ is simply a low-pass filter, i.e., it retains only frequency components

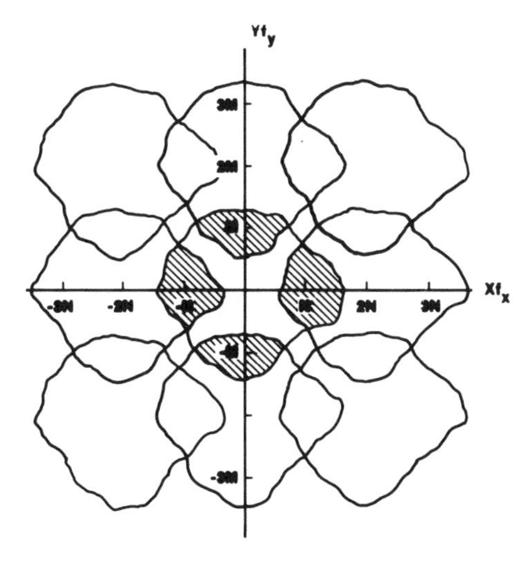

Fig. 1.5.4 Spectrum of the sampled image $B_S(x,y)$ when the original image $\overline{B}(x,y)$ contains frequency components above half the sampling frequency. Components in the crosshatched areas are corrupted by alias frequencies. This distortion is called *pre-aliasing* and is due to insufficient pre-filtering of $\overline{B}(x,y)$.

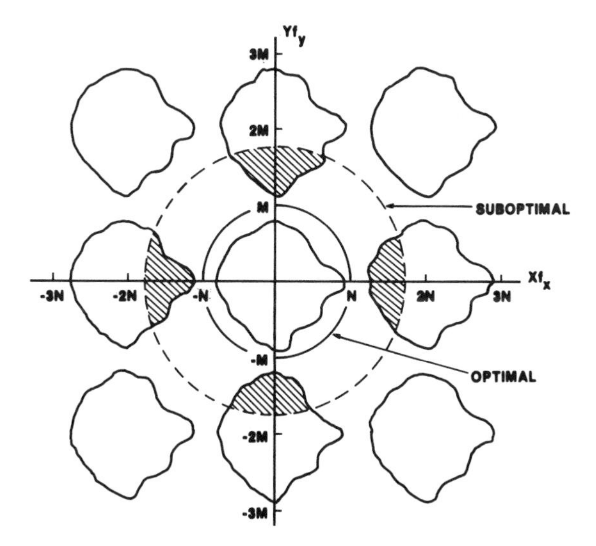

Fig. 1.5.5 Non ideal post-filtering. The dashed line indicates the frequency extent of a suboptimal post-filter $h_d(u,v)$. The solid line indicates an optimal filter. Alias frequency components in the crosshatched areas appear in the reconstructed image. This distortion is called *post-aliasing*, and is due to inadequate post-filtering of $B_S(x,y)$.

$$|f_x| < F_x = \frac{N+0.5}{X} \text{ cycles/length}$$

$$|f_y| < F_y = \frac{M+0.5}{Y} \text{ cycles/length} \tag{1.5.31}$$

Thus, in this *baseband* of frequencies the spectrum of $\bar{B}(x,y)$ is exactly the same as that of the sampled image $B_S(x,y)$. Equivalently, the frequency spectrum of $B_S(x,y)$ is simply the original $B(x,y)$ spectrum periodically replicated as shown in Fig. 1.5.3. Frequency components outside the baseband are called *alias* frequencies.

We are now in a position to describe some of the effects of these alias frequencies. Suppose the image $\bar{B}(x,y)$ that is sampled contains frequency components above half the sampling frequency. Then the frequency spectrum of the sampled image $B_S(x,y)$, shown in Fig. 1.5.4, contains some overlapping areas in which the original frequency components are corrupted by alias frequencies. The original image $\bar{B}(x,y)$ cannot be recovered from $B_S(x,y)$ by filtering. This distortion is called *pre-aliasing* and is due to insufficient pre-filtering of $\bar{B}(x,y)$. It is exactly the same kind of distortion that was discussed in Section 1.3.

Now suppose the sampling rate is high enough so that we have no pre-aliasing, and the spectrum of $B_S(x,y)$ is as shown in Fig. 1.5.3. However, suppose we cannot implement the ideal filter $\chi(u,v)$ of Eq. (1.5.30), but instead we employ a suboptimum reconstruction

$$\tilde{B}(x,y) = \int_0^X \int_0^Y B_S(w,z)\, h_d(x-w,y-z)\, dz dw \tag{1.5.32}$$

where the frequency extent of $h_d(u,v)$ is shown by the dashed curve of Fig. 1.5.5. In this case, many alias frequency components appear in $\tilde{B}(x,y)$ that were not in the original $\bar{B}(x,y)$. This distortion is *post-aliasing*, and is caused by inadequate post-filtering of $B_S(x,y)$. It is the same post-aliasing distortion that was discussed in Section 1.3.

1.5.5 Frequency Spectra of Raster Scanned Images

The raster scanned video signal $\bar{B}(t)$ defined by Eqs. (1.4.1)–(1.4.3) can also be described in the frequency domain. In this case, N_y is the number of scan lines, each of duration T_L seconds, and

$$B(f_m) = \frac{1}{N_y T_L} \int_0^{N_y T_L} \bar{B}(t) \exp(-2\pi\sqrt{-1}\, f_m t)\, dt \tag{1.5.33}$$

where

$$f_m = \frac{m}{N_y T_L} \quad Hz$$

$\bar{B}(t)$ can be filtered, sampled and reconstructed in the same way as described above, except that it is a one-dimensional signal. In fact, if filtering effects near $t=0$ and $t=N_y T_L$ can be ignored, the Integral Fourier Transform representation can also be used without much disadvantage.

$\bar{B}(t)$ has several characteristics peculiar to raster scanned signals.[1.5.3] First, discontinuities occur every T_L seconds due to the end of one scan line and the beginning of the next. If $\bar{B}(t)$ is filtered, these may cause undesirable "edge effects". Second, because of similarities in successive scan lines of the image, $\bar{B}(t)$ has strong periodicity at the line scan rate $1/T_L$, and, therefore, $B(f_m)$ will have large components at harmonics of the line rate, as shown in Fig. 1.5.6.

The bandwidth of $\bar{B}(t)$, i.e., of the envelope of $B(f_m)$, is determined almost exclusively by the horizontal spatial frequencies f_x (defined in Eqs. (1.5.1)–(1.5.4) and (1.5.7)) that remain after pre-filtering by the camera scanning spot. The vertical spatial frequencies f_y mainly affect the widths of the harmonic peaks shown in Fig. 1.5.6.

1.5.6 Discrete Fourier Transform (DFT)

In this section we assume that the image has already been spatially sampled, for example, with a rectangular array of $N_y \times N_x$ pels. We saw from Eq. (1.5.15) that for bandlimited images the Fourier frequency components could be found from the pel values by simple multiplication and summation, and without evaluating any integrals. However, for practical images, which may contain thousands or hundreds of thousands of pels, even this simplified procedure may be too complex and costly.

Thus, in most applications of Fourier Transforms to pictorial data, the sampled image is first subdivided into relatively small sections called *blocks*. Suppose we consider blocks of size $1 \times N$, that is, one pel vertically by N pels* horizontally. Further, suppose we denote the pel values of a block by $\{b_i, i=1,2,...,N\}$. Then by changing slightly the subscripts and normalization of Eq. (1.5.15), as well as setting $X = Y = 1$, we can write the Nth order *Discrete Fourier Transform* (DFT) coefficients as

* The block-size N should not be confused with the earlier usage of N to denote horizontal bandwidth.

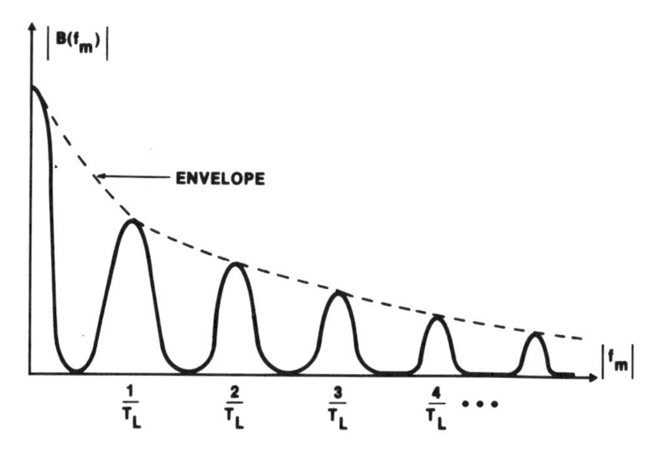

Fig. 1.5.6 Frequency domain representation of a raster scanned video signal. $|B(f_m)|$ is highly peaked at harmonics of the line scan rate $1/T_L$ and relatively small in between. The widths of the lobes are determined by the vertical spatial frequency components f_y. The overall bandwidth of $B(t)$ is determined by the horizontal spatial frequency components f_x.

$$c_m = \frac{1}{\sqrt{N}} \sum_{i=1}^{N} b_i \exp[-\frac{2\pi}{N}\sqrt{-1}(i-1)(m-1)] \qquad m = 1,\ldots,N$$

(1.5.34)

If the pels $\{b_i\}$ and the DFT coefficients $\{c_m\}$ are arranged into column matrices $\boldsymbol{b}$ and $\boldsymbol{c}$, respectively, then the DFT can be written as

$$\boldsymbol{c} = \boldsymbol{T}\boldsymbol{b} \tag{1.5.35}$$

where the square transform matrix $\boldsymbol{T}$ has (row m, column i) elements

$$t_{mi} = \frac{1}{\sqrt{N}} \exp[-\frac{2\pi}{N}\sqrt{-1}(i-1)(m-1)] \tag{1.5.36}$$

Let $\boldsymbol{T}'$ be the conjugate transpose of $\boldsymbol{T}$. If the mth column of $\boldsymbol{T}'$ is denoted by $\boldsymbol{t}_m$, then it is easily shown using formulas for sums of geometric series that

$$\boldsymbol{t}'_m \boldsymbol{t}_n = \delta_{mn} \tag{1.5.37}$$

This also implies that the transform $\boldsymbol{T}$ is *orthonormal* or *unitary*, i.e.,

$$\boldsymbol{T}^{-1} = \boldsymbol{T}' \tag{1.5.38}$$

and

$$\boldsymbol{b} = \boldsymbol{T}'\boldsymbol{c} \tag{1.5.39}$$

The vectors $\boldsymbol{t}_m$ are often called *orthonormal basis* vectors of the unitary transform $\boldsymbol{T}$. They are orthonormal because of Eq. (1.5.37). The term basis arises when Eq. (1.5.39) is written as

$$\boldsymbol{b} = \sum_{m=1}^{N} c_m \boldsymbol{t}_m \tag{1.5.40}$$

That is, any $\boldsymbol{b}$ is a linear combination of the basis vectors, each weighted by its corresponding transform coefficient.

Thus, the DFT gives another way of representing images. If the DFT coefficients are each quantized to finite accuracy, then the overall bit-rate is also finite.

1.6 Time Varying Images

If time variation of the visual information is to be represented, then $B(x,y,t)$ must be represented by a finite number of bits per second. The concepts described previously for two-dimensional $B(x,y)$ are easily extended[1.5.4] to the three-dimensional $B(x,y,t)$. Thus, spatio-temporal filtering of an image becomes:

$$\bar{B}(x,y,t) = \int\int\int B(w,z,\tau)\,h(x-w,y-z,t-\tau)\,dwdzd\tau \qquad (1.6.1)$$

and a frequency domain representation can be written for the time interval $[0 \le t \le T]$

$$B(x,y,t) = \sum_{m,n,p=-\infty}^{\infty} c_{mnp}\,\phi_{mnp}(x,y,t) \qquad (1.6.2)$$

where

$$\phi_{mnp}(x,y,t) = \frac{1}{\sqrt{XYT}} \exp 2\pi\sqrt{-1}\left[\frac{nx}{X} + \frac{my}{Y} + \frac{p\,t}{T}\right]$$

$$(1.6.3)$$

$$c_{mnp} = \int_0^T \int_0^X \int_0^Y B(x,y,t)\,\phi'_{mnp}(x,y,t)\,dydxdt$$

The quantity p/T is a temporal frequency,* and, as in Section 1.5, if the image is pre-filtered so that all temporal frequencies are smaller than, say, W hertz, then the bandlimited image can be completely specified by sampling, in time, at a rate of $2W$ hertz.

For applications involving human viewers, temporal frequency bandwidths of the order of 10 to 15 Hz are usually adequate, which implies a required temporal sampling, or frame rate of 20 to 30 Hz. Again, as in Section 1.5, the bandlimited image can be reconstructed from its time samples by temporal interpolation, i.e., post-filtering.

---

* Conceptually, we could set $T=\infty$ and use the Integral Fourier Transform representation for the time and frequency parameters in Eqs. (1.6.2) and (1.6.3). However, this requires finite energy in $B(x,y,t)$, i.e., that it be square integrable over all time — an assumption that is not especially useful for imaging applications.

In practice, some temporal pre-filtering takes place in the sensor itself. Prior to sampling, most imaging sensors simply accumulate light energy over some "exposure" time T_I so that the pre-filtered image becomes, conceptually,

$$\bar{B}(x,y,t) = \int_o^X \int_o^Y \int_{t-T_I}^t B(w,z,\tau)\,h(x-w,y-z)\,d\tau dz dw \qquad (1.6.4)$$

For example, a movie camera samples this pre-filtered image periodically[†] to give a sequence of *frames* $\{\bar{B}(x,y,kT_F), k=1,2,...\}$. The integration time T_I is less than or equal to the time between frames T_F, and is chosen on the basis of lighting conditions and the sensitivity of the particular film being used. Converting the sequence of frames to a finite data rate can then be done by applying the two-dimensional techniques described previously to each individual frame.

The temporal pre-filtering implied by the operation of Eq. (1.6.4) attenuates each temporal frequency component by

$$sinc(f_t T_I) \qquad (1.6.5)$$

where f_t (in hertz) is the temporal frequency of interest. We see that if the integration time T_I is very small, then very little temporal averaging occurs, and pre-aliasing effects such as stagecoach wheels rotating backward and jerky motion can be expected. However, even with T_I equal to the entire frame period T_F, the attenuation at half the temporal sampling frequency, i.e., $f_t = \frac{1}{2T_F}$, is only about 4 dB, which is not nearly enough to avoid pre-aliasing according to the sampling theorem.

Television cameras generally work somewhat differently. With *progressive scan* television, the pre-filtered image of Eq. (1.6.4) is raster scanned as described in Section 1.4. That is, T_L seconds between scan lines and N_y lines* per frame imply

$$T_F = N_y T_L \qquad (1.6.6)$$

as the time between frames. Television cameras have no shutters and, therefore, $T_I = T_F$. This scanning process produces a one dimensional waveform

† The sampling rate $1/T_F$ is typically 16 Hz for home movies and 24 Hz for commercial films.

* N_y usually includes several blanked lines at the bottom of the picture to allow time for the scanning beam to return to the first line of the next frame.

$$\bar{B}(t) = \bar{B}(x_t, y_t, t) \tag{1.6.7}$$

where during the ith line of the mth frame, i.e.,

$$(m-1)T_F + (i-1)T_L < t < (m-1)T_F + iT_L \tag{1.6.8}$$

we have

$$y_t = \frac{(i-0.5)Y}{N_y}$$

$$x_t = \left[t-(m-1)T_F-(i-1)T_L\right] \frac{X}{T_L} \tag{1.6.9}$$

The one-dimensional signal $\bar{B}(t)$ is then sampled, usually at some rate N_x / T_L where N_x is an integer, and quantized to give the desired finite bit-rate representation with $N_x N_y$ pels per frame.

The pictorial information produced by frame sequential television and raster scanned movie film are identical, except in one respect that is important in some applications. Movie cameras take a snapshot of the pre-filtered image, i.e., all spatial points x,y in a frame are sampled at the same time, whereas with most† television cameras there is a considerable time gap of about 1/60 second between sampled pels at the top of a frame and those at the bottom. This gives rise to a slight distortion of the shape of large moving objects if a single frame of television is examined. However, the shape of rapidly moving objects is also distorted or blurred by temporal pre-filtering, i.e., the camera integration of Eq. (1.6.4), and in most applications this overshadows any shape distortion due to raster scanning.

We have seen that reconstruction of a replica image $\tilde{B}(x,y,t)$ from a finite data-rate representation generally requires an interpolation or post-filtering operation. Temporally sampled images are no exception. However, due to the difficulty of implementation, temporal post-filtering is provided by only a few spatio-temporal imaging displays. Instead, such displays usually rely on the temporal post-filtering characteristics of the human eye and, typically, they display each frame briefly without any temporal interpolation or post-filtering whatsoever.

Temporal post-filtering by the eye will be treated in more detail in Chapter 4. Here, it suffices to mention two important characteristics. First, the ability of

---

† Charge coupled device (CCD) television cameras also take a snapshot, and store the image locally. The stored image is then raster scanned to produce a one-dimensional output signal.

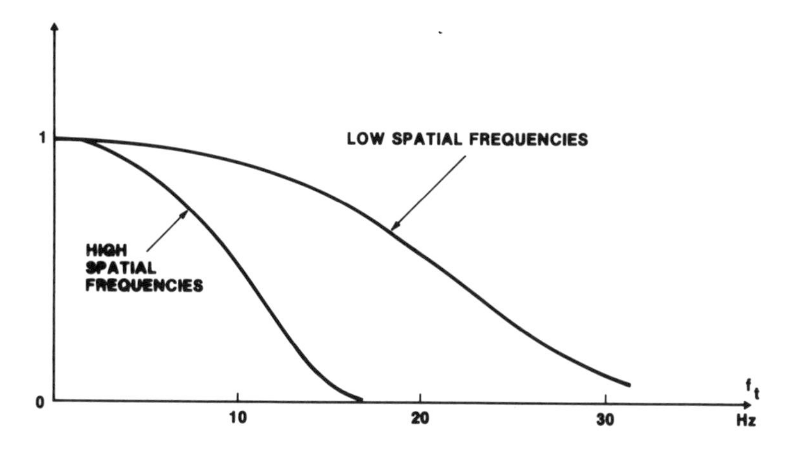

Fig. 1.6.1 Temporal post-filtering characteristic of the human eye. Filtering is more severe for high spatial frequencies than for low spatial frequencies. Filtering characteristics also change with room lighting, display brightness, etc. (See Section 4.3.1c for more details.)

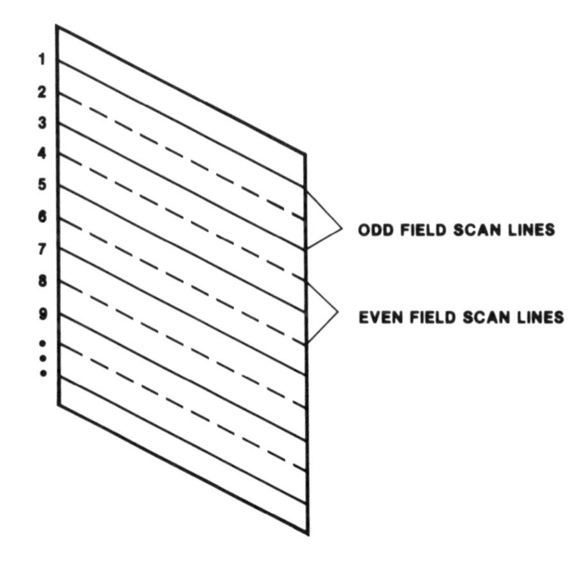

Fig. 1.6.2 With 2:1 interlace, odd numbered lines (field 1) are scanned and displayed first, followed by the even numbered lines (field 2). In this way, low spatial frequencies are displayed twice as often as high spatial frequencies, thus better matching the characteristics of human vision.

the eye to temporally interpolate between frames is inversely dependent on the ambient lighting level and the overall brightness of the displayed frames — the brighter the display, the higher the likelihood of visible flicker due to insufficient temporal post-filtering by the eye. Thus, motion pictures, which are normally displayed at relatively low light level, are able to achieve adequate rendition with a display rate of 32 or 48 Hz, while television, in a well lit room with a relatively high image brightness, requires a 50 to 60 Hz display rate for flicker free rendition. As already stated, for many applications the speed of movement in a scene is limited, so that temporal frequencies in $B(x,y,t)$ rarely exceed 10 to 15 Hz. Thus, a 50 to 60 Hz frame rate is often much higher than is needed for adequate motion rendition. This discrepancy is overcome in motion pictures by the projector displaying each frame two or three times. Thus, commercial movie cameras take pictures, typically, at a 24 Hz frame rate, which is adequate for the usual speeds of movement encountered. However, the projector then flashes each frame onto the screen two or three times to give a flashing rate of 48 or 72 Hz, which is high enough in dark theaters to avoid visible flicker.

Television cannot economically use this technique since a large amount of signal storage would be required at the receiver in order to display each frame a second time. Instead, television takes advantage of a second property of the human eye shown in Fig. 1.6.1 — namely that temporal post-filtering by the eye is much more effective for fine detail in an image, i.e., higher spatial frequencies, than it is for flat, low detail areas, i.e., lower spatial frequencies. Thus, efficient transmission can be achieved if low spatial frequencies are detected by the camera and displayed at a rate of 50 to 60 Hz, while high spatial frequencies are detected and displayed at a rate of only 25 to 30 Hz. This is exactly what is attempted by a technique called *interlace*. With 2-to-1 interlaced television, shown in Fig. 1.6.2, the camera first scans (and the receiver displays) the odd numbered lines $1,3,5,...,N_y$ of a frame. This set of lines is called *field 1*. Following this, the even numbered lines $2,4,6,...,N_y-1$ (called *field 2*) are scanned and displayed, which completes the processing of the frame. Virtually all of the world's television systems use 2-to-1 interlace in an effort to preserve efficiently high-detail visual information and, at the same time, avoid visible flicker at the display due to insufficient temporal post-filtering by the human eye.

A few effects of interlace deserve mention. First, *line flicker* becomes visible if viewers move too close to the display. This is due to the scan lines becoming too far apart relative to the viewer, i.e., too low a vertical spatial frequency, so that a 25–30 Hz display rate is no longer adequate for flicker-free rendition. Frame sequential television does not usually have this problem. Another effect of interlace occurs in scenes containing objects that are moving horizontally. If a single frame, i.e., two fields, is examined, then alternate scan lines on such objects appear displaced with respect to each other due to object motion between fields. In particular, edges of moving objects appear serrated, as shown in Fig. 1.6.3. Finally, in most TV cameras the raster scanning spot is

Fig. 1.6.3 Effect of interlace on moving objects. Moving objects are shifted in successive fields, which causes sharp moving edges to appear serrated.

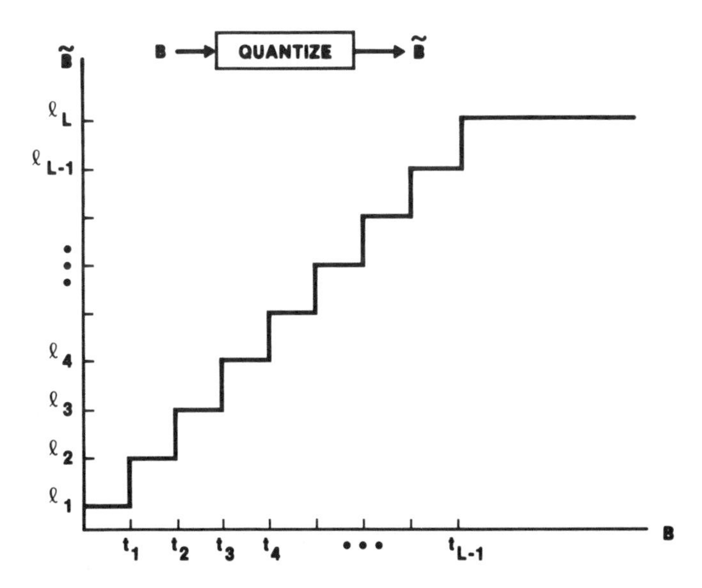

Fig. 1.7.1 Relationship between input and output of a quantizer. The input takes on a continuum of values. The output has only a finite number of levels L, which is often a power of two, e.g., 2^K for K bits.

actually two lines high. Thus, most of the image area is scanned in a single-field period, and the integration time $T_I \approx 1/2\ T_F$.

1.7 Quantization and Binary Word Assignment — PCM

One of the major steps in producing a finite bit-rate representation of visual information is amplitude quantization; that is, representing a given numerical value whose range is a *continuum* of numbers by an approximation chosen from a *finite* set of numbers. For example, voltages in the range 0–20 volts might be measured only to the nearest 0.1 volt. We may choose to quantize pel values, or we might wish to use quantized Fourier coefficients in our specification of an image. In any event, the resulting representation, illustrated in Fig. 1.7.1, is characterized by quantities that can take on only a finite number of values or levels, as opposed to the continuum of values possible in the original image.

Each member of the finite set of possible levels is assigned a unique binary word of finite length, i.e., a code is constructed. This code (as well as the set of approximating levels) may even change from time to time depending on certain circumstances discussed in later chapters. For example, it is inefficient to quantize and code all of the Fourier coefficients of an image according to the same rule.

Having constructed such a code, the procedure for producing a finite bit-rate specification of an image is simply to quantize the (time domain or Fourier domain) sampled values of the image and represent each quantized level by its associated code word. Since the samples occur at a finite rate and the code words are of finite bit-length, the resulting representation is of finite bit-rate. Coding of image intensity samples in this manner is called *pulse code modulation* and is abbreviated PCM.

Quantization necessarily results in loss of information since coded values are specified with less accuracy than in the original image. Minimizing this information loss will be discussed in Chapter 5. However, a few things can be said at this point regarding the very common situation where a replica image is to be reproduced for human viewing.

Replacing original pels by their quantized approximations, as in Fig. 1.7.1, causes errors in the reconstructed replica image B that, in a well designed PCM system, appear to the viewer as random noise or *snow*. For most applications involving photographic or cathode ray tube (CRT) displays, this noise can be made practically invisible with 8-bit quantization, i.e., quantization with 256 judiciously chosen representative levels.

Visibility of PCM quantization noise is also strongly dependent on how much noise there is in the original image prior to quantization. More will be said about this in Chapter 5. It suffices to point out here that most applications require only enough bits of quantization to insure that quantization noise is small compared to the noise in the original samples. Seven bit quantization (or even six) is, thus, sufficient for some applications where the original image is noisy.

1.8 Color Images

Light is a form of electromagnetic energy that can be completely specified at a point in the image plane by its wavelength distribution. Not all electromagnetic radiation is visible to the human eye. In fact, the entire visible portion of the radiation is only within the narrow wavelength band of 380 to 780 nanometers (nm). This radiation when incident on the eye produces a variety of sensations. In the previous sections, we were concerned mostly with light intensity, i.e., the sensation of *brightness* produced by the aggregate of various wavelengths. However, light of many wavelengths also produces another important visual sensation called *color*. Different spectral distributions generally, but not necessarily, have different perceived color. Thus, color is that aspect of visible radiant energy by which an observer may distinguish between different spectral compositions. A color stimulus therefore, is specified by visible radiant energy of a given intensity and spectral composition. Color is generally characterized by attaching names to the different stimuli, e.g., white, gray, black, red, green, blue. Color stimuli are generally more pleasing to the eye than "black and white" stimuli. Consequently, pictures with color are widespread in television, photography, and printing. Color is also used increasingly in computer graphics to add "spice" to the synthesized pictures. Coloring of black and white pictures by transforming intensity into colors (called pseudocolor) has been used extensively by artists and by workers in pattern recognition.

This section is concerned with questions of how to specify color, and how to reproduce it. In general color specification consists of three parts: (a) color matching, i.e., what are the characteristics of two colors that appear identical under given conditions of observations? (b) color differences, i.e., how different are two colors? (c) color appearance (or perceived color), i.e., what does a color look like? We will deal with the first of these questions in this chapter and defer the discussion of the other two questions to Chapter 4.

1.8.1 Representation of Color for Human Vision

Let us assume that the light emanating from a picture element of the image plane is specified by the spectral power distribution $S(\lambda)$ (in watts per square meter per unit wavelength), λ being the wavelength. The human retina contains predominantly three different color receptors (called cones) that are sensitive to three overlapping areas of the visible spectrum. [See Chapter 4 for more details on cones and other receptors (e.g. rods).] Fig. 1.8.1 shows the spectral sensitivities of individual primate cones. Although the data is quite noisy, it is seen that there are three types of receptors whose sensitivities peak at approximately 445 (called blue), 535 (called green), and 570 (called red) nanometers. Each type of receptor integrates the energy in the incident light at various wavelengths in proportion to their sensitivity to light at that wavelength. The three resulting numbers are primarily responsible for color sensation. This then is the basis for the trichromatic theory of color vision, which states that the color of light

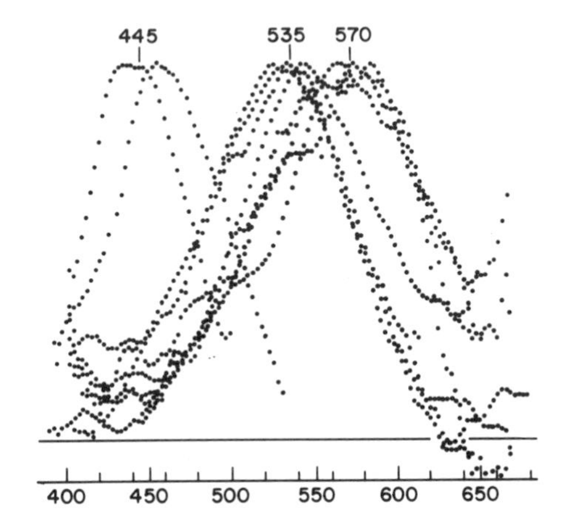

Fig. 1.8.1 Records of spectral sensitivity from individual cones. Some of these data come from monkeys and others come from human eyes (from Marks *et al.* [1.8.7]).

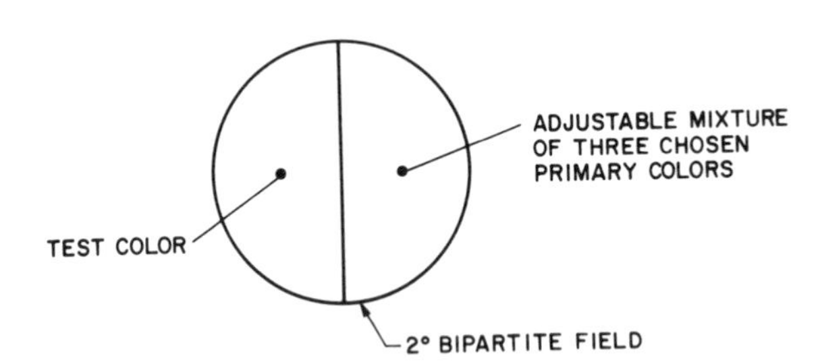

Fig. 1.8.2 The basic experiment in color matching. The bipartite field subtends an angle of 2° at the viewer's eye.

entering the eye may be specified by only three numbers, rather than a complete function of wavelengths over the visible range. It lends a significant economy in color specification and reproduction for human viewing. Much of the credit for this significant advance goes to Thomas Young.[1.8.1]

The counterpart to the trichromacy of vision is the trichromacy of *color mixture*. This important principle states that light of any color can be synthesized by an appropriate mixture of three properly chosen primary colors. Maxwell[1.8.2] demonstrated this in 1855 by using a three-color projection system, and several developments since that time have created a large body of knowledge known as *colorimetry*.[1.8.3] Although the trichromacy of color is based on subjective and physiological findings, there are precise measurements that can be made to examine color matches. An experiment using a *colorimeter* to examine color matches is shown in Fig. 1.8.2. A bipartite field contains the test color on the left and an adjustable mixture of three suitably chosen primary colors on the right. The entire bipartite field is viewed against a dark, neutral surround that subtends a 2° angle at the eye.

It is found experimentally that most test colors can be matched by a proper mixture of the three primary colors, as long as one primary is not a mixture of the other two. The primary colors are usually chosen to be either red, green, and blue, or red, green and violet. The *tristimulus values* of a test color are the amounts of the three primary colors required to give a match by additive mixture. Furthermore, they are unique within the accuracy of the experiment.

Much of the colorimetry is based on the above experimental results as well as other rules attributed to Grassman.[1.8.4] Two important rules, which appear to be valid over a large range of observing conditions, are *linearity* and *additivity*. They state that the color match between any two color stimuli holds even if the intensities of the stimuli are increased or decreased by the same multiplying factor, as long as their relative spectral distributions remain unchanged. As a consequence of linearity and additivity, if stimuli $S_1(\lambda)$ and $S_2(\lambda)$ match and stimuli $S_3(\lambda)$ and $S_4(\lambda)$ also match, then the additive mixtures $[S_1(\lambda) + S_3(\lambda)]$ and $[S_2(\lambda) + S_4(\lambda)]$ will also match. Another consequence of the above two rules and the trichromacy is that any four colors cannot be linearly independent, i.e., the tristimulus values of at least one of the four colors can be represented by a weighted linear sum of the tristimulus values of the other three colors. This amounts to representation of any color by a 3-component vector as specified by the tristimulus values (see Fig. 1.8.3). Thus, a color $\boldsymbol{C}$ is specified by its projection on three axes $\boldsymbol{R}, \boldsymbol{G}, \boldsymbol{B}$ corresponding to the chosen set of primaries. A mixture of two colors then follows the ordinary laws of vector addition, i.e., a mixture of two colors $\boldsymbol{S}_1$ and $\boldsymbol{S}_2$ is given by $\boldsymbol{S} = \boldsymbol{S}_1 + \boldsymbol{S}_2$. If $\boldsymbol{S}_1$ and $\boldsymbol{S}_2$ are specified by vectors of color components, i.e., tristimulus values $(R_{S_1}, G_{S_1}, B_{S_1})$ and $(R_{S_2}, G_{S_2}, B_{S_2})$, respectively, then the mixture $\boldsymbol{S}$ has color components $(R_{S_1} + R_{S_2}, G_{S_1} + G_{S_2}, B_{S_1} + B_{S_2})$.

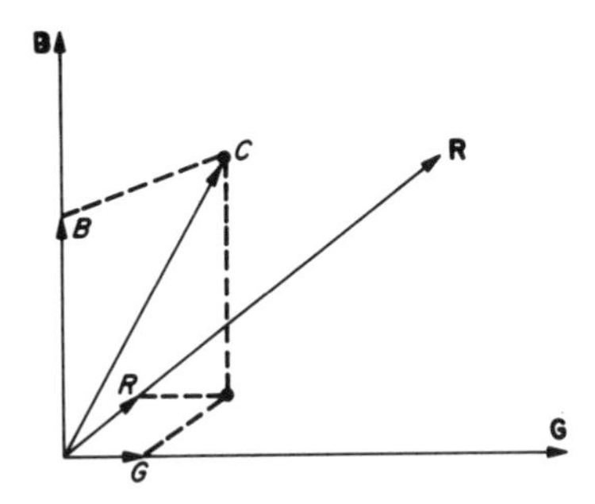

Fig. 1.8.3 R,G,B tristimulus space. A color $\boldsymbol{C}$ is specified by a vector in three-dimensional space with components R, G, and B (tristimulus values).

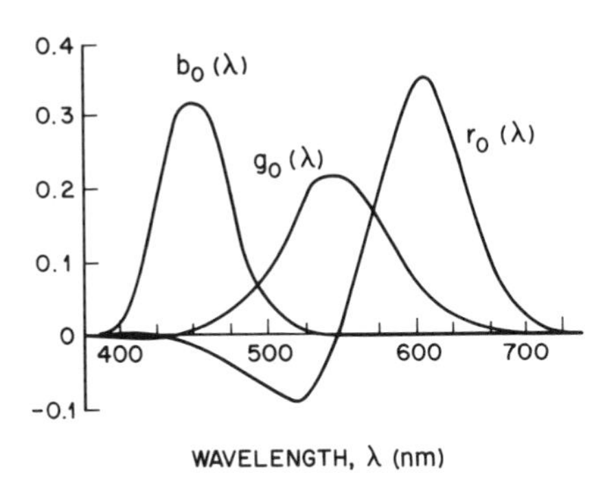

Fig. 1.8.4 The color-matching functions for the 2° Standard Observer, using primaries of wavelengths 700 (red), 546.1 (green), and 435.8 nm (blue), with units such that equal quantities of the three primaries are needed to match the equal energy white, E.

A constraint of the color matching experiment described so far is that only nonnegative amounts of primary colors can be added to match a given test color. However, in practice it is found that sometimes this is not sufficient to effect a match. In this case, since negative amounts of a primary cannot be produced, a match is made by simple transposition, i.e., by adding positive amounts of the primary to the test color. Thus, a test color $\boldsymbol{S}$ might be matched by

$$\boldsymbol{S} + 3\boldsymbol{G} = 2\boldsymbol{R} + \boldsymbol{B} \tag{1.8.1}$$

or

$$\boldsymbol{S} = 2\boldsymbol{R} - 3\boldsymbol{G} + \boldsymbol{B} \tag{1.8.2}$$

Thus, with this definition the negative tristimulus values (2, -3, 1) present no special problem.

By convention, tristimulus values are expressed in normalized form. This is done by a preliminary color matching experiment in which the left side of the split field of Fig. 1.8.2 is allowed to emit light of unit intensity whose spectral distribution is constant with respect to λ (called *equal energy white* $\boldsymbol{E}$). Then the amount of each primary required for a match is taken by definition as one unit. The amount of primaries for matching other test colors are then expressed in terms of this unit. In practice, primaries are usually chosen such that equal energy white $\boldsymbol{E}$ can be matched with positive amounts of each primary.

The tristimulus values of spectral colors (i.e., light of a single wavelength) with unit intensity are called *color matching functions*. Thus, color matching functions $\{r(\lambda), g(\lambda), b(\lambda)\}$ are the tristimulus values, with respect to three given primary colors, of unit intensity monochromatic light of wavelength λ. Fig. 1.8.4 shows color matching functions when the primary colors are chosen to be spectral colors of wavelengths $700.0(\boldsymbol{R}_o)$, $546.1(\boldsymbol{G}_o)$, $435.8(\boldsymbol{B}_o)$ nanometers.

Any color with spectral energy distribution $S(\lambda)$ can be thought of as an additive mixture of monochromatic components $\{S(\lambda)\,d\lambda\}$. If the tristimulus values of such monochromatic components are $\{R_S(\lambda)\,d\lambda\}$, $\{G_S(\lambda)\,d\lambda\}$ and $\{B_S(\lambda)\,d\lambda\}$ respectively, then by linearity and additivity, we get the tristimulus values of the spectral distribution $S(\lambda)$ in terms of tristimulus values of components as follows:

$$R_S = \int_\lambda R_S(\lambda)\, d\lambda$$

$$G_S = \int_\lambda G_S(\lambda)\, d\lambda \tag{1.8.3}$$

$$B_S = \int_\lambda B_S(\lambda)\, d\lambda$$

Since the tristimulus values of unit intensity monochromatic colors are the color matching functions, the above equations can be written in terms of the color matching functions as:

$$R_S = \int_\lambda S(\lambda) r(\lambda)\, d\lambda$$

$$G_S = \int_\lambda S(\lambda) g(\lambda)\, d\lambda \tag{1.8.4}$$

$$B_S = \int_\lambda S(\lambda) b(\lambda)\, d\lambda$$

We have thus found that for the purpose of color matching under specified viewing conditions, the tristimulus values of any test color $S(\lambda)$ can be found analytically using the color matching functions. This also implies that two colors with spectral distribution $S_1(\lambda)$ and $S_2(\lambda)$ match if and only if

$$R_1 = \int S_1(\lambda) r(\lambda)\, d\lambda = \int S_2(\lambda) r(\lambda)\, d\lambda = R_2$$

$$G_1 = \int S_1(\lambda) g(\lambda)\, d\lambda = \int S_2(\lambda) g(\lambda)\, d\lambda = G_2 \tag{1.8.5}$$

$$B_1 = \int S_1(\lambda) b(\lambda)\, d\lambda = \int S_2(\lambda) b(\lambda)\, d\lambda = B_2$$

where $\{R_1, G_1, B_1\}$ and $\{R_2, G_2, B_2\}$ are the tristimulus values of the two distributions $S_1(\lambda)$ and $S_2(\lambda)$, respectively. It is not necessary that $S_1(\lambda) = S_2(\lambda)$ for all wavelengths in the visible region 380–780 nm. This phenomenon of trichromatic matching is easily explained in terms of the trichromatic theory of color vision. For if all colors are analyzed by the retina and converted into only three different types of responses, the eye will be unable to detect any difference between two stimuli that give the same retinal response, no matter how different they are in spectral composition.

Color matching experiments have some important limitations and characteristics. First, the color matches normally depend only mildly on the conditions of observations and the previous exposures of the eye. Second, although the color matching functions are averaged for people with normal color vision, there are slight differences in the color matches made by different observers. The relative insensitivity of color matches to previous exposure of the eye to white or colored light and to the colors in the surrounding area is an experimentally observed fact and is referred to as the law of the *persistence of color match.* It holds for pre-exposures of moderate to fairly high brightness, but it breaks down for very high brightness. Its validity is important for the practical use of colorimetry. This does not imply, however, that the cognitive perception of the two matched stimuli remains unchanged under different viewing conditions. Indeed both the matching colors may be perceived as different colors than before, while they are still perceived as matching.

One consequence of the trichromacy of color vision is that there are many colors having different spectral distributions that nevertheless have matching tristimulus values. Such colors are called *metamers.* The spectral distributions of two metamers are shown in Fig. 1.8.5. Colors that have identical spectral distributions are called *isomers.* The condition of isomerism is obviously much stricter and not required in most applications.

One case that often arises is that of a light source with spectral distribution $S(\lambda)$, which is viewed after reflecting from surfaces having reflectance factors $\beta_1(\lambda)$ and $\beta_2(\lambda)$, respectively. Then the two surfaces will match in color if

$$\begin{aligned}\int S(\lambda)\beta_1(\lambda)r(\lambda)d\lambda &= \int S(\lambda)\beta_2(\lambda)r(\lambda)d\lambda \\ \int S(\lambda)\beta_1(\lambda)g(\lambda)d\lambda &= \int S(\lambda)\beta_2(\lambda)g(\lambda)d\lambda \\ \int S(\lambda)\beta_1(\lambda)b(\lambda)d\lambda &= \int S(\lambda)\beta_2(\lambda)b(\lambda)d\lambda\end{aligned} \tag{1.8.6}$$

However, this match may be affected if the conditions of observation are changed for example by changing the illuminant from $S(\lambda)$ to $S'(\lambda)$. The amount of mismatch will be a function of the differences between $\beta_1(\lambda)$ and $\beta_2(\lambda)$. If, on the other hand, $\beta_1(\lambda)$ and $\beta_2(\lambda)$ were isomeric, then the match would hold for any source distribution.

1.8.2 Color Coordinate Systems and Transformations

We have seen so far that most colors can be matched by a mixture of three suitably chosen primary colors. Instead of specifying a color by its tristimulus values R,G,B, colorimetrists often use normalized quantities called *chromaticity coordinates*. These are expressed by

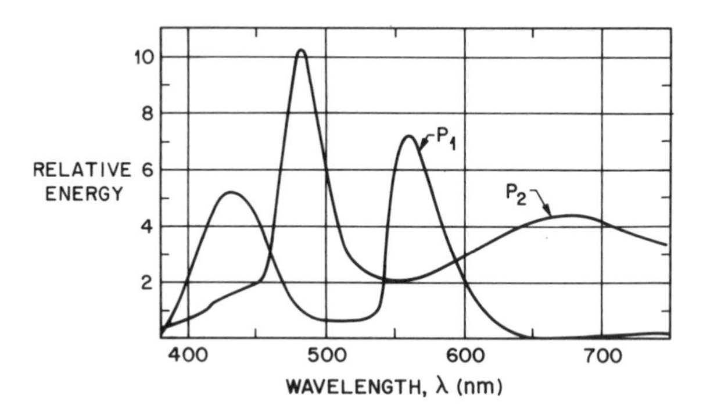

Fig. 1.8.5 Example of two spectral energy distributions P_1 and P_2 that are metameric with respect to each other, i.e., they look the same.

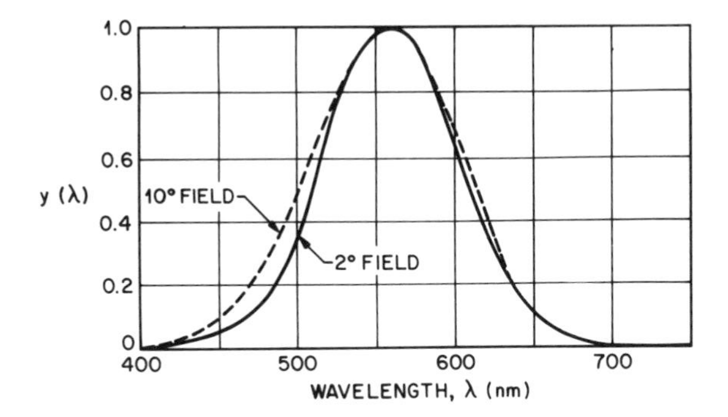

Fig. 1.8.6 The CIE relative luminous efficiency curve $y(\lambda)$ for the Standard Observer, for both 2° and 10° photopic (daylight) vision. (See Appendix for details on how radiometric quantities can be converted to photometric quantities using $y(\lambda)$.)

$$r = \frac{R}{R+G+B}$$

$$g = \frac{G}{R+G+B} \tag{1.8.7}$$

$$b = \frac{B}{R+G+B}$$

Of course, since $r+g+b=1$, two chromaticity coordinates are sufficient. This, however, leaves us with only two pieces of information.

The third dimension of color is called the *luminance* (Y), which may be obtained by a separate match. Luminance is an objective measure of that aspect of visible radiant energy that produces the sensation of *brightness*. Radiation of different wavelengths contributes differently to the sensation of brightness. Careful measurements have been made to determine the relative contribution of monochromatic radiation of a given wavelength to luminance, i.e., the brightness sensation. This relative response is termed the *relative luminous efficiency*, $y(\lambda)$. Since this is obtained by photometric matches (i.e., matching of brightnesses), it is again dependent on the conditions of observations. Fig. 1.8.6 shows the $y(\lambda)$ curve for 2° and 10° fields of view. Both these curves are normalized such that maximum $y(\lambda)$ is taken to be unity. The luminance of any given spectral distribution $S(\lambda)$ is then taken to be

$$Y = k_m \int S(\lambda)\; y(\lambda)\, d\lambda \tag{1.8.8}$$

By expressing k_m as 680 lumens/watt, luminance can be expressed in candelas/meter^2 (see Appendix for more details). As in color matches, a brightness match is observed between two spectral distributions $S_1(\lambda)$ and $S_2(\lambda)$ if

$$\int S_1(\lambda) y(\lambda)\, d\lambda = \int S_2(\lambda) y(\lambda)\, d\lambda \tag{1.8.9}$$

Also, using the definition of the luminance in Eq. (1.8.8), it is easy to see that the luminance of the sum of two spectral distributions is the sum of their luminances.

We will see below that a complete specification of color is given by the luminance and chromaticities. This specification is often used since it is very close to familiar concepts defining perceived color. However, there is another specification of color that is also popular. This was generated by an international body of color scientists called the CIE (Commission Internationale de L'Eclairage).

1.8.2a CIE System of Color Specification

The CIE has established several standards to help precise measurements, as well as cooperation and communication between many practicing colorimetrists. We will consider some of these here. Although a given color can be matched by an additive mixture of three primary colors, when different observers match the same color, there are slight differences in the tristimulus values that they require to effect a match. In 1931, the CIE defined a *standard observer* by averaging the color matching data of a large number of observers having normal color vision. This standard observed data consists of color-matching functions for primary stimuli of wavelengths $700(\boldsymbol{R}_0)$, $546.1(\boldsymbol{G}_0)$ and $435.8(\boldsymbol{B}_0)$ nm, with units normalized in the standard way; i.e., equal amounts of the three primaries are required to match the light from the equal energy illuminant $\boldsymbol{E}$ (see Fig. 1.8.4). Using these curves and given the spectral distribution of any color, one can calculate by Eq. (1.8.4) the tristimulus values required by the standard observer to match that color.

In addition to the standard observer, the CIE defined three new primaries $\boldsymbol{X}, \boldsymbol{Y}$, and $\boldsymbol{Z}$ in which standard-observer results could be expressed. As we shall see later, it is possible to calculate the amounts of $\boldsymbol{X}, \boldsymbol{Y}, \boldsymbol{Z}$ needed to match any color, given its tristimulus values corresponding to any other primaries, such as spectral colors $\boldsymbol{R}_0, \boldsymbol{G}_0, \boldsymbol{B}_0$. In order to do this we need to know the transformation equations relating the two primary systems. For example, the CIE has defined

$$\begin{aligned} \boldsymbol{X} &= 2.365\boldsymbol{R}_0 - 0.515\boldsymbol{G}_0 + 0.005\boldsymbol{B}_0 \\ \boldsymbol{Y} &= -.897\boldsymbol{R}_0 + 1.426\boldsymbol{G}_0 - 0.014\boldsymbol{B}_0 \\ \boldsymbol{Z} &= -0.468\boldsymbol{R}_0 + 0.089\boldsymbol{G}_0 + 1.009\boldsymbol{B}_0 \end{aligned} \tag{1.8.10}$$

By matrix inversion we obtain

$$\begin{aligned} \boldsymbol{R}_0 &= 0.490\,\boldsymbol{X} + 0.177\boldsymbol{Y} \\ \boldsymbol{G}_0 &= 0.310\,\boldsymbol{X} + 0.813\boldsymbol{Y} + 0.01\boldsymbol{Z} \\ \boldsymbol{B}_0 &= 0.200\,\boldsymbol{X} + 0.010\boldsymbol{Y} + 0.990\boldsymbol{Z} \end{aligned} \tag{1.8.11}$$

As in the case of the tristimulus values R, G, B, the tristimulus values X, Y, Z are also normalized to equal energy white. Two properties of the CIE coordinate system make it an interesting and useful choice. First, the $\boldsymbol{Y}$ tristimulus value corresponds to the definition of luminance normalized to equal energy white $\boldsymbol{E}$. That is, the color matching function for $\boldsymbol{Y}$ is proportional to the relative luminous efficiency of Fig. 1.8.6. Also, unlike an $\boldsymbol{R}, \boldsymbol{G}, \boldsymbol{B}$ system, where sometimes certain tristimulus values must be negative for a match, the tristimulus values

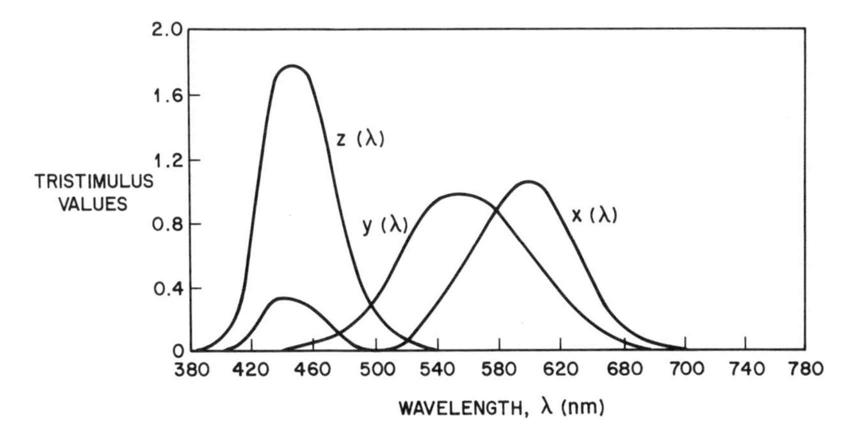

Fig. 1.8.7 Color matching functions $x(\lambda)$, $y(\lambda)$, $z(\lambda)$ for the 2° Standard Observer (from Wintringham [1.8.5]). Note that $y(\lambda)$ is the same as the 2° relative luminous efficiency curve of Fig. 1.8.6.

and the color matching functions in the CIE-XYZ system are always positive, as shown in Fig. 1.8.7. This positivity also makes the $\boldsymbol{X},\boldsymbol{Y},\boldsymbol{Z}$ primaries nonreal or imaginary; that is, they cannot be realized by any actual color stimuli. In the X, Y, Z tristimulus vector space, the primaries are represented by vectors outside the domain representing real colors. The reason for this will be clear in the next section.

Chromaticity coordinates can also be defined in the CIE-XYZ system. For the tristimulus values X,Y,Z the chromaticity coordinates are

$$x = \frac{X}{X+Y+Z}$$

$$y = \frac{Y}{X+Y+Z} \tag{1.8.12}$$

$$z = \frac{Z}{X+Y+Z}$$

Thus, a color can be specified by the two chromaticity coordinates x,y and the luminance Y.

It is useful to indicate qualitative color appearance in terms of this specification. A large value of x indicates a substantial amount of red light and can be matched by colors that appear orange, red, or reddish-purple. If y is large, the color appears green, bluish-green, or yellowish-green. A small value for both x and y indicates that chromaticity z is large, and such a color can be matched by colors appearing blue, violet or purple.

The color specification (x,y,Y) gives the luminance Y plus two chromaticity coordinates (x,y) that can be thought of as the color of the stimulus, devoid of brightness. Chromaticity coordinates can be plotted for the physical colors to form a chromaticity diagram. Two such diagrams are shown for the chromaticity coordinates (r_0,g_0) and (x,y) in Fig. 1.8.8. These chromaticity diagrams also show the chromaticity coordinates of each spectral color, i.e., light of single wavelength in the visible band of frequencies. These pure spectral colors are plotted on the elongated horseshoe-shaped curve called the *spectral locus*.

The straight line connecting the two extremes of the spectral locus is called the *line of purples*. It is worth noting that in the (r_o,g_o) chromaticity diagram the spectral locus extends outside the triangle formed by the three primaries $(\boldsymbol{R}_o,\boldsymbol{G}_o,\boldsymbol{B}_o)$, which are located at (0,0), (1,0) and (0,1). However, in the (x,y) chromaticity diagram the spectral locus is contained completely in the triangle formed by the primaries $\boldsymbol{X},\boldsymbol{Y},\boldsymbol{Z}$. The significance of this will be obvious in our discussion on color mixtures to follow. Chromaticity coordinates of the spectral colors can be plotted individually for each wavelength resulting in Fig. 1.8.9. These chromaticity diagrams are different from the color matching functions (see Fig. 1.8.4) and should not be confused with them. Note that for certain

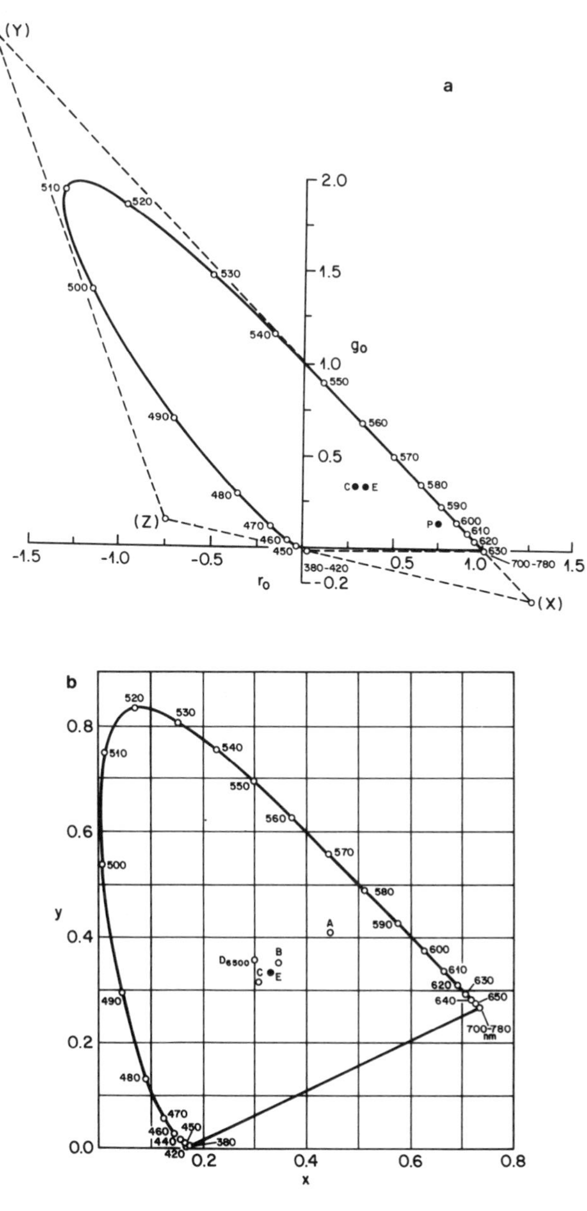

Fig. 1.8.8 (a) The (r_o,g_o) chromaticity diagram for the Standard Observer (from Wintringham [1.8.5]). The wavelengths (in nm) of the spectral colors appear on the horseshoe shaped locus. Point E represents equal-energy white, illuminant C is a standard bluish-white source, P represents a specific color sample irradiated by illuminant C and $(\mathbf{X},\mathbf{Y},\mathbf{Z})$ are the standard CIE primaries. (b) 1931 CIE-xy chromaticity diagram containing the spectral locus, line of purples, and the chromaticity locations A, B, C, D_{6500} and E for CIE standard illuminants (from Wyszecki and Stiles [1.8.3]).

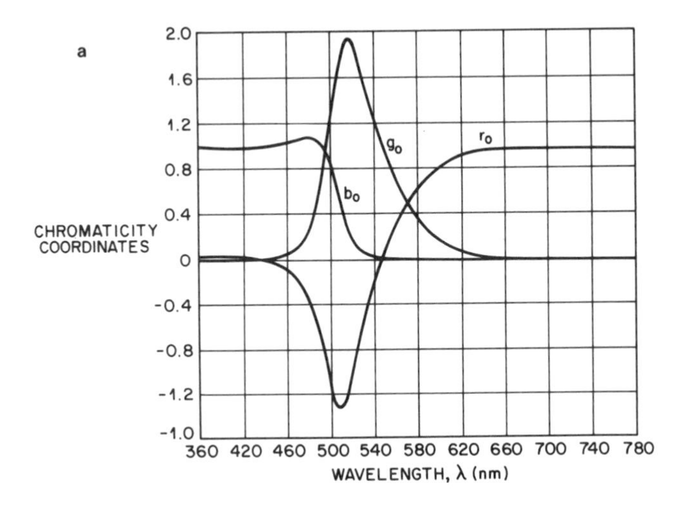

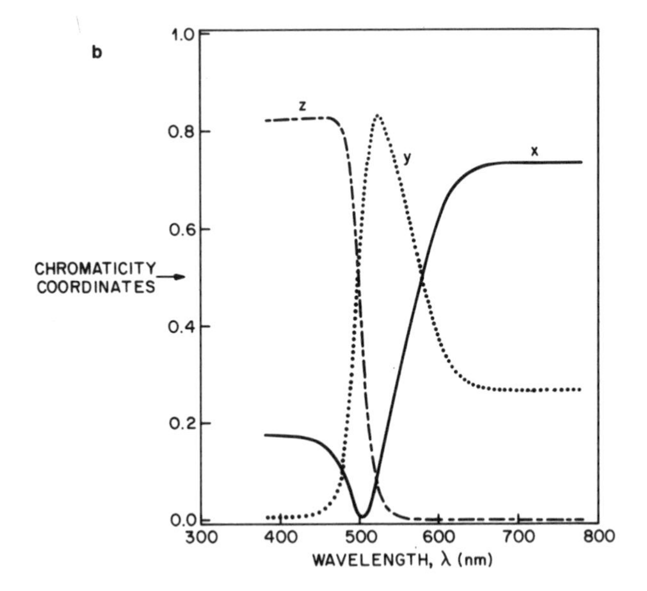

Fig. 1.8.9 (a) Chromaticity coordinates (r_o, g_o, b_o) of spectral colors for the 2° Standard Observer Primaries $\boldsymbol{R}_o$, $\boldsymbol{G}_o$, $\boldsymbol{B}_o$ at wavelengths 700.0 nm, 546.1 nm and 435.8 nm, respectively. Equal-energy white is used as the reference white. (b) Chromaticity coordinates (x,y,z) of spectral colors.

spectral colors (e.g., highly saturated green) the chromaticity coordinate r_o is negative whereas, the chromaticity coordinates (x,y,z) are always positive.

Besides specifying a color coordinate system in terms of primaries $\boldsymbol{X},\boldsymbol{Y},\boldsymbol{Z}$ the CIE has also specified certain sources of light energy that are important in experimental colorimetry. These sources are specified in terms of *color temperatures*. This arises from the fact that as the temperature of a blackbody radiator is increased, it begins to emit visible radiation whose spectral distribution changes from almost red to that containing a larger percentage of shorter wavelengths. Fig. 1.8.10 shows the locus of the chromaticity coordinates as the blackbody temperature is raised from 1000K (°Kelvin) to 10,000K. Several radiant sources on this locus are taken as standard sources by the CIE. Illuminant A represents light from a tungsten lamp, having temperature 2856K. Illuminants B and C are intended to represent direct sunlight (4870K) and average sunlight (6770K), respectively. Both of these are expected to be phased out, and illuminant D_{6500} will be used as a representative of average daylight. It belongs to a series of spectral power distributions that represent different phases of natural daylight (4000K to 25,000K) quite well. D_{6500} has a blackbody temperature 6500K. In addition to standard sources A,B,C,D_{6500}, source E is specified as having a flat spectral distribution and a temperature of 5500K. This is the normalizing reference source used for all the primaries. CIE xy chromaticity coordinates of these standard sources are given in Table 1.8.1

Table 1.8.1 Chromaticities of CIE standard sources.

Source	x	y
A	0.4476	0.4075
B	0.3484	0.3516
C	0.3101	0.3162
D_{6500}	0.3127	0.3291
E	0.3333	0.3333

1.8.2b Color Mixtures

One important objective of colorimetry is to be able to specify the color of a mixture in terms of the components of the mixture. We have already seen that the experimentally observed Grassman's laws state that the tristimulus values of a color mixture are obtained by vector addition of the tristimulus values of the components of the mixture. Thus, if colors $\boldsymbol{S}_1(=X_1,Y_1,Z_1)$ and $\boldsymbol{S}_2(=X_2,Y_2,Z_2)$ are mixed to obtain color $\boldsymbol{S}(=X,Y,Z)$, then

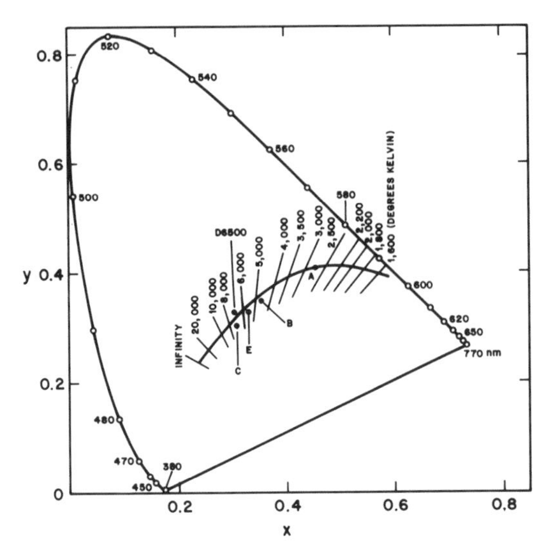

Fig. 1.8.10 Locus of chromaticity coordinates as a function of blackbody temperature. Also shown are light sources A, B, C, D_{6500} and the equal energy white E ($x = 0.33$, $y = 0.33$) (from Judd *et al.* [1.8.6]).

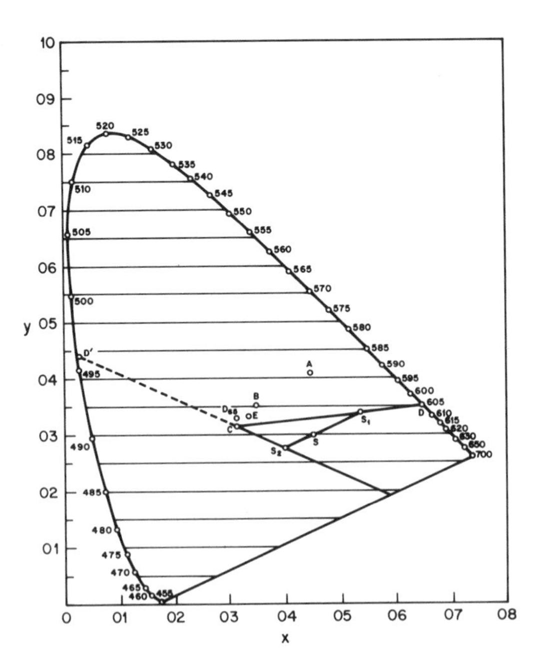

Fig. 1.8.11 Dominant wavelength (point D, 605 nm) for color $\boldsymbol{S}_1$ and complementary wavelength (point D', 496 nm) for color $\boldsymbol{S}_2$. Excitation purity of $\boldsymbol{S}_1$ is given by the ratio (CS_1/CD) when the reference color is CIE illuminant $\boldsymbol{C}$. Color $\boldsymbol{S}$ is a mixture of colors $\boldsymbol{S}_1$ and $\boldsymbol{S}_2$.

$$X = X_1 + X_2$$

$$Y = Y_1 + Y_2 \tag{1.8.13}$$

$$Z = Z_1 + Z_2$$

However, the chromaticities (x,y,z) do not simply add. Thus, we would like to obtain similar equations for the chromaticity-luminance coordinates of the mixture $\boldsymbol{S}(x,y,Y)$ obtained from the colors $\boldsymbol{S}_1(x_1,y_1,Y_1)$ and $\boldsymbol{S}_2(x_2,y_2,Y_2)$. Using the relationship between tristimulus values and chromaticity coordinates, we get

$$X_1 + Y_1 + Z_1 = Y_1/y_1 = X_1/x_1$$

$$X_2 + Y_2 + Z_2 = Y_2/y_2 = X_2/x_2 \tag{1.8.14}$$

Adding both sides and using Eq. (1.8.13)

$$X + Y + Z = Y_1/y_1 + Y_2/y_2$$

Thus,

$$x = \frac{X_1 + X_2}{X + Y + Z} = \frac{x_1(Y_1/y_1) + x_2(Y_2/y_2)}{(Y_1/y_1) + (Y_2/y_2)}$$

$$y = \frac{Y_1 + Y_2}{(Y_1/y_1) + (Y_2/y_2)} \tag{1.8.15}$$

and

$$Y = Y_1 + Y_2 \tag{1.8.16}$$

Similar results can be obtained for other sets of chromaticity coordinates such as (r_o,g_o).

We see that the chromaticity coordinates of the mixture are obtained by a weighted linear sum of the chromaticity coordinates of the components of the mixture. This means that if we mix two colors having chromaticities (x_1,y_1) and (x_2,y_2), then the chromaticity of the mixture will lie on the straight line connecting the points (x_1,y_1) and (x_2,y_2). The exact location on the straight line will be determined by the luminances Y_1 and Y_2. This is shown in Fig. 1.8.11, where the chromaticity of a mixture of colors S_1 and S_2 lies on the straight line $S_1\ S_2$.

Carrying this further, consider the three primary colors $\boldsymbol{R}_o$, $\boldsymbol{G}_o$, $\boldsymbol{B}_o$. We see that the chromaticities of all possible (positive) mixtures of the three primary colors $\boldsymbol{R}_o, \boldsymbol{G}_o, \boldsymbol{B}_o$, whose chromaticity coordinates are plotted in Fig. 1.8.8a at (0,0), (1,0), (0,1), will lie within the triangle having those vertices. We see, therefore, that there is a large range of spectral colors that cannot be synthesized by nonnegative amounts of primaries $\boldsymbol{R}_o, \boldsymbol{G}_o, \boldsymbol{B}_o$. On the other hand, in the (x,y) chromaticity diagram of Fig. 1.8.8b all the spectral colors, including the line of purples, lie within the triangle with vertices (0,0), (1,0) and (0,1) and, therefore, all real colors can be synthesized by positive amounts of the $\boldsymbol{X},\boldsymbol{Y},\boldsymbol{Z}$ primary colors. All three vertices are, however, outside the spectral locus, thus making them nonreal. This illustrates two of our previous statements: (a) Negative amounts of a primary in the $\boldsymbol{R}_o$, $\boldsymbol{G}_o$, $\boldsymbol{B}_o$ system may be required for a color match; (b) All real colors can be matched by positive amounts of primary colors $\boldsymbol{X},\boldsymbol{Y},\boldsymbol{Z}$. However, these primary colors themselves are nonreal.

Another way of representing the chromaticity of a color is by polar coordinates, with some preselected reference normalizing white, e.g., CIE illuminant $\boldsymbol{C}$, as the origin. The two polar coordinates are *Dominant Wavelength* (or Complementary Wavelength) and *Excitation Purity*. The Dominant wavelength of a color is the wavelength of the spectral color that, when additively mixed in suitable proportions with the specified reference white, yields a match with the color being considered. This is shown as point D for the color $\boldsymbol{S}_1$ of Fig. 1.8.11. For colors in the lower right-hand portion of the chromaticity diagram the straight line connecting the color under consideration with the reference white may have to be projected backwards in order to intersect the spectral locus. In such a case, this *complementary wavelength* is specified instead of the dominant wavelength. The complementary wavelength of a test color is the wavelength of the spectral color that when mixed in suitable proportions with the test color yields a match with the specified reference white. Fig. 1.8.11 shows the complementary wavelength (point D') for color $\mathbf{S}_2$. Every color has either a complementary wavelength or a dominant wavelength. Some colors have both.

The radial coordinate of this polar color specification after normalization is called the excitation purity. Excitation purity is the ratio of two lengths (CS_1/CD) on the chromaticity diagram. The first length CS_1 is the distance between the origin (reference white) and the color S_1 being considered; the second length CD is the distance along the same direction from the origin to the spectrum locus D (or the line of purples). Obviously, 100 percent excitation purity usually corresponds to a spectral color.

Qualitatively, the dominant wavelength of a color indicates what part of the spectrum has to be mixed with the reference neutral color to match the given color, and the excitation purity indicates how "far" the color lies from the reference neutral color. Dominant wavelength and excitation purity are related to the common specifications of perceived color, namely *hue* and *saturation* to be treated in Chapter 4. Thus, a dominant wavelength of 600 nm (orange) and purity 50 percent represents a color of orange hue of about half the saturation of spectral orange. In fact, such a color is perceived as more pinkish than this, but not enough to vitiate the usefulness of the designation. Colors of constant dominant wavelength are perceived to have nearly constant hues.

1.8.2c Color Transformations

Many situations arise in the application of colorimetry where it is required to compute the tristimulus values of a color in terms of a second set of primaries when these values are known in terms of a first set of primaries.[1.8.5] Such transformations allow comparison of results obtained with different colorimeters, for example. Also, in color television such transformations are very useful since the spectral characteristics of the camera may be quite different from the primaries used by the television receiver (see Chapter 2). The problem then can be stated as follows:

> Given the tristimulus values R_1, G_1 and B_1 of some specific color in terms of one set of primaries, derive the tristimulus values R_2, G_2, and B_2 of the same color in terms of the second set of primaries.

Obviously, in order to derive the transformation, the second set of primaries and the reference white must be completely specified in terms of the first set of primaries. We assume in the following that the reference white is the same equal energy white E for both sets of primaries. The general case of different reference whites is treated in reference [1.8.5]. As an example, we treat the problem of transforming between the two primary systems $\boldsymbol{R}_o$, $\boldsymbol{G}_o$, $\boldsymbol{B}_o$ and $\boldsymbol{X}$, $\boldsymbol{Y}$, $\boldsymbol{Z}$.

Let us represent the (x,y,z) chromaticity coordinates of primaries $\boldsymbol{R}_o$, $\boldsymbol{G}_o$, $\boldsymbol{B}_o$ by $(x_{R_o}, y_{R_o}, z_{R_o})$, $(x_{G_o}, y_{G_o}, z_{G_o})$ and $(x_{B_o}, y_{B_o}, z_{B_o})$, respectively. Also let

$$\boldsymbol{m} = \begin{bmatrix} x_{R_o} & y_{R_o} & z_{R_o} \\ x_{G_o} & y_{G_o} & z_{G_o} \\ x_{B_o} & y_{B_o} & z_{B_o} \end{bmatrix} \tag{1.8.17}$$

Similarly, let the (X,Y,Z) tristimulus values of the primaries $\boldsymbol{R}_o$, $\boldsymbol{G}_o$, $\boldsymbol{B}_o$ be represented by $(X_{R_o},Y_{R_o},Z_{R_o})$, $(X_{G_o},Y_{G_o},Z_{G_o})$ and $(X_{B_o},Y_{B_o},Z_{B_o})$, respectively. These are given by Eq. (1.8.11). Also, let

$$\boldsymbol{M} = \begin{bmatrix} X_{R_o} & Y_{R_o} & Z_{R_o} \\ X_{G_o} & Y_{G_o} & Z_{G_o} \\ X_{B_o} & Y_{B_o} & Z_{B_o} \end{bmatrix} \tag{1.8.18}$$

By definition of tristimulus values (see Eq. (1.8.11))

$$\begin{aligned} \boldsymbol{R}_o &= X_{R_o}\boldsymbol{X} + Y_{R_o}\boldsymbol{Y} + Z_{R_o}\boldsymbol{Z} \\ \boldsymbol{G}_o &= X_{G_o}\boldsymbol{X} + Y_{G_o}\boldsymbol{Y} + Z_{G_o}\boldsymbol{Z} \\ \boldsymbol{B}_o &= X_{B_o}\boldsymbol{X} + Y_{B_o}\boldsymbol{Y} + Z_{B_o}\boldsymbol{Z} \end{aligned} \tag{1.8.19}$$

or in matrix notation

$$\begin{pmatrix} \boldsymbol{R}_o \\ \boldsymbol{G}_o \\ \boldsymbol{B}_o \end{pmatrix} = \boldsymbol{M} \begin{pmatrix} \boldsymbol{X} \\ \boldsymbol{Y} \\ \boldsymbol{Z} \end{pmatrix} \tag{1.8.20}$$

For an arbitrary color stimulus $\boldsymbol{C}$, having tristimulus values (R_C, G_C, B_C) and (X_C, Y_C, Z_C) with respect to primaries $(\boldsymbol{R}_o, \boldsymbol{G}_o, \boldsymbol{B}_o)$ and $(\boldsymbol{X}, \boldsymbol{Y}, \boldsymbol{Z})$, respectively, the following is true:

$$\boldsymbol{C} = R_C\boldsymbol{R}_o + G_C\boldsymbol{G}_o + B_C\boldsymbol{B}_o = X_C\boldsymbol{X} + Y_C\boldsymbol{Y} + Z_C\boldsymbol{Z} \tag{1.8.21}$$

In matrix notation,

$$(R_C G_C B_C) \begin{pmatrix} \boldsymbol{R}_o \\ \mathbf{G}_o \\ \boldsymbol{B}_o \end{pmatrix} = (X_C Y_C Z_C) \begin{pmatrix} \boldsymbol{X} \\ \boldsymbol{Y} \\ \boldsymbol{Z} \end{pmatrix} \tag{1.8.22}$$

Using (1.8.20),

$$(R_C G_C B_C)\ \boldsymbol{M} \begin{pmatrix} \boldsymbol{X} \\ \boldsymbol{Y} \\ \boldsymbol{Z} \end{pmatrix} = (X_C Y_C Z_C) \begin{pmatrix} \boldsymbol{X} \\ \boldsymbol{Y} \\ \boldsymbol{Z} \end{pmatrix} \tag{1.8.23}$$

Therefore,

$$[R_C G_C B_C]\boldsymbol{M} = [X_C Y_C Z_C] \tag{1.8.24}$$

Thus, the tristimulus values can be transformed between the two primaries by knowing matrix $\boldsymbol{M}$. In particular, the color matching functions corresponding to the two sets of primaries are related by

$$[r_o(\lambda) g_o(\lambda) b_o(\lambda)]\boldsymbol{M} = [x(\lambda) y(\lambda) z(\lambda)] \tag{1.8.25}$$

The chromaticity coordinates of a color $\boldsymbol{C}$ can also be transformed between primary systems. For example, if we define sums of tristimulus values of $\boldsymbol{C}$ by

$$S_X = X_C + Y_C + Z_C \qquad S_R = R_C + G_C + B_C$$

then Eq. (1.8.24) can be written in terms of chromaticity coordinates as

$$S_X [x_C y_C z_C] \, \boldsymbol{M}^{-1} = S_R [r_C g_C b_C] \tag{1.8.26}$$

Rewriting and employing the matrix element summation operator Σ, we get

$$\Sigma \, \{ \, [x_C y_C z_C] \, \boldsymbol{M}^{-1} \} = \Sigma \left\{ \frac{S_R}{S_X} [r_C \; g_C \; b_C] \right\}$$

$$= \frac{S_R}{S_X}$$

Thus,

$$[r_C g_C b_C] = \frac{[x_C \; y_C \; z_C] \, \boldsymbol{M}^{-1}}{\Sigma \, \{ \, [x_C \; y_C \; z_C] \, \boldsymbol{M}^{-1} \}} \tag{1.8.27}$$

which gives the *rgb* chromaticity coordinates in terms of the *xyz* chromaticity coordinates and the transformation matrix $\boldsymbol{M}$. A similar relation can be derived for *xyz* in terms of *rgb*.

In some cases, only the matrix $\boldsymbol{m}$ of chromaticity coordinates relating the two primary systems is given. To handle this case we now derive a relation between $\boldsymbol{M}$ and $\boldsymbol{m}$. Let the sum of the tristimulus values be defined as below:

$$\begin{aligned} S_{R_o} &= X_{R_o} + Y_{R_o} + Z_{R_o} \\ S_{G_o} &= X_{G_o} + Y_{G_o} + Z_{G_o} \\ S_{B_o} &= X_{B_o} + Y_{B_o} + Z_{B_o} \end{aligned} \tag{1.8.28}$$

Substituting the relationships between chromaticity coordinates and tristimulus values (e.g., $x_{R_o} = X_{R_o} / S_{R_o}$) into Eq. (1.8.18), we get

$$M = \begin{bmatrix} x_{R_o} S_{R_o} & y_{R_o} S_{R_o} & z_{R_o} S_{R_o} \\ x_{G_o} S_{G_o} & y_{G_o} S_{G_o} & z_{G_o} S_{G_o} \\ x_{B_o} S_{B_o} & y_{B_o} S_{B_o} & z_{B_o} S_{B_o} \end{bmatrix} \tag{1.8.29}$$

Let $\boldsymbol{S} = \text{Diag}[S_{R_o} S_{G_o} S_{B_o}]$ be a 3×3 diagonal matrix. Then

$$\boldsymbol{M} = \boldsymbol{S}\ \boldsymbol{m} \tag{1.8.30}$$

Thus for an arbitrary color $\boldsymbol{C}$, substituting for $\boldsymbol{M}$ in Eq. (1.8.24)

$$[R_C\ G_C\ B_C]\boldsymbol{Sm} = [X_C\ Y_C\ Z_C] \tag{1.8.31}$$

For unit intensity equal energy white, all the tristimulus values are unity, and therefore we must have

$$\begin{aligned} [1\ 1\ 1]\boldsymbol{Sm} &= [1\ 1\ 1] \\ [S_{R_o}\ S_{G_o}\ S_{B_o}]\boldsymbol{m} &= [1\ 1\ 1] \\ [S_{R_o}\ S_{G_o}\ S_{B_o}] &= [1\ 1\ 1]\boldsymbol{m}^{-1} \end{aligned} \tag{1.8.32}$$

Thus, from Eq. (1.8.30) and the definition of $\boldsymbol{S}$

$$\boldsymbol{M} = Diag\left[[1\ 1\ 1]\boldsymbol{m}^{-1} \right] \boldsymbol{m} \tag{1.8.33}$$

and we have derived the relationship between matrices $\boldsymbol{m}$ and $\boldsymbol{M}$, which can then be used to convert arbitrary colors between the two primary systems.

1.9 Graphics

We have so far been concerned with images with many shades of gray and color. There is yet another class of images wherein each pel has either strictly or predominantly two levels of gray, i.e., binary.[1.9.1] Examples of such images are: business documents, typewritten letters, engineering drawings, newspapers, weather maps, fingerprint cards, etc. There are also many examples in which binary images are derived from gray-level or color images as an intermediate step. For example, a gray-level image containing K bits per pel can be thought of as K binary images, each containing one bit per pel. Also, in frame-to-frame coding, each frame is often partitioned into two segments: moving areas and non-moving background. Each pel is either in the moving area or in the background, and thus the partitioning information constitutes a two-level image.

Two characteristics distinguish such images from the rest. First, in most applications, they are still images (or single frame) rather than a sequence of images rendering motion. Second, since the picture content is mostly text or

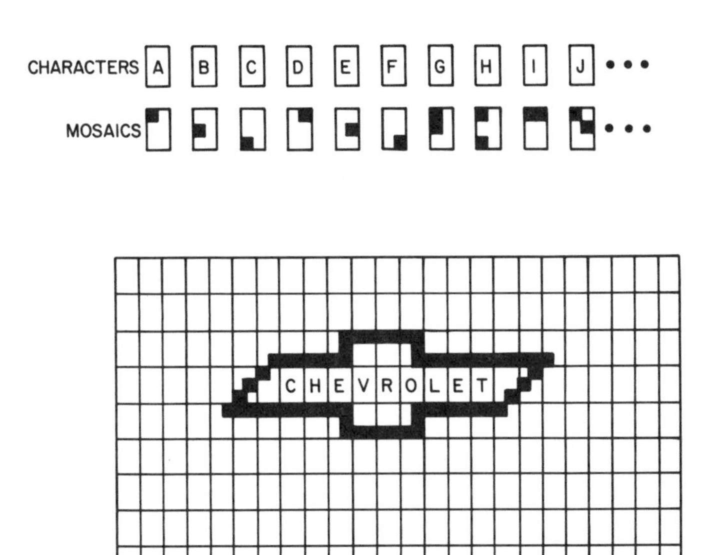

Fig. 1.9.1 Example of characters and mosaics used and a picture constructed from them. Rectangular cells are used.

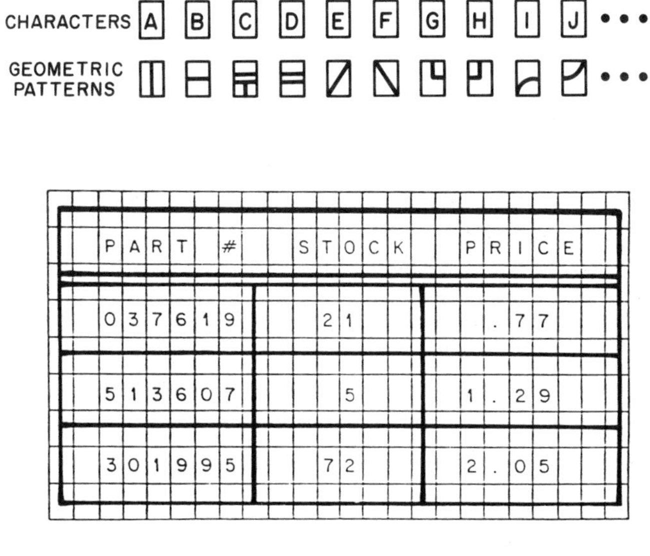

Fig. 1.9.2 An example of characters and geometric patterns used and a picture composed with them. A rectangular grid is also shown.

drawings, for clarity and crispness they require much higher spatial resolution (i.e., number of scan lines or pels per unit distance), as compared to standard television.[1.9.2] Thus, standard television, which was created mostly for entertainment, is adequate for displaying natural scenes but lacks the resolution required for high-quality graphics. Not all *graphics* is two-level; neither are all two-level images called graphics. The term graphics is used quite loosely in the literature. In this section, we will be dealing with representation of two-level still images. Broadly speaking, still images can be divided into three categories.

(a) Character-Mosaic

In this case, a picture is represented (or approximated) by small two-dimensional cells, each cell portraying either an alpha-numeric character or a pattern. Fig. 1.9.1 gives an example of character and mosaic patterns used plus a picture composed with these patterns. It is clear that the quality of the picture depends upon the number and size of the possible cells. If cells are large and therefore contain a large number of pels, then cell to cell resolution is small and therefore picture quality may suffer. Also, if the number of patterns is small, then the flexibility in constructing general pictures is reduced. Thus, economy achieved in storage and transmission, by the use of cells rather than individual pels, comes at the expense of reduced resolution and picture quality. Such representations are of most value in graphic systems where images are mostly alpha-numeric text plus simple patterns containing horizontal and vertical lines.

(b) Character-Geometric

The quality of character-mosaic images can be improved by specifying structured geometric shapes in terms of a set of predefined primitives, such as straight lines, curves of various lengths and shapes, etc. Fig. 1.9.2 shows an example of character and geometric patterns and their use in composing an image. The quality of such pictures is a function of both the type and the number of geometric patterns used. Compared to character-mosaic representation, a much richer set of geometric patterns is used in this representation. In general, at the slight expense of increased storage or transmission, a much better quality image can be obtained compared to character-mosaic representation. Both character-mosaic and character-geometric classes may not be only two-level. In fact, in VIDEOTEX (see Chapter 2) color is often used. However, two-level representation is common in document systems. There is an effort to standardize a set of character, mosaic, and geometric patterns used in VIDEOTEX sources. Details are given in Chapter 2.

(c) Photographic Binary Images

Images are usually specified by their gray level or color at every pel. Since we restrict ourselves to two-level images, photographic binary images are specified by a binary value at each pel. Thus, compared to character-mosaic or character-geometric images, photographic images are specified at each pel rather than cell by cell. Fig. 1.9.3 shows a map of Canada using a pel-by-pel binary representation (Fig. 1.9.3a) and its approximation using character and geometric patterns (Fig. 1.9.3b). A further subclass of two-level photographic images is *line drawings.* Line drawings generally contain interconnections of "thin" lines on a contrasting background. The thickness and color of the lines as well as the color and texture of the background are of little or, at most, symbolic significance. There are many graphics that are often recognizable as line drawings, for example, highway maps and graphs of statistical data. Also there are many images that are perceived as line drawings or can be converted to line drawings. Fig. 1.9.4 gives an example of a map that is readily perceived as a line drawing by following contours, whereas Fig. 1.9.5 gives an example of an image in which the original image (Fig. 1.9.5a) is *thinned* to make it into a line drawing. We will see later that line drawings are quite efficient in conveying information about shapes and offer great economy in storage and transmission. Also, there are many display systems, e.g., vector graphic displays, which allow manipulation and processing of line drawings. In Chapter 5 we shall consider methods of coding line drawings.

Most graphics systems use nonerasable hard-copy printing mechanisms of various types (e.g., impact matrix, ink jet, electrostatic) for output. However, there is a growing tendency to use *soft-copy* output, e.g., cathode ray tubes (CRT), as a result of the low cost of both CRT's and the digital frame buffers needed to refresh them. Other soft-copy output devices, such as plasma panels, are also used. Graphical images displayed on CRT's present different challenges and problems. Legibility of characters, user fatigue, etc., are some of the problems that need to be considered, whereas the benefits include the ability to erase and modify the graphical information.

1.9.1 Scanning and Resolution

One of the major characteristics of graphics information is its high sampling density, i.e., number of scan lines and samples per unit area, compared with pictorial images. The graphical document is first dissected into pels by the process of raster scanning and sampling as discussed in previous sections. The quality of the reproduced graphical image is determined, to a large extent, by the resolution. The vertical resolution is determined by the scanning aperture height and the scan line spacing (called *Pitch*). The horizontal resolution is determined by the scanning aperture width and the sampling distance. The amplitude of the picture signal is converted to two levels, and the inverse operation is performed at the display. However, there is usually no bandlimiting or post-filtering during reconstruction since the display has only two levels. Fig. 1.9.6 shows the effect

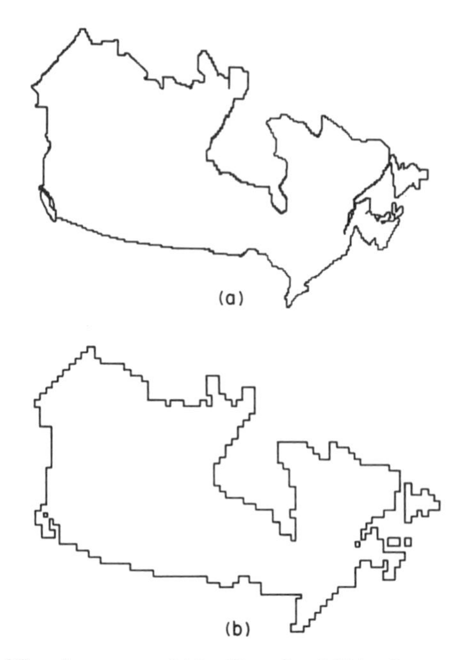

Fig. 1.9.3 Map of Canada represented (a) pel by pel and (b) by character-geometric patterns.

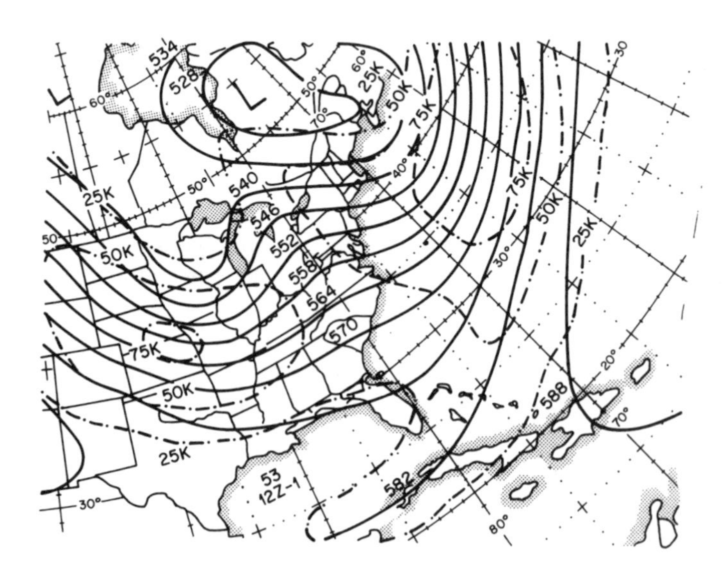

Fig. 1.9.4 Section of a typical weather map containing largely contour information that can be converted into a line drawing (from Huang *et al.* [1.9.7]).

Fig. 1.9.5 (a) Original binary image. (b) Image after thinning. Deleted pels are denoted by small dots.

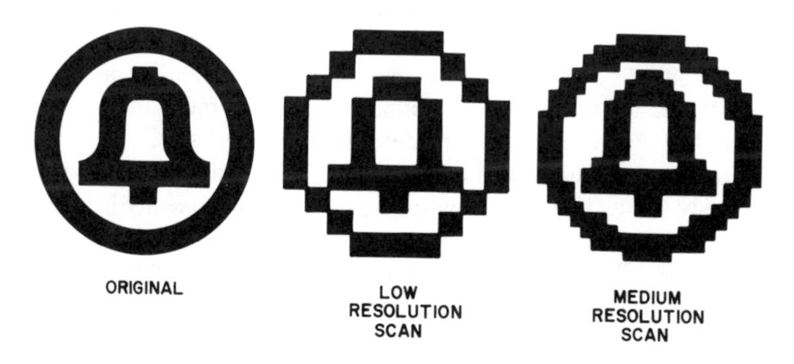

Fig. 1.9.6 Effect of spatial resolution on the image. Lower resolution results in staircase aliasing patterns.

of medium and low resolution scans on the BELL symbol when the reconstruction is accomplished by a rectangular aperture. It is clear that lower resolution has the effect of spatially distorting (i.e. post-aliasing) the shape into a staircase pattern. As a point of reference, standard broadcast television with a 52-centimeter picture [diagonally measured] has about 15 scan lines per centimeter.

There are various methods of specifying resolution. The IEEE facsimile chart (Fig. 1.9.7) contains a variety of test patterns for measuring the resolution of a system. At the top of the chart are vertical bar patterns ranging from 10 to 96 lines per inch. Below these are clusters of triple line groups ranging from 61 to 406 lines per inch in 12 discrete steps. The radial line pattern permits resolution measurement at an angle. It has three concentric circles with 50 scan lines per inch (outer circle) to 200 lines per inch (inner circle). The wedge pattern is calibrated in number of contiguous lines per inch. In addition to these patterns, there is a variety of type faces in the lower left quadrant of the chart. The character height is varied along with the font. This IEEE facsimile chart can be used to measure the resolution of the graphical signal at different angles and in different units.

There are several factors that determine the resolution of the graphical displays. The spot size of the scanning beam is a major factor limiting the resolution. The smaller the spot size, the larger the potential for high resolution. Sometimes the scanning lines of the graphical display system are permitted to overlap by increasing the height of the display spot. This makes the scanning structure invisible, whereas in a television system the display spot size is typically smaller than the pitch, and consequently the scanning structure is clearly visible. In existing facsimile equipment the spot size may vary from 1/20 cm to 1/400 cm. The shape of the scan aperture is generally a rectangle with its longer dimension along the scan line. Also, the scanning direction is the same as in standard TV, i.e., left to right, top to bottom. It has been found that the direction of scanning has very little effect on the resolution of the system.

The resolution requirements for different applications vary. Higher resolution obviously results in a better quality picture and consequently helps the task at hand. However, it puts a significant burden on the system in terms of storage and transmission. Precise studies to determine optimum resolutions required for different applications do not yet exist. Below we give a short summary of many scattered experiments reported in the literature.

1.9.1a Character Legibility

One method of determining the resolution is to measure the legibility of individual characters as distinct from the legibility of words.[1.9.3] In character legibility a random combination of characters with varying heights, widths and containing different numbers of scan lines per character is presented to an observer. The percentage of characters correctly identified is taken to be a measure of document quality. This obviously depends upon many other factors such

Fig. 1.9.7 The IEEE Facsimile Test Chart containing a variety of test patterns used for spatial resolution studies (IEEE Std. 167A-1980, Facsimile Test Chart).

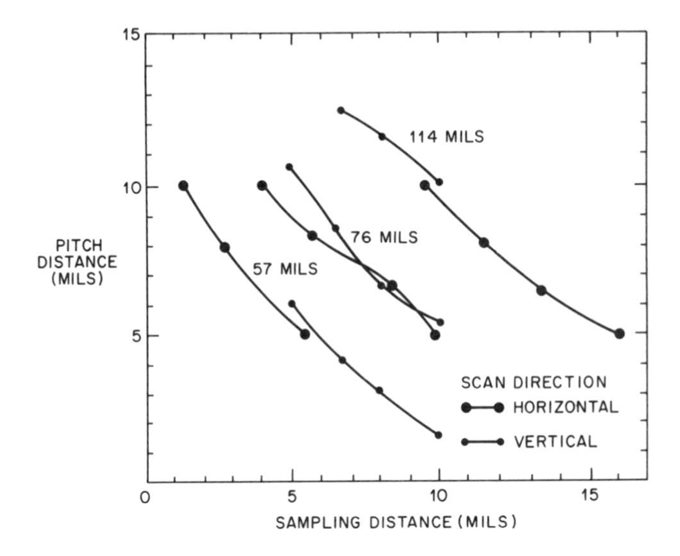

Fig. 1.9.8 Requirements for 97.5 percent legibility of individual characters that are mid-century type and upper case (from Arps *et al.* [1.9.8]).

A. 3885SEKHRN 2GADXZOCVU B. HHRKECOZUV B3S85XDAG2 C. GA2XD5B883 UZCVONKHRE D. UVZOCD2GXA HNREK8538S BLACK AND WHITE COPYING	a. ckoevutlie erxheenjbo b. njobeecvok aitluhexre c. tulalxrseh bjonekcvoa d. cherxojbne vcoakiutia he has a rolling gait
A. Z8HRNEB3KS OCVXDUGA25 B. BEKS3RH8ZN 2A5GUDVOCX C. A2G5UDXOCV HNZR8KBS3E D. CVOXD5G2AU 3KBSE8NHZR TAKE HOLD OF THE ROPE	a. ekloecvtbh arnejoxuis b. njreasuixo kleeotvhbc c. bvtchranje xsoiuhkaeo d. iuxosoaike hbtvcejnar he was a determined man
a. xoanjersou kvtbhacile b. lceiajnxao uersobhtvk c. esruotbkvh aleicoxajn d. vbthkclaie njoaxrseou his mouth dropped open	A. HCVS8EOBZ3 DRNX2GAU5K B. DXR2NK5AUG VCSH8OZB3E C. BZO3ENX2RD A5GKUS8HVC D. 5GUAK8CVHS ZO3EBXND2R THINK FOR YOURSELF

Fig. 1.9.9 A typical test document used for individual character legibility study (from Arps *et al.* [1.9.8]).

as contrast, viewing conditions, amount of time spent in recognizing a character and, of course, the font. Also, such measurements do not take into account the user fatigue resulting from prolonged viewing of documents. The results of a typical character legibility study are shown in Fig. 1.9.8 for the mid-century type of characters shown in Fig. 1.9.9. Fig. 1.9.8 shows the sampling and pitch resolution combination that gives 97.5% character legibility for upper case characters of three heights (57, 76, 114 mils). We see that for 97.5% character legibility a resolution of about 110×110 pels to the inch with each pel 7 mils square would be required for upper case letters with 100 mils in height. In general, legibility depends on pitch, sampling frequency, character height and stroke width. Upper case characters are more legible than the lower case. While such studies are adequate for individual character legibility, similar results for word recognition indicate a much lower required resolution. Fig. 1.9.10 shows the scan resolutions required for various minimal character heights, based on 5 lines per character to identify words and 10 lines to identify individual characters. For elite type, we see that 16 lines per centimeter would be sufficient for word identification whereas for character recognition this number must be doubled to 32 lines per centimeter (see Fig. 1.9.7).

The resolution requirement for handwriting is generally taken to be between 32 and 40 lines per centimeter for a variety of signature identification systems, whereas for fingerprints 80 lines per centimeter is generally used. It is an accepted practice to use about 40 lines per centimeter for the majority of weather maps. For Chinese characters with Mincho font a resolution of 24×24 pels per character is considered an absolute minimum. However, 32×32 pel resolution is normally used for characters of size 4mm×4mm.

Character legibility experiments performed on a cathode ray tube display indicate slightly different results. Characters displayed on CRT's are usually synthetically generated on a matrix of dots, where each dot position can be on or off. In character-oriented displays, coded characters (e.g., ASCII codes) are stored and fed to a character generator that generates an electrical waveform that drives the scanning beam of the display. Bit-map displays, on the other hand, use a frame buffer that contains binary information for each pel. Because of higher flexibility and the potential to produce a full gamut of pictures, bit map displays are gaining in use, despite their higher cost. Dot-matrix representation of characters is used with either type of display, and compared to raster scanned characters, much lower resolution per character is adequate.

Many studies have been made to optimize the parameters of dot-matrix displays.[1.9.4] Since dot-matrix displays are used extensively for a variety of applications, other criteria (e.g., random search and menu search) besides character legibility are used as measures of quality. Random search and menu search tasks are representative of situations in which the observer must locate a symbol or a group of symbols that are unrelated to other displayed information. This type of task is facilitated by characters that are much larger, have greater dot

MIN. CHARACTER HEIGHT	APPROX. EQUIVALENT TYPE SIZE	MIN SCAN LINES PER INCH (LP)	
		1.-IDENTIFY WORDS	2.-IDENTIFY INDIV. CHARACTERS
$\frac{1''}{64}$	6-7 pt.	107	214
$\frac{1''}{16}$	8-10 pt.	80	160
$\frac{3''}{32}$	11-12 pt.	53	106
$\frac{1''}{8}$	14-18 pt.	40	80
$\frac{5''}{32}$	20-22 pt.	32	64

Fig. 1.9.10 Scan resolutions required for various minimum character heights, based on five lines per character to identify words and 10 lines to identify individual characters.

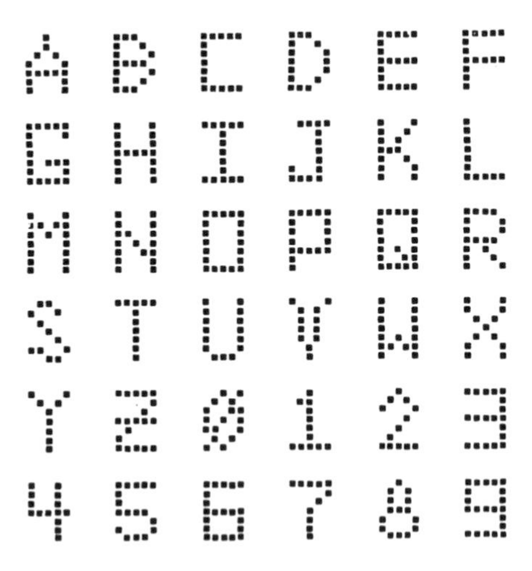

Fig. 1.9.11 An example of a dot-matrix character set in a 5×7 Lincoln/Mitre Font.

modulation (i.e., contrast ratio between lighted and dark dot) and in color. The following are conclusions from several experiments.

For single character recognition in a known display location, no further improvements are obtained beyond a character size of 11 minutes of arc (i.e., angle subtended at the eye) when the dot modulation of characters is high. If the modulation is reduced (e.g., 40%), then larger characters (e.g., 17 minutes of arc) are needed. Similar experiments for raster scan characters suggest characters as large as 36 minutes of arc. Smaller dot sizes (e.g., 0.76 mm) are better for reading contextual material, while larger dots (and therefore larger characters) are best for search tasks. Square dots are better than either circular or elongated dots for reading, search, and recognition. Recognition errors and search time decrease as the interdot spacing is decreased. A dot spacing/size ratio of 0.5 is superior to one of 1.0 or 1.5. Thus, the closer a dot-matrix character approximates a continuous stroke character, the better will be the observer's performance in recognition. Dot modulation above 80% is considered adequate. The font given in Fig. 1.9.11, called the Lincoln/Mitre font, is preferred over many other fonts. A 7×9 or 7×11 size matrix is required when both lower and upper case characters, symbols, sub- and superscripts need to be displayed, although a 5×7 size matrix is used quite widely (mostly for upper case letters). Based on the above, parameters of a good quality dot-matrix display are given in Fig. 1.9.12.

Most of the above discussion on character legibility for scanned printed characters, as well as dot-matrix type of characters on CRT's, are for nominal two-level pictures. For multilevel or color characters, comparable results are not yet known. Although color characters add to the aesthetic value of the pictures and are easy to locate among non-colored characters, it is not yet known whether color can be exploited to improve character legibility. Another advantage of characters with multiple shades of gray or color is that they tend to show less interline flicker on interlaced displays. This is due to the fact that gray level characters have fewer and more gradual changes in intensity in adjacent scan lines (i.e., lines of adjacent fields) compared with dot-matrix characters. Also visual effects of spatial distortion, which are evident as jaggedness at the edges, can be reduced by post-filtering and using gray levels or colors. This is referred to as *anti-aliasing* in the computer graphics literature.

1.9.2 Amplitude Quantization

We have seen the effect of spatial resolution on the quality of graphical documents. In this section, we look at the amplitude resolution appropriate for graphical documents. Studies have shown that for higher quality documents (clean originals), it is better to have a higher spatial resolution and only two shades of gray, i.e. black and white. However, for original documents that are not so clean, lower resolution and more shades of gray appear to give better quality reproduced picture.

VARIABLE	CONTEXTUAL DISPLAY	NONCONTEXTUAL DISPLAY
DOT SIZE	0.75 mm	1.2 TO 1.5 mm
DOT SHAPE	SQUARE	SQUARE
DOT SPACING/SIZE RATIO	≤0.5	≤0.5
MATRIX SIZE*	7×9	9×11
CHARACTER SIZE	16 TO 25 ARCMINUTES	1.0 TO 1.2 ARCDEGREE
DOT LUMINANCE	≥20 cd/m^2	≥30 cd/m^2
DOT MODULATION	≥75%	≥90%
AMBIENT ILLUMINANCE	≤125 LUX**	≤75 LUX
FONT*	LINCOLN/MITRE	LINCOLN/MITRE

*NUMERALS AND UPPER CASE LETTERS ONLY.
**LUX IS A UNIT OF ILLUMINANCE AND IS EXPLAINED IN APPENDIX A.

Fig. 1.9.12 Recommended design parameters for a good quality dot-matrix display.

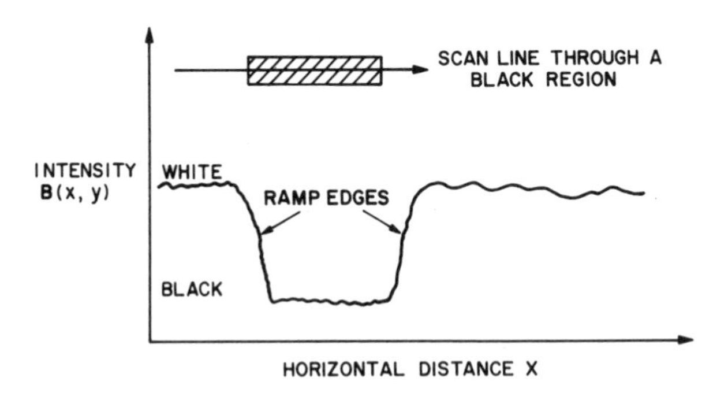

Fig. 1.9.13 Typical output from a facsimile scanner at a white-to-black transition.

In those cases where only two levels of amplitude quantization are used, the two levels are obtained normally by simply thresholding the analog voltage output of a document scanner. If $B(x,y)$ is the intensity at position (x,y), then thresholding approximates it by one of two levels B_b (for black) and B_w (for white). That is, the coded value $\tilde{B}(x,y)$ is given by

$$\tilde{B}(x,y) = \begin{cases} B_b & \text{if } B(x,y) < T(x,y) \\ B_w & \text{otherwise} \end{cases} \tag{1.9.1}$$

where $T(x,y)$ is, in general, a spatially varying threshold that is applied at each position (x,y). The performance of the thresholding operation depends obviously on the choice of the threshold function T. Although the problem of thresholding has been recognized for quite some time, it still remains something of an art.

A constant threshold can be applied globally to an image, or a variable threshold can be applied locally depending upon the local characteristics of the signal. If a fixed threshold is used throughout the document, then it does not accommodate many of the problems associated with poor quality originals, e.g., bad lighting, shading, etc., and it is frequently unsatisfactory. A major artifact associated with fixed thresholding is the amplification of low noise levels that occur at gray levels near the threshold value. Such noise gives rise to *false contours* that appear as black and white boundaries. Fixed thresholding combined with a low sampling rate also creates aliasing or *moiré* patterns. A good variable threshold technique is therefore desirable since it is much more effective in dealing with documents that may be impaired by low print contrast, imperfect characters, paper speckle and shading of the document. Also, proper thresholding should preserve as much as possible large areas of constant intensity, so that efficient coding can be performed. Also, it should make the task of character recognition based on two-level images much easier.

Fig. 1.9.13 shows the typical output of a document scanner at a white/black/white transition. It is clear that although in the white background the image intensity is approximately constant, it does vary somewhat as a result of aforementioned problems. The threshold can be applied by partitioning an image into nonoverlapping windows, each having its own separate threshold. Within each window, the distribution of amplitudes is computed to determine the threshold by using one of the techniques discussed below. The size of the window must be selected with care since too small a window would result in an unreliable amplitude distribution, while too large a window would reduce the threshold sensitivity to local intensity variations. Typically, square blocks containing 5 to 10 scan lines and pels are used as windows. Another concern that arises from the use of different thresholds is the possibility that a drastic change in threshold from one window to the next would cause distortions to appear for curves that cross several windows. Such effects can be decreased by defining the threshold for each window as a running average of the thresholds of neighboring windows.

1.9.2a Global Threshold Selection

One effective way of selecting the threshold is to examine the histogram of the gray-level occurrences of the image intensity. For a typical black and white document this histogram (Fig. 1.9.14) indicates the presence of a large number of white pels belonging to the background and a small number of black pels belonging to the graphics information. The occurrence of pels having intermediate gray levels is due to the paper noise or to the gray level of pels near black/white edges. When a large separation in intensity exists between the white and black areas (e.g., for clean originals) a single gray level selected in the valley between the peaks can be used as a suitable threshold value. The selected threshold value can be either the gray level of the minimum of the histogram between peaks or the average of the gray levels of the two peaks. It is desirable that the gray level separation between the two peaks be of the order of the contrast difference between the white and black areas. More sophisticated techniques have been developed based on the histograms observed for certain classes of documents.[1.9.1]

1.9.2b Local Threshold Selection

A large number of techniques exist that use the local gray level differences (or intensity gradients) for adjusting the threshold. Two of these are described below. A technique called *notchless* quantization adjusts the threshold $T(x,y)$ at each pel by raising or lowering it by a fixed amount depending upon the vertical and horizontal intensity differences at that pel. When a positive intensity difference above a certain magnitude is detected in either the vertical or horizontal direction, the threshold at that point is decreased by a predetermined number. Conversely, when a negative difference is encountered, the threshold is raised by the same predetermined number. The nominal value of the threshold is generally determined by conventional methods. The adjustment of the threshold in this manner decreases the effects of minor fluctuations in intensity over a character. Gaps in vertical or horizontal edges as well as pin holes in a character are also minimized. This technique does not, however, deal with the problems of large area shading if a constant nominal threshold is used.

Another technique allows the threshold to change continuously (instead of by finite increments), according to the value of intensity gradient at a pel. The gradient at each pel is computed as

$$g(x,y) = \frac{1}{2}\left[\Delta_x B(x,y) + \Delta_y B(x,y)\right] \tag{1.9.2}$$

where Δ_x and Δ_y are first order finite difference operators in the x and y directions, respectively. The threshold $T(x,y)$ is then taken as

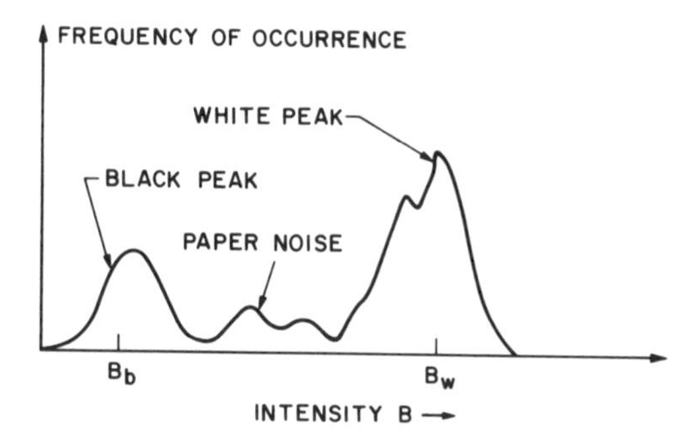

Fig. 1.9.14 Histogram of gray-levels from a black and white document. B_w and B_b are the average values of white and black pels, respectively. Even for a strictly binary document, noise and the scanning process may give rise to a variety of in-between gray levels.

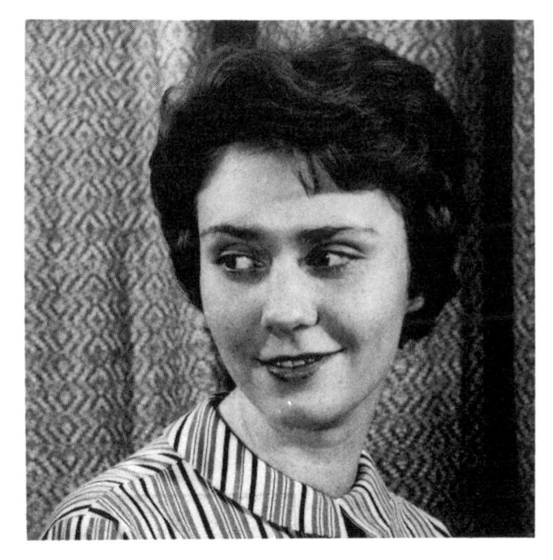

Fig. 1.9.15 An original 512×512 photograph where each pel is specified by 8 bits (from Jarvis *et al.* [1.9.5]).

$$T(x,y) = m \cdot g(x,y) + t_o \qquad (1.9.3)$$

where t_o is the nominal threshold. The quantity m is generally taken to be -1, and its function is to bias the decision so that a pel having a high negative intensity gradient would tend to be set to black. As in the previous technique, the nominal threshold t_o must be adjusted to handle shading problems. This technique, by being sensitive to edges, will work with low contrast images and is also capable of producing a cartoon-like black and white rendition of multilevel pictorial images.

In addition to these two techniques, there is much work on edge extraction that is relevant to this problem. Many of the edge extraction techniques, however, appear to be complex and have not yet found wide usage in document transmission/storage systems.

1.9.3 Multilevel Pictures on Two-Level Display

In the previous section, we considered methods of quantizing a predominantly two level black and white image into a binary signal by thresholding. These are important for display media that are basically bi-level in nature, e.g., liquid crystal displays, plasma panels, microfilm and ink on paper. Sometimes, however, it is important to present a continuous tone (i.e., gray level) image on these bi-level displays. In this section we consider techniques for processing a gray level picture so that when it is displayed on a bi-level display, it will appear to have multiple tones. These techniques are based on proper control of the spatial density of the bi-level display sites (or pels) so that the subjective effect of multiple tones can be obtained. This is done by transferring the artifacts associated with a two-level display to high spatial frequencies, and since high frequencies are less visible to the eye (see Chapter 4) the pictures look better.

There are many processing techniques[1.9.5,1.9.6] that give a bi-level representation that appears to have multiple shades of gray. We will consider two promising techniques because of their simplicity. We assume that the displayed intensity $\tilde{B}(x,y)$, corresponding to input intensity $B(x,y)$, can have values $\{B_w, B_b\}$ corresponding to the two display conditions *on* or *off*. The processing algorithms then give a mapping between $B(x,y)$ and $\tilde{B}(x,y)$.

1.9.3a Ordered Dither

The ordered dither technique creates a bi-level representation by comparing the image intensity $B(x,y)$ to a position dependent threshold that is one of the elements of a $n \times n$ dither matrix $\boldsymbol{D}^n$. The particular matrix element $\boldsymbol{D}^n(i,j)$ to be used as a threshold depends only on the coordinates x,y:

$$i = y \text{ (modulo } n)$$
$$j = x \text{ (modulo } n) \tag{1.9.4}$$

The threshold* at (x,y) is then given by $T(x,y) = \boldsymbol{D}^n(i,j)$, and the decision of whether or not to turn on the cell at position (x,y) is governed by:

$$\tilde{B}(x,y) = \begin{cases} B_w & , \text{ if } B(x,y) > D^n(i,j) \\ B_b & , \quad \text{otherwise} \end{cases} \tag{1.9.5}$$

Thus, we see from Eq. (1.9.4) that the dither threshold at each pel in the image is obtained by repeating the dither matrix in a checkerboard fashion over the entire image. The quality of the picture obviously depends upon the content of the dither matrix, its size and, most importantly, on the resolution of the display. Dither matrices of various sizes can be synthesized, for example, by the following recursive relationship:

$$\boldsymbol{D}^n = \begin{bmatrix} 4\boldsymbol{D}^{n/2} + \boldsymbol{D}^2(0,0)\,\boldsymbol{U}^{n/2} & 4\boldsymbol{D}^{n/2} + \boldsymbol{D}^2(1,0)\,\boldsymbol{U}^{n/2} \\ 4\boldsymbol{D}^{n/2} + \boldsymbol{D}^2(0,1)\,\boldsymbol{U}^{n/2} & 4\boldsymbol{D}^{n/2} + \boldsymbol{D}^2(1,1)\,\boldsymbol{U}^{n/2} \end{bmatrix} \tag{1.9.6}$$

where

$$\boldsymbol{D}^2 = \begin{bmatrix} 0 & 2 \\ 3 & 1 \end{bmatrix} \tag{1.9.7}$$

$$\boldsymbol{U}^n = \begin{bmatrix} 1 & 1 & \cdots & 1 \\ 1 & 1 & & \\ \cdot & \cdot & & \\ \cdot & \cdot & & \\ 1 & 1 & \cdots & 1 \end{bmatrix} \tag{1.9.8}$$

---

* In practice, the dither matrix values need to be scaled properly to obtain the threshold values. For example, a 2×2 dither matrix Eq. (1.9.7) uses 4 equally spaced thresholds of {32, 96, 160, 224} corresponding to the matrix values {0, 1, 2, 3} for an 8 bit input.

and $D^n(i,j)$ is the (i,j)th elment of the dither matrix D^n. As an example, a 4×4 dither matrix generated in this manner is given by

$$D^4 = \begin{bmatrix} 0 & 8 & 2 & 10 \\ 12 & 4 & 14 & 6 \\ 3 & 11 & 1 & 9 \\ 15 & 7 & 13 & 5 \end{bmatrix} \tag{1.9.9}$$

The corresponding matrix of threshold values after proper scaling for an 8 bit input would be

$$\begin{bmatrix} 8 & 136 & 40 & 168 \\ 200 & 72 & 232 & 104 \\ 56 & 184 & 24 & 152 \\ 248 & 120 & 216 & 88 \end{bmatrix} \tag{1.9.10}$$

A preliminary goal in the design of dither matrices is to provide a sufficient number of distinct dither thresholds so that false contouring is minimized and to distribute them spatially so that the spectral energy of the output dots is as high as possible. This implies that similar thresholds should be separated from each other by as large a distance as possible. With these goals in mind, several studies have been performed to optimize the matrices. Eq. (1.9.6) shows a matrix with some interesting properties. First, (n^2+1) shades of gray can be produced by a matrix D^n since elements include each integer value from 0 to (n^2-1) exactly once. The displayed image does not necessarily lose spatial resolution as n is increased, even though the intensity resolution does increase. Fig. 1.9.15 shows an original 8-bit per pel image of a head and shoulders view, and Fig. 1.9.16 shows a dithered image using an 8×8 dither matrix. Dither matrices smaller than 8×8 tend to show contouring, while larger ones (e.g., D^{16}) do not generate a noticeably different output than the matrix D^8.

Implementation of ordered dither requires memory to store the dither matrix as well as a comparator. No additional memory is needed for image data. Since the size of the dither matrix is in powers of two, the modulo operation of Eq. (1.9.4) is performed simply by taking the low order bits of the x and y positions.

Fig. 1.9.16 Dithered image from the original picture of Fig. 1.9.15. A 8×8 ordered dither matrix is used (from Jarvis *et al.* [1.9.5]).

Fig. 1.9.17 Image of Fig. 1.9.15 processed by the error diffusion technique. Matrix $\{\alpha_{ij}\}$ of Eq. (1.9.13) is used.

1.9.3b Error Diffusion Technique

This technique is also called the minimized average error scheme. It attempts to choose the pattern of lighted display cells in such a way as to minimize the average error between the input $B(x, y)$ and the displayed intensity $\tilde{B}(x, y)$. If the error $E(x, y)$ is defined by

$$E(x, y) = B(x, y) - \tilde{B}(x, y) \tag{1.9.11}$$

then a modified value of input intensity $B'(x, y)$ can be computed from the previous (in the sense of scanning direction) errors and current pel intensity as

$$B'(x, y) = B(x, y) + \frac{\sum \alpha_{ij} E(x_j, y_i)}{\sum \alpha_{ij}} \tag{1.9.12}$$

Coefficients $\{\alpha_{ij}\}$ define the relative contributions of the previous errors to the corrected intensity. Indices $\{x_j, y_i\}$ vary over a small neighborhood above and to the left of (x, y). One typical configuration of matrix $\{\alpha_{ij}\}$ is

$$\{\alpha_{ij}\} = \begin{bmatrix} 1 & 3 & 5 & 3 & 1 \\ 3 & 5 & 7 & 5 & 3 \\ 5 & 7 & * & & \end{bmatrix} \tag{1.9.13}$$

where * is the current pel location. As in the dither threshold case, the corrected intensity is compared to a threshold $T(x, y)$ to determine whether the pel at (x, y) should be turned on. Thus,

$$\tilde{B}(x, y) = \begin{cases} B_w , & \text{if } B'(x, y) > T(x, y) \\ B_b , & \text{otherwise} \end{cases} \tag{1.9.14}$$

Typically $T(x, y)$ is a constant, independent of (x, y). Often the error Eq. (1.9.11) is modified using the corrected intensity $B'(x, y)$ values.

The quality of the pictures generated by this technique depends to some extent on the characteristics of matrix $\{\alpha_{ij}\}$. In most cases, microscopic patterns are observed in the resulting image that are not contained in the original image. However, the rendition of edges and small detail is excellent. Fig. 1.9.17 shows a processed image demonstrating these effects. It is clear that compared to ordered dither, the amount of processing is significantly increased with some increase of quality in certain areas of the picture.

1.9.4 Multilevel Halftoning of Color Images

In section 1.8, we saw that a high quality representation of color images requires 8 bits using PCM quantization of red, green and blue components.

Thus, at each pel, 24 bits of amplitude quantization is required, allowing specification of 2^{24} = 16777216 distinguishable colors. Many imaging systems use a frame-buffer that refreshes the color CRT. For a system requiring 1000×1000 spatial resolution, the size of the frame buffer would be 24 Mbits. The cost of such a high speed memory to support display of 2^{24} colors may be too high for some applications. This cost can be decreased by allowing only a limited number of colors at each pel, (e.g., 6 bits per pel giving 2^6 = 64 colors/pel). However, since natural images normally contain a large number of distinguishable colors, displaying such images with a limited number of colors produces the usual quantization artifacts, e.g., false colors and contours. Such artifacts can be reduced by first choosing the *best* set of colors (called color palette) to map the input image colors into, and then associating each pel of the image with a color from this palette to yield the best quality image. Thus, the problem can be separated into two steps: a) Design of the color palette, b) Mapping pixels into the color palette. These two steps are described in detail below.

1.9.4a Color Palette Design

A color palette can be chosen independent of the statistical distribution of colors in an image by simply dividing the space of possible colors into 2^m (=M) representative colors by simple space partitioning. Within each of the 2^m partitions, the center point of that partition is chosen as the representative color. Cubic partitions of three dimensional color space are the simplest to use. Unfortunately, since most images do not contain a uniform distribution of colors (see section 3.3), many partitions are either empty or sparse, thereby wasting valuable representative colors. This can be avoided by taking into account the statistics of the image. In such a situation, a preprocessing step first computes the color palette. This palette is then associated with the image for any further operations such as storage, transmission or display. One method[1.9.9, 1.9.10] of using the statistics, is to construct an initial palette by centering colors at peaks of the histogram of color values. To avoid concentrating too many colors at one peak, the histogram values are reduced in a region surrounding each assigned color after the color is selected. Using this initial palette, one can design an iterative process (similar to that of the next section) to refine the palette so as to converge to the palette that minimizes the mean square error over the distribution of colors.

A standard method [1.9.11] of iteratively designing the color palette for minimizing the mean-square error, called the LBG algorithm appears in many other contexts as well (see section 5.5.1). Let $\{\mathbf{C}_n\}$, n=1,...,N ($N \leq 2^{24}$) be the set of possible colors in an image, and $\{\mathbf{P}_n\}$ n=1,...,M ($M \ll N$) be the colors in a palette. Any representation of colors may be used, but as we will see later, color coordinates that are better matched perceptually (see section 4.4) lead to better image quality. The LBG algorithm proceeds as follows:

Step 1 : Choose an initial color palette $\{\mathbf{P}_n\}^1_{n=1...M}$

Step 2 : Form M clusters, $\{CL_i\}_{i=1,...,M}$ using the rule

$$\{CL_k\} = \{\mathbf{C}_j : ||\mathbf{C}_j - \mathbf{P}_k|| \leq || \mathbf{C}_j - \mathbf{P}_l||, l=1,...,M\ j=1,...,N\}$$

Thus in this step each color from the input colors is assigned to the *closest* representative color of the palette and a new set of clusters is created. $||\cdot||$ above denotes distance between two colors. The simplest distance is the Euclidean distance for the 3-component color vector. However, perceptually matched distance measures of section 4.4 give better picture quality.[1.9.12]

Step 3 : Revise the color palette using

$$\mathbf{P}^2_n = \sum_{C_i \in CL_n} \mathbf{C}_i \,/\, |CL_n| \qquad n=1,...,M$$

Where $|CL_n|$ is the number of pels in cluster CL_n.

Step 4 : Iterate between steps 2 and 3 until the difference between $\mathbf{P}^k_n$ and $\mathbf{P}^{k+1}_n$ becomes small.

Each iteration of the above algorithm causes the following mean square error to either decrease or remain the same,

$$Mean\ Square\ Error = \sum_{k=1,...,M} \sum_{C_i \in CL_k} || C_i - P_k||^2 \qquad (1.9.15)$$

Therefore, the algorithm is guaranteed to converge to a minimum. Unfortunately, in most such problems, there can be many local minima, and the algorithm converges to one of them depending on the initial color palette. Many techniques are described in the literature to overcome this including the one in which several different initial palettes are used. However, computational cost then becomes an issue. A more structured palette (e.g., tree structured) can be used to reduce computations substantially[1.9.13].

1.9.4b Pel Mapping

Having selected a color palette, the color of each pel can be mapped into the representative color of the palette that minimizes the color distance. Such a strategy for pel mapping minimizes the mean square error. However, for palettes of small size (i.e., small M) the usual quantization distortions such as false contours and colors are seen. These can be reduced by different pel mapping techniques that match the local averages of the input and the displayed color, analogous to the case of monochrome pictures discussed in section 1.9.3. In particular the techniques of ordered dither and error diffusion described in that section are directly applicable, and, as before, they reduce the error power at lower frequencies while increasing the error power at high frequencies where the artifacts are less visible.

As explained in section 1.9.3, ordered dither can be thought of as adding* a pseudo-random noise pattern $\{n_{ij}\}_{i,j=1,...,L}$ to the pels before quantizing. Thus, input color C_{ij} at location (ij) is modified

$$\mathbf{C}_{ij} = \mathbf{C}_{ij} + \mathbf{N}_{i\ modL,j\ modL} \tag{1.9.16}$$

As before, the dither matrix $\{\mathbf{N}_{ij}\}_{i,j=1,...,L}$ is chosen such that its mean value is approximately equal to the average value of the pel colors. Unlike in the monochrome case, here each N_{ij} is a 3-component vector. For simplicity, each color component can be dealt with separately, thereby, treating the above equation as 3 scalar equations, one for each color component. The amplitudes of the dither matrix are chosen such that any area of constant color will get quantized into a variety of colors breaking the false contours, but averaging to the input color constant. If the color palette is designed using the LBG algorithm on the image, the distance between the nearby representative colors of the palette could vary dramatically. Therefore, it may not be possible to design a dither matrix with amplitudes that will sufficiently dither all areas of constant color. Dither matrices that depend not only on the spatial position, but also on the values of nearby input colors can only partially overcome this problem. For this reason, the error diffusion method is preferred.

The error diffusion method of section 1.9.3 can be extended simply by treating brightness B as one of the three color components. Thus, as before, the average error between the input color $\mathbf{C}_{ij}$ and the displayed output color $\mathbf{O}_{ij}$ is used to modify the input color after appropriate filtering. Thus, the modified color is given by

$$\mathbf{C}_{ij} = \mathbf{C}_{ij} + \frac{\sum \alpha_{i-k,\, j-\ell}\, (\mathbf{C}_{ij} - \mathbf{O}_{ij})}{\sum \alpha_{k\ell}} \tag{1.9.17}$$

The filter coefficients define the relative contributions of the previous errors to the corrected color. Although not shown in the above equation, different color components may use different filter coefficients $\{\alpha_{ij}\}$. Section 1.9.3 gives typical values for these coefficients. Since it is desirable to suppress the low frequency error components, the filter is usually high pass. The modified color is then simply represented as one of the colors in the palette that is the closest.

Since 24 Mbits frame buffers continue to be expensive, many practical systems have implemented different variations of the above color halftoning techniques. In general, image independent palettes perform rather poorly.

---

* Alternatively, the input image can be compared to pseudorandom thresholds for quantization as in section 1.9.3.

Substantial improvement in picture quality is obtained by using the image color distribution in the LBG algorithm with a perceptually matched color metric. The improvement is particularly significant for palettes of smaller size. As in the case of monochrome images, error diffusion performs better than ordered dither, although not in all situations. In the case of color images, difference between error diffusion and ordered dither is even higher due to a higher degree of filter control for individual components. Natural images displayed on a 1000x1000 display look quite pleasant with 128 to 256 size color palettes. However, from a close distance, contouring noise is visible and can only be reduced by increasing the palette size beyond 1024.

1.9.5 Resolution Reduction of Graphical Images

Many applications of graphical imagery require viewing or manipulation at different resolution. One such application is progressive transmission in which a coarse (or low resolution) image is transmitted first, and then is gradually refined. For browsing an image database, this works better than scan oriented transmission, since the decision to further view the image can be made faster by looking at the coarse image. In the case of gray level or color images, resolution is reduced simply by first filtering the image to avoid aliasing and then subsampling at an appropriate rate. In the case of bilevel or halftone monochrome or color images, resolution reduction must be done differently. Resolution reduction by simple subsampling can eliminate (or break) thin lines from bilevel images containing text or drawings. In halftoned images, the appearance of grayness or constant color could be lost if alternate samples are removed due to subsampling. This could be particularly bad if the subsampling phase coincides with the periodic dither matrix, in the case of ordered dither. For these reasons, a different procedure has been developed for reducing resolution. It accepts a high resolution image and creates an image with half as many rows and columns. The procedure works well with black and white, dithered as well as error-diffused images. It is developed as part of the JBIG standard[1.9.14] for lossless progressive compression of still pictures (see chapter 7). Figure 1.9.18 shows a configuration of pels. $\{h_{ij}\}$ are the high resolution pels and $\{l_{ij}\}$ are the low resolution pels computed in the scan line order. Thus l_{00}, l_{01}, l_{10} are already computed, and l_{11} is the next one to be determined. The value of any low resolution pel (e.g., l_{11}) is determined by the color of the nine high resolution pels that surround it (h_{ij}, i=1,...,3, j=1,...,3) and the three low resolution pels (l_{00}, l_{01}, l_{10}) in the neighborhood. The resolution reduction algorithm works on the principle of preserving the density of pels via the use of filtering. Specifically, the following difference is formed

$$4h_{22} + 2(h_{23} + h_{32}) + h_{33} + (h_{11} - l_{00}) + 2(h_{21} + l_{10}) + (h_{31} - l_{10}) + 2(h_{12} - l_{01}) + (h_{13} - l_{01}) \tag{1.9.18}$$

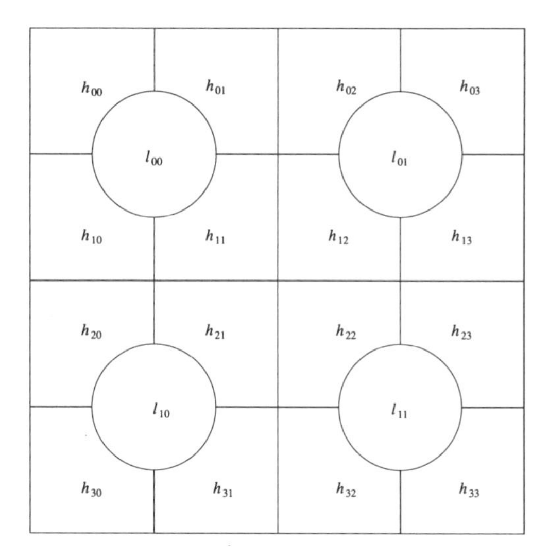

Fig. 1.9.18 Configuration of Pels for Resolution Reduction. $\{h_{ij}\}$ are higher resolution pels and $\{l_{ij}\}$ are lower resolution pels.

Table 1.9.1 The value of the Low Resolution Pel is a function of the index composed of surrounding pels $\{l_{00}, l_{01}, l_{10}, h_{11}, h_{12}, h_{13}, h_{21}, h_{22}, h_{23}, h_{31}, h_{32}, h_{33}\}$ in Fig. 1.9.18. The complete table is given in the JBIG standard specification[1.9.16]

No	Index	Value of l_{11}
1	0 0 0 0 0 0 0 0 0 0 0 0 0	0
2	0 0 0 0 0 0 0 0 0 0 0 0 1	0
3	0 0 0 0 0 0 0 0 0 0 0 1 0	0
4	0 0 0 0 0 0 0 0 0 0 0 1 1	1
5	0 0 0 0 0 0 0 0 0 0 1 0 0	0
⋮	⋮	⋮
4095	1 1 1 1 1 1 1 1 1 1 1 1 0	1
4096	1 1 1 1 1 1 1 1 1 1 1 1	1

This quantity explicitly considers differences between h_{11} and l_{00}, h_{21} and l_{10}, etc. It has an average value of 4.5 for a neighborhood in which black (=0) and white (=1) are equally likely and statistically independent. The pel l_{11} is decided to be a 1 if and only if the above expression is greater than 4.5. This rule can be formulated as a table which is indexed by a string

$$l_{00}, l_{01}, l_{10}, h_{11}, h_{12}, h_{13}, h_{21}, h_{22}, h_{23}, h_{31}, h_{32}, h_{33}. \tag{1.9.19}$$

Since each entry above is either 0 or a 1 and there are 12 entries, the total number of indices is $2^{12} = 4096$. For each index, Table 1.9.1 gives the value of the low resolution pel l_{11}. While the rule given above does an excellent job most of the time, some modifications have been incorporated in the table to improve the picture quality. These are:

a. Edge Preservation: This modification keeps edges straight instead of becoming zig-zag due to subsampling. These are 132 indices that need this modification as determined empirically.

b. Line Preservation: This modification preserves linkages along slanted lines for 420 indices, which are determined empirically.

c. Periodic Pattern Preservation: Periodic patterns in the high resolution picture are already preserved in the low resolution picture by the general rule above, except at the boundary between periodic and aperiodic regions. Ten indices are modified to handle such boundary regions.

d. Dither Pattern Preservation: These modifications preserve very low or very high density dithering that arise from isolated dark or bright areas in the picture. There are 12 indices which are modified this way.

As remarked earlier, Table 1.9.1 has been used quite extensively to reduce resolution by factors of 2 in both horizontal and vertical dimensions with excellent picture quality due to preservation of much of the edge and local average brightness information.

Appendix

PHOTOMETRIC QUANTITIES AND UNITS

In this Appendix, we familiarize the reader with photometric quantities and their units. This Appendix is necessitated by the fact that many photometric quantities (e.g. intensity) are used rather loosely in the literature (and in this book) and each quantity has many units that are used in different sections. Since light is the visible portion of the electromagnetic radiation, there is considerable similarity between radiometric and photometric quantities and their units. Therefore, we start with some well-known radiometric quantities and relate them to photometric quantities.

Radiant energy, Q, is the energy propagating in the form of electromagnetic radiation and is measured in **joules**. *Radiant flux*, Φ, measured in **watts** (**joules**/sec), is the radiant power or the time rate of flow of radiant energy. Radiant energy may be generated by a source (e.g. a body raised to some temperature) or may be reflected from a surface. *Radiant exitance, M*, which includes flux emitted from, transmitted through, or reflected by a surface, is the density of radiant flux and is therefore measured in **watts**/meter^2. *Irradiance, E*, is a restricted case of exitance and denotes density of radiant flux incident on a surface. *Radiant intensity, I*, is the radiant flux per unit solid angle in a given direction. It is measured in **watts**/steradian. *Radiance, L*, is the radiant flux per unit of projected area and unit solid angle either leaving a point in a given direction, or arriving at a given point in a given direction. Its unit is **watts**/steradian-meter^2. Each of these quantities can be restricted to a narrow bandwidth of $\Delta\lambda$ centered around a wavelength of λ. In such a case, the wavelength specific nature of the quantity is denoted, for example $\Phi(\lambda)$. These radiometric quantities have exact analogs in photometry, except they have different units.

The capacity of radiant flux $\Phi(\lambda)$ to create a sensation of light can be evaluated by multiplying it by the *relative luminous efficiency*, $y(\lambda)$. The luminous flux at wavelength λ is thus given by $\Phi(\lambda)\,y(\lambda)$ and the total luminous flux Φ_v is obtained by integrating this over the visible spectrum. Thus,

$$\Phi_v = \int_{380}^{780} \Phi(\lambda)\,y(\lambda)\,d\lambda \qquad (1.10.1)$$

Since $y(\lambda)$ is dimensionless, Φ_v is measured in the same units as $\Phi(\lambda)$, i.e. **watts**. An alternative unit of luminous flux is **lumens**. To convert Φ_v into lumens, a factor k_m is used:

$$\Phi_v = k_m \int_{380}^{780} \Phi(\lambda) y(\lambda) d\lambda \tag{1.10.2}$$

where $k_m = 680$ **lumens/watt**. Similarly *luminous intensity*, I_v (= visible radiant intensity), and *luminance*, L_v (= visible radiance), are given by

$$I_v = k_m \int_{380}^{780} I(\lambda) y(\lambda) d\lambda \text{ in } \textbf{candelas}$$

$$L_v = k_m \int_{380}^{780} L(\lambda) y(\lambda) d\lambda \text{ in } \textbf{candelas}/\text{square meter} \tag{1.10.3}$$

Luminous energy, Q_v, or the quantity of light is measured in **lumen**-seconds, also denoted as **talbots**. As in the case of radiometric intensity, *luminous intensity*, I_v, is the density of luminous flux per unit solid angle and therefore

$$I_v(\lambda) = \frac{\Delta\Phi_v(\lambda)}{\Delta\omega} \tag{1.10.4}$$

where $\Delta\omega$ is an element of solid angle. Thus,

$$\Phi_v(\text{in } \textbf{lumens}) = \text{intensity (in } \textbf{candelas}) \cdot \text{solid angle (in steradians)} \tag{1.10.5}$$

Illumination, $E(\lambda)$, is the density of luminous flux incident upon a surface. Thus,

$$E(\lambda) = \frac{\Phi_v(\lambda)}{A} \tag{1.10.6}$$

$$E(\lambda)(\text{in } \textbf{lux}) = \Phi_v(\lambda) \text{ (in } \textbf{lumens})/\text{area (in square meters)}$$

or

$$E(\lambda) \text{ (in } \textbf{foot candles}) = \Phi_v(\lambda) \text{ (in } \textbf{lumens})/\text{area (in square feet)}$$

with

$$1 \textbf{ foot candle} = 10.76 \textbf{ lux} \tag{1.10.7}$$

Luminance, L_v, or photometric brightness, is the luminous flux per unit solid angle, per unit projected area, or luminous intensity per unit projected area, whereas *Luminous exitance* is the density of the luminous flux emitted from a surface. Therefore luminance is directional and determines the ability of a luminous object to produce an effect at a given point in space. Luminous exitance, on the other hand, only specifies density of flux over an area with no regard to direction. A common unit of luminous exitance is the **lambert**, which is defined by

$$\text{Luminous exitance (in } \textbf{lamberts}) = \text{Luminous flux (in } \textbf{lumens})/\text{cm}^2 \tag{1.10.8}$$

The table below summarizes these quantities and their units. For more details, see *Handbook of Optics*, W. G. Driscoll and W. Vaughn (editors), McGraw-Hill 1978.

QUANTITY	Radiant	Luminous
Energy: Q	joule	lumen-second (= talbot)
Flux Φ: Q/time	watt	lumen
Exitance: Φ/area	watt/meter^2	lumen/meter^2
Intensity		
I: Φ/solid angle	watt/steradian	lumen/steradian = candela
E: Φ/area	Irradiance watt/meter^2	Illumination lumen/meter^2 = lux lumen/ft^2 = foot candle
L: Φ/area · solid angle	Radiance watt/steradian-meter^2	Luminance candela/meter^2 = nit candela/π cm^2 = lambert

References

1.3.1 N. S. Jayant and P. Noll, *Digital Coding of Waveforms*, Chapter 4, Prentice Hall, New York, 1984.

1.3.2 L. D. Hammon and B. Julesz, "Masking in Visual Recognition: Effects of 2-D Filtered Noise," Science *180*, 1194-97.

1.5.1 A. Papoulis, *The Fourier Integral and Its Applications*, McGraw-Hill, New York, 1962.

1.5.2 L. R. Rabiner and B. Gold, *Theory and Application of Digital Signal Processing*, Prentice-Hall, Englewood Cliffs, New Jersey, 1975.

1.5.3 P. Mertz and F. Gray, "A Theory of Scanning and Its Relation to the Characteristics of the Transmitted Signal in Telephotography and Television," *BSTJ*, Vol. 13, 1934, pp. 464-515.

1.5.4 D. E. Pearson, *Transmission and Display of Pictorial Information*, Pentech Press, London, 1975.

1.5.5 H. S. Hou and H. C. Andrews, "Cubic Splines for Interpolation and Filtering," IEEE Trans. ASSP, v. ASSP-26, No. 6, 1978, pp. 508-517.

1.8.1 T. Young, "On the Theory of Light and Colors," Philosophical Transactions of the Royal Society of London; Vol. 92, 1802, pp. 20-71.

1.8.2 D. L. MacAdam, *Sources of Color Science,* Cambridge, MA, MIT Press 1970.

1.8.3 G. Wyszecki and W. S. Stiles, *Color Science,* John Wiley, New York, 1967.

1.8.4 H. G. Grassman, "Theory of Compound Colors," Philosophical Magazine, Vol. 4, No. 7, 1854, pp. 254-264.

1.8.5 W. T. Wintringham, "Color Television and Colorimetry," Proc. of IRE Vol. 39, No. 10, 1951, pp. 1135-1173.

1.8.6 D. B. Judd and G. Wyszecki, *Color in Business, Science, and Industry*, 2nd Edition, John Wiley & Sons, Inc., 1963.

1.8.7 W. B. Marks, W. H. Dobelle, and E. F. MacNihol, Jr., "Visual Pigments of Single Primate Cones," Science, Vol. 143, 1964, pp. 1181-1183.

1.9.1 A. N. Netravali (Editor), *Digital Encoding of Graphics*, Special Issue, Proceedings of IEEE, March 1980.

1.9.2 D. M. Costigan, *Electronic Delivery of Documents and Graphics,* Van Nostrand Reinhold, 1978.

1.9.3 D. Y. Coraog, and F. C. Rose, Legibility of Alphanumeric Characters and Other Symbols: II. A Reference Handbook, U. S. Government Printing Office, National Bureau of Standard Miscellaneous Publication 262-2.

1.9.4 H. L. Snyder and M. E. Maddox, *Information Transfer from Computer Generated Dot-Matrix Displays*, Report, EFL-78-3/ARO-78-1, U. S. Army Research Office, North Carolina, 1978.

1.9.5 J. F. Jarvis, C. N. Judice, and W. H. Ninke, "A Survey of Techniques for the Display of Continuous Tone Pictures on Bilevel Displays," Computer Graphics and Image Processing, Vol. 5, 1976, pp. 13-40.

1.9.6 J. C. Stoffel and J. F. Moreland, "A Survey of Techniques for Pictorial Reproduction," IEEE Trans. on Communications, Vol. COM-29, No. 12, December, 1981, pp. 1898-1925.

1.9.7 *Picture Bandwidth Compression,* ed. by T. S. Huang and O. J. Tretiak, Gordon & Breach, 1972.

1.9.8 R. B. Arps, R. L. Erdmann, A. S. Neal, and C. L. Schlaepfer, "Character Legibility Versus Resolution in Image Processing of Printed Matter," IEEE Transactions on Man-Machine Systems, Vol. MMS-10, No. 3, pp. 66-71, 1969.

1.9.9 P. Heckbert, "Color Image Quantization for Frame Buffer Display", Computer Graphics, Vol. 16, No. 3, July 1982, pp. 297-307.

1.9.10 G. Braudauray, "A Procedure for Optimum Choice of a Small Number of Colors from a Large Color Palette for Color Imaging", Electronic Imaging, 1987.

1.9.11 Y. Linde, A. Buzo, and R. Gray, "An Algorithm for Vector Quantizer Design", IEEE Trans. Communication, Vol. Com-28, No. 1, Jan. 1980, pp. 84-95.

1.9.12 R. S. Gentile, E. Walowit, and J. P. Allebach, "Quantization and Multilevel Halftoning of Color Images for Near-Original Image Quality", Journal of Optical Society of America, Vol. 7, No. 6, June 1990, pp. 1019-1026.

1.9.13 M. Orchard and C. A. Bouman, "Color Quantization of Images", IEEE Trans. on Signal Processing, Vol. 39, No. 12, Dec. 1991, pp. 2677-2690.

1.9.14 H. Hampel, et al., "Technical Features of the JBIG Standard for Progressive Bilevel Image Compression", Signal Processing: Image Communication, Vol. 4, 1992, pp. 103-111. Also, the ISO-JBIG Standard, ISO 11544.

Questions for Understanding

1.1 Assume a rectangular array of sampling points with $\Delta_x = \Delta_y = 1$. If bilinear interpolation is used for postfiltering, derive the equation for $\tilde{B}(x,y)$ between the sample points (0,0), (0,1), (1,0) and (1,1).

1.2 Show that the cosine series basis functions of Eq. (1.5.6) are orthonormal.

1.3 Prove Eq. (1.5.14).

1.4 Prove Eq. (1.5.21).

1.5 Derive Eq. (1.5.26).

1.6 Derive Eq. (1.5.30).

1.7 Prove Eq. (1.5.37).

1.8 Explain trichromacy of vision and color mixture.

1.9 Explain Primaries, Tristiulus Values, Color Matching Functions and Chromaticity Coordinates.

1.10 Explain Spectral Locus and Line of Purples.

1.11 What are the x,y Chromaticity Coordinates (approximately) of a color with Dominant Wavelength 700 nm and Exitation Purity 0.5? What is its approximate Complementary Wavelength, if any?

1.12 A color has x,y chromaticities (0.4, 0.5) and luminance 5. What are its R_0, G_0, B_0 chromaticities and tristimulus values?

1.13 We wish to scan 8.5×11 inch pages containing mid century, upper case characters that are 76 mils high. How many pels per page are needed for 97.5% legibility?

1.14 From Eq. (1.9.6) derive D^8 scaled for 8-bit input.

2

Common Picture Communication Systems

In this chapter we describe the operation of several picture communication systems in order to understand better the fundamentals and applications of image coding methods to be described later. We begin with monochrome television, followed by the three major color TV systems, NTSC, PAL and SECAM.

As of this writing, transmission and recording of television signals is still largely analog. For example, over-the-air and coaxial cable transmission to consumers is by amplitude modulation (AM) of RF carriers in the VHF or UHF band. Satellite and terrestrial microwave transmission is by frequency modulation (FM), as is the recording on most magnetic tape or disk. A large percentage of optical fiber transmission and almost all optical video disk recording is carried out using pulse frequency modulation, i.e., the frequency of a train of optical pulses is modulated by the analog video signal.

Digital television signals can be found in a wide variety of TV studio signal processing equipment. For example, time-base correctors remove the variation in scan line duration due to stretching of magnetic tape in video recorders. Frame synchronizers can completely synchronize two video signals so that switching from one to the other is possible without disruption of raster scanning in receivers. And special effect generators are able to combine, reshape, slow down, speed up, etc., a wide variety of picture and graphics material. However, digital *transmission* of broadcast television is still in its infancy.

After describing broadcast TV systems, we discuss other television systems such as videotelephone and high definition television (HDTV). Videotelephone and videoconferencing are growing in popularity, especially in the business community, both as an adjunct to voice communication and as a substitute for time-consuming and expensive travel. HDTV is still in its infancy and attempts to provide picture resolution approaching that of motion picture film.

Finally, we describe several graphics communication systems that send text or line drawings for hard copy output or computer terminal display. For example, facsimile services are growing rapidly since they are a quicker and more convenient alternative to sending letters and documents by mail. Teletext and videotex, aimed more at the consumer market, are able to make a wide variety of information services available through television or telephone data channels.

2.1 Monochrome Television

A typical television communication system is shown in Fig. 2.1.1. Virtually all monochrome (i.e., black and white) and color television employs raster scanning with 2:1 line interlace, as described in Chapter 1. Several international standards exist, and some of these are shown in Table 2.1.1. The ratio of visible or *active* picture width to picture height is called the *Image Aspect Ratio* (*IAR*). In all of the systems of Table 2.1.1 it is 4:3. However, for high definition television (HDTV) a higher *IAR* = 16:9 is subjectively more pleasing for most scenes.

Many TV cameras and virtually all cathode ray tube (CRT) displays have an inherent nonlinear relationship between signal voltage and light intensity. A power law is usually assumed of the form

$$B = cv^{\gamma} + B_o \tag{2.1.1}$$

where B is light intensity, v is voltage above cutoff, c is a constant gain factor, B_o is the cutoff (camera) or black level (CRT) light intensity, and γ is in the range 1.0 to 3. For example, the vidicon camera tube has $\gamma \approx 1.7$, whereas for plumbicons and image orthicons $\gamma \approx 1$.* Cathode ray tube displays in use today have gammas in the range 2.2 to 3.

In order to avoid correction circuitry inside each of millions of TV receivers, *gamma correction* is usually done at the camera prior to transmission. That is, if γ_c is the gamma of the camera, and a standard γ_d (e.g., 2.5) is assumed, for the CRT display, then the camera voltage v_c is *gamma corrected* prior to output by the relation

$$v_d = v_c^{\gamma_c/\gamma_d} \tag{2.1.2}$$

to obtain a voltage v_d that can be transmitted and applied directly to the display CRT. The gamma corrected voltage v_d is also less susceptible to transmission noise, as we shall see later in Chapter 4.

Approximately 17 percent of each scan line is blanked to allow for horizontal retrace of the scanning spot to the beginning of the next line. Similarly, 6 to 10 percent of each field is blanked to allow for vertical retrace to the top of the

---

* Some authors use $1/\gamma$ as the gamma of the camera.

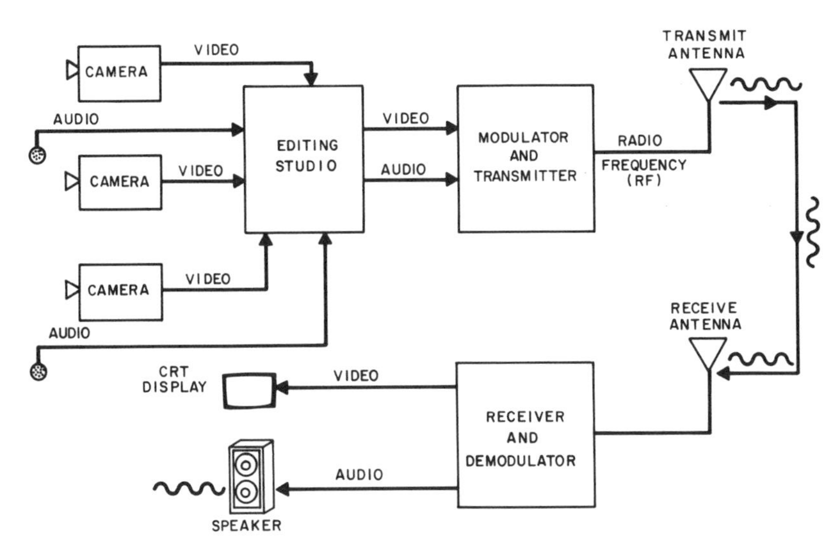

Fig. 2.1.1 Typical television communication system. Several sources of video and audio go to the studio where they are switched, edited and possibly recorded for later transmission. The modulator multiplexes the audio and video onto the assigned radio frequency whereupon it is transmitted via an antenna. The TV receiver selects the desired channel and demultiplexes the video and audio, which then pass to a display cathode ray tube (CRT) and a loudspeaker, respectively.

Table 2.1.1 International standards for monochrome television. System M is used in North America and Japan; system I is used in the United Kingdom; France uses systems E and L; the rest of Western Europe uses B, C, G and H; and Eastern Europe uses D and K. High Definition Television (HDTV) has more than 1000 lines/frame; however, international standards have yet to be set.

Standard	A	M	B,C,G,H	I	D,K,L	E
Lines/Frame	405	525	625	625	625	819
Fields/Second	50	60	50	50	50	50
Frames/Second	25	30	25	25	25	25
Lines/Second	10125	15750	15625	15625	15625	20475
Bandwidth (MHz)	3.0	4.2	5.0	5.5	6.0	10.0
Bit-Rate Mbs (8-bit PCM)	48	67.2	80	88	96	160

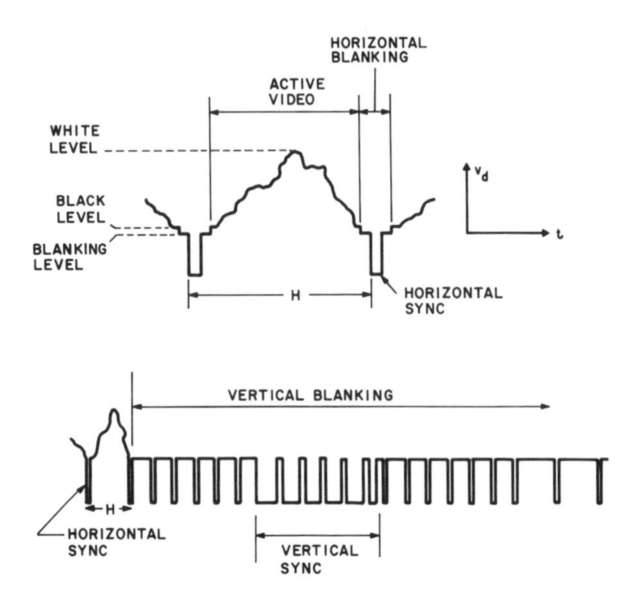

Fig. 2.1.2 Synchronization pulses are added to the blanked regions of a raster scanned video signal. Between the *active* regions of each scan line, a horizontal sync pulse resides. Between fields a vertical sync pulse is inserted.

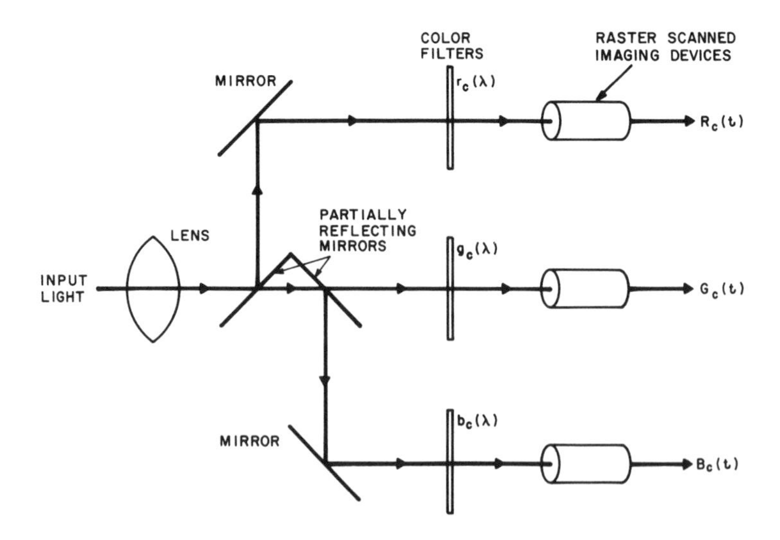

Fig. 2.2.1 Color camera arrangement in which the incoming light is split into three paths, filtered and raster scanned to produce three tristimulus signals $R_c(t), G_c(t), B_c(t)$.

next field. Prior to transmission, synchronization pulses are inserted into these blanked portions of the signal, as shown in Fig. 2.1.2, in order to provide timing information for the horizontal and vertical scanning circuitry of the display.

Monochrome video signals can be sampled at their Nyquist rate, i.e., twice the bandwidth, and the samples quantized to about 8-bit accuracy, in order to adequately represent the visual information for most applications of interest. For example, such PCM coding of standard television produces the bit rates shown in Table 2.1.1.

If the visual information is destined for human viewers, then better results are usually obtained if the gamma corrected voltage v_d of Eq. (2.1.2) is PCM coded instead of the actual brightness B or the camera tube voltage [unless, of course, $v_c \approx v_d$]. This is due to the fact, as we shall see in Chapter 4, that errors in intensity are much less visible at high light levels, where gamma correction causes compression of the voltage v_d, than at low light levels.

2.2 Color Television

In Chapter 1 we saw how monochrome visual information can be approximated by raster scanning, and how color information destined for human viewers can be approximated by a summation of light intensities in three *primary* wavelength bands. Each of the three primary intensity distributions can be thought of as a monochrome picture that can be raster scanned to give three tristimulus waveforms. For example, these tristimuli might be designated $R(t)$, $G(t)$ and $B(t)$ if the primaries lie in the red, green and blue wavelength bands, respectively. In principle, any three colors can be used as primaries as long as one is not a mixture of the other two. For some colors this means that negative tristimulus values may be needed, as described in Chapter 1. If the CIE, nonphysical primaries $\boldsymbol{X},\boldsymbol{Y},\boldsymbol{Z}$ are used, then negative tristimulus values are never required.

However, real-world color displays are only capable of producing *positive* amounts of *physical* primaries, such as $\boldsymbol{R},\boldsymbol{G},\boldsymbol{B}$. Thus, there will always be some colors that cannot be shown on practical, three-primary displays. Moreover, display primaries must be chosen to produce adequate *luminance*, which places a further restriction on the gamut of colors that can be portrayed.

In order for a color TV camera to measure three tristimulus values at every point in the image, three weighted averages of the wavelengths must be computed as indicated in Eq. (1.8.4). Fig. 2.2.1 shows an arrangement for doing this in which the incoming light is first split into three paths. It then passes through three color filters $r_c(\lambda), g_c(\lambda), b_c(\lambda)$ which perform the wavelength averaging and then to three monochrome, raster scanned imaging devices that produce signals $R_c(t), G_c(t), B_c(t)$.

We saw in Chapter 1 that the color matching or weighting functions $r(\lambda), g(\lambda), b(\lambda)$ corresponding to primaries $\boldsymbol{R},\boldsymbol{G},\boldsymbol{B}$ will in general be negative for some wavelengths. This precludes their direct use in television cameras since optical filters with negative transmission are not possible. Conceptually, we could use a camera that has positive color matching functions $x(\lambda), y(\lambda), z(\lambda)$ corresponding to the CIE primaries $\boldsymbol{X}, \boldsymbol{Y}, \boldsymbol{Z}$. The resulting tristimulus signals $X(t), Y(t), Z(t)$ could then be converted by matrix transformation to the tristimulus signals $R(t)$, $G(t)$, $B(t)$ corresponding to primaries $\boldsymbol{R}, \boldsymbol{G}, \boldsymbol{B}$, as we saw in Chapter 1. However, this is not done for several reasons. First, certain colors in some scenes will produce negative R, G or B tristimulus values, which cannot be handled by real-world displays. Also, it may be difficult to manufacture $x(\lambda), y(\lambda), z(\lambda)$ optical filters in large quantity. But most importantly, viewing conditions vary widely, both at the camera and at the display.[2.2.1] Original scene illumination and viewing room lighting are rarely the same. Also, the luminance of the scene may be very different from the luminance of the display. Under these circumstances, correct reproduction of *perceived* colors may not be achieved by correct reproduction of chromaticities and tristimulus values. Instead, other criteria, such as accurate display of reference white, are often used in designing and adjusting color TV cameras. We will return to the subject of perceived color in Chapter 4.

Practical TV cameras have nonnegative matching or weighting functions $r_c(\lambda), g_c(\lambda), b_c(\lambda)$ that may not correspond to *any* primary, physical or nonphysical. The resulting camera "tristimulus" waveforms $R_c(t), G_c(t), B_c(t)$ are linearly transformed to correspond to the primaries $\boldsymbol{R}, \boldsymbol{G}, \boldsymbol{B}$ of the standard display via the matrix equation

$$\begin{bmatrix} R(t) \\ G(t) \\ B(t) \end{bmatrix} = \boldsymbol{T} \begin{bmatrix} R_c(t) \\ G_c(t) \\ B_c(t) \end{bmatrix} \tag{2.2.1}$$

The matrix $\boldsymbol{T}$ is then chosen to optimize either the *average* error in perceived color or perhaps the reproduction of certain key colors such as white and human skin tones. In some cases, the camera may be designed to conform to nonphysical display primaries that lie outside the gamut of reproducible colors. As long as these nonphysical primaries have the same dominant wavelength as the actual display primaries $\boldsymbol{R}, \boldsymbol{G}, \boldsymbol{B}$, acceptable pictures are produced.

According to basic principles of colorimetry, various other linear combinations of the R, G, B tristimulus values can be formed to yield alternate primary representations that may have more useful properties than the original red, green and blue signals. For color television we denote these new representations by $Y(t)$, $C_1(t)$ and $C_2(t)$, where in practically all cases of interest $Y(t)$ is chosen to approximate the luminance. The remaining two waveforms $C_1(t)$ and $C_2(t)$ contain *chrominance*, i.e., color information. The form of the chrominance varies depending on the particular application.

There are two main advantages in using Y, C_1 and C_2 for color television. The first is compatibility: $Y(t)$ can be used directly by monochrome television receivers without further processing. The second is power and bandwidth saving: by proper choice of chrominance representation, $C_1(t)$ and $C_2(t)$ can be bandlimited to frequencies significantly below those of $Y(t)$ without seriously affecting the perceived quality of the reproduced color picture. Subjective aspects of color vision will be treated in more detail in Chapter 4.

In most of today's color television systems, practical considerations have led to certain compromises and departures from the above idealized situation of simply computing the tristimulus waveforms $R(t)$, $G(t)$ and $B(t)$ at a camera, transmitting $Y(t)$, $C_1(t)$ and $C_2(t)$, and reproducing $R(t)$, $G(t)$ and $B(t)$ at a receiver/display. In the first place, it is highly desirable that color signal transmission take no more bandwidth than monochrome signals. Thus Y, C_1 and C_2 must be multiplexed together into the same bandwidth that would normally be reserved for Y, the luminance signal alone. Furthermore, this multiplexing must produce a signal that can be displayed on a monochrome receiver without undue picture degradation.

Another compromise is to perform the camera gamma correction described in Section 2.1 directly on the R,G,B tristimulus waveforms, so that no additional processing is necessary at the receiver to maintain amplitude accuracy. However, when $Y(t)$ is calculated as a linear combination of gamma corrected tristimulus waveforms, the result is not a gamma corrected luminance signal. Indeed $Y(t)$ may contain significant chrominance information, and $C_1(t)$ and $C_2(t)$ may contain luminance information. This also causes a certain amount of luminance distortion in monochrome displays. Further compromises in color TV signal definition are also made in order to utilize transmission bandwidth efficiently and to simplify circuitry in TV receivers. These are discussed in more detail below.

2.2.1 NTSC

The National Television System Committee (NTSC) in 1952 defined the color television[2.2.2-4] system that is currently in use in North America and Japan. The scanning raster is almost the same as with monochrome television in these countries (see Table 2.1.1, system M). The defined standard display phosphors and color primaries are defined in terms of CIE *xyz* chromaticity coordinates as

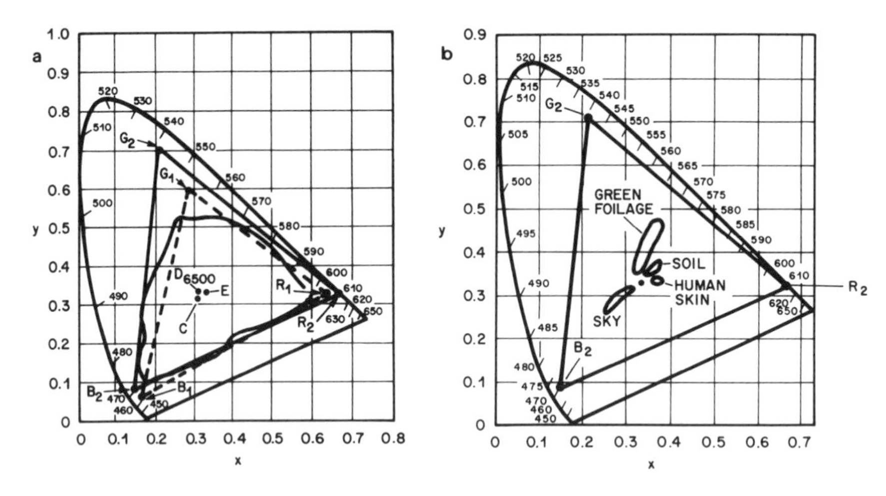

Fig. 2.2.2 Primaries used for the PAL ($\boldsymbol{R}_1,\boldsymbol{G}_1,\boldsymbol{B}_1$) and the NTSC ($\boldsymbol{R}_2,\boldsymbol{G}_2,\boldsymbol{B}_2$) systems of color television. (a) The range of achievable colors is the interior of the triangle defined by the primaries. The irregular curve outlines the gamut of inks, pigments and dyes. (b) Most natural colors are pastel and occur near the center of the triangles.

Table 2.2.1 International standards for color television. South America uses the video signal parameters of System M, but the color modulation methods of PAL.

	NTSC	PAL	SECAM
TV System	M	G, I	L
Field Rate (Hz)	59.94	50	50
Lines/Frame	525	625	625
Lines/Second	15734	15625	15625
Audio Carrier (MHz)	4.5	5.5, 6.0	6.5
Gamma	2.2	2.8	2.8
Reference White	C	D_{6500}	D_{6500}
Color Subcarrier (Hz)	3579545*	4433618	4250000 (+U) 4406500 (-V)
Color Modulation	QAM	QAM	FM
Luminance Bandwidth (MHz)	4.2	5.0, 5.5	6.0
Chrominance Bandwidth (MHz)	1.3 (I) 0.6 (Q)	1.3 (U_t) 1.3 (V_t)	>1.0 (U_t) >1.0 (V_t)

* Multistandard tape players and monitors sometimes use a Modified NTSC (M-NTSC) in which the color subcarrier frequency is the same as PAL.

$$
\begin{array}{cccc}
 & x & y & z \\
\boldsymbol{R}: & 0.67 & 0.33 & 0.00 \\
\boldsymbol{G}: & 0.21 & 0.71 & 0.08 \\
\boldsymbol{B}: & 0.14 & 0.08 & 0.78
\end{array}
\tag{2.2.2}
$$

Fig. 2.2.2a shows the locations of these NTSC primaries on the CIE xy chromaticity diagram along with the color triangle they define. It might seem from this diagram that there are a lot of physical colors that cannot be displayed. However, we will see in Chapter 4 that the area on the CIE xy diagram does not represent color gamut very well. Also, most real-world scenes have rather pastel and unsaturated colors, as shown in Fig. 2.2.2b, and only the seldom occurring saturated hues suffer from using the NTSC primaries. Moreover, at the time the NTSC system was defined, these primaries could be produced with higher luminance than could more saturated primaries. Modern color displays often utilize nonstandard phosphors having higher color saturation as well as higher output luminance.

Using the relationships described in Section 1.8.2 we can compute the XYZ tristimulus matrix $\boldsymbol{M}$ for the primaries $\boldsymbol{RGB}$ using the xyz chromaticities of Eq. (2.2.2). Thus, we obtain for the NTSC primaries from Eq. (1.8.30)

$$
\boldsymbol{M} = \begin{bmatrix} 0.661 & 0.326 & 0.0 \\ 0.171 & 0.578 & 0.065 \\ 0.168 & 0.096 & 0.935 \end{bmatrix} \tag{2.2.3}
$$

$$
\boldsymbol{M}^{-1} = \begin{bmatrix} 1.754 & -0.998 & 0.070 \\ -0.489 & 2.027 & -0.141 \\ -0.265 & -0.029 & 1.071 \end{bmatrix} \tag{2.2.4}
$$

For an arbitrary color $\boldsymbol{A}$, the transformation of tristimulus values between the CIE XYZ system and the NTSC primaries is then carried out via the equations

$$
[X_A \; Y_A \; Z_A] = [R_A \; G_A \; B_A] \; \boldsymbol{M} \tag{2.2.5}
$$

$$
[R_A \; G_A \; B_A] = [X_A \; Y_A \; Z_A] \; \boldsymbol{M}^{-1} \tag{2.2.6}
$$

In particular, the luminance of a color (dropping the A subscript) is given by

$$Y = 0.326R + 0.578G + 0.096B \tag{2.2.7}$$

where R, G, B are the NTSC tristimulus values normalized to equal energy white (standard illuminant $\boldsymbol{E}$) as described in Chapter 1. However, reference white for the NTSC system is chosen to approximate average daylight and is defined as illuminant $\boldsymbol{C}$, with xyz chromaticity coordinates

$$\begin{aligned} x_C &= 0.3101 \\ y_C &= 0.3162 \\ z_C &= 0.3737 \end{aligned} \tag{2.2.8}$$

The relative amounts of primaries $\boldsymbol{R}$, $\boldsymbol{G}$ and $\boldsymbol{B}$ necessary to give reference white $\boldsymbol{C}$ are the rgb chromaticity coordinates of $\boldsymbol{C}$, found from Eq. (1.8.27)

$$\begin{aligned} r_C &= 0.2937 \\ g_C &= 0.3244 \\ b_C &= 0.3819 \end{aligned} \tag{2.2.9}$$

The tristimulus signals R, G and B are first divided by these chromaticities so that equal amplitudes produce $\boldsymbol{C}$, the reference white. They are further normalized so that (denoting normalized quantities by tildes) $\tilde{R} = \tilde{G} = \tilde{B} = 1$ corresponds to reference white $\boldsymbol{C}$ of unity luminance. The luminance of a color in terms of *normalized** waveforms then becomes from Eq. (2.2.7)

$$\begin{aligned} Y &= \frac{0.326 r_C \tilde{R} + 0.578 g_C \tilde{G} + 0.096 b_C \tilde{B}}{0.326 r_C + 0.578 g_C + 0.096 b_C} \\ &= 0.299\tilde{R} + 0.587\tilde{G} + 0.114\tilde{B} \end{aligned} \tag{2.2.10}$$

---

* Many color cameras have circuitry for "white correction," which ensures that white objects are displayed as white even under a variety of lighting conditions at the camera. Also, the camera output component signals $\tilde{R}, \tilde{G}, \tilde{B}$ are gamma corrected as described earlier. Thus, Y, C_1, C_2 are not true linear combinations of the tristimulus values R, G, B.

where the actual camera output component signals (tildes) are calculated from the NTSC tristimulus values by

$$\tilde{R} = 1.088R$$
$$\tilde{G} = 0.987G \qquad (2.2.11)$$
$$\tilde{B} = 0.837B$$

followed by gamma correction using a gamma of 2.2. Thus, if $\tilde{R}$,$\tilde{G}$ and $\tilde{B}$ are limited to the range 0 to 1 volts, then so is the luminance signal Y.

Having designated the luminance, we must now define two chrominance waveforms in order to specify the three-dimensional color information. The first step in this is to shift the origin of the color axes to reference white **C**, and this is done by calculating the *color-difference* chrominance signals

$$U_t = \frac{\tilde{B} - Y}{2.03}$$
$$V_t = \frac{\tilde{R} - Y}{1.14} \qquad (2.2.12)$$

with attenuation factors to reduce their amplitudes. Note that reference white, i.e., $\tilde{R} = \tilde{G} = \tilde{B} = Y$, produces zero chrominance. Thus, since most scenes consist of pastel colors, the chrominance signals will be much smaller than the luminance, on the average. Next, a 33° axis rotation* is carried out, as shown in Fig. 2.2.3, resulting in the final NTSC chrominance signals

$$I = V_t \cos 33° - U_t \sin 33°$$
$$Q = V_t \sin 33° + U_t \cos 33° \qquad (2.2.13)$$

or in matrix notation

---

* The rotation angle is chosen to minimize the required bandwidth for Q. See Chapter 4.

$$\begin{bmatrix} Y \\ I \\ Q \end{bmatrix} = \begin{bmatrix} 0.299 & 0.587 & 0.114 \\ 0.596 & -0.274 & -0.322 \\ 0.211 & -0.523 & 0.311 \end{bmatrix} \begin{bmatrix} \tilde{R} \\ \tilde{G} \\ \tilde{B} \end{bmatrix} \tag{2.2.14}$$

$$\begin{bmatrix} \tilde{R} \\ \tilde{G} \\ \tilde{B} \end{bmatrix} = \begin{bmatrix} 1.0 & 0.956 & 0.621 \\ 1.0 & -0.272 & -0.649 \\ 1.0 & -1.106 & 1.703 \end{bmatrix} \begin{bmatrix} Y \\ I \\ Q \end{bmatrix}$$

where $\tilde{R}$, $\tilde{G}$, $\tilde{B}$ are the normalized camera output component signals. As mentioned previously, $\tilde{R}$, $\tilde{G}$ and $\tilde{B}$ are all gamma corrected *prior* to matrixing via Eq. (2.2.14). This means that Y is *not* the true gamma corrected luminance, nor are I and Q gamma corrected chrominances. In general, with this procedure Y contains some chrominance information, and I and Q contain some luminance information although for reference white signal purity is maintained. The effects of this are relatively small, however, and usually not noticeable by the average viewer.

The chrominance signals $I(t)$ and $Q(t)$ have the virtue of not requiring the full 4.2 MHz bandwidth for subjectively acceptable picture quality in most cases. This is due to viewers being relatively insensitive to chrominance signal transitions in high detail areas of a picture and at edges of objects, where large luminance transitions tend to mask the chrominance. More will be said about subjective aspects of color vision in Chapter 4. Very good picture rendition can be obtained in most cases with I and Q bandlimited* to about 1.5 MHz and 0.6 MHz, respectively. This is normally done following the *matrixing* of Eq. (2.2.14).

We now describe how the Y, I and Q signals are multiplexed into a 4.2 MHz bandwidth, even though Y itself has frequency components as high as 4.2 MHz. We saw in Chapter 1 that successive scan lines of a television picture are, on average, very much alike, i.e., highly *correlated.* This is true even with interlace, with the result that signals obtained by raster scanning contain strong periodicities at the line-scan-rate and its harmonics. Moreover, for most pictures there is very little spectral energy *between* the line-rate harmonics in both the luminance and chrominance components, as shown in Fig. 1.5.6. Thus, by interleaving the chrominance and luminance frequencies it is possible to transmit both in the allotted bandwidth without excessive interference of one with the other.

---

* Bandlimiting of I and Q normally reduces only the horizontal chrominance resolution. Ideally, vertical filtering should also be carried out.

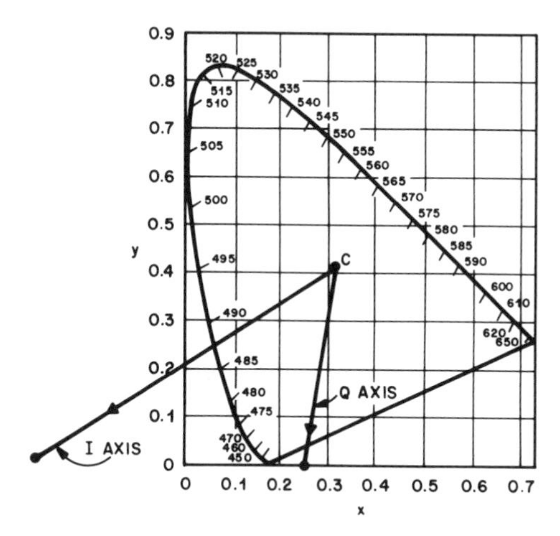

Fig. 2.2.3 Color axes used for the NTSC color television system. Reference white is ***C*** for NTSC. The color difference axes U_t and V_t are rotated to give the final NTSC chrominance signals I and Q. In polar coordinates, hue is approximately the angle $\tan^{-1}(Q/I)$, and the radial distance $\sqrt{Q^2+I^2}$ is approximately proportional to saturation.

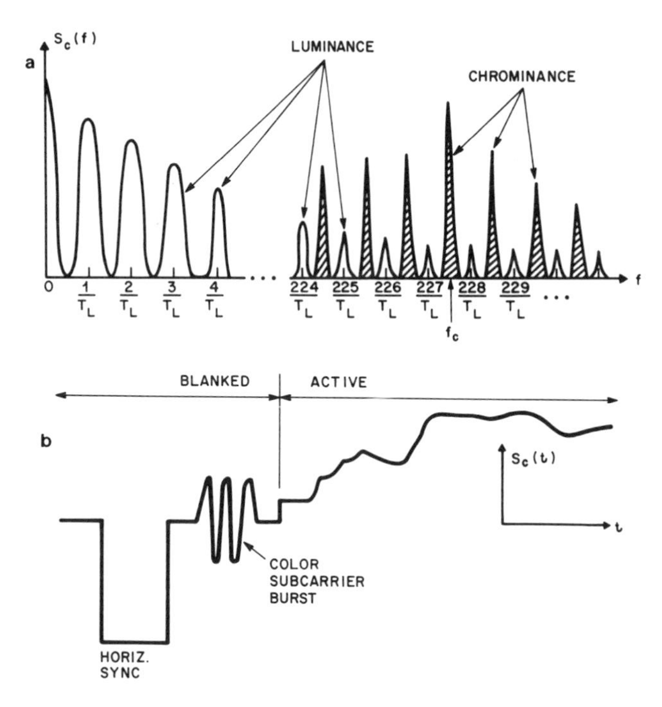

Fig. 2.2.4 NTSC composite color video signal. a) The color subcarrier lies halfway between two line-scan-rate harmonics. Ideally this enables frequency interleaving of luminance and chrominance signals. However, some pictures do cause luminance/chrominance overlap. b) A burst of several cycles of color subcarrier is inserted into the horizontal blanking interval to enable quadrature demodulation of chrominance at the receiver.

This is accomplished by first defining a color sub-carrier frequency that is exactly mid-way between two line-rate harmonics. For NTSC,

$$f_c \triangleq \frac{455}{2} f_H \triangleq 3579545 \text{ Hz} \tag{2.2.15}$$

where f_H is the line-scan-rate ($\approx$ 15734 Hz). The color sub-carrier frequency is chosen high enough so it is hard to see on monochrome receivers, yet low enough to be able to carry chrominance on two side-bands.

The two chrominance signals I and Q are then quadrature amplitude modulated (QAM) onto this carrier via the relation

$$C(t) = I(t)\cos(2\pi f_c t + 33°) + Q(t)\sin(2\pi f_c t + 33°) \tag{2.2.16}$$

The envelope of this QAM signal relative to the luminance, namely

$$\frac{\sqrt{I^2 + Q^2}}{Y} \tag{2.2.17}$$

is approximately the *saturation* of the color with respect to reference white, and the phase of $C(t)$, namely

$$\tan^{-1} \frac{Q(t)}{I(t)} \tag{2.2.18}$$

is approximately* the *hue*. Additive transmission noise affects mainly the envelope of the signal. Thus, this arrangement takes advantage of viewers' relative insensitivity to saturation distortion as compared with hue distortion. More will be said about hue and saturation in Chapter 4.

The chrominance and luminance signals are then added together to form the composite video signal shown in Fig. 2.2.4a

$$S_c(t) = Y(t) + C(t) \tag{2.2.19}$$

It can be shown that even though $\tilde{R}$, $\tilde{G}$, $\tilde{B}$ are restricted say to the range 0 to 1 volt, S_c can be negative and can also be as large as 1.2 volts positive.

---

* The terms hue and saturation are often (and incorrectly) used to describe dominant wavelength and excitation purity.

A short burst of several cycles of color sub-carrier

$$\sin 2\pi f_c t \tag{2.2.20}$$

plus a horizontal synchronization pulse are transmitted in the horizontal blanking time of the video signal, as shown in Fig. 2.2.4b. The phase difference of $33°$ between the sinusoidal carriers of Eqs. (2.2.16) and (2.2.20) facilitates demodulation in some receiver designs.

Eq. (2.2.16) defines a double sideband QAM signal $C(t)$ centered at $f_c \approx 3.58$ MHz. The modulated $Q(t)$ signal extends from 2.98 MHz to 4.18 MHz since its baseband width is 0.6 MHz. However, the modulated $I(t)$ signal extends from 2.08 MHz to 5.08 MHz since its baseband width is 1.5 MHz. Thus, the modulated $I(t)$ extends above the allocated total bandwidth, and when the composite signal is bandlimited to 4.2 MHz by filtering prior to transmission, $I(t)$ loses part of its upper sideband. Therefore, the high frequencies of $I(t)$ end up being transmitted via single sideband modulation (SSB).

Audio is usually frequency modulated (FM) on a carrier above 4.2 MHz. For example, over-the-air and cable transmissions to consumers use an audio carrier frequency of 4.5 MHz.

Demodulation, shown in Fig. 2.2.5, ideally involves a comb filtering* of $S_c(t)$ to separate $Y(t)$ and $C(t)$, followed by synchronous demodulation of $C(t)$ to recover $I(t)$ and $Q(t)$, followed by equalization of $I(t)$ to recover the missing portion of its sidebands. However, for reasons of economy, simple bandpass filters are often used for luminance-chrominance separation, and equalization is disregarded, resulting in bandwidths of about 3 MHz for the reproduced luminance and 0.5 MHz for the I and Q chrominance signals. Suboptimum demodulation results in visible, but not often objectionable degradation in the reproduced picture. For example, chrominance-to-luminance crosstalk is due to chrominance components below 3 MHz being included in the luminance signal. In the displayed picture it shows up mainly as serrations on edges of colored objects. Luminance-to-chrominance crosstalk is due to luminance components near the color subcarrier frequency. It shows up mainly as colored patterns in high detail areas of the picture, e.g., black/white stripes may erroneously produce color in a received picture.

In theory, digitization of the composite NTSC signal via PCM may be accomplished by sampling slightly above the Nyquist rate of 8.4 MHz. However, in reality, at this sampling frequency fine quantization plus very stringent and impractical pre/post-filters are needed to avoid color subcarrier aliasing

---

* Comb filtering requires both horizontal and vertical filtering.

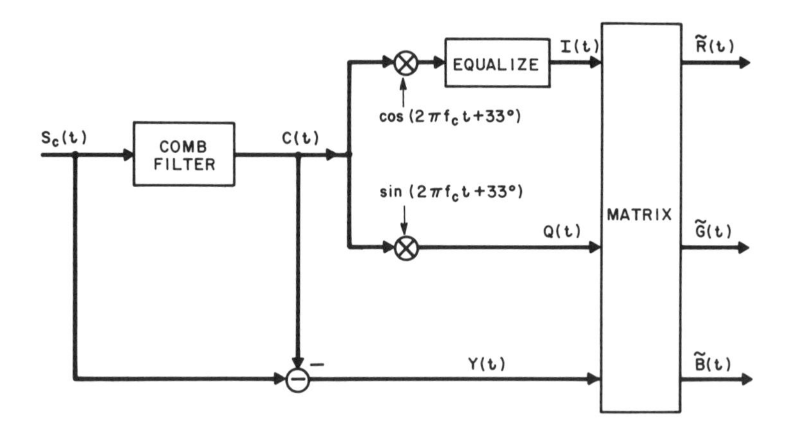

Fig. 2.2.5 Quadrature demodulation of NTSC composite signal $S_c(t)$. A comb filter with nulls at line-rate harmonics separates the chrominance $C(t)$ and luminance $Y(t)$. Quadrature mixing followed by equalization produce I and Q. Matrixing then produces $\tilde{R}$, $\tilde{G}$, and $\tilde{B}$.

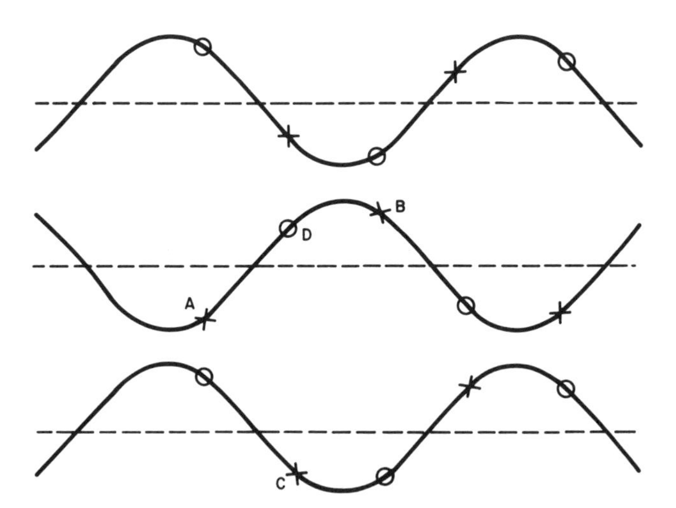

Fig. 2.2.6 Two successive lines of NTSC TV sampled at the sub-Nyquist rate of $2f_c$. The X pels are the original samples. The O pels must be interpolated for proper reconstruction.

distortions, which can appear as extraneous moving patterns in the reproduced picture.

The filtering requirements can be eased and the visibility of such aliasing patterns reduced by increasing the sampling rate and locking it to the color sub-carrier frequency. Sampling and quantization errors are then stationary with respect to the scanning raster and much less visible in the reproduced picture. A convenient sampling rate is $3f_c \approx 10.7$ MHz. However, $3f_c$ is not a multiple of the line-scan frequency, which means that samples are not aligned vertically* from line to line or from frame to frame and many digital signal processing operations cannot be easily implemented, e.g., comb filtering and digital demodulation. A sampling rate of $4f_c \approx 14.3$ MHz is a multiple of the line-scan rate and is usually used when high quality digital signal processing is contemplated.

TV studio signal processing often employs 9-bit quantization (because of the increased dynamic range of the composite waveform) and a sampling rate of $4f_c$. This results in a bit-rate of about 129 Mbs which, although economically acceptable for short distance transmission, may present a severe economic problem for digital recording or long distance transmission.

Experiments have been carried out with a *Sub-Nyquist* sampling rate of $2f_c$[2.2.5] shown by the *X* pels in Fig. 2.2.6. Reconstruction generally requires an interpolation in order to estimate the signal value between the samples. Moreover, this interpolation uses samples from adjacent lines above and/or below. For example, Fig. 2.2.6 shows two adjacent lines of a field sampled at $2f_c$. Pels *A*, *B* and *C* are known, while pel *D* is to be interpolated. The luminance of *D* is

$$Y_D \approx \frac{1}{2}(A+B) \tag{2.2.21}$$

The chrominance of *D* is then

$$C_D \approx Y_D - C \tag{2.2.22}$$

The interpolated value of *D* then becomes

$$D \approx Y_D + C_D = \frac{1}{2}(A+B) - C \tag{2.2.23}$$

* If the phase of the sampling clock is shifted appropriately every line period, vertical alignment of pels can be achieved.[2.2.8]

2.2.2 PAL

The *Phase Alternation Line* (PAL) color television system[2.2.6–7] is currently used in most of Western Europe. Scanning raster specifications are, as with NTSC, close to older monochrome standards (see Table 2.1.1, systems G, I and K).

PAL was developed some time after NTSC. Thus, the display primaries used in PAL are more in accord with the high luminance phosphors used in modern CRT displays. They are defined in terms of CIE chromaticity coordinates as

$$\begin{array}{cccc} & x & y & z \\ \boldsymbol{R}: & 0.64 & 0.33 & 0.03 \\ \boldsymbol{G}: & 0.29 & 0.60 & 0.11 \\ \boldsymbol{B}: & 0.15 & 0.06 & 0.79 \end{array} \tag{2.2.24}$$

Fig. 2.2.2a shows their locations on the CIE xy chromaticity diagram, along with the color triangle they define.

Using the relationships described in Section 1.8.2 we can compute the XYZ tristimulus matrix for the PAL primaries $\boldsymbol{RGB}$ using the xyz chromaticities $\boldsymbol{m}$ of Eq. (2.2.24). For the PAL primaries

$$\boldsymbol{M} = \begin{bmatrix} 0.514 & 0.265 & 0.024 \\ 0.324 & 0.670 & 0.123 \\ 0.162 & 0.065 & 0.853 \end{bmatrix} \tag{2.2.25}$$

$$\boldsymbol{M}^{-1} = \begin{bmatrix} 2.565 & -1.022 & 0.975 \\ -1.167 & 1.978 & -0.252 \\ -0.398 & 0.044 & 1.177 \end{bmatrix} \tag{2.2.26}$$

Thus, the luminance of a color is

$$Y = 0.265R + 0.670G + 0.065B \tag{2.2.27}$$

where R, G, B are the tristimulus values, normalized to equal energy white. However, reference white for PAL is chosen as D_{6500}, defined in Chapter 1. It has CIE xy chromaticity coordinates

$$x_D = 0.3127$$
$$y_D = 0.3291 \tag{2.2.28}$$
$$z_D = 0.3582$$

The relative amounts of primaries R, G, B necessary to give reference white D_{6500} are the rgb chromaticity coordinates of D_{6500}, found from Eq. (1.8.27)

$$r_D = 0.2797$$
$$g_D = 0.3525 \tag{2.2.29}$$
$$b_D = 0.3678$$

When the tristimulus values R, G, B of a color are normalized as in Eq. (2.2.10) so that $\tilde{R} = \tilde{G} = \tilde{B} = 1$ corresponds to reference white of unity luminance, the true luminance becomes

$$Y = 0.222\tilde{R} + 0.707\tilde{G} + 0.071\tilde{B} \tag{2.2.30}$$

where

$$\tilde{R} = 1.190\ R$$
$$\tilde{G} = 0.494\ G \tag{2.2.31}$$
$$\tilde{B} = 0.911\ B$$

followed by gamma correction using a gamma of 2.8.

For historical reasons, the three linear combinations of $\tilde{R}$, $\tilde{G}$, $\tilde{B}$ that are transmitted by PAL systems are given by

$$Y = 0.299\tilde{R} + 0.587\tilde{G} + 0.114\tilde{B}$$
$$U_t = \frac{\tilde{B}-Y}{2.03} \tag{2.2.32}$$
$$V_t = \frac{\tilde{R}-Y}{1.14}$$

or in matrix form

Table 2.2.2 Image Aspect Ratios of current systems.

	IAR
NTSC, PAL (4:3)	1.33
35mm and 16mm Movie Film	1.33
35mm Slides	1.48
EDTV, HDTV (16:9)	1.78
70mm Movie Film	2.10

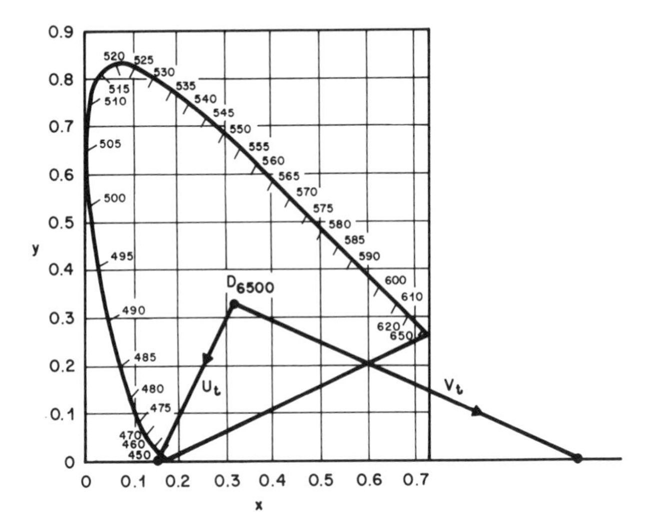

Fig. 2.2.7 Color axes used by the PAL and SECAM color television systems. Reference white is D_{6500}.

$$\begin{bmatrix} Y \\ U_t \\ V_t \end{bmatrix} = \begin{bmatrix} 0.299 & 0.587 & 0.114 \\ -0.147 & -0.289 & 0.436 \\ 0.615 & -0.515 & -0.100 \end{bmatrix} \begin{bmatrix} \tilde{R} \\ \tilde{G} \\ \tilde{B} \end{bmatrix}$$

$$\begin{bmatrix} \tilde{R} \\ \tilde{G} \\ \tilde{B} \end{bmatrix} = \begin{bmatrix} 1.0 & 0.0 & 1.140 \\ 1.0 & -0.394 & -0.581 \\ 1.0 & 2.030 & 0.0 \end{bmatrix} \begin{bmatrix} Y \\ U_t \\ V_t \end{bmatrix}$$

where the color-difference chrominance axes are shown in Fig. 2.2.7. $\tilde{R}$, $\tilde{G}$, $\tilde{B}$ are the gamma corrected, normalized values as described earlier for NTSC. Note that these are the same as Eqs. (2.2.10) and (2.2.12) for NTSC, but with different *RGB* primaries.

With PAL, the U_t and V_t chrominance signals of Eq. (2.2.32) are both transmitted with bandwidths of about 1.5 MHz. A color subcarrier is modulated with U_t and V_t via QAM, and the composite signal $S_c(t)$ is bandlimited to the allocated frequency band, thus truncating part of the upper sideband of the QAM signal $C(t)$, as with NTSC. However, unlike NTSC, the two chrominance signals U_t and V_t are of the same bandwidth, which results in their both losing part of their upper sideband. Thus the high frequencies of both the U_t and V_t chrominance signals are transmitted via single sideband (SSB). Since simple SSB can only transmit one signal at a time, further processing is required.

This problem is circumvented by taking advantage of the fact that chrominance signals do not, on average, change very much in successive scan lines of the picture. Thus, PAL alternately transmits the V_t chrominance signals as $+V_t$ and $-V_t$ in successive scan lines. That is,

$$C(t) = U_t \cos 2\pi f_c t \pm V_t \sin 2\pi f_c t$$

$$S_c(t) = Y(t) + C(t) \tag{2.2.33}$$

Demodulation of the QAM signal $C(t)$ is similar to NTSC. With *simple PAL* only the low frequency chrominance is reproduced, and simple QAM demodulation suffices. However, with *standard PAL* the high chrominance frequencies are received as $U_t \pm V_t$. In order to fully recover the high chrominance frequencies, successive demodulated scan lines must be averaged together (with proper signs) to obtain the chrominance signals* U_t and V_t. Equalization of both U_t and V_t is then required to restore the truncated sidebands to proper amplitude.

An unfortunate side effect of alternating the sign of V_t is that its frequency spectrum is shifted by one-half the line rate and occurs halfway between the peaks of the luminance spectrum.[2.2.1] This means that the color subcarrier cannot be chosen half-way between two line-scan-rate harmonics as it is in NTSC. Instead, it is chosen to be one quarter of the way as shown in Fig. 2.2.8. A small additional offset is also added to further reduce visibility on monochrome receivers and to reduce cross-color interference. The resulting color subcarrier value is

$$f_c \triangleq 283.75\, f_H + 25 \approx 4.43 \text{ MHz} \tag{2.2.34}$$

where $f_H \triangleq 15625$ Hz.

A fortunate result of sign alternation is that phase errors, i.e., hue errors, due to transmission are canceled somewhat by the line-to-line averaging during demodulation. In fact, the cancellation is good enough so that hue and tint controls are not required on standard PAL receivers as they are on NTSC receivers.

Comparison of PAL and NTSC is hampered by the different monochrome standards on which they are based. PAL has higher spatial resolution so that pictures look sharper, but NTSC has higher temporal resolution so that large area flicker visibility is reduced. However, PAL does have an advantage insofar as immunity to certain analog transmission degradations is concerned.

Digitization of the composite PAL signal via PCM suffers the same problems as with NTSC. Unless the sampling rate is a multiple of the color subcarrier frequency, anomalous interference in the reproduced picture may be difficult to remove, and certain signal processing operations may be difficult to implement digitally. Even with synchronous sampling, the pels are not vertically aligned. Sampling at a rate of $4f_c \approx 17.7$ MHz and quantizing to nine-bit accuracy results in a bit-rate of about 160 Mbs. Experiments have also been carried out with a sampling rate of $2f_c$.[2.2.10]

2.2.3 SECAM

The SECAM color television system[2.2.7] is used in France, Russia, the Middle East and Eastern Europe (see Table 2.1.1, systems D, K, L). The name is derived from *Sequentiel Couleur avec Memoire,* which refers to the fact that

---

* The color resolution in the vertical direction is thus reduced. However, it is still larger than the color resolution in the horizontal direction. A delay of one scan line is required to perform the averaging.

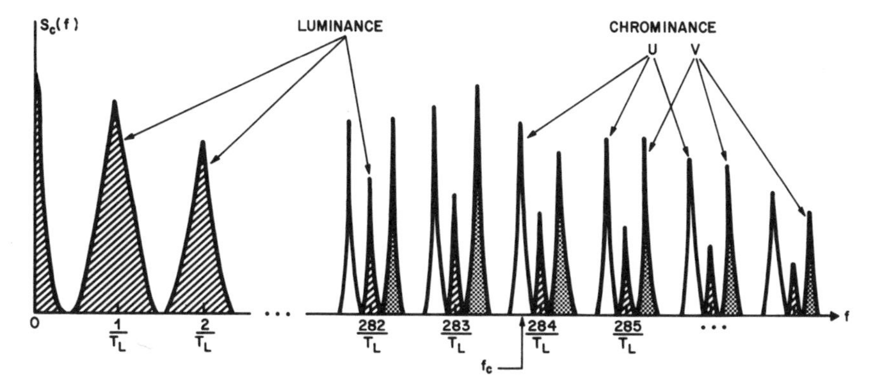

Fig. 2.2.8 PAL composite color video signal. The color subcarrier f_c and the U_t chrominance signal lie one-quarter of the way between two line-scan-rate harmonics. The V_t chrominance signal lies three-quarters of the way between.

the U_t and V_t chrominance signals are transmitted alternately on successive scan lines and that a delay element (memory) of one scan line duration is used to make the chrominance signal of the previous line also available for producing the present line. Since only one of the two chrominance signals U_t and V_t is transmitted on any scan line, there is no possibility of interference between them.

In SECAM the raster scan specifications, as well as the luminance and chrominance definitions, are the same as for PAL, except the video bandwidth is 6 MHz. The chrominance signals $+U_t$ and $-V_t$ are transmitted on alternate lines by frequency modulation (FM) of the two color subcarriers:

$$f_U = 272\ f_H \approx 4.25\ \text{MHz}$$

$$f_V = 282\ f_H \approx 4.41\ \text{MHz} \tag{2.2.35}$$

As with most FM transmissions, the baseband signals are *pre-emphasized* prior to modulation. However, unlike other systems, the carrier frequency is amplitude modulated by a small amount.[2.2.4] In order to further reduce visibility on monochrome receivers, the phase of the subcarriers is reversed 180° every third line.

With an allocated composite signal bandwidth of 6 MHz, very little truncation of chrominance sidebands occurs, with the result that high quality demodulation is simpler than with the QAM used in PAL and NTSC. FM is also much less susceptible to gain distortion, although phase distortion causes about the same degradation as with PAL.

Digitization of SECAM is slightly more difficult than with PAL or NTSC. Since there is no stable color subcarrier signal for locking the sampling rate, it must be locked to the line-scan rate.[2.2.10] Also, the FM chrominance is not very amenable to signal processing; e.g., even a simple picture fade-out requires chrominance demodulation in its implementation since zero luminance does not make the FM chrominance signals zero. Sampling at a rate of $4f_V$ and quantizing to nine-bit accuracy results in a bit-rate of about 159 Mbs.

2.2.4 Standard Conversion

A technical summary of the NTSC, PAL and SECAM color television systems is given in Table 2.2.1. Conversion of a television signal from one standard to another is no trivial task. The most straightforward approach is the so called "optical scan converter" where a receiver/monitor designed according to one standard takes the input signal and produces a color image on its CRT display. A color camera operating according to the desired output signal standard is then pointed at the CRT display, thus producing the output video waveform.

Far better results are obtainable using digital standards converters.[2.2.13] Basically, these operate by first decomposing the input composite signal into color components, e.g., Y, $R-Y$ and $B-Y$. The field rates and number of lines per frame of these components are then changed by first digitizing and then using spatial and temporal interpolation, i.e., digital filtering. Finally, the components are modulated and combined to form an output waveform according to the desired standard.

Another form of standards conversion occurs when movie film is to be scanned for television broadcast. For PAL it is straightforward, since 24 frame/sec film can be speeded up to PAL's 25 frames/sec without much degradation. However, a speedup to the 29.97 frames/sec of NTSC would be intolerable.

An easy solution for NTSC is to scan alternate frames of film using three fields instead of the customary two. This process is called "3:2 pulldown". The film is advanced at the rate of 23.97 fps, while the video scans at 59.94 fields per second. For those film frames that use three fields, the first and third field scan the same film area; however, their composite signal is not the same because of the color carrier phase shift between NTSC frames.

2.2.5 Component Digitization

Another approach to digitizing broadcast color television is to first demodulate the composite signal into its luminance and chrominance components, or, alternatively, to start with the components themselves right at the camera. This avoids the distortion problems that are due to intermodulation with the color subcarrier and, in some cases, enable lower bit-rate coding. The $\tilde{R}$, $\tilde{G}$, $\tilde{B}$ signals themselves could be PCM coded directly, thus avoiding the need for matrixing. However, coding the luminance and chrominance signals Y, C_1, C_2 results in a lower bit-rate because of the much lower bandwidths of the chrominance,[2.2.11] and is preferable in most applications.

For example, with NTSC signals the Y, I and Q components ideally require only Nyquist rate sampling. Using 8-bit quantization for luminance and 7-bits for chrominance results in a bit-rate of about 97 Mbs, which is the same as the $3f_c$, 9-bit PCM rate for composite signals. If line-to-line similarity of chrominance signals is exploited so that I and Q are transmitted on alternate lines (both sampled at 3 MHz), the component bit-rate falls to about 88 Mbits/sec.

If slightly less picture quality is acceptable, the I and Q chrominance signals can both be bandlimited to 0.6 MHz and quantized to 6-bit accuracy. The two digital chrominance signals can then be time compressed and transmitted alternately during the horizontal blanking time of each scan line, resulting in a bit-rate of about 67 Mbs.

If a further reduction in picture quality is tolerable, e.g., only slightly annoying degradation comparable to home cassette videotape quality, then the luminance can be bandlimited to 3 MHz, PCM coded to 7-bit accuracy, and all components transmitted as above. The resulting bit-rate is about 42 Mbs, which is less than half the above studio quality bit-rate of 97 Mbs. Component coding of PAL and SECAM video can be accomplished in similar fashion, albeit at higher bit-rates because of the larger bandwidths involved.

CCIR 601

In an attempt to produce a digital signal that is more compatible with NTSC, PAL and SECAM, an international standard for component video coding has been established by the ITU-R*, formerly known as the CCIR.[2.2.12] The CCIR 601 standard comes in two flavors. One is referred to as 525/60 and has the same scanning format as NTSC. The other is 625/50 and has the same scanning format as PAL.

The luminance and chrominance components used for CCIR 601 are linear combinations of gamma corrected $\tilde{R}$, $\tilde{G}$, $\tilde{B}$ normalized either to NTSC's reference white $\boldsymbol{C}$ or PAL's and SECAM's reference white $\boldsymbol{D}_{6500}$. All three components are quantized with 8-bits, which gives a nominal digital range of 0-255. However, the maximum and minimum values are reserved for synchronization, which leaves the range 1–254 actually available for video.

If $\tilde{R}$, $\tilde{G}$, $\tilde{B}$ are in the range 0–1 volt, then, as before, the luminance

$$Y = 0.299\,\tilde{R} + 0.587\,\tilde{G} + 0.114\,\tilde{B} \tag{2.2.36}$$

is also in the range 0–1. In order to provide working margins for coding, filtering, etc., the full digital video range is not used for the initial digitization of the component signals. Instead the digital luminance is defined as

$$Y_d = 219\,Y + 16 \tag{2.2.37}$$

rounded to the nearest integer, which gives a digital luminance range of 16–235, i.e., 220 levels starting at 16.

Similarly, the standard digital color-difference signals are defined by

$$\begin{aligned} C_B &= \frac{112(\tilde{B} - Y)}{0.886} + 128 \\ C_R &= \frac{112(\tilde{R} - Y)}{0.701} + 128 \end{aligned} \tag{2.2.38}$$

---

* International Telecommunications Union - Radio Sector.

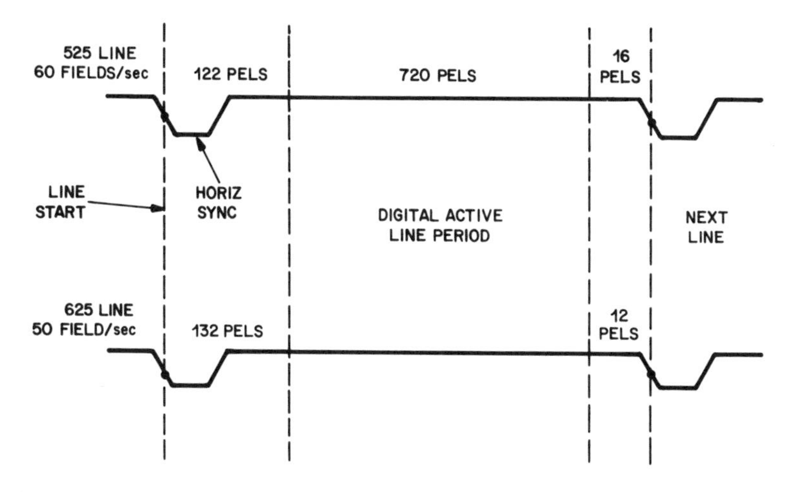

Fig. 2.2.9 CCIR 601 standard luminance sampling structure for 525/60 and 625/50 local video formats. The start-of-line is defined by the leading edge of the horizontal sync. The standard specifies 720 active luminance pels per line and 360 active chrominance pels. Only the active pels need be transmitted.

rounded to the nearest integer, which gives a digital color-difference range of 16–240, i.e., 225 levels centered at 128. Maximum and minimum values of C_R occur for red ($\tilde{R}=1,\ \tilde{G}=\tilde{B}=0$) and cyan ($\tilde{R}=0,\ \tilde{G}=\tilde{B}=1$), respectively. Maximum and minimum values of C_B occur for blue ($\tilde{R}=\tilde{G}=0,\ \tilde{B}=1$) and yellow ($\tilde{R}=\tilde{G}=1,\ \tilde{B}=0$), respectively.

If we define the 8-bit digital variables $R_d = 255\tilde{R}$, $G_d = 255\tilde{G}$ and $B_d = 255\tilde{B}$ then we can write the CCIR 601 color transformations as:

$$\begin{bmatrix} Y_d \\ C_B \\ C_R \end{bmatrix} = \begin{bmatrix} 0.257 & 0.504 & 0.098 \\ -0.148 & -0.291 & 0.439 \\ 0.439 & -0.368 & -0.071 \end{bmatrix} \begin{bmatrix} R_d \\ G_d \\ B_d \end{bmatrix} + \begin{bmatrix} 16 \\ 128 \\ 128 \end{bmatrix} \tag{2.2.38a}$$

$$\begin{bmatrix} R_d \\ G_d \\ B_d \end{bmatrix} = \begin{bmatrix} 1.164 & 0.0 & 1.596 \\ 1.164 & -0.392 & -0.813 \\ 1.164 & 2.017 & 0.0 \end{bmatrix} \begin{bmatrix} Y_d - 16 \\ G_d - 128 \\ B_d - 128 \end{bmatrix}$$

The sampling rates for the luminance and chrominance components are, respectively,

$$\begin{aligned} f_y &= 858 f_{NTSC} \text{ or } 864 f_{PAL} \\ &\approx 13.5 \text{ MHz} \\ f_c &= \frac{1}{2} f_y \approx 6.7 \text{ MHz} \end{aligned} \tag{2.2.39}$$

where f_{NTSC} and f_{PAL} are the respective line scan frequencies for NTSC and PAL. The sampling structure should be orthogonal and static, spatially and temporally, and color pels are co-sited with the 1st, 3rd, etc., luminance pels of each line. The number of active luminance pels in each line is 720, as shown in Fig. 2.2.9. The number of active lines per frame is 486 for NTSC and 576 for PAL (the top and bottom active lines may contain only a half line of active pels).

Care should be taken when dealing with *frames*, i.e., two interlaced fields, of CCIR 601 data. According to current standards, the field containing the top line of the frame should occur temporarily *later* than the field containing the bottom line. However, not all implementations adhere to this convention. Also, when live video is interspersed with 3:2 pull down film source, the convention may be violated.

Since all three components (Y_d, C_B, C_R) are quantized with 8-bits, the total bit-rate for this standard is about 216 Mbs, which is relatively large for purposes of digital recording or transmission, but which may be useful for studio digital signal processing and standards conversion.

In deciding whether to employ component coding or composite signal coding, several factors must be taken into account. The first consideration is whether the components are already available, or whether they have to be obtained by demodulating the composite signal. Demodulation entails extra expense and often some loss in picture quality. The second consideration is the type of signal processing to be done. For example, standards conversion is better done with components. However, time-base-correction may be easier with a composite signal since all that is required is a time stretching or time compression of the scan lines. A final consideration is bit-rate reduction. As we shall see later, component coding is often more amenable to redundancy reducing techniques than is composite signal coding.

Image and Pel Aspect Ratio

The Image Aspect Ratio (*IAR*) defined as the ratio of active width to height is shown for several image formats in Table 2.2.2. Related to *IAR* is the concept of Pel Aspect Ratio (*PAR*). We saw in Chapter 1 that pels have no intrinsic shape of their own. However, in the very restrictive case of nearest-neighbor reconstruction, shown in Figs. 1.3.4c and 1.3.5a, we can think of pels as adjacent rectangles completely filling the picture. *PAR* is defined as the ratio of width to height of one of these rectangles.* CCIR 601 has *PAR* values of 0.900 for NTSC 525/60 and 1.067 for PAL 625/50.

For a digitized picture of width W, height H, N_x pels horizontally and N_y pels vertically, we have

$$PAR = \frac{W/N_x}{H/N_y} = \frac{N_y}{N_x} IAR \tag{2.2.40}$$

PAR is important in editing, cropping, picture insertion, etc. It is also a factor to be considered when stored or transmitted images must be reproduced on a variety of displays. In all such applications, the *PAR* must be preserved to within an accuracy of a few percent. Otherwise, objects will change shape, circles become ellipses and thin/fat people become fat/thin people.

If the *PARs* of image and display do not match, further processing is required. Changing an image *PAR* is possible by increasing or decreasing the number of pels per line N_x. Sometimes, simply replicating or deleting a certain percentage of randomly chosen pels is sufficient. Otherwise, more sophisticated methods of interpolation and decimation may be required.[2.2.14]

* Some authors define IAR and PAR as height/width.

2.3 Videoconferencing

Videoconferencing systems are applicable where two or more groups of people at widely separated locations wish to communicate in real time, both visually and orally, i.e., hold a conference. The objective is to save the time and costs that would be incurred by physically bringing all of the conferees to the same location. Cameras are usually fixed, and participants are usually seated. Thus, the amount of motion in a typical videoconference scene is considerably less than with broadcast television.

In this situation the number of viewers per transmission path is much less than with broadcast television. Thus, costs of implementation and transmission are extremely important in determining whether or not such a service is viable. For this reason, picture quality constraints are usually much less stringent than for broadcast television. Also, the application is different — conferencing being much more task oriented and broadcast television being much more entertainment oriented.

A standard-rate broadcast television picture can adequately display only a few people at a time for conferencing purposes. One way of accommodating more conferees is to use several cameras, each one focussed on a different subgroup, and to switch between cameras depending on who is talking as shown in Fig. 2.3.1a. This has the advantage of producing only one video signal for transmission, using standard audio conferencing equipment and standard video components. However, it has the disadvantage of never showing people who remain silent, and also introducing a certain amount of discontinuity into the interpersonal communication. Human factors tests have shown that conferees would like to see all participants, if possible.

More people can also be accommodated by the use of a split frame video signal, using two cameras at each location. With this technique one camera is aimed to show one group of people in the bottom half of its frame; the other is aimed to show another group in the top half of its frame. The video output is then switched from one camera to the other halfway through each frame, thus transmitting pictures of two groups of people in one video signal. At the receiving end, the switching process is undone to show one group on each monitor in their proper locations, or, alternatively, both groups of people could be shown on one monitor by simply displaying the received signal as is shown in Fig. 2.3.1b.

Of course, multiple conferees can always be portrayed by simply sending multiple video signals and bearing the extra costs entailed. However, if a lower resolution picture and/or lower frame rate can be tolerated, then frequency multiplexing or time multiplexing techniques can reduce transmission costs considerably. For example, three cameras might be used at each location, each one focused on a different subgroup of the participants. If only one field out of three is transmitted from each camera, then by interleaving the fields (time division multiplexing) a standard TV signal can be produced for transmission. At the receiving end, each field is routed to the appropriate one of three monitors and

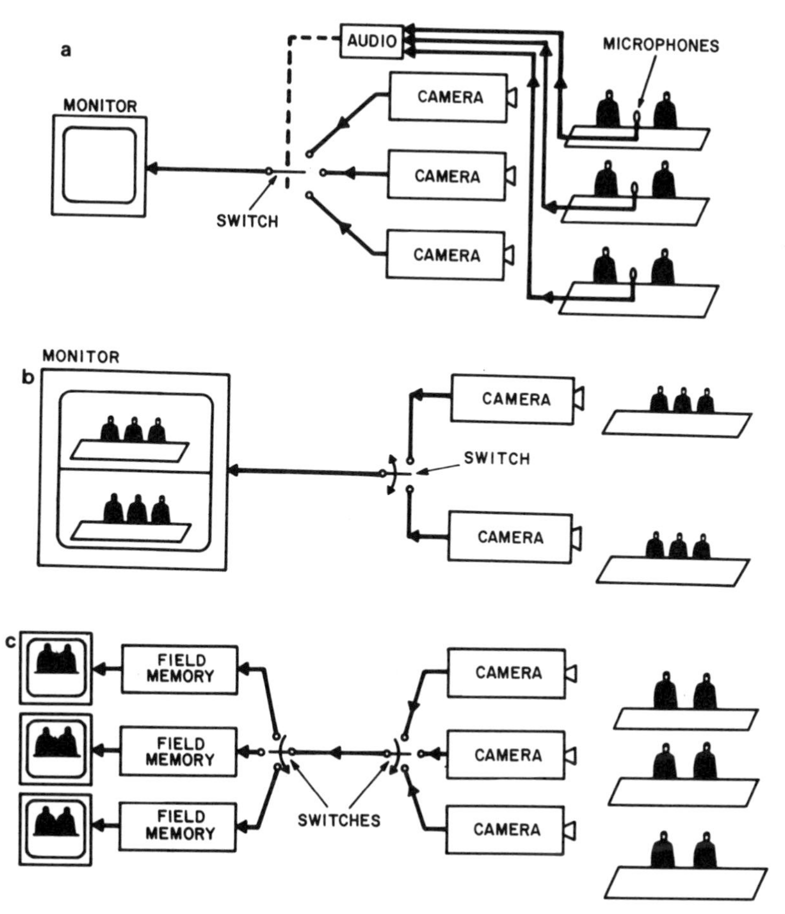

Fig. 2.3.1 Videoconferencing arrangements for sending multiple pictures using only one video channel. (a) Three cameras are pointed at three groups, each having a separate microphone. The strongest audio signal causes the switch to move to the corresponding camera video output, thus displaying the speaker at the receiving end. (b) The switch toggles from one camera to the other at the start and the middle of each field. The received picture shows all participants. The image could also be split at the receiver to show participants side by side on two separate monitors. (c) The switches move to the next position at the start of each field. At the receiver, field memories enable each field to be displayed three times. Thus, all participants are shown, but at a reduced temporal resolution.

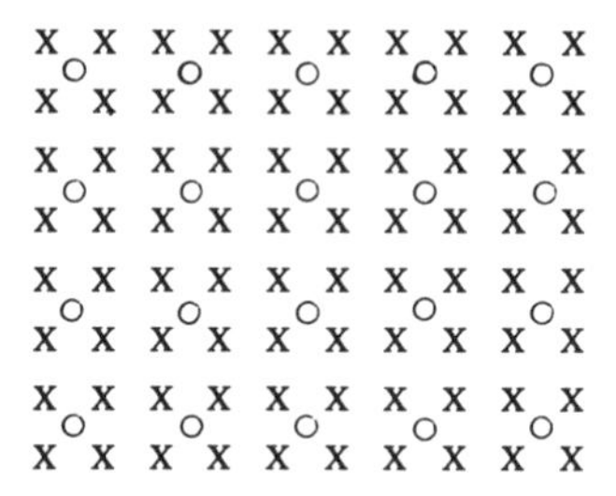

Fig. 2.3.2 Positioning of luminance and chrominance samples in P *64 video.

Table 2.3.1 Video parameters for P*64, active area.

	CIF	QCIF
Luminance Pels per Line	352	176
Chrominance Pels per Line	176	88
Luminance Lines per Frame	288	144
Chrominance Lines per Frame	144	72
Maximum Frames per Second	29.97	29.97
Interlace	1:1	1:1
Color Components	YC_BC_R	YC_BC_R
Luminance Range	16-235	16-235
Chrominance Range	16-240	16-240
Zero Color Difference Level	128	128
Bits per Pel	8	8
Pel Aspect Ratio (PAR)	1.067	1.067
Image Aspect Ratio (IAR)	4:3	4:3

displayed repetitively three times as shown in Fig. 2.3.1c.

As described so far, transmission could be either digital or analog. However, digital coding offers the possibility of bit-rate reduction techniques and less expensive transmission. At present, the digital transmission facilities of greatest use for videoconferencing are the so-called *primary rates* of 1.544 Mbits/sec in North America and Japan, and 2.048 Mbits/sec in Western Europe, plus the Integrated Services Digital Networks (ISDN), which are being designed and operated according to international standards. The most useful ISDN bit-rates for videoconferencing services are probably 384 kbits/sec and multiples thereof, up to and including the primary rates.

However, standardized bit-rates are not the only thing needed for internationally compatible videoconferencing service. For this reason, ISDN standard video coding methods have been defined. They are commonly referred to as the P∗64 video coding standards,[2.3.1] which indicates that they operate on multiples of 64 kbits/sec ISDN B-channels. Values of P can range from 1 to 30.

P∗64 defines a standard video input called the Common Intermediate Format (CIF). It uses a frame rate of 29.97 Hz (same as NTSC), no interlace, 288 active lines per frame (one active field of PAL) and 352 active pels per line aligned vertically from line to line. Color sampling is at half the rate of luminance, both horizontally and vertically as shown in Fig. 2.3.2.

For reasons of interoperability and low cost, a lower resolution format, called Quarter Common Intermediate Format (QCIF) has also been defined in $P*64$. It has half the resolution of CIF, both horizontally and vertically, and for compatibility it is required to be supported in all P∗64 terminals. Parameters of CIF and QCIF are shown in Table 2.3.1.

Input video bit rates at maximum frame rate for CIF and QCIF are 36.5 Mbits/sec and 9.1 Mbits/sec, respectively. Decoders are allowed to signal that they cannot operate at maximum frame rate. However, they must be able to handle at least 7.49 frames per second.

2.4 Videotelephone

Interpersonal video communication between two seated individuals does not usually require the full resolution that is provided by the Common Intermediate Format. Specifications for the early U. S. Picturephone®, shown in Fig. 2.4.1, called for 271 lines per frame, 30 frames per second, 2:1 line interlace and about 1.0 MHz of video bandwidth. This corresponds to roughly half the CIF of P∗64 videoconferencing. The lower resolution of QCIF is also quite suitable for videophone, especially if transmission is via modems on the analog

Fig. 2.4.1 Early U. S. Picturephone®, circa 1968.

Table 2.5.1 HDTV parameters for SMPTE 240M.

Parameter	Value
Lines/Frame (total)	1125
Lines/Frame (active)	1035/1080
Interlace	2:1
Image Aspect Ratio (IAR)	16:9
Field Rate	60.00 Hz
Line Rate	$\approx$ 33750 Hz
Gamma (normal light levels)	2.2
(low light levels)	1.0
Reference White	D_{6500}
Bandwidth ($\tilde{R}$, $\tilde{G}$, $\tilde{B}$, Y)	30MHz
Bandwidth (P_B, P_R)	15MHz

Public Switched Telephone Network (PSTN). The ITU-T* (old CCITT) is standardizing video coding techniques for PSTN that are small modifications of the P*64 methods. A major objective of these standards is low cost. However, high compression is also required since the PSTN can support, at the very most, about 40 kbits/sec, including video and audio. See Chapter 8 for more details.

2.5 High Definition Television (HDTV)

HDTV is generally thought of as high fidelity color television. It is raster scanned with at least twice the vertical resolution and twice the horizontal resolution as standard color television. Thus, bandwidths and bit-rates are at least four times those of standard TV. HDTV is especially beneficial in projection television systems where viewers may be located as close as three times picture height. By comparison, standard viewing distance for today's TV is assumed to be six to eight times picture height.

It has been found from subjective tests that viewer appreciation of image quality is increased if the image aspect ratio of HDTV is increased from the 4:3 of standard TV to 16:9. This is due to the fact that most natural scenes benefit far more from an increase in width than from an increase in height. If horizontal resolution is increased to match the increased aspect ratio, then the overall HDTV signal bandwidth may be more than five times that of standard TV.

HDTV signal standards are nearly complete after a long period of negotiation and study. As in the case of color TV, the question of compatibility between HDTV and standard TV is an important consideration. However, it is not certain that such compatibility is economically feasible or subjectively acceptable. The situation would be eased somewhat if HDTV were to have exactly twice the number of scan lines as standard TV. In this case, alternate lines could be stretched in time and sent in different frequency bands, and one of the bands would be at least compatible in line scan rate with standard TV. Additional processing would probably be necessary to construct standard composite color waveforms having the standard aspect ratio. Naturally, if converters must be provided to enable standard TV sets to receive HDTV, then these should be as simple and inexpensive as possible.

A standard for HDTV production has been established by the Society of Motion Picture and television Engineers.[2.5.2] It is called SMPTE 240M, and has the parameters shown in Table 2.5.1. The standard color primaries are defined in terms of CIE *xyz* chromaticity coordinates as:

---

* International Telecommunications Union - Telecommunication Sector.

$$
\begin{array}{cccc}
 & x & y & z \\
\boldsymbol{R}: & 0.630 & 0.340 & 0.030 \\
\boldsymbol{G}: & 0.310 & 0.595 & 0.095 \\
\boldsymbol{B}: & 0.155 & 0.070 & 0.775
\end{array}
\tag{2.5.1}
$$

Using these chromaticities plus those of D_{6500} and normalizing as before we obtain

$$
Y = 0.212\tilde{R} + 0.701\tilde{G} + 0.087\tilde{B} \tag{2.5.2}
$$

where $\tilde{R}$, $\tilde{G}$ and $\tilde{B}$ are gamma corrected components. In order that all components have the same peak-to-peak amplitude, the color differences $\tilde{B} - Y$ and $\tilde{R} - Y$ are divided by 1.826 and 1.576, respectively, to give

$$
\begin{bmatrix} Y \\ P_B \\ P_R \end{bmatrix} = \begin{bmatrix} 0.212 & 0.701 & 0.087 \\ -0.116 & -0.384 & 0.500 \\ 0.500 & -0.455 & -0.055 \end{bmatrix} \begin{bmatrix} \tilde{R} \\ \tilde{G} \\ \tilde{B} \end{bmatrix}
$$

(2.5.3)

$$
\begin{bmatrix} \tilde{R} \\ \tilde{G} \\ \tilde{B} \end{bmatrix} = \begin{bmatrix} 1.000 & 0.0 & 1.576 \\ 1.000 & -0.227 & -0.477 \\ 1.000 & 1.826 & 0.0 \end{bmatrix} \begin{bmatrix} Y \\ P_B \\ P_R \end{bmatrix}
$$

An alternative to HDTV in improving television resolution is to maintain the standard number of scan lines per frame, but increase the allowable bandwidth, and perhaps the image aspect ratio, of the video signals both in luminance and chrominance. This is termed *Extended Definition Television* (EDTV). If no further steps were taken, the result would be an increase in horizontal resolution, but none in vertical resolution. In order to improve vertical resolution, several approaches have been proposed.[2.5.1] For example, one suggestion is to simply operate the television camera in a noninterlaced mode at the standard frame rate. By means of digital memories at the receiver, each frame is then displayed twice, and in this way the reduction in vertical resolution due to interlace (see Section 1.6) is removed. In fact, with digital memory at the camera, the odd numbered lines of a frame can be transmitted first, followed by the even numbered lines, thus maintaining interlace compatibility with standard receivers. With this proposal, EDTV receivers would still display each frame twice in a noninterlaced mode of operation, however.

2.6 Graphics

Several systems have been invented for communication of graphical signals. Although a variety of graphical signals were described in Chapter 1, accepted standards and compatible systems have not yet emerged for many cases. However, for document facsimile systems considerable progress has been made towards standardizing the signal and transmission format so that communication between machines made by different manufacturers would be possible. Much of this has been done by the CCITT* It is for this reason that we describe the document facsimile system in detail.

2.6.1 Facsimile Systems

A block diagram of a Facsimile system is shown in Fig. 2.6.1. An original document is scanned and the light is converted into an electrical signal. In the photoelectric conversion block, older drum scanners in which the document is wrapped around a drum and scanned are being replaced by flat-bed scanners in which scanning is done while the document remains flat. The electrical signal is then processed and encoded for transmission over a variety of channels. The public switched telephone network (either analog or digital using modems) is the most popular transmission channel, but more and more use of public data networks is also occurring, e.g., ISDN (Integrated Services Digital Network).

Fig. 2.6.2 shows the frequency characteristics of a typical analog telephone channel. The characteristics indicate a reasonably flat portion between 300 to 2500 Hz and a sharp fall on both sides of this band. Analog transmission of facsimile signals over private telephone lines generally uses amplitude modulation (AM) of a carrier by the scanner output signal. For a switched public telephone network, because of the higher noise levels involved, frequency modulation (FM) of a carrier is used. In order to allow for increased scanning speeds and, therefore, decrease the transmission time, Vestigial Side Band (VSB) analog modulation is sometimes used. Also, since many documents such as business letters and engineering drawings can be satisfactorily reproduced when quantized in only two-tones (see Chapter 1), modern facsimile systems deal only with bi-level signals and allow a reduction in transmission time by data compression techniques.

Data transmission using modems over the telephone network is typically at 2400, 4800, 9600 or 14400 bits/sec. Newer data networks offer higher transmission speeds (e.g. 56 kbits/sec) and error free reception. Also services such as store-and-forwarding of documents are possible using packet-switched data

---

* Comité Consultatif International Télégraphique et Téléphonique. The CCITT has since renamed itself ITU-T.

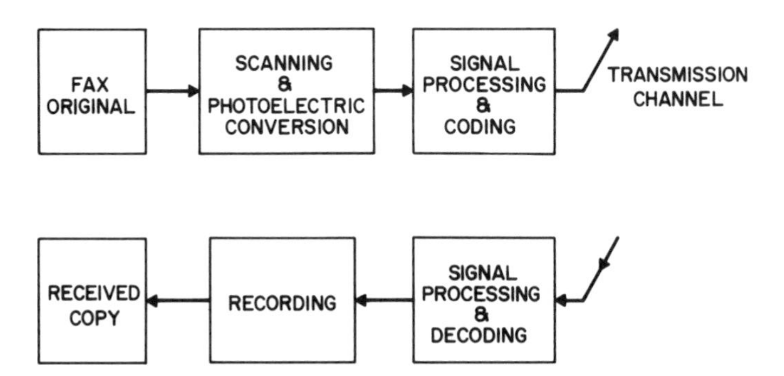

Fig. 2.6.1 Block diagram of a document facsimile system.

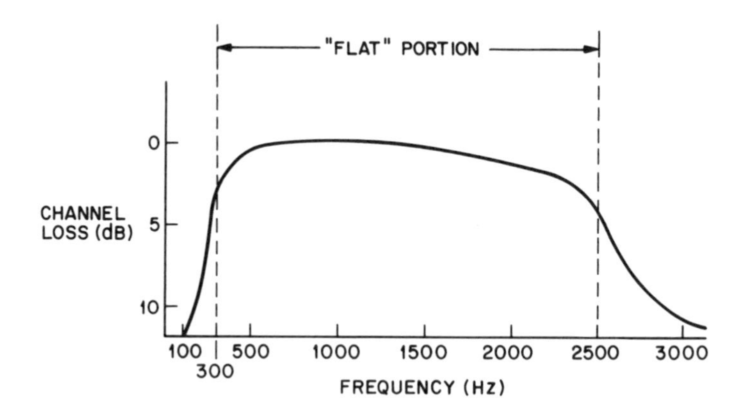

Fig. 2.6.2 Frequency characteristic of a typical telephone circuit.

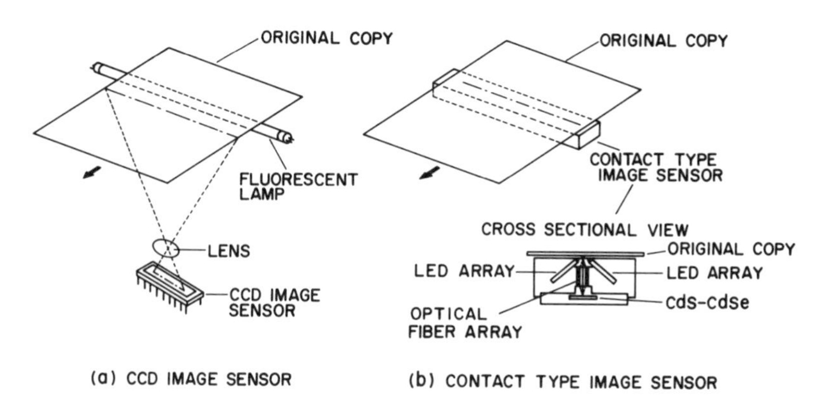

Fig. 2.6.3 Two popular methods for photoelectric conversion in a facsimile scanner.

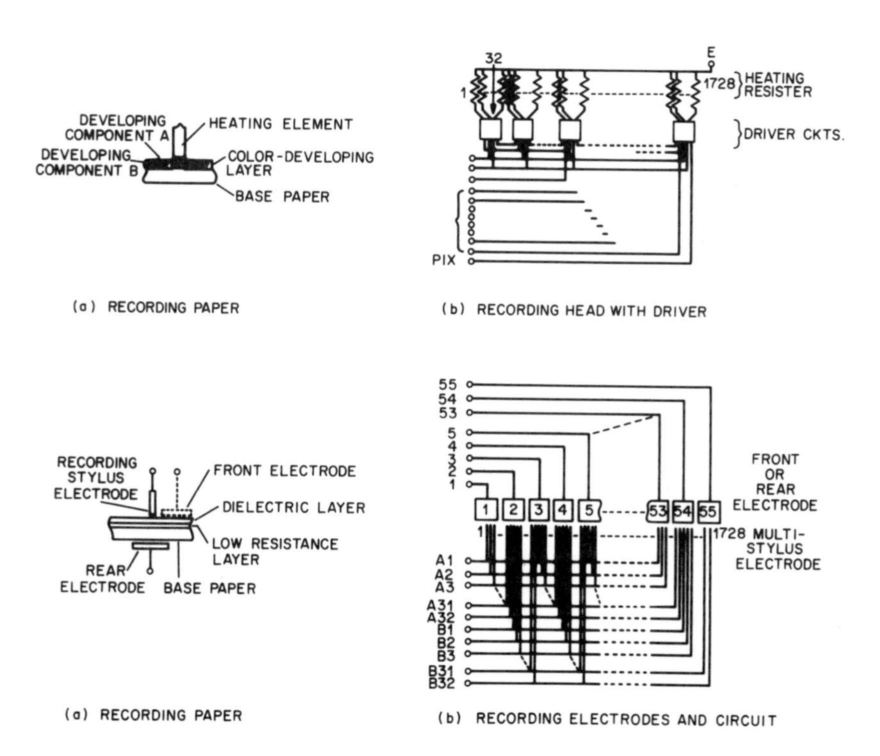

Fig. 2.6.4 Thermal (top) and electrostatic (bottom) methods of recording used in facsimile receivers.

networks. At the receiver, signal processing is used to convert the received, coded signal into an analog signal that drives the machine to record on paper.

Fig. 2.6.3 shows two popular methods of scanning. Fig. 2.6.3a shows a projection system in which a fluorescent lamp is used as a light source and a reduced image of the original document is formed by a lens on to an image sensor. A charge coupled device (CCD) array with about 2,000 to 5,000 photodiodes arranged linearly is generally used as the sensor. Since a lens system is required to focus the image onto the small CCD array, a long optical path is necessary resulting in a bulky scanner. This is obviated by a contact type scanner that allows image formation and photoelectric conversion to be performed at the same width as the original copy as shown in Fig. 2.6.3b. This Figure also shows a light emitting diode (LED) array used as a light source and a photo sensitive material using CdS-CdSe. In both these systems, due to the reduction of the light transmission at the periphery of the lens and uneven illumination from the light source, the output signal within a single scanning period is not uniform. This phenomenon, called *shading*, has to be compensated by proper signal processing.

At the facsimile receiver, many different kinds of recording or writing methods have been used such as electrostatic, ink jet, thermal recording, and electro photographing. Of these, thermal and electrostatic recording are the most widely used. Fig. 2.6.4 (top) shows the thermal recording method. The recording paper contains a layer of color-developing emulsion created on a base paper. As shown in the figure, the emulsion contains a dispersion of two components A and B that chemically react when heat is applied, thus developing the image.

The thermal recording head contains a row of heating resistors spaced at small intervals. The density of heating elements can be as high as 12 elements/mm in practical systems. Fig. 2.6.4 (bottom) shows the electrostatic recording method. Here the recording paper contains a low resistance layer applied to a paper base with a dielectric layer on top. This paper is inserted between two electrodes, one on the recording stylus and one in the rear. By applying a high voltage across the electrodes, an electrostatic latent image is formed and then developed on paper. The density of electrodes on the stylus can be as high as 6 electrodes/mm. Compared to thermal recording, which requires no developing, electrostatic recording is bulkier. However, electrostatic recording is capable of high speed and resolution. Thermal recording is widely used for smaller facsimile machines in offices, whereas electrostatic recording is used in larger machines that transmit documents in high volume.

Both analog and digital transmission systems have been used. The older facsimile machines modulate the baseband waveform from the scanner to a higher frequency band so that efficient transmission is possible on analog telephone networks. Traditional modulation methods are amplitude modulation (AM) and frequency modulation (FM). Table 2.6.1 shows Group 1 and Group 2 CCITT standards for analog transmission systems. In Group 1 a simple

Table 2.6.1 CCITT standard for document facsimile systems. Three transmission networks that could be used are: PSTN (public switched telephone network), PDN (public data network) and the ISDN (integrated services digital networks). Information on protocols and control procedures (e.g. T.30) can be obtained from the appropriate CCITT recommendations. Coding schemes (one-dimensional modified Huffman code) MH, and (two-dimensional modified read code) MR, are described in Chapter 6.

APPARATUS / PARAMETER				GROUP 4 ***		
	GROUP 1	GROUP 2	GROUP 3	CLASS 1	CLASS 2	CLASS 3
APPARATUS RECOMMENDATION	T.2	T.3	T.4	T.5	T.5	T.5
NETWORK	PTN	PTN	PTN	PDN (PTN.ISDN)****	PDN (PTN.ISDN)****	PDN (PTN.ISDN)****
TRANSMISSION TIME/A4(min.)	6	3	APPROX. 1	—	—	—
NUMBER OF PELS ALONG A SCAN LINE	—	—	1728	1728. 2074* 2592*. 3456*	1728. 2074** 2592. 3456*	1728. 2074** 2592. 3456*
SCANNING DENSITY	3.85 (PEL/mm)	3.85 (PEL/mm)	3.85. 7.7* (PEL/mm)	200. 240** 300*. 400* (PEL/INCH)	200. 240** 300. 400* (PEL/INCH)	200. 240** 300. 400* (PEL/INCH)
MODEM	FM (1700±400Hz)	AM-PM-VSB (fc: 2100Hz)	PM (V.27ter). AM-PM (V.29)*	—	—	—
DATA RATE (k b/s)	—	—	2.4. 4.8. 7.2*. 9.6*	2.4. 4.8. 9.6. 48	2.4. 4.8. 9.6. 48	2.4.4.8. 9.6. 48
CODING SCHEME	—	—	MH MR*	MODIFIED MR (T.6)	MODIFIED MR (T.6)	MODIFIED MR (T.6)
CONTROL PROCEDURE PROTOCOL, RECOMMENDATION	T.30	T.30	T.30	T.62, T.70 T.71, T.73	T.62, T.70, T.71, T.72, T.73	T.62, T.70 T.71, T.72, T.73
REMARKS					RECEPTION ONLY FOR TELETEX AND MIXED-MODE	TRANSMISSION AND RECEPTION FOR TELETEX AND MIXED-MODE

*OPTION: **REQUIRED FOR TELETEX AND MIXED-MODE RECEPTION: ***WAS RECOMMENDED END OF 1984: ****FURTHER STUDY. (AS OF 1986)

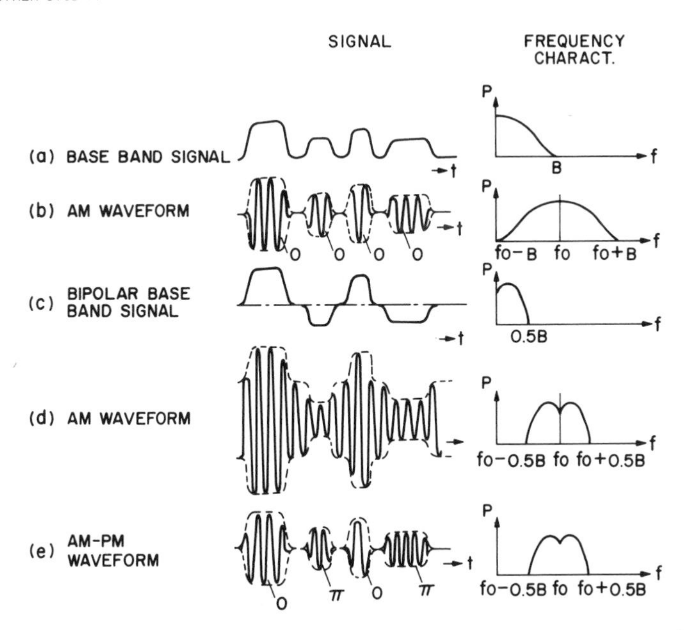

Fig. 2.6.5 Baseband, AM, and AM-PM waveforms used in analog facsimile systems and their frequency spectra.

frequency modulation in the 1700 ± 400 Hz band is used. In Group 2, a more sophisticated modulation is used. This is shown in Fig. 2.6.5. Here the baseband signal is first converted to a bipolar signal. This is then used to modulate the amplitude of a carrier signal. Since such a modulation has symmetric upper and lower sidebands, only one of them is transmitted (called AM-PM-VSB). This scheme is the Group 2 standard and reduces the transmission time for a standard size document from 6 to 3 minutes.

Today's facsimile machines use digital conversion and coding for transmission. These are standardized in Group 3 and 4. Group 3 contains two schemes: the first one utilizes one dimensional correlation in the signal for data compression, whereas the second one uses two dimensional correlation as well. In the first one, the encoder codes runs of black or white pels. A variable length coder is then used to give shorter codes to frequently occurring runs, and this achieves significant compression. Each encoded scan line is followed by a unique end of line (EOL) codeword so that resynchronization can occur even in the presence of transmission errors. In the second scheme, two dimensional correlation is exploited by encoding the shifts of runs from one line to the next. More details of the Group 3 coding scheme are given in Chapter 6. The compression efficiency depends on the document being transmitted. Fig. 2.6.6 shows 8 documents used by the CCITT to optimize the code tables used by the schemes. These documents have either 3.86 pels/mm or 7.7 pels/mm and are standard A4 size (210×298 mm) resulting in 1728×2376 pels per document. As shown in Table 2.6.1, transmission time is reduced to about one minute. Group 4 systems are divided into 3 classes depending on whether they are capable of working with a teletext terminal or with a mixed-mode terminal (teletext and facsimile). Also Group 4 systems allow for a wide range of scanning density, but use the same type of compression scheme as in Group 3.

Facsimile systems are currently used for communication in homes and offices over the telephone network. However, it is expected that such systems will grow in the future and use public data networks for their speed. Also they will be combined with other systems such as optical character recognition and character oriented systems such as telex.

2.7 Interactive Picture Communication Systems

In addition to the facsimile systems described in Section 2.6, there are other picture communication systems that have come into existence recently. Two of them, called Teletext and Videotex, use a narrow band transmission medium and display still frames of information from a remote data base on a home or office TV set. In general they are meant to retrieve information (text and pictures) from a data base. Thus, communication from the user is of much lower bandwidth compared with communication to the user.

In teletext, as shown in Fig. 2.7.1, data signals replace the normal video signals in some of the scan lines of a broadcast TV signal. The data is arranged

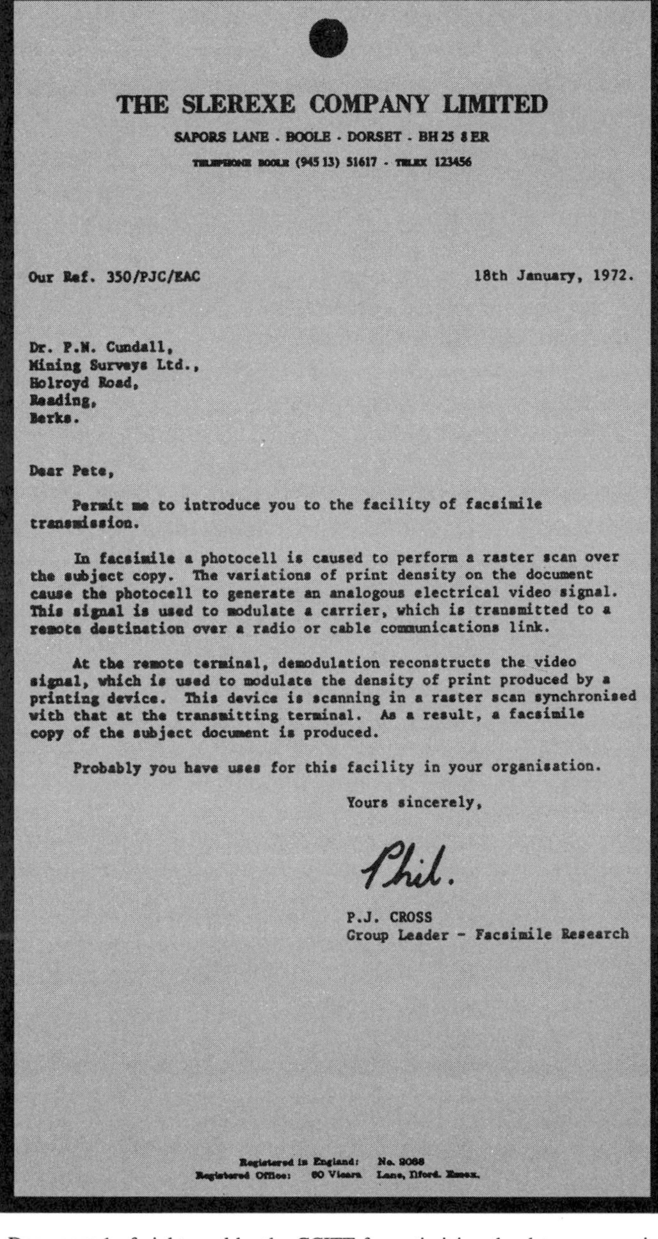

THE SLEREXE COMPANY LIMITED

SAPORS LANE . BOOLE . DORSET . BH 25 8 ER

TELEPHONE BOOLE (945 13) 51617 - TELEX 123456

Our Ref. 350/PJC/EAC

18th January, 1972.

Dr. P.N. Cundall,
Mining Surveys Ltd.,
Holroyd Road,
Reading,
Berks.

Dear Pete,

Permit me to introduce you to the facility of facsimile transmission.

In facsimile a photocell is caused to perform a raster scan over the subject copy. The variations of print density on the document cause the photocell to generate an analogous electrical video signal. This signal is used to modulate a carrier, which is transmitted to a remote destination over a radio or cable communications link.

At the remote terminal, demodulation reconstructs the video signal, which is used to modulate the density of print produced by a printing device. This device is scanning in a raster scan synchronised with that at the transmitting terminal. As a result, a facsimile copy of the subject document is produced.

Probably you have uses for this facility in your organisation.

Yours sincerely,

Phil.

P.J. CROSS
Group Leader - Facsimile Research

Registered in England: No. 2038
Registered Office: 60 Vicars Lane, Ilford. Essex.

Fig. 2.6.6 Document 1 of eight used by the CCITT for optimizing the data compression code for bilevel document facsimile.

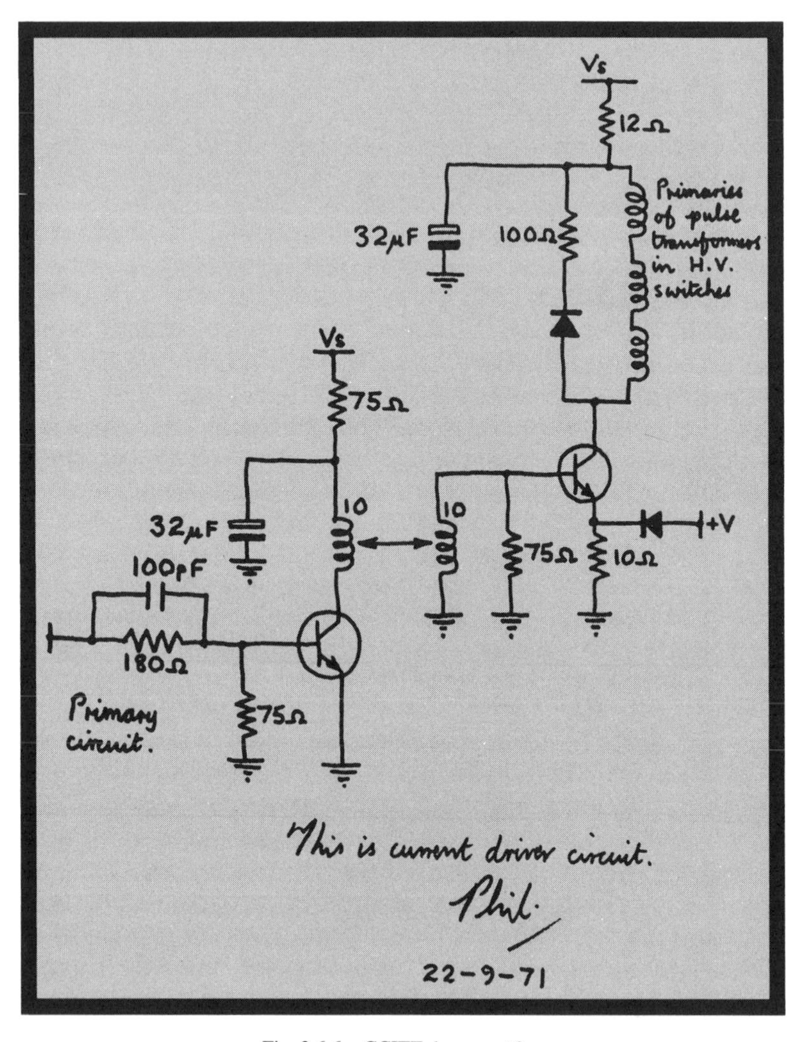

Fig. 2.6.6 CCITT document 2.

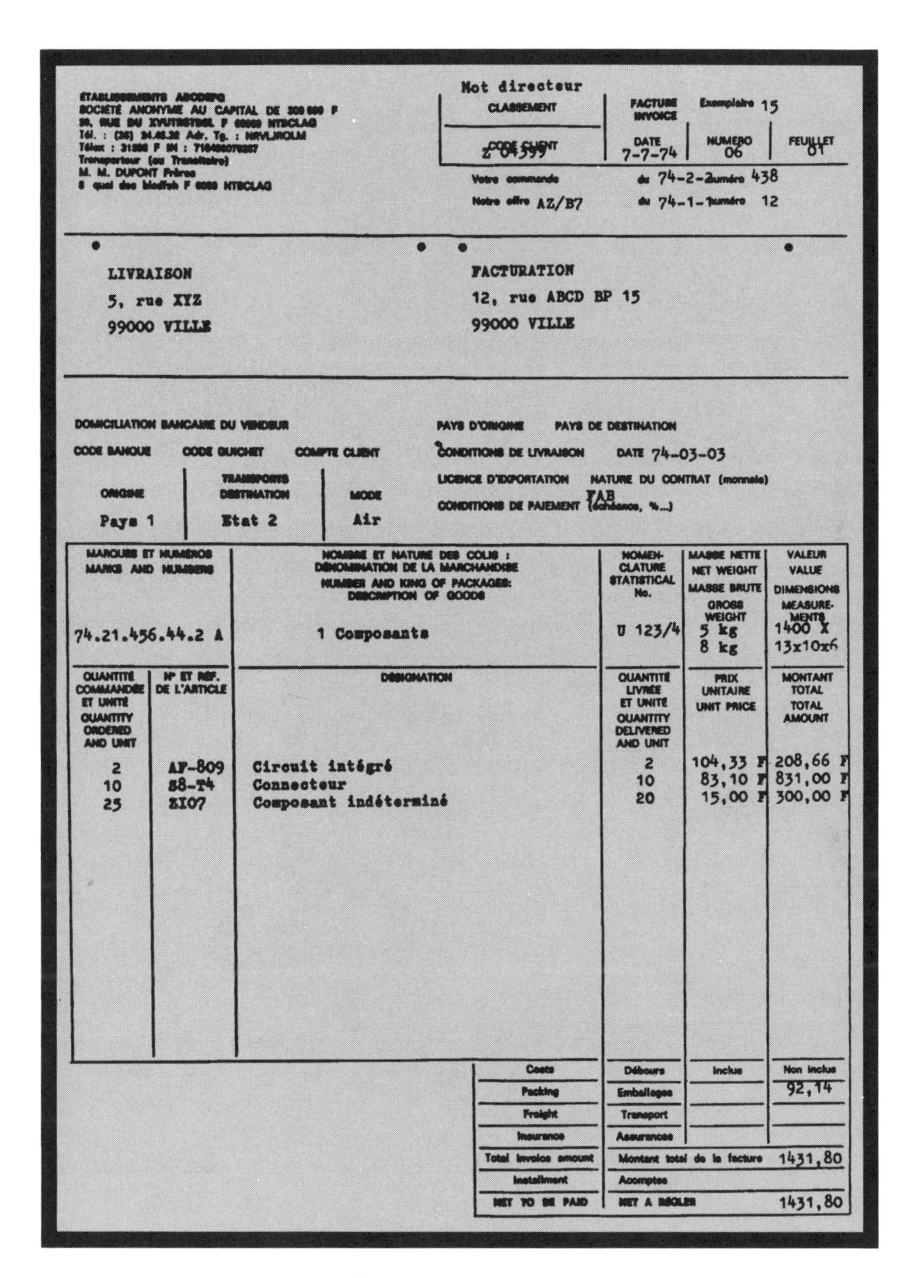

ÉTABLISSEMENTS ABCDEFG
SOCIÉTÉ ANONYME AU CAPITAL DE
Transporteur (ou Transitaire)
M. M. DUPONT Frères

Mot directeur

CLASSEMENT	FACTURE INVOICE	Exemplaire 15	
CODE CLIENT Z 04399	DATE 7-7-74	NUMERO 06	FEUILLET 01

Votre commande du 74-2-2 numéro 438
Notre offre AZ/B7 du 74-1-1 numéro 12

LIVRAISON
5, rue XYZ
99000 VILLE

FACTURATION
12, rue ABCD BP 15
99000 VILLE

DOMICILIATION BANCAIRE DU VENDEUR
CODE BANQUE CODE GUICHET COMPTE CLIENT

ORIGINE	TRANSPORTS DESTINATION	MODE
Pays 1	Etat 2	Air

PAYS D'ORIGINE PAYS DE DESTINATION
CONDITIONS DE LIVRAISON DATE 74-03-03
LICENCE D'EXPORTATION NATURE DU CONTRAT (monnaie)
CONDITIONS DE PAIEMENT (échéance, %...) FAB

MARQUES ET NUMÉROS MARKS AND NUMBERS	NOMBRE ET NATURE DES COLIS : DÉNOMINATION DE LA MARCHANDISE NUMBER AND KIND OF PACKAGES: DESCRIPTION OF GOODS	NOMENCLATURE STATISTICAL No.	MASSE NETTE NET WEIGHT MASSE BRUTE GROSS WEIGHT	VALEUR VALUE DIMENSIONS MEASUREMENTS
74.21.456.44.2 A	1 Composants	U 123/4	5 kg 8 kg	1400 X 13x10x6

QUANTITÉ COMMANDÉE ET UNITÉ QUANTITY ORDERED AND UNIT	N° ET RÉF. DE L'ARTICLE	DÉSIGNATION	QUANTITÉ LIVRÉE ET UNITÉ QUANTITY DELIVERED AND UNIT	PRIX UNITAIRE UNIT PRICE	MONTANT TOTAL TOTAL AMOUNT
2	AF-809	Circuit intégré	2	104,33 F	208,66 F
10	S8-T4	Connecteur	10	83,10 F	831,00 F
25	ZI07	Composant indéterminé	20	15,00 F	300,00 F

Costs	Débours	Inclus	Non inclus
Packing	Emballages		92,14
Freight	Transport		
Insurance	Assurances		
Total invoice amount	Montant total de la facture		1431,80
Instalment	Acomptes		
NET TO BE PAID	NET A RÉGLER		1431,80

Fig. 2.6.6 CCITT document 3.

- 34 -

L'ordre de lancement et de réalisation des applications fait l'objet de décisions au plus haut niveau de la Direction Générale des Télécommunications. Il n'est certes pas question de construire ce système intégré "en bloc" mais bien au contraire de procéder par étapes, par paliers successifs. Certaines applications, dont la rentabilité ne pourra être assurée, ne seront pas entreprises. Actuellement, sur trente applications qui ont pu être globalement définies, six en sont au stade de l'exploitation, six autres se sont vu donner la priorité pour leur réalisation.
Chaque application est confiée à un "chef de projet", responsable successivement de sa conception, de son analyse-programmation et de sa mise en oeuvre dans une région-pilote. La généralisation ultérieure de l'application réalisée dans cette région-pilote dépend des résultats obtenus et fait l'objet d'une décision de la Direction Générale. Néanmoins, le chef de projet doit dès le départ considérer que son activité a une vocation nationale donc refuser tout particularisme régional. Il est aidé d'une équipe d'analystes-programmeurs et entouré d'un "groupe de conception" chargé de rédiger le document de "définition des objectifs globaux" puis le "cahier des charges" de l'application, qui sont adressés pour avis à tous les services utilisateurs potentiels et aux chefs de projet des autres applications. Le groupe de conception comprend 6 à 10 personnes représentant les services les plus divers concernés par le projet,et comporte obligatoirement un bon analyste attaché à l'application.

II - L'IMPLANTATION GEOGRAPHIQUE D'UN RESEAU INFORMATIQUE PERFORMANT

L'organisation de l'entreprise française des télécommunications repose sur l'existence de 20 régions. Des calculateurs ont été implantés dans le passé au moins dans toutes les plus importantes. On trouve ainsi des machines Bull Gamma 30 à Lyon et Marseille, des GE 425 à Lille, Bordeaux, Toulouse et Montpellier, un GE 437 à Massy, enfin quelques machines Bull 300 TI à programmes câblés étaient récemment ou sont encore en service dans les régions de Nancy, Nantes, Limoges, Poitiers et Rouen ; ce parc est essentiellement utilisé pour la comptabilité téléphonique.
A l'avenir, si la plupart des fichiers nécessaires aux applications décrites plus haut peuvent être gérés en temps différé, un certain nombre d'entre eux devront nécessairement être accessibles, voire mis à jour en temps réel : parmi ces derniers le fichier commercial des abonnés, le fichier des renseignements, le fichier des circuits, le fichier technique des abonnés contiendront des quantités considérables d'informations.
Le volume total de caractères à gérer en phase finale sur un ordinateur ayant en charge quelques 500 000 abonnés a été estimé à un milliard de caractères au moins. Au moins le tiers des données seront concernées par des traitements en temps réel.
Aucun des calculateurs énumérés plus haut ne permettait d'envisager de tels traitements.
L'intégration progressive de toutes les applications suppose la création d'un support commun pour toutes les informations, une véritable "Banque de données", répartie sur des moyens de traitement nationaux et régionaux, et qui devra rester alimentée, mise à jour en permanence, à partir de la base de l'entreprise, c'est-à-dire les chantiers, les magasins, les guichets des services d'abonnement, les services de personnel etc.
L'étude des différents fichiers à constituer a donc permis de définir les principales caractéristiques du réseau d'ordinateurs nouveaux à mettre en place pour aborder la réalisation du système informatif. L'obligation de faire appel à des ordinateurs de troisième génération, très puissants et dotés de volumineuses mémoires de masse, a conduit à en réduire substantiellement le nombre.
L'implantation de sept centres de calcul interrégionaux constituera un compromis entre : d'une part le désir de réduire le coût économique de l'ensemble, de faciliter la coordination des équipes d'informaticiens; et d'autre part le refus de créer des centres trop importants difficiles à gérer et à diriger,et posant des problèmes délicats de sécurité. Le regroupement des traitements relatifs à plusieurs régions sur chacun de ces sept centres permettra de leur donner une taille relativement homogène. Chaque centre "gèrera" environ un million d'abonnés à la fin du VIème Plan.
La mise en place de ces centres a débuté au début de l'année 1971 : un ordinateur IRIS 50 de la Compagnie Internationale pour l'Informatique a été installé à Toulouse en février ; la même machine vient d'être mise en service au centre de calcul interrégional de Bordeaux.

Photo n° 1 - Document très dense lettre 1,5mm de haut -
Restitution photo n° 9

Fig. 2.6.6 CCITT document 4.

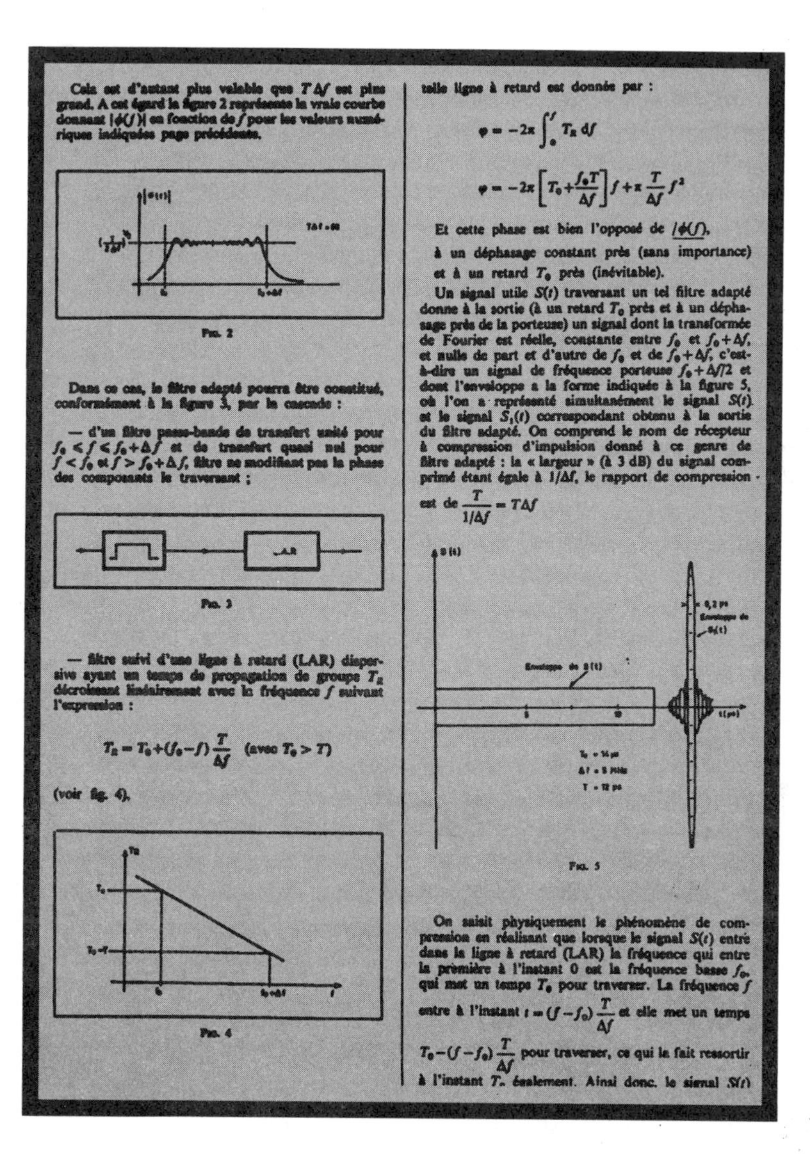

Cela est d'autant plus valable que $T\Delta f$ est plus grand. A cet égard la figure 2 représente la vraie courbe donnant $|\phi(f)|$ en fonction de f pour les valeurs numériques indiquées page précédente.

FIG. 2

Dans ce cas, le filtre adapté pourra être constitué, conformément à la figure 3, par la cascade :

— d'un filtre passe-bande de transfert unité pour $f_0 < f < f_0 + \Delta f$ et de transfert quasi nul pour $f < f_0$ et $f > f_0 + \Delta f$, filtre ne modifiant pas la phase des composants le traversant ;

FIG. 3

— filtre suivi d'une ligne à retard (LAR) dispersive ayant un temps de propagation de groupe T_R décroissant linéairement avec la fréquence f suivant l'expression :

$$T_R = T_0 + (f_0 - f)\frac{T}{\Delta f} \quad (\text{avec } T_0 > T)$$

(voir fig. 4).

FIG. 4

telle ligne à retard est donnée par :

$$\varphi = -2\pi \int_0^f T_R \, df$$

$$\varphi = -2\pi \left[T_0 + \frac{f_0 T}{\Delta f}\right] f + \pi \frac{T}{\Delta f} f^2$$

Et cette phase est bien l'opposé de $/\phi(f)$, à un déphasage constant près (sans importance) et à un retard T_0 près (inévitable).

Un signal utile $S(t)$ traversant un tel filtre adapté donne à la sortie (à un retard T_0 près et à un déphasage près de la porteuse) un signal dont la transformée de Fourier est réelle, constante entre f_0 et $f_0 + \Delta f$, et nulle de part et d'autre de f_0 et de $f_0 + \Delta f$, c'est-à-dire un signal de fréquence porteuse $f_0 + \Delta f/2$ et dont l'enveloppe a la forme indiquée à la figure 5, où l'on a représenté simultanément le signal $S(t)$ et le signal $S_1(t)$ correspondant obtenu à la sortie du filtre adapté. On comprend le nom de récepteur à compression d'impulsion donné à ce genre de filtre adapté : la « largeur » (à 3 dB) du signal comprimé étant égale à $1/\Delta f$, le rapport de compression est de $\dfrac{T}{1/\Delta f} = T\Delta f$

FIG. 5

On saisit physiquement le phénomène de compression en réalisant que lorsque le signal $S(t)$ entré dans la ligne à retard (LAR) la fréquence qui entre la première à l'instant 0 est la fréquence basse f_0, qui met un temps T_0 pour traverser. La fréquence f entre à l'instant $t = (f - f_0)\dfrac{T}{\Delta f}$ et elle met un temps $T_0 - (f - f_0)\dfrac{T}{\Delta f}$ pour traverser, ce qui la fait ressortir à l'instant T_0 également. Ainsi donc, le signal $S(t)$

Fig. 2.6.6 CCITT document 5.

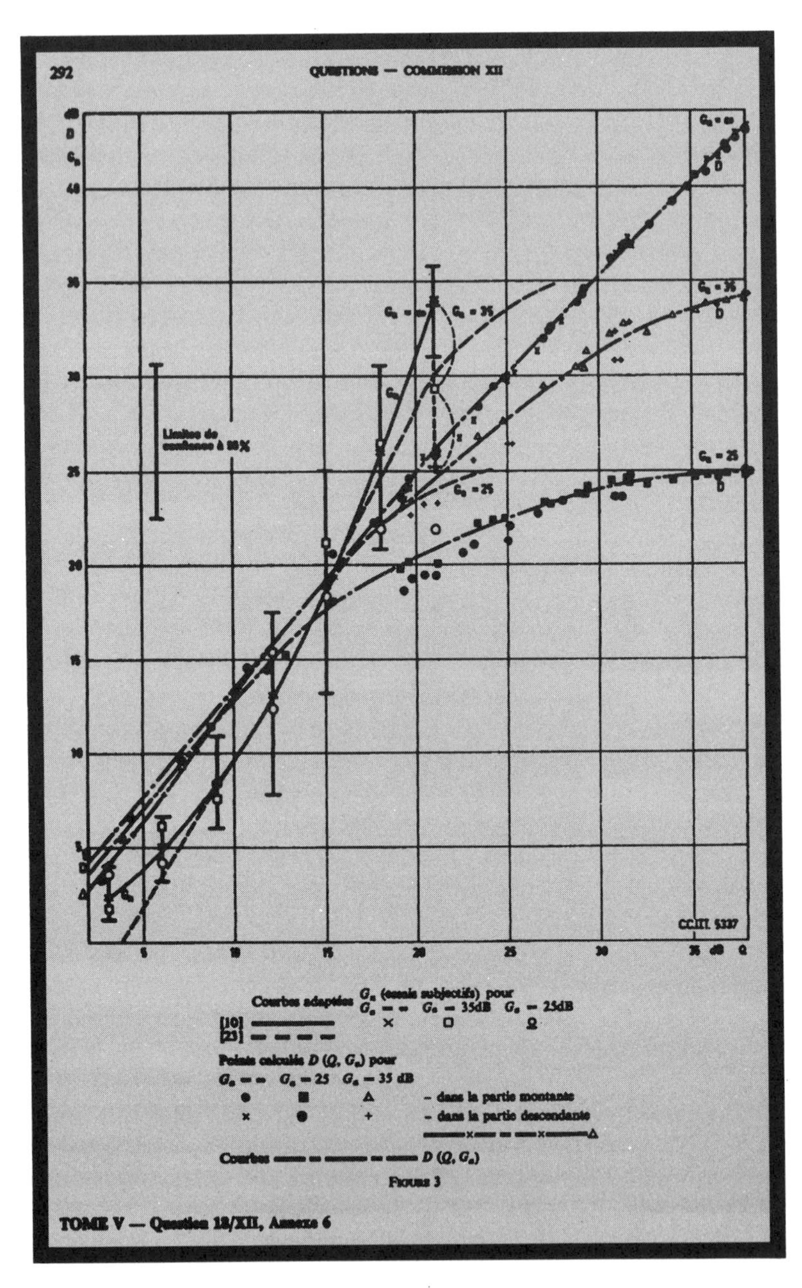

Fig. 2.6.6 CCITT document 6.

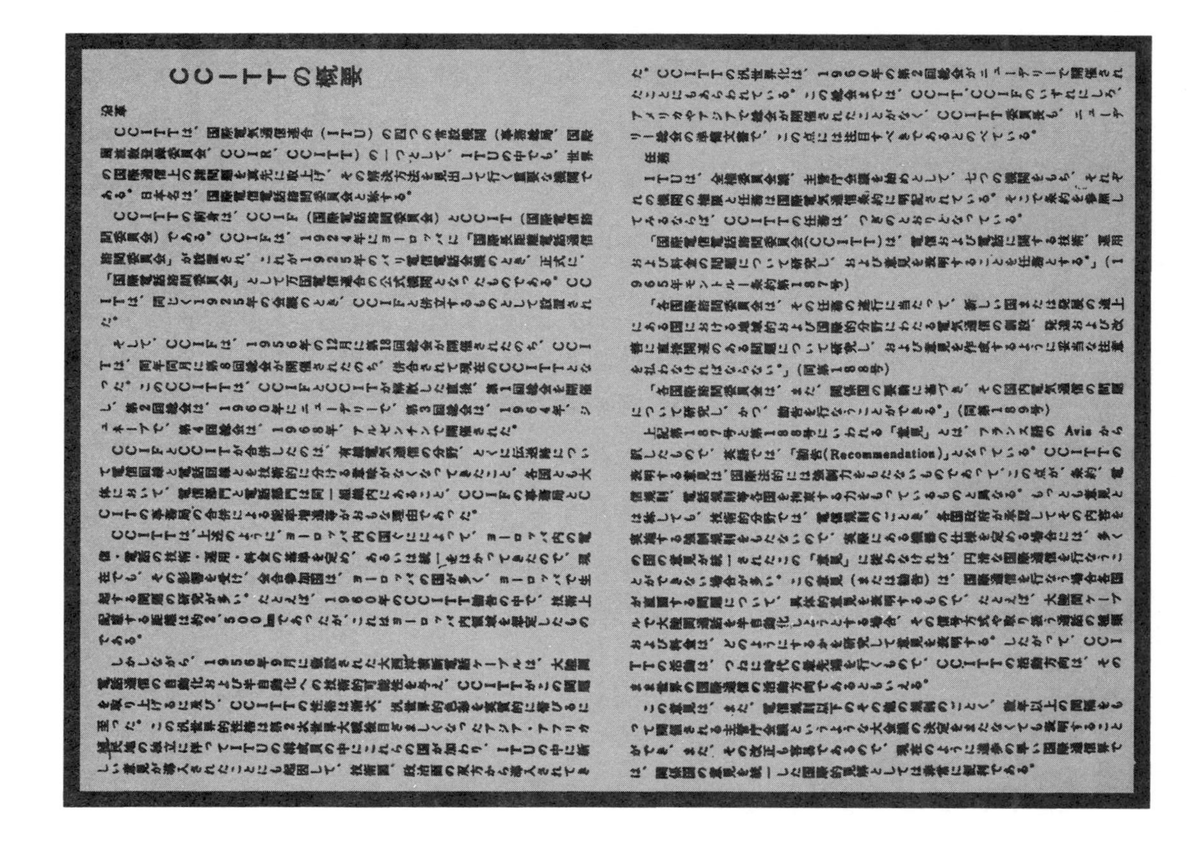

CCITTの概要

沿革

CCITTは、国際電気通信連合（ITU）の四つの常設機関（事務総局、国際周波数登録委員会、CCIR、CCITT）の一つとして、ITUの中でも、世界の国際通信上の諸問題を真先に取上げ、その解決方法を見出して行く重要な機関である。日本名は、国際電信電話諮問委員会と称する。

CCITTの前身は、CCIF（国際電話諮問委員会）とCCIT（国際電信諮問委員会）である。CCIFは、1924年にヨーロッパに「国際長距離電話通信諮問委員会」が設置され、これが1925年のパリ電信電話会議のとき、正式に、「国際電話諮問委員会」として万国電信連合の公式機関となったものである。CCITは、同じく1925年の会議のとき、CCIFと併立するものとして設置された。

そして、CCIFは、1956年の12月に第18回総会が開催されたのち、CCITは、同年同月に第8回総会が開催されたのち、併合されて現在のCCITTとなった。このCCITTは、CCIFとCCITが解散した直後、第1回総会を開催し、第2回総会は、1960年にニューデリーで、第3回総会は、1964年、ジュネーブで、第4回総会は、1968年、アルゼンチンで開催された。

CCIFとCCITが合併したのは、有線電気通信の分野、とくに伝送路について電信回線と電話回線とを技術的に分ける意味がなくなってきたこと、各国とも大体において、電信部門と電話部門は同一組織内にあること、CCIFの事務局とCCITの事務局の合併による能率増進等がおもな理由であった。

CCITTは、上述のように、ヨーロッパ内の国々によって、ヨーロッパ内の電信・電話の技術・運用・料金の基準を定め、あるいは統一をはかってきたので、現在でも、その影響を受け、会合参加国は、ヨーロッパの国が多く、ヨーロッパで生起する問題の研究が多い。たとえば、1960年のCCITT勧告の中で、技術上配慮する距離は約2,500kmであったが、これはヨーロッパ内領域を想定したものである。

しかしながら、1956年9月に敷設された大西洋横断電話ケーブルは、大陸間電話通信の自動化および半自動化への技術的可能性を与え、CCITTがこの問題を取り上げるに及び、CCITTの性格は漸次、汎世界的色彩を実質的に帯びるに至った。この汎世界的性格は第2次世界大戦後目ざましくなったアジア・アフリカ植民地の独立に伴ってITUの構成員の中にこれらの国が加わり、ITUの中に新しい意見が導入されたことにも起因して、技術面、政治面の双方から導入されてきた。CCITTの汎世界化は、1960年の第2回総会がニューデリーで開催されたことにもあらわれている。この総会までは、CCIT、CCIFのいずれにしろ、アメリカやアジアで総会が開催されたことがなく、CCITT委員長も、ニューデリー総会の準備文書で、この点には注目すべきであるとのべている。

任務

ITUは、全権委員会議、主管庁会議を始めとして、七つの機関をもち、それぞれの機関の権限と任務は国際電気通信条約に明記されている。そこで条約を参照してみるならば、CCITTの任務は、つぎのとおりとなっている。

「国際電信電話諮問委員会(CCITT)は、電信および電話に関する技術、運用および料金の問題について研究し、および意見を表明することを任務とする。」（1965年モントルー条約第187号）

「各国際諮問委員会は、その任務の遂行に当たって、新しい国または発展の途上にある国における地域的および国際的分野にわたる電気通信の創設、発達および改善に直接関連のある問題について研究し、および意見を作成するように妥当な注意を払わなければならない。」（同第188号）

「各国際諮問委員会は、また、関係国の要請に基づき、その国内電気通信の問題について研究し、かつ、勧告を行なうことができる。」（同第189号）

上記第187号と第188号にいわれる「意見」とは、フランス語の Avis から訳したもので、英語では、「勧告(Recommendation)」となっている。CCITTの表明する意見は、国際法的には強制力をもたないものであって、この点が、条約、電信規則、電話規則等各国を拘束する力をもっているものと異なる。もっとも意見とは称しても、技術的分野では、電信規則のごとき、各国政府が承認してその内容を実施する強制規則をもたないので、実際にある機器の仕様を定める場合には、多くの国の意見が統一されたこの「意見」に従わなければ、円滑な国際通信を行なうことができない場合が多い。この意見（または勧告）は、国際通信を行なう場合各国が直面する問題について、具体的意見を表明するもので、たとえば、大陸間ケーブルで大陸間通話を半自動化しようとする場合、その信号方式や取り扱う通話の種類および料金は、どのようにするかを研究して意見を表明する。したがって、CCITTの活動は、つねに時代の最先端を行くもので、CCITTの活動方向は、そのまま世界の国際通信の活動方向であるともいえる。

この意見は、また、電信規則以下のその他の規制のごとく、数年以上の間隔をもって開催される主管庁会議というような大会議の決定をまたなくても表明することができ、また、その改正も容易であるので、現在のように進歩の早い国際通信界では、関係国の意見を統一した国際的見解としては非常に便利である。

Fig. 2.6.6 CCITT document 7.

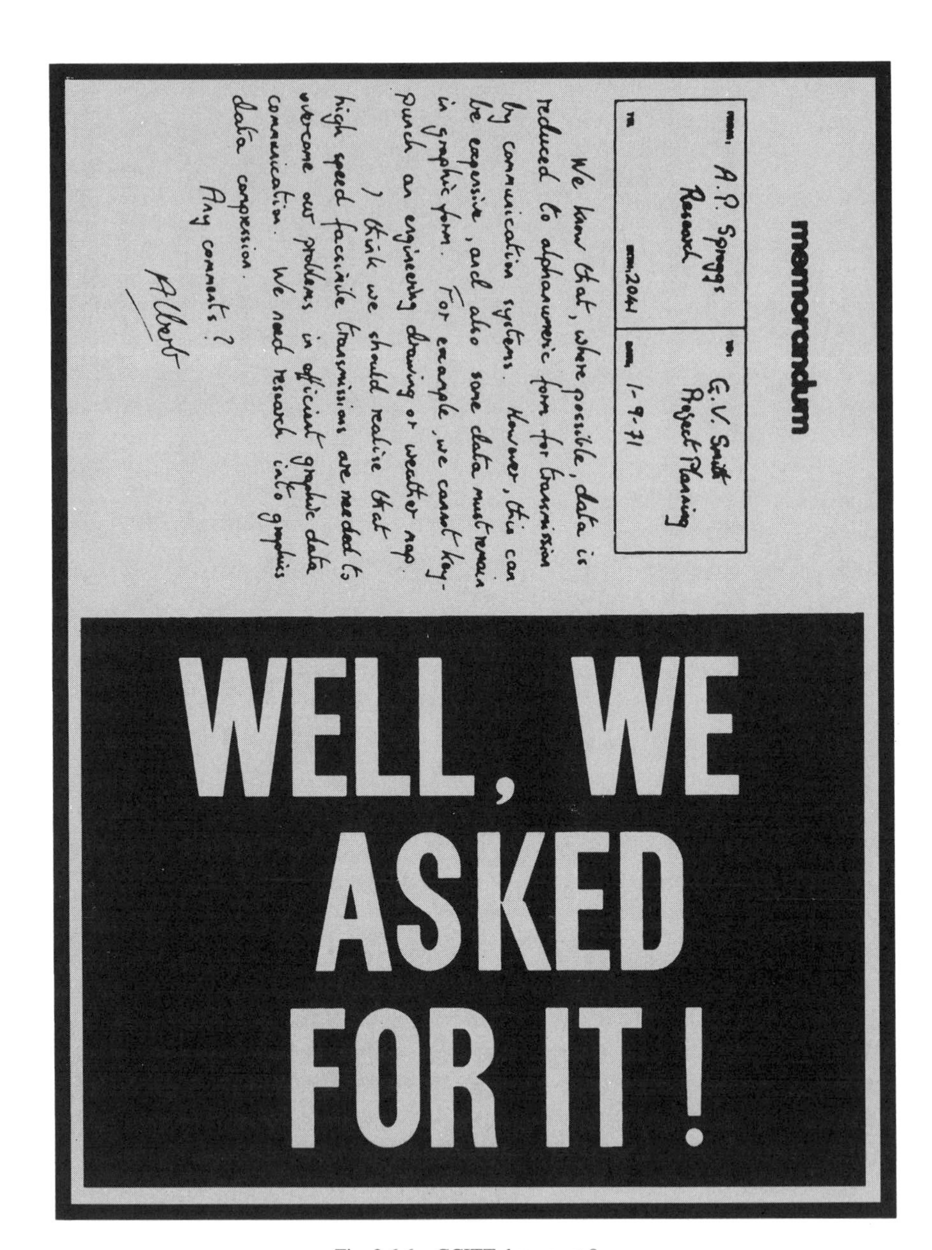

memorandum

A.P. Spriggs
Research

2041

G.V. Smith
Project Planning

1-9-71

We know that, where possible, data is reduced to alphanumeric form for transmission by communication systems. However, this can be expensive, and also some data must remain in graphic form. For example, we cannot key-punch an engineering drawing or weather map.

I think we should realise that high speed facsimile transmissions are needed to overcome our problems in efficient graphic data communication. We need research into graphics data compression.

Any comments?

Albert

WELL, WE ASKED FOR IT!

Fig. 2.6.6 CCITT document 8.

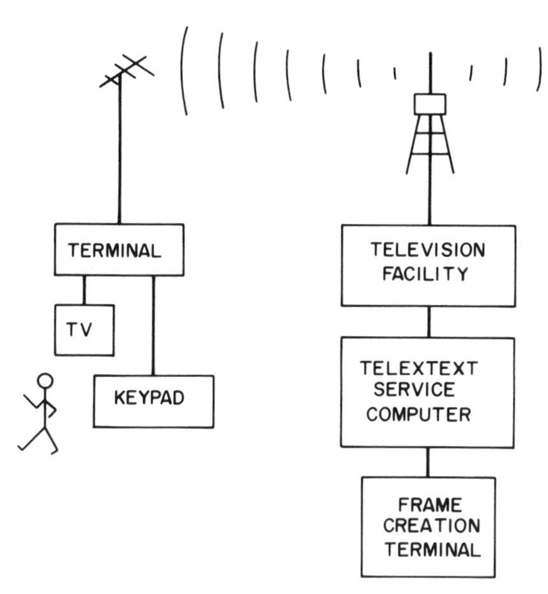

Fig. 2.7.1 Block Diagram of a Teletext System. Coded information replaces some of the scan lines of the video signals. Video signal may be broadcast or may be delivered on a cable (from Ninke [2.7.4]).

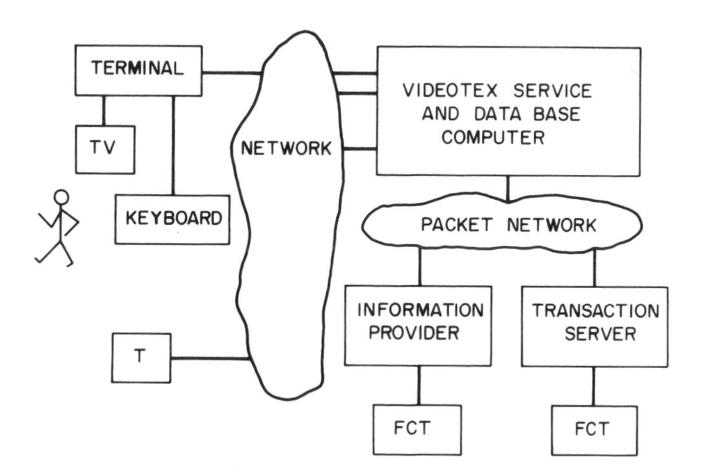

Fig. 2.7.2 Block Diagram of Videotex System. The Frame Creation Terminal (FCT) is used by the information provider to create frames which are then stored in the data base. Frames can be accessed over the switched telephone network using a keyboard and a telephone (T) (from Ninke [2.7.4]).

into a magazine format that is cycled through repetitively. A customer uses a keypad to indicate first the magazine number and then the frame number in that magazine. When this frame is broadcast again, its description is captured by the terminal and displayed. One of the frames of the magazine contains an index to the remaining frames. If the data is sent in only a few scan lines (normally 2 to 4) during the Vertical Blanking Interval (VBI), the system is called VBI-Teletext. In such a system the normal TV picture is not disturbed, and data can be mixed with the normal picture to give additional functionality (e.g., subtitles or news flashes). If data is sent for every scan line, the system is called Full-Channel Teletext. VBI-Teletext is normally used with broadcast television, whereas Full-Channel Teletext is more common in the cable TV environment.

In videotex, as shown in Fig. 2.7.2, digital transmission on a switched telephone network is used. A customer dials a data base (located perhaps in a telephone central office), and requests a frame of information to be transmitted to his terminal. As in teletext, certain frames in the data base serve as an index, and the desired frame can be accessed by requesting more and more specialized indexes in a tree-structured fashion. In videotex, the data transmission is usually at 1200 bits/sec on a switched telephone line resulting in a typical response time of 10 secs to receive a frame. Thus, the major difference between teletext and videotex is that in teletext, a customer grabs one of the many frames that are transmitted in a round-robin fashion. Therefore, the response time (or the ability to grab the required frame) grows with the size of the data base. In videotex, however, since a customer has a two-way connection to the data base, the number of accessible frames can be much larger while still maintaining a reasonable response time. Moreover, the database is usually arranged in a tree-structured hierarchy so that access time is decreased. Teletext frames are essentially free of charge once the terminal is purchased or rented. However, for videotex a fee is charged for each frame in addition to the terminal rental and the telephone line.

Two major standards for videotex and teletext have been adopted in North America. Both are based on the International Standards Organization (ISO) seven-layer model for Open System Interconnection (OSI).[2.7.1] One standard, North American Presentation Level Protocol Syntax (NAPLPS) specifies the data syntax to be used in both videotex and teletext.[2.7.2] The other, North American Broadcast Teletext Specification (NABTS) specifies all the protocol layers for the teletext, using NAPLPS for the presentation layer data syntax.[2.7.3] We will describe the NAPLPS in some detail and leave the rest of the protocol specification to the above references.

NAPLPS is an agreement on the syntax for representing data describing images at the presentation layer. It is used for encoding text, graphics and images in color. In NAPLPS a character-geometric representation is used. It contains the following graphical objects: (a) characters and mosaics (upward compatible from ASCII), (b) points, (c) lines, (d) arcs, (e) rectangles,

				b7	0	0	0	0	1	1	1	1
				b6	0	0	1	1	0	0	1	1
				b5	0	1	0	1	0	1	0	1
				COLUMN	0	1	2	3	4	5	6	7
b4	b3	b2	b1	ROW								
0	0	0	0	0	NUL	DLE	SP	0	@	P	'	p
0	0	0	1	1	SOH	DC1	!	1	A	Q	a	q
0	0	1	0	2	STX	DC2	"	2	B	R	b	r
0	0	1	1	3	ETX	DC3	#	3	C	S	c	s
0	1	0	0	4	EOT	DC4	$	4	D	T	d	t
0	1	0	1	5	ENQ	NAK	%	5	E	U	e	u
0	1	1	0	6	ACK	SYN	&	6	F	V	f	v
0	1	1	1	7	BEL	ETB	'	7	G	W	g	w
1	0	0	0	8	APB (BS)	CAN	(	8	H	X	h	x
1	0	0	1	9	APF (HT)	SS2	)	9	I	Y	i	y
1	0	1	0	10	APD (LF)	SDC	*	:	J	Z	j	z
1	0	1	1	11	APU (VT)	ESC	+	;	K	[	k	{
1	1	0	0	12	CS (FF)	APS	,	<	L	\	l	|
1	1	0	1	13	APR (CR)	SS3	-	=	M	]	m	}
1	1	1	0	14	SO	APH	.	>	N	^	n	~
1	1	1	1	15	SI	NSR	/	?	O	—	o	DEL

C SET (columns 0–1) G SET (columns 2–7)

Fig. 2.7.3 The control and graphics set of the initial (or default) code table for NAPLPS (from Ninke [2.7.5]).

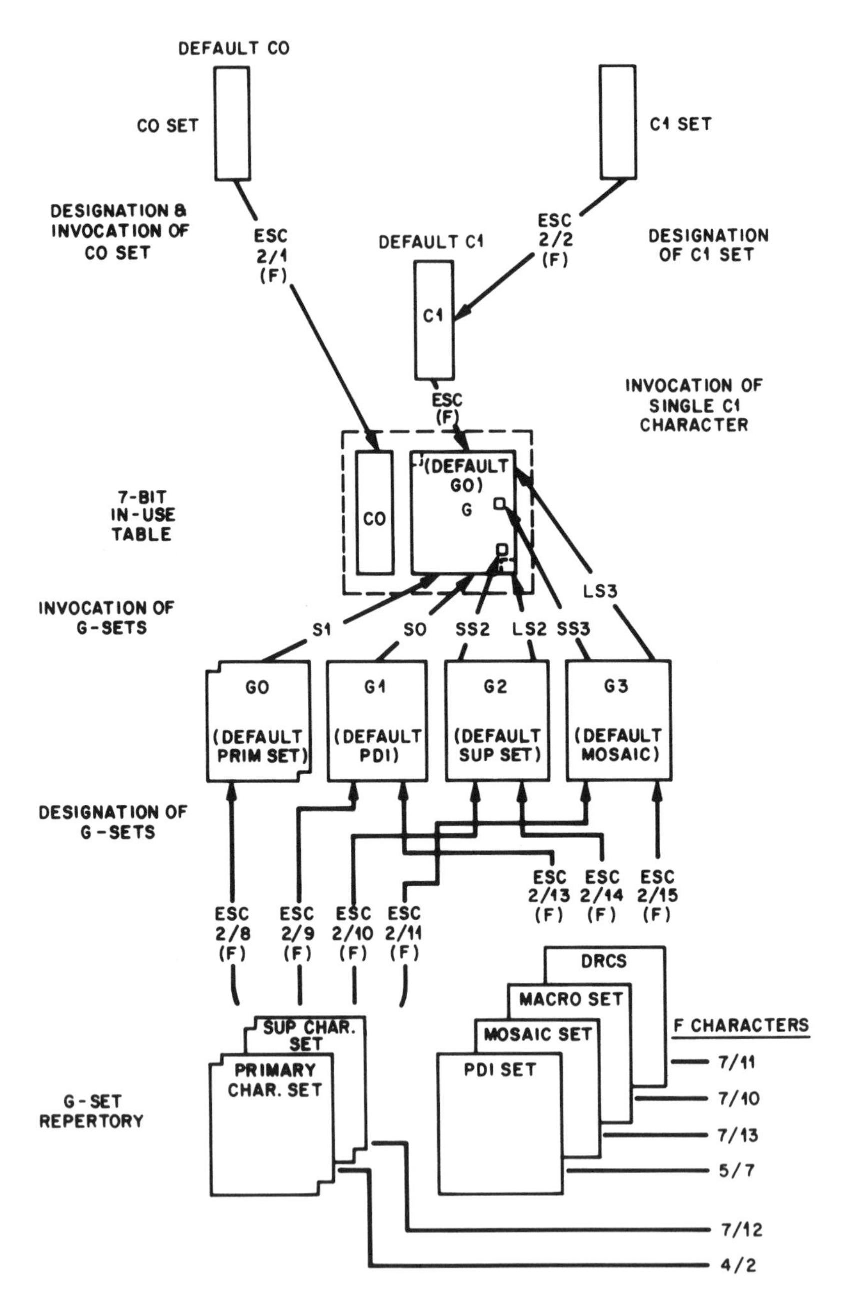

Fig. 2.7.4 Code extensions for the NAPLP. It also shows various steps necessary to access a particular code table (from Ninke [2.7.5]).

(f) polygons, and (g) incrementals. NAPLPS interpreter contains several code tables arranged in a hierarchical fashion. The string of data bits received by the terminal are interpreted first to bring into use the appropriate code table, and then to interpret the bytes to form the displayed frame. The initial (default) code table is shown in Fig. 2.7.3. Bits $b7$, $b6$ and $b5$ address one of the 8 columns, and bits $b4$, $b3$, $b2$ and $b1$ address one of the 16 rows. As an example, bit pattern 1000011 indicates the entry in column 4 and row 3, which represents C. The code table is divided into 2 portions: a control set (called C set) containing 32 entries in columns 0 and 1; and a graphics set (called G set) containing 96 entries in columns 2,3,...,7. The initial G set contains upper- and lower-case characters, numbers, and punctuation marks, and is identical to the ASCII character set. The initial C set contains the ASCII C set with some modified code functions.

If a feature is desired that is not available in the initial code table, the new code table containing the feature must first be brought into use and then the feature specified. In NAPLPS, there are two C sets, C0 and C1, and six G sets: primary character sets, picture description instruction (PDI) set, the supplementary character set, mosaic set, dynamically redefinable character set (DRCS), and a macro set. The sets can be brought into use by the sequence of steps shown in Fig. 2.7.4. As an example, the PDI set can be brought into use by the SO control character.

The supplementary character set is necessary for non-English text. It contains accents, diacritical marks, and special characters for Latin-based alphabets. The mosaic set, shown in Fig. 2.7.5, contains 64 different 2×3 block mosaic characters. Mosaics can be combined to form approximations to a variety of graphical objects.

The picture description instruction set (PDI) is used to specify six types of geometric graphical objects shown in Fig. 2.7.6. It is also used for coloring and controlling the appearance of these graphical objects. The PDI set operates somewhat differently compared to the other G sets. The PDI table (Fig. 2.7.6) has two parts. Columns 2 and 3 contain details of the operation to be performed, and columns 4, 5, 6 and 7 specify the data to be used for these operations. The operations fall into two groups: drawing codes that form geometric objects and control codes that supply parameters to be applied to the geometric objects. In general, control codes are employed to set up colors, textures, and other parameters, and then geometric object codes are specified to draw using these parameters. The six types of PDI geometric graphical objects are shown in Fig. 2.7.7. In the drawing of the graphical object the concept of the logical pel is used. The size of the logical pel is specified by the PDI. A large logical pel, therefore, gives fat lines and outlines, and a small logical pel gives thinner lines.

The C1 control set contains the codes through which the entries in the macro and DRCS graphics sets are defined. It also can be used to protect and unprotect the active field and to define textures.

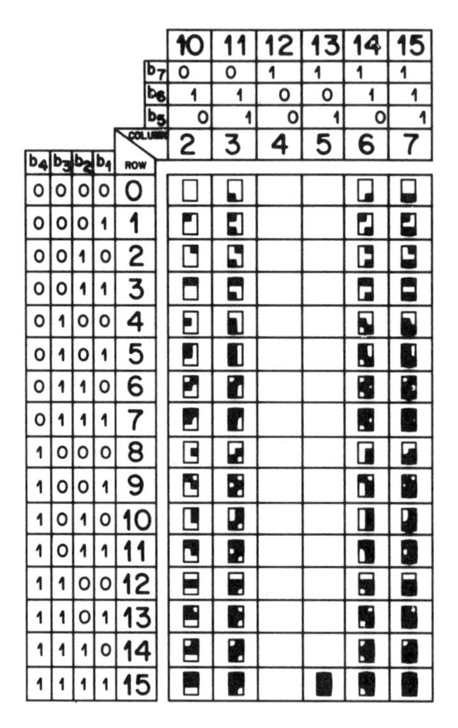

Fig. 2.7.5 Mosaic set used in the NAPLPS (from Ninke [2.7.5]).

					10	11	12	13	14	15
				b_7	0	0	1	1	1	1
				b_6	1	1	0	0	1	1
				b_5	0	1	0	1	0	1
b_4	b_3	b_2	b_1	ROW \ COLUMN	2	3	4	5	6	7
0	0	0	0	0	RESET	RECT (OUT. LINED)	NUMERIC DATA			
0	0	0	1	1	DOMAIN	RECT (FILLED)				
0	0	1	0	2	TEXT	SET & RECT (OUT. LINED)				
0	0	1	1	3	TEXTURE	SET & RECT (FILLED)				
0	1	0	0	4	POINT SET (ABS)	POLY (OUT. LINED)				
0	1	0	1	5	POINT SET (MEL)	POLY (FILLED)				
0	1	1	0	6	POINT (ABS)	SET & POLY (OUT. LINED)				
0	1	1	1	7	POINT (MEL)	SET & POLY (FILLED)				
1	0	0	0	8	LINE (ABS)	FIELD				
1	0	0	1	9	LINE (MEL)	MCR POINT				
1	0	1	0	10	SET & LINE (ABS)	MCR LINE				
1	0	1	1	11	SET & LINE (MEL)	MCR POLY (FILLED)				
1	1	0	0	12	ARC (OUT. LINED)	SET COLOR				
1	1	0	1	13	ARC (FILLED)	WA/7				
1	1	1	0	14	SET & ARC (OUT. LINED)	SELECT COLOR				
1	1	1	1	15	SET & ARC (FILLED)	BLANK				

Fig. 2.7.6 Picture Description Instruction (PDI) Table in the NAPLPS (from Ninke [2.7.5]).

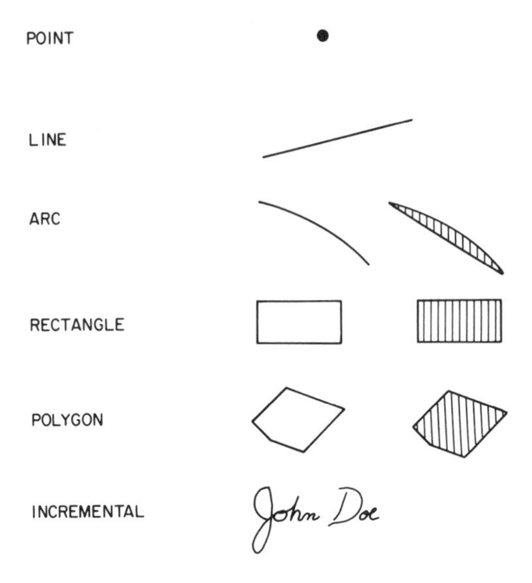

Fig. 2.7.7 Drawing Primitives used in Picture Description Instructions (PDI) (from Ninke [2.7.5]).

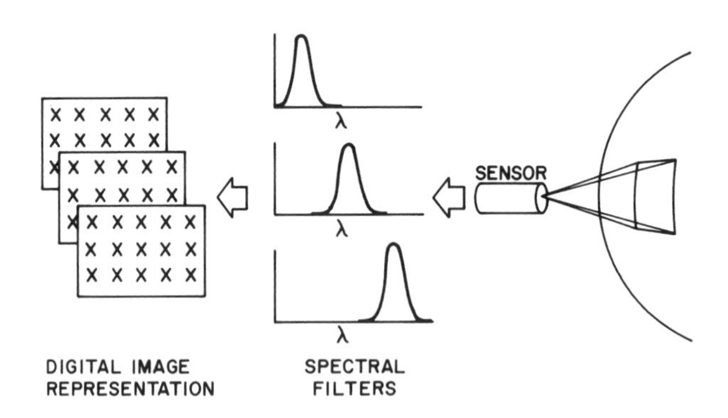

Fig. 2.8.1 Multispectral Image Acquisition (from Green [2.8.1]).

The dynamically redefinable character set (DRCS) allows downloading of custom character shapes. The downloaded shape is confined to the same aspect ratio as the character field, and the color information is not included in the downloading. The macro set is also a downloaded set of shapes. A sequence of presentation level codes is used for each macro. A macro once downloaded is displayed by designating and invoking the macro set into the table currently being used and specifying its G set code. Thus, the first occurrence of a macro may take a long time, but its subsequent use can be quite fast. There are four custom texture patterns that can also be downloaded. To reduce complexity, only bilevel patterns are used.

In summary, NAPLPS incorporates considerably more functionality than can be specified by a 7- or 8-bit code table, and therefore several code tables (six G sets and two C sets) are used with escape sequences to bring the desired code table into use. The interpreter at the terminal that interprets the NAPLPS syntax performs three major functions: (a) activate the appropriate code table, (b) update and store parameters, attributes and controls, (c) execute the drawing and coloring commands.

A number of developments are currently underway to extend the basic teletext and videotex concepts in many exciting directions. Thus, instead of just retrieving previously composed frames, customers may have the ability to enter information so that services such as catalog ordering or bill paying become possible. Direct information exchange between customers is possible for services such as electronic mail or teleconferencing. Quality of the material may be enriched by including character-photographic pictures, animated graphics and full-motion video. The concept of a public picture information service to homes is being amended to include private picture services to specialized businesses. In the next decade or two, we will know which of these is profitable and receives public acceptance.

2.8 Earth Resources Imagery

The images dealt with so far have been from a single range of the electromagnetic spectrum (in the case of monochrome) or in three ranges of the spectrum for color. There are other applications of image processing and communication that require *multispectral* imagery. Fig. 2.8.1 shows a sensor acquiring images of the same scene in different parts of the electromagnetic spectrum. The same area is viewed through three filters that separate the image into different spectral components. Each separate component may then be treated as a separate image of the scene.

A popular application of multispectral images is in remote sensing which involves the use of imaging sensors and other instruments mounted on aircraft, spacecraft, undersea vehicles and platforms designed to measure and monitor the earth's characteristics. The Earth Resources Survey program of the National Aeronautics and Space Agency (NASA) has used a series of LANDSAT

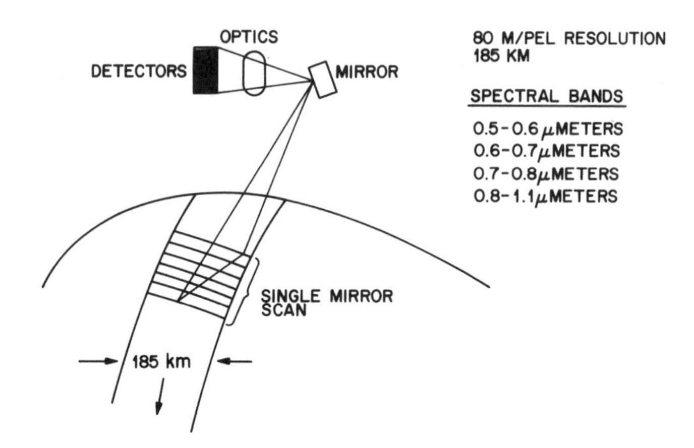

Fig. 2.8.2 LANDSAT Image Acquisition Using a Multispectral Scanner (from Green [2.8.1]).

Table 2.8.1 Four spectral bands used by the LANDSAT Multispectral Scanner.

MSS band	Wavelength μm	Spectral Region
4	0.5 to 0.6	Green
5	0.6 to 0.7	Red
6	0.7 to 0.8	Near Infrared
7	0.8 to 1.1	Near Infrared

Fig. 2.8.3 The spectral component of a LANDSAT scene of New Orleans in the band of wavelength 0.5 - 0.6 μm (from Green [2.8.1]).

satellites to acquire repetitive multispectral imagery of the earth's surface. The LANDSAT Multispectral Scanner (MSS) acquires images in four spectral bands denoted as 4, 5, 6, and 7 (Fig. 2.8.2). The spectral range of each band is given in Table 2.8.1.

Multispectral imagery has been used in geological applications to study the composition and dynamics of earth surfaces in cartography, mineral exploration, analysis of land utilization, weather and climate analysis, as well as agriculture and forestry applications. Fig. 2.8.3, for example, shows one of the spectral components (band 4) of a LANDSAT scene over the city of New Orleans. The low spectral response of water bodies in the infrared spectral bands, as well as the variation in the response of vegetation observed in band 4, is quite clear. In most systems the image acquired by the sensor is transmitted to a processing center on earth. Significant research has taken place in techniques for compression of such images, as well as processing to recognize patterns of interest. In addition to using correlation between the different pels of the image in one spectral band, techniques have also been invented to utilize the cross correlation between the images in different spectral bands.

References

2.2.1 D. E. Pearson, *Transmission and Display of Pictorial Information*, Pentech Press, London, 1975.

2.2.2 D. Fink, ed., *Television Engineering Handbook,* University Microfilms, Ann Arbor, Michigan.

2.2.3 R. S. O'Brien, ed., *Color Television*, Society of Motion Picture and Television Engineers, New York, 1970.

2.2.4 T. Rzeszewski, ed., *Color Television*, IEEE Press, New York, 1983.

2.2.5 J. Ouellet and E. Dubois, "Sampling and Reconstruction of NTSC Video Signals at Twice the Color Subcarrier Frequency," *IEEE Trans. Communications,* v. COM-29, no. 12, December 1981, pp. 1823-1832.

2.2.6 G. B. Townsend, *PAL Colour Television*, Cambridge University Press, 1970.

2.2.7 P. G. Carnt and G. B. Townsend, *Colour Television, Volume 2*, Iliffe, London, 1969.

2.2.8 J. Rossi, "Color Decoding a PCM NTSC Television Signal," *J. SMPTE.*, v. 83 no. 6, June 1974, pp. 489-495.

2.2.9 V. G. Devereux, "Digital Video: Sub-Nyquist Sampling of PAL Colour Signals," British Broadcasting Corp. Report 1975/4, January 1975.

2.2.10 V. G. Devereux, "Comparison of Picture Impairments Caused by Digital Coding of PAL and SECAM Video Signals," British Broadcasting Corp. Report 1974/16, April 1974.

2.2.11 J. O. Limb, C. B. Rubinstein and J. E. Thompson, "Digital Coding of Color Video Signals — A Review," *IEEE Trans. Communications,* v. COM-25 no. 11, November 1977, pp. 1349-1385.

2.2.12 A. H. Jones, "Digital Video Coding Standards: Factors Influencing the Choice in Europe," *J. SMPTE,* March 1982, pp. 260-265. Also, CCIR Rec. 601 and 601-1.

2.2.13 J. Baldwin, "Digital Standards Conversion," in *Digital Video*, SMPTE, Scarsdale, New York, 1977.

2.2.14 N. Jayant ad P. Noll, *"Digital Coding of Waveforms*", Prentice-Hall, Englewood Cliffs, New Jersey, 1984.

2.3.1 CCITT Rec. H. 261

2.5.1 T. Rzeszewski, ed., *Television Technology Today*, IEEE Press, New York, 1985.

2.5.2 SMPTE Standard 240M, *J. SMPTE,* Sept. 1989, pp. 723-725.

2.6.1 D. Costigan, *Electronic Delivery of Documents and Graphics*, Van Nostrand Co., 1978.

2.6.2 R. Hunter and A. H. Robinson, "International Digital Facsimile Coding Standards", *Proceedings of the IEEE*, Vol. 68, No. 7, July, 1980, pp. 854-867.

2.7.1 Basic Reference Model for Open System Interconnection, International Standards Organization, 150 DIS 7498–1983.

2.7.2 VIDEOTEX/TELETEXT Presentation Level Protocol Syntax, North American PLPS, ANSI X3.110 — 1983 and CSA T500 — 1983, American National Standards Institute, Inc. and Canadian Standards Organization, 1983.

2.7.3 North American Broadcast Teletext Specification (NABTS) Electronic Industries Association and Canadian Videotex Consultative Committee, 1983.

2.7.4 W. H. Ninke, "Guest Editorial, Interactive Picture Information Systems — Where From? Where To?" *IEEE Journal on Selected Areas in Communications*, Vol. SAC-1, No. 2, February 1983.

2.7.5 W. H. Ninke, "Design Considerations of the Data Syntax for Videotex and Teletext in North America," *Proceedings of the IEEE*, Vol. 73, No. 3, 1985.

2.8.1 W. B. Green, *Digital Image Processing, A Systems Approach*, New York: Van Nostrand Reinhold Company, 1983.

Questions for Understanding

2.1 How many pels per frame are required for Nyquist sampling of System M monochrome television, including blanking information?

2.2 Derive the matrix values in Eqs. (2.2.3) and (2.2.4).

2.3 Why don't TV systems use the XYZ primaries?

2.4 Derive the x,y chromaticities of the Y, I and Q primaries.

2.5 Why is the NTSC Y signal not the true luminance?

2.6 How many samples per frame are needed for four times color subcarrier sampling of NTSC? PAL?

2.7 Explain the differences between the color modulations of NTSC, PAL and SECAM.

2.8 Suppose a CCIR 601 receiver performs matrixing followed by clipping to recover R, G, B in the range 0 to 255. What color is produced by $Y = C_R = C_B = 0$?

2.9 How many luminance samples per frame are needed for 60 Hz SMPTE 240 television?

2.10 How many pels per 8.5×11 inch page are needed by the high resolution option of G3 fax?

3

Redundancy-Statistics-Models

In the first two chapters we discussed primarily how to represent visual information by a finite amount of digital data, or in the case of time-varying images, by a finite data rate. However, the data produced by these techniques normally contains a considerable amount of superfluous information that can be removed by methods to be discussed here and in later chapters.

Generally, we speak of two kinds of superfluous information. The first is *statistical redundancy*, having to do with the similarities, correlation and predictability of the data. Statistical redundancy has the property that it can be removed from the data without destroying any information whatsoever. This means that the original data can be recovered, if need be, subsequent to the removal of statistical redundancy. The second type of superfluous information is *subjective redundancy*, which has to do with data characteristics that can be removed without complaint by a human observer. Unlike statistical redundancy, the removal of subjective redundancy is irreversible. The original data cannot be recovered following the removal of subjective redundancy.

In order to study these two types of redundancy, we need to understand both the statistics of visual information as well as the characteristics of human vision. The former we treat in some detail in this chapter. The latter is covered in Chapter 4.

In this chapter we generally assume the images to have known or measurable statistical properties, in spite of the fact that this is rarely completely true as we shall see later. Thus, a pel value is a *random variable*. In the early sections we will sometimes need to distinguish between a random variable and its set of possible values. Thus, where necessary we will denote the random variable by upper case, e.g., B, one of its possible values by lower case, e.g., b or sometimes $b(i)$ when we need to denote the ith possible value, and the set of all possible values by braces, e.g., $\{b\}$. We will represent the histogram or probability distribution (PD) of B by $\{P(b)\}$.

We will also utilize joint probability distributions of two or more random variables. We denote values of several (usually adjacent) pels by N-tuples, e.g., $(b_1 b_2 ... b_N)$. Also, we sometimes refer to them as N-length vectors $\boldsymbol{b}$. The corresponding random vector is $\boldsymbol{B}$, or sometimes $(B_1 B_2 ... B_N)$, and its PD is denoted by $\{P(\boldsymbol{b})\}$, or sometimes $\{P(b_1 b_2 ... b_N)\}$.

In most of this chapter we will assume *stationary* statistics, i.e., the statistics of pels are the same in all parts of the picture and do not change with time. This assumption also is rarely completely true.

3.1 Redundancy in the Sampled Data—Information Theory

Suppose we have a monochrome image that has been raster scanned, sampled to produce MN pels and quantized to K-bit accuracy according to the techniques of Chapter 1. Thus, PCM would produce MNK bits for representation of the image. Also, suppose (as is usually the case) that the quantized luminance values are not all equally likely, having, for example, the histogram of Fig. 3.1.1. In this case, a reduction is possible in the number of bits required to specify the image if, instead of assigning K-bit words to each of the possible 2^K luminance levels, we assign words of various lengths. We can reduce the average number of bits per word if luminance levels having high probability are assigned short code-words and levels having lower probability are assigned longer code words. This method is called variable word-length coding, or sometimes *entropy coding* for reasons that will soon be apparent.

For example, suppose that quantized luminance level b has probability of occurrence $P(b)$ and that it is assigned a code word of length $L(b)$ bits. Then the average code word length for the image under consideration would be

$$\bar{L} = \sum_b L(b) P(b) \quad \text{bits/pel} \tag{3.1.1}$$

where summation is over the 2^K possible pel values. We would like to find codes for which $\bar{L}$ is as small as possible.

The average word-length $\bar{L}$ cannot be made arbitrarily small, however, and still have correct decoding by a receiver. In fact, a lower bound on $\bar{L}$ has been derived from information theory.[3.1.1] This lower bound is called the *entropy* of the random variable B, and is given by

$$H(B) = -\sum_b P(b) \log_2 P(b) \quad \text{bits/pel} \tag{3.1.2}$$

Entropy is never negative since $P(b)$ lies in the range [0,1]. As stated above

$$H(B) \leq \bar{L} \tag{3.1.3}$$

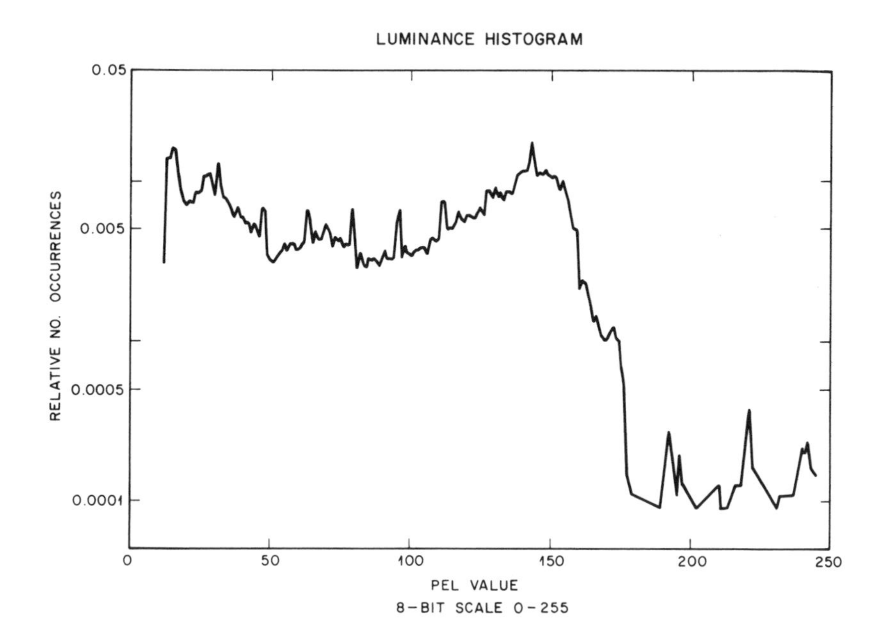

Fig. 3.1.1 Example of a luminance histogram. Luminances are quantized to 8-bits and range from 0 to 255.

P(0)	P(1)	P(2)	P(3)	P(4)	P(5)	P(6)	P(7)	Entropy (bits/pel)
1.0	0	0	0	0	0	0	0	0.00
0	0	0.5	0.5	0	0	0	0	1.00
0	0	0.25	0.25	0.25	0.25	0	0	2.00
0.06	0.23	0.30	0.15	0.08	0.06	0.06	0.06	2.68
0.125	0.125	0.125	0.125	0.125	0.125	0.125	0.125	3.00

Fig. 3.1.2 Five probability distributions and their corresponding entropies. For 8-level quantization entropies range from 0 to 3 bits/pel.

for all decodable, variable word-length codes that code pels independently of one another.

The entropy $H(B)$ is a measure of the amount of information carried by random variable B. For example, if the value of B is highly predictable,† e.g., $P(b) = \delta_{bb_o}$, then the entropy is zero. On the other hand, if all values of B are equally likely, i.e., $P(b) = 2^{-K}$, then the entropy is maximized and $H(B) = K$ bits/pel. Note that in the latter example, there is no benefit in variable word-length coding.

Thus, as $\{P(b)\}$ becomes more highly concentrated, the entropy becomes smaller, and the advantage of variable word-length coding is greater. Several PD's and their corresponding entropies are shown in Fig. 3.1.2.

3.1.1 Huffman Codes

If the quantized luminance for each sample is coded with variable word lengths, and the resulting code words are concatenated to form a stream of binary digits (bits) for storage or transmission, then correct decoding by a receiver (which, of course, knows the code) requires that every combination of concatenated code words be uniquely decipherable. A sufficient condition for this is that the code satisfy the so called *prefix rule*, which states that no code word may be the prefix of any other code word. Such codes are also *tree codes*, as we shall see.

A variable word-length code that minimizes the average bit-rate in Eq. (3.1.1) is the *Huffman code*.[3.1.2] It satisfies the prefix rule, and its average word length $\bar{L}_1$ is in the range

$$H(B) \le \bar{L}_1 \le H(B) + 1 \qquad \text{bits/pel} \tag{3.1.4}$$

That is, it performs within one bit/pel of the entropy.

Construction of a Huffman code involves the use of a *tree* as shown in Fig. 3.1.3a for 3-bit, i.e., 8-level* quantization. Levels $\{b(i)\}$ to be coded are first arranged in Section I of the tree according to decreasing order of probability. Construction of the tree then proceeds right-to-left as follows: the two bottom-most branches of Section I are combined via a *node* and their probabilities added to form a new branch in Section II of the tree. The branches of Section II are then rearranged according to decreasing order of probability. The procedure is repeated using the two bottom-most branches of Section II to form the branches of Section III that are then reordered according to decreasing

---

† $\delta_{bb_o} = 1$ if $b = b_o$. $\delta_{bb_o} = 0$ otherwise.

* The Huffman code does not require that the number of levels be a power of two. However, PCM quantization usually does produce 2^K levels.

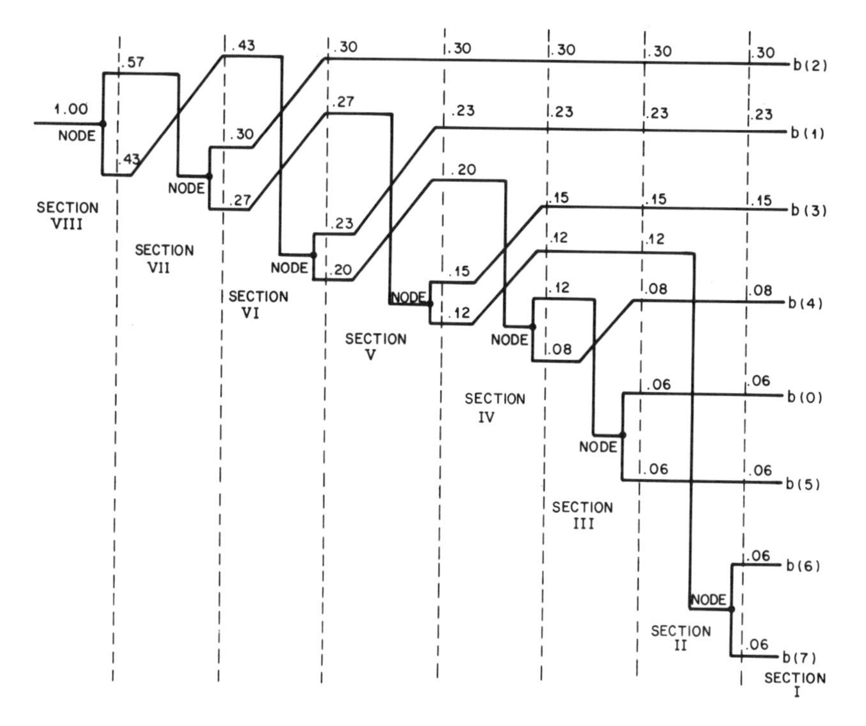

Fig. 3.1.3a Construction of a binary Huffman code proceeds from right to left. At each section the two bottom-most branches are combined to form a node and followed by a reordering of probabilities into descending order. These probabilities are then used to start the next section.

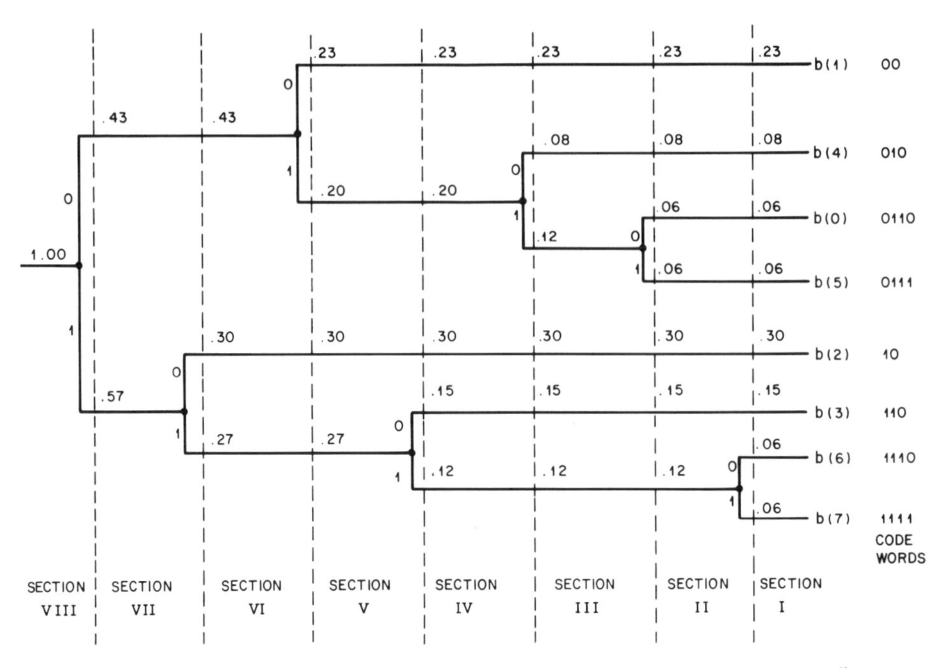

Fig. 3.1.3b After code construction, the tree is rearranged to eliminate crossovers, and coding proceeds from left to right. At each node a step-up produces a zero and a step-down a one. Resulting code words are shown at the right for each of the eight levels. Note that no code word is a prefix of any other code word.

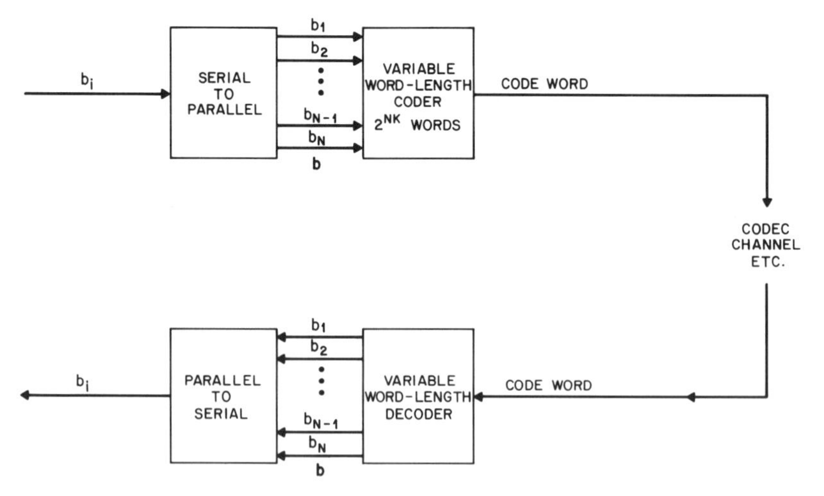

Fig. 3.1.4 Variable word-length coding of blocks of N pels. For K-bit quantization 2^{NK} code words are required. Decoding is accomplished via the tree of Fig. 3.1.3b.

probability, as with previous sections. The procedure continues until in Section VIII only one branch remains whose probability is one.

The entire tree may then be reordered to eliminate crossovers as shown in Fig. 3.1.3b. Coding is then accomplished for each sample by moving from the left of the tree toward the level $b(i)$ to be coded and, at each node, transmitting a binary 0 if an upward step is required and a binary 1 to indicate a downward step. The receiver decodes using the same tree in exactly the same way, i.e., moving upward for a received 0 and downward for a received 1. Resulting code words are shown at the right of Fig. 3.1.3b. The average bit-rate $\bar{L}$ is 2.71 bits/pel for the particular probabilities shown; the entropy is 2.68 bits/pel as seen in Fig. 3.1.2.

It can be shown that the Huffman code achieves entropy if all the probabilities $\{P(b)\}$ are a negative power of two. However, its average word-length is also constrained by

$$\bar{L}_1 \geq 1 \text{ bit/pel} \tag{3.1.5}$$

regardless of how small the entropy is.

Although the Huffman code minimizes the average word-length, in practice other codes are also used. For example, codes having a small number of word-lengths are often easier to implement than Huffman codes. Also, codes having very long word-lengths are usually avoided.

Although fixed word-length codes require more bits, they have the advantage that in case of a transmission bit error, the beginning of the next code word is easily found. Variable word-length codes, on the other hand, and Huffman codes in particular generally lose synchronization in case of transmission bit errors. That is, several code words may be decoded incorrectly before correct decoding is reestablished. Also, during this period of nonsynchronization, pels may be deleted or extra pels added. In video and facsimile systems, synchronization bits are often inserted at the end of each scan line in an attempt to limit the effect of transmission bit errors to one scan line.

3.1.2 Block Coding

If pels are coded and transmitted in blocks of N samples instead of one at a time, further bit-rate reductions are possible that may be significant depending on the particular application. We denote an N-tuple block of pels by

$$\boldsymbol{b} = (b_1 b_2 b_3 \ldots b_N) \tag{3.1.6}$$

where the components of the N-tuple, or vector, are K-bit quantized values as before. Thus, there are 2^{NK} possible values for $\boldsymbol{b}$. In this context, we speak of the random vector $\boldsymbol{B}$ whose possible values are $\{\boldsymbol{b}\}$ with probability distribution $\{P(\boldsymbol{b})\}$.

The *Nth order entropy* is defined as

$$H(\boldsymbol{B}) = -\sum_{\boldsymbol{b}} P(\boldsymbol{b})\log_2 P(\boldsymbol{b}) \quad \text{bits}/N\text{-tuple} \tag{3.1.7}$$

where the summation is over all the 2^{NK} possible N-tuples. The Nth order or *block* entropy $H(\boldsymbol{B})$ is a measure of the information carried by the random vector $\boldsymbol{B}$. If the pels of the block were completely independent of one another, then we would expect $\boldsymbol{B}$ to divulge N times as much information as B. However, if the pels of the block are highly dependent, i.e., their values are very similar on the average, then knowing a few pels tells us a lot about the remaining ones, and we would expect the information divulged by $\boldsymbol{B}$ to be much less. This is in fact true, and it can be shown rigorously that

$$\frac{1}{N}\, H(\boldsymbol{B}) \leq H(B) \quad \text{bits/pel} \tag{3.1.8}$$

with equality if and only if the pels are statistically independent, i.e.,

$$P(\boldsymbol{b}) = P(b_1)P(b_2)\ \dots\ P(b_N) \tag{3.1.9}$$

for all $\boldsymbol{b}$. The greater the statistical dependence between the pels of $\boldsymbol{B}$, the smaller the block entropy $H(\boldsymbol{B})$. However, the block entropy is also constrained by

$$H(\boldsymbol{B}) \geq H(B) \tag{3.1.10}$$

regardless of how much dependence there is.

Now, a Huffman code can be constructed for the N-tuples $\{\boldsymbol{b}\}$ according to the probability distribution $\{P(\boldsymbol{b})\}$ resulting in 2^{NK} code words as shown in Fig. 3.1.4. The average Huffman code-word length $\bar{L}_N$ in bits/N-tuple will then be within one bit of $H(\boldsymbol{B})$, as in Eq. (3.1.4). More specifically, in terms of bits per pel and using Inequality (3.1.8)

$$\frac{1}{N}\, H(\boldsymbol{B}) \leq \frac{1}{N}\, \bar{L}_N \leq \frac{1}{N}\, H(\boldsymbol{B}) + \frac{1}{N} \leq H(B) + \frac{1}{N} \qquad \text{bits/pel} \tag{3.1.11}$$

Thus, for sufficiently large N it is always possible, although it may not be practical, to block code at a bit-rate (in bits/pel) near or below the first order entropy $H(B)$. Moreover, if there is large statistical dependence between pel values, then the achievable bit-rate could be much lower than the single-pel entropy

b_1 \ b_2	0	1	2	3	4	5	6	7
0	0.004	0.014	0.018	0.009	0.005	0.004	0.004	0.004
1	0.014	0.053	0.069	0.034	0.018	0.014	0.014	0.014
2	0.018	0.069	0.090	0.045	0.024	0.018	0.018	0.018
3	0.009	0.034	0.045	0.023	0.012	0.009	0.009	0.009
4	0.005	0.018	0.024	0.012	0.006	0.005	0.005	0.005
5	0.004	0.014	0.018	0.009	0.005	0.004	0.004	0.004
6	0.004	0.014	0.018	0.009	0.005	0.004	0.004	0.004
7	0.004	0.014	0.018	0.009	0.005	0.004	0.004	0.004

$P(b_1 b_2)$

$H(B_1) = H(B_2) = 2.68$ bits/pel $\qquad \bar{L}_1 = 2.71$ bits/pel

$\frac{1}{2} H(B_1 B_2) = 2.68$ bits/pel $\qquad \frac{1}{2} \bar{L}_2 = 2.70$ bits/pel

$H(B_2|B_1) = 2.68$ bits/pel $\qquad \bar{L}_c = 2.71$ bits/pel

Fig. 3.1.5a Joint probability distribution (PD) of two successive pels. In this case, the pels are statistically independent, and entropies are equal. $\{P(b_1)\} = \{P(b_2)\}$ is the same as in Fig. 3.1.3.

b_1 \ b_2	0	1	2	3	4	5	6	7
0	0.050	0.010	0.000	0.000	0.000	0.000	0.000	0.000
1	0.010	0.180	0.020	0.020	0.000	0.000	0.000	0.000
2	0.000	0.020	0.210	0.040	0.020	0.010	0.000	0.000
3	0.000	0.020	0.040	0.070	0.010	0.010	0.000	0.000
4	0.000	0.000	0.020	0.010	0.030	0.010	0.010	0.010
5	0.000	0.000	0.010	0.010	0.010	0.020	0.010	0.000
6	0.000	0.000	0.000	0.000	0.010	0.010	0.030	0.010
7	0.000	0.000	0.000	0.000	0.000	0.000	0.010	0.050

$P(b_1 b_2)$

$H(B_1) = H(B_2) = 2.68$ bits/pel $\qquad \bar{L}_1 = 2.71$ bits/pel

$\frac{1}{2} H(B_1 B_2) = 2.07$ bits/pel $\qquad \frac{1}{2} \bar{L}_2 = 2.11$ bits/pel

$H(B_2|B_1) = 1.46$ bits/pel $\qquad \bar{L}_c = 1.70$ bits/pel

Fig. 3.1.5b In this case the pels are partially dependent. $\{P(b_1)\} = \{P(b_2)\}$ is the same as above.

		b_2							
		0	1	2	3	4	5	6	7
	0	0.060	0.000	0.000	0.000	0.000	0.000	0.000	0.000
	1	0.000	0.230	0.000	0.000	0.000	0.000	0.000	0.000
	2	0.000	0.000	0.300	0.000	0.000	0.000	0.000	0.000
	3	0.000	0.000	0.000	0.150	0.000	0.000	0.000	0.000
b_1	4	0.000	0.000	0.000	0.000	0.080	0.000	0.000	0.000
	5	0.000	0.000	0.000	0.000	0.000	0.060	0.000	0.000
	6	0.000	0.000	0.000	0.000	0.000	0.000	0.060	0.000
	7	0.000	0.000	0.000	0.000	0.000	0.000	0.000	0.060

$P(b_1 b_2)$

$$H(B_1) = H(B_2) = 2.68 \text{ bits/pel} \qquad \bar{L}_1 = 2.71 \text{ bits/pel}$$

$$\frac{1}{2} H(B_1 B_2) = 1.34 \text{ bits/pel} \qquad \frac{1}{2} \bar{L}_2 = 1.38 \text{ bits/pel}$$

$$H(B_2 | B_1) = 0.00 \text{ bits/pel} \qquad \bar{L}_c = 1.00 \text{ bits/pel}$$

Fig. 3.1.5c In this case pels are completely dependent. $\{P(b_1)\} = \{P(b_2)\}$ is the same as above.

		b_2							
		0	1	2	3	4	5	6	7
	0	0.930	0.000	0.000	0.000	0.000	0.000	0.000	0.000
	1	0.000	0.010	0.000	0.000	0.000	0.000	0.000	0.000
	2	0.000	0.000	0.010	0.000	0.000	0.000	0.000	0.000
	3	0.000	0.000	0.000	0.010	0.000	0.000	0.000	0.000
b_1	4	0.000	0.000	0.000	0.000	0.010	0.000	0.000	0.000
	5	0.000	0.000	0.000	0.000	0.000	0.010	0.000	0.000
	6	0.000	0.000	0.000	0.000	0.000	0.000	0.010	0.000
	7	0.000	0.000	0.000	0.000	0.000	0.000	0.000	0.010

$P(b_1 b_2)$

$$H(B_1) = H(B_2) = 0.56 \text{ bits/pel} \qquad \bar{L}_1 = 1.20 \text{ bits/pel}$$

$$\frac{1}{2} H(B_1 B_2) = 0.28 \text{ bits/pel} \qquad \frac{1}{2} \bar{L}_2 = 0.60 \text{ bits/pel}$$

$$H(B_2 | B_1) = 0.00 \text{ bits/pel} \qquad \bar{L}_c = 1.00 \text{ bits/pel}$$

Fig. 3.1.5d In this case pels are completely dependent and per pel entropies are small compared with unity. Note that the block coding bit-rate is lower than with conditional coding even though the entropy is higher.

$H(B)$. However, as with all variable word-length codes, we also have the constraint

$$\bar{L}_N \geq 1 \text{ bit}/N\text{-tuple} \tag{3.1.12}$$

Fig. 3.1.5 shows entropies and Huffman code bit-rates for several probability distributions of 2-tuples with 3-bit quantization. Note that the higher the statistical dependence, the lower the block coding bit-rate.

3.1.3 Conditional Coding

A difficulty with block coding is the often large size of the codes. A lookup table or tree containing 2^{NK} code words must be provided in order to code N-tuples of pels quantized to K-bits each. This problem is partially overcome by *conditional* coding and its extensions. Moreover, as we shall see, conditional coding generally produces a lower bit-rate than block coding. Thus, it is of great interest both from a practical and a performance standpoint.

With conditional coding, as shown in Fig. 3.1.6, we assume that the components $b_1 b_2 \ldots b_{N-1}$ of N-tuple $\boldsymbol{b}$ have already been transmitted and are known at the receiver. Component b_N can then be coded more efficiently by utilizing this information (assuming there is statistical dependence between pels). Specifically, for a particular $(N-1)$-tuple $(b_1 b_2 \ldots b_{N-1})$, a Huffman code for b_N can be constructed based on the conditional PD

$$\{P(b_N | b_1 b_2 \ldots b_{N-1})\} \tag{3.1.13}$$

The code will have 2^K words, and the average word length in bits/pel (for this particular $(N-1)$-tuple) will be within one bit of

$$H(B_N | b_1 b_2 \ldots b_{N-1}) = -\sum_{b_N} P(b_N | b_1 \ldots b_{N-1}) \log_2 P(b_N | b_1 \ldots b_{N-1}) \text{ bits/pel} \tag{3.1.14}$$

If such a code is constructed for each possible $(N-1)$-tuple, then $2^{K(N-1)}$ codes result, each having 2^K code words. Thus, the size of each code is reduced considerably compared with block coding; however, we now have $2^{K(N-1)}$ codes. The overall conditional coding bit-rate $\bar{L}_c$ in bits/pel is then approximated to within one bit by averaging Eq. (3.1.14) over all of the $(N-1)$-tuples. The result is called the *conditional entropy* and is given by

$$H(B_N | B_1 B_2 \ldots B_{N-1}) = \sum_{b_1 \ldots b_{N-1}} P(b_1 \ldots b_{N-1}) H(B_N | b_1 \ldots b_{N-1}) \tag{3.1.15}$$

$$= -\sum_{\boldsymbol{b}} P(\boldsymbol{b}) \log_2 P(b_N | b_1 b_2 \ldots b_{N-1}) \text{ bits/pel}$$

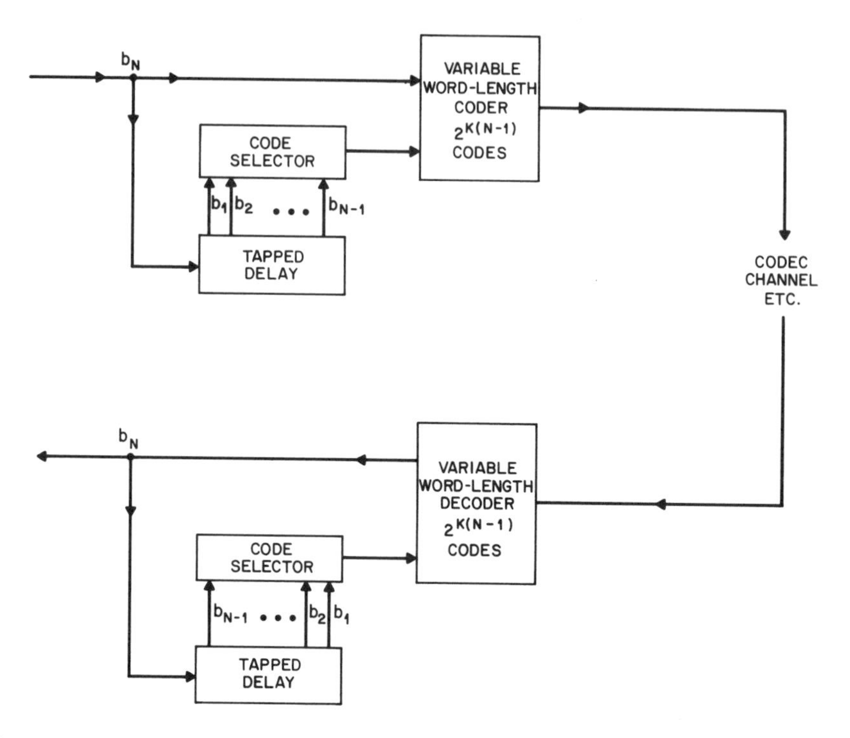

Fig. 3.1.6 Conditional coding. Each pel is coded differently depending on the value of the previous $N-1$ pels. The system requires $2^{K(N-1)}$ variable word-length codes. Each code has 2^K code words. The previous $N-1$ pels must be stored at both the coder and decoder.

More specifically, the overall average bit-rate $\bar{L}_c$ of conditional coding satisfies

$$H(B_N|B_1B_2...B_{N-1}) \;\le\; \bar{L}_c \;\le\; H(B_N|B_1B_2...B_{N-1}) + 1 \quad \text{bits/pel} \tag{3.1.16}$$

However, as with all variable word-length codes the average word-length is also constrained by

$$\bar{L}_c \ge 1 \text{ bit/pel} \tag{3.1.17}$$

Fig. 3.1.5 shows conditional entropies and Huffman code bit-rates $\bar{L}_c$ for several probability distributions of 2-tuples with $K = 3$. Note that with complete dependence the conditional entropy is zero, whereas the block entropy is bounded below by Inequality (3.1.10).

Since the complexity of idealized conditional coding, as described above, and block coding is about the same, we naturally wonder what the relative performances are. To examine this question we make use of the relation

$$\begin{aligned} H(\boldsymbol{B}) &\triangleq H(B_1B_2...B_N) \quad \text{bits}/N\text{-tuple} \\ &= H(B_N|B_1...B_{N-1}) + H(B_1B_2...B_{N-1}) \end{aligned} \tag{3.1.18}$$

which follows easily from Eqs. (3.1.7) and (3.1.15). This says that the entropy of the N-tuple is equal to the entropy of the $(N-1)$-tuple plus the conditional entropy. Similarly,

$$H(B_1...B_{N-1}) = H(B_{N-1}|B_1...B_{N-2}) + H(B_1...B_{N-2}) \tag{3.1.19}$$

etc., until $H(\boldsymbol{B})$ is decomposed into $N-1$ conditional entropies plus $H(B_1)$. Now, assuming all pels have the same statistics, i.e., the statistics are *stationary*, conditional entropies and $H(B_1)$ are all lower bounded by $H(B_N|B_1...B_{N-1})$.* Thus,

---

* That is, for stationary statistics $H(B_N|B_1...B_{N-1}) \le H(B_{N-1}|B_1...B_{N-2})$, etc. since less information is available for the right hand side conditional entropies.

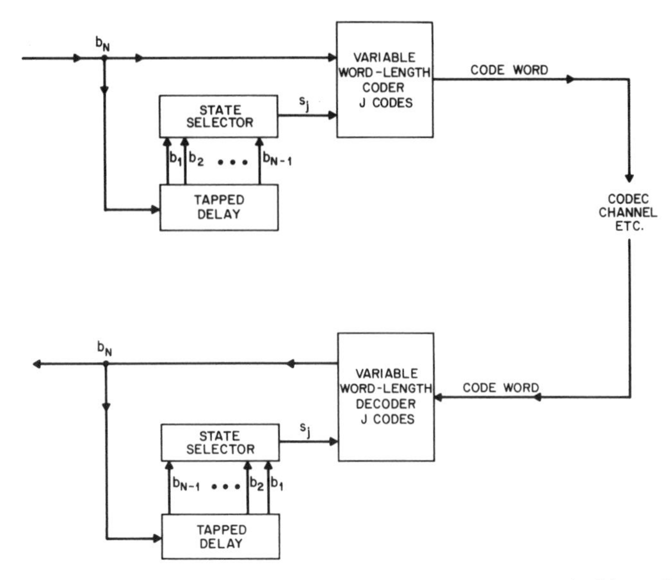

Fig. 3.1.7 State space conditional coding. Here the number of codes required is much less than with conditional coding. The previous $N-1$ pels are stored and used to compute the state s_j. Then the $j-th$ variable word-length code is used to code b_N.

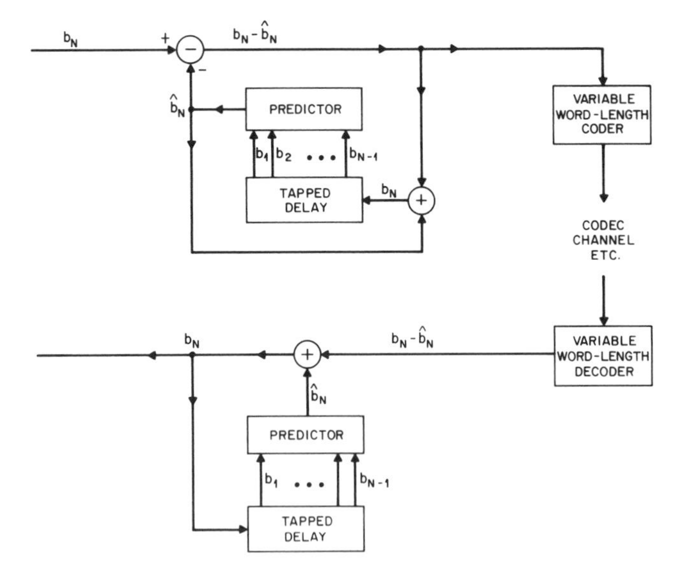

Fig. 3.1.8 Predictive coding. Here the previous $N-1$ pels are used to compute a prediction $\hat{b}_N$ of pel b_N both at the coder and decoder. The differential signal $b_N - \hat{b}_N$ is then coded usually with a single variable word-length code.

$$\frac{1}{N} H(\boldsymbol{B}) \geq H(B_N | B_1 \ldots B_{N-1}) \text{ bits/pel} \tag{3.1.20}$$

and we see that the conditional entropy is always less than the per pel block entropy. Thus, conditional coding is usually better than block coding. However, if the conditional and per pel block entropies are much less than one bit/pel, then block coding with variable word-lengths may perform better than conditional coding, since the conditional coding average word-length $\overline{L}_c$ is constrained to operate above 1 bit/pel. See Inequalities (3.1.17) and (3.1.12) and also Fig. 3.1.5.

3.1.4 State-Space Conditional Coding

The major impediment to using conditional coding is the number of variable word-length code sets that must be generated and made available during coding and decoding. As mentioned above, this number is $2^{K(N-1)}$, which is often too large to allow for practical implementation of conditional coding.

One way of reducing the required number of codes utilizes the concept of state space, which is closely related to the theory of Markov random processes. With this approach the set of possible $(N-1)$-tuples

$$\{ b_1 b_2 \ldots b_{N-1} \} \tag{3.1.24}$$

is partitioned into J subsets or states $\{ s_j,\ j=1 \ldots J \}$ in such a way that the conditional PD

$$\{ P(b_N | b_1 b_2 \ldots b_{N-1}) \} \tag{3.1.25}$$

is nearly invariant for $(N-1)$-tuples within each subset. Following this, J variable word-length codes are constructed based on the J conditional PD's

$$\{ P(b_N | s_j) \} \tag{3.1.26}$$

Each of the J codes has 2^K code words for a total of $J2^K$ code words, which, with appropriate partitioning, is considerably smaller than the 2^{NK} code words required for block or conditional coding. State space conditional coding is shown in Fig. 3.1.7.

The average bit-rate $\overline{L}_j$ in bits/pel for each state will be within one bit of the *state entropy*

$$H(B_N | s_j) = - \sum_{b_N} P(b_N | s_j) \log_2 P(b_N | s_j) \text{ bits/pel} \tag{3.1.27}$$

and the overall bit-rate will be within one bit of

$$H(B_N|S) = \sum_j P(s_j)H(B_N|s_j) \text{ bits/pel}$$

$$= -\sum_b P(b)\,log_2 P(b_N|s_j) \tag{3.1.28}$$

This is approximately equal to the conditional entropy $H(B_N|B_1B_2...B_{N-1})$ if $P(b_N|b_1b_2...b_{N-1})$ does not vary much for the $(N-1)$-tuples within a given state.

Now, it may happen (especially with graphics) that some of the state entropies $H(B_N|s_j)$ are very small, and the benefits of block coding should be utilized, whereas for other states, block coding may not yield much saving. This can be accomplished by a technique known as *reordering* wherein instead of coding and transmitting pels in the order in which they occur, they are reordered so that pels having state s_1 are coded and transmitted together, followed by pels having state s_2, etc. until finally pels having state s_J are coded and transmitted. In this way, block coding can be carried out for some states, as desired, but not for the remaining states. As we shall see later, reordering finds most of its application in graphics coding.

3.1.5 Predictive Coding

Predictive coding[3.1.4] is very closely related to conditional coding, the main difference being that instead of coding and transmitting b_N, we transmit another quantity $b_N - \hat{b}_N$ as shown in Fig. 3.1.8. We call $\hat{b}_N$ the *prediction* of b_N, and it is usually a fixed, deterministic function of the previously transmitted pels $b_1b_2...b_{N-1}$. We call $b_N - \hat{b}_N$ the *differential** signal, or sometimes the *Predication Error*. Since $\hat{b}_N$ can be made available also at the receiver, b_N is recovered simply by adding the prediction to the received differential signal.

In theory, nothing is lost or gained by conditional coding, block coding, etc. of differential signals instead of the pels themselves. Conditional entropies are unchanged, i.e.,

$$H(B_N - \hat{B}_N|B_1...B_{N-1}) = H(B_N|B_1...B_{N-1}) \text{ bits/pel} \tag{3.1.29}$$

as long as $\hat{b}_N$ depends only on $b_1...b_{N-1}$. Block coding can easily be applied to the differential signal values, as can state-space coding. However, in the latter case, $\hat{b}_N$ is still usually a function of the pels $b_1...b_{N-1}$ themselves, rather than being a function only of the state s_j.

---

* Predictive coding is also known as differential PCM or DPCM.

An advantage of predictive coding over conditional coding is that the conditional PD of the differential signal

$$\{P(b_N - \hat{b}_N | b_1 \ldots b_{N-1})\} \tag{3.1.30}$$

usually shows much less variation with $\{b_1 \ldots b_{N-1}\}$ than does the conditional PD of b_N itself. Thus, with state-space predictive coding, the number of states can be reduced significantly compared with state-space conditional coding. In fact, in an overwhelming number of applications, only one state is used, i.e., no conditional coding, in which case only one variable word-length code need be designed based on the simple PD

$$\{P(b_N - \hat{b}_N)\} \tag{3.1.31}$$

Another advantage of a well-designed predictive coding is that successive differential signals usually have much less statistical dependency, i.e., less redundancy than successive pels. That is*

$$\frac{1}{N} H(B_1 - \hat{B}_1, B_2 - \hat{B}_2 \ldots B_N - \hat{B}_N) \approx H(B - \hat{B}) \text{ bits/pel} \tag{3.1.32}$$

In fact, a completely optimum predictive coder has statistically independent† outputs. Thus, block predictive coding is of little benefit unless $H(B - \hat{B})$ is small compared to unity (as for example in graphics). However, state-space predictive coding with reordering may benefit from block coding if for some states the conditional entropy

$$H(B_N - \hat{B}_N | s_j) \text{ bits/pel} \tag{3.1.33}$$

is small compared with unity.

A major objective of predictive coding is to produce a differential signal $b_N - \hat{b}_N$ that is small on the average and large only occasionally, i.e., it has a highly peaked PD and a small entropy. Thus, the better the predictor, the smaller the entropy and the lower the bit-rate required for coding. For example, a *maximum likelihood* predictor chooses $\hat{b}_N$ to be the value of b_N that maximizes the conditional probability $P(b_N | b_1 \ldots b_{N-1})$. A *conditional mean*

---

* Again we implicitly assume stationarity, i.e., that the first order entropy $H(B_N - \hat{B}_N)$ is independent of N.

† Statistical independence is necessary, but not sufficient, for optimality. For example, simple encryption gives statistical independence, but no bit-rate reduction.

predictor chooses

$$\hat{b}_N = \sum_{b_N} b_N \; P(b_N | b_1 \ldots b_{N-1}) \tag{3.1.34}$$

which can be shown to minimize the mean square prediction error# $E(b_N - \hat{b}_N)^2$. Thus, Eq. (3.1.34) also defines the *minimum mean square* predictor.[3.1.5]

A *linear* predictor chooses

$$\hat{b}_N = \sum_{i=1}^{N-1} \alpha_i b_i \tag{3.1.35}$$

where the weighting coefficients $\{\alpha_i\}$ are chosen for ease of implementation and/or according to some performance criterion. A minimum mean square *linear* predictor can easily be derived by differentiating $E(b_N - \hat{b}_N)^2$ with respect to each α_j and setting the result to zero. This yields

$$-2E\left[b_N b_j - \sum_{i=1}^{N-1} \alpha_i b_i b_j\right] = 0 \qquad j=1 \ldots N-1 \tag{3.1.36}$$

Letting $d_j = E(b_N b_j)$ and $r_{ij} = E(b_i b_j)$, and using column matrices $\boldsymbol{d}$, $\boldsymbol{\alpha}$ and square matrix $\boldsymbol{R}$, Eqs. (3.1.36) become

$$\boldsymbol{d} - \boldsymbol{R\alpha} = 0 \tag{3.1.37}$$

The square matrix $\boldsymbol{R}$ is known as the *correlation matrix* of the pels $b_1 \ldots b_{N-1}$, and r_{ij} is the *correlation* between pels b_i and b_j. If the correlation matrix is nonsingular, then

$$\boldsymbol{\alpha} = \boldsymbol{R}^{-1}\boldsymbol{d} \tag{3.1.38}$$

is the *unique* solution of Eq. (3.1.37). Otherwise, many solutions exist, each having the same mean square prediction error $E(b_N - \hat{b}_N)^2$.

The mean square prediction error can be reduced somewhat if in Eqs. (3.1.35) through (3.1.38) each pel has its mean value subtracted prior to

---

\# E is the statistical averaging or expectation operator. For a function f of the random variable b, $E[f(b)] \triangleq \sum_b P(b) f(b)$. From now on we no longer distinguish between a random variable and its possible values.

calculation, e.g., Eq. (3.1.35) becomes

$$\hat{b}_N - Eb_N = \sum_{i=1}^{N-1} \alpha_i (b_i - Eb_i) \tag{3.1.39}$$

However, the improvement is small if, as is the usual case, adjacent pels are highly correlated.

3.1.6 Discrete Transform Coding

Suppose a reversible transformation[3.1.6] is carried out on N-tuple blocks $\boldsymbol{b}$ of PCM coded pels prior to transmission as shown in Fig. 3.1.9. We denote the transformation (and inverse) by

$$\boldsymbol{c} = T(\boldsymbol{b})$$

$$\boldsymbol{b} = T^{-1}(\boldsymbol{c}) \tag{3.1.40}$$

and refer to $\boldsymbol{c}$ as the block (usually an N-tuple) of *transform coefficients*. A major objective of transform coding is to make as many transform coefficients as possible small enough so that they are insignificant and need not be coded for transmission. A parallel objective is to minimize statistical dependence, i.e., redundancy between the transform coefficients.

If the transformation can be represented as a (possibly complex) matrix $\boldsymbol{T}$ operating on column vectors $\boldsymbol{b}$ and $\boldsymbol{c}$, i.e.,

$$\boldsymbol{c} = \boldsymbol{Tb}$$

$$\boldsymbol{b} = \boldsymbol{T}^{-1}\boldsymbol{c} \tag{3.1.41}$$

then T is called a *linear* transform. If in addition*

$$\boldsymbol{T}^{-1} = \boldsymbol{T}' \tag{3.1.42}$$

then $\boldsymbol{T}$ is an *orthonormal* or *unitary* transform. For example, we saw in Section 1.5.6 that the Discrete Fourier Transform (DFT) for which $\boldsymbol{T}$ is defined by

---

* Prime denotes conjugate transpose.

$$t_{mi} = \frac{1}{\sqrt{N}} \exp\left[-\frac{2\pi}{N}\sqrt{-1}\,(i-1)(m-1)\right]$$

$$i,m = 1 \ldots N \qquad (3.1.43)$$

is a unitary transform.

Reiterating briefly from Chapter 1, if we denote the mth column of matrix $\boldsymbol{T}'$ by the column vector $\boldsymbol{t}_m$, then Eq. (3.1.42) is equivalent to

$$\boldsymbol{t}'_m \, \boldsymbol{t}_n = \delta_{mn} \qquad (3.1.44)$$

The vectors $\boldsymbol{t}_m$ are called *orthonormal basis* vectors for the linear transform $\boldsymbol{T}$. This terminology arises when Eq. (3.1.41) is written as

$$\boldsymbol{b} = \sum_{m=1}^{N} c_m \boldsymbol{t}_m \qquad (3.1.45)$$

i.e., $\boldsymbol{b}$ is represented as a weighted sum of the basis vectors. Unitary transforms preserve signal energy, i.e.,

$$\sum_{i=1}^{N} b_i^2 = \sum_{m=1}^{N} |c_m|^2 \qquad (3.1.46)$$

This is known classically as Parseval's theorem, and is easily seen from

$$\boldsymbol{b}'\boldsymbol{b} = (\boldsymbol{T}'\boldsymbol{c})'(\boldsymbol{T}'\boldsymbol{c}) = \boldsymbol{c}'\boldsymbol{T}\boldsymbol{T}'\boldsymbol{c} = \boldsymbol{c}'\boldsymbol{c} \qquad (3.1.47)$$

The DFT suffers from undesirable effects due to discontinuities in the periodic extension of the block $\boldsymbol{b}$ of N pels. This is known classically as the *Gibbs phenomenon.* It can be reduced by constructing a $2N$ point even extension of $\boldsymbol{b}$ given by

$$\boldsymbol{b}_e = [b_1 b_2 \ldots b_N \; b_N \ldots b_2 b_1]^t \qquad (3.1.48)$$

and taking a $2N$ point DFT. Moreover, if instead of the DFT of Eq. (3.1.43), we use the time shifted, $2N$ point DFT

$$t_{mi} = \frac{1}{\sqrt{2N}} \exp\left[-\frac{2\pi}{2N}\sqrt{-1}\left(i - \frac{1}{2}\right)(m-1)\right] \qquad (3.1.49)$$

then the transform coefficients will be real. After orthonormalization we obtain the *Discrete Cosine Transform* or DCT.[3.2.3] Elements of $\boldsymbol{T}$ are given by

$$t_{mi} = \sqrt{\frac{2-\delta_{m1}}{N}} \cos\left[\frac{\pi}{N}\left(i - \frac{1}{2}\right)(m-1)\right] \qquad i,m = 1\ldots N \tag{3.1.50}$$

Another of the early proposed transforms was the *Walsh-Hadamard Transform* (WHT) for N equal to a power of two. Its (symmetric) transform matrix can be described recursively. Let

$$\boldsymbol{H}_2 = \begin{bmatrix} 1 & 1 \\ 1 & -1 \end{bmatrix} \tag{3.1.51}$$

Then for N equal to a power of two

$$\boldsymbol{H}_{2N} = \begin{bmatrix} \boldsymbol{H}_N & \boldsymbol{H}_N \\ \boldsymbol{H}_N & -\boldsymbol{H}_N \end{bmatrix} \tag{3.1.52}$$

The corresponding orthonormal transform matrix is then given by

$$\boldsymbol{T} = \frac{1}{\sqrt{N}} \boldsymbol{H}_N = \boldsymbol{T}' \tag{3.1.53}$$

An advantage of the WHT is that most of the computation requires only addition and subtraction, as opposed to most other transforms that require multiplication as well.

Since information about $\boldsymbol{b}$ is neither lost nor gained by a reversible transformation, entropies are unchanged. That is,

$$H(\boldsymbol{b}) = H(\boldsymbol{c}) \text{ bits/block} \tag{3.1.54}$$

In addition, if a transform can be found that *removes statistical dependence*, i.e., produces transform coefficients $\{c_1 c_2 \ldots c_N\}$ that are statistically independent, then

$$H(\boldsymbol{c}) = \sum_{m=1}^{N} H(c_m) \text{ bits/block} \tag{3.1.55}$$

In this case, block coding of the coefficients is of relatively little value (except possibly for coefficients with $H(c_m)$ small compared with unity). Thus, efficient coding and transmission of transform coefficients is possible with much less complexity than block coding of the pels themselves would entail.

Thus far, no transformation has been found for pictorial information that completely removes statistical *dependence* between the transform coefficients. However, pairwise statistical *correlation* can be removed by use of an orthonormal linear transformation known as the *Karhunen-Loeve Transform* (KLT).[3.1.6] That is, the KLT transform coefficients satisfy (for $m \neq n$)

$$\sum_{c_m c_n} c_m c_n P(c_m, c_n) = \sum_{c_m} c_m P(c_m) \cdot \sum_{c_n} c_n P(c_n) \qquad (3.1.56)$$

where summation is over all possible values. This is usually written, using the statistical averaging operator E, as

$$E(c_m c_n) = Ec_m \cdot Ec_n \quad m \neq n \ , \quad \text{or}$$

$$E[(c_m - Ec_m)(c_n - Ec_n)] = \lambda_m \delta_{mn} \ \forall \ m,n \qquad (3.1.57)$$

where λ_m is the variance of c_m. Note that statistical independence implies uncorrelation, but the reverse is generally not* true.

The KLT can be derived by assuming that pel vectors are first biased to have zero mean, that is

$$E\boldsymbol{b} = \boldsymbol{0} \text{ and therefore } E\boldsymbol{c} = \boldsymbol{0} \qquad (3.1.58)$$

The $N \times N$ correlation matrix of pel values then becomes

$$\boldsymbol{R} = E(\boldsymbol{bb}')$$

$$= \boldsymbol{E}(\boldsymbol{T}'\boldsymbol{cc}'\boldsymbol{T}) = \boldsymbol{T}'E(\boldsymbol{cc}')\boldsymbol{T} \qquad (3.1.59)$$

from Eqs. (3.1.41) and (3.1.42). From the orthonormality condition of Eq. (3.1.42)

$$\boldsymbol{RT}' = \boldsymbol{T}'E(\boldsymbol{cc}') \qquad (3.1.60)$$

From Eqs. (3.1.57) and (3.1.58), $E(\boldsymbol{cc}')$ is a diagonal $N \times N$ matrix with the variances of c_m along the main diagonal. Thus, in terms of column vectors of $\boldsymbol{T}'$, Eq. (3.1.60) becomes

* It is true for jointly Gaussian random variables, however.

$$\boldsymbol{R}\boldsymbol{t}_m = \lambda_m \boldsymbol{t}_m \quad m = 1,\ldots,N \tag{3.1.61}$$

and we see that the basis vectors of the KLT are the eigenvectors of the correlation matrix $\boldsymbol{R}$, orthonormalized (by suitable linear combinations of eigenvectors) to satisfy Eq. (3.1.44). The variances $\{\lambda_m\}$ of the KLT coefficients are the eigenvalues of $\boldsymbol{R}$, and since $\boldsymbol{R}$ is symmetric and positive definite, they are real and positive. The KLT basis vectors and transform coefficients are also real.

Besides decorrelating transform coefficients, the KLT has another useful property: it maximizes the number of coefficients that are small enough so that they are insignificant. For example, suppose the KLT coefficients $(c_1 \ldots c_N)$ are ordered according to decreasing variance, i.e., λ_1 largest, λ_N smallest. Also, suppose that for reasons of economy we transmit only the first pN coefficients where $0<p<1$, as shown in Fig. 3.1.9. The receiver then uses the truncated column vector

$$\tilde{\boldsymbol{c}} = (c_1 \ldots c_{pN}, 0, \ldots 0)^t \tag{3.1.62}$$

to form the reconstructed pel values

$$\tilde{\boldsymbol{b}} = \boldsymbol{T}'\tilde{\boldsymbol{c}} \tag{3.1.63}$$

The *mean squared error (MSE)* between the original pels $\boldsymbol{b}$ and the reconstructed pels $\tilde{\boldsymbol{b}}$ is then

$$\begin{aligned} E\,\frac{1}{N}\sum_{i=1}^{N}(b_i - \tilde{b}_i)^2 &= \frac{1}{N}\,E[|\boldsymbol{b}-\tilde{\boldsymbol{b}}|^2] \\ &= \frac{1}{N}\,E[(\boldsymbol{b}-\tilde{\boldsymbol{b}})'(\boldsymbol{b}-\tilde{\boldsymbol{b}})] \\ &= \frac{1}{N}\,E[(\boldsymbol{c}'\boldsymbol{T}-\tilde{\boldsymbol{c}}'\boldsymbol{T})(\boldsymbol{T}'\boldsymbol{c}-\boldsymbol{T}'\tilde{\boldsymbol{c}})] \\ &= \frac{1}{N}\,E[|\boldsymbol{c}-\tilde{\boldsymbol{c}}|^2] \\ &= \frac{1}{N}\sum_{pN+1}^{N}\lambda_m \end{aligned} \tag{3.1.64}$$

It is shown in the Appendix that the KLT minimizes the MSE due to truncation.

The treatment here of transform coding has assumed digitization, i.e., sampling and quantization, prior to transformation. The order of these operations can, of course, be reversed as discussed in Chapter 1. From an information

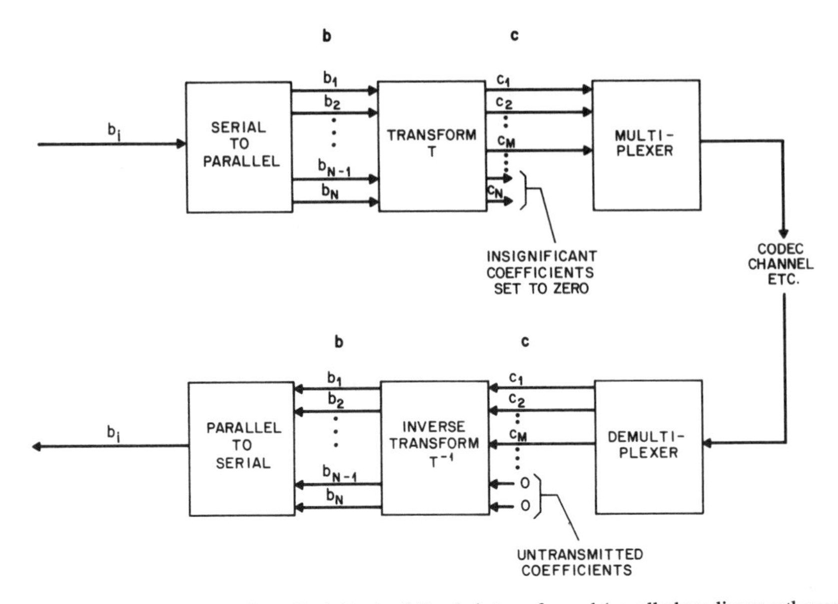

Fig. 3.1.9 Transform coding. Each block of N pels is transformed (usually by a linear orthonormal matrix) into a block of transform coefficients, and the insignificant coefficients are discarded. The remaining $M = pN$ $(p < 1)$ coefficients are then coded and transmitted to the receiver where the inverse operation takes place.

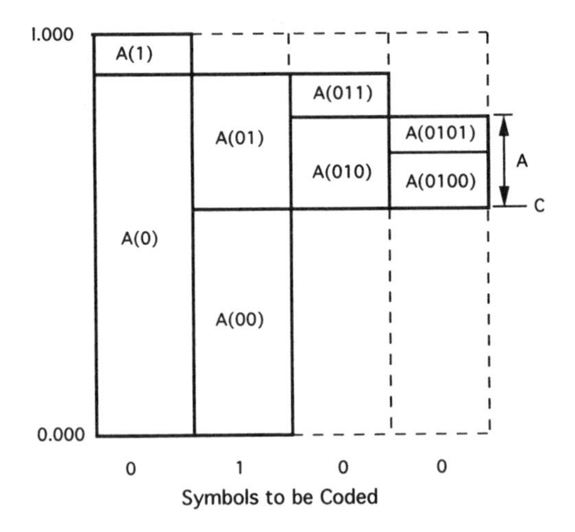

Fig. 3.1.10 Example of interval subdivision for Arithmetic Coding. Four symbols have been input so far.

theoretic point of view, however, filtering and sampling can always be carried out with negligible distortion, followed by whatever transformation, coding and transmission operations one wishes to use. Since there is no appreciable loss of visual information, either approach yields essentially the same results.

3.1.7 Universal Coding

A number of data compression methods have been suggested for cases where the statistics of the data are not initially known by the coder. Basically, these techniques, called Universal Coding Algorithms, attempt to measure the statistics during the actual coding operation, and adapt themselves thereto in order to maximize the compression. Typically, they are fairly adept at exploiting the redundancy in repetitive pel patterns, but are less capable of removing correlations between pel values in adjacent lines, fields or frames. They are able to adapt initially to a variety of statistics, but they cannot handle very well changes in statistics within the data. They work better when the pels are initially quantized to a relatively small number of levels. Also, implementation often utilizes large random access memories with complex logical operations, which may require a general purpose digital computer.

One such Universal Coding method is known as the Lempel-Ziv-Welch algorithm.[3.1.3] With this technique, coding is accomplished by means of a string table, which contains up to say 2^J strings. The string table is initialized simply with the set of 2^K possible pel values, i.e., all strings of length one. Best performance results if $K \ll J$, depending on the amount of redundancy in the data.

Coding starts by defining the *current string* S to be simply the first pel of the image. Note that S is a member of the string table. Coding then continues as follows:

1. If there are no more pels (end-of-image), output the J-bit code for S and quit. Otherwise,
2. Input the next pel P and append to S to get the string SP.
3. If SP is in the string table, reset S to equal SP, and continue with step 1. Otherwise,
4. Output the J-bit code for S.
5. Add SP to the string table if there is still room left.
6. Reset S to equal P, and continue with step 1.

Countless variations on this algorithm exist that increase compression performance by sometimes significant amounts. For example, the table entries could be coded using variable word-lengths, which leads to better compression during the earlier part of the coding. A running measure could be taken of compression performance. When it falls below a certain value, the string table could be reinitialized and rebuilt, to enable tracking of changes in statistics.

Another method of universal coding is called Arithmetic Coding.[3.1.11] Conceptually, Arithmetic Coding maps the string of symbols to be coded into a single real number x, where $0 \leq x < 1$. The particular value of x to which an input sequence maps is determined by the recursive probability interval subdivision of the Elias coder.[3.1.12] Fig. 3.1.10 shows, for binary symbols, an example of such interval subdivision through an initial sequence 0100 to be coded.

The portion of the unit interval on which x is known to lie after coding an initial sequence of symbols is known as the current coding interval. For each new binary input symbol the current coding interval is divided into two sub-intervals with sizes proportional to estimates of the conditional probabilities of symbol-value occurrences. The new current coding interval then becomes that portion of the old coding interval associated with the new symbol value actually occurring. Conceptually, this process continues until there are no more input symbols, after which a binary fraction representing some x within the last coding interval is transmitted.

Practically, of course, this process cannot be easily implemented as described, especially for long symbol strings. The accuracy required for the arithmetic simply becomes too large. However, certain approximations can be made to render the algorithm practical without seriously degrading performance. These will be described in Chapter 5.

3.1.8 Coding with Approximation—Rate Distortion Theory

In this chapter we have assumed, up to now, that the visual information is sampled and quantized with adequate fidelity, and that the resulting digital picture elements are coded, transmitted and decoded without changing their values. We have discussed mainly the removal of statistical redundancy from the visual information. We now consider not only reducing the statistical redundancy, but also reducing the *subjective* redundancy, i.e., that part of the visual information that the user is either unable to perceive, is content to be without or cannot afford.

Removing subjective redundancy entails coding with approximation. This means reducing the fidelity or information content of the digital data *prior* to transmission in order to achieve a practical system implementation, i.e., for economic reasons we do not want the system to be over designed.

An information theoretic approach to removing subjective redundancy, i.e., coding with a fidelity criterion, was put forth by Shannon in 1948 in his famous treatise on Information Theory.[3.1.7] Although useful quantitative results are often hard to obtain using this theory, considerable insight can be gained from the qualitative mathematical treatment of the problem.

We start with the classical communication system representation, shown in Fig. 3.1.11, consisting of source, coders, channel, decoders and destination. The

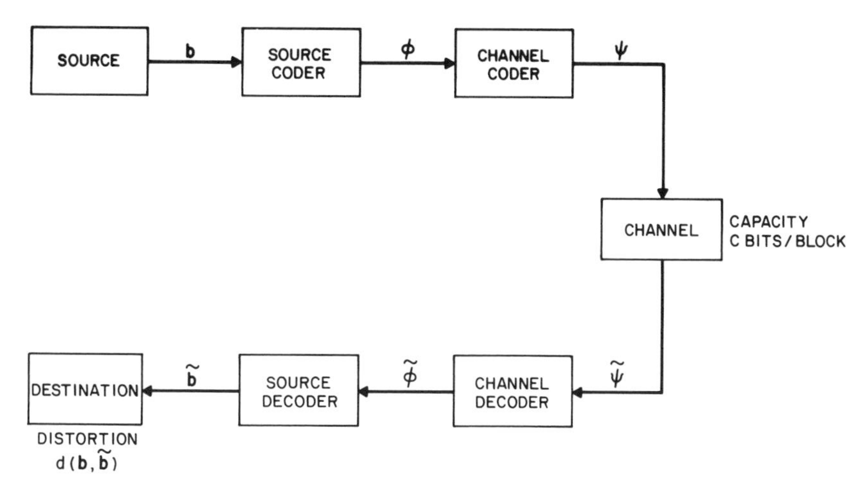

Fig. 3.1.11 Classical communication system representation. The source produces block $\boldsymbol{b}$ of N pels that are converted by the source and channel coders into symbols ϕ and ψ, respectively. The channel produces (possibly corrupted) symbols $\tilde{\psi}$ that are decoded to finally yield the received block of pels $\tilde{\boldsymbol{b}}$.

source produces N-tuples, or blocks, of quantized pels $\boldsymbol{b}$ that the source coder uses to produce symbols ϕ. The channel coder converts the ϕ into channel symbols ψ that the channel transmits as best it can producing $\tilde{\psi}$ at its output. The channel decoder converts the $\tilde{\psi}$ to $\tilde{\phi}$, which the destination decoder uses to produce blocks $\tilde{\boldsymbol{b}}$ that are then made available for use at the destination. In practice, the symbols $\boldsymbol{b}$, ϕ and ψ (with and without tildes) are often represented as binary words and transmitted as such.

The channel is characterized as having *capacity* C bits per pel or NC bits per channel symbol. Capacity is a fundamental information theoretic property of communication channels. It is the maximum bit-rate that can be accommodated by the channel with negligible error. Equivalently, for any bit-rate below C the bit-error-rate can be made negligible by appropriate channel coding.

More precisely, suppose the conditional probability $P(\tilde{\psi}|\psi)$ of the channel is independent of previously transmitted channel symbols, i.e., the channel is memoryless. This is usually a reasonable assumption. For a given $P(\psi)$, the *Mutual Information* is then defined as

$$I(\psi,\tilde{\psi}) = \sum_{\psi,\tilde{\psi}} P(\psi)P(\tilde{\psi}|\psi)\log_2 \frac{P(\tilde{\psi}|\psi)}{P(\tilde{\psi})} \quad \text{bits/symbol} \tag{3.1.65}$$

$I(\psi,\tilde{\psi})$ is a measure of the amount of information about ψ that is conveyed to $\tilde{\psi}$ by the transmission channel. Shannon has shown[3.1.7] that the channel capacity as defined above can be found from

$$C = \frac{1}{N} \underset{P(\psi)}{Max}\; I(\psi,\tilde{\psi}) \quad \text{bits/pel} \tag{3.1.66}$$

where the maximization is over all possible probability distributions on the channel symbols ψ.

As an example, suppose the channel is noiseless, i.e.,

$$P(\tilde{\psi}|\psi) = \begin{cases} 1 & \tilde{\psi} = \psi \\ 0 & \text{otherwise} \end{cases} \tag{3.1.67}$$

Then,

$$I(\psi,\tilde{\psi}) = H(\psi) \quad \text{bits/symbol} \tag{3.1.68}$$

That is, the channel conveys all of the information about ψ to $\tilde{\psi}$. In this case $I(\psi,\tilde{\psi})$ is maximized if all ψ are equally likely. If the set $\{\psi\}$ of possible ψ has J members, then

$$C = \frac{1}{N} \; Max \; H(\psi) = \frac{1}{N} \log_2 J \text{ bits/pel} \tag{3.1.69}$$

or

$$J = 2^{NC} \tag{3.1.70}$$

Thus, the set $\{\psi\}$ can be used to send binary words of length NC bits, and transmission, indeed, does take place at a bit-rate of C bits per pel.

If the channel is noisy, computation of C is more difficult. Transmission at a bit-rate near C is still possible. However, error-correction-coding[3.1.1] must then be incorporated into the channel coder. Discussion of such coding is beyond our scope, except to say that, generally, the more powerful and sophisticated the error-correction-code, the greater the transmission delay between input and output of the channel codec. See [3.1.1] for more details on channel coding.

For *distortionless* source coding, we have seen in previous sections that the source coder output bit-rate $H(\phi)$ can be no smaller than the source entropy $H(\boldsymbol{b})$ in bits/block. Also, for reliable transmission it must be less than the channel capacity. Thus, for *distortionless* communication, i.e., $\tilde{\boldsymbol{b}} = \boldsymbol{b}$, we must have

$$NC \geq H(\phi) \geq H(\boldsymbol{b}) \text{ bits/block} \tag{3.1.71}$$

However, in many situations the requirement $\tilde{\boldsymbol{b}} = \boldsymbol{b}$ is neither necessary nor economical. In such cases, the source coder must make approximations and thereby produce symbols ϕ having entropy considerably smaller than $H(\boldsymbol{b})$. Since the channel capacity need only satisfy

$$NC \geq H(\phi) \; , \tag{3.1.72}$$

considerable savings may be possible depending on circumstances.

If $H(\phi) < H(\boldsymbol{b})$, the source coding is irreversible, information is irretrievably lost, and the received visual information $\tilde{\boldsymbol{b}}$ cannot always be equal to the source information $\boldsymbol{b}$. Thus, we define a *distortion* measure

$$d(\boldsymbol{b},\tilde{\boldsymbol{b}}) \tag{3.1.73}$$

as a non-negative measure of the penalty incurred when $\boldsymbol{b}$ is output by the source, but $\tilde{\boldsymbol{b}}$ is received at the destination. Furthermore, we assume that the *average* distortion

$$\bar{d} = E\{d(\boldsymbol{b},\tilde{\boldsymbol{b}})\} \tag{3.1.74}$$

is a meaningful measure of the overall performance of the communication system. For example, a simple and widely used measure of distortion is mean square error (MSE), i.e., coding noise power. For the case of N-tuples of pels this becomes

$$d(\boldsymbol{b},\tilde{\boldsymbol{b}}) = \frac{1}{N} |\boldsymbol{b}-\tilde{\boldsymbol{b}}|^2 = \frac{1}{N} \sum_{i=1}^{N} (b_i - \tilde{b}_i)^2 \tag{3.1.75}$$

Because of channel errors, a given $\boldsymbol{b}$ may not always give rise to the same $\tilde{\boldsymbol{b}}$. This also may occur if the source coder accumulates several inputs before producing an output in an attempt, possibly, to have ϕ with nearly equal binary word lengths. In any event, we assume that all systems give rise to a conditional probability distribution

$$P(\tilde{\boldsymbol{b}}|\boldsymbol{b}) \tag{3.1.76}$$

between input and output. The mutual information between input and output is then given by

$$I(\boldsymbol{b},\tilde{\boldsymbol{b}}) = \sum_{\boldsymbol{b},\tilde{\boldsymbol{b}}} P(\boldsymbol{b})P(\tilde{\boldsymbol{b}}|\boldsymbol{b}) \log_2 \frac{P(\tilde{\boldsymbol{b}}|\boldsymbol{b})}{P(\tilde{\boldsymbol{b}})} \tag{3.1.77}$$

Now consider only those systems having an average distortion less than or equal to some acceptable value D, i.e.,

$$\bar{d} = \sum_{\boldsymbol{b},\tilde{\boldsymbol{b}}} P(\boldsymbol{b})P(\tilde{\boldsymbol{b}}|\boldsymbol{b}) d(\boldsymbol{b},\tilde{\boldsymbol{b}}) \le D \tag{3.1.78}$$

The *rate-distortion function* is then defined[3.1.8] as

$$R(D) = \frac{1}{N} \underset{P(\tilde{\boldsymbol{b}}|\boldsymbol{b})}{MIN} \; I(\boldsymbol{b},\tilde{\boldsymbol{b}}) \text{ bits/pel} \tag{3.1.79}$$

where the minimization is over all $P(\tilde{\boldsymbol{b}}|\boldsymbol{b})$ satisfying Eq. (3.1.78), i.e., $\bar{d} \le D$.

$R(D)$ gives a lower bound on the channel capacity required to achieve $\bar{d} \le D$. This is easily seen from the *Data Processing Theorem*[3.1.1], which states that mutual information between random variables cannot be increased by operations on either random variable. For example, $\tilde{\phi}$ is produced from $\tilde{\psi}$. Therefore, the mutual information with ψ cannot increase, i.e.,

$$I(\psi,\tilde{\psi}) \ge I(\psi,\tilde{\phi}) \tag{3.1.80}$$

Thus, for any system having $\bar{d} \le D$,

$$\begin{aligned} NR(D) &\le I(\boldsymbol{b},\tilde{\boldsymbol{b}}) && \text{from (3.1.79)} \\ &\le I(\phi,\tilde{\phi}) && \text{from (3.1.80)} \\ &\le I(\psi,\tilde{\psi}) && \\ &\le NC && \text{from (3.1.66)} \end{aligned}$$

Thus, $C \ge R(D)$ is a requirement on channel capacity in order to achieve an average distortion $\overline{d} \le D$.

Shannon has also shown[3.1.7] that C can be made arbitrarily close to $R(D)$ with appropriate source and channel coding. The proof of this is beyond our scope except to say that the greater the coding efficiency, the greater the system complexity and the greater the delay between system input and output. Eventually, a point of diminishing returns is reached, beyond which additional coding is more expensive than the additional channel capacity.

Unlike noiseless coding theory, rate-distortion theory is also valid for continuous amplitude, unquantized vectors $\boldsymbol{b}$ and $\tilde{\boldsymbol{b}}$. $I(\boldsymbol{b},\tilde{\boldsymbol{b}})$ and $\overline{d}$ are defined by using probability density functions instead of probability distributions and by replacing summations with integrals. In some cases, the theory can also be extended to continuous time, unsampled signals $b(t)$ and $B(x,y)$.[3.1.8]

$R(D)$ is usually very difficult to compute[3.1.9] and can usually be found only approximately through the use of various bounds. The best known of these is the *Shannon Lower Bound*, which for continuous amplitude $\boldsymbol{b}$ and $\tilde{\boldsymbol{b}}$ is given as follows:

If the distortion criterion depends only on the difference $z = \tilde{\boldsymbol{b}} - \boldsymbol{b}$, i.e.,

$$d(\boldsymbol{b},\tilde{\boldsymbol{b}}) = d(\tilde{\boldsymbol{b}} - \boldsymbol{b}) \triangleq d(z) \tag{3.1.81}$$

then*

$$NR(D) \ge H(\boldsymbol{b}) - H(z) \text{ bits/block} \tag{3.1.82}$$

where the random vector z has probability density function

$$p(z) = c_1 \exp\left[-c_2 d(z)\right] \tag{3.1.83}$$

* For a continuous amplitude random variable x, the integral entropy is defined as $H(x) \triangleq = \int p(x)\log_2 p(x)\,dx$. The actual entropy of a continuous amplitude random variable is, of course, infinite.

the constants c_1 and c_2 being chosen such that $p(z)$ has unity probability mass and $E[d(z)] = D$.

For the MSE criterion of Eq. (3.1.75), $p(z)$ is Gaussian, and the Shannon lower bound becomes

$$NR(D) \geq H(\boldsymbol{b}) - \frac{N}{2} \log_2 (2\pi e D) \quad \text{bits/block} \tag{3.1.84}$$

Note that, like many information theoretic results, rate-distortion theory does not indicate how to design optimum systems. However, it conceivably could provide a benchmark against which the performance of practical systems can be compared.

There are three major difficulties in trying to apply rate-distortion theory to the coding of visual information. The first is that, as we have seen, $R(D)$ is usually very difficult to compute. The second is that source statistics are often hard to come by, they are decidedly non-Gaussian and, more importantly, they vary considerably from picture to picture and from region to region within a picture, i.e., they are nonstationary. This implies that minimum bit-rates can only be achieved if coders adapt themselves to the local statistics at hand, a situation not covered by traditional rate-distortion theory. Several approaches to statistical modeling of picture sources have been suggested to deal with this situation, and they will be discussed in later sections. For example, one approach is to partition the picture into regions having similar statistics. This is akin to state-space coding as described earlier in this chapter. Another approach is to decompose the picture into components[3.1.10] that have more stationary statistics and which can be coded separately. This is the approach of transform coding, as mentioned above and about which more will be said in later sections.

The third, and probably the greatest difficulty of rate-distortion theory is finding a distortion criterion that is meaningful and, at the same time, mathematically tractable. In fact, Chapter 4 is devoted entirely to this topic. The MSE criterion has enjoyed considerable mathematical success. However, its application to images is questionable in many situations. In the first place the visibility of different error patterns $\tilde{\boldsymbol{b}} - \boldsymbol{b}$ varies considerably, even though the error energy $|\tilde{\boldsymbol{b}} - \boldsymbol{b}|^2$ may be the same. Another characteristic of image distortion is that a given error pattern $\tilde{\boldsymbol{b}} - \boldsymbol{b}$ has varying visibility depending on the image itself. For example, as we shall see in Chapter 4, errors are much more visible in flat, low-detail areas of a picture than they are near edges of objects or in high-detail areas. This property cannot be taken into account by *any* difference distortion measure of the form $d(\tilde{\boldsymbol{b}} - \boldsymbol{b})$, thus increasing the difficulty of applying rate-distortion theory.

In the case where images are to be coded, transmitted and reproduced without visible distortion (beyond the original PCM coding), a distortion criterion might be derived by a lengthy series of subjective tests. For example, suppose the goal is 95% viewer acceptability. Then a meaningful criterion

Fig. 3.2.1 The image "Stripes" used to produce some of the statistical data of this section. Pels are obtained by sequential raster scanning.

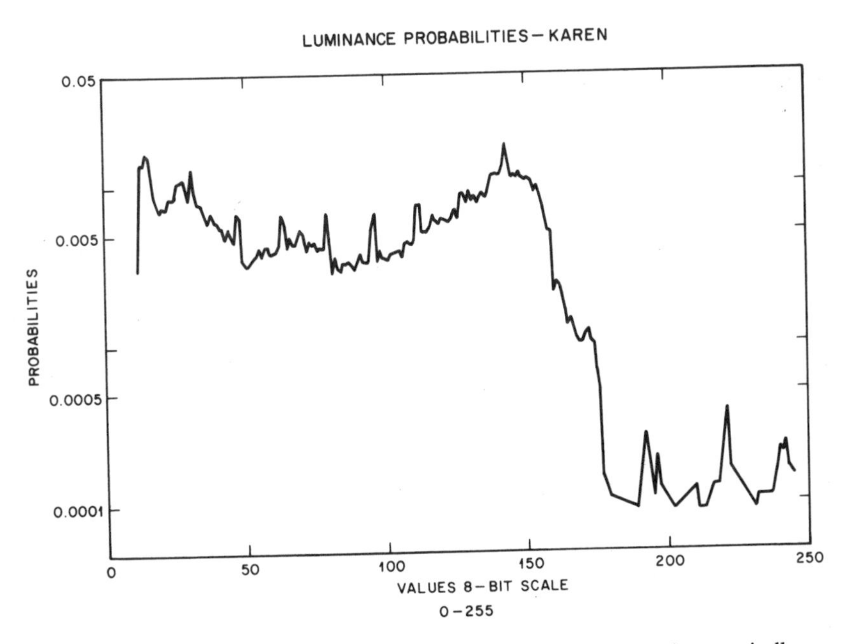

Fig. 3.2.2 Histogram of luminance for "Stripes". For 8-bit PCM, pel values nominally range from 0 to 255. Entropy = 7.2 bits/pel.

would be

$$d(\boldsymbol{b},\tilde{\boldsymbol{b}}) = \begin{cases} 0\ , & \text{if distortion is invisible to} \\ & \text{at least 95\% of viewers} \\ 1\ , & \text{otherwise.} \end{cases} \tag{3.1.85}$$

Coding at an average distortion $D \approx 0$ would achieve system goals, and the rate-distortion $R(D)$ would indicate the optimum bit-rate.

3.2 Monochrome Luminance Statistics

In this section we present some measured statistics from actual pictures. The purpose is not to provide absolute numerical data from which a system design can be derived, since this is impossible due to the nonstationarity and variability of images. Instead we hope to convey an understanding of the broad overall statistical behavior of some of the parameters that are important in picture coding. These include luminance, transform coefficients, differential signals, time-varying properties, etc.

We first concentrate on monochrome gray-level images because in most cases the preponderance of visual information resides in the luminance signal. Chrominance and two-level graphics statistics will be covered in subsequent sections.

3.2.1 First Order Statistics

For a given picture, the probability distribution (PD) of luminance can be very nonuniform as illustrated in Fig. 3.2.2 where a histogram of 8-bit PCM luminance values {0–255} is plotted for the picture "Stripes" shown in Fig. 3.2.1. The entropy of this distribution is 7.4 bits/pel. However, if luminance is averaged over many pictures, the resulting probability distribution is somewhat more uniform as shown in Fig. 3.2.3. This distribution has an entropy of 7.8 bits/pel, which is only slightly less than the 8 bits/pel produced by the original PCM coding.

In general, luminance PD's vary greatly from picture to picture, i.e., statistics are nonstationary with entropies ranging from relatively small values to near maximum. Thus, in order to reduce bit-rates by variable word-length coding of luminance, it is necessary to use different codes for different pictures. This requires that some *side information* be transmitted so that the decoder knows which code is being used on a particular picture. The quantity of side information and the saving in bits accrued from adaptive coding depend on many factors, e.g., the number of codes that can be accommodated, the statistics of the class of pictures that are to be coded, etc. More will be said about adaptive variable word-length coding techniques in later sections. Here it suffices to say that variable word-length coding that relies solely on nonuniform luminance PD's,

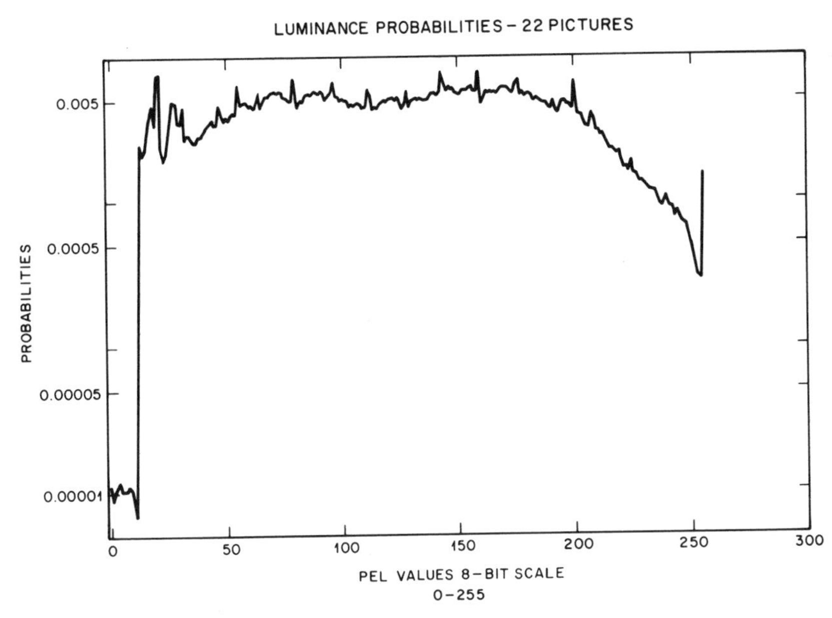

Fig. 3.2.3 Histogram of luminance averaged over 22 different images. Entropy = 7.8 bits/pel.

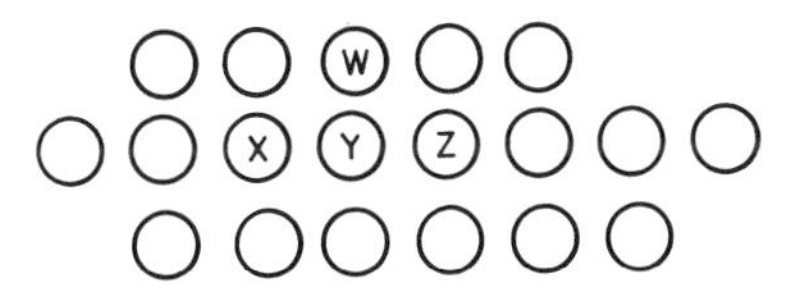

Fig. 3.2.4 Pel configuration used for entropy data. Pels X,Y,Z are on one scan line. Pel W is on the previous scan line.

and nothing else, has found very little application in gray-level picture coding systems.

3.2.2 Block Statistics of Single, Still Pictures

As we saw in Section 3.1 statistical dependencies between adjacent picture elements can be quantified by higher order entropies. Consider the pel configuration of Fig. 3.2.4, which shows three successive pels *XYZ* on one scan line and one pel *W* from the previous line. Various higher order entropies for these pels (normalized to bits/pel) are shown in Fig. 3.2.5 for the picture "Stripes".

Other pictures will, of course, have different entropies. However, certain generalizations can be stated regarding the relative values of the different entropies. For examples:

- As the block size increases, the entropy (per pel) decreases, as expected.
- For small block sizes the conditional entropies are significantly less than the block entropies. As block size increases, this disparity diminishes.
- Conditional entropies decrease as the number of conditional pels increases. However, the decrease is relatively small for conditional pels beyond the immediate vicinity of the unconditioned pel.
- Statistical dependence in television pictures is greater in the vertical direction than in the horizontal direction. This property is rather exaggerated in "Stripes" because of the background. However, most television scanning rasters have a smaller vertical pel spacing (within a frame) than horizontal pel spacing, which gives rise to a greater vertical statistical dependence with most pictures.

At this point a word or two should be said about what we mean exactly by *statistics* of a single picture such as "Stripes". For example, starting with 8-bit PCM, as we do in this case, the 3-tuple (X,Y,Z) has 256^3 possible values. However, the picture has only 256^2 pels, and therefore, good statistics for (X,Y,Z) cannot be obtained for all values. What we really want are statistics for pictures *similar* to the one of interest, and *similar* has to be defined for each application. For example, in order to obtain Fig. 3.2.5 we constructed *similar* pictures by simply adding different brightness levels to the original picture. We then smoothed the resulting PD's on the assumption that the least significant 2 to 3 bits are mostly random noise and that similar pel values should not have markedly different probabilities of occurrence.

In other contexts, sets of similar images can be obtained more readily. For example, chest x-rays are very much alike, and reliable statistics can be generated fairly easily. Bi-level text images are another class for which good statistics can be measured.

H(Y)	=	7.4	bits/pel
H(XY)/2	=	6.6	
H(Y|X)	=	5.8	
H(WY)/2	=	5.5	
H(Y|W)	=	3.6	
H(XZ)/2	=	7.0	
H(XYZ)/3	=	5.9	
H(Y|XZ)	=	3.7	
H(Z|XY)	=	4.5	
H(XW)/2	=	6.7	
H(WXY)/3	=	5.5	
H(Y|XW)	=	3.1	

Fig. 3.2.5 Entropies in bits/pel for "Stripes". Original data is 8-bit PCM.

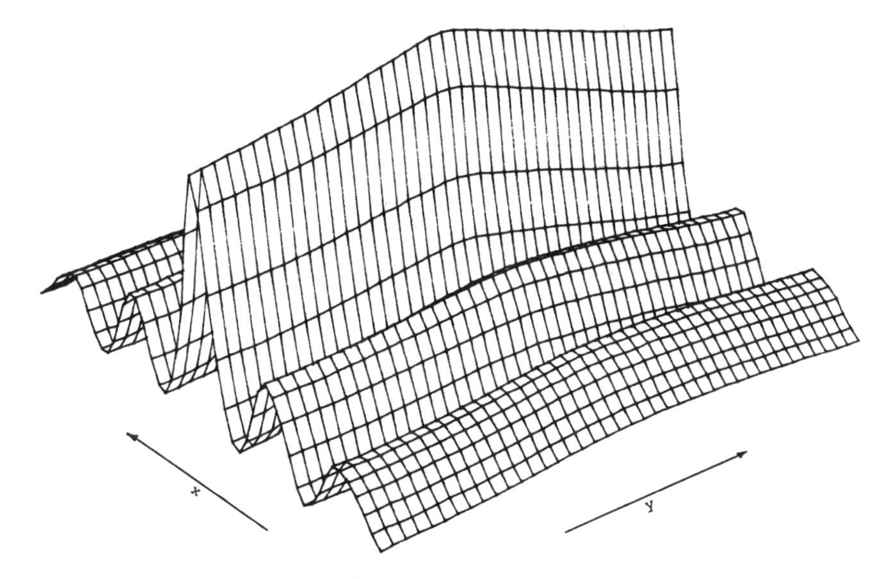

Fig. 3.2.6 Normalized autocovariance function $\rho(x,y)$ for "Stripes". It is much less peaked in the vertical y direction than in the horizontal x direction due to the high vertical correlation of this particular image.

3.2.3 Second Order Statistics of Single, Still Pictures

A very widely used measure of statistical dependence between adjacent picture elements in an image is the *autocorrelation function*

$$R(x,y) \triangleq E(bb^*) \tag{3.2.1}$$

where pel b^* is x units to the right and y units below pel b in the picture. The autocorrelation function is required in many mathematical formulations, e.g., the Karhunen-Loeve transform. It is also much easier to measure and calculate than higher-order entropies. However, it does have the drawback that statistical stationarity is tacitly assumed even though, in reality, such is rarely the case. Some properties of the autocorrelation are:

$$\begin{aligned} R(0,0) &\geq R(x,y) \\ R(x,y) &= R(-x,-y) \end{aligned} \tag{3.2.2}$$

The *normalized autocovariance function*

$$\rho(x,y) = \frac{R(x,y) - (\bar{b})^2}{E(b^2) - (\bar{b})^2} \tag{3.2.3}$$

where $\bar{b} \triangleq E(b)$, gives a useful measure of statistical dependence between pels. If $\rho(x,y)$ is highly peaked at the origin, then statistical dependence between pels is less than if $\rho(x,y)$ has a broad, relatively flat peak indicating pel-to-pel correlation over a large area. Fig. 3.2.6 shows $\rho(x,y)$ for the picture "Stripes" for a 21×21 pel area. For this picture the vertical correlation is much larger than the horizontal correlation as evidenced by the considerable disparity in sharpness of $\rho(x,y)$ in the horizontal and vertical directions.

Autocorrelation functions vary considerably from picture to picture and even within the same picture, thus illustrating once again the extreme nonstationarity of picture statistics.

3.2.4 Differential Signal Statistics of Single, Still Pictures

Statistics of the differential signals typically generated by predictive coders exhibit less variation from picture to picture and within the same picture than do block statistics or second-order statistics. Their PD's are almost always approximately Laplacian in shape with a variance that depends on the particular picture and on the prediction algorithm utilized. For example, Fig. 3.2.7 shows for "Stripes" differential signal PD's for three different two-dimensional predictors that (see Fig. 3.2.4) predict pel Y, given previous pels W and X. Fig. 3.2.8 shows

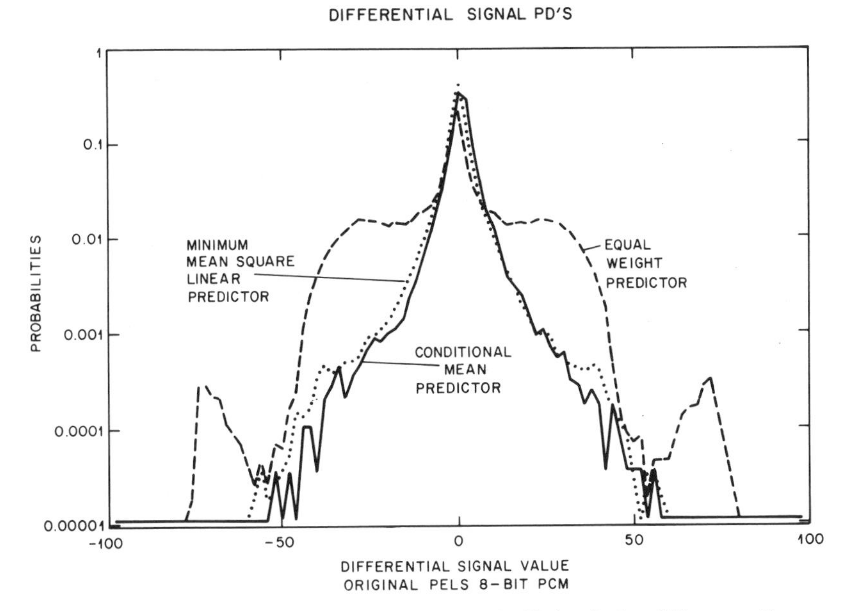

Fig. 3.2.7 Differential signal probability distributions for "Stripes" when different predictors are used. See Section 3.1.6 for definition of minimum mean square linear and conditional mean predictors. The equal weight predictor is $\hat{Y} = (X+W)/2$, where pels are defined in Fig. 3.2.4.

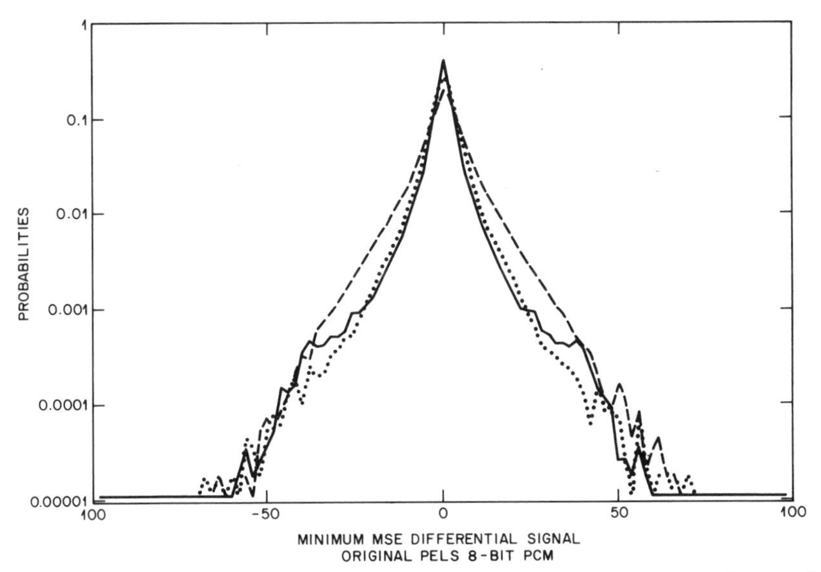

Fig. 3.2.8 Differential signal probability distribution for the minimum mean square linear predictor for three different images. Note the similarity in shapes.

Fig. 3.2.9 Typical differential signal magnitude for "Stripes".

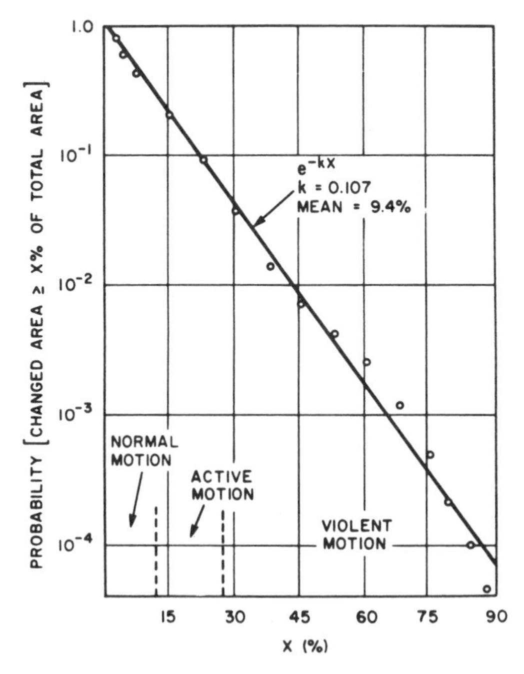

Fig. 3.2.10 Statistics of the amount of motion in a typical videoconference scene consisting of a person seated.

differential signal PD's for three different pictures using the same predictor, namely the minimum mean square linear predictor described in Section 3.1.5. Fig. 3.2.9 shows a typical prediction error magnitude for "Stripes". Note that large error occurs mainly at edges in the picture.

The entropy of the differential signal and, therefore the average bit-rate of a Huffman code, depends almost entirely on the shape of the PD near the origin. Thus, for a given predictor a single variable word-length code very often performs near optimum for a wide variety of pictures.

3.2.5 Temporal Luminance Statistics

With television or motion pictures, the statistics of successive frames change with time depending on the scene being viewed and on camera motion. However, even within a single frame there are important statistical distinctions between single pictures of nonmoving objects and television or motion pictures. For example, the exposure time of an individual frame is often quite large for the speeds of motion encountered. In fact, television cameras typically have no shutters at all, so that the frame exposure time is the time between frames. For such large exposure times, moving objects are blurred considerably when viewed in a single frame. If the camera is moving then the entire scene is blurred.

A distinguishing property of television is that most systems use 2:1 line interlace to reduce the visibility of flicker in the display. With interlace, odd-numbered scan-lines of a frame are transmitted during field 1 followed by the even numbered lines during field 2. Thus, adjacent scan lines of a frame are separated in time by a field period, and during this time significant movement within the scene or by the TV camera can occur. This causes moving objects in a single frame of television to appear with serrated edges (see Fig. 1.6.3). It also tends to reduce the amount of statistical dependency between adjacent lines of a television frame.

Frame rates normally used by motion picture cameras (18–24 Hz) and television cameras (25–30 Hz) are not high enough to portray accurately all of the movements that can occur in a scene. Stated in terms of sampling theory, significant energy can occur at temporal frequencies higher than half the frame rate, and therefore temporal aliasing can occur. Nevertheless, for a large percent of the time, motion in a scene is relatively slow, which gives rise to significant frame-to-frame statistical dependence and correlation. In fact, in many applications, large portions of the picture remain practically unchanged between frames.

For example, Fig. 3.2.10 shows the probability of more than $x\%$ of the picture changing significantly between frames for a video telephone or conference TV application where the camera is stationary.[3.2.1] Here a change is deemed significant if it exceeds 1.5% of maximum signal amplitude. On the average only 9 percent of the picture changes from frame to frame, and the statistics are well represented by a one-sided exponential distribution.

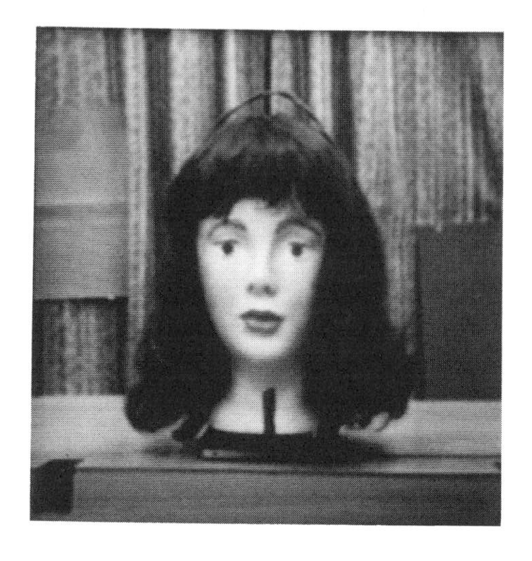

Fig. 3.2.11 Images used for measuring interframe differential signal entropies.

For other applications such as broadcast television, the shape of the curve remains about the same. However, the mean value changes considerably depending on the scene content, amount of camera motion, etc. For noisy pictures such as old movies or low quality videotape, the curve is shifted to the right and rounded off at the top.

In order to exploit temporal correlation and remove frame-to-frame redundancy, one or more previous frames must also be made available. For real-time applications this means that a sizable amount of memory must be incorporated into the system in order to have access to pels in adjacent lines, fields or frames that are highly correlated with the pel or pels being coded. Having accomplished this, all of the basic redundancy reduction techniques defined previously for single pictures are also applicable to frame-to-frame coding (also known as *interframe* coding).

For example, predictive coding using the previous frame pel as a prediction of the pel to be coded produces a differential signal that is essentially zero in the nonchanging parts of the picture and nonzero only in the *moving area*. In this case, we need to transmit differential signal values only for moving-areas of the picture. This technique is called *conditional frame replenishment*.

PD's of moving-area differential signals all have typically a Laplacian shape. However, their entropies vary considerably depending on the amount of movement, the amount of picture detail and the particular prediction being used. Fig. 3.2.12 shows entropies of various moving-area differential signals for scenes such as Fig. 3.2.11 containing a moving mannequin's head as well as scenes containing live subjects. The camera is stationary. In this case, speed is measured in pel spacings per frame period. Note that entropies increase with speed for predictors that employ pels from the previous field or frame. However, if the moving-area predictor uses only pels from the present field, then entropy decreases as a function of speed. This is due to the fact that blurring of moving areas increases with speed, and such blurring tends to increase the correlation between intrafield pels.

Frame-to-frame correlation of moving-area pels generally decreases with speed of movement, giving rise to the aforementioned increase of entropy for interframe predictors. However, if the velocity of a moving area is known or can be found, then in principle at least, its position in the previous frame can also be determined. Then, instead of using a fixed set of pels in the previous frame as a prediction, a translated set can be used to improve the prediction. This situation is shown in Fig. 3.2.13 where the previous frame pels used to calculate the linear prediction are shifted by an amount corresponding to the frame-to-frame displacement of the moving area. Velocity measurement and displaced prediction are discussed in detail in Section 5.2.3.

Unfortunately, this idealized situation does not always exist in practice. Objects do not necessarily move in pure translation, and accurate estimates of velocity are difficult to obtain from a real-time television signal. Interlace

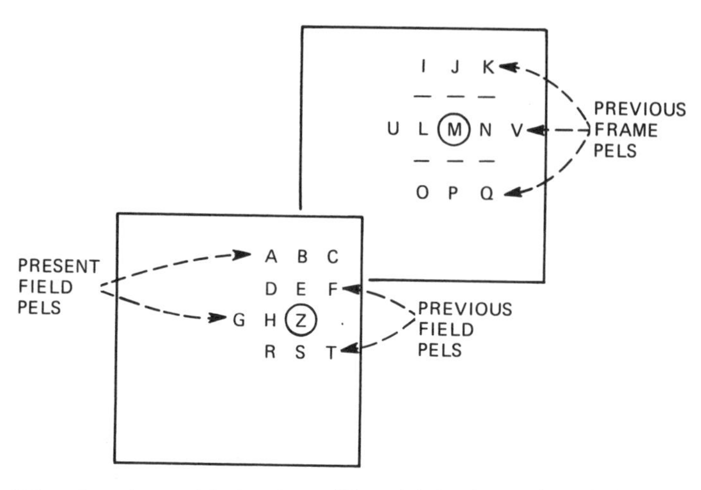

Fig. 3.2.12a Pel configuration used for interframe differential signal entropies. Moving-area pel Z is to be predicted from the remaining pels, which are from three successive fields. Pels Z and M are separated by exactly one frame period.

Transmitted Signal $Z - \hat{Z}$	Prediction $\hat{Z}$	Entropies in Bits Per Moving-Area Pel (35 level quantization)
Frame Difference	M	≈ 2.1 - 3.9
Element-Difference	H	≈ 2.0 - 3.7
Element-Difference of Frame-Difference	M + H − L	≈ 1.8 - 3.1
Line-Difference of Frame-Difference	M + B − J	≈ 1.5 - 3.5
Field-Difference	(E + S)/2	≈ 1.8 - 3.2
Element-Difference of Field-Difference	H + (E + S)/2 − (D + R)/2	≈ 1.5 - 2.5

Fig. 3.2.12b Typical differential signal entropies of moving-area pels for some fixed linear predictors when the differential signal is quantized to 35 levels.

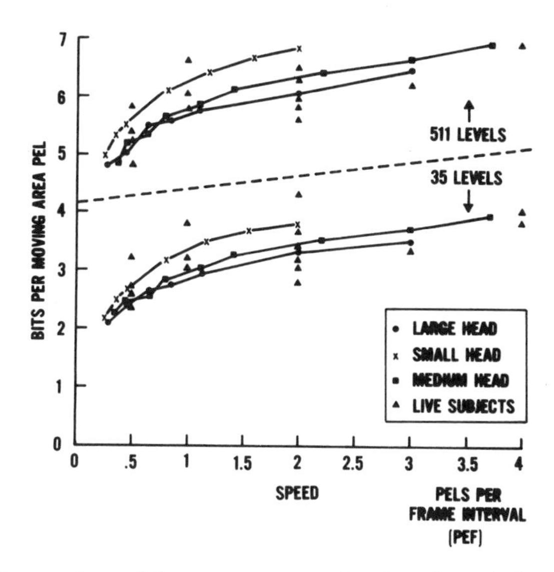

Fig. 3.2.12c Moving-area frame-difference entropy as a function of speed of movement. Two quantization scales are shown.

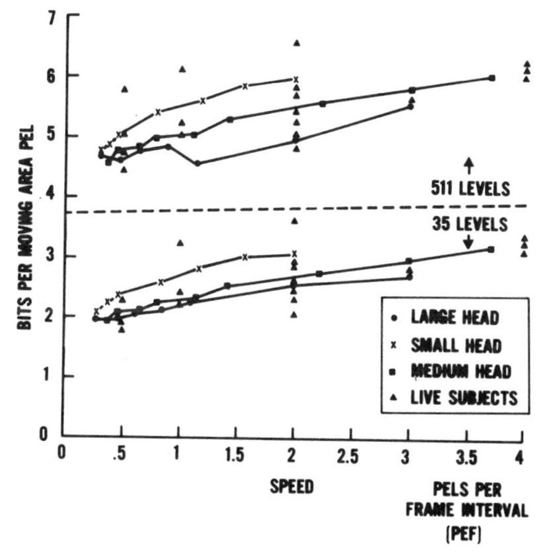

Fig. 3.2.12d Moving-area field-difference entropy as a function of speed of movement. Two quantization scales are shown.

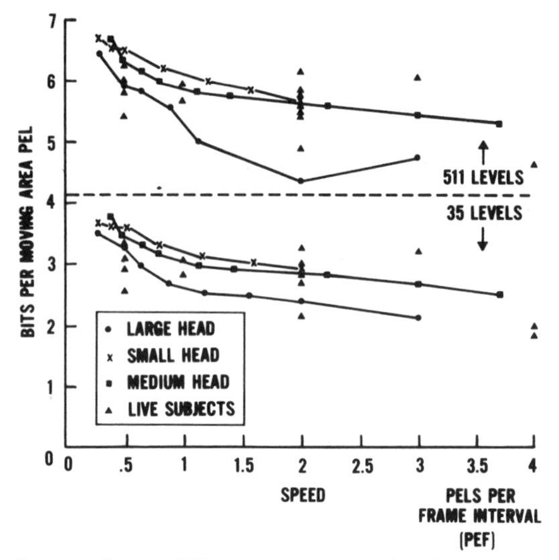

Fig. 3.2.12e Moving-area element-difference entropy as a function of speed of movement. Two quantization scales are shown. Note that for this intrafield predictor, entropy decreases as speed increases.

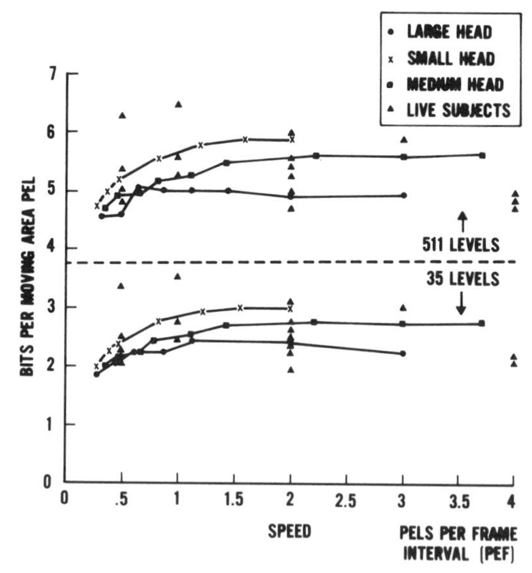

Fig. 3.2.12f Moving-area entropy of element-difference-of-frame-difference as a function of speed of movement. Two quantization scales are shown.

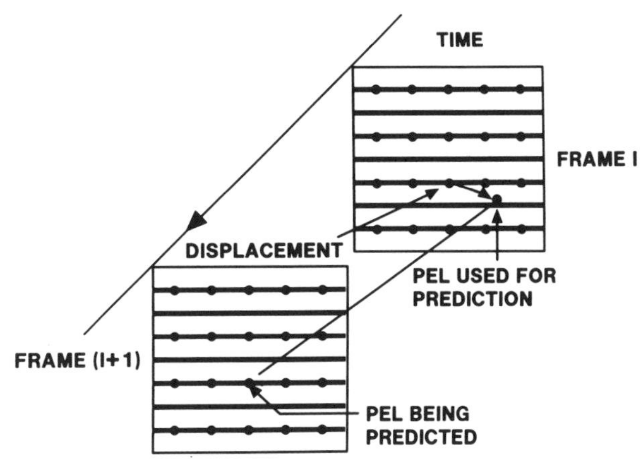

Fig. 3.2.13 If the velocity (speed and direction) of a moving area is known, then the prediction can be improved by using displaced pels in the previous frame.

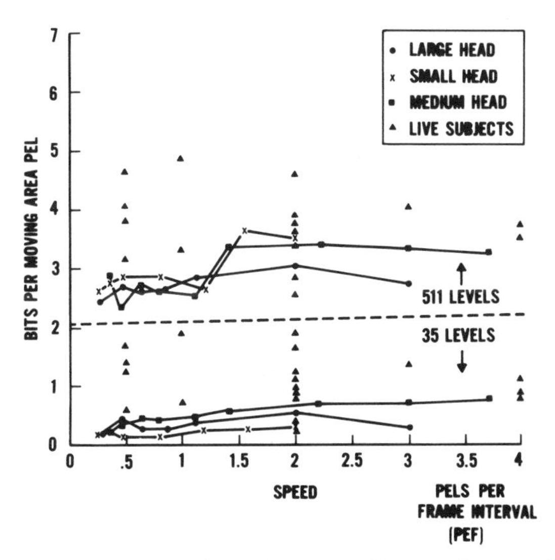

Fig. 3.2.14 Moving-area entropy of minimum mean square linear predictor as a function of speed of movement. Two quantization scales are shown. Entropies are considerably smaller than those of fixed linear predictors.

further confuses the issue as does camera integration and camera motion. Also, objects usually do not move by an integral number of pel spacings in a frame period. Thus, a single translated pel from the previous frame will rarely be a good prediction of a given pel in the present frame. Instead, an interpolation or linear combination of adjacent pels in the previous frame is required to give a good prediction. The weighting coefficients $\{\alpha_i\}$ of the linear prediction must therefore adapt to the direction and speed of movement of objects within the scene, and the algorithm thus becomes an *adaptive* linear predictive coder. With such schemes there is no need to limit the predictor to only the pels in the previous frame. Previously transmitted pels in the present and previous fields can also be included in the linear combination in order to exploit whatever intraframe redundancy may exist, e.g., for the pels of Fig. 3.2.12a the prediction of pel Z would be

$$\hat{Z} = \alpha_1 A + \alpha_2 B + \cdots + \alpha_{21} U + \alpha_{22} V \tag{3.2.4}$$

Fig. 3.2.14 shows the entropy of the moving-area differential signal for such an adaptive linear predictor using the 22 pels of Fig. 3.2.12a. In this case the weighting coefficients were chosen to minimize the mean squared differential signal over the moving area of each field.

We see that even for the live subjects, the adaptive linear predictor entropies are considerably below those of the fixed linear predictors. However, these entropies do not include the bits needed to send the weighting coefficients. In Chapter 5 we will consider more practical methods for such "motion compensation" predictive coding of television and videotelephone pictures.

3.2.6 Statistical Multiplexing

A major problem with interframe coders is that, although the average data rate per field may be relatively small, the peak data rate can be quite large. Moreover, when a high data rate does occur, due to large areas of changing luminance in the picture, it is likely to last for more than a short time. This is because rapid motion of objects in a scene usually lasts at least for many fields. For example, Fig. 3.2.15 shows for video telephone scenes the normalized autocorrelation coefficient of the number of significantly changed pels in successive fields. Notice that the amount of changed area in a picture is very highly correlated from field to field, and that there is still more than 50% correlation even in fields that are one second apart.

It is usually impractical to achieve very much smoothing of this highly irregular data for two reasons. First, an extremely large amount of storage or buffering would be required, making such systems inordinately expensive. Second, the large transmission delay that this would entail would be unacceptable in many applications, e.g., videoconferencing. Thus, excess channel capacity must normally be provided to accommodate these peaks in the data rate, and

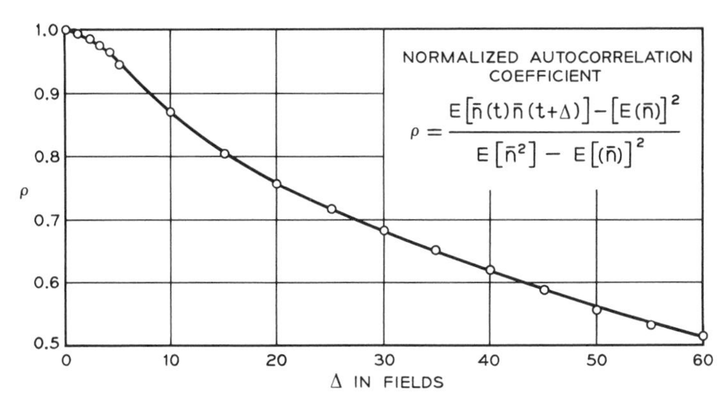

Fig. 3.2.15 Normalized autocorrelation coefficient of the number of changes per field. There is still a 50-percent correlation between fields that are 1 second apart.

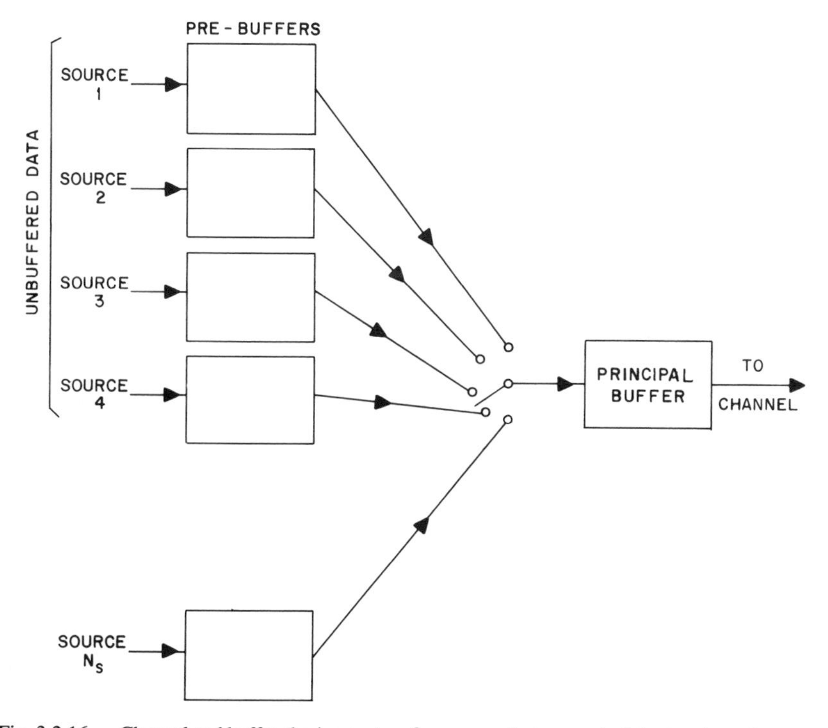

Fig. 3.2.16 Channel and buffer sharing system for sources that generate data at an irregular rate.

this means that during the intervals between data peaks the channel may have little or nothing to transmit.

An obvious method to utilize the channel more efficiently is to transmit the data from several interframe coders over a shared channel, allocating more capacity and buffer space to those sources that happen to require it at a given moment and less to those that are producing relatively little data. Since separate conversations can be assumed to be independent, the combined data on a per-source basis should be much less peaked than that from a single source, or in statistical terms, if the data from N_s independent, identically distributed sources is averaged, then the variance is reduced by a factor of N_s (the mean is unchanged). This *statistical multiplexing* will not only reduce the required channel capacity, but it will also reduce the required buffering.

A number of schemes have been proposed. Most of them are variations of the system shown in Fig. 3.2.16. Unbuffered data from the various sources is first stored in individual prebuffers, then transferred to the principal buffer via the multiplexing switch, and finally transmitted over a single high-capacity data channel to the receiver where the inverse operation takes place. The switch may rotate either sequentially or nonsequentially and either at a constant rate or at a variable rate. In any event, some source labeling information must be sent so that at the receiver a given block of data is identified with the correct source.

Fig. 3.2.17 is the same as Fig. 3.2.10, but expressed in number of changed pels per field for medium resolution videotelephone when a single video source uses the channel. Figs. 3.2.18–21 show the corresponding curves when data streams from more than one source are statistically multiplexed prior to transmission. To obtain these curves, simultaneous occurrence of several videotelephone conversations was simulated by starting each conversation at a different point in the single conversation data. We see that as the number of sources sharing the channel increases, the standard deviation of the data decreases, indicating that the data is becoming more and more uniform.

Not only do the data peaks become smaller, they also have shorter duration. Figs. 3.2.22–26 show the average durations of peaks in the number of changes per field per source, as the number of sources is increased. We see a marked decrease, indicating further the efficiencies possible by many interframe coders sharing the same communication channel.

3.2.7 Discrete Transform Statistics

We saw in Section 3.1.6 that discrete transform coding has the potential of reducing redundancy in two ways. First, many of the transform coefficients are small enough to be insignificant and, therefore, need not be transmitted. Second, the remaining coefficients can be made relatively uncorrelated with each other. Thus far, only *linear* transforms of blocks $\boldsymbol{b}$ of original quantized pels into transform coefficients $\boldsymbol{c}$ have been studied. In fact, most effort has centered on unitary transforms. In matrix notation using column vectors $\boldsymbol{b}$ and $\boldsymbol{c}$

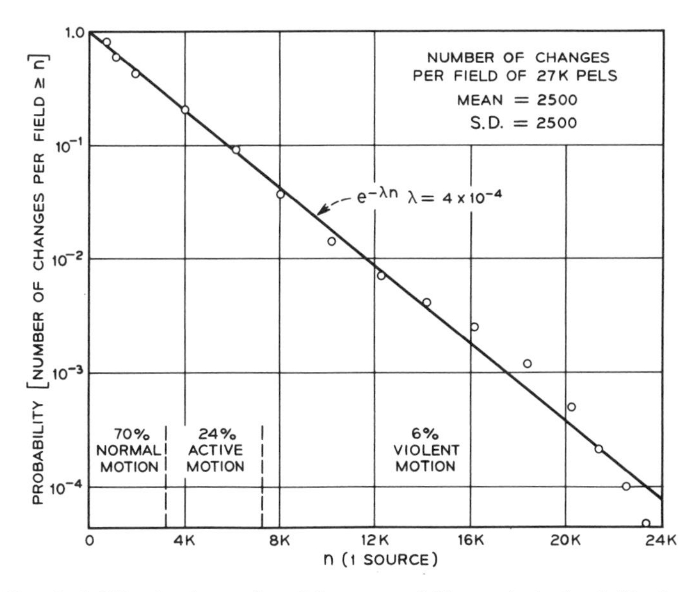

Fig. 3.2.17 Probability that the number of changes per field exceeds the threshold value n. Points lie very close to the exponential $e^{-\lambda n}$ with $\lambda = 4 \times 10^{-4}$.

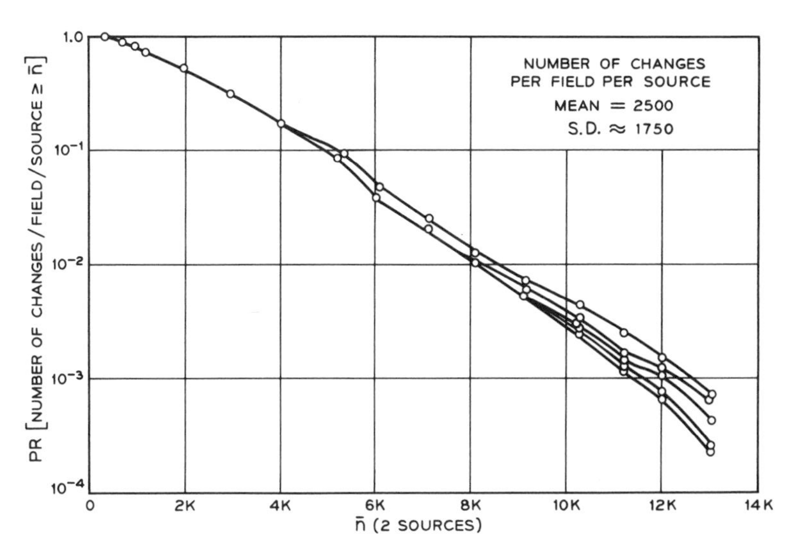

Fig. 3.2.18 Probability that the number of changes per field per source exceeds $\bar{n}$ when data from *two sources* are combined. Multiplicity of curves results from different starting positions in the data.

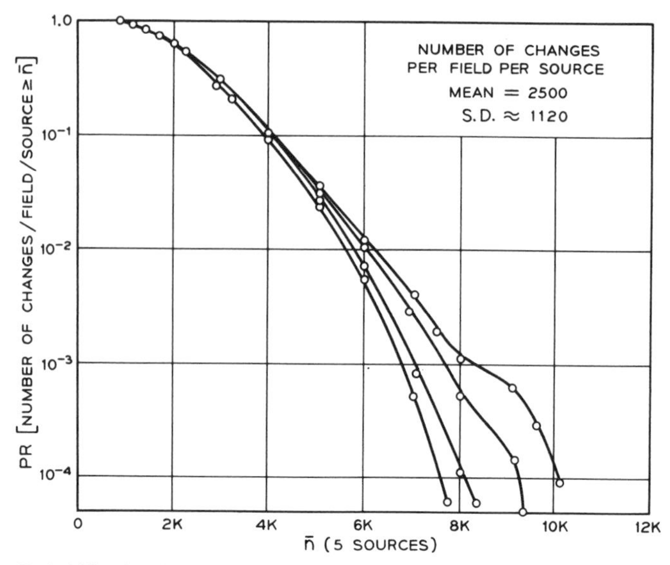

Fig. 3.2.20 Probability that the number of changes per field per source exceeds $\bar{n}$ when data from *twelve sources* are combined.

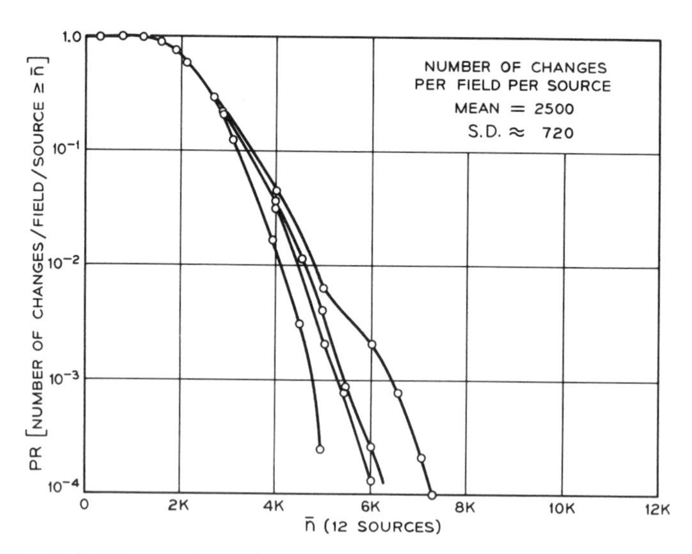

Fig. 3.2.19 Probability that the number of changes per field per source exceeds $\bar{n}$ when data from *five sources* are combined.

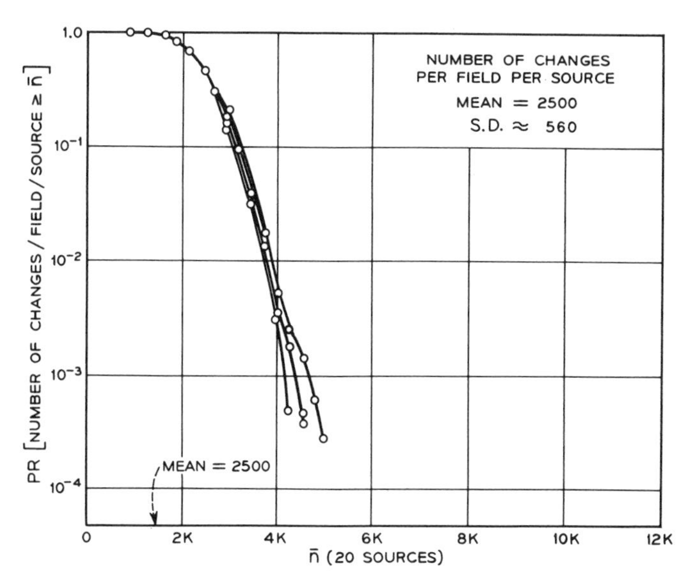

Fig. 3.2.21 Probability that the number of changes per field per source exceeds $\bar{n}$ when data from *twenty sources* are combined.

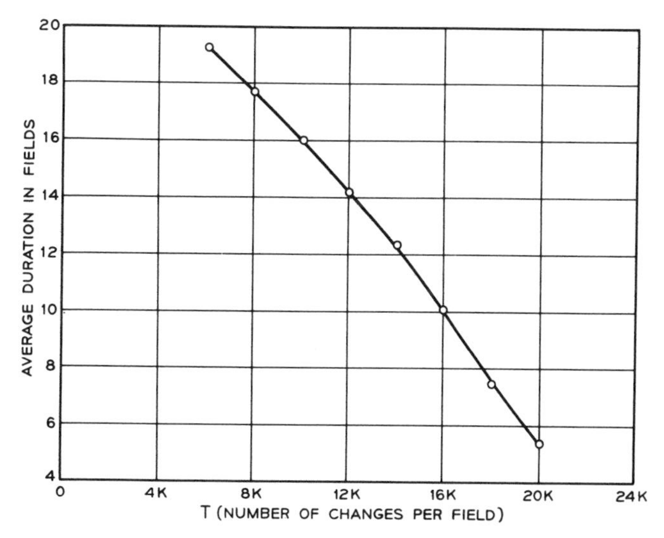

Fig. 3.2.22 Average duration of peaks in the number of changes per field. A peak starts when n rises above T and ends when n falls back below T.

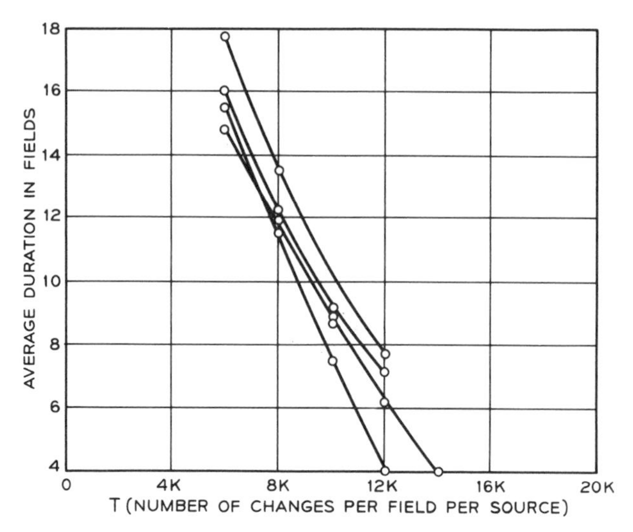

Fig. 3.2.23 Average duration of peaks in the number of changes per field per source when data from *two sources* are combined.

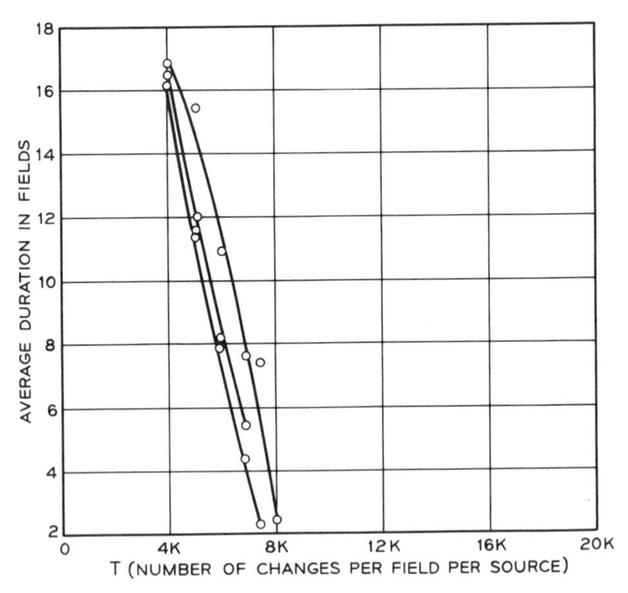

Fig. 3.2.24 Average duration of peaks in the number of changes per field per source when data from *five sources* are combined.

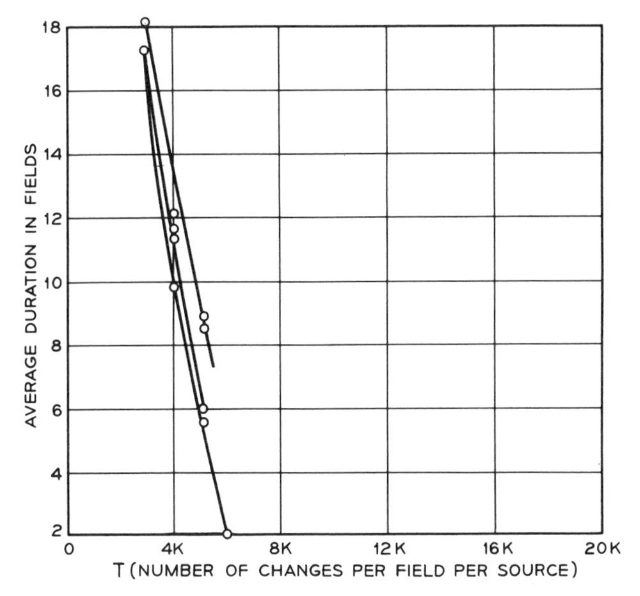

Fig. 3.2.25 Average duration of peaks in the number of changes per field per source when data from *twelve sources* are combined.

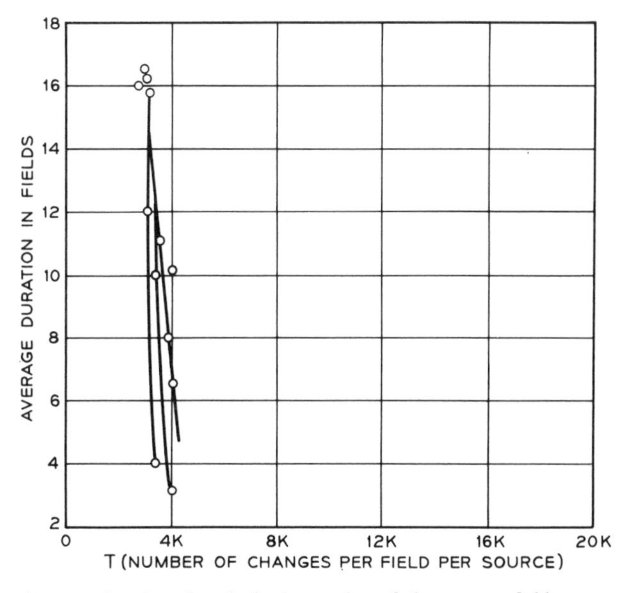

Fig. 3.2.26 Average duration of peaks in the number of changes per field per source when data from *twenty sources* are combined.

1.000	1.000	1.000	1.000	1.000	1.000	1.000	1.000
1.000	.707	.000	-.707	-1.000	-.707	.000	.707
1.000	.000	-1.000	.000	1.000	.000	-1.000	.000
1.000	-.707	.000	.707	-1.000	.707	.000	-.707
1.000	-1.000	1.000	-1.000	1.000	-1.000	1.000	-1.000
1.000	-.707	.000	.707	-1.000	.707	.000	-.707
1.000	.000	-1.000	.000	1.000	.000	-1.000	.000
1.000	.707	.000	-.707	-1.000	-.707	.000	.707

.000	.000	.000	.000	.000	.000	.000	.000
.000	-.707	-1.000	-.707	.000	.707	1.000	.707
.000	-1.000	.000	1.000	.000	-1.000	.000	1.000
.000	-.707	1.000	-.707	.000	.707	-1.000	.707
.000	.000	.000	.000	.000	.000	.000	.000
.000	.707	-1.000	.707	.000	-.707	1.000	-.707
.000	1.000	.000	-1.000	.000	1.000	.000	-1.000
.000	.707	1.000	.707	.000	-.707	-1.000	-.707

Fig. 3.2.27 Transform matrix for the DFT with $N = 8$. Top is real part, bottom is imaginary part.

$$c = Tb$$

$$b = T'c = T^{-1}c \tag{3.2.5}$$

where prime denotes conjugate transpose.

The columns of T' are the *basis* vectors t_m of the transform. In Section 3.1.6 we described the Discrete Fourier Transform (DFT). For $N = 8$, the DFT transform matrix T is shown in Fig. 3.2.27. For real b, the complex coefficients c have conjugate symmetry. For example, for N even, c_1 and $c_{\frac{N}{2}+1}$ are both real, and

$$c_{2+j} = c_{N-j} \qquad j=0...N-2 \tag{3.2.6}$$

Thus, reconstruction of the N pels of b, in fact, requires only the transmission of at most N real transform coefficient values. An advantage of the DFT is that a fast computation algorithm called the FFT exists as we shall see in a later chapter.

The Walsh-Hadamard Transform (WHT) was described in Section 3.1.6 and also has a symmetric transform matrix. For $N=4$ and 8, WHT matrices T are shown in Fig. 3.2.28.

The Karhunen-Loeve Transform (KLT) matrix T is derived from the correlation matrix

$$R \triangleq E(bb') \tag{3.2.7}$$

The basis vectors t_m are the orthonormalized eigenvectors of R. The corresponding eigenvalues λ_m, customarily arranged in decreasing order, are the mean square values of the transform coefficients, i.e., $Ec_m^2 = \lambda_m$. Although the derivation of the KLT basis vectors in Sec. 3.1.6 assumed that pels were biased to have zero mean, actual calculation of the KLT coefficients is often performed without such a bias. Thus,

$$c_m = t'_m b \tag{3.2.8}$$

$$Ec_m = t'_m (Eb) \neq 0$$

However, it usually turns out that only the first few of the transform coefficients are affected very much by $Eb \neq \mathbf{0}$.

The discrete cosine transform (DCT) has transform matrix elements

$$\frac{1}{2}\begin{bmatrix} 1 & 1 & 1 & 1 \\ 1 & -1 & 1 & -1 \\ 1 & 1 & -1 & -1 \\ 1 & -1 & -1 & 1 \end{bmatrix}$$

$$\frac{1}{\sqrt{8}}\begin{bmatrix} 1 & 1 & 1 & 1 & 1 & 1 & 1 & 1 \\ 1 & -1 & 1 & -1 & 1 & -1 & 1 & -1 \\ 1 & 1 & -1 & -1 & 1 & 1 & -1 & -1 \\ 1 & -1 & -1 & 1 & 1 & -1 & -1 & 1 \\ 1 & 1 & 1 & 1 & -1 & -1 & -1 & -1 \\ 1 & -1 & 1 & -1 & -1 & 1 & -1 & 1 \\ 1 & 1 & -1 & -1 & -1 & -1 & 1 & 1 \\ 1 & -1 & -1 & 1 & -1 & 1 & 1 & -1 \end{bmatrix}$$

Fig. 3.2.28 Transform matrices for the WHT with $N = 4$ and 8.

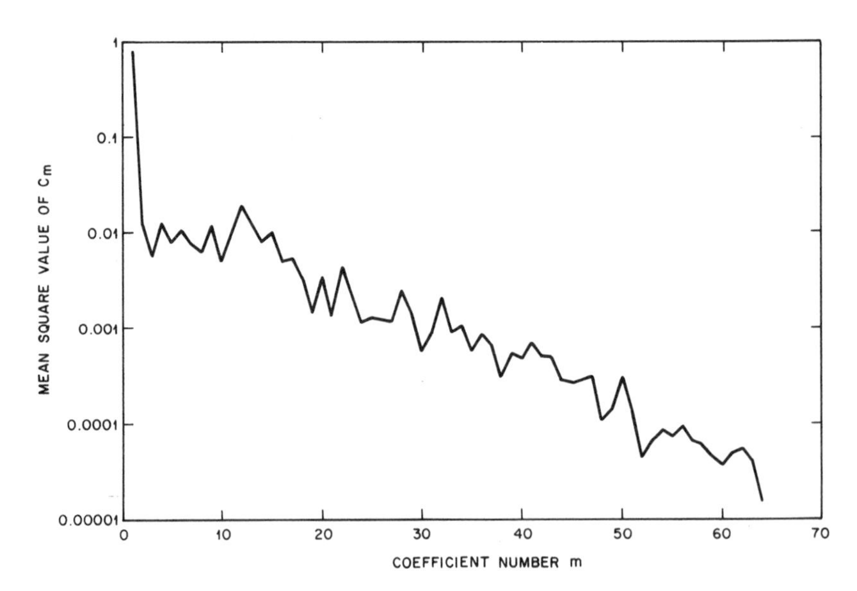

Fig. 3.2.29 Normalized mean square values (MSV) of the DCT coefficients (64-pel blocks) for "Stripes". Pel data is 8-bit PCM.

$$t_{mi} = \sqrt{\frac{2-\delta_{m1}}{N}} \cos\left[\frac{\pi}{N}\left(i - \frac{1}{2}\right)(m-1)\right] \quad i,m = 1 \ldots N \tag{3.2.9}$$

The DCT basis vectors $\boldsymbol{t}_m$ are sinusoids with frequency indexed by m. Fig. 3.2.29 shows the mean square values of DCT coefficients c_m (64 pel blocks) for the picture "Stripes". We see that, in general, the coefficient amplitudes are largest for small values of m, i.e., for small frequencies.

A major objective of transform coding is to produce statistically independent transform coefficients, thus reducing the need for block coding in order to achieve good efficiency. A simultaneous objective is *energy compaction,* which, loosely speaking, means making as many transform coefficients as possible small enough that they need not be transmitted. It is conjectured that achieving one objective implies the other. Indeed, most of the unitary transforms in use today approach both objectives reasonably well. However, rigorous proof of this conjecture is available only for the KLT.

Fig. 3.2.30 shows mean square truncation error results for four different transforms (64 pel blocks) applied to "Stripes" when only pN of the N = 64 transform coefficients are transmitted ($0<p<1$). In each case the pN coefficients having the largest mean square values are chosen. We see that in this case, the KLT is by far the best performing transform in terms of energy compaction.

For all of the unitary transforms, the magnitudes of the transform coefficients depend very strongly on the statistics of the pels themselves.

$$\begin{aligned} E[|c_m|^2] &= E\, c_m c_m' \\ &= E\, \boldsymbol{t}_m' \boldsymbol{b}(\boldsymbol{t}_m' \boldsymbol{b})' \\ &= \boldsymbol{t}_m'\, E(\boldsymbol{b}\boldsymbol{b}')\, \boldsymbol{t}_m \\ &= \boldsymbol{t}_m' \boldsymbol{R} \boldsymbol{t}_m \end{aligned} \tag{3.2.10}$$

where $\boldsymbol{R}$ is the correlation matrix of the pel blocks. Thus, if $\boldsymbol{t}_m$ is of a certain spatial frequency, then $\boldsymbol{c}_m$ will be sizable if $\boldsymbol{R}$ contains periodicities of that spatial frequency, i.e., the pels have correlations corresponding to that periodicity.

The KLT has basis vectors $\boldsymbol{t}_m$ that are eigenvectors of $\boldsymbol{R}$. Thus, they are chosen to match the periodicities in $\boldsymbol{R}$ in the best way possible.

The transform coefficients themselves are simply weighted sums and differences of pel values. If pel statistics were stationary (which they are not) and if the block size N were very large, then by the central limit theorem the transform coefficients would have Gaussian probability distributions. However, for more

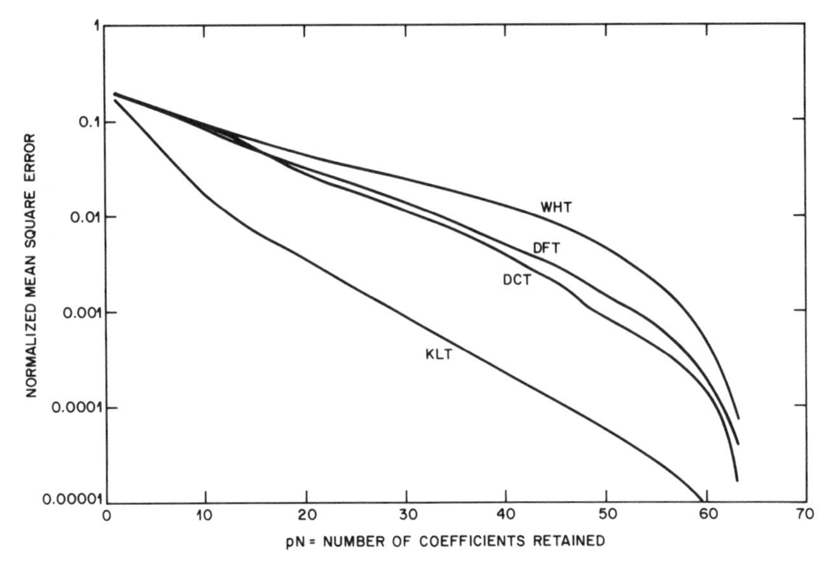

Fig. 3.2.30 Normalized mean square truncation error when percentage p of the $N = 64$ coefficients are retained using four transforms on "Stripes".

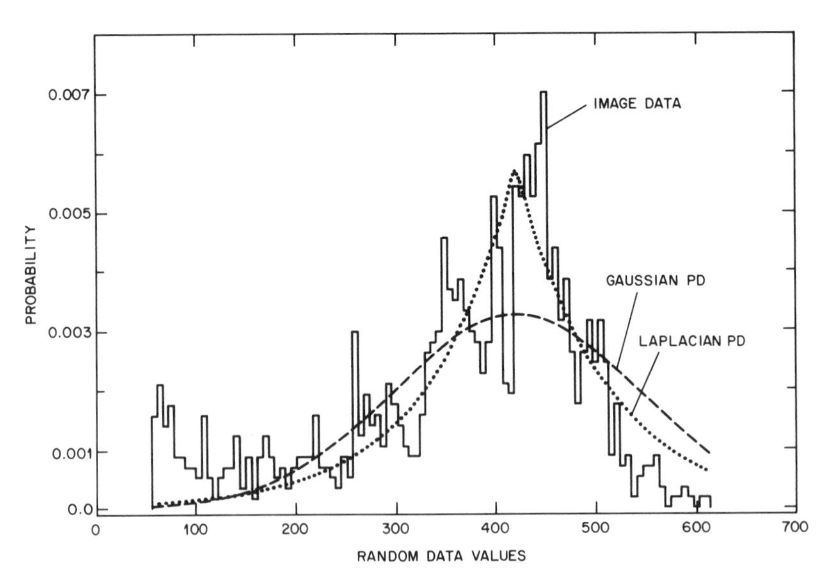

Fig. 3.2.31a Histogram for the first DCT coefficient (64-pel blocks) for "Stripes". Laplacian and Gaussian PD's with the same mean and variance are also shown.

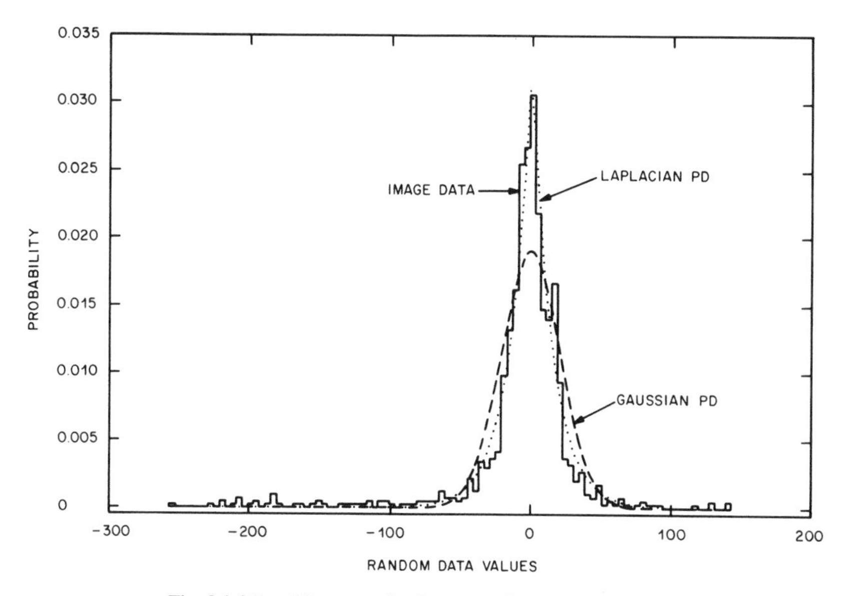

Fig. 3.2.31b Histogram for the second DCT coefficient.

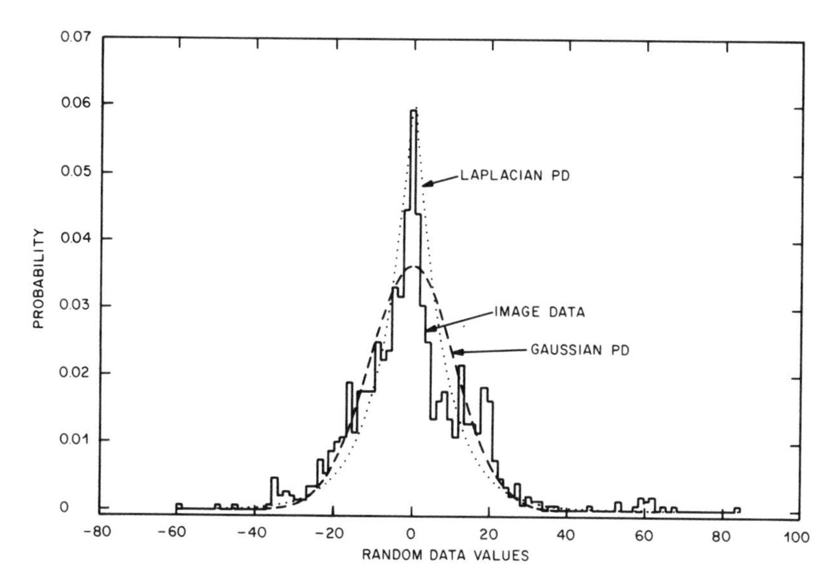

Fig. 3.2.31c Histogram for the third DCT coefficient.

practical block sizes the statistics tend to be more Laplacian in nature. For example, Fig. 3.2.31 shows histograms for the first three DCT coefficients computed from "Stripes" using 64-pel blocks of pels. Also shown for reference are the Gaussian and Laplacian distributions having the same mean and variance as the respective coefficient distributions. We see that the Laplacian distribution is a fairly good representation of the true* statistics.

The actual mechanisms of transform coding, including fast computation and optimum quantization of coefficients, will be covered in detail in Chapter 5.

3.2.8 Statistical Models and Parameters

A simple, accurate statistical model for visual information is desirable for many reasons. For example, with a reliable statistical model plus a distortion measure, rate-distortion theory could be invoked to reveal the minimum bit-rate at which coding could take place. For differential PCM systems, predictors could be optimized, and efficient Huffman codes matched to statistics could be generated. Alternatively, optimum transform coding could be carried out, or at the very least, the Karhunen-Loeve transform could be derived *a priori* without having to measure correlation functions in real time.

Alas, these objectives are far from being realized. In the first place, we have seen that the statistics of pictorial information are highly nonstationary. In addition to nonstationarity, picture data and waveforms do not lend themselves to a simple, concise mathematical description. Pictures, after all, consist of objects with fairly well defined edges, whose locations and orientations within a scene may not be completely random. For example, in many scenes edges oriented vertically and horizontally occur in much greater proportion than edges with other orientations (see Fig. 3.2.32). The interior of some objects may contain texture that is more or less random, while in other objects the interior may have a very definite structure. In short, the situation is hardly conducive to a simple mathematical representation. Even the verbal description above is not without ambiguity. For example, how do we tell the difference between many small objects in a picture and textural structure within a single large object? How do we define an object when its edges are blurred or otherwise difficult to determine?

In spite of these hurdles, many statistical models have been suggested and studied for representation of imaging data.[3.2.6] Although none of these are completely satisfactory, some have proved useful in forming a mathematical basis for certain experimentally observed phenomena. Most of the models assume pels having continuous amplitudes.

---

* For low detail pictures, low order coefficients are somewhat better represented by a Gamma distribution, whereas high order coefficients are approximately Gaussian [3.2.5]. For high detail pictures, statistics are much less stationary.

3.2.8a Gaussian Model

A very tempting model because of its mathematical tractability is the stationary, Gaussian random vector model. It would be realistic only for images with predominant grays, and very little bright white or dark black. If $\boldsymbol{b}$ is a column vector of N pels with correlation matrix*

$$\boldsymbol{R} = E[\boldsymbol{bb}'] \quad \text{and} \quad |\boldsymbol{R}| = \text{determinant } of\ \boldsymbol{R}, \tag{3.2.11}$$

then the joint Gaussian probability *density* function (PDF) of the block $\boldsymbol{b}$ of continuous amplitude pels is given by

$$p(\boldsymbol{b}) = \frac{1}{\sqrt{|\boldsymbol{R}|(2\pi)^N}} \exp\left[-\frac{1}{2} \boldsymbol{b}\boldsymbol{R}^{-1}\boldsymbol{b}' \right] \tag{3.2.12}$$

The rate-distortion function for jointly Gaussian random vectors $\boldsymbol{b}$ and mean square error distortion criterion is known exactly.[3.1.8] In terms of a parameter $\theta \geq 0$ and the orthonormal eigenvalues $\{\lambda_m\}$ of the correlation matrix $\boldsymbol{R}$,

$$D = \frac{1}{N} \sum_{m=1}^{N} \min(\theta, \lambda_m) \tag{3.2.13}$$

$$NR(D) = \sum_{m=1}^{N} \max\left[0, \frac{1}{2} \log_2 \left(\frac{\lambda_m}{\theta}\right)\right] \quad \text{bits/block}$$

For very small values of D, i.e., $\theta \leq \lambda_m \ \forall m$, we have $D = \theta$ and

$$R(D) = \frac{1}{2N} \log_2 \prod_{1}^{N} \frac{\lambda_m}{D} = \frac{1}{2} \log_2 \frac{|\boldsymbol{R}|^{\frac{1}{N}}}{D} \quad \text{bits/pel} \tag{3.2.14}$$

---

* Assuming pels are biased so that $E\boldsymbol{b} = \boldsymbol{0}$.

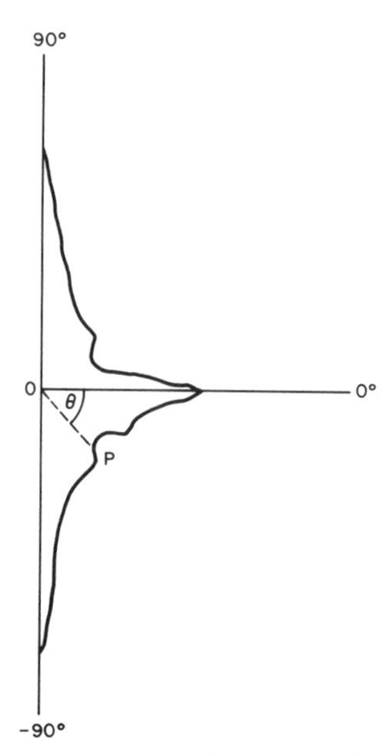

Fig. 3.2.32 Probability distribution of the angle (with respect to horizontal) of edges for an outdoor scene containing man-made structures. The distance OP is the probability of angle θ. Vertical and horizontal edges occur with higher frequency than edges at other angles (from F. Kretz).

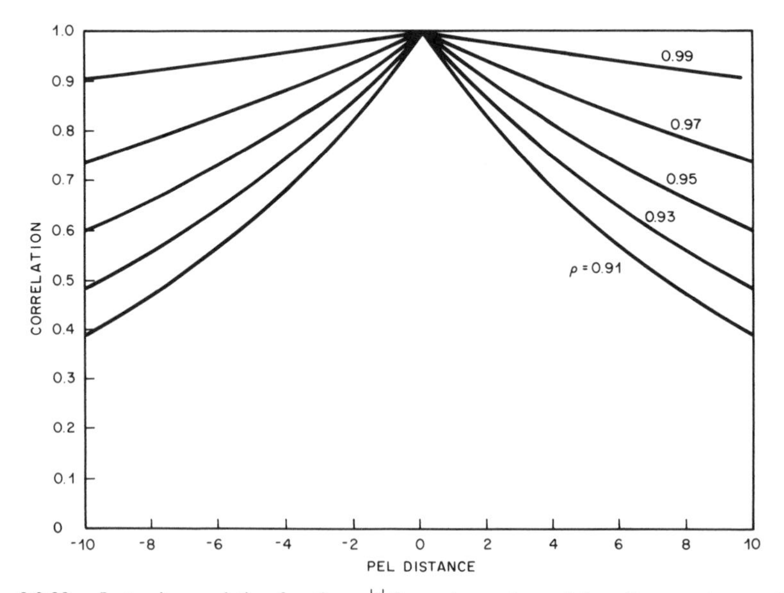

Fig. 3.2.33 Isotropic correlation function $\rho^{|x|}$ for various values of the adjacent pel normalized correlation ρ.

For jointly Gaussian $\boldsymbol{b}$ the integral entropy is given by

$$H(\boldsymbol{b}) = \frac{N}{2} \log_2 (2\pi e |\boldsymbol{R}|^{\frac{1}{N}}) \quad (3.2.15)$$

and we see by comparing Eqs. (3.2.14) and (3.1.84) that for small D and Gaussian random vectors, $R(D)$ is equal to the MSE Shannon lower bound.

Shannon has also shown that Eq. (3.2.13) is an *upper* bound on the MSE rate-distortion function for *any* random vector $\boldsymbol{b}$ having correlation matrix $\boldsymbol{R}$. Furthermore, for small D it converges to the Shannon lower bound. Thus, for MSE at least, rate-distortion functions can be approximated reasonably well for low distortion, which is the case of predominant practical interest.

Application of the Karhunen-Loeve transform to the Gaussian model results in Gaussian transform coefficients that are not only uncorrelated, but also statistically independent. If each coefficient is then optimally encoded* with MSE and rate given by the summands of Eq. (3.2.13), then the overall encoding is optimal and achieves the rate-distortion function. Similar results can be shown for a frequency weighted MSE distortion criterion.[3.2.8]

These results agree well with the observed performance of the KLT on real pictures. However, numerically synthesized data having Gaussian statistics do not look very much like images since edges of objects are usually missing. Also, statistics of predictive coding differential signals and transform coefficients are not well represented by the Gaussian model.

3.2.8b Correlation Models

Very often image statistics are assumed to be isotropic, and the correlations (for pels biased to zero mean)

$$r(i,j) \triangleq E[b_i b_j] \quad (3.2.16)$$

are modeled as some decreasing function of the geometric distance $d(i,j)$ between pels b_i and b_j. For example,

$$r(i,j) = \sigma^2 \rho^{d(i,j)} \quad (3.2.17)$$

where $\rho < 1$, gives reasonably good agreement with experimental data. For natural images, ρ usually exceeds 0.90.

---

* This generally requires block coding of the coefficients.

If vectors $\boldsymbol{b}$ are formed from 1-by-N blocks of pels, then Eq. (3.2.17) becomes

$$r(i,j) = \sigma^2 \rho^{|i-j|} \tag{3.2.18}$$

which is shown in Fig. 3.2.33 for various ρ. For example, using this model we can estimate the MSE of unitary transform coding when only a fraction $p<1$ of the transform coefficients are transmitted. From Eq. (3.1.64)

$$\begin{aligned} MSE &= \frac{1}{N} \sum_{m>pN}^{N} E\,|c_m|^2 \\ &= \frac{1}{N} \sum_{m>pN} E\left[\boldsymbol{t}'_m \boldsymbol{b}\boldsymbol{b}'\boldsymbol{t}_m\right] \\ &= \frac{1}{N} \sum_{m>pN} \boldsymbol{t}'_m \boldsymbol{R}\boldsymbol{t}_m \\ &= \frac{1}{N} \sum_{m>pN} \sum_{i,j=1}^{N} r(i,j)\, t'_{im} t_{jm} \end{aligned} \tag{3.2.19}$$

Fig. 3.2.34 shows the relative performance of several transforms using this model.

In some cases a *separable* correlation is used. For example, if $d_x(i,j)$ and $d_y(i,j)$ are the horizontal and vertical distances, respectively, between pels b_i and b_j, then a separable correlation might be

$$r(i,j) = \sigma^2 \rho_x^{d_x(i,j)} \rho_y^{d_y(i,j)} \tag{3.2.20}$$

where ρ_x, $\rho_y < 1$ are the horizontal and vertical adjacent pel correlations.

If the correlation matrix is to be estimated from real picture data, then Eq. (3.2.16) can be evaluated as a simple arithmetic average over all blocks in the picture or pictures. If the block size is large, then overlapping blocks may be used in the averaging process in order to obtain a more realistic estimate. In either case, the resulting $\boldsymbol{R}$ will be positive definite, i.e., a legitimate correlation matrix.

One might be tempted to construct a function $f[d]$ such that

$$r(i,j) \approx f[d(i,j)] \tag{3.2.21}$$

as in Eq. (3.2.17) where we assume $f(d) = \sigma^2 \rho^d$. The function $f(d)$ would presumably be approximated by averaging the product of all pels that are separated geometrically by distance d. The difficulty with this approach is that the image statistics may not be circularly symmetric (see Fig. 3.2.6); and when the estimated function f is used in Eq. (3.2.21), the resulting correlation matrix $\boldsymbol{R}$ may not be positive definite. It will thus possess negative eigenvalues and give bizarre mathematical results when used to model image statistics.

3.2.8c Uniform Model

Sometimes the best or only probability density function (PDF) that is suitable, e.g., see Fig. 3.2.3, is the uniform distribution

$$p(x) = \begin{cases} a^{-1} & 0<x<a \\ 0 & \text{elsewhere} \end{cases} \tag{3.2.22}$$

If the range $(0,a)$ of a uniform random variable is quantized uniformly using L levels, then the length of a single quantization interval is

$$\Delta = \frac{a}{L} \tag{3.2.23}$$

Values of the random variable lying within a quantization interval are all represented by the midpoint value of the interval. Thus, the MSE due to quantization is the same within each interval, and the total MSE is

$$MSE = L \int_0^{\Delta} \left[x - \frac{\Delta}{2} \right]^2 p(x)\, dx \tag{3.2.24}$$

$$= \frac{\Delta^2}{12}$$

The root mean square error then becomes

$$RMSE = \frac{\Delta}{\sqrt{12}} \tag{3.2.25}$$

For video applications, the amount of quantization or other noise is often written as the ratio of peak-to-peak signal to RMSE expressed in decibels (PSNR). For L level quantization and PCM coding using constant binary word lengths, the bit-rate is

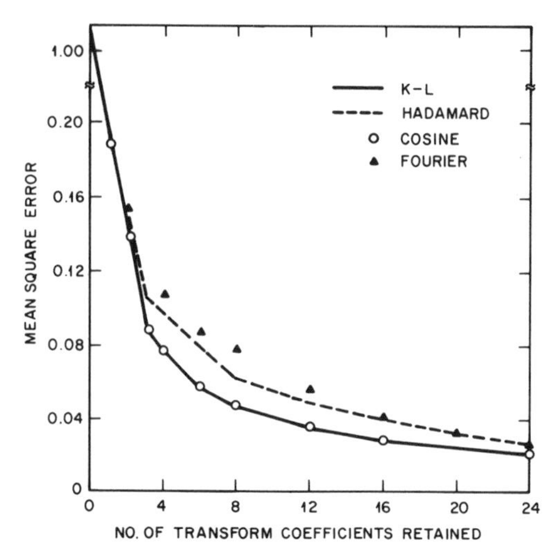

Fig. 3.2.34 Normalized MSE when only a fraction of the $N = 64$ coefficients are retained using various transforms and the correlation of Eq. (3.2.17). $\sigma^2 = 1$ and adjacent pel correlation $\rho = 0.95$.

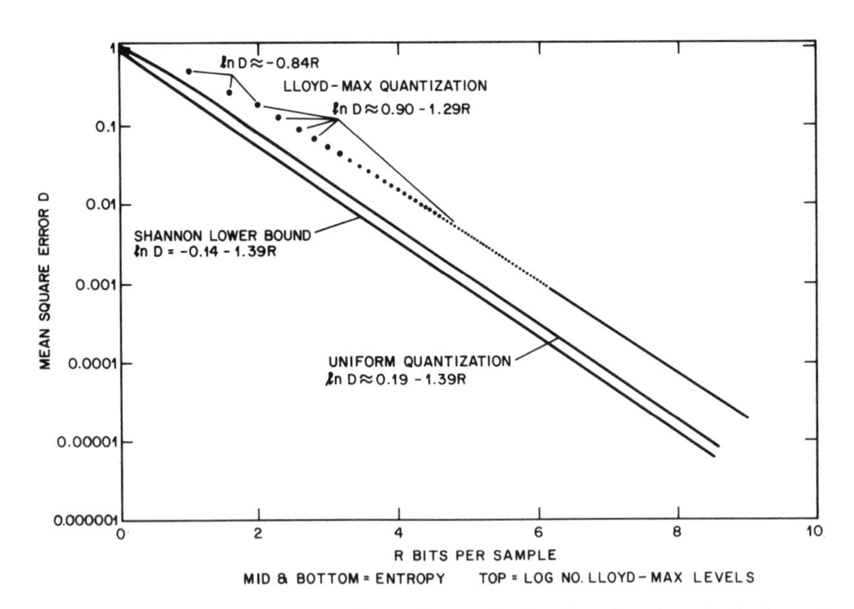

Fig. 3.2.35 Mean square error versus bit-rate for quantization of a unit-variance Laplacian random variable. Lloyd-Max and uniform quantizers were chosen to be mid-step quantizers. Also, shown is the Shannon lower bound on the Rate-Distortion function.

$$R = \log_2 L = \frac{\log_{10} L}{\log_{10} 2} \quad \text{bits/pel} \tag{3.2.26}$$

Thus, we have

$$\begin{aligned} PSNR &= 20 \log_{10} \frac{a\sqrt{12}}{\Delta} \\ &\approx 6R + 10.8 \text{ dB} \end{aligned} \tag{3.2.27}$$

and we see that each additional bit of PCM quantization yields a 6 dB improvement in PSNR.

3.2.8d Laplacian Model

The Laplacian PDF

$$p(x) = \frac{\lambda}{2} e^{-\lambda|x|} \tag{3.2.28}$$

having mean zero and variance $2/\lambda^2$, closely matches the measured PDF's of most differential signals. It also approximates the PDF's of transform coefficients, except for the first one which is usually a simple arithmetic average of the pels in the block (see Fig. 3.2.31).

Thus, the Laplacian model can be useful, for example, in designing Huffman codes for the transmission of these quantities. With an appropriate distortion criterion, it can also be used to derive quantization strategies. For example, the optimum MSE quantizer having a fixed number of output values is derived as follows:

Let $\{\ell_k\}$ be a monotonically increasing set of L quantizer representative values with $\{t_k\}$ the corresponding set of thresholds, i.e., for input x, if $t_{k-1} < x \le t_k$, then the quantizer output is ℓ_k (see Fig. 1.7.1). For random variables like the Laplacian, which have negative as well as positive values, if zero is one of the representative levels ℓ_k, then the quantizer is called a *mid-step* quantizer. If zero is one of the threshold values t_k, the quantizer is called a *mid-riser* quantizer. The overall MSE due to quantization is then

$$MSE = \sum_{k=1}^{L} \int_{t_{k-1}}^{t_k} (x-\ell_k)^2 p(x)\, dx \tag{3.2.29}$$

For fixed L, two necessary conditions for minimizing this error are

$$t_k = \begin{cases} -\infty, \infty & k = 0, L \text{ respectively} \\ \frac{1}{2}(\ell_k + \ell_{k+1}) & \text{otherwise} \end{cases} \tag{3.2.30}$$

and by simple differentiation

$$\int_{t_{k-1}}^{t_k} (x - \ell_k) p(x) \, dx = 0 \qquad \forall k$$

This relationship defines a *Lloyd-Max* quantizer.[3.2.7] In general, these equations must be solved numerically. An estimate of the resulting optimum MSE (assuming L is large and $p(x)$ is symmetric about 0) is then given by Bennet's integral

$$MSE \approx \frac{1}{12 \, L^2} \left[\int_{t_{L-1}}^{t_L} p^{\frac{1}{3}}(x) \, dx \right]^3 \tag{3.2.31}$$

Fig. 3.2.35 shows the result of Lloyd-Max mid-step quantization of a unit variance Laplacian random variable. Mean square error D is plotted versus the bit-rate $R = \log_2 L$, and we see that the curve is very well approximated by the relation

$$\ln D = \begin{cases} -0.84R & R \leq 2 \\ 0.90 - 1.29R & R \geq 2 \end{cases} \tag{3.2.32}$$

Also shown is the result of uniform mid-step quantization with entropy coding. It can be shown that with entropy coding of Laplacian random variables, uniform mid-step quantization is optimum.[3.2.7a] We see that for this case the curve is well approximated in the range $R \geq 1$ by

$$\ln D = 0.19 - 1.39R \tag{3.2.33}$$

Finally, Fig. 3.2.35 shows the Shannon lower bound on the Rate-Distortion function $R(D)$. From Eq. (3.2.28) the integral entropy for a Laplacian random variable is

$$H(x) \triangleq -\int p(x) \log_2 p(x)\,dx$$
$$= \log_2 \frac{2e}{\lambda} \tag{3.2.34}$$

Then from Eq. (3.1.84) the Shannon lower bound is

$$R(D) \geq \frac{1}{2} \log_2 \frac{2e}{\pi D \lambda^2} \tag{3.2.35}$$

which for unit variance becomes

$$\ln D \geq -0.14 - 1.39R \tag{3.2.36}$$

Thus, we see that uniform quantization with entropy coding is quite efficient, performing within 0.25 bits of the Rate-Distortion function.

Fig. 3.2.36 shows the quantization interval Δ versus entropy for uniform mid-step quantization of a unit variance Laplacian random variable. We see that for $R \geq 1$

$$1.39R \approx 2.69 - 2\ln\Delta \tag{3.2.37}$$

This can also be derived analytically for small Δ by assuming a constant PDF within each quantization interval. Combining Eqs. (3.2.37) and (3.2.33) results in

$$D \approx \frac{\Delta^2}{12} \tag{3.2.38}$$

which is the same MSE as in Eq. (3.2.24) for uniform random variables. Thus, we see that for bit-rates even as small as 1 bit/sample, the MSE due to uniform mid-step quantization and entropy coding is mostly dependent on step size Δ, and relatively independent of PDF.

3.2.8e Autoregressive Models

A plausible model for one-dimensional picture data (at least away from the image boundaries) is the so called pth order causal autoregressive process[3.1.8]

$$b_i = \sum_{k=1}^{p} a_k b_{i-k} + \varepsilon_i \tag{3.2.39}$$

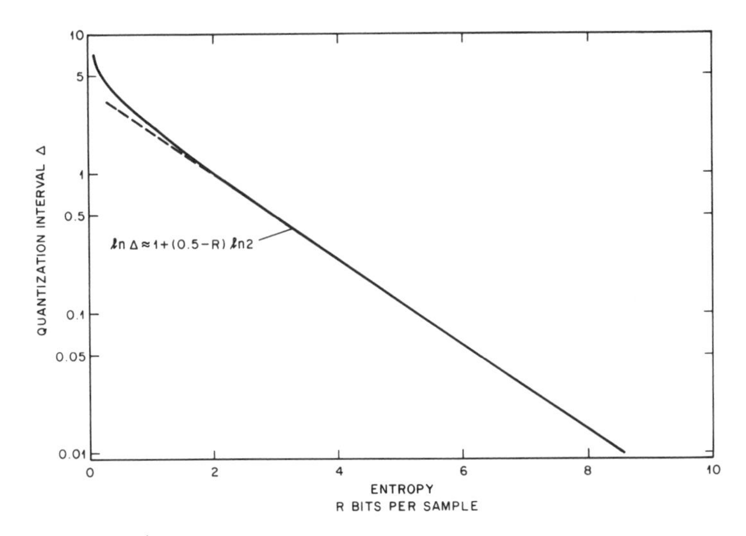

Fig. 3.2.36 Quantization interval Δ versus entropy for uniform mid-step quantization of a unit variance Laplacian random variable.

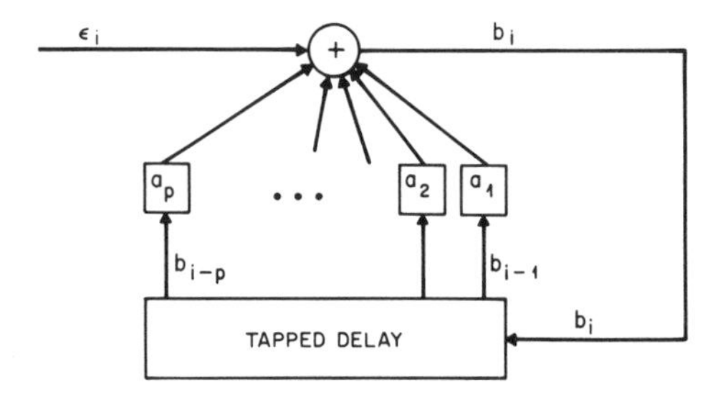

Fig. 3.2.37 Casual autoregressive source model. The generator is driven by independent identically distributed random variables ε_i, also known as white noise.

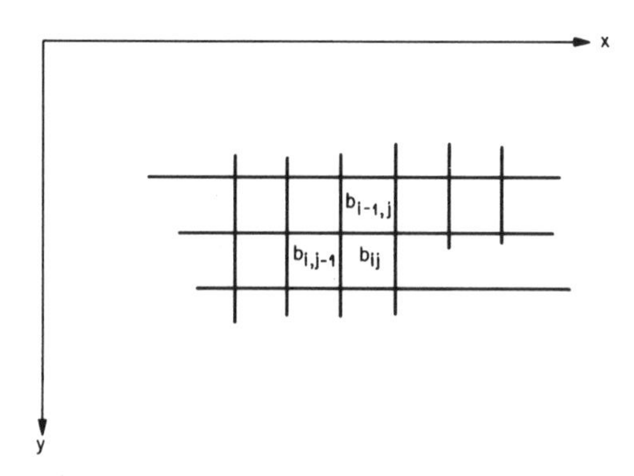

Fig. 3.2.38 Pel configuration for two-dimensional causal autoregressive source model. Pel b_{ij} is produced by adding ε_{ij} to a linear combination of previously generated pels.

where the ε_i are independent, identically distributed, zero mean* random variables. Such a model is shown in Fig. 3.2.37. The process is stationary and stable, i.e., useful for image modeling, only if the roots of

$$1 - \sum_{k=1}^{p} a_k x^{-k} \tag{3.2.40}$$

lie inside the unit circle. If such is the case, then the correlations $r(i,j) = E(b_i b_j)$ can be found fairly easily.[3.2.6] Conversely, given a set of desired correlations, an autoregressive source can be found that produces these correlations to any desired degree of accuracy.[3.2.6]

In particular, the first order causal autoregressive process has received much attention because of its analytical and computational tractability. In this case $(p = 1,\ 0 < a_1 < 1)$

$$\begin{aligned} E\varepsilon_i^2 &= \left(1 - a_1^2\right)\sigma^2 \\ Eb_i^2 &= \sigma^2 \ , \end{aligned} \tag{3.2.41}$$

and the correlation matrix is given by

$$r(i,j) \triangleq E(b_i b_j) = \sigma^2 a_1^{|i-j|} \tag{3.2.42}$$

If the ε_i are Gaussian,† then so are the b_i, and in this case the MSE rate distortion function is known. For small D and large block size, it is given by[3.1.8]

$$R(D) = \frac{1}{2} \log_2 \frac{\sigma^2(1-a_1^2)}{D} \ \text{bits/pel} \tag{3.2.43}$$

$$\text{where } D < \frac{\sigma^2(1-a_1)}{(1+a_1)}$$

It has also been shown[3.2.9] that for this model, the Discrete Cosine Transform (DCT) performs very close to the Karhunen-Loeve Transform (KLT) for the

---

* The model produces pels with zero mean. Thus, the image is assumed to be biased to zero mean. The causal autoregressive process is also pth order Markov.

† In this case, the process is also known as first-order Gauss-Markov.

range $0.5 < a_1 < 1$ that most often occurs in practice. The model has also been used to prove the conjecture that DPCM and KLT coding perform comparably for low mean square distortions, e.g., $D \leq 0.01\sigma^2$, whereas for larger distortions the KLT is superior.[3.2.9]

Eq. (3.2.39) has been extended to raster scanning of a two dimensional image (for pels away from the boundaries) by using double subscripts, j for the horizontal dimension and i for the vertical. The causal, autoregressive process then becomes

$$b_{ij} = \sum_{k=1}^{p} a_{k0} b_{i,j-k} + \sum_{k=-p}^{p} \sum_{\ell=1}^{p} a_{k\ell} b_{i-\ell,j-k} + \varepsilon_{ij} \tag{3.2.44}$$

Note that pels on the right side of this equation all occur before pel b_{ij} in the scanning, as shown in Fig. 3.2.38. In both of the models of Eqs. (3.2.39) and (3.2.44), the minimum MSE predictor [see Eq. (3.1.34)] is given by the right hand side of the equations without the ε's.

A noncausal autoregressive model can be written simply by

$$\sum_{k,\ell=-p}^{p} a_{k\ell} b_{i-k,j-\ell} = -\varepsilon_{ij} \tag{3.2.45}$$

with $a_{00} = -1$. Such a model (although not very usable for predictive coding) can be used to accurately represent correlation matrices of real pictures and to estimate the MSE performance of transform coders. If the coefficients $a_{k\ell}$ vary with time, then nonstationarity is produced.

3.2.8f Object Based Models

We have already stated that statistical image models are often unsatisfactory. More realistic, but less mathematically tractable, models have also been suggested. Some of the proposals involve decomposition of the original image into two or more parts: that is, one part that can be described loosely as *edge information* giving locations and boundaries of objects in the scene, and the other parts or *residual information* that describe the interior color and texture of those objects.[3.2.10] With this method, one set of statistics would control the edge behavior in the resulting images, while other statistics would be used to produce interior texture that resembles that of real images.[3.2.11]

Another approach involves computer modeling of shapes, positions and motion of relatively simple objects within a scene. Each object must be specified completely in three dimensions, including surface texture and reflectivity. Objects are then placed in an idealized scene with the chosen incident light and observer viewing position. The resulting two dimensional images can then be calculated and used as input for various processing algorithms. More discussion of these points is found in Chapter 11.

Fig. 3.3.1 A color picture containing a head and shoulders view (from Limb et al. [3.3.5]).

The methods above can be described as *scene synthesis*, and can be useful in testing certain algorithms. However, very often real picture data can also be used without much additional expense or inconvenience.

3.3 Color Picture Statistics

Statistics of color picture signals are highly dependent on the color components being considered (e.g., R, G, B or Y, I, Q). Also, the statistics depend very much on the ensemble of pictures used for computing them. Often, the statistics are computed by averaging over a very large number of pictures. However, the resulting probability density functions for R,G and B may turn out to be nearly uniform, as with the luminance signal for monochrome pictures. Fig. 3.3.2 shows the probability densities for the gamma-corrected tristimulus values Red, Green and Blue for the head-and-shoulders type of picture shown in Fig. 3.3.1.

There is considerable correlation between the Red, Green and Blue components, especially for natural pictures. This is due to the fact that most natural pictures do not contain large areas with saturated colors. Thus, similar amplitudes of the Red, Green and Blue signals occur most frequently. Colors having the highest probability of occurrence are the different shades of gray. For example, Fig. 3.3.3 shows, for a test picture, the relative frequency of occurrence of the joint amplitudes of Red and Green, Blue and Green, and Red and Blue signals. The three joint distributions are very different from each other, but show remarkable correlation between the components. Moreover, the correlation increases as the gamut of colors decreases.[3.3.1] The correlation is especially large for pictures containing mostly gray color for which $R \approx G \approx B$.

Much of the amplitude correlation between the color components can be eliminated by linear transformation. In fact, the Karhunen-Loeve transformation can be computed for *RGB* 3-tuples so that most of the signal energy is in one of the transformed components, and very little is in the other two. Although such a transform, in general, varies from picture to picture, the first component usually consists of approximately equal amounts of Red, Green and Blue signals.[3.3.2−3] Table 3.3.1 shows the energy compaction ability of the Karhunen-Loeve transform (coefficients K_1, K_2 and K_3) compared to the familiar *Y,I,Q* transformation. Compared to the Red, Green and Blue signals, both the K_1, K_2, K_3 and the *Y,I,Q* signals provide high energy compaction. The *Y,I,Q* system has the advantage that in addition to providing a compatible system with monochrome television, its energy compaction property comes close to the K-L transform. Because of this, we will now deal only with the component signals *Y,I,Q* and Y, U_t, V_t that are used in broadcast television systems.

Most natural scenes cover only a relatively small range of chromaticities compared to the gamut enclosed by the spectral locus. We saw in Section 2.2 that the range of chromaticities that can be reproduced by a television system depends on the chromaticities of the phosphors used. Furthermore, this range is

Table 3.3.1 Comparison of the energy compaction ability of three different component representations of two pictures.

PICTURE	COORDINATE SYSTEM	POWER (%) 1st COMP.	POWER (%) 2nd COMP.	POWER (%) 3rd COMP.
WOMAN	RGB	45.14	35.41	19.45
	YIQ	78.32	17.54	4.14
	K1, K2, K3	85.84	12.10	2.06
COUPLE	RGB	51.55	31.09	17.36
	YIQ	84.84	13.81	1.35
	K1, K2, K3	92.75	6.46	0.79

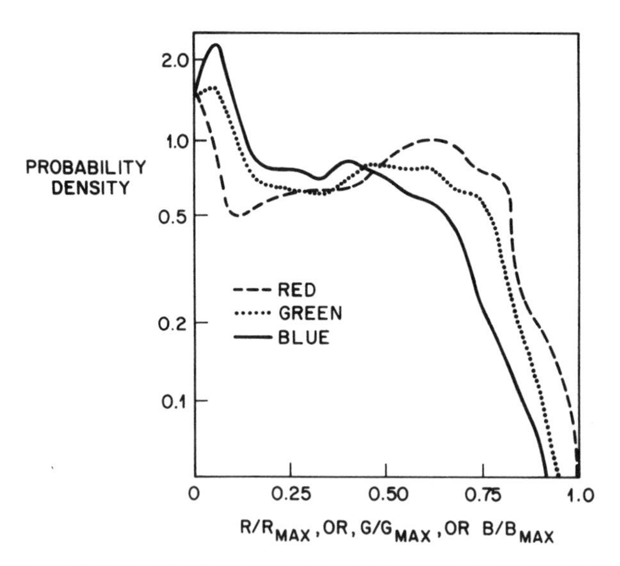

Fig. 3.3.2 Probability density functions of the Red, Green and Blue signals for the picture shown in Fig. 3.3.1 (from Limb et al. [3.3.5]).

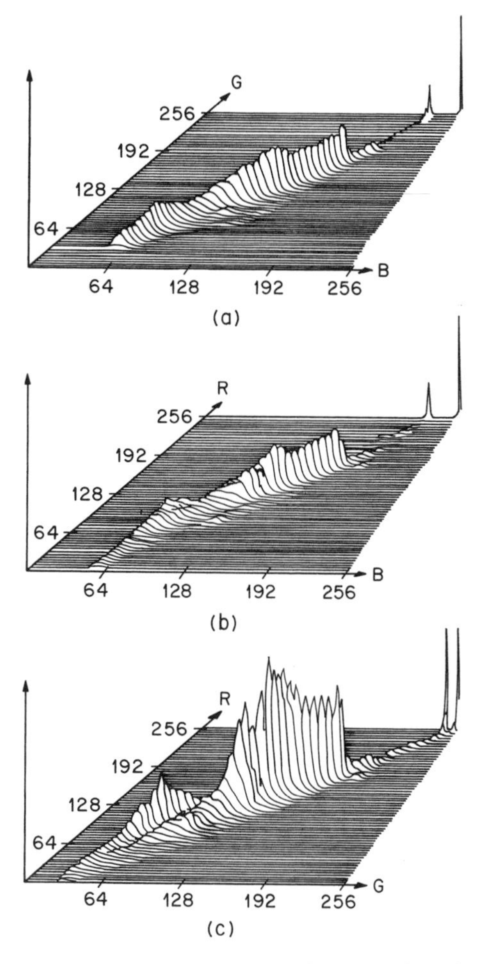

Fig. 3.3.3 Two-dimensional joint probability densities of the three-dimensional *RGB* signal, for a test picture, showing the strong correlation between components. (a) *B* vs. *G*, (b) *R* vs. *B*, and (c) *R* vs. *G* (from Frei et al. [3.3.9]).

completely defined in a CIE chromaticity diagram by the triangle formed by the chromaticities of the three phosphors. Fig. 3.3.4 shows for NTSC primaries that the gamut of colors enclosed by this triangle is quite large and covers the chromaticities of most of the pigments, dyes and inks.[3.3.4] Also, as Fig. 3.3.5 shows for two typical pictures, the range of chromaticities is usually very small compared to the allowable range.[3.3.5] In this figure, the brightness of the location in the CIE-*uv* diagram* indicates the logarithm of the frequency of occurrence of that particular chromaticity.

Since the output of the camera bounds the Red, Green and Blue signals (suppose, they are in the range [0,1]), it is not possible at all luminance levels to obtain all the chromaticities within the triangle formed by the receiver phosphors. As the luminance increases, the range of attainable chromaticities decreases, and for a luminance value of 1 unit, the range of chromaticities shrinks to a point centered on the reference white. Fig. 3.3.6 shows the range of achievable chromaticities for different values of luminance, assuming a linear system, i.e., gamma = 1.

Similar results are obtained for other color components. For example, Fig. 3.3.7 shows typical probability density functions of the Y, U_t and V_t signals used in the PAL system.[3.3.6] Here, just as with the components used in the NTSC system, the reduction of signal energy as a result of transforming the signal from R,G,B to Y,U_t,V_t is clearly evident.

Statistical measurements have also been made to evaluate the probability densities of spatial and temporal DPCM differential signals for color components. The PDF's of DPCM differential signals for the chrominance components are quite similar to those of the luminance components, i.e., close to being Laplacian, but with smaller variance and therefore a more rapid fall off than with the luminance signal. This is, again, a result of the smaller dynamic range of the chrominance signal for most natural scenes. Measurements also indicate that there is very little statistical correlation between the DPCM differential signals of the different components.[3.3.3,3.3.6]

However, the amplitudes of the DPCM differential signals for the color components exhibit an interesting type of spatial correlation. It is found that almost always, pel locations in the picture that represent large transitions in the chrominance signal also represent large transitions in the luminance signal.[3.3.7–8] This is equivalent to the statement that strong chrominance edges are accompanied by strong luminance edges. It has also been found that the reverse of this is not true, i.e., there are many luminance edges that are not

---

* The CIE-*uv* diagram is a chromaticity diagram like the CIE-*xy* diagram. CIE-*uv* chromaticity coordinates are more closely related to perceptual attributes of color and are discussed in detail in Chapter 4.

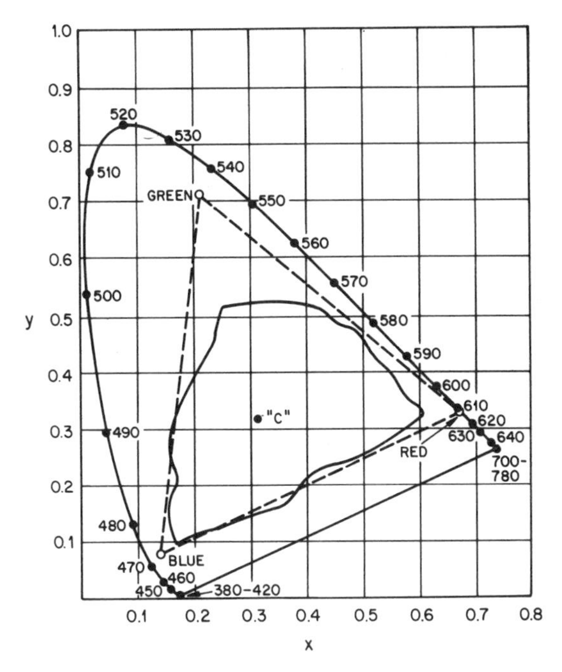

Fig. 3.3.4 CIE-xy diagram comparing the gamut of colors reproducible by the NTSC receiver (area lying within the dashed triangle) and the extreme purities of pigments, dyes and inks under Illuminant C (from Fink [3.3.4]).

Fig. 3.3.5 Log probability density plot of CIE-uv chromaticity for the picture of Fig. 3.3.1 (from Limb et al. [3.3.5]).

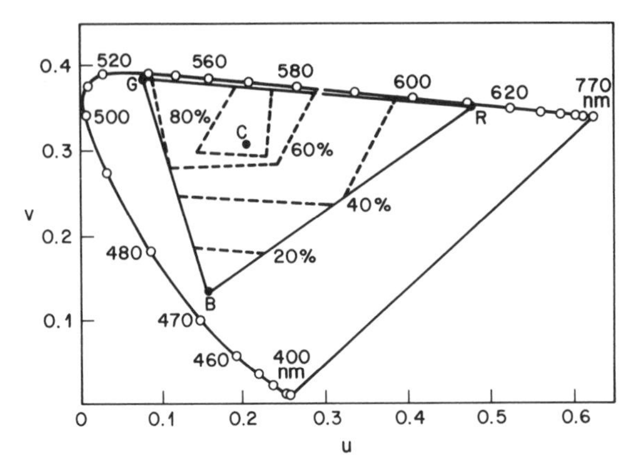

Fig. 3.3.6 The range of chromaticities possible with a television signal at a given luminance. The dashed contour lines indicate the range at 20, 40, 60 and 80 percent of maximum luminance. Below a luminance of 11 percent, all chromaticities within the triangle are achievable (from Limb et al. [3.3.5]).

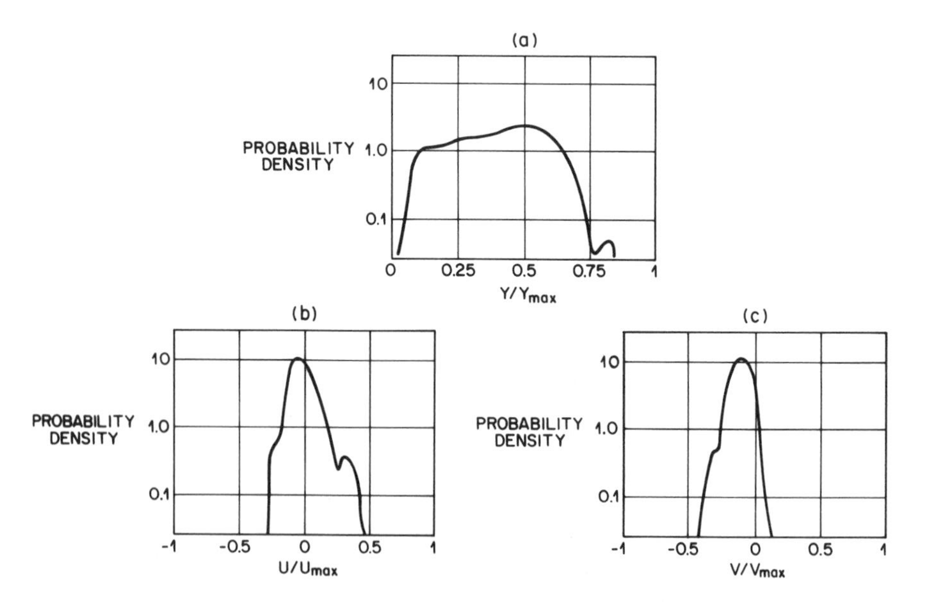

Fig. 3.3.7 Probability density functions of Y, U_t, V_t signals averaged across six different test pictures. U_t and V_t have reduced dynamic range compared to Y (from Pirsch et al. [3.3.6]).

accompanied by chrominance edges. Such correlation has been used for efficient coding of the chrominance components [see Chapter 5].

3.4 Statistics of Graphical Signals

As we mentioned in Section 2.6, graphical signals generally require very high spatial resolution, but only two shades of gray, i.e., black and white. This helps images to have sharply defined edges improving the overall quality. Spatial resolutions of 4 or 8 pels per mm in both the horizontal and vertical dimensions is common. Because of this, statistical properties of graphical signals and the models used for their characterization are quite different from those used for monochrome and color pictures. Here, we study three types of statistics and models that have been found useful for graphics: (1) *N*th Order Markov; (2) Run-Length and Line-Differential of Run lengths; and (3) Block.

3.4.1 *N*th Order Markov

The output of a scanner that generates a two-level graphical signal can be described mathematically as a random sequence of black (B or 0) and white (W or 1) pels with certain statistical properties. Several models can be formulated based on how the "present" pel depends statistically on the "previous" pels. If the present pel X_o depends statistically only on the horizontally previous pel X_1, as in Fig. 3.4.1, then a one dimensional first order Markov model results. In this case, the dependence is expressed mathematically by the conditional probabilities $P(X_0|X_1)$, and the conditional entropy (in bits/pel) is given by

$$H(X_0|X_1) = -\sum_{X_0}\sum_{X_1} P(X_0X_1)\cdot\log_2 P(X_0|X_1) \tag{3.4.1}$$

where the summation is over the four possible values of two adjacent pels.

Virtually all graphical signals also show two-dimensional correlation, and therefore a two-dimensional Markov model is more appropriate than a one-dimensional model. In this case, the present pel X_0 depends upon N previous pels, $X_1X_2...X_N$, as shown in Fig. 3.4.2. This dependence is expressed by the conditional probabilities $P(X_0|X_1X_2...X_N)$, and the conditional entropy (in bits/pel) is given by

$$H(X_0|X_1X_2...X_N) = -\sum_{X_0}\cdots\sum_{X_N} P(X_0X_1...X_N)\cdot\log_2 P(X_0|X_1...X_N) \tag{3.4.2}$$

Fig. 3.4.3 shows measurements for four documents that are scanned at a resolution of 4 pel/mm.[3.4.1] It is clear that by including more pels in the model, a better exploitation of statistical correlation is possible. Although not shown in the figure, the advantages of the two-dimensional model are even higher (in

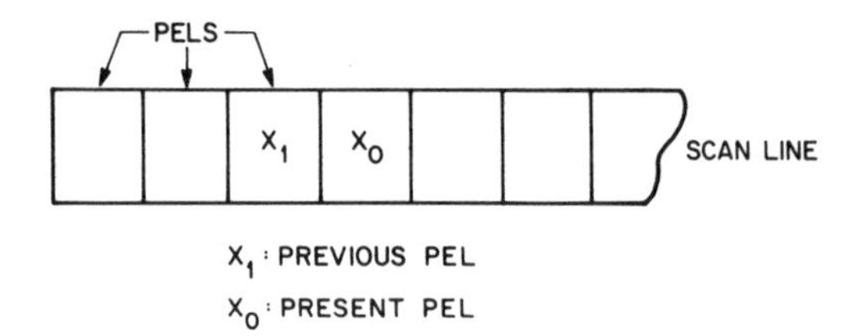

Fig. 3.4.1 One dimensional Markov Model. The color of Pel X_o depends statistically only on the color of pel X_1.

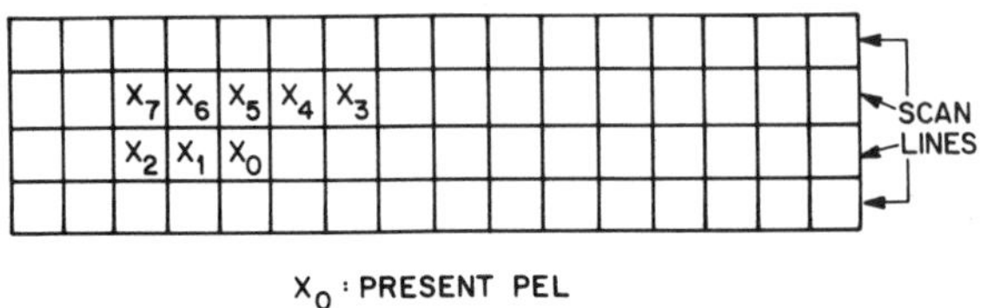

Fig. 3.4.2 Two dimensional Markov Model. Pel X_0 depends statistically only on pel $X_1, X_2, ... X_n$.

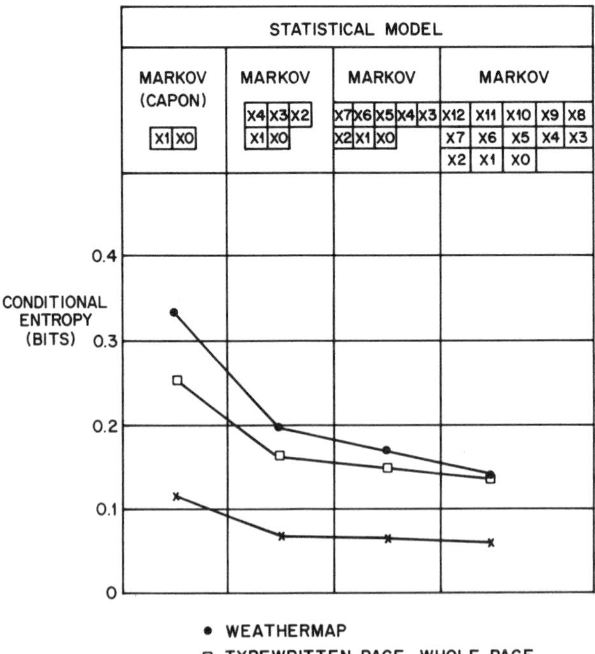

Fig. 3.4.3 Conditional entropies for different statistical models (from Preuss [3.4.1]).

bits/pel) if documents are sampled at 8 pel/mm. However, the bits per document generally increase with the sampling density.

3.4.2 Run-Length Model

A special case of the one-dimensional Markov model is the *run-length model*.[3.4.2] A sequence of pels of the same color along a scanning line is called a *run*. Obviously, such runs alternate in color. For a document containing black information on white background, runs of black pels are much shorter than those of white pels. The sparser the document, the larger is the average run length. Thus, documents containing engineering drawings have larger average white run length compared to a document containing densely-typed, single-spaced text.

Statistics of the run lengths can be calculated for the one-dimensional Markov model. Let the probability that a white run has length r be denoted by $P_W(r)$. For such a run the first white pel is followed by $(r-1)$ successive white pels followed by a transition from a white pel to a black pel. Thus,

$$P_W(r) = [P(W/W)]^{r-1} \cdot P(B/W) \tag{3.4.3}$$

where $W \triangleq$ white, $B \triangleq$ black, and $P(W/W)$ and $P(B/W)$ are the Markov conditional probabilities. Similarly, the black run-length distribution is given by

$$P_B(r) = [P(B/B)]^{r-1} \cdot P(W/B) \tag{3.4.4}$$

The average run lengths are then given by

$$\bar{r}_W = \sum_{r=1}^{\infty} r \cdot P(W/W)^{r-1} \cdot P(B/W) = \frac{1}{P(B/W)} \tag{3.4.5}$$

$$\bar{r}_B = \sum_{r=1}^{\infty} r \cdot P(B/B)^{r-1} \cdot P(W/B) = \frac{1}{P(W/B)} \tag{3.4.6}$$

The entropies of the black and white run lengths (in bits/run) are given by

$$H_W = -\sum_{r=1}^{\infty} P_W(r) \cdot \log_2 P_W(r) = -\log_2 \left[\frac{P(B/W)}{P(W/W)}\right] - \frac{\log_2(P(W/W))}{P(B/W)} \tag{3.4.7}$$

and

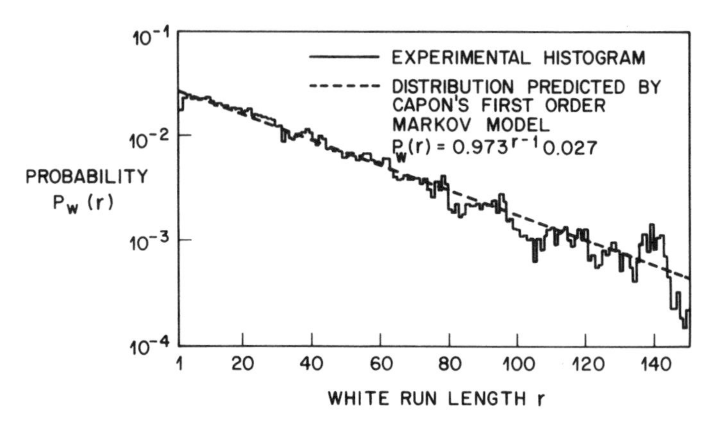

Fig. 3.4.4 Theoretical and experimentally measured white run-length distributions for a weather map. The agreement between the model and the experimental data is quite good (from Kunt [3.4.4]).

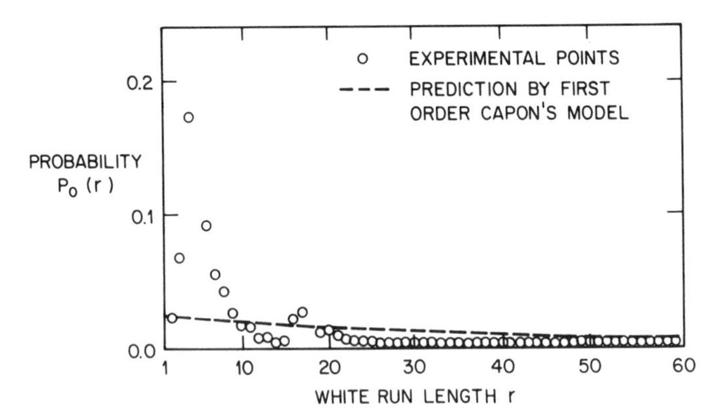

Fig. 3.4.5 Theoretical and experimentally measured white run-length distributions for a typewritten page. Unlike the data for weather maps (Fig. 3.4.4), the agreement is not good (from Kunt [3.4.4]).

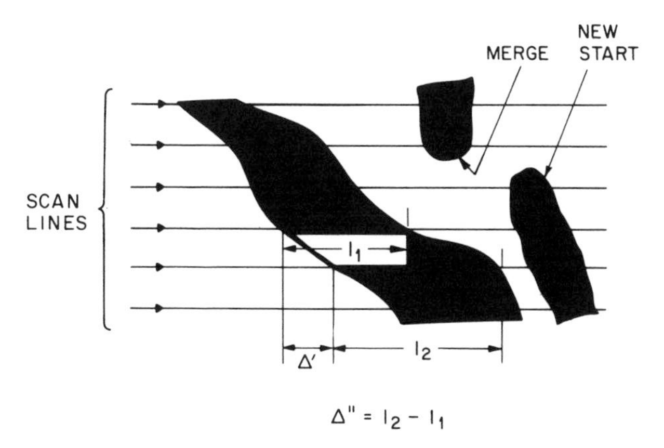

Fig. 3.4.6 Illustration of similarities between runs of successive lines. The quantity Δ' indicates the direction of a line, while Δ'' shows how the run width changes from line to line (from Huang et al. [3.4.7]).

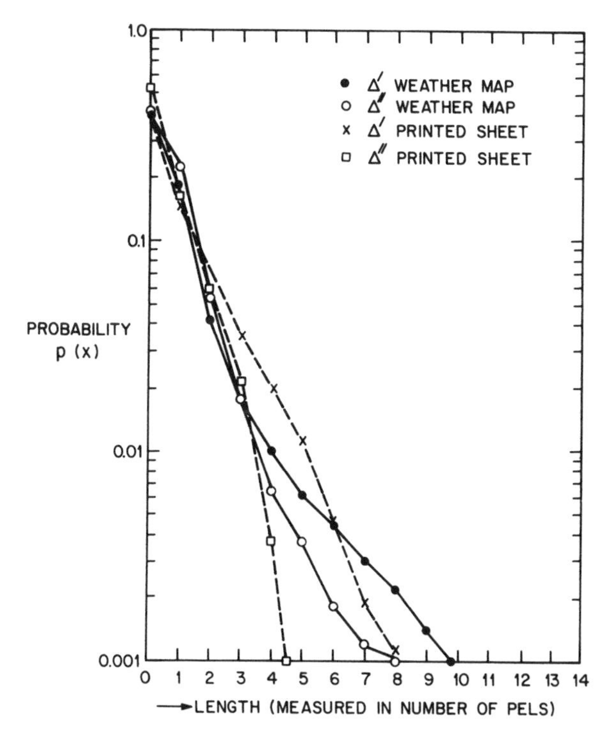

Fig. 3.4.7 Experimentally measured Probability Distribution of line differentials of starting positions of runs (Δ') and run lengths (Δ'') (from Huang et al. [3.4.7]).

$$H_B = -\sum_{r=1}^{\infty} P_B(r) \log_2 P_B(r) = -\log_2 \left[\frac{P(W/B)}{P(W/W)}\right] - \frac{\log_2 P(B/B)}{P(W/B)} \tag{3.4.8}$$

The above entropies can be converted to a more traditional measure of entropy, namely bits/pel by

$$H = \frac{H_W + H_B}{\bar{r}_W + \bar{r}_B} \text{ bits/pel} \tag{3.4.9}$$

since the number of black and white runs in a document are practically equal. This model performs reasonably well for most pictures that do not contain any particular structure. For example, Fig. 3.4.4 shows run-length distributions both experimentally measured as well as computed by the above model. It is easy to see that the agreement is reasonably good.

Structured pictures, such as typed text, have certain regular geometrical characteristics, e.g., character size, spacing between character, words, margins, etc. These have a strong effect on the run-length probability distribution. For example, Fig. 3.4.5 shows that for such pictures there is considerable disagreement between the experimentally measured distribution and that predicted by the above model. Alternative models exist that better predict the experimental data.[3.4.3–4]

The run-length statistics described above were derived from the one-dimensional Markov model. Considerable two-dimensional correlation also exists that can be exploited by considering similarities between corresponding run lengths of successive scan lines. One measure of similarity of runs from one line to the next is the shift of runs (Δ',Δ''). Fig. 3.4.6 shows the definition of quantities Δ' and Δ'', that are relevant for coding schemes based on run-length similarities (see Chapter 5). Fig. 3.4.7 shows experimentally measured probability distributions for Δ' and Δ'' for two pictures. It is clear that considerable correlation between starting positions of runs exists from line-to-line. Also, from the distribution of Δ'' it can be concluded that similar correlation exists from line-to-line in the terminating positions of runs.

Much of the experimental statistics depends upon the spatial resolution with which the pictures are scanned. Fig. 3.4.8 shows the effect of resolution on run-lengths as well as the line differentials of runs. It is seen that the entropy in bits per run increases more or less logarithmically with resolution, while the entropy of Δ' remains approximately constant. A generalization can be made from Fig. 3.4.8 regarding the increase of bit-rate (or entropy) as a function of resolution. Coding schemes become more efficient in terms of compression

PICTURE	RUN-LENGTH CODING			LINE DIFFERENTIALS		
	AVE RUN-LENGTH (IN PELS)	ENTROPY PER RUN (BITS)	ENTROPY PER PEL (BITS)	ENTROPY PER Δ' (BITS)	ENTROPY PER Δ" (BITS)	ENTROPY PER PEL (BITS)
WEATHER MAP						
33 LINES PER INCH	13	4.5	0.35	2.88	2.65	0.35
50	12	4.6	0.38	2.82	3.0	0.36
100	15.5	5.0	0.33	2.57	2.92	0.24
200	27	5.6	0.20	2.55	2.9	0.12
300	43	6	0.14	2.52	2.88	0.08
PRINTED SHEET						
6.25 LINES/ LETTER HEIGHT	10.4	3.0	0.29	2.3	3.0	0.49
8.3	9.0	2.92	0.33	2.06	2.9	0.52
12.5	10	3.3	0.34	1.87	2.97	0.43
25	18	4	0.26	1.94	3.15	0.23
50	30	5	0.18	2.07	3.07	0.12

Fig. 3.4.8 Dependence of statistics of run lengths and line differential on resolution.

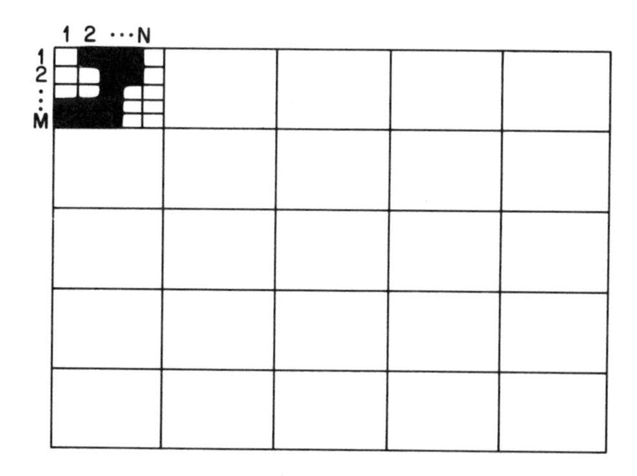

Fig. 3.4.9 The picture is divided into 5×5 blocks, each with $M \times N$ pels. Since each block contains NM pels, there are 2^{NM} types of blocks.

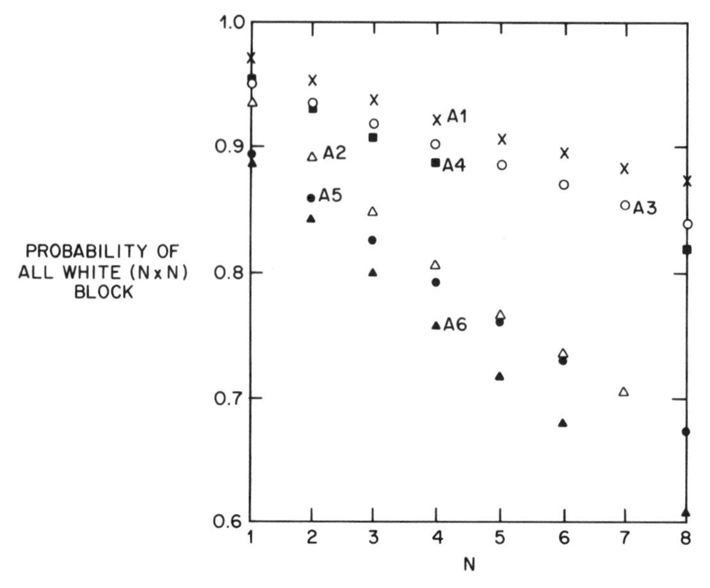

Fig. 3.4.10 Probability of all white square blocks as a function of block size $N \times N$ for six different documents ($A_1 A_2 ... A_6$ in Fig. 2.6.6) at a resolution of 5 pels/mm, containing drawing, text and handwritten material (from Kunt et al.[3.4.5]).

ratio, i.e., bits/pel as resolution is increased. However, as resolution is increased equally in horizontal and vertical directions, the total number of pels increases as the square of the resolution. The net effect of these two factors, in general, is that the compressed number of bits per picture increases more or less linearly with resolution.

3.4.3 Block Models

In the block models, a picture may be viewed as a set of juxtaposed rectangular blocks of size N pels horizontally by M pels vertically (Fig. 3.4.9). Since each pel can be either black or white, the number of different possibilities for blocks is 2^{NM}. The probability of occurrence of each one of these blocks can be experimentally measured, and it is found that the all-white block (assuming the document is black information on a white background) occurs most frequently. Many coding schemes are based on this fact.[3.4.5–6]

Fig. 3.4.10 shows the experimentally measured probabilities of the all white block, which as expected, occurs most frequently. However, its frequency of occurrence drops as the block size is increased. Although the probabilities of other blocks are not shown in this figure, they can be measured and used to compute the entropy as a function of block size (Fig. 3.4.11). It is clear that as block size is increased, the entropy decreases. However, the drop becomes less significant as the block size becomes larger than 4 by 4 (at a resolution of approximately 4 pels/mm). It should be observed that these entropies are higher than those for other two-dimensional coding schemes such as line-to-line differences of the runs (or differential run lengths). Thus statistics favor coding schemes based on line differences of runs, and these are discussed in detail in Chapter 5.

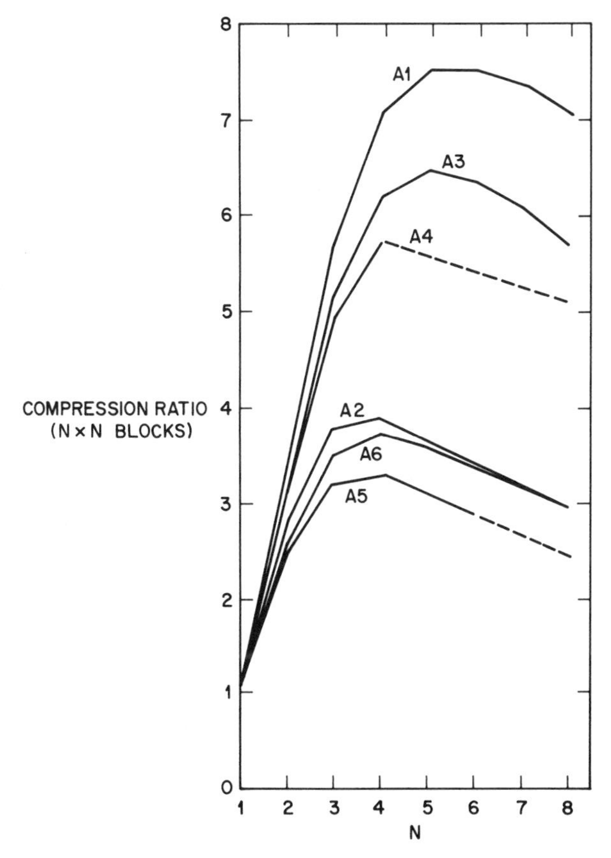

Fig. 3.4.11 Experimentally measured compression ratios (1/entropy) as a function of N for square blocks of size $N \times N$ for six documents ($A_1 A_2 ... A_6$).

Appendix

We will now show that of all possible linear transforms operating on length-N pel vectors having stationary statistics, the KLT minimizes the MSE due to truncation.[3.1.5] Let $\boldsymbol{U}$ be some other linear transform having unit length basis vectors $\{\boldsymbol{u}_m\}$. First we show heuristically that the truncation error is minimum if the basis vectors $\{\boldsymbol{u}_m\}$ are orthogonalized, using for example the well known Gram-Schmidt procedure. Let S_T be the vector space spanned by the "transmitted" vectors

$$\{\boldsymbol{u}_m, \quad 1 \le m \le pN\} \tag{3A.1}$$

and let S_E be the space spanned by the remaining vectors. Then, as shown in Fig. 3A.1 any vector $\boldsymbol{P}$ can be represented by a vector $\boldsymbol{P}_T \in \boldsymbol{S}_T$ plus a vector $\boldsymbol{P}_E \; \varepsilon \; \boldsymbol{S}_E$. If only $\boldsymbol{P}_T$ is transmitted, then the MSE is the squared length of the error vector $\boldsymbol{P}_E$, which is clearly minimized if $\boldsymbol{P}_E$ is orthogonal to $\boldsymbol{P}_T$, i.e., $\boldsymbol{S}_E$ is orthogonal to $\boldsymbol{S}_T$. Further orthogonalization of the basis vectors within $\boldsymbol{S}_T$, similarly $\boldsymbol{S}_E$, does not affect the MSE.

Thus, we assume $\boldsymbol{U}$ is unitary. Its transform coefficients are then given by

$$c_n = \boldsymbol{u}_n' \boldsymbol{b}, \qquad n = 1 \ldots N \tag{3A.2}$$

The mean square truncation error is, as in Eq. (3.1.64),

$$\begin{aligned} MSE &= \frac{1}{N} \sum_{n > pN} E|c_n|^2 \\ &= \frac{1}{N} \sum_{n > pN} E\left[\boldsymbol{u}_n' \boldsymbol{b}\boldsymbol{b}' \boldsymbol{u}_n\right] \\ &= \frac{1}{N} \sum_{n > pN} \boldsymbol{u}_n' \boldsymbol{R} \boldsymbol{u}_n \end{aligned} \tag{3A.3}$$

Now suppose we represent the vectors $\{\boldsymbol{u}_n\}$ in terms of the KLT basis vectors $\{\boldsymbol{t}_m\}$, i.e.,

$$\boldsymbol{u}_n = \sum_{m=1}^{N} w_{mn} \boldsymbol{t}_m$$

where

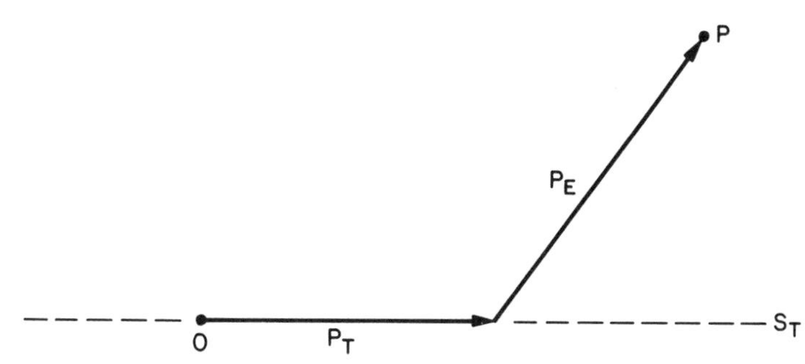

Fig. 3A.1 If S_T is the space spanned by the transmitted basis vectors, then the truncation error is the length of the error vector $\boldsymbol{P}_E$. The error is minimized if $\boldsymbol{P}_E$ is orthogonal to S_T.

$$w_{mn} = \boldsymbol{t}'_m \boldsymbol{u}_n \tag{3A.4}$$

Since $\boldsymbol{U}$ is unitary

$$|\boldsymbol{u}_n|^2 = \sum_{m=1}^{N} |w_{mn}|^2 = 1 \tag{3A.5}$$

From Eq. (3A.4)

$$w'_{mn} = \boldsymbol{u}'_n \boldsymbol{t}_m \tag{3A.6}$$

and therefore $\boldsymbol{t}_m$ can be written in terms of $\{\boldsymbol{u}_n\}$ as

$$\boldsymbol{t}_m = \sum_{n=1}^{N} w'_{mn} \boldsymbol{u}_n \tag{3A.7}$$

and since the KLT is unitary

$$|\boldsymbol{t}_m|^2 = \sum_{n=1}^{N} |w_{mn}|^2 = 1 \tag{3A.8}$$

We may now write

$$\boldsymbol{u}'_n \boldsymbol{R} \boldsymbol{u}_n = \boldsymbol{u}'_n \ \boldsymbol{R} \sum_{m=1}^{N} w_{mn} \boldsymbol{t}_m$$

$$= \boldsymbol{u}'_n \sum_{m=1}^{N} w_{mn} \lambda_m \boldsymbol{t}_m \tag{3A.9}$$

which from Eq. (3A.6)

$$= \sum_{m=1}^{N} |w_{mn}|^2 \lambda_m \tag{3A.10}$$

which from Eq. (3A.5)

$$= \lambda_{pN} + \sum_{m=1}^{N} |w_{mn}|^2 (\lambda_m - \lambda_{pN}) \tag{3A.11}$$

$$= \lambda_{pN} + \sum_{m=1}^{pN} \cdots + \sum_{m>pN} \cdots$$

$$\geq \lambda_{pN} + \sum_{m>pN} |w_{mn}|^2 (\lambda_m - \lambda_{pN}) \tag{3A.12}$$

since $\lambda_m \geq \lambda_{pN}$ for $m \leq pN$. Thus, from Eq. (3A.3) we have for the $\boldsymbol{U}$ transform

$$MSE \geq \frac{1}{N} \sum_{n>pN}^{N} \left[\lambda_{pN} - \sum_{m>pN} |w_{mn}|^2 (\lambda_{pN} - \lambda_m) \right] \tag{3A.13}$$

$$= \frac{N-pN}{N} \lambda_{pN} - \frac{1}{N} \sum_{m>pN} (\lambda_{pN} - \lambda_m) \sum_{n>pN} |w_{mn}|^2$$

The second summation is ≤ 1 from Eq. (3A.8), and since $(\lambda_{pN} - \lambda_m) \geq 0$ for $m > pN$, we have

$$MSE \geq (1-p)\lambda_{pN} - \frac{1}{N} \sum_{m>pN} (\lambda_{pN} - \lambda_m)$$

$$= \frac{1}{N} \sum_{m>pN} \lambda_m = MSE \text{ for KLT} \tag{3A.14}$$

Thus, the MSE for any linear transform U exceeds that of the KLT.

References

3.1.1 R. G. Gallager, *Information Theory and Reliable Communication,* John Wiley and Sons, Inc., New York, 1968.

3.1.2 D. A. Huffman, "A Method for the Construction of Minimum Redundancy Codes," *Proceedings IRE,* v. 40 (1962), pp. 1098-1101.

3.1.3 T. A. Welch, "A Technique for High-Performance Data Compression", *IEEE Computer,* v. 17, June 1984, pp. 8-19.

3.1.4 C. C. Cutler, "Differential Quantization of Communication Signals," U. S. Patent 2,605,361, July 1952.

3.1.5 J. L. Brown, Jr., "Mean Square Truncation Error in Series Expansions of Random Functions", J. Soc. Indust. Appl. Math., v. 8, March 1960, pp. 28-32.

3.1.6 P. A. Wintz, "Transform Picture Coding," *Proceedings IEEE,* v. 60, July 1972, pp. 809-820.

3.1.7 C. E. Shannon, "A Mathematical Theory of Communication," *Bell Syst. Tech. J.,* v. 27 (1948), pp. 379-423 and 623-656. Also, "Coding Theorems for a Discrete Source with a Fidelity Criterion", IRE Nat. Conv. Record, part 4, pp. 142-163.

3.1.8 T. Berger, *Rate Distortion Theory— A Mathematical Basis for Data Compression,* Prentice-Hall, Englewood Cliffs, New Jersey, 1971.

3.1.9 B. G. Haskell, "Computation and Bounding of Rate-Distortion Functions", *IEEE Trans. Information Theory,* v. IT-15, Sept. 1969, pp. 525-531. Also, R. E. Blahut, "Computation of Channel Capacity and Rate-Distortion Functions," *IEEE Trans. Information Theory,* v. IT-18, July 1972, pp. 460-473.

3.1.10 J. K. Yan and D. J. Sakrison, "Encoding of Images Based on a Two-Component Source Model," *IEEE Trans. Communications,* v. COM-25, Nov. 1977, pp. 1315-1322.

3.1.11 T. C. Bell, J. G. Cleary and I. H. Witten, *Text Compression*, Prentice Hall, 1990.

3.1.12 N. Abramson, *Information Theory and Coding*, pp. 61-62, McGraw Hill, 1963.

3.2.1 B. G. Haskell, F. W. Mounts and J. C. Candy, "Interframe Coding of Videotelephone Pictures," *Proc. IEEE,* v. 60, July 1972, pp. 971-800.

3.2.2 H. C. Andrews and W. K. Pratt, "Fourier Transform Coding of Images," Hawaii Int. Conf. on System Sciences, Jan. 1968, pp. 677-679.

3.2.3 N. Ahmed and K. R. Rao, *Orthogonal Transforms for Digital Signal Processing,* Springer-Verlag, New York, 1975.

3.2.4 A. Habibi, "Hybrid Coding of Pictorial Data," *IEEE Trans. Communications*, v. COM-22, May 1974, pp. 614-624.

3.2.5 R. J. Clarke, *Transform Coding of Images,* Academic Press, London, 1985.

3.2.6 A. K. Jain, "Advances in Mathematical Models for Image Processing," *Proc. IEEE,* v. 69, May 1981, pp. 502-528.

3.2.7 J. Max, "Quantization for Minimum Distortion," *IRE Trans. Information Theory,* v. IT-6, Mar. 1960, pp. 7-12. Also, R. Lloyd, "Least Squares Quantization in PCM," *IEEE Trans. Info. Theory,* March 1982, pp. 129-137.

3.2.7a T. Berger, "Optimum Quantizers and Permutation Codes," *IEEE Trans. Information Theory* v. IT-18, no. 6, Nov. 1972, pp. 759-765.

3.2.8 D. J. Sakrison, "Image Coding Applications of Vision Models," Chapter 2 of *Image Transmission Techniques,* W. K. Pratt ed., Academic Press, New York 1979.

3.2.9 A. K. Jain, "Image Data Compression: A Review," *Proc. IEEE,* v. 69, no. 3, March 1981, pp. 349-389.

3.2.10 J. K. Yan and D. J. Sakrison, "Encoding of Images Based on a Two-Component Source Model," *IEEE Trans. Communications,* v. COM-25, no. 11, November 1977, pp. 1315-1322.

3.2.11 W. K. Pratt et al., "Applications of Stochastic Texture Field Models to Image Processing," *Proc. IEEE,* v. 69, no. 3, March 1981, pp. 542-551.

3.3.1 W. Frei and P. A. Jaeger, "Some Basic Considerations for the Source Coding of Color Pictures," Proc. of International Conference on Communications, Seattle, Wash. 1973, pp. 48.26-48.29.

3.3.2 W. K. Pratt, "Spatial Transform Coding of Color Images," *IEEE Trans. Communications Technology,* v. COM-19, Dec. 1971, pp. 980-992.

3.3.3 C. B. Rubinstein and J. O. Limb, "Statistical Dependence Between Components of a Differentially Quantized Color Signals," *IEEE Trans. on Communications Technology,* v. COM-20, October 1972, pp. 890-899.

3.3.4 D. G. Fink (Editor), Color Television Standards: Selected Papers and Records of the National Television Systems Committee, McGraw-Hill, 1955.

3.3.5 J. O. Limb, C. B. Rubinstein, and J. E. Thompson, "Digital Coding of Color Video Signals - A Review," *IEEE Trans. on Communications,* v. COM-25, Nov. 1977, pp. 1349-1385.

3.3.6 P. Pirsch and L. Stenger, "Statistical Analysis and Coding of Color Video Signals," *Acta Electronica,* Vol. 19, 1976, pp. 277-287.

3.3.7 J. O. Limb and C. B. Rubinstein, "Plateau Coding of the Chrominance Components of Color Picture Signals," *IEEE Trans. on Communications,* v. COM-22, July 1974, pp. 812-820.

3.3.8 A. N. Netravali and C. B. Rubinstein, "Luminance Adaptive Coding of Chrominance Signals," *IEEE Trans. on Communications,* v. COM-27, April 1979, pp. 703-710.

3.3.9 W. Frei and P. A. Jaeger, "Some Basic Considerations for the Source Coding of Color Pictures," Proc. Int. Conf. Commun., Seattle, Wash., June 11-13, 1973, pp. 48.26-48.29.

3.4.1 D. Preuss, "Two Dimensional Facsimile Source Coding Based on a Markov Model," NTZ, Vol. 28, Oct. 1975, pp. 358-363.

3.4.2 J. Capon, "A Probabilistic Model for Run-Length Coding of Pictures," IRE Trans. on Information Theory, Vol. IT-5, 1959, pp. 157-163.

3.4.3 R. B. Arps, "The Statistical Dependence of Run-Lengths in Printed Matter," Nachr.-tech Z. 28 (1975) H.10, S.358-363.

3.4.4 M. Kunt, "Statistical Models and Information Measurements for Two-Level Digital Facsimile," Information and Control, Vol. 33, April 1977, pp. 333-350.

3.4.5 M. Kunt and O. Johnsen, "Block Coding of Graphics: A Tutorial Review," *Proc. of IEEE,* Vol. 68, July 1980, pp. 770-786.

3.4.6 T. S. Huang, A. B. Shahid-Hussain, "Facsimile Coding by Skipping White," *IEEE Trans. on Communications,* Vol. COM-23, 1975, pp. 1452-1460.

3.4.7 T. S. Huang and O. J. Tretiak, eds., *Picture Bandwidth Compression,* Gordon and Breach, 1972.

Questions for Understanding

3.1 How large can the entropy be for a random variable having 1000 possible values?

3.2 Construct a Huffman code for a random variable having six values with probabilities 0.1, 0.2, 0.25, 0.15, 0.3 and 0. How close is its average length to the entropy?

3.3 What are the relative advantages and disadvantages of block, conditional, state-space, transform and predictive coding?

3.4 How is a correlation matrix used in predictive coding?

3.5 Show that the DCT basis vectors are orthonormal.

3.6 What are the advantages and disadvantages of the KLT?

3.7 Explain statistical and subjective redundancy.

3.8 Explain Channel Capacity and Rate Distortion Function.

3.9 Calculate the Shannon Lower Bound on R(D) for the source of problem 2 above, and the probability of error distortion measure. How much can the bit rate be reduced below the entropy (approximately) if an error probability of 0.1 is allowed?

3.10 Explain the advantages and disadvantages of minimum MSE Linear Predictive Coding.

3.11 Explain channel and buffer sharing.

3.12 How are correlation models used in image coding?

3.13 Derive Eq. (3.2.24).

3.14 What is usually a good approximation of MSE vs. quantization step size? Under what assumptions?

3.15 Why are the color components of Fig. 3.3.3 so highly correlated?

3.16 Why are some chromaticities not possible for a given luminance?

3.17 What is an Nth order Markov bilevel source?

3.18 Which model would be more suitable for a dithered bilevel image, run-length or Markov? Why?

3.19 Derive Eq. (3.4.9).

4

Visual Psychophysics

One of the most important objectives in the design of visual communication systems is that it only represent, transmit and display the information that the human eye can see. To transmit and display characteristics of images that a human observer cannot perceive is a waste of channel resources and display media. Therefore, we must understand how we can represent pictures economically and transmit them with the minimum accuracy required by the human eye. In this chapter we study those properties of human vision that are helpful in evaluating the quality of a coded picture and which thereby help us in optimizing the coder to achieve the lowest transmission rate for a given picture quality.

In a practical communication system, the goal of making the reproduced picture identical to the original is beset by many difficulties and is not realistic. Systematic distortions occur, for example, in representing a live scene by a television system. The contrast ratio in a natural sunlit scene (the ratio of the highest to the lowest luminance) can frequently be 200:1 or greater, whereas for most television displays, a contrast ratio of even 50:1 is difficult to obtain. Also, color television systems use three primary colors to reproduce the approximate chromaticities of the original scene (i.e. metameric matches). Television does not reproduce a scene with the same spectral distribution as the original. In addition, the scanning standards (e.g., 525 lines in NTSC) introduce loss of resolution that cannot be overcome. We seem to have accepted these approximations, even though a direct comparison between the original scene and its reproduction reveals quality degradations easily. Hopefully, in the future some of these approximations and compromises that were made based on cost considerations may be alleviated to give better picture quality (e.g., High Definition Television).

An important consideration in determining the required picture quality is the task for which the pictures are used. A photo-interpreter would probably attach greater importance to sharpness of the picture than to tonal reproduction, whereas a broadcaster would be concerned with the tones, particularly the skin-tones. Since the relationship between required picture quality and a specific task is extremely complicated, we concern ourselves mainly with an average television viewer who is performing no specific tasks related to the image structure.

We start with a discussion of psychophysical measurements and their interpretation. Next we look at some analog transmission impairments (e.g., echoes, random noise) and their effect on picture quality. Digital coding introduces distortions that are so numerous and have such varied appearance that simple categorization and quantification of these distortions (as is done for analog systems) is extremely difficult. We therefore look at the fundamentals of processing by the human visual system and derive therefrom models of picture quality. A word of caution is relevant here. Although our understanding of the human visual system has increased significantly over the past century, it has also revealed many complexities of the visual system, and we are far from developing a simple and yet powerful model for visual processing that can be used for optimizing coding systems. It is for this reason that subjective tests are often necessary to evaluate the picture quality, despite the associated time and expense. Besides allowing one to optimize the coding system for best picture quality, subjective tests also provide a means for monitoring and maintaining an appropriate grade of service when the system is in operation. Thus, waveform testing (to be discussed in the sequel) is generally used only to monitor and maintain analog television transmission channels.

4.1 Subjective Testing

The field of study concerned with finding reliable and meaningful methods of testing for quality responses of humans to different stimuli is called *Psychometrics* or *Scaling*.[4.1,4.2] Two broad measurement techniques exist. In *primary* or *explicit* measurements of picture quality, a group of subjects examine a set of pictures and make subjective decisions about their quality. In *secondary* or *implicit* measurements, characteristics of standardized waveforms are measured objectively, and the results are then converted to quality measures through previously established relations (e.g., waveform testing used in analog transmission links). Primary methods are more useful if the distortions introduced by processing are complex in appearance, as in some methods of digital coding.

Both the primary and secondary methods assume that quality can be represented on a linear, unidimensional scale. Multidimensional scaling,[4.4] which has been successfully applied in scaling speech quality, has only recently been applied to television images[4.5]. Much remains to be done in this important area. Primary subjective evaluations are based on two broad methods, *Category-Judgment* methods and *Comparison* methods.[4.3] We will discuss each of these below.

Table 4.1 Quality and Impairment Ratings Commonly Used.

(a)	(b)	(c)
5 Excellent	5 Imperceptible	3 Much better
4 Good	4 Perceptible but not annoying	2 Better
3 Fair	3 Slightly annoying	1 Slightly better
2 Poor	2 Annoying	0 Same
1 Bad	1 Very annoying	–1 Slightly worse
		–2 Worse
		–3 Much worse

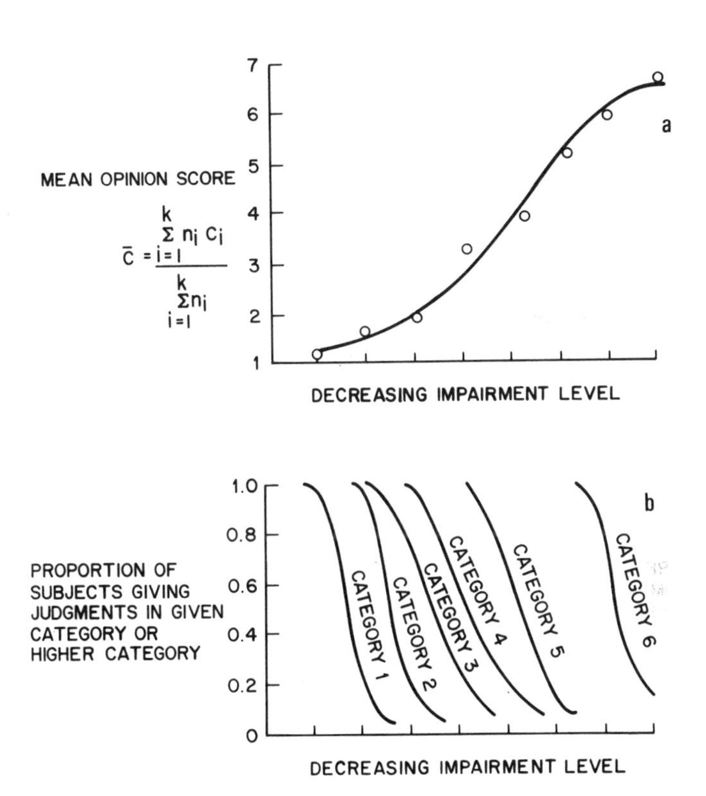

Fig. 4.1.1 (a-top) Typical plot of Mean Opinion Score $\overline{C}$ versus decreasing impairment level. (b-bottom) Cumulative method of plotting category-judgement results.

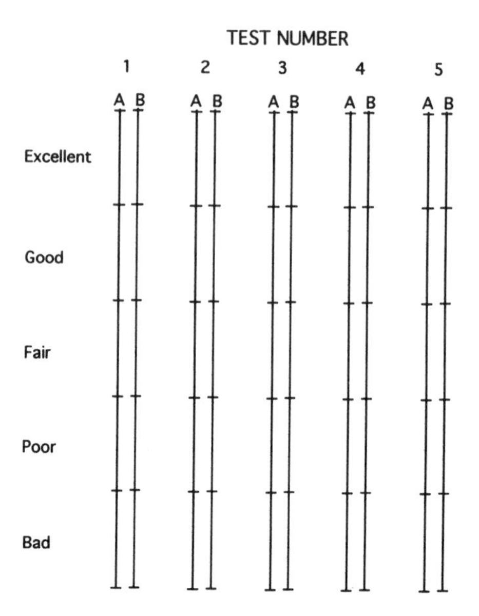

Fig. 4.1.1c Scoring sheet for Double-Stimulus Continuous-Quality-Scale test. For each test the subject writes two marks, one on vertical line A corresponding to the score for picture A, and the other on line B corresponding to the score for picture B.

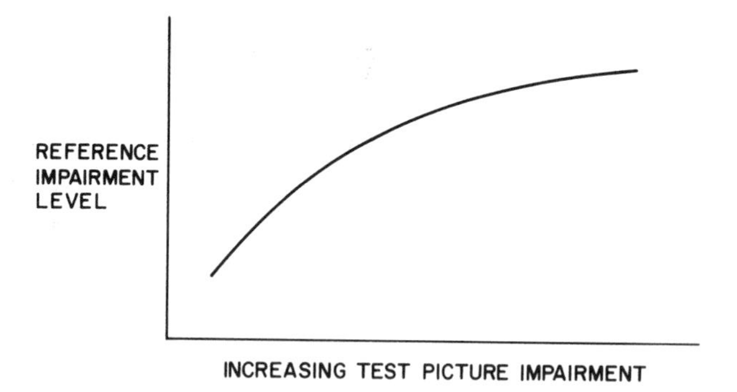

Fig. 4.1.2 Subjective equivalence of test and reference impairments obtained in a comparison test, averaged over all subjects.

4.1.1 Category-Judgment Methods

These are also referred to as *Rating-Scale* methods. Here, subjects view a sequence of pictures processed in a variety of predetermined ways under prescribed normal viewing conditions, and they assign each picture (or sequence) to one of several given categories. The categories may be based either on overall quality or on visibility of impairments as in Table 4.1 (a) and (b), respectively.

Subject response usually depends upon many factors. For example, the highlight luminance, contrast ratio, ambient room light, picture size, viewing distance, experience and motivation of the subjects, and the range of the picture material all affect the test results to a varying degree. Proper experimental design is necessary to avoid biases in the results due to such factors as the order of presentation of the processed pictures. Also, variability among subjects may be minimized by using "expert" subjects who are familiar with principles of television, particularly the visual appearance of impairments. However, experts are also more sensitive to imperfections in pictures and may not be representative of general viewers. Many of the above variables and their effect on subjective assessment have been explored in depth, and international standardization has taken place with recommendation CCIR 500.[4.6] This should make it possible to compare the results obtained by different workers at different times.

The results of the category-judgment procedure are normally presented by computing a Mean Opinion Score (MOS).

$$\text{MOS} = \frac{\sum_{i=1}^{k} n_i C_i}{\sum_{i=1}^{k} n_i} \tag{4.1.1}$$

where C_i is the numerical value corresponding to category i, n_i is the number of judgments in that category and k is the number of categories in the scale. For a scale containing seven categories, a typical MOS-plot as a function of impairment level is shown in Fig. 4.1.1(a). Category-judgment methods are popular in broadcast television, since they can be used for setting up and maintaining an appropriate grade of service.* Thus, in Fig. 4.1.1(b) the proportion of subjects giving a judgment in a given category or higher is plotted as a function of the impairment level. Impairment levels for a given grade of service can then be derived from such plots.

---

* Grade of service is a measure of customer satisfaction. Obviously, translation of MOS into a grade of service is a difficult task, since customer satisfaction also depends upon many other factors such as his past experience of watching television, what he is paying for the service and other available services.

CCIR 500 also specifies another test called the Double-Stimulus Continuous-Quality-Scale method. With this technique the subject is shown an unimpaired reference picture and a test picture in random order A/B and asked to score each on a continuous scale as shown, for example, in Fig. 4.1.1c. Many such tests using a variety of pictures, impairments and subjects produce a set of scores to which statistical analysis can be applied in order to remove biases, aberations, etc. In this way, a fairly accurate measure can be obtained of the subjective effects of image compression techniques.

4.1.2 Comparison Methods

The other method for primary subjective quality measurement is called the *comparison* method. Here, the subject compares an impaired or distorted test picture with a reference picture to which impairment of a standard type (e.g., white noise) has been added. The comparison may be on two monitors arranged side by side or on one monitor where both pictures are displayed sequentially in time. Impairment is added to the reference picture until both pictures appear to the subject to be of equal quality. The amount of this added impairment can be either under the subject's control, or alternatively pictures with variable amounts of impairments can be precomputed, stored and displayed in a given sequence.

The comparison between the test picture and the reference picture can usually be done accurately when the two types of distortions are visually similar, e.g., additive noise of different spectral characteristics. The distortion of the test picture can be assigned a quality (or category) by utilizing the previous category-judgment tests on the impaired reference pictures. The results of such tests are usually presented by a graph of the form given in Fig. 4.1.2.

A popular variation of this method is where the subject uses a comparison rating scale, e.g., Table 4.1(c), to compare test pictures having various levels of a distortion with a reference picture. The subject is then responding to the question "how much better or worse is the test picture compared to the reference picture?" The resulting data is then processed to determine the "point of subjective equality" between the distorted test and impaired reference pictures. One of the drawbacks of comparison methods is that transitivity among impairments may not always hold, especially when different impairments have very different appearance. Thus, having established separately the subjective equivalence of x amount of impairment 1 with y amount of impairment 2, and y amount of impairment 2 with z amount of impairment 3, we cannot always conclude that x amount of impairment 1 is subjectively equivalent to z amount of impairment 3.

4.1.3 Waveform Testing

Waveform testing is a secondary or implicit method of determining quality. It is used mainly as a quick test of the quality of analog point-to-point TV transmission links and consists of inserting special test waveforms (mostly

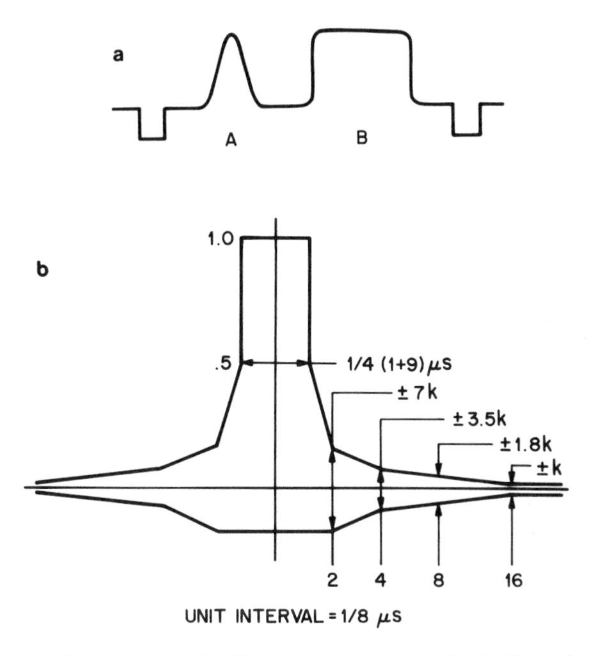

Fig. 4.1.3 (a-top) Test signal used for K-rating measurement method. Signal A is a sine-squared pulse of half-amplitude duration $2T$, where $2T$ equals the reciprocal of the system bandwidth. Signal B is a bar signal of width approximately half the duration of a horizontal scanning line. (b-bottom) K-rating graticule for the $2T$ pulse for the NTSC system.

during the field blanking period) so that in-service monitoring is possible. Test waveforms are chosen for their similarity to parts of actual picture signals.[4.7] Pulse and bar waveforms such as shown in Fig. 4.1.3 are popularly used and have a numerical channel evaluation procedure associated with them. The pulse is shown inside a tolerance window in Fig. 4.1.3. It has a sine-squared shape and a half-amplitude width of $1/W$, where W is the system bandwidth. The bar is a rectangular pulse approximately half the duration of a scan line, and shaped by the ideal transmission link. The quality of the channel is measured by how well it can keep the flat portion of the bar constant and how well it reproduces rapid changes present in the pulse. To establish a precise rating, the received pulse and bar waveforms are checked against tolerance windows. As an example, the window for the pulse is shown in Fig. 4.1.3b with all dimensions expressed as multiples of a single number K. The smallest value of K for both windows, which will still accommodate the responses, is then quoted as the K rating. Typically, a television link may have a K-rating of 0.03.

The relationship between distortion of the test waveforms (or the K-rating) and the quality of impaired pictures during actual transmissions is sometimes tenuous. Although waveform testing is a good engineering solution to a real need for quantifying analog channel quality, care is needed to confine its application to specific, well understood forms of distortion. Explicit testing methods must then be used to check results.

4.1.4 Kell Factor

Just as the K-rating is an engineering approach to specifying the quality of analog point-to-point communication links, the Kell factor is an approach to specifying imperfections in the vertical sampling process. One of the desirable criteria in specifying scanning and transmission standards is the provision for equal horizontal and vertical resolution. The basis for this comes from the fact that objects are, in general, randomly oriented in a picture, and acuity of vision is approximately equal in both the vertical and horizontal directions. In television, since the horizontal direction is continuous, and the vertical direction is scanned or sampled, this leads to the question of what is the equivalent resolution of a scanned signal compared to the continuous signal. Based on the sampling theorem, a raster containing n scan lines can only represent vertical sinusoidal components of frequencies below $n/2$ cycles per picture height. However, the apparent resolution as measured by the maximum number of resolvable black-white line pairs in a picture transmitted by television, is somewhat less because of the quantization of scan lines. This loss of vertical resolution is expressed as a numerical factor k, $0 \leq k \leq 1$, called the Kell factor. In other words, the Kell factor is that number k such that an image rendered in n scan lines is equal in vertical resolution to the original unscanned image filtered to a bandwidth of $nk/2$ cycles per picture height. Equivalently, k is the loss of vertical resolution in a scanned picture, relative to the original picture filtered to a bandwidth of $n/2$ cycles per picture height. In a well designed digital system, k is approximately equal to[†] PAR^{-1}, i.e., the ratio of the vertical to horizontal pel spacing. However, usually it is less than this ratio.

The Kell factor has also been used in predicting the maximum video frequency required to provide the same resolution in the horizontal direction as is offered by the scan line structure in the vertical direction. If a raster of n scan lines yields a resolution of $nk/2$ cycles per picture height, and if the image aspect ratio (IAR) is $R : 1$, then for equal resolution in horizontal and vertical direction, no horizontal signal components higher than $Rnk/2$ cycles per picture width* need to be reproduced. This implies that there are at most $Rn^2k/2$ cycles/frame. If the time between successive frames is T_F, then, the maximum video frequency would be

$$f_{max} = \frac{Rn^2k}{2T_F} \tag{4.1.2}$$

This formula has been used to define and measure k if f_{max} is given by some experimental or theoretical consideration. Most television systems have a Kell factor of around 0.6–0.8. Interlaced scanning results in a lower Kell factor compared to noninterlaced scanning, especially in scenes containing motion. A lower value implies a larger imperfection of the vertical spatial sampling process. Better spot shapes or higher sampling densities can raise the Kell factor. The Kell factor has not been too useful because it has many weaknesses. It is not clearly defined, resulting in many different definitions. Its value depends on many factors such as the specific television system used, viewing conditions, and scanning parameters. Also, with some of the modern sensors (e.g., CCD sensors) that employ two-dimensional sampling, the usefulness of the Kell factor has decreased. One important thing to remember concerning the Kell factor is that practical vertical resolution, regardless of how one measures it, falls short of the limit prescribed by the sampling theorem.

4.2 The Human Eye

We present in this section a functional description of the human eye as a background for constructing a phenomenological model of the visual processing consistent with physiology. The treatment is not sufficiently detailed to correlate precisely the places in the visual system with the equations of the model. A detailed description of the visual system appears in references [4.8–9]. A major

---

† Some authors actually define the Kell factor itself as $k \triangleq PAR^{-1}$.

* A picture having maximum horizontal resolution of $Rm/2$ cycles per picture width is said to have *m lines* of horizontal resolution. Thus, for equal vertical and horizontal resolution, $m \approx nk$.

source of the information to be presented is from physiological studies on vertebrates such as cats and monkeys whose visual systems are similar to ours. Since these studies are not made on the human species and since they involve measurements in limited regions of the visual system, they cannot provide a complete model of visual processing. Psychophysical experiments are necessary where responses of humans to given stimuli are measured.

Fig. 4.2.1 illustrates the principal components of the human eye. Light from an external object is focussed by the cornea and lens to form an image of the object on the retina at the back of the eye ball. Refraction takes place at the cornea and is also affected by the varying thickness of the lens. Since the eye is not a perfect optical system, a certain amount of image spreading and consequent degradation takes place at the retina. Another source of image degradation is eye movement. Of course, voluntary eye movements are necessary and enable us to track moving objects or shift our gaze from one object to another. However, involuntary eye movements also occur, even during steady fixation, and introduce a certain temporal variation to any image. These involuntary movements consist of slow drifts from the point of fixation, corrective flicks (called *saccades*) at time intervals of about 0.3 to 0.7 seconds, as well as high frequency tremors. Although involuntary eye movements degrade the image in general, they are important in maintaining continuous visibility of the visual field, since an image that is stationary on the retina fades and eventually disappears.

A schematic diagram of the retina is shown in Fig. 4.2.2. The retina consists of a layer of photoreceptors and connecting nerve cells. The photoreceptors are at the point of the layer that is farthest from the center of the eyeball, and therefore light energy must pass through the layer of nerve cells before reaching the receptors. The receptors contain photosensitive pigments that are capable of absorbing light and initiating the neural response.

The photoreceptors are of two kinds: rods and cones. In the region surrounding the fovea, only cones are present and they are densely packed. The density of cones decreases rapidly as we move away from the fovea, whereas the density of rods increases. Cones are responsible for spatial acuity and color vision at normal daylight levels (called the *photopic* range), while rods are responsible for low light vision (called the *scotopic* range). At light levels between the photopic and scotopic range, (called the *mesopic* region), both cones and rods provide vision. Light absorbed by the receptors initiates chemical reactions that bleach the photosensitive pigment, which reduces the light-sensitivity in proportion to the fraction of pigment bleached. A change in ambient illumination causes the amount of bleached pigment to rise or fall to a new equilibrium level, and this provides a mechanism for adapting to different light levels.

As seen from Fig. 4.2.2, photoreceptors first make an outer synaptic (chemical) contact with the bipolar cells that extend through the nerve layers of the retina closer to the lens. A second inner synapse then connects the bipolar cells to the ganglion cells. Lateral interactions also take place by means of horizontal

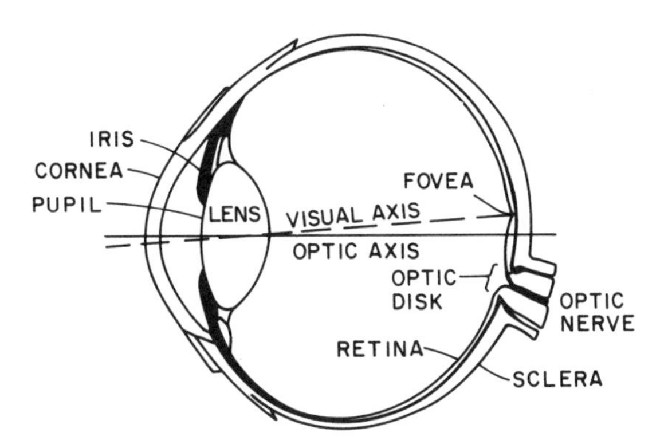

Fig. 4.2.1 Diagram of a cross-section of the human eye.

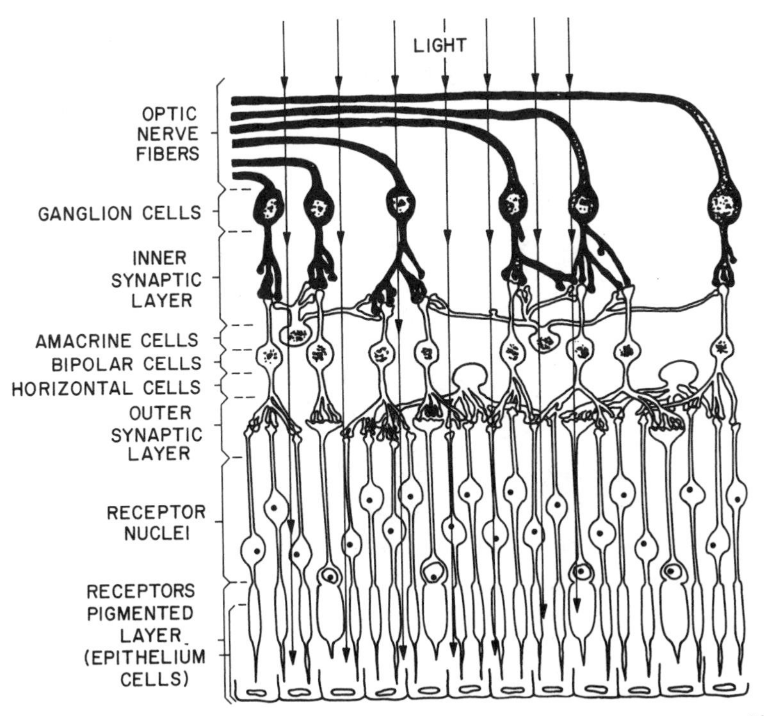

Fig. 4.2.2 Schematic diagram of the retina showing interconnections between receptors and bipolar, ganglion, horizontal and amacrine cells (from Dowling *et al.* [4.31]).

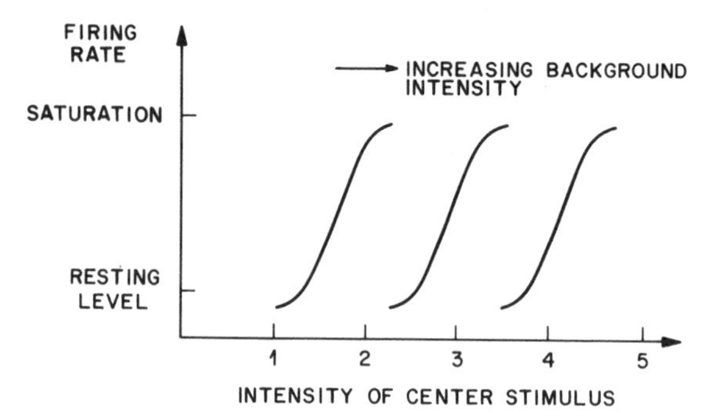

Fig. 4.2.3 Firing Rate as a function of stimulus intensity for several background intensity levels.

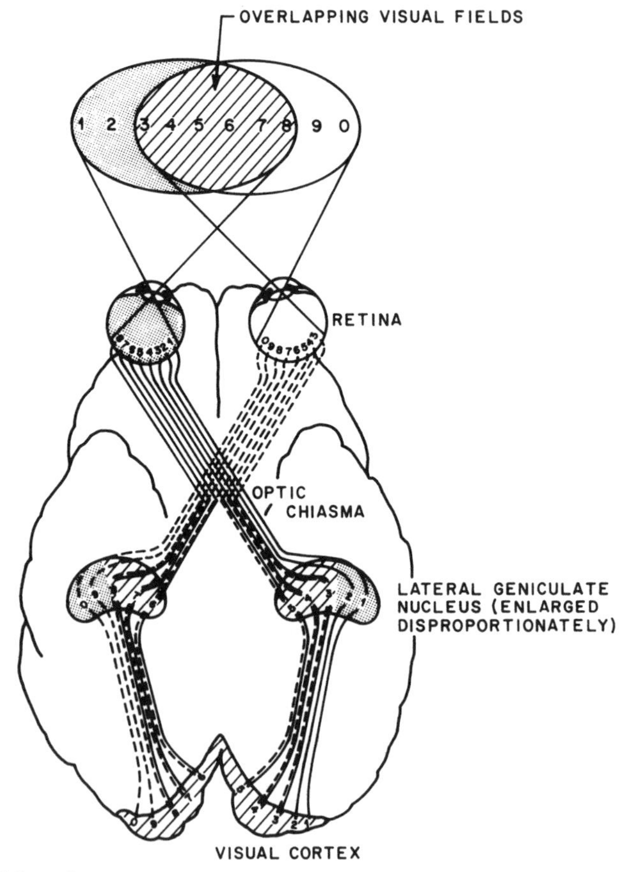

Fig. 4.2.4 Schematic representation of the visual pathways from each eye to the visual cortex, via the optic chiasma and the lateral geniculate nucleus (from Lukas [4.28]).

and amacrine cells. The axons of the ganglion cells form the fibers of the optic nerve by which the signal is transmitted to the brain. Because of this, the area around the optic nerve inside the center of the retina does not contain any photoreceptors and results in a blind spot. Lateral connections made by horizontal and amacrine cells are responsible for amplitude companding and spatial frequency preemphasis of the visual signal by mediating the sensitivity of the ganglion cells to light. This effect, called *lateral inhibition*, results in a reduction of the signal from a cell when the neighboring cells are also carrying a signal or are illuminated.

Although the signal transmission from the photoreceptors to the cells is by chemical reaction and conduction of certain chemicals, electrical potentials provide transmission within the photoreceptors, horizontal, amacrine and bipolar cells. Thus at the ganglion cells, electric potentials control the rate of generation of electrical pulses (firing) that propagate through the optic nerve. It is believed that the information coding is by the firing rate of these pulses. Fig. 4.2.3 shows the firing rate as a function of intensity of the light stimulus for three different levels of light adaptation. Two types of ganglion cells exist: those that provide a transient response, since their firing rate varies only if the light stimulus varies temporally; and those that provide a sustained response to a steady stimulus. Fig. 4.2.3 shows the effect of different background intensities on the firing rate of the ganglion cells. The firing rate varies from a threshold level (zero to tens of pulses/sec) to saturation (200–300 pulses/sec). Different ambient background light can be accommodated by the adaptation mechanism and results mostly in the shift of the active response region. In the active region, the firing rate is almost linearly related to the log of the intensity, and virtually independent of the background illumination.

The lateral connections made by the horizontal and amacrine cells result, for each ganglion cell, in receptive fields with an excitatory region in the center surrounded by the inhibitory region. Thus, a light stimulus exciting a ganglion cell raises its firing rate, but any stimulus exciting the surrounding ganglion cells inhibits its firing rate. This implies that the ganglion produces the highest response if it is excited by a light pattern with a high intensity center spot surrounded by a low intensity. This center-on surround-off pattern, called the receptive field, is generally circularly symmetric. Since the density of the receptors is the highest at and around the fovea, the size of the receptive field near the fovea is the smallest, and it increases with distance from the fovea.

Fig. 4.2.4 shows a schematic representation of the visual pathway beyond the retina. Significant spatial processing takes place beyond the retina; however, it is not clearly understood. The fibers of each optic nerve split at the optic chiasma according to the half-retina from which they originate. Fibers from the right half-retina go to the lateral geniculate nucleus (LGN) on the right side of the brain, and fibers from the left half-retina go to the LGN on the left side of the brain. At the LGN, the fibers map out the appropriate half-retina from the two eyes in a regular manner and then terminate in the appropriate area of the visual cortex. Cortical processing provides the basic functions needed for visual perception.

The above review of visual processing, although incomplete, points out the complexities of the visual system. Much of the psychophysical data to be presented below is necessary to develop models of visual processing. However, the consistency of these models with the physiological facts makes the models applicable in a wider variety of situations.

4.3 Psychophysics of Vision

In psychophysics, a systems approach to vision is taken by describing the visual system as an input-output system, with visual stimuli as input and prescribed sensations as the output. Obviously, the transfer function of such a "black box" completely defines the visual system. In applications of psychophysics to picture coding, we are most concerned with the *visibility* of coding impairments. Such stimuli are generally small, and may not be detected by the eye on every occasion. This is especially true for high-quality encoding and reproduction (as in the coding of broadcast network television). In such cases, the *visibility threshold* of a stimulus is an important quantity. It is defined as the magnitude of the stimulus at which it becomes just visible or just invisible, i.e., the probability of detection by a human viewer is 50%. Much work has been done in psychophysics to determine visibility thresholds in a variety of situations. We will review some of the work where the stimuli are somewhat pseudo random and therefore resemble coding errors in appearance.

We are also obviously interested in "how the coded pictures look" and how large is the subjective effect of visible coding distortions. However, it is difficult to specify the subjective magnitude of distortion that is clearly visible. It is much easier for subjects to specify when the distortion is barely visible. Thus, the visibility threshold can be determined reliably and consistently. However, there are many situations where, based on cost considerations only, lower quality encoding is desirable. In such cases, the subjective magnitude of the coding distortion is quite large and considerably above the visibility threshold i.e., it is *suprathreshold.* Unfortunately, psychophysical data on suprathreshold stimuli is so far quite limited.

4.3.1 Threshold Vision

In this section we shall describe briefly experiments in *threshold vision,* which are performed to evaluate the visibility thresholds of some useful stimuli. These are spatio-temporal stimuli and may be described either in the space-domain or in the frequency-domain, via the Fourier transform. There is no general agreement among psychophysicists about which domain is more suitable. Known physiological facts can be interpreted to support partially either of the two domains. We will start with the space-domain description of threshold

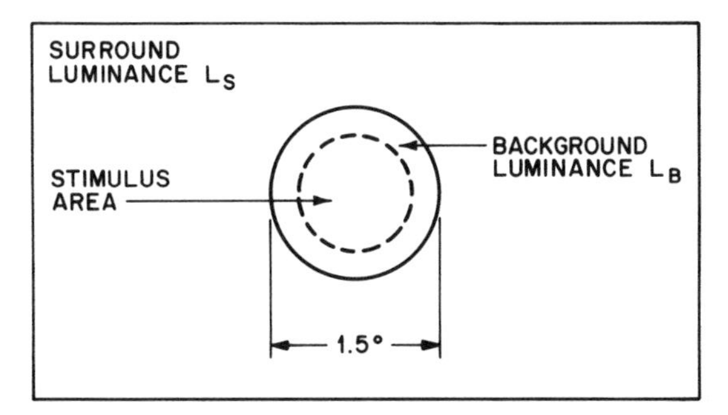

Fig. 4.3.1 Display used for experiments to validate Weber's law. Luminance of the stimulus area (inside the dotted circle) is perturbed and the visual threshold of this perturbation is determined by subjective tests.

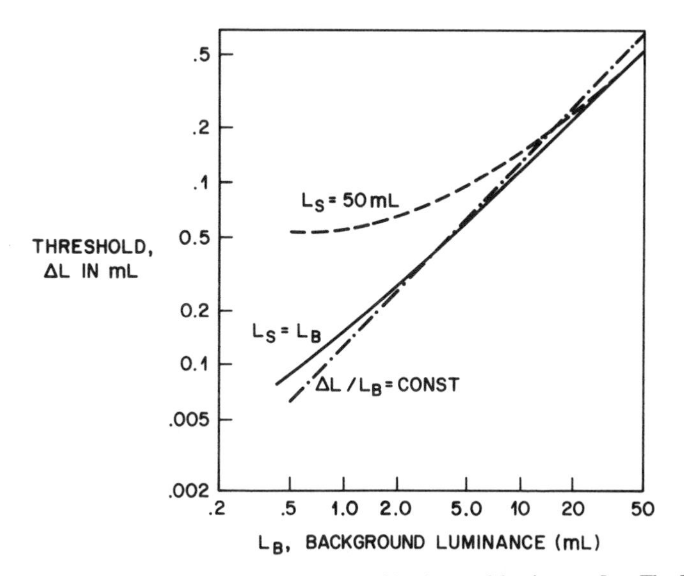

Fig. 4.3.2 Visibility threshold ΔL as a function of background luminance, L_B. The Weber law relation ($\Delta L / L_B$ = constant) is shown by the straight dash line. The short dashed curve is for surround luminance, L_s, maintained at a maximum value, while the full curve gives the result when the surround luminance is maintained equal to the background luminance (from Newell *et al.* [4.32]).

vision and then describe extensions to the frequency domain.

4.3.1a Space-Domain Threshold Vision

The visibility threshold of a particular stimulus depends on many factors. If we restrict ourselves to achromatic stimuli, the principal factors are:

(a) Average background (spatially and temporally constant) luminance level against which the stimulus is presented.

(b) Suprathreshold luminance changes (in space and time) adjacent to the test stimulus in space and time.

(c) Spatial shape and temporal variation of the stimulus.

We will consider each of these factors separately, although in reality, these factors are not independent at all. Indeed, in real-life complex images, the visibility threshold of an additive perturbation (or stimulus) would be a complicated function of all three factors. For simplicity, however, we will assume independence of these factors.

In experiments for determining the effects of the first factor, a constant *background* luminance of L_B subtends an angle of about 1.5° as shown in Fig. 4.3.1. The area outside this background region is called the *surround*, which has luminance L_S. The luminance of the dotted circled area is perturbed and subjects determine the magnitude of this perturbation, ΔL, at which it becomes just visible. ΔL is then the visibility threshold. It depends on background and surround luminances, as well as on the size of the background and surround areas. Experiments[4.10,4.11] show that the primary dependence of the visibility threshold is on the luminance of the background as shown in Fig. 4.3.2. In the special case, when $L_B = L_S$, the threshold ΔL increases almost linearly with L_B, i.e., $\Delta L / L_B$ is a constant. This is known popularly as *Weber's* law.

The *Weber fraction* $(\Delta L / L_B)$ remains approximately constant at high luminances. However, as the luminance L_B is decreased, $(\Delta L / L_B)$ begins to increase. If L_S is held at a high value and L_B is decreased, then ΔL decreases less than linearly and bottoms out at a value several times larger than for the condition $L_S = L_B$. This phenomenon is a result mainly of the smallness of the size of the background (1.5°) used in the experiment. Such a small background allows the edge between the background and the surround to contribute to the threshold (called visual masking, see Section 4.3.1b). Indeed, if the background size were reduced to 0.5°, then the Weber fraction would be even higher.

The threshold effects described above are global since the surround, background and the test stimuli span a large number of pels in the display. These effects are often used in choosing the number of quantization levels for PCM analog to digital conversion of the luminance portion of the video signal. One must exercise care, however, since the displayed luminance (L) and the electrical voltage v applied to a cathode ray tube (CRT) are nonlinearly related by a relationship of the form (called the gamma characteristic):

$$L = cv^{\gamma} + L_o \tag{4.3.1}$$

where c, L_o and γ are positive constants, and v is the voltage above cutoff. γ is in the range 2 to 3 for most CRTs (see Chapter 2). This implies that small variations ΔL are related to small variations Δv and L by

$$\Delta L = \gamma c \left(\frac{L - L_o}{c}\right)^{1-1/\gamma} \Delta v \tag{4.3.2}$$

Since γ is greater than 1, ΔL increases with L. Thus, part of the increase in ΔL with respect to L, that is considered desirable by Weber's law, is already provided by the nonlinear characteristics of the CRT.

In summary, the visibility threshold ΔL depends on many global factors. It shows a strong dependence on the neighboring background luminance, and a weaker dependence on the surrounding luminance that is further removed spatially from the stimulus. Although we have not discussed it, randomness (both spatial and temporal) of the stimulus as well as the background also has an effect. The gamma characteristics of the display tube partially compensate for the Weber's law effects. In practice, application of these effects for coding is further complicated by the fact that the ambient illumination falling on the display often plays a major role. In general, high visibility thresholds will occur in the regions of pictures that are either very dark or very bright, and lower thresholds will occur in medium to dark-gray regions.

One of the popular models for processing of visual signals in the early stages of human vision incorporates a logarithmic transformation of intensity. This has originated from the fact that light sensitive neurons fire at a rate approximately proportional to the logarithm of the incident intensity. A simple experiment that is often cited supporting this is to adjust a controllable light patch until it is just noticeably brighter or darker than a reference patch. If one were to do this by stepping through the entire gamut of intensities and plot the step numbers as a function of intensity of the reference patch, the resulting curve would be approximately logarithmic over several orders of magnitude of intensity. Such experiments indicate that there is a nonlinear transformation of intensity in the early stages of the visual system. This is followed by spatio-temporal filtering and masking, as we shall see later in this chapter.

4.3.1b Visual Masking

Most pictures contain a complex rather than a uniform luminance background. Therefore, it is important to know how the visibility threshold of a test stimulus changes when it is viewed in the vicinity of large visible (suprathreshold) spatial and temporal changes in the luminance. It is known that there is a

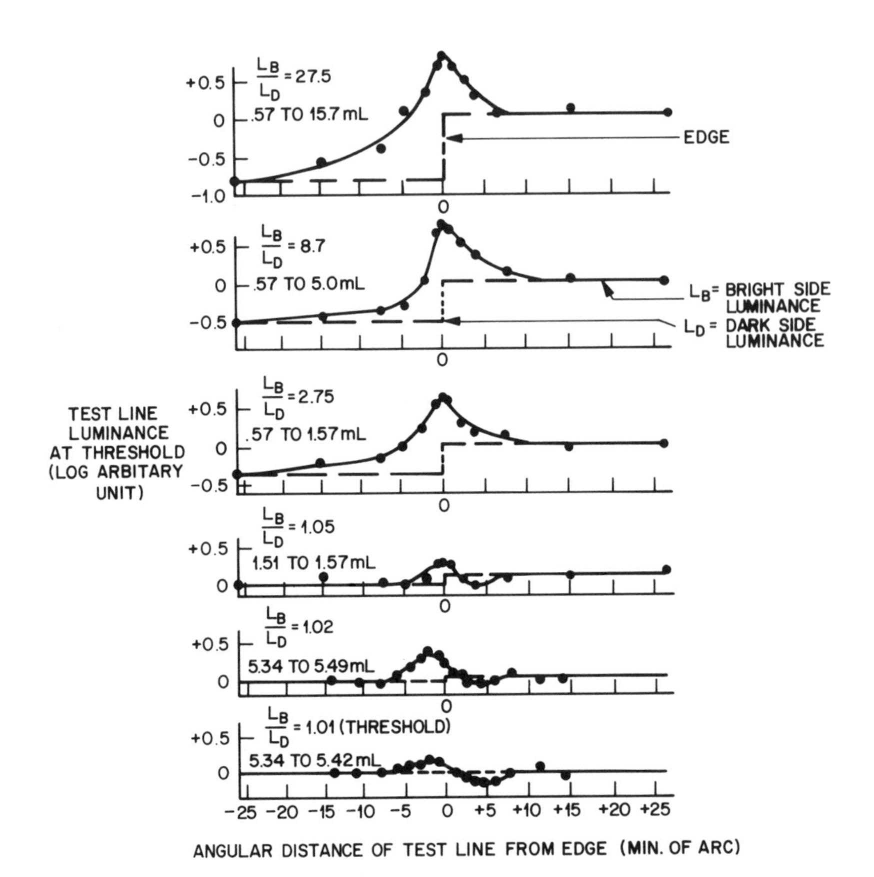

Fig. 4.3.3 Line thresholds in log units as a function of distance from luminance edges of different contrast ratios L_B/L_D (from Fiorentini, *et al.* [4.33]). The distance is measured in terms of angle subtended at the eye, and the luminance is measured in milliLamberts (mL).

reduction in the visibility of stimuli, i.e., increase in the visibility threshold, by spatial or temporal nonuniformity of the background. This is referred to as *masking* of the test stimulus by a nonuniform background. The test stimulus is generally a small, near threshold stimulus, whereas the masking pattern is well above the threshold of visibility. *Spatial masking*, i.e., reduced visibility of a test stimulus on both sides of a large change in the background luminance, has been known for some time[4.12,4.13]. Fig. 4.3.3 and Fig. 4.3.4 show results of a typical experiment in which a thin vertical line of light (stimulus) is presented at various locations relative to a continuously presented horizontal discontinuity in the luminance background, i.e., a vertical edge. The visibility threshold of the line stimulus (i.e., when the line is just barely visible) is measured and plotted as a function of distance from the vertical edge. For edges of both low and high contrast, the threshold rises as the edge is approached from either side. For edges of low contrast, the masking effect is substantially lower. Also, as shown in Fig. 4.3.4, the masking effect is very local; the typical spread is of the order of only $5'$ of subtended arc (approximately 4 picture elements at a viewing distance of 8 times the picture height for broadcast television).

Spatial masking should not be confused with the *Mach* effect.[4.14] The Mach effect refers to a change in *perceived* brightness at an edge, and as shown in Fig. 4.3.5, there is an apparent increase in the brightness on the light side and a decrease on the dark side. Thus, there is a perceived contrast enhancement of the luminance transitions, an overshoot in the bright side and an undershoot on the dark side. It is believed that the lateral interactions of the human eye (see Section 4.2) are mainly responsible for this.

Although spatial masking has been under investigation for quite some time, and despite the fact that its dependence on many factors is already known, no comprehensive model exists for it. Some of the relevant experimental results are as follows: The visibility threshold is not elevated significantly if the background masking edge is presented only briefly. This indicates that at least some of spatial masking is related to involuntary eye movements. Indeed, recent experiments show that the masking effect is considerably decreased if the image of the edge is stabilized (i.e., image is at the same position on the retina, despite eye motion) on the retina so that involuntary eye movements are eliminated. Another experiment shows that a longer length edge in the vicinity of the test stimulus tends to increase the elevation of the visibility threshold.

Spatial masking has received considerable attention in the field of picture coding. It has been recognized that a DPCM* (Differential Pulse Code Modulation) quantizer with unequal quantization steps relies on the masking of errors by

* See Chapters 3 and 5 for details.

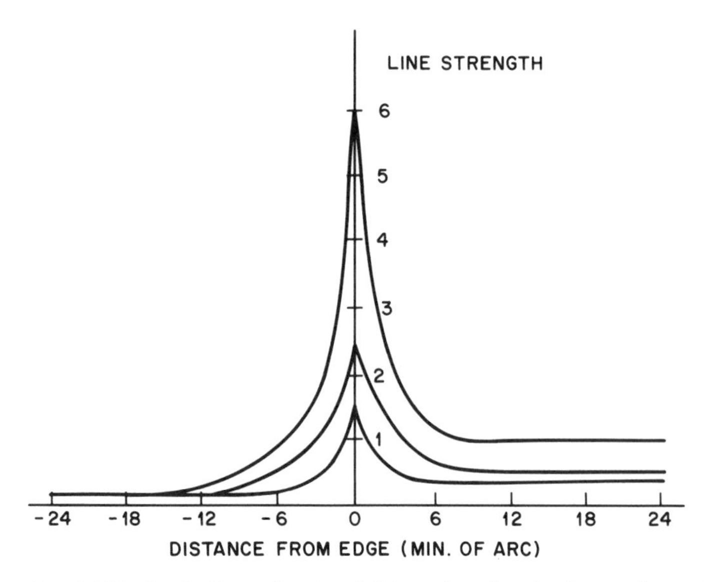

Fig. 4.3.4 Line visibility thresholds as a function of distance from three luminance edges; contrast ratios $L_B/L_D = 27.5, 8.7, 2.75$ (from Fiorentini *et al.* [4.33]).

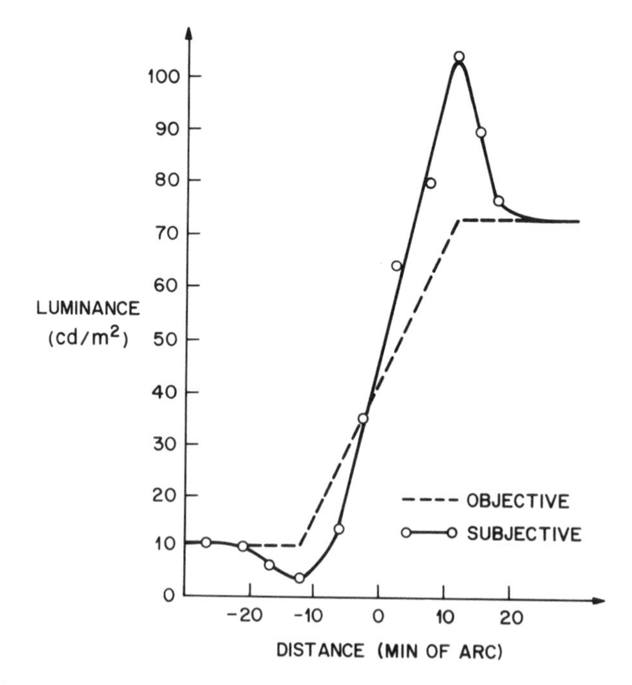

Fig. 4.3.5 Subjective brightness distribution across Mach bands at a graded luminance edge (from Lowry *et al.* [4.34]).

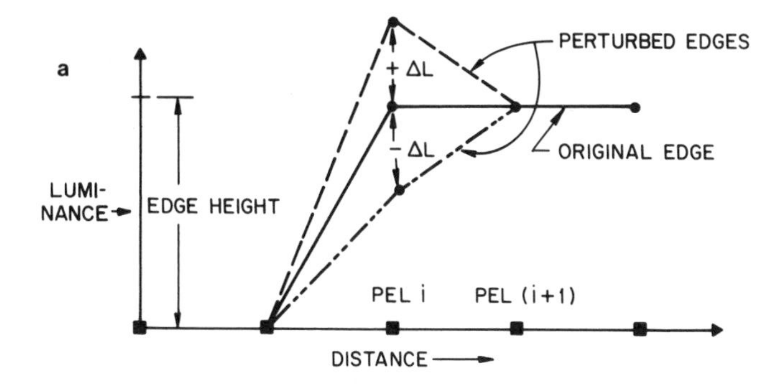

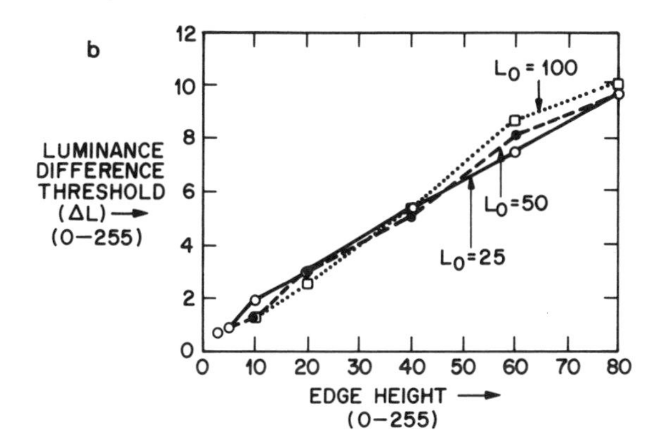

Fig. 4.3.6 (a-top) Stimuli for the determination of visibility thresholds. Angle subtended by adjacent picture elements was 2–2.5 minutes of arc. (b-bottom) Element difference visibility thresholds at different background luminance levels, L_0 (shown on a 0–255 scale).

spatial transitions in luminance.[4.15] Several psychovisual experiments[4.16] especially tuned to DPCM coding have been performed. Fig. 4.3.6 shows results of one such experiment, in which subjects were shown both a perturbed (with varying amount of perturbation ΔL) and an unperturbed edge. The perturbation was varied until it became just detectable. In this way, element difference visibility thresholds were obtained as a function of the amplitude of the edge. These results can be used to design DPCM quantizers (see Chapter 5).

Another procedure must be used for evaluating the effect of masking on suprathreshold perturbations. This procedure[4.16] is based on the comparison methods. It attempts to measure the subjective magnitude of the test stimulus (e.g., quantization noise) when it is above threshold. Here, at every picture element (called the central pel) a *spatial activity function*, consisting of weighted sums of the magnitudes of the horizontal and vertical gradients of luminance at several surrounding picture elements, is evaluated. The weights are a decreasing function of the distance between the neighboring pel and the central pel. Test conditions are first set up by adding random noise (of a known power) only to picture elements where the magnitude of the spatial activity lies in a certain narrow range. The subjective value of this noise is then determined by comparison with a reference picture in which white noise is added over the whole picture and varied in power until it appears equal in quality to the test picture.

The *noise visibility function* $f(x)$, is then defined as the ratio of the white noise power in the subjectively equivalent reference picture to the power of the noise that is added to the test picture only at pels where the spatial activity lies in an incremental range around x. An example of the noise visibility function, when the spatial activity function is the horizontal slope of the luminance signal (i.e., the magnitude of the element difference), is shown in Fig. 4.3.7. Although, in principle, noise visibility functions so determined apply only to the specific pictures used in the subjective tests, they appear to be useful for any other picture in the same general class (e.g., head and shoulders view). Noise visibility functions decrease with respect to their argument. This is a result of two factors: the decrease in noise visibility near spatial detail, and fewer elements having high spatial detail in most pictures. Fig. 4.3.7 shows both the noise visibility function and the probability of occurrence of pels with given horizontal slope. So far, it has not been possible to derive a relationship between the two that would be valid for any picture and not just for the picture used in tests.

4.3.1c Temporal Vision

Temporal masking and perception of temporally changing stimuli are extremely important in interframe coding [see Chapter 5]. However, temporal masking is complicated by at least two facts: 1) television cameras integrate the image of any object on the target and, thus, there is motion-related blurring and resolution loss; 2) perception of a moving object depends heavily on whether or not the object is tracked by the eye. The psychophysical literature[4.17] contains

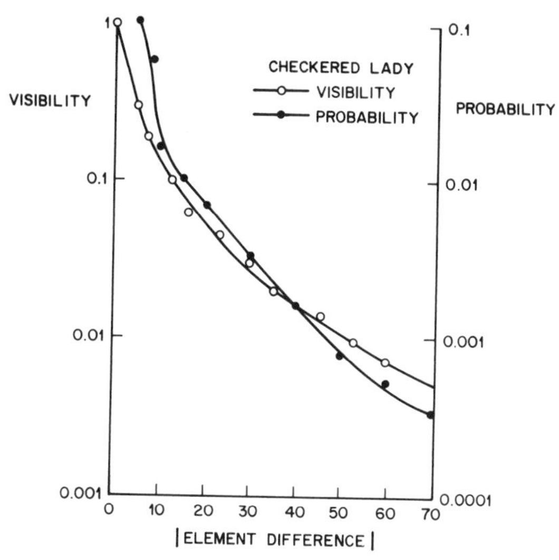

Fig. 4.3.7 Noise visibility function and probability density for a typical head and shoulders view type of picture (picture of checkered lady is shown in Fig. 5.1.3).

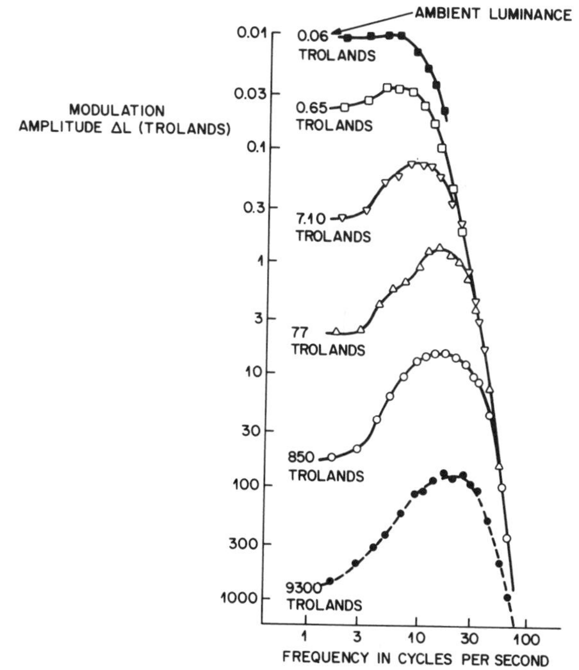

Fig. 4.3.8 Modulation amplitude in trolands needed to keep a flickering light of different frequencies at flicker fusion threshold, for six different average luminance levels (from Kelly [4.35]).

many facts about the perception of temporally changing stimuli. However, their application to coding is still in its infancy. Instead, several applied studies have attempted to evaluate the loss of perceived resolution (spatial and amplitude) as a result of movement in scenes. If movement is drastic, such as with a scene change (when TV cameras are switched), the perceived spatial resolution is reduced significantly immediately after the scene change. In fact, the perceived spatial resolution of the new scene may be reduced down to only one-tenth of normal without detection provided that full resolution is restored gradually within about half a second.[4.18] Experiments[4.19,4.20] also show that if a moving object is tracked by the eye, then perceived resolution due to camera integration dominates any reduction in resolution introduced by the visual system. However, when erratically moving objects are not tracked, the loss of perceived spatial resolution due to the visual system is significant. In practical television viewing, most displayed movement is not easily tracked. However, we have no quantitative data to tell when a viewer tracks an object and how accurately he tracks it. Also, since in many visual communication systems a transmitted picture may be viewed by many observers (e.g. broadcast TV), it is not clear how the resolution loss of nontracked objects can be used to improve the coding efficiency.

A form of temporal threshold that is relevant in many practical situations is the threshold of *flicker fusion*. If a test stimulus is presented repetitively in some temporal sequence, then there will be, in general, for a stimulus and viewing conditions, a particular frequency of repetition at which the presentation cannot be differentiated from a steady nonflickering field. The lowest such frequency is commonly known as the *Critical Flicker Fusion Frequency*. A large number of studies have been carried out to determine the dependence of this frequency on parameters of the test stimulus and viewing conditions. If the stimulus is spatially constant, but varies temporally according to $L + \Delta L \cos(2\pi f t)$, then data is normally expressed as a relationship of the modulation amplitude ΔL, to the critical flicker fusion frequency. Fig. 4.3.8 shows such experimental data. Both the modulation amplitude ΔL and the average luminance L are expressed in trolands, which is a unit of retinal illuminance. A troland is defined as the retinal illuminance when a surface of illuminance 1 candela (cd) per square meter (see Chapter 1 Appendix) is viewed through a pupil at the eye of area 1 square millimeter.[4.8] As a rough approximation, for display viewing 100 trolands is equivalent to about 10 cd/m^2 in the display.

We see that the eye is noticeably more sensitive to flicker at high luminances than at low luminances. The 9,300 troland curve is appropriate for the highlights of bright CRT displays, for which the flicker sensitivity peaks at about 15 Hz. Cut-off is reached around 70 Hz and there is a substantial fall-off at low frequencies. The 0.65 troland curve, appropriate for low light displays, shows a much reduced sensitivity to flicker, lower bandwidth and much less fall-off at low frequencies. Other factors affect the sensitivity to flicker as well.

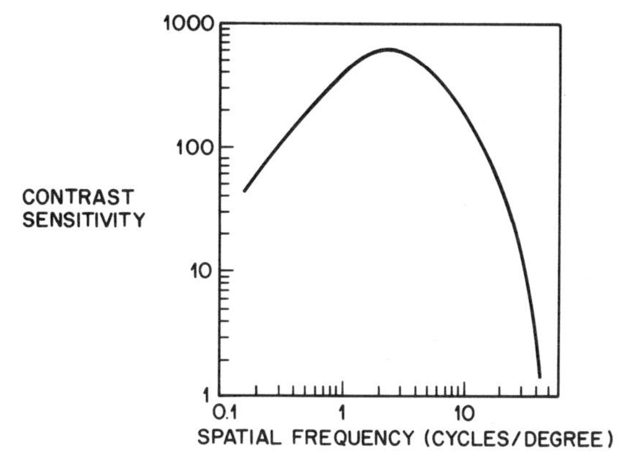

Fig. 4.3.9 Contrast sensitivity (inversely related to the visibility threshold) for sinusoidal gratings at luminance of 500 cd/m^2.

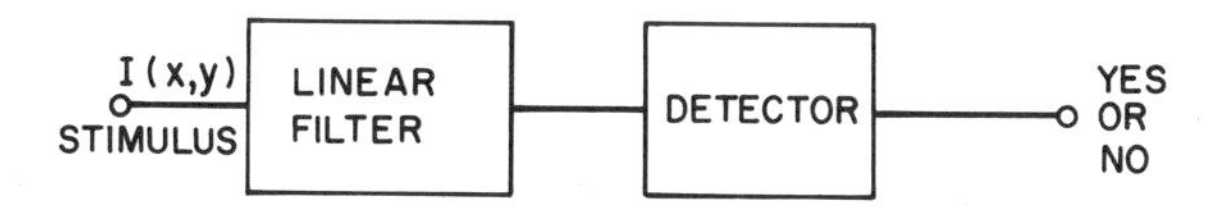

Fig. 4.3.10 Simple single-channel model of visual threshold for a stimulus presented against a plain background.

For example, the sensitivity increases gradually with increasing stimulus area; it is less for peripheral vision than for foveal vision.

4.3.1d Effect of Stimulus Shape

Under the assumption of linearity of the visual system, the visibility threshold of arbitrarily shaped test stimuli can be calculated if the transfer function or the impulse response of the linear system is known. The eye acts as a filter to the test stimulus. If the test stimulus is a spatial sinusoid with no temporal variation, then the visibility threshold can be plotted as a function of the spatial frequency, as in Fig. 4.3.9.* This function is called the *contrast sensitivity* function. It has a characteristic shape, having a maximum value (i.e., minimum contrast threshold) for spatial frequencies around 3–4.5 cycles/degree and falling off both at higher and lower frequencies.

Such a response can be predicted adequately by the simple model shown in Fig. 4.3.10, in which the test stimulus is linearly filtered first, and then if the resulting filtered output exceeds a threshold, the stimulus is considered visible. This model is successful for some stimuli, but fails for some others. However, the model can be extended to take care of some of these other situations. For example, a *multichannel model*, which utilizes a bank of linear filters, can predict responses to stimuli that are a mixture of several sine waves (see next section). Another extension, consistent with the spatial inhomogeneity of the retinal processing, is to allow the filter to vary spatially. This helps bring about more accurate prediction of the visibility of small distortions, but it has not yet been applied to any coding situation, principally because the registration of the image on the retina has to be known before the filter function can be applied. The filter function can also be extended to consider temporally varying stimuli. We will consider this in the next section.

4.3.1e Frequency Domain Threshold Vision

Threshold vision deals with small amplitude test stimuli, where the visual system behaves essentially as a linear system. Frequency domain methods, in particular the visual response to spatial and spatio-temporal sinusoidal gratings, have been used to characterize the linear system. Although gratings are structured and not pseudo random, they are still useful since they can be used as basis functions for the representation of pseudo random stimuli. Spatial gratings are

---

* Instead of plotting the visibility threshold as a function of spatial frequency, it is traditional as in Fig. 4.3.9 to plot the contrast sensitivity as a function of spatial frequency. Contrast sensitivity is defined as the ratio of the average background intensity to the peak amplitude of the spatial sinusoid at threshold that is used as the test stimulus. Thus, the visibility threshold is inversely related to the contrast sensitivity.

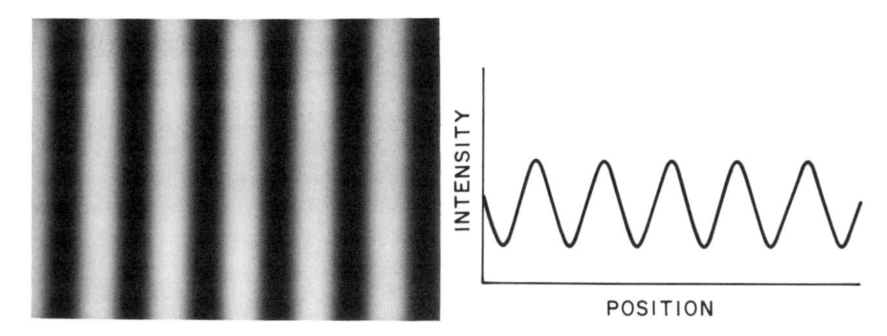

Fig. 4.3.11a Picture of a sine wave grating and the plot of corresponding intensity as a function of position (from Cornsweet [4.14]).

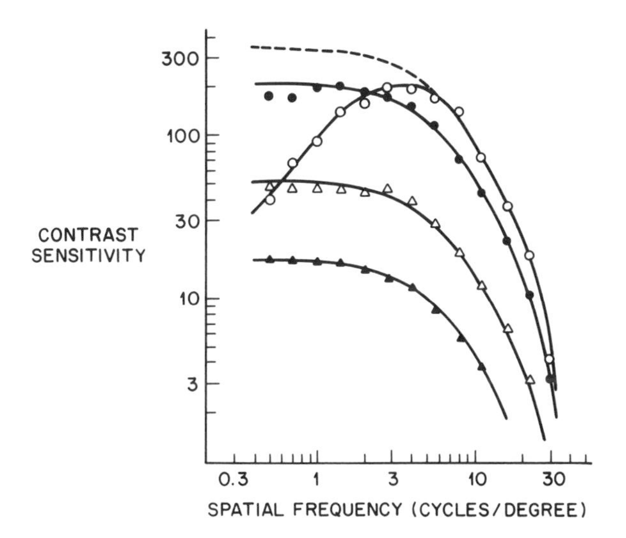

Fig. 4.3.11b Contrast sensitivity functions of the visual system. Spatial frequency response curves for different temporal frequencies f_T (○ 1 Hz, • 6 Hz, Δ 16 Hz, Δ 22 Hz).

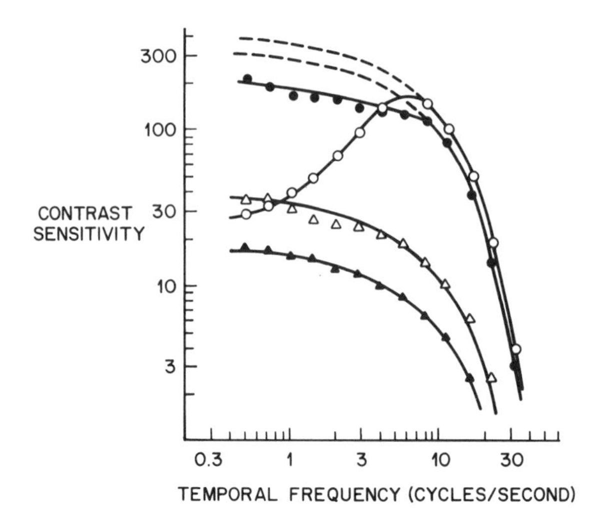

Fig. 4.3.11c Contrast sensitivity functions of the visual system. Temporal frequency response curves for different spatial frequencies f_o (○ 5 cpd, • 4 cpd, Δ 16 cpd, Δ 22 cpd) (from Robson [4.36]).

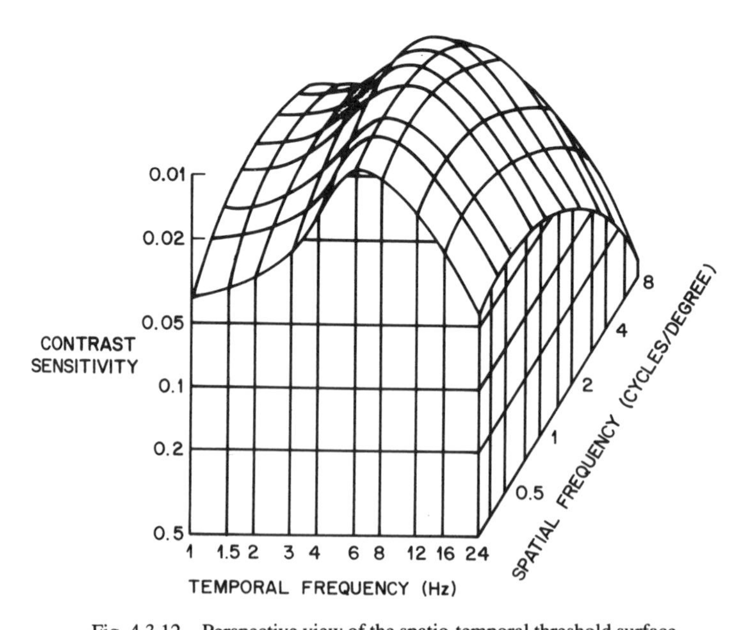

Fig. 4.3.12 Perspective view of the spatio-temporal threshold surface.

generated by:

$$B(x,y) = B_o + k \cos(2\pi f_o(x \cos\theta - y \sin\theta)) \tag{4.3.3}$$

The gratings thus generated are displayed against a background intensity B_0, and are bars of fundamental spatial frequency f_o oriented at an angle θ to the vertical axis. k is the modulation level that is adjusted to measure the visibility threshold for various f_o, B_o and grating angle θ. Fig. 4.3.11a shows a picture of such a grating. Important results from these experiments are (see Fig. 4.3.11b)

(i) At photopic illumination levels, visibility threshold of the grating for any fixed value of f_o and θ, depends primarily on the contrast k/B_o of the grating, and not separately on k and B_o.

(ii) The contrast sensitivity, i.e., value of $(k/B_o)^{-1}$ at the visibility threshold level, increases with f_o nearly linearly from low spatial frequencies up to a maximum in the mid-frequency range, and then falls off rapidly with increasing frequency. The frequency of maximum sensitivity is around 3 to 4.5 cycles/degree. The drop-off of contrast sensitivity at high frequency is a result of the resolution limits of the optical processing in the lens, the spatial density of the photoreceptors, and the subsequent neural processing. As examined in Section 4.2, the lateral connections between the cells cause neural interactions, leading to inhibition that spreads over a wide area and has longer time duration. This implies that the low frequency roll-off of the contrast sensitivity is a result of lateral inhibitions. Fig. 4.3.11b shows the typical response.

(iii) Contrast sensitivity also depends on θ, the grating angle. It is maximum for horizontal or vertical gratings and decreases with the angle from either axis, to approximately 3 dB at an angle of 45°.

(iv) Measurements of spatial sinusoidal gratings have been broadened to include spatio-temporal gratings. Stimuli generated by the following equation have been used.

$$B(x,y) = B_o + k \cos[2\pi f_o(x \cos\theta - y \sin\theta)] \cdot \cos 2\pi f_T t \tag{4.3.4}$$

where f_T is the temporal frequency. The response to such gratings is shown in Fig. 4.3.11c. This figure represents a measured cross section of a complete spatio-temporal response surface shown in perspective in Fig. 4.3.12. In both spatial and temporal directions, the eye acts as bandpass/low pass filter. Since the shapes of all cross sections of this surface are not similar, spatial and temporal properties of vision are not independent of each other, especially at low frequencies. At high frequencies, the separability of temporal and spatial components of the response appears to be more valid.

(v) Measurements have also been carried out using periodic gratings of various nonsinusoidal wave shapes and complex gratings consisting of two sinusoids of different frequencies. Experimental results suggest that frequency components separated by about an octave are detected by the observer independently. This implies that the level at which one frequency component is detected is unaffected by the presence and level of the second subthreshold component. This leads to the hypothesis that the visual system is composed of a number of independent parallel detection mechanisms called *spatial channels*, each tuned to a different spatial frequency and orientation angle. We will expand on this model in Section 4.5.

(vi) Spatial masking also occurs in the frequency domain. Experiments have been performed[4.21,4.22] to determine how the detectability of a stimulus of a given frequency near threshold is masked by background suprathreshold images within a narrow spatial frequency band. Results show that, indeed, there is a reduction in contrast sensitivity caused by the background, and although the greatest reduction takes place near the center frequency of the background, loss of sensitivities occurs in a broad range of frequencies. This may be due to the fact that spatial masking is a very local effect and therefore a frequency domain representation is not very appropriate.

4.3.2 Visibility of Random Noise

One of the major factors affecting the quality of analog television transmission is random wideband noise. Over the years, television engineers have developed formulas quantifying visibility of such random noise. Most of these formulas compute a weighted noise power (*WNP*) and relate it to a rating scale:

$$WNP = \int_0^B N(f) \cdot W(f)\, df \tag{4.3.5}$$

where B is the video bandwidth, $N(f)$ is the noise power spectral density and $W(f)$ is the weighting function. For monochrome television system M (see Table 2.1.1), it is given by

$$W(f) = \frac{[1 + (f/f_3)^2]}{[1 + (f/f_1)^2][1 + (f/f_2)^2]} \tag{4.3.6}$$

where $f_1 = 0.270$ MHz, $f_2 = 1.37$ MHz, $f_3 = 0.39$ MHz. The weighting function thus has a low-pass characteristic with asymptotic decay of 6 dB per octave. It is determined experimentally and incorporates visual effects of variations in the shape of the noise power spectrum.

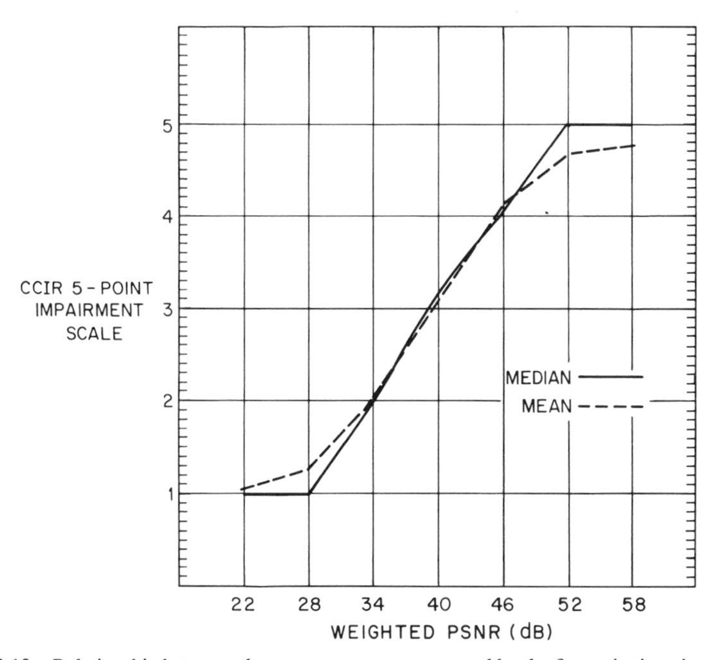

Fig. 4.3.13 Relationship between observer response as measured by the five-point impairment scale of Table 4.1(b) and the weighted peak signal to noise ratio WPSNR of Eq. (4.3.7).

The noise power by itself has little meaning unless it is related to the image waveform amplitude. The quantity commonly used by television engineers is the ratio of peak-to-peak luminance P to weighted rms noise, expressed in dB, i.e.,

$$WPSNR = 10 \log \frac{P^2}{WNP} \tag{4.3.7}$$

Although WPSNR was originally derived for monochrome television, it continues to be used for composite color TV waveforms, mainly because nothing better has been agreed upon so far by the television engineers.

The subjective rating attached to the weighted noise power depends upon many factors such as viewing distance, illumination, display luminance, etc. However, curves are available for many typical and standard viewing situations. As an example, Fig. 4.3.13 shows, for the NTSC signal, a relationship between the five-point impairment scale and the weighted signal-to-noise ratio expressed in dB. It is clear that above 52 dB of signal-to-noise ratio, the random noise cannot be perceived, and below 26 dB, it is very annoying and for most applications unusable. Also, the median of the observer response is almost linear with respect to the signal-to-noise ratio. One needs to exercise care in applying such curves to situations other than random noise. For example, quantization noise produced by a predictive encoder (see Chapters 3 and 5) tends to be correlated with the image itself, and therefore its visibility cannot be judged accurately by the above procedure. The major utility of the weighted noise power is in quick rough estimation of picture quality for a specific type of distortion.

4.4 Psychophysics of Color Vision

In Chapter 1, we described how colors are represented and the characteristics of two colors that attain a visual match under given conditions of observation. We saw that for the purposes of visual match, a color could be specified by only three values. But, how different are two colors that have different luminances and chromaticities, and what is the relationship between the *subjective* appearance of a color and its *objective* specification such as with chromaticity coordinates? What are the characteristics of visibility thresholds for color? These are some of the questions that will be considered in this section.

4.4.1 Perceived Color

In Chapter 1, we addressed the first part of colorimetry, namely color matching under given (standardized) conditions of observation. It is important to note that the chromaticity diagrams that we have discussed pertain to color in an objective (or psychophysical) sense. They are based on color matches, and not necessarily on the subjective color appearance. In this section we will be concerned with color appearance: not on which colors match, but what colors look like to a human observer. Although color appearance has been a subject of much research, standardized procedures for measuring color appearance do not yet exist. This may be a result of the complexity of the problem due to the subjective nature of color appearance. The ultimate goal, of course, is prediction of color appearance of stimuli presented in complex surroundings, and not just the restricted observing conditions of the CIE.

Color appearance is classified into two modes: the *object color* mode and the *light source* mode. Depending on the mode, different terminology is used in the specification of color appearance. In the object color mode visual objects do not appear to be emitting light, but rather reflecting or transmitting light. In this case, luminances of the surrounding are quite similar to those of the object that is being viewed. In the light source mode the surrounding field is usually relatively dark. These colors arise from light sources such as sun, lamps, TV displays, etc. The domains of object and light source colors cannot be made completely coincident. That is, there are certain perceived colors in one mode that can never be perceived in the other mode. For example, an object color such as olive brown cannot be perceived in the light-source mode.

The perceived appearance of a color stimulus depends upon many factors. Some of the more important ones are: (a) spectral power distribution of the source of light illuminating the field of view; (b) for object colors, the spectral reflectance or transmittance of both the object to which the observer's attention is directed and of all the other objects that are in the field of view; (c) spatial arrangement, sizes and shapes of the objects in the field of view; (d) spectral response of the observer as it applies to the entire field of view with his attention directed at the object.

It is difficult to quantify all of these factors, and one quickly sees the infinite number of possibilities. One may contemplate calibration and quantification of stimuli constituting the object and its surrounding. However, this is a mammoth task in which color appearance for each particular condition must be obtained from subjective tests. Such a procedure is clearly impractical.

The difficulty is compounded by its dependence upon not only the stimuli incident on the retina, but also on the state of the visual system as determined by previous exposures and other simultaneous exposures. A large body of literature deals with the prediction of visual response under different conditions. This area of study is called *chromatic adaptation*. However, as yet, there are no simple and comprehensive models for chromatic adaptation.

Experimental evidence indicates that perception of colors for each mode has three attributes: lightness, hue, and saturation for object colors; and brightness, hue, and saturation for light source colors. Lightness is that attribute of visual sensation according to which an area appears to reflect or transmit a smaller or greater fraction of incident light, whereas brightness is that attribute of visual sensation according to which an area appears to emit more or less light. Lightness is then a relative brightness. These two terms refer to the same characteristic that we discussed earlier as luminance. In common terms, they tell us how bright or dim an object appears to be.

Hue is that attribute of visual sensation wherein a stimulus appears to be similar to the perceived colors red, yellow, orange, green, blue, purple, etc. *Saturation* is that attribute of perceived color that tells us how different that color is from an achromatic [white or gray] color. Thus, spectral colors have high saturation, and pastel colors have low saturation.

The measurement of perceived color consists of an attempt to record our impressions of hue, brightness (or lightness), and saturation. We describe below two techniques used for this measurement. The first technique, called *binocular matching,* is similar to the brightness matching technique described earlier. The second technique is called *associative color memory*.

For binocular matching a subject gives his impression of a test color by relating it to another color seen under standard viewing conditions. The test color is viewed with one eye, and the reference color is seen with the other. The reference color is suitably synthesized by a mixture of three primaries. A shortcoming of this method is that the state of adaptation of the two eyes could be quite different, due to varied viewing conditions and previous exposures of the two eyes. The outcome of this experiment is an expression of the perceived test color in terms of objective color specifications, e.g., CIE coordinates x,y,Y.

With the associative memory method of measuring perceived color, the subject gives a name to each test color by reference to his memory of a previously perceived color. This obviously depends upon how well an observer is previously trained and, therefore, may show a large variability between observers. It is possible, however, to train most color normal observers within a short time to classify color perceptions consistently by showing them standard colors labeled in a suitable language. It is generally an accepted practice to use a 2° angular subtense with a neutral surround and an ordered collection of colored chips illuminated by a standard illuminant, e.g., D_{6500}.

The *Munsell renotation system*[4.23] specifies the color of such a collection of the chips by hue, lightness (called value) and saturation (called chroma). There are ten hues ranging from red to red-purple: R, YR, Y, GY, G, BG, B, PB, P, RP with ten subdivisions of each, of which four are commonly used, e.g.: 2.5R, 5R, 7.5R, 10R for red. There are ten steps of value, zero being dark and ten being pure white light, and over 20 steps of chroma (zero being for neutral color and 20 or more being for a color of high saturation). For example, the response of a subject could be 2.5R, 8/10, which means that the Munsell chip corresponding to a hue of 2.5R, value of 8 and chroma of 10 is perceived to have the same color as the test color.

The particular color chips of the Munsell System have been calibrated in terms of chromaticity coordinates with the standard source C representing daylight. In the 1931 CIE-*xy* diagram this leads to a network of points that, for a fixed Munsell value, shows lines of constant chroma and constant hue.

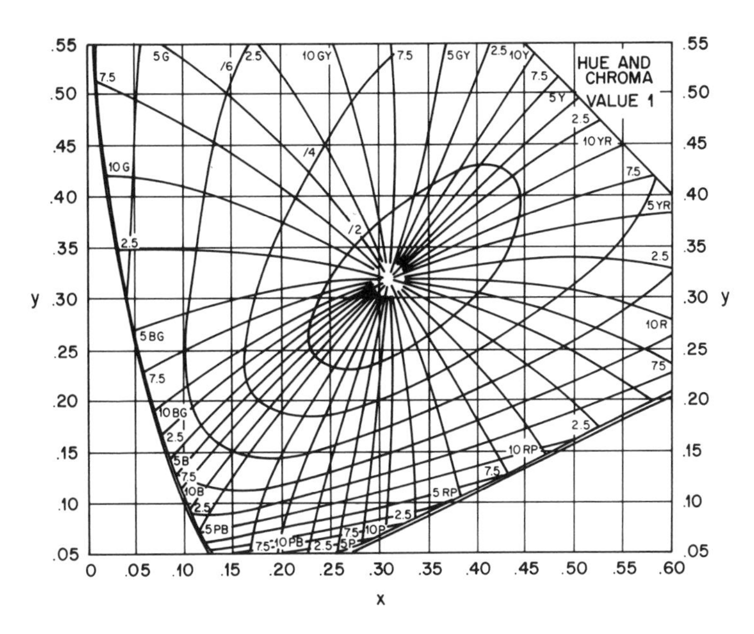

Fig. 4.4.1a 1931 CIE-xy Chromaticity diagram showing loci of constant hue and constant chroma at value 1/. of the Munsell renotation system.

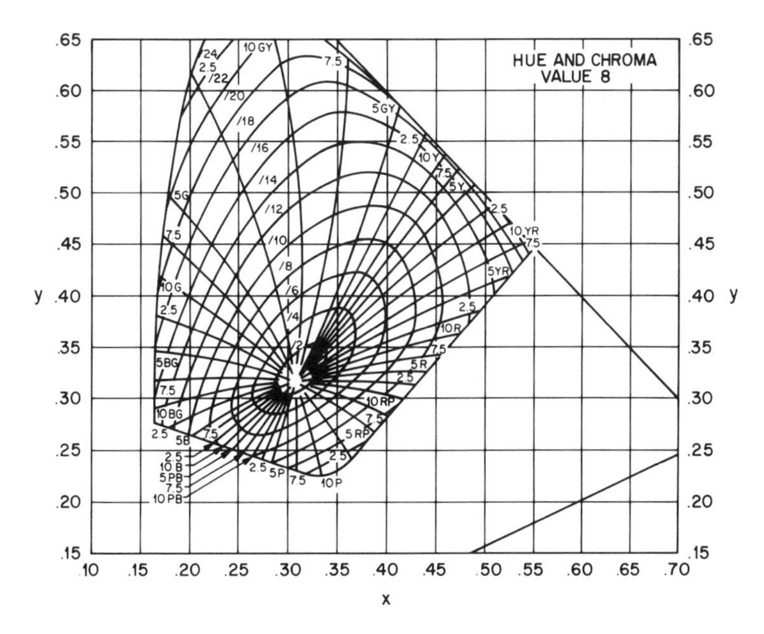

Fig. 4.4.1b 1931 CIE-xy Chromaticity diagram showing loci of constant hue and constant chroma at value 8/. of the Munsell renotation system (from Wyszecki *et al.* [4.37]).

Fig. 4.4.1 shows the loci of constant hue and constant chroma at values of 1/ and 8/ of the Munsell renotation system.

It is seen that the lines of constant hue radiate from the point corresponding to reference C, whereas points of constant chroma are nonintersecting ovoids. As expected, the network of points of constant hue or chroma does not exhibit uniformity. However, analytical expressions can be found that transform the CIE $[x,y,Y]$-space into a new space, which renders the network of points into a nearly uniform network. In this new space the lines of constant chroma closely resemble concentric circles, and lines of constant hue are nearly straight lines radiating from the central point corresponding to reference illuminant C. One such transformation is the CIE-uv coordinate system [also known as the 1960 CIE-UCS coordinate system] defined from the tristimulus values, X,Y,Z as

$$U = \frac{2}{3} X$$

$$V = Y \tag{4.4.1}$$

$$W = -\frac{1}{2} X + \frac{3}{2} Y + \frac{1}{2} Z$$

with the inverse transformation

$$X = \frac{3}{2} U$$

$$Y = V \tag{4.4.2}$$

$$Z = \frac{3}{2} U - 3V + 2W$$

Note that U, V and W are not true tristimulus values as defined in Chapter 1, since $U = V = W = 1$ does not correspond to equal energy white. The chromaticity coordinates (u,v) for this new system can then be derived as:

$$u = \frac{4x}{-2x + 12y + 3}$$

and

$$v = \frac{6y}{-2x + 12y + 3} \tag{4.4.3}$$

with the inverse transformation given by

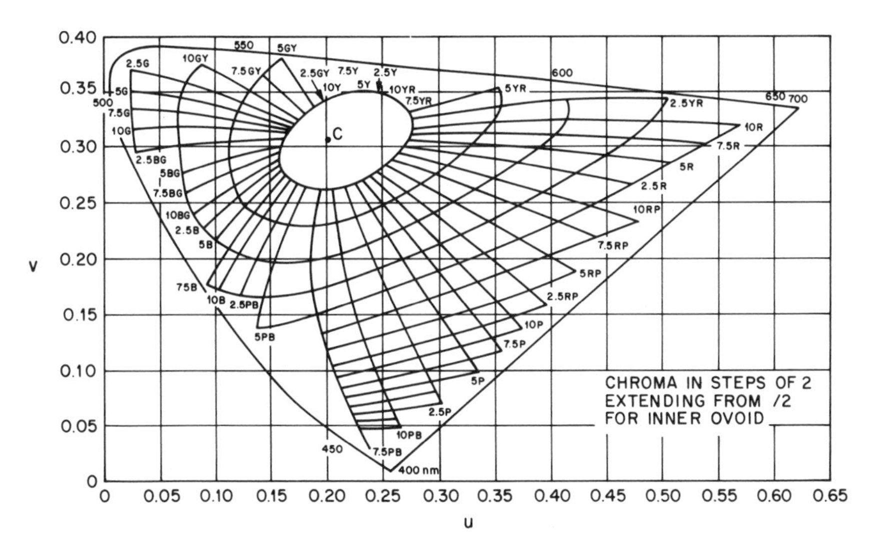

Fig. 4.4.2 Munsell Renotation System for illuminant C, Value 1/. in the 1960 CIE-uv diagram (from Pearson [4.39]).

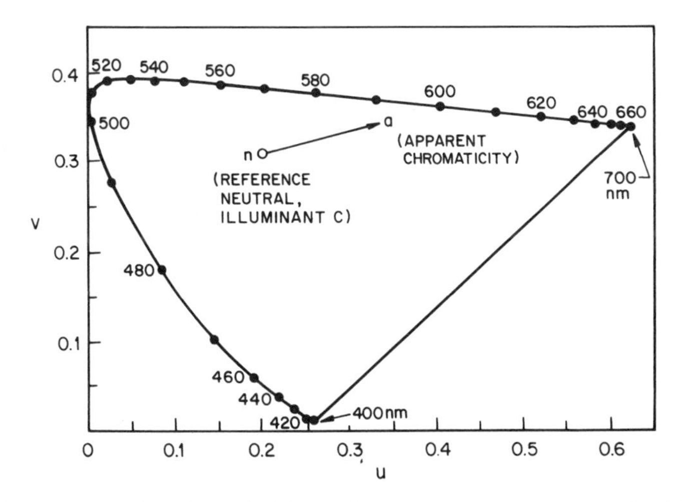

Fig. 4.4.3 Representation of perceived hue and saturation as a vector na on the 1960 CIE-uv chromaticity diagram. The origin of the vector n is the reference neutral color (here illuminant C). The arrowhead a is the apparent chromaticity.

$$x = \frac{\frac{3}{2}u}{u - 4v + 2} \tag{4.4.4}$$

$$y = \frac{v}{u - 4v + 2}$$

The lines of constant Munsell hue and chroma in the 1960 CIE-*uv* diagram are shown for the Munsell value of 1/. in Fig. 4.4.2. A comparison of Fig. 4.4.1(a) and Fig. 4.4.2 indicates that the 1960 CIE-*uv* coordinate system does in fact make lines of constant hue and chroma more uniformly spaced (although not entirely). Often, a vector representation is used to show perceived hue and saturation of a color (see Fig. 4.4.3). Also, several formulas have been experimentally obtained that specify color appearance in ways that are closely related to the perceptual attributes involved.[4.23]

4.4.2 Color Metrics

So far, we have considered the representation of color by the matching of two colors under given conditions of observation and perception of color. In this section we will be concerned with another part of colorimetry — namely how different are two colors? Obviously, we would like the quantity denoting the difference between two colors to be closely related to their perceptual difference. Determination of such a quantity depends on the ability of an observer to judge relative magnitudes of a color difference when viewing two color stimuli. As in the measurement of perceived color, conditions of observation such as sizes, shapes, luminances and spectral content of the test stimuli and surround affect the color differences. An approach towards measurement of color difference, called the color metric (or *line element*) has been developed using the mathematical notion of Riemannian spaces.[4.23] Most color metrics are characterized by the hypothesis that the perceived colors can be represented by points in three dimensional space. The problem, however, is to measure distances in this space. A Riemannian form of distance is assumed, which results in a positive definite quadratic form for the *Just Noticeable Color Difference*, Δs_0. Thus, if the two colors are specified by tristimulus values (U_1, U_2, U_3) and $(U_1 + \Delta U_1, U_2 + \Delta U_2, U_3 + \Delta U_3)$, where ΔU_1, ΔU_2 and ΔU_3 are infinitesimal increments in U_1, U_2 and U_3, respectively, then the distance between the two colors is taken to be

$$(\Delta s)^2 = g_{11}(\Delta U_1)^2 + 2\,g_{12}(\Delta U_1)\,(\Delta U_2) + g_{22}(\Delta U_2)^2 \tag{4.4.5}$$

$$+ 2\,g_{23}(\Delta U_2)\,(\Delta U_3) + g_{33}(\Delta U_3)^2 + 2\,g_{31}(\Delta U_1)\,(\Delta U_3)$$

where the coefficients $\{g_{ij}\}_{i,j=1,2,3}$ are continuous functions of U_1, U_2, U_3 that make $(\Delta s)^2 > 0$ for any choice of $\Delta U_1, \Delta U_2, \Delta U_3, U_1, U_2$, and U_3. The application of the above metric is made by hypothesizing that two neighboring colors (U_1, U_2, U_3) and $(U_1 + \Delta U_1, U_2 + \Delta U_2, U_3 + \Delta U_3)$ are *just-noticeably* or *just-perceptibly* different if and only if $\Delta s = \Delta s_0$. For two colors S_1 and S_2 that differ by more than Δs_0, the magnitude of the perceptible difference can be evaluated by counting the number of just perceptible steps (i.e., Δs_0) along a path from S_1 to S_2 that results in a minimum number of such steps. This corresponds to integrating (Δs) along the *Geodesic Line* between colors S_1 and S_2.

Having defined a difference metric between two colors, the next problem is to determine the coefficients $\{g_{ij}\}$ of the metric. Two methods are used.[4.26] One method, called the *inductive method*, is based on theoretical considerations regarding the human visual system, coupled with certain visual threshold data. In the other method, called the *empirical method*, coefficients are derived by an empirical analysis of a large number of measurements of threshold differences, obtained for colors covering an extended domain of space.

Color difference formulas given in Eq. (4.4.5) are usually too complex to use. A simplification is made by nonlinearly transforming the tristimulus values $\{U,V,W\}$ to other variables $\{U',V',W'\}$ in such a way that equally perceptible small differences in this transformed space can be defined by the simple Euclidean distance, i.e.,

$$(\Delta s)^2 = (\Delta U')^2 + (\Delta V')^2 + (\Delta W')^2 \tag{4.4.6}$$

where $\Delta U'$, $\Delta V'$ and $\Delta W'$, are the infinitesimal increments in the transformed tristimulus values. Many such color transformation formulas have been developed, each for a specific situation. None of them has yet found general agreement among all workers. For this reason we give below only one color formula recommended by the CIE in 1976 called the 1976 CIE-$L^*u^*v^*$ formula.

4.4.3 1976 CIE-L*u*v* Formula

In this system, the coordinates corresponding to lightness (L^*) and chromaticity (u^*, v^*) are given by

$$L^* = \begin{cases} 116(Y/Y_o)^{1/3} - 16, & \text{for } Y/Y_o > 0.01 \\ 903(Y/Y_o), & \text{otherwise} \end{cases}$$

$$u^* = 13\, L^*(u' - u'_o) \tag{4.4.7}$$

$$v^* = 13\, L^*(v' - v'_o)$$

where

$$
\begin{aligned}
u' &= \frac{4X}{X+15Y+3Z} \\
v' &= \frac{9Y}{X+15Y+3Z} \\
u'_o &= \frac{4\ X_o}{X_o+15Y_o+3Z_o} \\
v'_o &= \frac{9\ Y_o}{X_o+15Y_o+3Z_o}
\end{aligned}
\tag{4.4.8}
$$

and (X_o, Y_o, Z_o) are the tristimulus values of the reference white object-color stimulus. The color difference is then given by the Euclidean distance in the three-dimensional space $L^*\ u^*\ v^*$, i.e.,

$$
\Delta s = \{(\Delta L^*)^2 + (\Delta u^*)^2 + (\Delta v^*)^2\}^{1/2} \tag{4.4.9}
$$

Fig. 4.4.4 shows a plot of u' vs. v' with associated colors. A plot of u^* vs. v^* can be derived easily from this once the coordinates u'_o and v'_o for reference white are known. An additive mixture of two colors lies on the straight line joining them in the u' vs. v' plot, but not on the u^* vs. v^* plot due to the nonlinearity in its definition.

One application of color transformation formulas is to compute loci of constant Δs around given points on any chromaticity diagram, for a constant level of luminance. Such loci are ellipses since the color difference formulas are usually quadratics. MacAdam has experimentally determined loci of chromaticities that are equally noticeably different from each of 25 representative colors for a constant level of luminance. Such loci closely approximate ellipses, whose size, shape and orientation depend upon the representative color. These results are plotted in the 1931 CIE-xy diagram, as well as in the 1976 CIE-$L^*u^*v^*$ diagram in Fig. 4.4.5.

It is clear that the variation of size of the ellipses is rather large in the 1931 CIE-xy space compared with the 1976 CIE-$L^*u^*v^*$ space. However, even in the 1976 CIE-$L^*u^*v^*$ space the loci are not circles, indicating that the Euclidean metric in $L^*u^*v^*$ space is not quite uniform. Fig. 4.4.6 shows another aspect of the 1976 CIE-$L^*u^*v^*$ space in terms of perceived color. Munsell loci of constant hue and chroma are plotted (u^* vs. v^*) for a value of 5/. If the 1976 CIE-$L^*u^*v^*$ space were uniform in terms of perceived color, these loci would be straight, equally spaced radial lines (hue) and concentric, equally spaced circles (chroma). Although the Munsell data represent color differences very much larger than threshold, and therefore are not necessarily suitable for comparing color-difference formulas that are near-threshold, it is clear that the 1976 CIE-$L^*u^*v^*$ space is far from uniform.

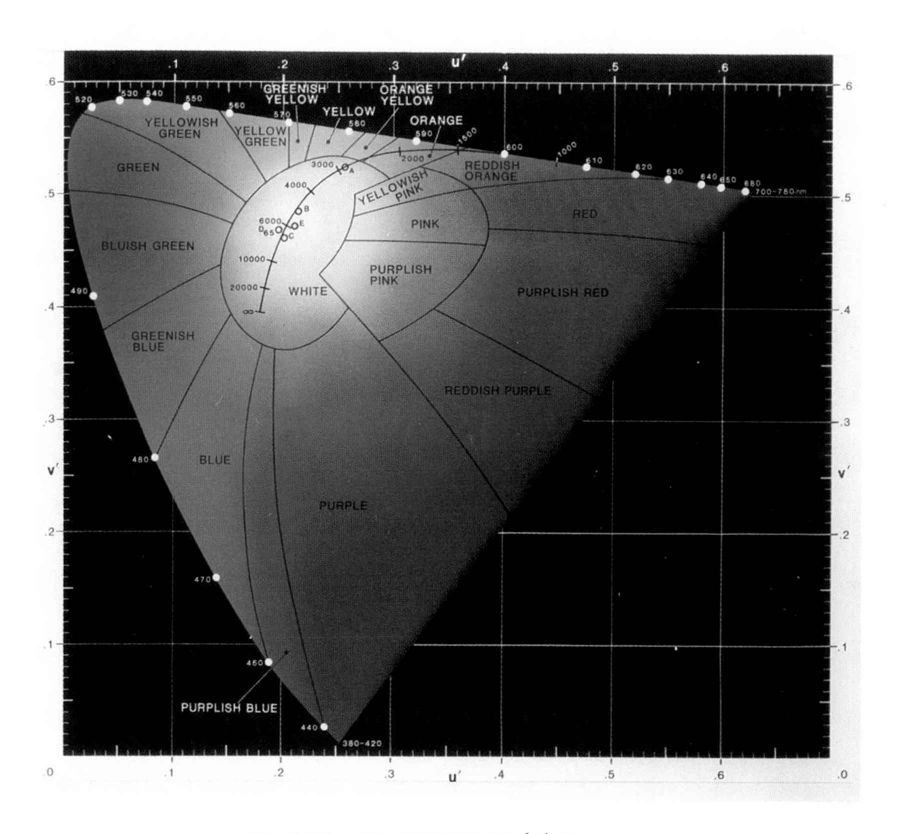

Fig. 4.4.4 The 1976 CIE-$L^*u'v'$ diagram.

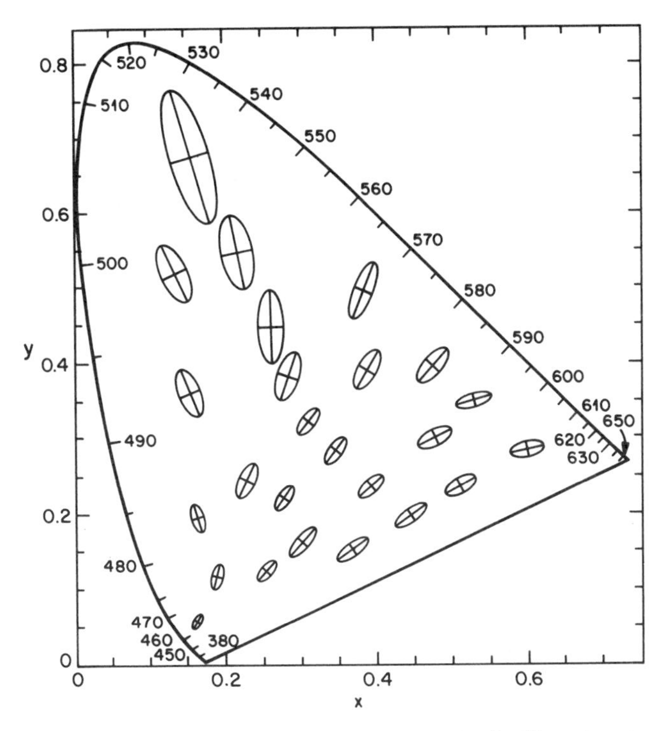

Fig. 4.4.5a 1931 CIE-xy chromaticity diagram showing MacAdam's ellipses (ten times enlarged) (from Wyszecki *et al.* [4.37]).

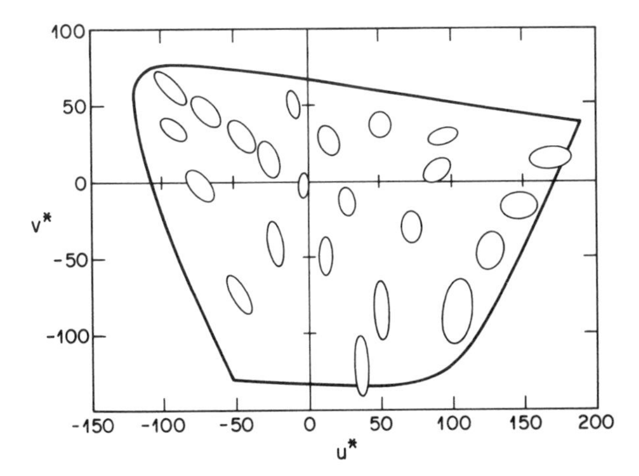

Fig. 4.4.5b MacAdam ellipses plotted in the 1976 CIE-$L^*u^*v^*$ diagram (from Robertson [4.41]).

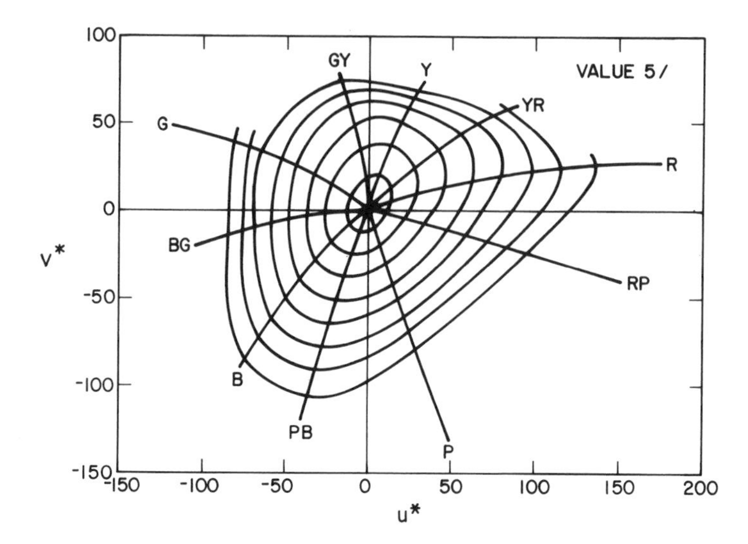

Fig. 4.4.6 Munsell Renotation System, Value 5/. plotted in the 1976 CIE-$L^*u^*v^*$ diagram (from Robertson [4.41]).

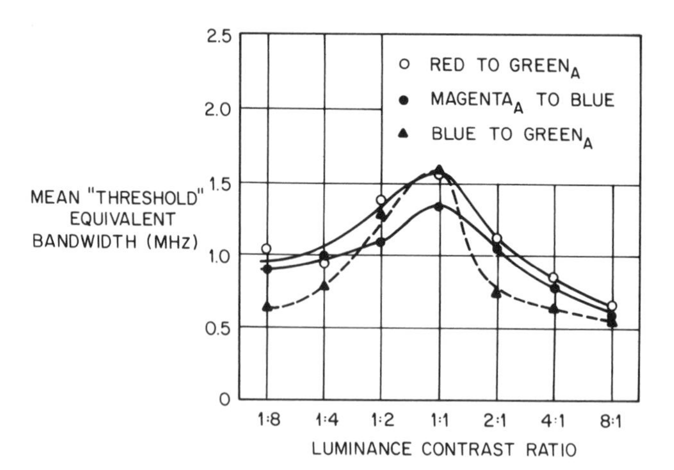

Fig. 4.4.7a Bandwidth versus luminance contrast ratio (expressed in terms of the first color for each pair) for transitions between saturated colors.

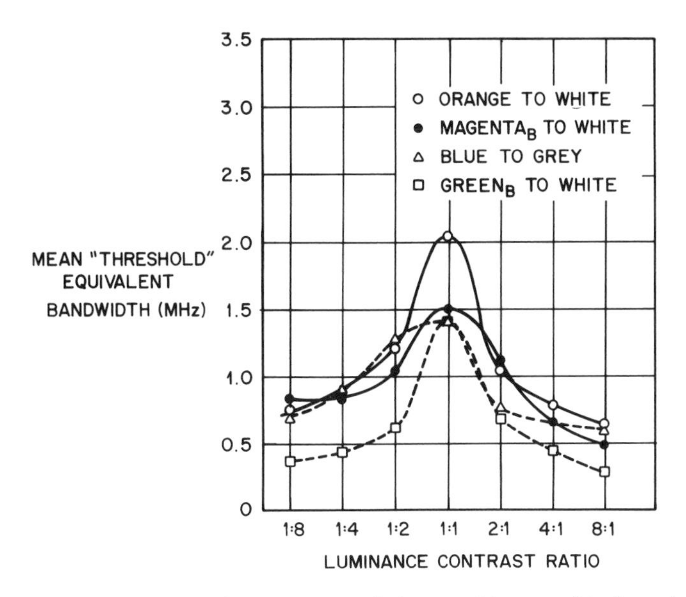

Fig. 4.4.7b Bandwidth versus luminance contrast ratio (expressed in terms of the first color for each pair) for transitions between saturated and "white" colors.

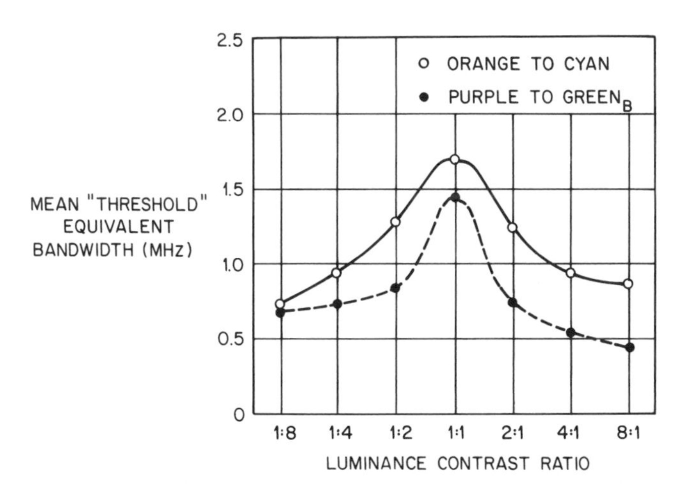

Fig. 4.4.7c Bandwidth versus luminance contrast ratio (expressed in terms of the first color for each pair) for transitions between orange to cyan and purple to green (from Hacking [4.24]).

4.4.4 Spatial Effects in Color Vision

We have so far been concerned with specification, appearance and discrimination of "large" patches of color, or large areas of uniform color. Real life pictures contain spatial and temporal variation of color. Thus, we are naturally interested also in specification, appearance and discrimination of small patches of color, when such patches are surrounded spatially and temporally by other colors. Color vision in such a complex case is not well understood. A large body of experiments exists, but no unifying theory has yet received wide acceptance. For this reason, we describe a few experiments that are relevant to coding applications, instead of simply describing theories.

We have already considered the monochrome properties of spatial vision (e.g., masking in the luminance signal). We will now consider masking of the chrominance signal by the luminance signal and the masking of the chrominance signal by itself. Masking of the chrominance signal by the luminance signal takes place in many ways. However, two types of masking are particularly relevant for the coding problem. In one type, measurements are made of the just perceptible degradation in the sharpness of the chrominance components at a boundary between two colors, as a function of the luminance contrast ratio across the boundary. In the second type of masking, experiments are made to determine changes in required amplitude resolution of the chrominance signal (near a boundary) as a function of luminance contrast.

Experiments of the first type contain a field of view having a reference pair of colors in the upper half and the same pair of colors in the lower half.[4.24] Starting with identically sharp transitions in the upper and lower half of the field, the sharpness of the chromaticity transition in the lower half is then gradually reduced until the subject can just perceive a degradation or blur at the boundary between the two lower colors. The test is repeated for different values of sharpness of the transition in the upper half and a threshold bandwidth is obtained from the settings. Fig. 4.4.7 shows results of such investigations for nine color pairs. As seen from these data, the visibility of blur changes rapidly between luminance contrasts of 1:1 and 2:1 (or 1:2) and changes much more slowly when luminance contrasts exceed 4:1 (or 1:4). The luminance contrast ratio has a large influence on the threshold as does the choice of the color pair. At a luminance contrast of 8:1 (or 1:8), the bandwidth requirement (which is inversely proportional to blur) decreases by a factor of about 3 from the 1:1 case. Thus the luminance contrast partially masks the ability of the subject to detect a decrease of color sharpness at the transition boundary.

Luminance masking also affects the required chrominance amplitude resolution at sharp luminance boundaries.[4.25] Such masking is determined by experiments that relate visibility of quantization noise in each (or both) chrominance components to a measure of spatial detail of the luminance signal. These relationships are called the noise visibility functions. Fig. 4.4.8 shows a typical visibility function for a portrait type picture when the horizontal slope of the

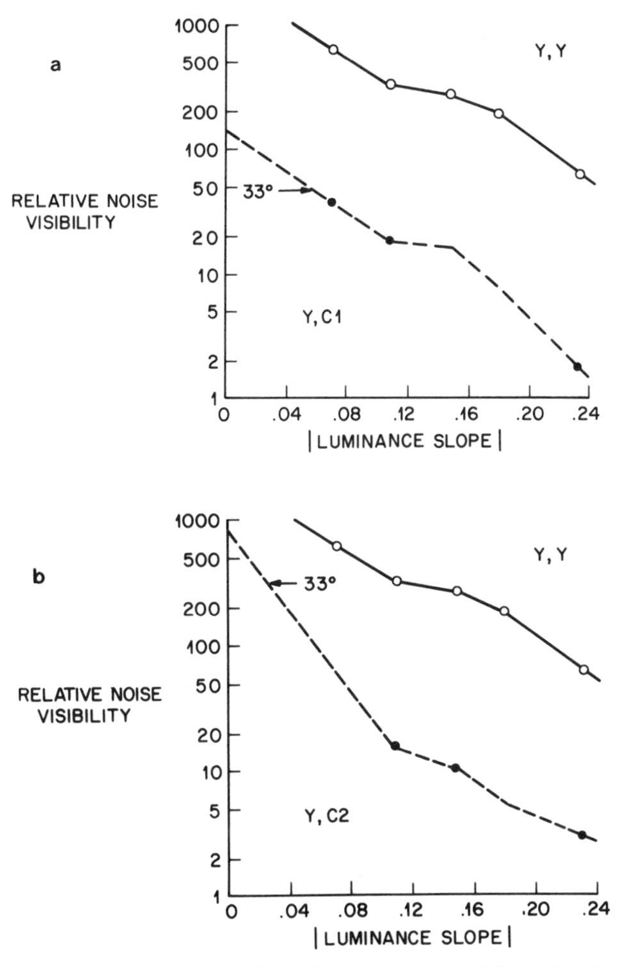

Fig. 4.4.8 Noise visibility functions with masking due to horizontal slope of the luminance signal. $Y-Y$, $Y-C1$, $Y-C2$ curves refer to masking of the luminance and the two chrominance signals, respectively. Luminance slope is measured in rate of change of luminance voltage per microseconds for a 256 line, 1 MHz videotelephone type of system with NTSC color primaries (33°).

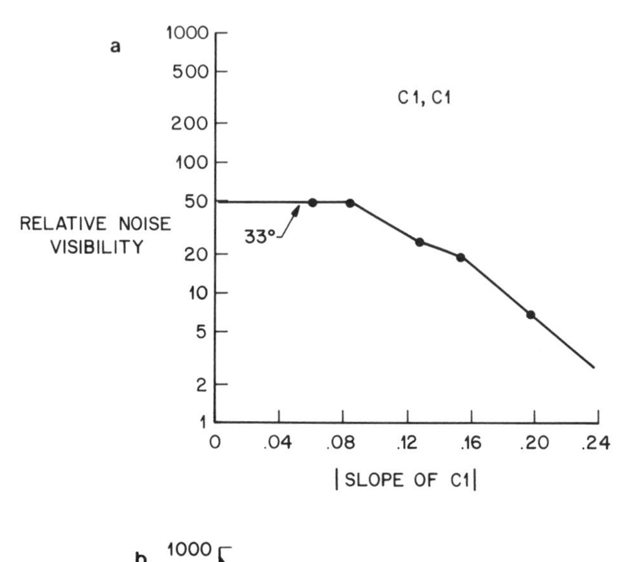

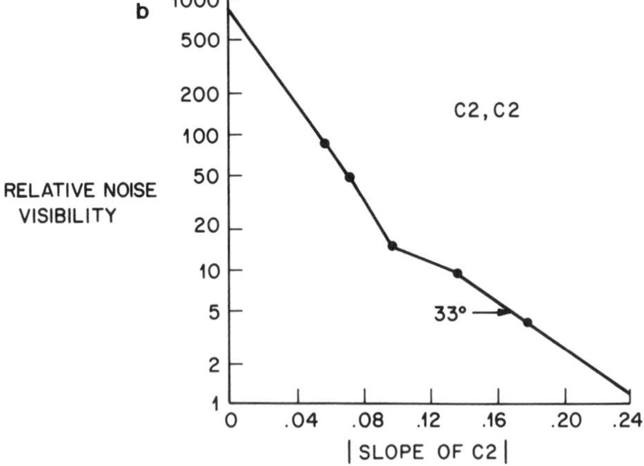

Fig. 4.4.9 Noise visibility functions with masking due to horizontal slope of the two chrominance signals.

luminance signal is used as a measure of spatial detail. We see that due to masking, the visibility of noise decreases rapidly as a function of luminance slope. Of course, the same experiments could be performed by using the horizontal slope of the chrominance signal as a measure of spatial detail. In this case, masking of the quantization noise in the chrominance signal is evaluated as a function of the slope of the chrominance signal. Data from such experiments is shown in Fig. 4.4.9. Again, the visibility of noise decreases with the increase of the chrominance slope. However, masking by luminance detail is much greater than masking by chrominance detail in most pictures. This is a result of several factors: perceptual, larger number of luminance boundaries and larger bandwidth of the luminance signal.

4.4.5 Opponent-Color Theory

Our discussion so far indicates that "Red", "Green", and "Blue" cones are truly the "mechanisms" of color vision. Yet the achromatic pattern vision that we experience when looking at a black-and-white photograph must be served by the same receptors, since there are no other cones. Probing of the monkey cortex with microelectrodes reveals cells that appear to carry chromatic information; yet at the same time many of these same units respond selectively to nonspectral characteristics of the stimulus. It is therefore plausible that the trichromacy, which is mediated by three types of cones at the initial stage of vision, needs subsequent recoding into two chromatic channels and one achromatic channel through which luminance information is transmitted.

Fig. 4.4.10 shows a schematic representation of an "Opponent-Color" model of chromatic vision.[4.26] Outputs from the R and G cones add and pass out of the retina as an achromatic message delivered through the "luminance channel". Note that unlike the luminance channel used in television, in this model B cones do not contribute to the luminance channel. This is still controversial and not well accepted. The three triangles at the bottom of the Fig. 4.4.10 are intended to represent the same cones, but they are regarded as initial elements of the chromatic channels. For this purpose, outputs of the R and G cones are subtracted and an $r-g$ opponent channel is created. The second chromatic pathway, called the $y-b$ opponent channel, carries the signal $R+G-B$. Of course, to mediate spatial vision it is necessary that millions of cones of the required type be organized, so that they feed into thousands of pathways of the luminance and opponent-color type. For each of these pathways, the receptive field consists of a center excitatory spot surrounded by a toroidal inhibitory region.

Earlier sections of this chapter dealt with the spatial properties of the achromatic channel. Compared to achromatic channels, the receptive fields of the chromatic channels are rather broad in spatial extent, and therefore they are much less sensitive to high spatial frequencies. Subjectively, observers looking at a grating of uniform luminance, but alternating in color, report a uniform field

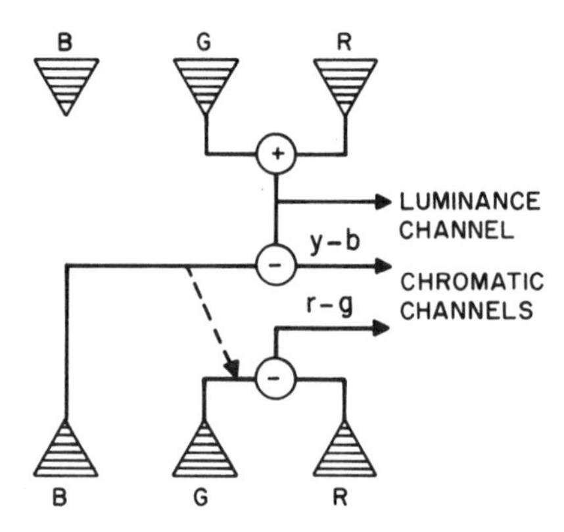

Fig. 4.4.10 An opponent-color model of human color vision. The B, G, and R cones at the top are intended to represent the same three cones that are also at the bottom. The achromatic (broad-band, nonopponent) pathway is activated by the summated output of R and G cones. The $r-g$ opponent pathway is activated by the difference in output of these same cones. The $y-b$ opponent pathway receives a signal that is the difference between the output of B cones and that of the luminance channel (from Boynton [4.26]).

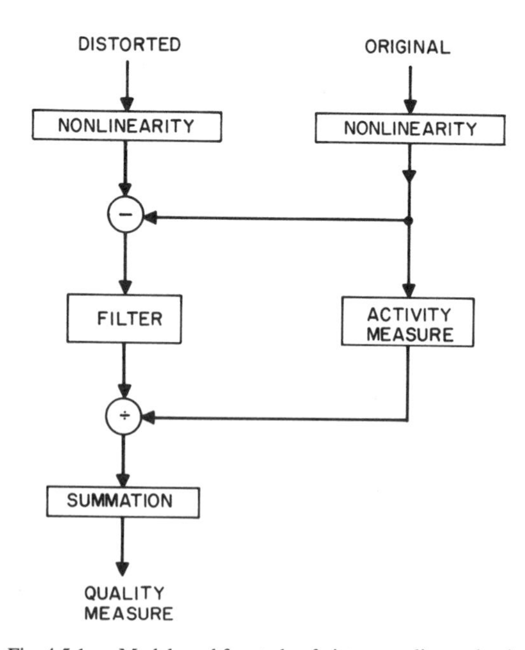

Fig. 4.5.1 Model used for study of picture quality evaluation.

of color for frequencies that would be perceptible if there were alternations of luminance as well as chrominance. Extending this to the spatio-temporal domain, experiments also indicate that the sensitivity to spatio-temporal changes in green is the highest, followed by red and then blue. Much work needs to be done before this can be quantified.

4.5 Models for Picture Quality

The complexities of visual processing are not yet fully understood, and therefore models for predicting picture quality are not completely satisfactory and are in a state of evolution. In addition, there is continuing debate about whether space-domain or frequency-domain modeling is more appropriate. We will consider below two recent models: one in the space-domain and the other in frequency-domain. These are both consistent with many physiological facts and psychophysical experiments. However, neither of these models can be uniquely derived based on experimental data. It is our belief that both the models, although simplified to a large extent, are too complex to be directly useful at present for optimizing coding algorithms.

A quality measure that is commonly used is the mean square error (MSE or equivalently PSNR) between the original and the distorted picture. The measure is usually applied after a nonlinear conversion of the intensity (similar to gamma correction) to offset the Weber's law effects. Sometimes, to improve the performance, frequency weighting (using a shape of the type in Fig. 4.3.9) of the error waveform is performed before computing the MSE. However, such a measure is signal independent and therefore does not reflect the effect of masking. The chief appeal of MSE is its simplicity, and therefore it has found wide use, particularly in mathematical studies.

4.5.1 Space-Domain Model

A block diagram of a space-domain model[4.27–4.28] for picture quality evaluation is shown in Fig. 4.5.1. As might be expected, it incorporates the operations of nonlinearity, filtering of the error, masking and averaging of the resultant error. Obviously, based on the psychophysical experiments of the previous sections, the parameters of various blocks can be specified only qualitatively. However, more experiments with coded pictures containing different types of distortions must be done in order to curve-fit the models to the experimental data.

The model normally uses a logarithmic type of nonlinearity based on the global threshold effects (Weber's law, etc) of Section 4.3.1. The error is then filtered by a spatio-temporal filter that emphasizes the mid-frequency range, both in the temporal and spatial domain. This is suggested by the frequency response given in Fig. 4.3.11. To incorporate the masking action, a measure of the amount of local change in the signal in a small neighborhood close to the picture element being processed is evaluated. This measure is then used to normalize

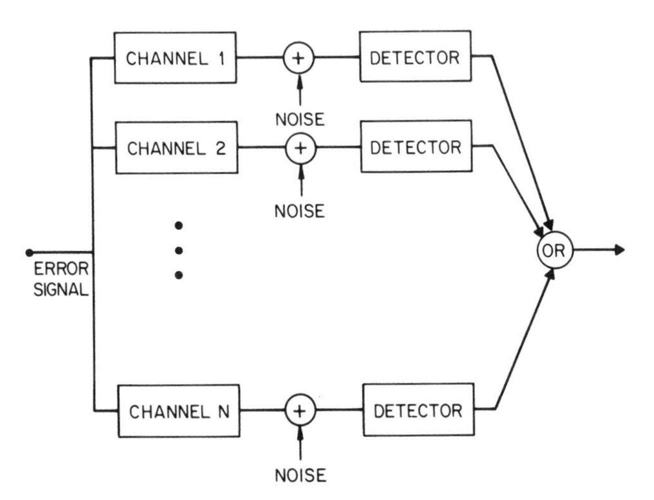

Fig. 4.5.2 A multiple-channel model for filtering the error (= original picture – coded picture) signal. The detection mechanism is random, i.e., simulated by adding noise to the filter output and comparing it to a threshold.

the error signal, and the resulting normalized error signal is averaged over an area.

In contrast to the mean square error measure, which is popular due to its mathematical tractability, it appears that observers base their estimates of picture quality on a few worst case local areas. Of course, these may be determined by context, background and other higher level functions. However, to a first degree of approximation, this situation can be modeled by summing the error signal over small local areas, and using the maximum of these for a prediction of picture quality.[4.27,4.28]

4.5.2 Frequency Domain Model

An alternative to the space domain model is the frequency domain model, which takes into consideration the variation of the sensitivity of the human eye to different spatial and temporal frequencies.[4.29,4.30] Here again, we start with a logarithmic type of nonlinearity of the intensity.* The log error waveform is then processed by a bank of filters to simulate the multichannel behavior.

Fig. 4.5.2 shows a block diagram of the overall structure. Each channel in this figure produces an output that is a random perturbation of a functional of the form

$$r_k = \int\int \left[\frac{w_k(x,y) \cdot v_k(x,y)}{m(x,y)} \right]^6 dxdy \tag{4.5.1}$$

where $v_k(x,y)$ is the output of a linear bandpass filter in the kth channel and the integration is performed over the entire image. The transfer function of these bandpass filters is given by

$$H_k(f_r,\theta) = H_{r_k}(f_r) \cdot H_{a_k}(\theta) \tag{4.5.2}$$

where $H_{r_k}(\cdot)$ and $H_{a_k}(\cdot)$ are the kth radial and angular transfer functions, respectively. They are given by

$$H_{r_k}(f_r) = \left[1 + \left(\frac{f_r - f_k}{2.5} \right)^2 \right]^{-\frac{1}{2}} \tag{4.5.3}$$

and

---

* The cube root function is sometimes used instead of the logarithmic function.

$$H_{a_k}(\theta) = \exp\left\{-\frac{1}{2}\left[\frac{\theta - \theta_k}{1.0}\right]^2\right\} + \exp\left\{-\frac{1}{2}\left[\frac{\theta - \theta_k - \pi}{b}\right]^2\right\} \quad (4.5.4)$$

where f_k and θ_k are radial and angular center frequencies, respectively. The gains of these filters depend on f_k, and are proportional to the contrast sensitivity function of Fig. 4.3.9 evaluated for a specific f_k. The radial bandwidth of 2.5 cycles/degree and the angular bandwidth of 10° are assumed. Although, data does not exist on the total number of channels, a rough estimate can be made to span the two-dimensional spatial frequency plane with radial frequencies from 1 cycle/degree to 20-25 cycles/degree and with angular frequencies from 0° to 180°. One possibility is four channel filters with radial passbands of frequency ranges (1-3), (2-6), (4-12), and (8-32) cycles/degree. In the angular dimension, spacings of every 10° near the horizontal and vertical axes and slightly larger spacing near 45° (visual acuity decreases off the horizontal and vertical axes) may be adequate, yielding 12-18 channels to cover the complete angular plane. Thus, complete coverage is possible with approximately 50-100 channels.

The function $w_k(x,y)$ provides spatial weighting to incorporate the sensitivity change of receptors with different locations in the visual field and to allow spatial summation over an area for a given channel. A raised cosine function is often used to represent $w_k(x,y)$; thus

$$w_k(x,y) = \begin{cases} \frac{1}{2}[1 + \cos(\frac{\pi\sqrt{x^2+y^2}}{r_0})] & , \sqrt{x^2+y^2} \le r_0 \\ 0 & , \sqrt{x^2+y^2} > r_0 \end{cases} \quad (4.5.5)$$

with $r_0 = 1.6°$ of visual angle.

The function $m(x,y)$ is a measure of reduced sensitivity as a result of visual masking. As discussed in Section 4.3.16, it depends upon the amount of detail present in the picture at location (x,y) and its surroundings. The integrand is raised to the sixth power based on experimental observations to emphasize the effects of errors with larger magnitudes. The resulting output from each of these channels is then connected to separate detectors whose outputs are OR'ed together. Detection of the error thus occurs whenever the signal magnitude in any of the independent channels exceeds its threshold level. Since each channel acts independently and makes an independent random decision, there is probability summation across the channels. Although explicit formulas are given above for the processing carried out by each channel, it is important to remember that they are based on models that are not unique, nor does there exist sufficient data to give us the required confidence in them. However, the models do give a reasonable fit to a multitude of data obtained from experiments.

Another restriction that is applicable to these models is that they use threshold data and should only be used for evaluating whether the error is just visible. Complete data does not yet exist for suprathreshold errors. Also for time varying images, filtering and masking ought to be spatio-temporal; however, here again, more data needs to be collected.

4.5.3 Color Images

For color images, much experimental data still needs to be collected. One reasonable distortion measure could be the sum of three terms, each of the terms being functionally the same as the distortion measures of Sections 4.5.1 (space domain) or 4.5.2 (frequency domain), but operating, respectively, on the filtered versions of the three channels of the opponent-color model (Fig. 4.4.8). The filters would be the contrast sensitivity functions for the achromatic and the two chromatic channels. Such independent processing of the three channels obviously does not take into account the interaction (particularly the masking) between the channels.

We close this chapter by pointing out that a large body of literature exists on visual psychophysics. It reveals many complexities of the human visual system and large gaps in knowledge. We are still far from a satisfactory model that can relate quantitatively the coding degradations to picture quality.

References

4.1 J. P. Guilford, *Psychometric Methods*, McGraw-Hill, New York, 1954.

4.2 W. S. Torgerson, *Theory and Methods of Scaling*, John Wiley, 1158.

4.3 D. E. Pearson, "Methods for Scaling Television Picture Quality: A Survey," in *Picture Bandwidth Compression,* Editors Huang and Tretiak, Gordon Breach, 1975, pp. 47-95.

4.4 R. N. Shepard, A. K. Romney and S. B. Nerlove, *Multidimensional Scaling* vol. 1: Theory, vol. 2: Applications, Seminar Press, New York, 1972.

4.5 D. E. Pearson, "A Three-Stage Process for the Evaluation of Image Quality," Proceedings of the SID, vol. 21/3, 1980, pp. 271-278.

4.6 C.C.I.R., "Method for Subjective Assessment of the Quality of Television Pictures," 13th Plenary Assembly, Recommendation 500, vol. 11, 1974, pp. 65-68.

4.7 NTC Report No. 7, "Video Facility Testing, An Engineering Report of the Network Transmission Committee," The Public Broadcasting Service, June 1975.

4.8 M. H. Pirenne, *Vision and the Eye*, Science Paperbacks, Chapman and Hall, 1967.

4.9 G. S. Brindley, *Physiology of the Retina and Visual Pathway*, Williams and Wilkins, 1970.

4.10 P. Moon and D. E. Spencer, "The Visual Effect of Nonuniform Surrounds," Journal of Optical Society of America, vol. 35, March 1945, pp. 233-248.

4.11 P. Mertz, "Perception of Television Random Noise," J. of Society of Motion Picture & Television Engineers, vol. 54, Jan. 1950, pp. 8-34.

4.12 A. Fiorentini, M. Jeanne, and G. T. Defranchi, "Measurements of Differential Threshold in the Presence of a Spatial Illumination Gradient," Atti Ford. Ronchi, vol. 10, pp. 371-379.

4.13 D. Y. Teller, "The Influence of Borders on Increment Threshold," Ph.D. Dissertation, University of California, Berkeley, 1965.

4.14 T. N. Cornsweet, *Visual Perception*, New York: Academic Press 1970, p. 276.

4.15 R. E. Graham, "Predictive Quantizing of Television Signals," IRE Wescon Conv. Record, Part IV, 1958, pp. 147-157.

4.16 A. N. Netravali, and B. Prasada, "Adaptive Quantization of Picture Signals Using Spatial Masking," Proc. of IEEE vol. 65, 1977, pp. 536-548.

4.17 N. Weinstein, "Metacontrast," in *Handbook of Sensory Physiology*, D. Jameson and L. M. Hurvich, (Editors), Springer-Verlag, 1972.

4.18 A. J. Seyler and Z. L. Budrikis, "Detail Perception after Scene Changes in a Television Image Presentations," IEEE Trans. on Information Theory, vol. 1T-11, Jan 1965, pp. 31-43.

4.19 D. J. Connor, and J. E. Berrang, "Resolution Loss in Video Images," National Telecommunications Conference, Dec. 1974, pp. 54-60.

4.20 M. Miyahara, "Analysis of Perception of Motion in Television Signals and its Application to Bandwidth Compression," IEEE Trans. Communications, vol. COM-23, July 1975, pp. 761-766.

4.21 H. Mostafavi and D. J. Sakrison, "Structure and Properties of a Single Channel in the Human Visual System," Vision Research, vol. 16, 1967, pp. 957-968.

4.22 C. F. Stromeyer and B. Julesz, "Spatial Frequency Masking in Vision: Critical Bands and Spread of Masking," Journal of Optical Society of America, vol. 62, Octo. 1972, pp. 1221-1232.

4.23 R. W. Hunt *"The Reproduction of Color"* (3rd Ed.) Wiley, New York, 1975.

4.24 K. Hacking "The Choice of Chrominance Axes for Color Television", Acta Electronica, Vol. 2, 1957, pp. 87-94.

4.25 A. N. Netravali and C. B. Rubinstein, "Quantization of Color Signals", Proceedings of IEEE, July, 1977.

4.26 R. M. Boynton, *Human Color Vision* New York, Holt, Rinchart, Winston, 1979.

4.27 J. O. Limb, "Distortion Criteria of the Human Viewer," IEEE Trans. on Systems, Man Cybernetics, Dec. 1979.

4.28 F. X. J. Lukas, "Picture Quality Prediction Based on a Visual Model," Ph. D. Dissertation, University of Western Australia, 1980.

4.29 D. J. Sakrison, "On the Role of the Observer and a Distortion Measure in Image Transmission," IEEE Trans. on Communications Nov. 1977, pp. 1251-1267.

4.30 D. Sakrison, M. Halter and H. Mostafavi, "Properties of the Human Visual System as Related to Encoding of Images," New Directions in Signal Processing in Communication and Control, J. K. Skwiyzyinski (Editor), Noordhoff and Leyden 1975, pp. 159-177.

4.31 J. E. Dowling and B. B. Boycott, "Organization of the Primate Retina: Electron Microscopy," Proc. Roy. Soc. **B,** 166, pp. 80-111, 1966.

4.32 G. F. Newell and W. K. E. Geddes, "The visibility of small luminance perturbations in television displays," Res. Dep., British Broadcasting Corp., Rep. No. T. 106, 1963.

4.33 A. Fiorentini and M. T. Zoli, "Detection of a Target Superimposed to a Step Pattern of Illumination. II. Effects of a Just-Perceptible Illumination Step," Atti. Fond. G. Ronchi, **22,** pp. 207-217, 1967.

4.34 E. M. Lowry and J. J. De Palma, "Sine-Wave Response of the Visual System. I. The Mach Phenomenon," J. Opt. Soc. Am., **51,** pp. 740-746, 1961.

4.35 D. H. Kelly, "Visual responses to time-dependent stimuli," J. Opt. Soc. Am., **51,** pp. 422-429, 1961.

4.36 J. G. Robson, "Spatial and Temporal Contrast Sensitivity Functions of the Visual System," J. Opt. Soc. Am., **56,** pp. 1141-1142, 1966.

4.37 G. Wyszecki and W. S. Stiles, *Color Science,* John Wiley & Sons, Inc., 1967.

4.38 A. N. Netravali and J. O. Limb, "Picture Coding: A Review," Proceedings of IEEE, v. 68, no. 3, pp. 366-406, March, 1980.

4.39 D. E. Pearson, *Transmission and Display of Pictorial Information,* Pentech Press Limited, 1975.

4.40 K. Miller, "Progress Toward a Three Dimensional Standard," *Photonics Spectra,* Feb. 1985, pp. 75-82.

4.41 A. R. Robertson, "The CIE 1976 Color-Difference Formulae", *Color Research and Application,* v.2, no.1, Spring 1977.

Questions for Understanding

4.1 A picture format having IAR=16:9 and 1050 scan lines is to be designed for a Kell factor of 0.7. How many pels per line are needed? How many lines of horizontal resolution? How many pels per line for square pels?

4.2 Explain photopic, mesoptic and scotopic vision. Explain lateral inhibition.

4.3 Explain visibility threshold and Weber's law. How is it related to gamma?

4.4 Explain spatial masking. Also, the Mach effect.

4.5 Explain spatial activity and noise visibility functions.

4.6 What are flicker fusion and critical flicker fusion frequency?

4.7 Explain the contrast sensitivity function.

4.8 What are PSNR and WPSNR?

4.9 Explain object color mode and light source mode.

4.10 What are hue and saturation? Also, the Munsell system?

4.11 What is the CIE-uv system? Why is it useful?

4.12 Explain the Space-Domain and Frequency-Domain models for image quality.

5

Basic Compression Techniques

We have so far described the nature and properties of picture signals such as the television signal and the facsimile signal. In particular, we considered statistical properties of pictures and the properties of the human viewer that are relevant to the coding problem. In this chapter we will describe many of the basic coding approaches that have been successfully used for digital picture communication. Emphasis will be on general principles, and how they are related or derived from the picture statistics and psychophysics of vision. We start with a classification of coding schemes and then describe them in some detail outlining procedures for optimizing their parameters.

The classification of coding schemes is given in Table 5.1.1. General waveform coding can be classified into four major categories: Pulse Code Modulation (PCM), Predictive Coding, Subband/Transform Coding and Interpolative/Extrapolative Coding. Besides these four classes, there are other schemes that may not fall into any of these four classes, but which are tailored for certain types of pictures. As an example, run length coding is popular and efficient for black and white facsimile pictures. Each of these classes can be further divided based on whether the parameters of the coder are fixed or whether they change as a function of the type of data that is being coded (adaptive). In practice, a coder may use a mixture of these approaches in a compatible manner to achieve proper cost/performance trade-offs. Examples of complete coding algorithms are given in Chapter 6.

In PCM, also known as analog to digital or A/D conversion, a time discrete, amplitude discrete representation of picture elements (pels) is made without removing much statistical or perceptual redundancy from the signal. The time discreteness is provided by sampling the signal generally at the Nyquist rate, whereas the amplitude discreteness is provided by using a sufficient number of quantization levels so that degradation due to quantization errors is tolerable.

Table 5.1.1 Classification of Picture Waveform Coding Algorithms.

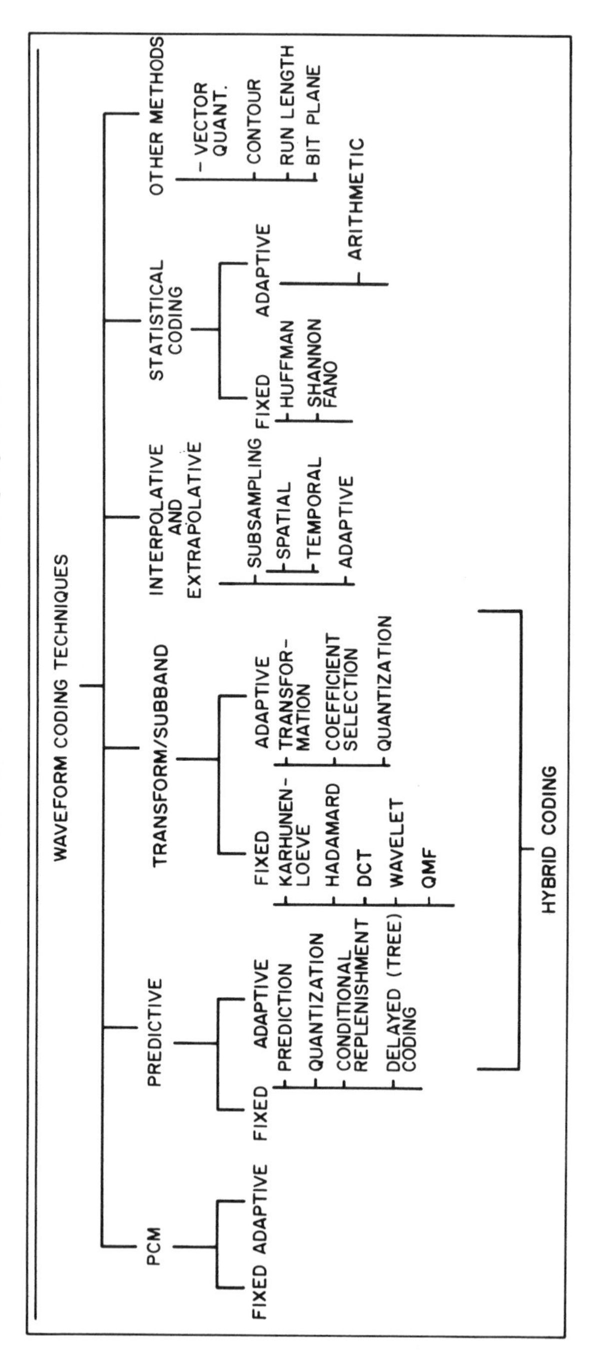

In predictive coding, also known as the Differential Pulse Code Modulation (DPCM), an attempt is made to predict the pel to be encoded. The prediction is made using the encoded values of the "previously" transmitted pels (or already encoded pels), and only the prediction error (differential signal) is quantized for transmission. Such an approach can be made adaptive by changing the prediction based on local picture statistics, or by varying the coarseness of the quantizer based on visual criteria, or by not transmitting the prediction error whenever it is below a certain threshold, as in conditional replenishment. A further possibility is to delay the encoding of a pel until the "future trend" of the signal can be observed, and then code to take advantage of this trend. This is called Delayed Coding.

In transform coding, instead of coding the image as discrete intensity values on a set of sampling points, an alternative representation is made first by linearly transforming the pels into blocks of data called coefficients and then quantizing the coefficients that are selected for transmission. Several transformations (such as the simple Hadamard or the fairly complex Karhunen-Loeve) have been used. Cosine transforms have become more popular in recent years because they appear to be well matched to the statistics of the picture signal. One, two and three dimensional blocks (i.e., two spatial dimensions and one time dimension) have been studied for transformation. Recently, however, two dimensional transforms have been used almost exclusively. Adaptation of transform coders is possible either by changing the transformation in order to match picture statistics or by changing the criteria for selection and quantization of the coefficients in order to match the subjective quality requirements.

With subband coding, the entire picture is passed through a bank of bandpass filters, thus decomposing the waveform into a number of subband pictures. Each subband picture can then be subsampled and coded using, for example, DPCM or run length coding.

Interpolative and extrapolative coding techniques work on a different principle. They attempt to send a subset of the pels to the receiver, which then either extrapolates or interpolates to obtain the untransmitted pels. These techniques are heavily used for interframe systems in conjunction with predictive coding. Adaptation of these systems consists of varying the criteria for selection of the samples to be sent as well as the strategy for interpolating or extrapolating the remaining samples.

A practical coding system may be a combination of many of the above schemes. One interesting combination is called hybrid transform coding in which predictive coding based on previously transmitted adjacent (spatially or temporally) pels is followed by linear transformation of the prediction error. Other methods include run-length coding and its two dimensional extension for bilevel signals. In run-length coding, runs of consecutive pels of the same color (0 or 1) are assigned a codeword. This can be extended to multilevel signals by coding the runs in several *bit planes*. Contour coding is a scheme in which a

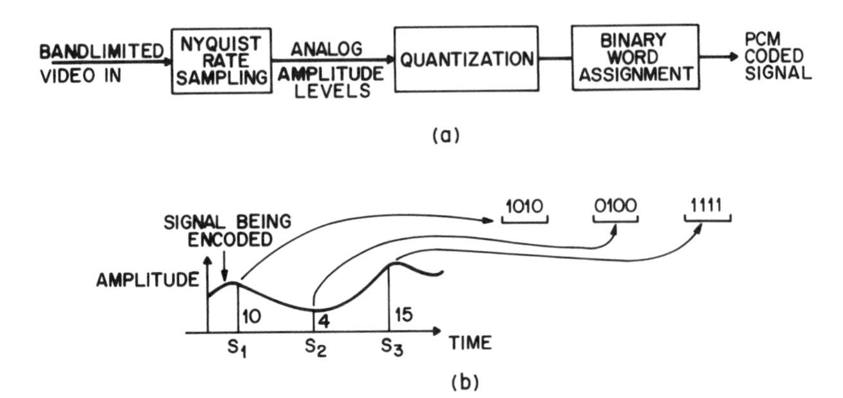

Fig. 5.1.1 PCM Encoding. (a) Components of a PCM encoder. (b) Four-bit binary representation of amplitude levels between 0 to 15.

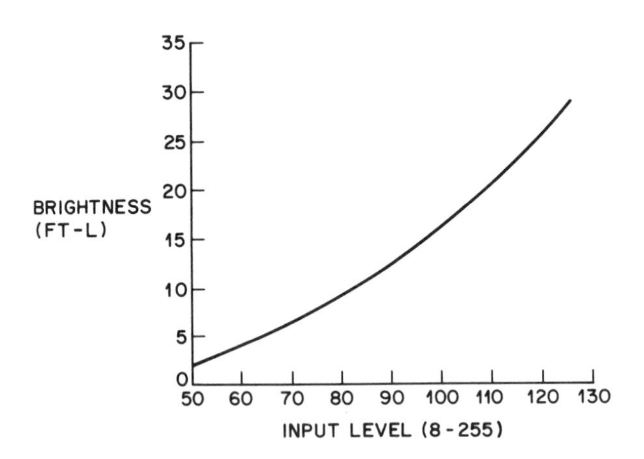

Fig. 5.1.2 Nonlinear input-output characteristic (nonunity *gamma*) of a cathode-ray tube. Input is voltage above cut-off. Brightness is in foot-lamberts.

picture is separated into 1) high contrast boundaries (or contours) and 2) all the rest. Since a picture of contours is a two-level picture, techniques such as run length coding are applicable. The rest of the picture contains only low frequency and texture information and is coded by predictive or transform techniques.

The methods mentioned so far, (except those for two level pictures) generally result in information loss, i.e., the original picture cannot be reconstructed exactly from the coded picture data. Statistical coding techniques, on the other hand, are generally information-preserving. They are used to assign bits to the quantizer outputs of DPCM or transform coded signals to minimize the average bit rate. An efficient comma-free code (i.e., a code in which the boundaries between adjacent code-words are not marked, but can be automatically deduced from the transmitted sequence of bits), is given by the Huffman code (see Chapter 3). Here, the code assignment depends upon the probability of occurrence of the source symbols; therefore, this assignment can be fixed based on long-term average statistics, or adaptive based on the short term statistics. In the remainder of the chapter we describe principles and optimization procedures for most of the above schemes.

5.1 Pulse Code Modulation (PCM)

Pulse code modulation is a discrete time (or space), discrete amplitude representation of visual information.[5.1.1] Although conceptually straightforward, its first application for television was not until 1951 by Goodall.[5.1.2] This is perhaps due to the fact that electronics technology up to that time was not able to handle the high sampling speeds required for a television signal. Since then, PCM has been used as a video digitizing scheme for the purposes of storage and processing, as well as for digitizing prior to the application of other more sophisticated coding techniques.

As shown in Fig. 5.1.1, basic PCM consists of sampling a one-dimensional raster scanned waveform (usually at the Nyquist rate) and quantizing each sample using 2^K levels. Although, not explicitly shown in this figure, as described in Section 1.5, an appropriate prefilter is used before sampling so that the sampling rate is approximately Nyquist and the aliasing distortions are avoided. Each level is represented by a binary word containing K bits. Usually, the binary code word is related monotonically to the level so that further arithmetic can be done with ease. At the decoder, these binary words are converted to a time sequence of discrete amplitude levels, which is then low-pass filtered. Basic PCM exhibits a conceptual simplicity uncommon to most other coders, but suffers from inefficiency since it does not fully utilize redundancy present in the picture signal.

Placement of quantizing levels for PCM coding can be based on psychovisual criteria. For gray-level pictures the principal effect used is Weber's law (see Chapter 4), which states that the visibility threshold of a perturbation in luminance, ΔL, increases almost linearly with increasing background luminance. This would imply that the visibility of a given amount of quantization noise decreases with the luminance level, and therefore coarseness of the PCM quantizer can be increased with the luminance level. Such a nonlinear *companding* of the quantization levels does improve the quality of PCM coded pictures for a given number of bits per sample, and in linear display systems with unity *gamma*, advantages can be realized. However, in most television systems, where cameras and cathode ray tubes (CRT's) with nonunity gamma are used, improvements are not significant. Typical nonlinear CRT characteristics are shown in Fig. 5.1.2. It increases the amplitude of the quantization error at larger values of the luminance to some extent, and therefore, partially compensates for the Weber's law effects. However, since in practice, viewers may look at a picture on displays with different brightness and contrast settings, it is difficult to make full use of Weber's law. Thus, in most applications uniform PCM on a gamma corrected camera signal is used. PCM coding systems for monochrome video, in general, require 128 to 256 levels (7 or 8 bits) per pel for good quality pictures under most viewing conditions. For monochrome television with a sampling rate of 8 MHz, this amounts to a bit rate of 56 to 64 Mbits/sec. For systems using still frames, the quantization noise does not have any time variation as in television and is therefore frozen on the screen. This reduces its visibility and therefore, only 6 or 7 bits per pel may be sufficient.

Quantization noise that may be visible due to the coarseness of the quantizer can be reduced in many ways. One technique is to put pre- and post filters around the quantizer of Fig. 5.1.1. Approximating the quantizer as an additive noise source, filters that are optimum in the mean-square error sense can be calculated. Computer simulations show that, by using these filters, one can obtain pictures essentially free of artificial contours with fewer than 5-6 bits per pel. However, such filtering also reduces the resolution of the reproduced images.

We have seen in Chapter 4 that the human eye is more sensitive to noise or distortions that have strong structure than to random noise. For high quality original pictures sampled at a sufficiently high rate, as the number of quantization levels is decreased, quantization error is seen as *contouring*, i.e., as false contours in low-detail areas of the pictures. This is shown in Fig. 5.1.3a where an analog picture with PSNR of 50 dB was encoded with 5 bit/pel PCM. As seen in the picture, the quantization contours are visible in the sky. Visibility of this quantization noise can be decreased by adding some high-frequency noise (called *Dither*) to the original signal before quantizing. This noise causes the coded signal to oscillate between the quantizing levels thereby increasing the frequency content of the quantizing noise. The contouring patterns whose visibility is high are broken and since the higher frequency noise is less visible, such quantizing noise is less visible. This is seen in Fig. 5.1.3b where the

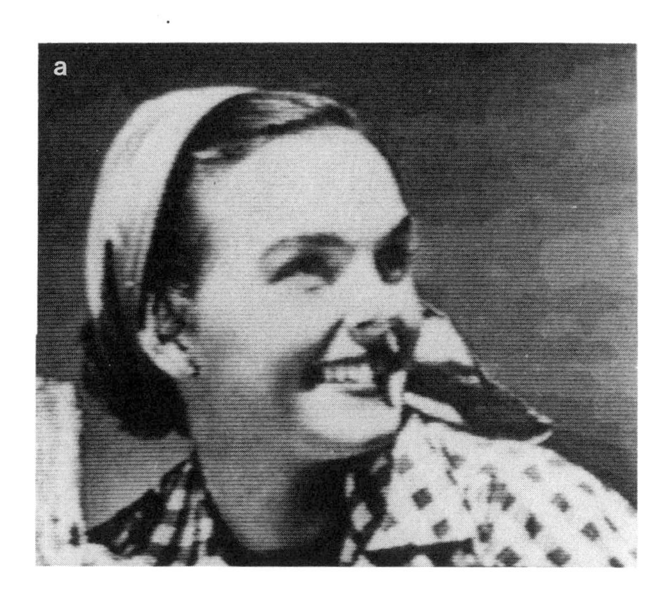

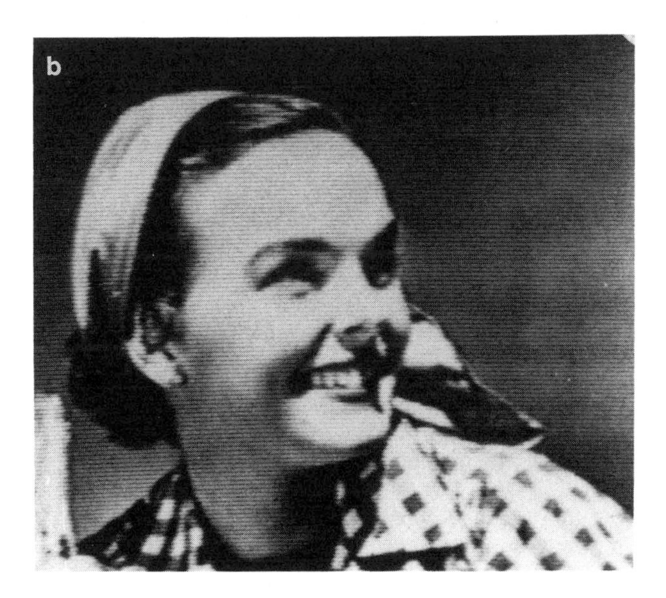

Fig. 5.1.3 Effect of coarse quantization in PCM coded picture. (a) Contouring is visible in this high input signal-to-noise ratio, five-bit PCM picture. (b) Dither reduces the visibility of contours.

quantizing contours are no longer visible. Dithering is thus a technique of adding pseudo random noise to the picture signal before quantization, and later at the receiver subtracting the same noise from the quantized picture.[5.1.3] Such techniques have also been useful in converting a gray level picture into a two-level picture (see Section 1.9.3). Additional improvements are possible by optimizing the parameters of the pseudo random noise, e.g., amplitude, spatial and temporal characteristics.

5.1.1 PCM Encoding of Component Color

A color camera usually has available red (R), green (G) and blue (B) tristimulus values for each pel. However, PCM encoding may not always utilize these R, G and B signals. Instead, these waveforms may be transformed to another color space before digitization, e.g., the NTSC components Y, I and Q. The digitization may also use a quality criterion based on measurements in another color space that may be psychovisually more appropriate (see Chapter 4). Also, instead of independent quantization of each component, multidimensional quantization by simultaneous treatment of more than one component is possible. We will consider each of these cases below.

The simplest components to use for digitization are R, G, B. However, it is not necessary to quantize each of them with the same accuracy, since quantization noise is not equally visible in each of these components. As an example, experiments on still natural pictures show that additive noise (which approximates quantization noise) in the blue signal is 10 dB less visible than in the red signal, and 20 dB less visible than noise in the green signal.[5.1.4] Thus, if all three components R, G, B have the same bandwidth, then fewer bits can be used for quantization of the red and blue signals compared to the green signal.

The chief drawback of coding the R, G, B components is that relatively high spatial resolution is required for each component. Although some studies indicate that for most natural scenes, slightly lower bandwidth can be assigned to the red and blue signals with respect to the green signal, much more of a bandwidth reduction is possible when other color components (e.g., luminance and chrominance) are used. In practice, however, PCM digitization of R, G, B using the full bandwidth for each component is common. Such PCM signals can then be used easily for further processing such as matrixing, filtering, compression, storage, etc.

Luminance and chrominance signals, e.g., Y, I, Q, may also be digitized directly. In this case, sampling rates are adjusted to correspond to the lower allowable bandwidths of the I and Q signals. Several studies have been made to determine the required accuracy of quantization for the NTSC I and Q signals, the PAL U_t and V_t signals, as well as the hue and saturation type of chromaticity signal. In these studies, an attempt is made to relate the quantizer level spacing in the chrominance component domain to another domain that is perceptually more meaningful. As an example, Fig. 5.1.4 shows results of a

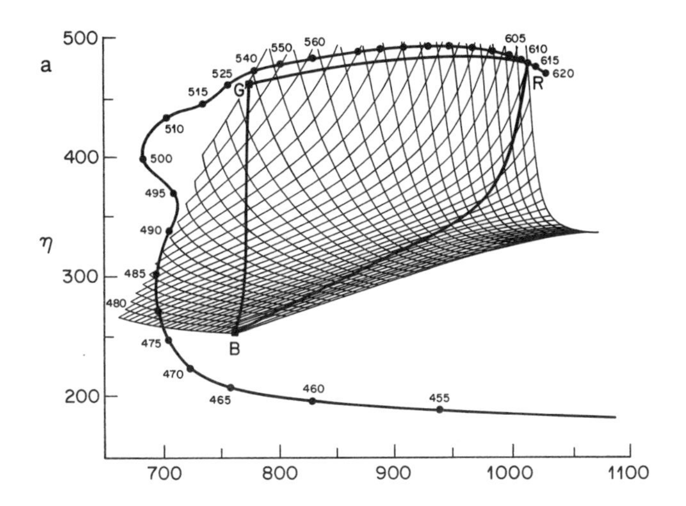

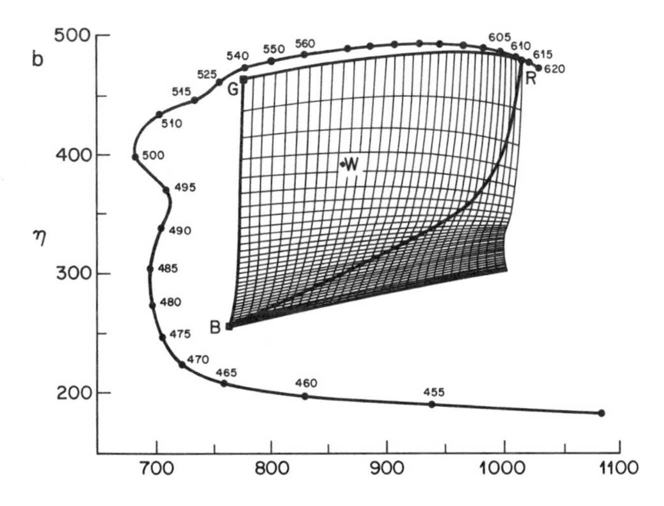

Fig. 5.1.4 Plots of 5-bit uniform quantization grids mapped from three different color spaces onto MacAdam's geodesic (See Section 4.4.2) diagram: (a) NTSC I and Q signals. (b) PAL U_t and V_t signals.

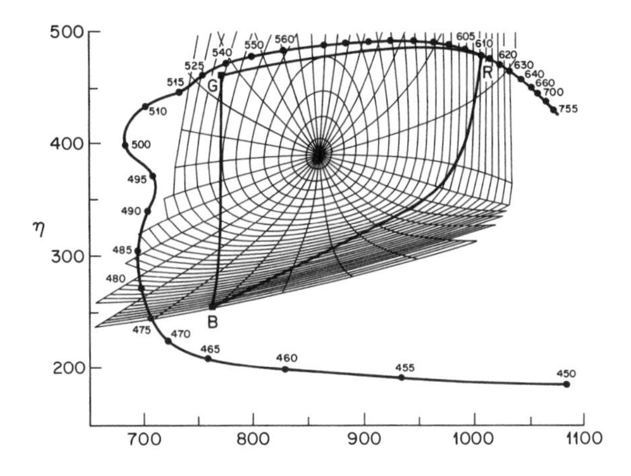

Fig. 5.1.4c Hue and saturation signals derived from U_t and V_t signals. (from Frei et al. [5.1.5])

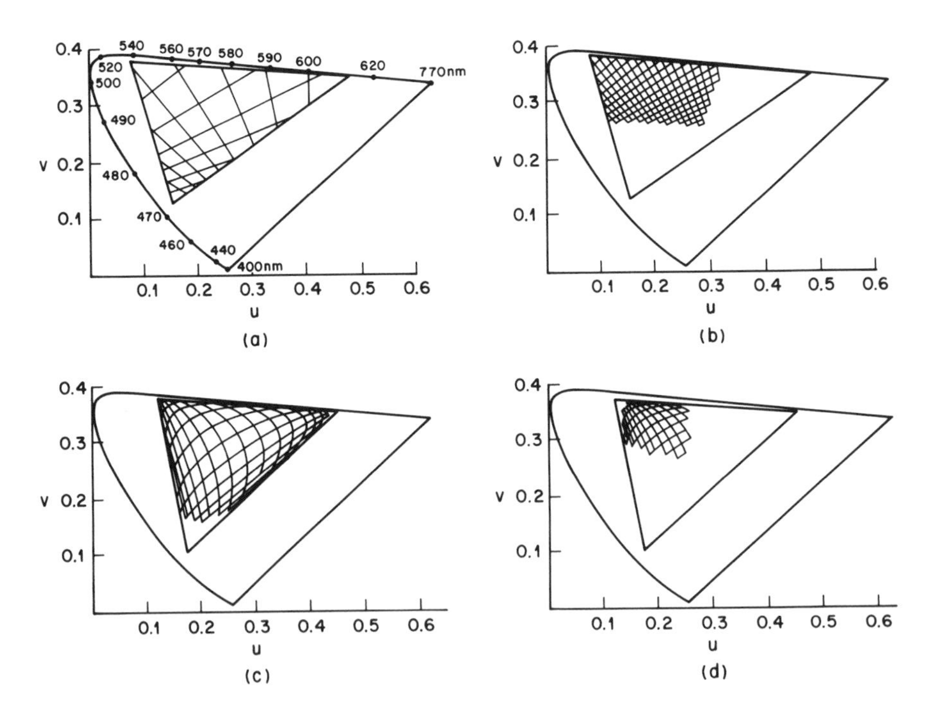

Fig. 5.1.5 Plots of the uniform quantization grids of the NTSC I and Q signals mapped on the 1960 CIE-uv diagram. Quantization is performed with 20 levels. (a) and (b) use gamma corrected luminance of 10 percent and 50 percent of the maximum value, respectively. (c) and (d) use gamma corrected R,G,B signals. Only the possible squares of the quantization grid are shown. (from Marti [5.1.6])

study in which a rectangular grid caused by uniform quantization of both components is mapped into a space in which equal distances represent more nearly uniform changes in perceived chromaticity.[5.1.5] Here, MacAdam's Geodesic space (see Section 4.4.2) is used. Plots are given for three types of chrominance components and for one value of luminance, without any gamma correction. From these plots we see that the quantization grid of the components considered in the study is, at least, locally uniform in the MacAdam's Geodesic diagrams. However, in all cases there is considerable compression of the grid in the blue and purple area. Also, there are many squares in the grid that are outside the color triangle for hue and saturation. If the quantization grid for I and Q is mapped onto the 1960 CIE-*uv* diagram (see Fig. 5.1.5a-b) rather than MacAdam's Geodesic diagram, the grid is again nonuniform with significant compression in the blue area.[5.1.6] At a high value of luminance, a smaller subset of the *uv*-space is occupied, and quantization is perceptually more uniform (see Fig. 5.1.5b).

In practice the nonlinear operation of gamma correction affects this quantization process significantly as shown in Fig. 5.1.5(c-d). With gamma correction of the R, G, B signals, the grid is less dependent on luminance; however, it is still quite nonuniform. One technique for perceptually more uniform quantization is to use logarithmic mapping of R, G, and B signals in order to reduce dependence of the grid size on luminance, followed by nonlinear pre- and post-processing of I and Q signals before and after quantization, i.e., companded quantization. Such a nonlinear mapping is discussed in Section 4.5.

Independent quantization of the components leads to several inefficiencies. First, the sensitivity of the human observer to quantization noise in chrominance components varies throughout the three dimensional color space, and it is these interactions between the chrominance components that cannot be exploited by independent quantization. Also, as we saw in Chapter 3, the range of possible values of the chrominance signal depends on the value of the luminance signal. For example, at high and low values of the luminance signal there is a considerable reduction in the possible range of the chrominance components.

Two equivalent approaches to digitization may be used to exploit this. In the first approach, signal components are first converted by a nonlinear transformation to a new three-dimensional space that approximates a uniform chromaticity space. Then uniform quantization is performed in this new space. In this new three-dimensional space, uniform quantization is accomplished by subdividing the space into cubes and giving each cube an output value. As an example, by scaling of the chrominance signals U_t and V_t as a function of the luminance, so that the scaled chrominance signals completely fill the quantizing range, it is experimentally observed that only four-bit quantization of the chromaticity components produces pictures without objectionable chrominance contouring or color errors. However, such an approach requires a nonlinear mapping before quantization. This is more difficult to implement than the following method.

In the second approach, a perceptually uniform color space is chosen first, and uniform quantization of this space is mapped back to quantization of the original chrominance signal space for digitization. Thus instead of nonlinear mapping of the chrominance signals followed by uniform quantization, nonlinear quantization or companding is performed directly on the chrominance signals. Such nonlinear quantization can be easily implemented in the form of look-up tables using read only memories (ROM's). In one example, [5.1.7] U_t and V_t signals of the PAL system were quantized by constructing volumes having approximately the same size in 1976 CIE-$L^*u^*v^*$ space. Dependence on luminance was included by using four different quantizations for four different luminance ranges. Only 1000 quantizer values (or regions), requiring 10 bits, were found to give a good quality picture.

In summary, simultaneous quantization of the chrominance components using perceptual redundancy and constraints on the possible range, can lead to significant reduction in the PCM bit rate for a required quality. The total bits required for PCM quantization of the chrominance signals is thus substantially smaller than the bits required for coding of the luminance component. This is a combined result of smaller bandwidth (lower bandwidth in both horizontal and vertical direction) and fewer quantization levels for the chrominance components. In general, chrominance bits take up about 10–20% of the total bits.

5.1.1a Color Maps

In many applications, either for economical reasons or because the pictures have inherently a limited number of colors in them (e.g., layouts of integrated circuits), it is necessary to map the space of colors into a small number of representative colors. Such maps are also called color tables, especially in the computer graphics literature. If 8 bits are used for each color component before mapping, then the color space has 2^{24} distinct colors.* However, if only a small number (say 8) of colors are to be used to represent this space, then a mapping is necessary from 2^{24} to eight colors. This is usually done by plotting a histogram in the 3-dimensional color space for a given picture (or a class of pictures) and choosing those representative colors that minimize the overall representation error. Since the 1976 CIE-$L^*u^*v^*$ space is perceptionally more uniform, it can be used to quantify the representation error. The maps or tables derived by such a procedure obviously depend upon the pictures used for computing the histogram. In many applications, it is possible to first load such tables for the

* They are not all visually distinct, however.

appropriate picture and then use them for that picture. Such techniques are similar to vector quantization discussed in detail in Section 5.5.

5.1.2 PCM Encoding of Composite Color Waveform

In Chapter 2, we saw that color television signals of today are broadcast in a frequency multiplexed composite form. Sampling and quantization of such a composite signal requires special considerations. Considerable freedom in the choice of sampling frequency exists for coding of the color components. Sampling frequency for components need only satisfy the Nyquist criterion, and be an integer harmonic of the scan line repetition rate. Quantization distortion is then less visible since the sampling sites are at the same spatial location on the raster from frame to frame. However, with composite coding, in order to avoid intermodulation distortion between the sampling and color subcarrier frequencies that occurs in the nonlinear process of A/D conversion, sampling should be a simple multiple of the subcarrier frequency (f_c). In addition, to minimize quantization noise patterns, sampling should be at a harmonic of the line frequency for NTSC, and an odd harmonic of half the line rate for PAL. Other reasons for sampling at a multiple of subcarrier frequency are: (i) ease of conversion from the composite to the component domain and vice versa; (ii) simplification of subsequent processing.

Two sampling frequencies are commonly utilized: $3f_c$ and $4f_c$. $3f_c$ is the lowest multiple of the subcarrier frequency that is above the Nyquist rate. Sampling at $4f_c$ generates more samples than sampling at $3f_c$, but allows considerable simplification of further processing such as filtering and coding. At $4f_c$, one chrominance component can be excluded from each sample by choosing the sampling phase appropriately. For NTSC, this results in a time division multiplexed signal consisting of a linear combination of luminance and chrominance signals (e.g., $Y+I$, $Y-Q$, $Y-I$, $Y+Q$). Similar remarks can be made about PAL television signals. However, due to frequency modulation, SECAM signals behave quite differently.

If uniform quantization with 8 bits/pel and gamma correction are used, quantization distortions are not generally visible provided the A/D converter is ideal. However, nonideal operation of the A/D converter, particularly the sample and hold circuit, may result in considerable visible patterning. Thus, it is generally agreed that due to the increased dynamic range of the composite signal (compared to the component signals) 9 bits per pel are required for PCM encoding of the composite signal. Although this might be considered overspecification in some instances, it helps overcome effects of subsequent processing, e.g., multiple encodings, digital filtering and other effects of finite word lengths.

In component encoding, different perceptual amplitude and spatial resolution requirements of the different components could be exploited to reduce the required quantization levels. Such an opportunity is not easily available in PCM quantization of the composite signal. Of course, to some extent, differing requirements of spatial bandwidth of the different components are already utilized in constructing the composite signal; but some of this advantage is lost due to the oversampling (at $3f_c$ or $4f_c$) performed to avoid visible distortion patterns.

5.2 Predictive Coding

PCM systems transmit quantized amplitudes of each pel. In Chapter 3 we saw that there is a strong correlation between adjacent pels that are spatially, as well as temporally, close to each other. Predictive coding exploits this correlation. In basic predictive coding systems,[5.2.1] an approximate prediction of the sample to be encoded is made from previously coded information that has been transmitted (Fig. 5.2.1). The error (or *differential* signal) resulting from the subtraction of the prediction the actual value of the pel is quantized into a set of L discrete amplitude levels. These levels are then represented as binary words of fixed or variable word-lengths and sent to the channel coder for transmission.

Thus, the predictive coder has three basic components: 1) Predictor, 2) Quantizer, 3) Code Assigner. Depending upon the number of levels of the quantizer, a distinction is often made between delta modulation[5.2.2] (DM), which has $L = 2$, and Differential Pulse Code Modulation (DPCM), which has L greater than 2. Since only two levels are used in DM, in order to get an adequate picture quality, the sampling rate has to be several times the Nyquist rate. Although DM has been used extensively in encoding other waveforms (e.g., speech), it has not found great use in encoding of pictures, due perhaps to the high sampling rates required. Consequently, we will give only a short description of DM coders, followed by considerable detail of the DPCM encoders.

5.2.1 Delta Modulation

A block diagram of a simple delta modulator is shown in Fig. 5.2.2. The relevant waveforms of the analog input signal, digital coded output and reconstructed waveforms are shown in Fig. 5.2.3. The comparator compares the analog input signal $y(t)$ with $Q(t)$, which is a reconstructed version of $y(t)$ at the transmitter. In the absence of any errors in transmission, $Q(t)$ is also very close to the reconstructed waveform at the receiver. The polarity of the difference signal $I(t)$ is calculated for each new sample by the latch. The comparator and latch combine the functions of sampling, holding and quantizing, and transmit a logical 1 if the error signal $I(t) > 0$ or a logical 0 if $I(t) \leq 0$. The output of the latch is combined with a delayed and inverted clock pulse to drive an integrator whose output is negatively fed back to the comparator.

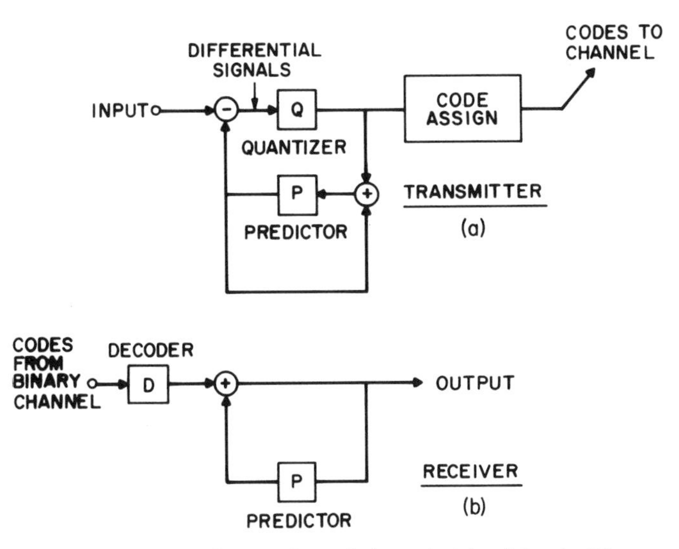

Fig. 5.2.1 Block diagram of a predictive coder (a) and decoder (b).

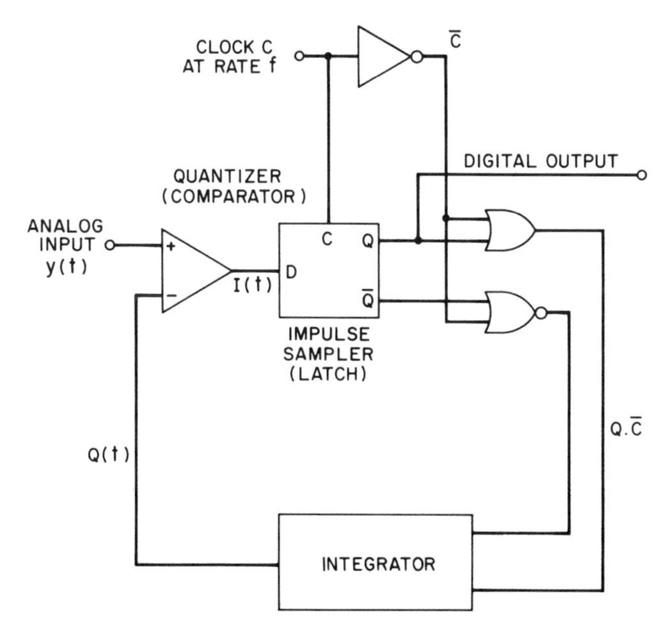

Fig. 5.2.2 Delta modulator encoder block diagram.

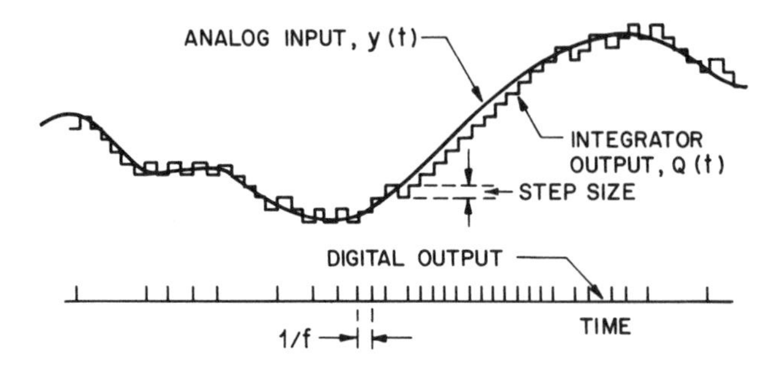

Fig. 5.2.3 Waveforms for typical analog input $y(t)$, integrator output $Q(t)$, and corresponding digital output from a delta modulation encoder. The sampling frequency is f.

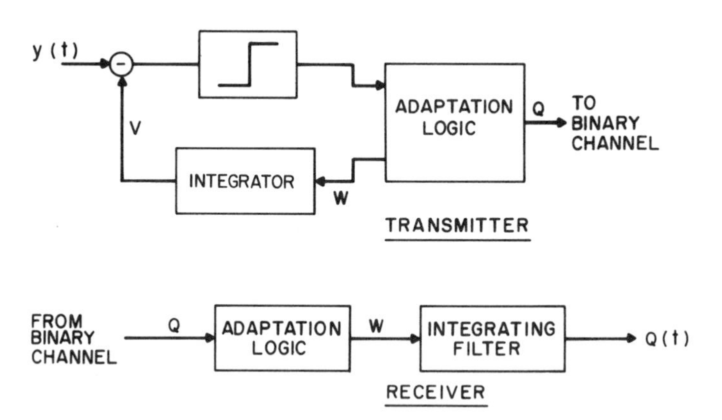

Fig. 5.2.4 Block diagram of companded or adaptive delta modulator. $y(t)$ = input waveform; $Q(t)$ = processed output waveform; W = weighting sequence; V is integrator output; Q = code sequence.

Thus the basic DM is extremely simple and, in addition, has the following advantages. Since DM generates a code that is exactly one bit per sample, the beginning and end of a code word are easily known. Also, the one-bit code can be easily converted to a wide variety of other digital codes suitable for transmission and storage. Performance of a DM depends generally on only two quantities, the quantizer step size and the sampling rate.

A larger step size gives the coder the ability to follow large transitions in the signal, but results in higher quantization noise in flat areas (low detail) of the picture. A smaller step-size, on the other hand, results in better reproduction of flat areas, but reduces the ability of the coder to follow sharp edges in the picture, which results in blurring of the picture. Thus the choice of step size is a trade-off between noise in the flat areas versus sharpness of the picture. Increasing the sampling rate gives improved picture quality at the cost of higher bit rate. Most DM coders need to sample the input signal at a rate several times higher than the Nyquist rate in order to achieve good quality. Sampling at a higher rate also has the beneficial effect that the quantization noise is then of higher frequency and is therefore less visible.

For a given sampling rate (or bit rate), the performance of the DM coder can be improved by adapting (or *companding*) the step size as a function of the local properties of the picture signal. In *syllabic* companding, which is used very successfully for speech coding, the step size is dependent on local averaging of some property of the signal. Such averaging is usually done over several samples and can be used to track slow or long term variations in statistics or other properties of the signal. In *instantaneous* companding, on the other hand, the step size is dependent on local signal values in a very small neighborhood, such as a few previously coded bits or the slope of the picture signal. Fig. 5.2.4 shows a block diagram of a companded delta modulator. The output code can be used to detect transitions in the signal (e.g., edges in pictures) and thereby change the step size. For example, three successive coded bits being the same indicates a sharp transition, and therefore the step size for the third sample could be made double that of the second sample. If two successive bits are different, indicating a flat area of the picture, the step size of the second sample could be half that of the first, thereby reducing the quantization noise.

Such adaptation improves performance, but if carried too far can result in instabilities. Some of these instabilities can introduce large oscillations and quantization noise that is within band and therefore more visible. Also, performance in the presence of transmission errors may suffer. However, if some of the future samples can be used in the adaptation process, the trend of the signal can be better estimated and instabilities can be avoided. This is done in *Delayed Coding*. In order to explain delayed coding, consider Fig. 5.2.5. At every sample a delta modulator allows two possible encodings (or two branches in a tree at each node). Each of these paths branch to give four options for the next sample, eight for two samples into the future, etc. Thus, from the present

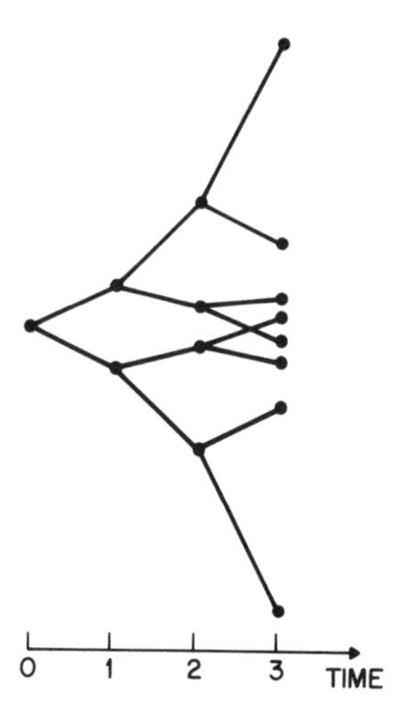

Fig. 5.2.5 Encoding tree for companded delta modulator with delayed encoding.

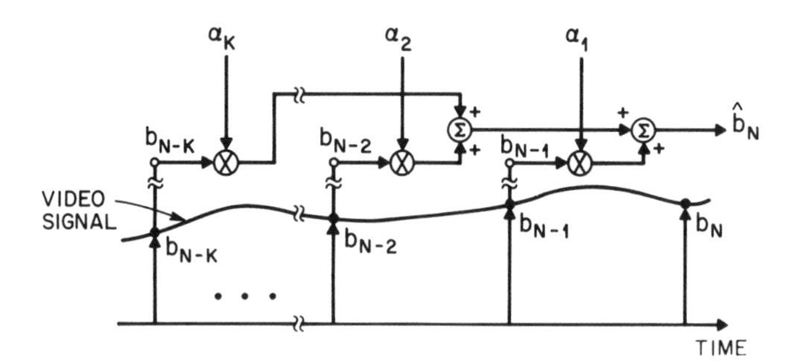

Fig. 5.2.6 An outline of linear prediction used for predictive coding. Note that quantization effects are neglected. The prediction, therefore, is made by using a weighted sum of K previous samples of the original signal, with $\{\alpha_i,\ i=1,\ldots,K\}$ as weights.

sample to the $n-th$ sample in the future, there are 2^n possible paths corresponding to every possible bit pattern that can be transmitted. The resulting decoded signal from each of these possible paths is compared sample by sample with the actual input signal, and a measure of encoded quality is evaluated (e.g., a sum of weighted difference). The encoding option, up (1) or down (0), for the present pel that lies on the path for which the encoded quality is the highest is then selected and transmitted. In the process of encoding the next sample, the window of $n+1$ samples is shifted forward by one sample, and the whole process is repeated. Experimental results indicate that the signal to noise ratio can be improved by about 3 dB by delayed encoding. Moreover, pictures appear a lot sharper due to better reproduction of edges. The original appeal of delta modulation was its simplicity. However, with a sophisticated adaptation algorithm and delayed coding, the DM coder becomes just as complex as the DPCM coders, which are discussed in the next section.

5.2.2 DPCM Coding

In DPCM coding, the analog signal is first sampled at or around the Nyquist rate. In its simplest form, DPCM uses the coded value of the horizontally previous pel as a prediction. This is equivalent to quantizing an approximation of the horizontal gradient of the signal. More sophisticated predictors make better use of the correlation by using more elements in the present field (including present and previous lines) as well as previous fields or frames of information. These are called *intrafield* and *interframe* predictors, respectively. Coding systems based on such predictors have been traditionally called intrafield and interframe coders. Interframe coders require a field or frame memory and are generally more complex than intrafield coders. However, as the cost of memory chips continues to drop, the distinction between intrafield and interframe coders is becoming less important. In the next few sections we will see how the three components of the predictive coder, namely predictor, quantizer and the code assigner, can be optimized for coding of different types of picture signals.

5.2.3 Predictors (General)

Predictors for DPCM coding can be classified as linear or nonlinear depending upon whether the prediction is a linear or a nonlinear function of the previously transmitted pel values. A further division can be made depending upon the location of previous pels used: *one-dimensional* predictors use previous pels in the same line as the pel being predicted; *two-dimensional* predictors use pels in the previous line(s) as well, whereas interframe predictors use pels also from the previously transmitted fields or frames. Predictors can be fixed or adaptive. Adaptive predictors change their characteristics as a function of the data, whereas fixed predictors maintain the same characteristics independent of the data.

5.2.3a Linear Predictors

Linear predictors for monochrome images have been studied using the general theory of linear prediction described in Chapter 3[5.2.3–5.2.4]. If $\{b_1,\ldots b_N\}$ is a block of zero mean pels for which $b_1,\ldots b_{N-1}$ have already been transmitted, (and therefore can be used to predict b_N) then a linear predictor for b_N can be written* as

$$\hat{b}_N = \sum_{i=1}^{N-1} \alpha_i \, b_{N-i} \tag{5.2.1}$$

This is shown diagrammatically in Fig. 5.2.6. The coefficients $\{\alpha_i\}$ can be obtained by minimizing the mean square prediction error (MSPE), $E(b_N - \hat{b}_N)^2$ as in Section 3.1.5. Using the optimum coefficients [see Eq. (3.1.38)], the mean square prediction error is given by

$$\text{optimum (MSPE)} = \sigma^2 - \sum_{i=1}^{N-1} \alpha_i d_i \tag{5.2.2}$$

where $d_i = E(b_N \cdot b_{N-i})$, and pels are assumed to be identically distributed with zero mean and variance σ^2. Thus the MSPE of the input to the DPCM quantizer is reduced by $\sum_{i=1}^{N-1} \alpha_i d_i$ from σ^2, the mean square of the input to a PCM quantizer. It is found in most cases that the sum of the coefficients $\{\alpha_i\}$ is close to one, and therefore Eq. (5.2.1) is typically used with original pel values, i.e., without first subtracting the mean value of the pels.

The above analysis assumes stationarity and identical statistics of the sequence of pels $\{b_n\}$. Although the theory of linear prediction can treat signals having a wide variety of stationary statistics, its application to image coding has not been very successful. The principal reasons are that there are, as yet, no satisfactory statistical models that accurately describe the picture signal and secondly, while minimizing mean square prediction error is important, it is not equivalent to minimization of bit-rate or optimization of coded picture quality. Also, the above analysis neglects the effects of quantization in the DPCM coder. In a real DPCM coder, the prediction of pel b_N can only be made using previously encoded representations of the past samples $\tilde{b}_{N-1}, \tilde{b}_{N-2},\ldots$, and not by using the original uncoded pel values. This is necessary to allow the receiver to also compute the prediction. For a coder that produces high quality pictures,

---

* If $\hat{b}_N$ is outside the range of original pel values, it is usually clipped. Also, if correlations are high enough, zero mean is unnecessary.

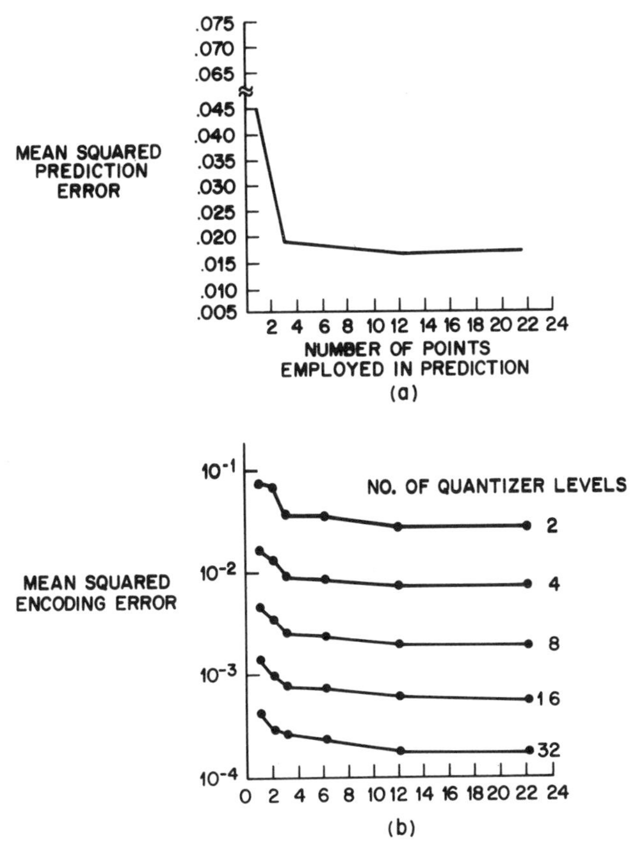

Fig. 5.2.7 Performance of different linear predictors. (a) MSPE is plotted as a function of the order of the predictor (number of points employed in the predictor) without a quantizer in the loop. (b) Mean square encoding error is plotted with a quantizer having 2,4,8,16 and 32 levels. (from Habibi [5.2.5])

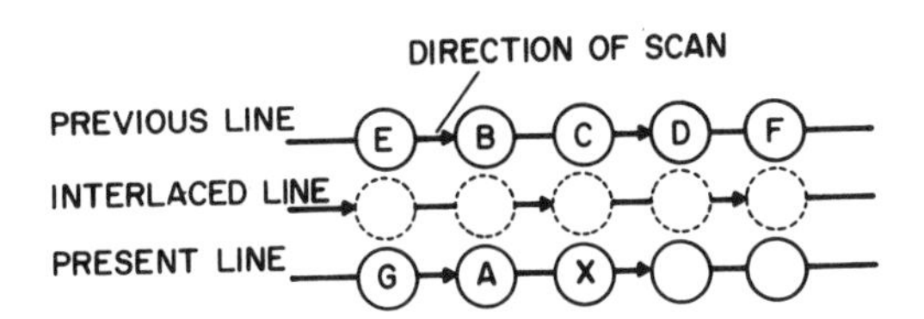

Fig. 5.2.8 Position and naming of picture elements used in Table 5.2.1 for prediction of pel X. Present and previous lines are from the same field.

however, the effects of quantization are usually small and may often be neglected.

The minimum mean square prediction error (MSPE) can be determined experimentally for linear predictors that use different numbers of picture elements within a field.[5.2.5] Results of such calculations are plotted in Fig. 5.2.7a. For these results the statistics were first precomputed for each given picture and then the predictor weighting coefficients were calculated [as in Eq. (3.1.38)] to match the statistics. In such a case, there is a dramatic decrease in the MSPE by using up to three pels, but further decrease by using more pels is small. If the statistics of the picture are not precisely known then the predictor coefficients may not be well matched to the picture, and the decrease in the prediction error by using three pels may not be very significant compared to using only the adjacent previous pel. Also the percent improvement in mean square encoding error as a function of the number of samples used changes little with respect to the coarseness of the quantizer used in the DPCM loop, as shown in Fig. 5.2.7b.

5.2.3b Practical Intrafield Predictors

In practice, two-dimensional predictors are often used. Although the improvement in entropy of the prediction error may not be substantial (see Chapter 3) by using two dimensional prediction, there can be considerable decrease in the peak prediction error. Moreover, subjective evaluations of quantized images indicate that the rendition of edges of various orientations using two-dimensional prediction is significantly improved. Also, by the proper choice of coefficients, it is possible to get improved prediction as well as quick decay of the effects of transmission bit errors in the reconstructed picture (see Section 5.2.3.j). In general, since correlations are usually high and the mean of the picture signal does not change dramatically within a few samples, the sum of the optimum MSPE predictor coefficients $\{\alpha_i\}$ is usually close to one. However, to ensure that the effect of any perturbation (e.g., quantization or round-off errors) in the DPCM loop would die off as a function of time (i.e., the loop would be stable), the worst case gain $\sum_{i=1}^{N-1} |\alpha_i|$ should be as small as possible. This implies that it is better to use a larger number of picture elements, each with a small positive coefficient, in the prediction process.

Another important consideration for ease of implementation is that the predictor coefficients $\{\alpha_i\}$ should be powers of (2^{-1}). Table 5.2.1 lists several predictors and their performance (in terms of mean square prediction error) for the luminance and chrominance components of a color television signal. The relative positions of pels are shown in Fig. 5.2.8. This data was experimentally collected for a set of pictures that were sampled at 10 MHz for luminance and 2 MHz for both the chrominance signals $(R-Y)$ and $(B-Y)$ and uniformly quantized to 8 bits. It is seen that one of the two-dimensional predictors with

Table 5.2.1 Prediction Coefficients and Resulting Variances of Prediction Errors for Luminance and Chrominance Signals. Pel positions are shown in Fig. 5.2.8.

	PREDICTION COEFFICIENTS			VARIANCE
VIDEO SIGNAL	α_A	α_B	α_C	σ_e^2
Y	1	—	—	53.1
	1	−1/2	1/2	29.8
	3/4	−1/2	3/4	27.9
	7/8	−5/8	3/4	26.3
R − Y	1	—	—	22.6
	—	—	1	6.8
	1/2	−1/2	1	4.9
	5/8	−1/2	7/8	4.7
B − Y	1	—	—	13.3
	—	—	1	3.2
	1/2	−1/2	1	2.5
	3/8	−1/4	7/8	2.5

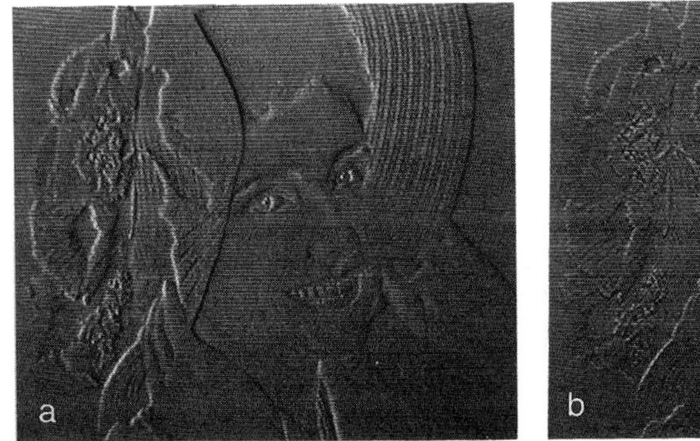

Fig. 5.2.9 Prediction error of a television picture processed with a 4-bit DPCM system using (a) one-dimensional prediction ($\alpha_A = 1.0$) and (b) two dimensional prediction ($\alpha_A = 0.75, \alpha_B = -0.5, \alpha_C = 0.75$). (from Musmann [5.2.6])

simple coefficients reduces the MSPE by a factor of 2, compared to the simple previous-element predictor.

Fig. 5.2.9 shows pictures of luminance prediction errors for one- and two-dimensional predictors with a four-bit quantizer in the DPCM loop. Positive and negative errors are represented by white and black elements, respectively, while zero error is shown as mid gray. The two-dimensional prediction reduces the error on vertical edges, but increases the error on horizontal edges as compared to one-dimensional prediction. Thus a particular choice of predictor coefficients and samples used in the predictor affects the ability of the predictor to handle correlation in a particular direction. Excellent predictors can be designed if the direction of maximum correlation in the local neighborhood is known. Such predictors are treated in Section 5.2.3d.

For the chrominance signals, due to lower sampling rates, in Fig. 5.2.8 the distance between pels A and X is much greater than between X and C. This results in sample C being a much better predictor than sample A by almost a factor of 4 using MSPE as the criterion (Table 5.2.1).

5.2.3c Intrafield Predictors for Composite Television

In Section 5.1.2 we discussed some special considerations necessary in the sampling of signals bearing subcarriers such as the composite TV signal. Special problems are also created in the prediction process, since much of the correlation in the adjacent samples is lost due to the presence of modulated chrominance components. It is possible, however, to overcome to a large extent the effect of the color subcarrier by proper design of the predictor. In Section 5.1.2 we saw the advantages of sampling the composite signal at an integral multiple of the subcarrier frequency. With such sampling, the effects of the subcarrier might be accommodated by using only samples of the same phase for prediction. Thus, if the sampling rate is nf_{sc}, (n being an integer), then the n-th previous sample along the same scan line can be used as the prediction. Since the n-th previous sample has the same phase, the effect of the subcarrier is minimized. However, such a sample is also further away resulting in less correlation than if the signal were a baseband color component.

More complex predictors can also be designed in order to match the bimodal spectrum of the composite signal (see Fig. 2.2.4a) so that simultaneous prediction of the baseband luminance and modulated chrominance is almost as efficient as could be obtained by separate prediction of the individual color components. We illustrate some of these predictors for an NTSC signal sampled at $4f_{sc}$. Similar predictors are possible, with some modifications, for PAL signals as well as for $3f_{sc}$ sampling. A $4f_{sc}$ sampled signal is shown in Fig. 5.2.10 for a portion of a scan line where the chrominance and luminance are approximately constant. Fig. 5.2.11 shows adjacent lines in the present and previous fields along with pels denoted by A, B, C, D. The pel to be encoded is denoted as X (having the same phase of the color subcarrier as pel D). If the

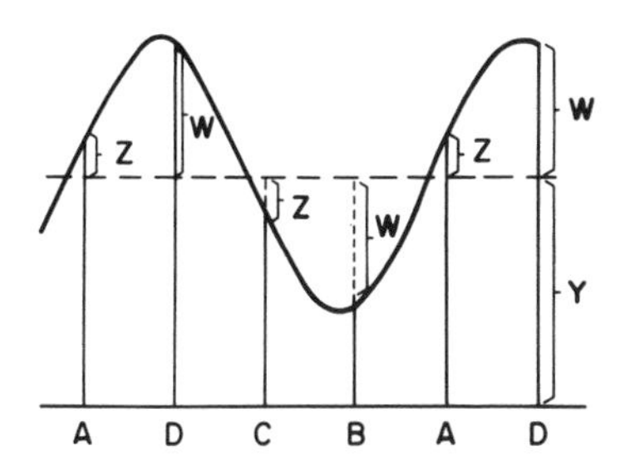

Fig. 5.2.10 Sampling of the composite signal at four times the subcarrier frequency.

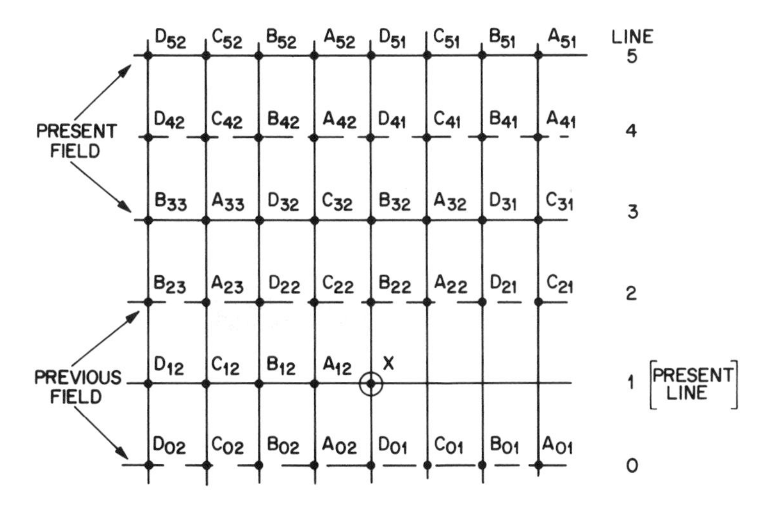

Fig. 5.2.11 Position of the pels surrounding the predicted pel X in NTSC signals sampled at $4f_c$.

sampling covers some area of the picture where both the luminance and the chrominance are constant, then the pel amplitudes are:

$$\begin{aligned} A &= Y + Z \\ B &= Y - W \\ C &= Y - Z \\ X = D &= Y + W \end{aligned} \tag{5.2.3}$$

where Z and W are instantaneous color carrier amplitudes. In terms of pel amplitudes, they can be written as:

$$\begin{aligned} W &= D - \frac{1}{2}(A+C) \\ Z &= A - \frac{1}{2}(D+B) \end{aligned} \tag{5.2.4}$$

Since the color phase of X and D is the same, any linear predictor $\hat{X}$ for X can be written as

$$\hat{X} = \hat{Y} + \hat{W} \tag{5.2.5}$$

where $\hat{Y}$ predicts the luminance at X and $\hat{W}$ the instantaneous color carrier value. Two possible estimates of Y are:

$$\hat{Y}_1 = \frac{1}{2}(A+C) \tag{5.2.6}$$

and

$$\hat{Y}_2 = \frac{1}{2}(B+D) \tag{5.2.7}$$

Also three possible predictors for the color carrier component W are:

$$\hat{W}_1 = \frac{1}{2}(A+C)-B \tag{5.2.8}$$

$$\hat{W}_2 = \frac{1}{2}(D-B) \tag{5.2.9}$$

$$\hat{W}_3 = D - \frac{1}{2}(A+C) \tag{5.2.10}$$

The pels A, B, C, D can be selected from any of those shown in Fig. 5.2.11. However, pels closer to X would be more correlated with it and therefore more desirable. From Eqs. (5.2.6-5.2.10), we see that there are two elementary predictors for X, namely

$$\hat{X}_1 = D \tag{5.2.11}$$

and

$$\hat{X}_2 = (A-B+C) \tag{5.2.12}$$

Since the elements $\{A, B, C, D\}$ can be chosen in a number of different ways and formed into linear combinations to give more complex predictors that satisfy Eq. (5.2.5), a more general predictor is given by

$$\hat{X} = \sum_{i,j} \alpha_{ij} D_{ij} + \sum_{i,j} \beta_{ij}(A_{ij}-B_{ij}+C_{ij}) \tag{5.2.13}$$

Also, since the range of the signal does not change from sample to sample, the coefficients should add up to one, i.e.,

$$\sum_{i,j} \alpha_{ij} + \sum_{i,j} \beta_{ij} = 1 \tag{5.2.14}$$

Also, it is possible to add any number of null predictors that are zero in an area of constant luminance and chrominance, but which may give useful information about the slope of luminance and chrominance signals in the area of the change. This makes the following generalization possible.

$$\hat{X} = \sum_{i,j} \alpha_{ij} D_{ij} + \beta_{ij}(A_{ij}-B_{ij}+C_{ij}) + \gamma_{ij}(A_{ij}-A'_{ij}) + \delta_{ij}(B_{ij}-B'_{ij}) + \varepsilon_{ij}(C_{ij}-C'_{ij}) \tag{5.2.15}$$

with the same constraint as Eq. (5.2.14) for unity gain. The notation A'_{ij}, B'_{ij}, C'_{ij}

indicates that these pels are chosen to be in the same neighborhood as A_{ij}, B_{ij}, and C_{ij}, and have the same color carrier phase.

Although the above expressions might indicate considerable flexibility, in practice based on cost and performance, only a few predictors have been used. As with predictors for component signals, the best pels are those that are close to X in space and time, in a combination that gives reasonable amplification of edges of various orientation. With $4f_{sc}$ sampling, adjacent pel distance along a scan line is small. Calculation shows that B_{32} is about 2.3 times further away from X than is A_{12}. Although the above techniques are optimum for signals containing phase modulated subcarriers where the required prediction distance is constant in areas of uniform color, they can also be applied to SECAM signals. As an example, if the SECAM signal is sampled at $3f_{sc}$, the third previous element predictor is found to be quite efficient on most picture material.

5.2.3d Adaptive Intrafield Predictors

We saw in Chapter 3 that a picture signal is highly nonstationary. Therefore, it is advantageous to change the prediction based on the local properties of the picture signal. If the calculation of the local properties uses some yet untransmitted pels, then side information may have to be transmitted so that the receiver can make the appropriate change in the prediction. If only previously transmitted information is used in computing the predictor, then the receiver can perform the same calculation, and no side information need be sent. For the intrafield situation, one commonly used method is to compute some measure of directional correlation based on the local neighboring pels that have already been transmitted and use this measure to select a predictor along the direction of maximum correlation. The set of predictors from which a predictor is selected are usually linear and are chosen such that each one of them will give small prediction error if the signal is correlated in a certain direction. For example, in Graham's predictor[5.2.7] either the previous line (pel C in Fig. 5.2.8) or the previous pel (pel A) is used for prediction, and the switching between them is done by the following rule:

$$\hat{X} = \text{predictor for element } X$$

$$= \begin{cases} A & , \text{ if } |B-C| < |A-B| \\ C & , \text{ otherwise} \end{cases} \tag{5.2.16}$$

Compared to the previous pel prediction, this prediction reduces slope overload on vertical edges and in general results in sharper pictures. However, it does not improve the edges at an angle (e.g., diagonal edges) substantially.

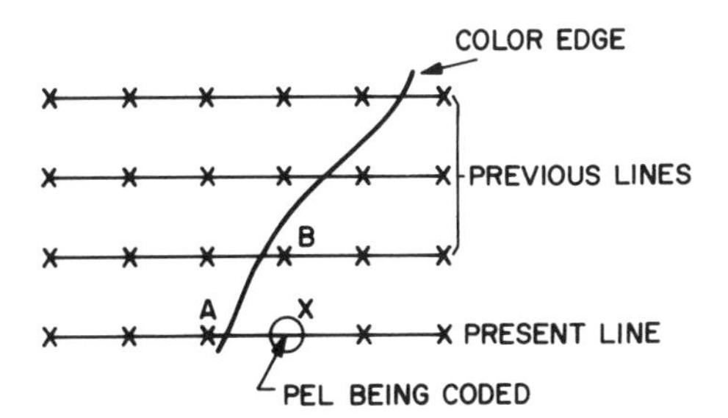

Fig. 5.2.12 Adaptive Prediction of Chrominance Components. Due to the color edge, correlation between pels *X* and *B* is high (compared to the correlation between *X* and *A*) for all the color components.

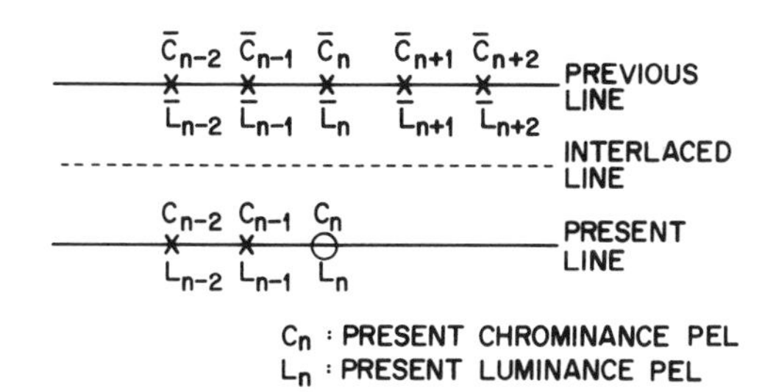

Fig. 5.2.13 Pel Configuration for Adaptive Prediction. It is assumed that luminance and chrominance samples are available at same location. As is common practice, if chrominance is subsampled with respect to luminance, then luminance samples at subsampled chrominance locations may be used for adaptive prediction.

An extension of this concept is the so called *contour* prediction. Here, if we assume that the direction of a contour (i.e., large spatial change in intensity) crossing a scan line does not change significantly, then the contour direction can be determined at pel A by searching for that pel among E, B, C, or G that gives the minimum difference $|A-E|$, $|A-B|$, $|A-C|$, or $|A-G|$. (See Fig. 5.2.8) Then, the neighboring pel to the right can be taken as a prediction of pel X. Contour prediction improves picture quality whenever the direction of the contour can be predicted reliably. However, in practice contour direction cannot always be predicted accurately, which often results in artifacts. Another variation in adaptive prediction is to use a weighted sum of several predictors, where the weights are switched from pel to pel and are chosen by observing certain characteristics of already transmitted neighboring pels. However, experimental evidence shows that such adaptive intrafield prediction does not improve the mean square prediction error or its entropy significantly, unless the edges in the picture are very sharp. In some cases, however, the subjective rendition of certain types of quantized edges can be remarkably improved.

In the case of component color coding, adaptation of the luminance prediction may be done in one of the ways discussed above. Having coded the luminance component, it is desirable to derive from it as much useful information about coding of the chrominance components as possible. In most natural color pictures, the majority of large spatial changes in the chrominance signal are also accompanied by large spatial changes in the luminance signal (see Section 3.3). In Fig. 5.2.12, due to a color edge passing between the present pel X and the previous coded pel A, pel B provides a better prediction of X for both the luminance and chrominance. Thus, in this case, knowing the better predictor for the luminance of pel X makes it possible to derive a better predictor for the chrominance. This can be applied as in Fig. 5.2.13, where L_n, C_n are the luminance and chrominance of the present pel, respectively. It is assumed that the luminance pels L_n, L_{n-1}, $\bar{L}_{n-1}$, $\bar{L}_n$, $\bar{L}_{n+1}$ and the chrominance pels C_{n-1}, $\bar{C}_n$, $\bar{C}_{n+1}$ are available to the transmitter and receiver in coded form. First, several predictions of L_n are made using values of various previous pels and their combinations. Knowing L_n, the prediction error is computed for each predictor, and the predictor that gives the least absolute prediction error is selected. It is then assumed that the best representation of the present chrominance sample is also provided by the same predictor. Thus, if L_{n-1} was the best predictor for L_n, then C_{n-1} is the best predictor for C_n. Experimental results indicate that the entropy of the prediction error of the chrominance signals is decreased by about 15 to 20 percent due to such adaptation. In this scheme, it is assumed that both the luminance and chrominance samples are available at the same location in the image. In practice, however chrominance components are subsampled to a rate below that of the luminance. In such a case, luminance samples at subsampled chrominance locations may be used in the same way as above, but then the reduction in entropy is less.

5.2.3e Interframe Prediction

Interframe prediction, in general, uses a combination of pels from the present field as well as previous fields (see Section 3.2.5). For scenes with low detail and small motion, frame difference prediction appears to be the best. In scenes with higher detail and motion, field difference prediction does better than frame difference prediction.[5.2.8] As the motion in the scene is increased further, intrafield predictors do better. This is largely due to two reasons: 1) for higher motion, there is less correlation between the present pel and either the previous field or previous frame pel, and 2) due to the time integration of the signal in the video camera, the spatial correlation of the television signal in the direction of motion is increased. For the same reasons, predictors that produce an element or line difference of frame or field differences also perform better for higher motion. Recapping the results of Sec. 3.2.5, a typical variation of entropy of the prediction error for elements that have changed significantly since the previous frame (called the *moving area pels*) is shown in Fig. 3.2.12a–f as a function of different predictors and motion of the object in the scene.

Adaptive frame-to-frame prediction is based on an approach similar to intrafield prediction. As an example, the extension of Graham's predictor into three dimensions would select either the previous frame or an intrafield predictor depending on surrounding information. Thus, in Fig. 5.2.14, pel Z is predicted by either pel H (previous element), pel B (previous line in the same field), or pel M (previous frame) depending on which of the differences $|H-G|$ (element difference), $|H-A|$ (line difference), or $|H-L|$ (frame difference) is the smallest. Such switching implies that the coded values of samples, G, H, B,..., are available. However, in some interframe coders, during coder overload, these pels may not be available due to subsampling. In such cases, the predictor has to be suitably modified. However, better performing adaptive predictors for frame-to-frame coding are the ones that take into account the speed and direction of motion of objects, which is the subject of next section.

5.2.3f Motion Estimation

If a television scene contains moving objects and if an estimate of their translation is available, then more efficient prediction can be performed using elements in the previous frame that are appropriately spatially displaced. Such prediction is called *motion compensated* prediction. In real scenes, motion can be a complex combination of translation and rotation. Such motion is difficult to estimate and may require large amounts of processing. However, translational motion is easily estimated and has been used successfully for motion compensated coding. Its success depends upon the amount of translational motion in the scene and the ability of an algorithm to estimate translation with the accuracy that is necessary for good prediction. The crucial problem is the algorithm used for motion estimation. We give below a brief review of algorithms for motion estimation that have been successfully used in coding applications.

Most of the algorithms for motion estimation in interframe coding make the following assumptions:* (i) objects move in translation in a plane that is parallel to the camera plane, i.e., the effects of camera zoom, and object rotation are not considered, (ii) illumination is spatially and temporally uniform, (iii) occlusion of one object by another, and uncovered background are neglected.

Under these assumptions the monochrome intensities $b(\mathbf{z}, t)$ and $b(\mathbf{z}, t-\tau)$ of two consecutive frames are related by

$$b(\mathbf{z}, t) = b(\mathbf{z}-\mathbf{D}, t-\tau) \tag{5.2.17}$$

where τ is the time between two frames, $\mathbf{D}$ is the two dimensional translation vector of the object during the time interval $[t-\tau, t]$ and $\mathbf{z}$ is the two dimensional vector $[x,y]'$ of spatial position. Thus, in real scenes a very good prediction of $b(\mathbf{z}, t)$ is $\hat{b} = b(\mathbf{z}-\mathbf{D}, t-\tau)$. The problem then is to estimate $\mathbf{D}$ from the intensities of the present and previous frame. Two broad types of methods have been used: (a) Block Matching Methods, in which $\mathbf{D}$ is generally coded and transmitted as side information, and (b) Recursive Methods, in which $\mathbf{D}$ is not transmitted, but instead estimated from previously transmitted pels. It is also possible to combine these two methods in a variety of ways.

5.2.3g Block Matching Methods

With block methods, it is assumed that the object displacement is constant within a small two-dimensional block B of pels. This assumption presents difficulties in scenes with multiple moving objects or in scenes in which different parts of the same object move with different displacements. If the size of the block is decreased then this assumption become more valid; however, then the overhead of computation and transmission of displacement information increases.

The displacement $\mathbf{D}$ can be estimated by correlation or matching techniques. Thus, $\mathbf{D}$ can be chosen such that it minimizes some measure of the prediction error such as

$$PE(\mathbf{D}) = \sum G(b(\mathbf{z},t) - b(\mathbf{z}-\mathbf{D},t-\tau)) \tag{5.2.18}$$

where $G(\cdot)$ is a distance metric such as the magnitude or the square function.

---

* However, several algorithms described in this section have been shown experimentally to be robust even when these assumptions are not met fully.

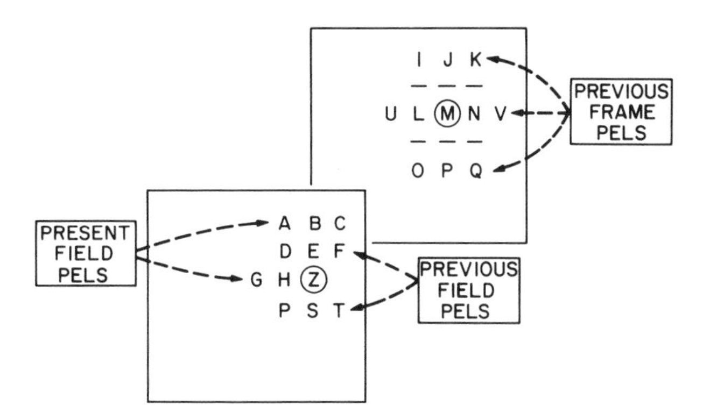

Fig. 5.2.14 Pel configuration for interframe predictors. The current pel Z is to be predicted from pels using previous scan lines, field and frame as well as previous pels in the same scan line as pel Z.

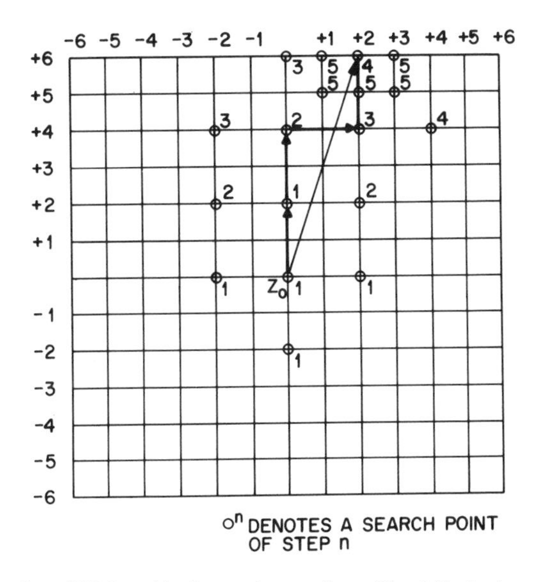

Fig. 5.2.15 Illustration of 2D-logarithmic search procedure. The shifts in the search area of the previous frame are shown with respect to a pel (z_0) in the present frame. Here the approximated displacement vectors $(0,2)'$, $(0,4)'$, $(2,4)'$, $(2,5)'$, $(2,6)'$ are found in steps 1, 2, 3, 4 and 5. d_{max} is 6 pels. (from Musmann et al. [5.2.9])

Having found the best **D**, all pels $z \in B$ are predicted, coded and transmitted using that value of **D**.

As an example consider a block of $M \times N$ pels centered around pel $\mathbf{z}_0$ in the current frame at time t. Assuming a maximum horizontal and vertical displacement of d_{max} pels, the region in the previous frame where the minimum of Eq. (5.2.18) has to be found would be given by*

$$\mathbf{z}_0 \pm \begin{bmatrix} \frac{1}{2}(M-1) + d_{max} \\ \frac{1}{2}(N-1) + d_{max} \end{bmatrix} \tag{5.2.19}$$

The number of pels in this region would be given by

$$(M + 2d_{max})(N + 2d_{max}) \tag{5.2.20}$$

Thus if the block size is 9×9 and a maximum displacement of $d_{max} = 10$ is used, the search region in the previous frame would be an area containing 29×29 pels. An exhaustive search method would be to evaluate Eq. (5.2.18) for every pel shift in the horizontal and vertical direction and choose the minimum. This would require $(2d_{max} + 1)^2$ evaluations of Eq. (5.2.18). Moreover unless shifts by fractional pel distances are also included, the accuracy with which **D** could be obtained is limited to one pel. Several simplifications have been suggested in the literature. One type of simplification is to use a simple distance criterion G for Eq. (5.2.18). Examples are:

$$PE(\mathbf{z}_0, i, j) = \frac{1}{MN} \sum_{|m| \le \frac{M}{2}} \sum_{|n| \le \frac{N}{2}} [b(\mathbf{z}_{mn}, t) - b(\mathbf{z}_{m+i,\, n+j}, t-\tau)]^2 \tag{5.2.21}$$

or

$$PE(\mathbf{z}_0, i, j) = \frac{1}{MN} \sum_{|m| \le \frac{M}{2}} \sum_{|n| \le \frac{N}{2}} | b(\mathbf{z}_{mn}, t) - b(\mathbf{z}_{m+i,\, n+j}, t-\tau) | \tag{5.2.22}$$

where

---

* In these equations, we assume M,N to be odd. A similar result for even M,N follows easily.

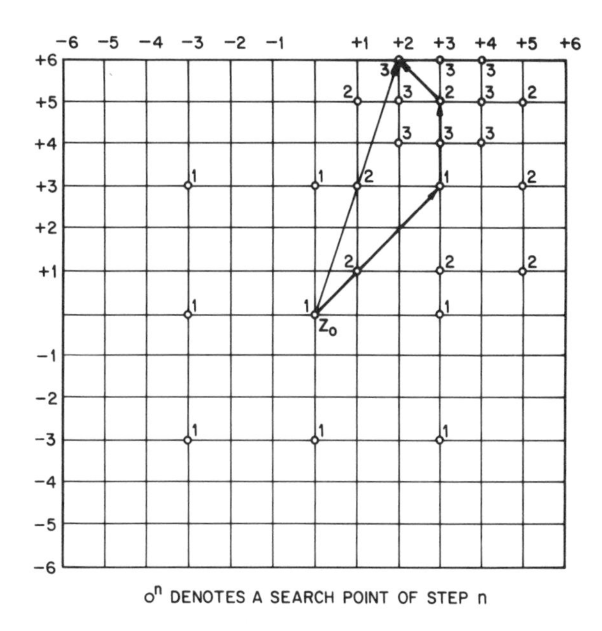

Fig. 5.2.16 Illustration of the three step search procedure. Here, $(3,3)'$, $(3,5)'$ and $(2,6)'$ are the approximate displacement vectors found in steps 1, 2 and 3. d_{max} is 6 pels. (from Musmann et al. [5.2.9])

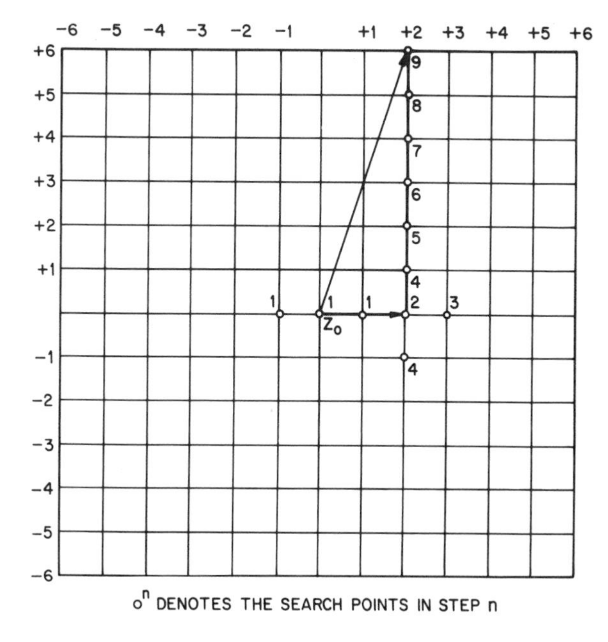

Fig. 5.2.17 Illustration of a simplified conjugate direction search method. Here, $(2,6)'$ is the displacement vector found in step 9. d_{max} is 6 pels. (from Musmann et al. [5.2.9])

$$-d_{\max} \leq i,\ j \leq +d_{\max}$$

$$\mathbf{z}_{mn} = \mathbf{z}_0 + [m,\ n]' \tag{5.2.23}$$

The second definition of $PE(\cdot)$ in Eq. (5.2.22) has the advantage that no multiplications are required. Also, it is experimentally observed that the precise definition of *PE* does not have a significant effect on the amount of searching or the accuracy of estimation of **D** and, therefore due to simplicity, the criterion of Eq. (5.2.22) is generally preferred.

In addition to the simplification of the matching criterion, several methods for simplifying the search procedure have been investigated. We describe three promising ones: (a) 2D-logarithmic search, (b) three step search, and (c) modified conjugate direction search. In all these, the goal is to require as few shifts as possible and therefore evaluate *PE* as few times as possible (each shift requires one calculation of *PE*). In reducing the number of shifts the assumption that is commonly made is that $PE(\mathbf{z}_0,\ i,\ j)$ increases monotonically as the shift $(i,\ j)$ moves away from the direction of minimum distortion.

In a 2D-logarithmic search, the algorithm follows the direction of minimum distortion. At each step, five shifts are checked as shown in Fig. 5.2.15. The distance between the search points is decreased if the minimum is at the center of search locations, or at the boundary of the search area. For the example shown in Fig. 5.2.15, five steps were required to get the displacement vector at the point $(i,j) = (2,6)$.

In the three step search procedure, the center position at $\mathbf{z}_0$, as well as eight coarsely spaced positions around the center position, are tested in the first step. In the second step, again eight positions are used around the position of minimum *PE* found from the first step, but in this step spacing is finer than the first step. Fig. 5.2.16 shows points $\mathbf{z}_0 + (3,3)'$ and $\mathbf{z}_0 + (3,5)'$ as first and second approximations, respectively. This process is continued for yet another step, resulting in the final displacement vector at position $\mathbf{z}_0 + (2,6)'$.

The conjugate direction search, in its simple form, searches for the direction of minimum $PE(\mathbf{z}_0,\ i,\ j)$. It is a two step procedure. In the first step, a search is performed along the horizontal direction. First the minimum of $PE(\mathbf{z}_0, -1, 0)$, $PE(\mathbf{z}_0,\ 0,\ 0)$, and $PE(\mathbf{z}_0,\ 1,\ 0)$ is computed. If $PE(\mathbf{z}_0,\ 1,\ 0)$ is the smallest, then $PE(\mathbf{z}_0,\ 2,\ 0)$ is also computed and compared. Proceeding in this way, the algorithm determines the minimum in the horizontal direction, when it is located between two higher values for its neighboring positions. In the second step, the identical procedure is used to compute the minimum in the vertical direction. Fig. 5.2.17 shows an example of the conjugate direction method in which the results of the two steps are positions $\mathbf{z}_0 + (2,\ 0)'$ and $\mathbf{z}_0 + (2,\ 6)'$ respectively.

It is obvious that all the three search methods are based on some heuristics and may experience difficulties in certain cases. Limited simulations show promise, but more experience is necessary before definitive statements can be made about their efficiency and usefulness. As mentioned before, computing complexity of the search procedure is related to the number of shifts for which PE needs to be evaluated. Fig. 5.2.18 shows a comparison of the number of search shifts and sequential steps required by each of the above search procedures. It should be noted that the displacement estimation accuracy of the block-methods is generally limited to 0.5 pels. It could, in principle, be increased by including interpolated values of the pels and thereby making the grids shown in Figs. 5.2.15–17 finer. However, the complexity is usually prohibitive.

One of the characteristics of the block methods is that, since pels that are not yet transmitted to the receiver are used in the displacement calculation, displacement values need to be transmitted separately as overhead. This increases the bit-rate, but also has the potential of better motion estimates. As an example, if a 5×5 block is used, and if d_{max} is ± 8 pels, then for a displacement accuracy of ½ pel, for every block, 12 bits specifying the motion vector have to be transmitted, resulting in a bit rate of 0.48 bits/pel. Of course, many techniques can be used to reduce this.

For example, since there is significant block-to-block correlation of motion vectors, predictive coding of motion vectors is more efficient than coding each $\mathbf{D}$ independently. Also, since the PDF of the resulting differential motion vectors is highly peaked around zero, Huffman coding is far more efficient than constant word lengths.

Simpler displacement estimation techniques utilize the relationship between spatial and temporal changes of the intensity. Using Eq. (5.2.17), we can write the frame difference signal, $FDIF(\mathbf{z}, t)$ as

$$FDIF(\mathbf{z}, t) \triangleq b(\mathbf{z}, t) - b(\mathbf{z}, t-\tau) \tag{5.2.24}$$

$$= b(\mathbf{z}, t) - b(\mathbf{z}+\mathbf{D}, t) \tag{5.2.25}$$

which can be written* for small $\mathbf{D}$ using Taylor's expansion about $\mathbf{z}$ as

$$FDIF(\mathbf{z}, t) = -\mathbf{D}' \, \nabla_{\mathbf{z}} b(\mathbf{z}, t) + \text{Higher Order Terms in } \mathbf{D} \tag{5.2.26}$$

where $\nabla_{\mathbf{z}}$ is the spatial gradient with respect to $\mathbf{z}$. If the translation of the object is constant over some moving area $\boldsymbol{R}$ (except for uncovered background) and if the higher order terms in $\mathbf{D}$ can be neglected, then the above equation can be

* We now consider $\mathbf{D}$, $\mathbf{z}$, and ∇ to be column vectors of size (2×1). Prime denotes their transpose.

SEARCH PROCEDURE	REQUIRED NUMBER OF SEARCH POINTS		REQUIRED NUMBER OF SEQUENTIAL STEPS	
	a	b	a	b
2D-LOGARITHMIC	18	21	5	7
THREE STEP	25	25	3	3
CONJUGATE DIRECTION (SIMPLIFIED)	12	15	9	12

a) FOR A SPATIAL DISPLACEMENT VECTOR (2, -6)′.
b) FOR A WORST CASE SITUATION.

Fig. 5.2.18 Required number of search points and sequential steps for various search procedures and a search area corresponding to a maximum displacement of 6 pels per frame. Total number of search points is $Q = 169$. (from Musmann et al. [5.2.9])

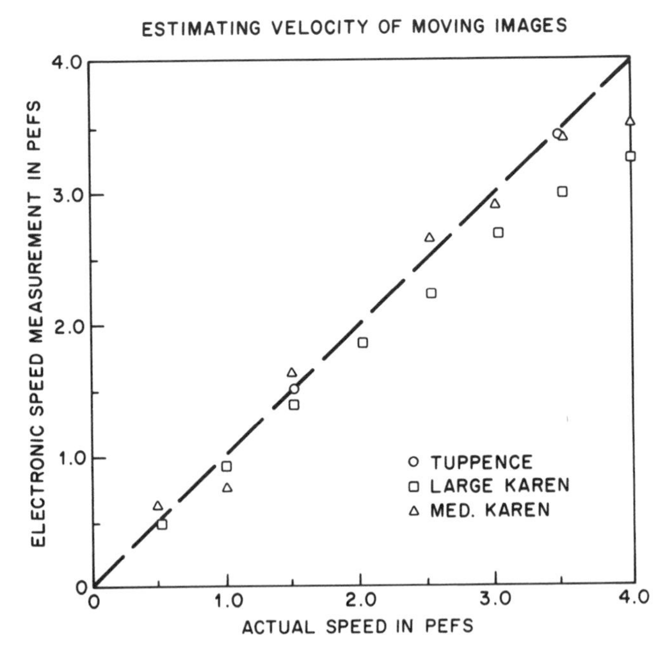

Fig. 5.2.19 Results of a speed measurement system. Measured speed in pels per frame (PEF) as a function of actual speed for three different targets. Moving object "Tuppence" is approximately 16 times the area of moving object "Medium Karen." The algorithm of Eq. (5.2.34) was used.

used at every pel in the entire moving area to obtain an estimate for translation. Moreover using multidimensional linear regression, $\hat{\mathbf{D}}$, the minimum mean square estimate of $\mathbf{D}$ can be obtained. One way to do this is to move the first term on the right hand side of Eq. (5.2.26) to the left side and obtain $\mathbf{D}$ that minimizes

$$\sum_{R} \{FDIF(\mathbf{z}, t) + \mathbf{D}'\nabla_{\mathbf{z}} b(\mathbf{z}, t)\}^2 \tag{5.2.27}$$

with respect to $\mathbf{D}$. Since this is a quadratic in $\mathbf{D}$, by simple differentiation the optimum $\mathbf{D}$ is given by

$$\hat{\mathbf{D}} = -\left[\sum_{R} \nabla_{\mathbf{z}} b(\mathbf{z}, t) \cdot \nabla'_{\mathbf{z}} b(\mathbf{z}, t)\right]^{-1} \left[\sum_{R} FDIF(\mathbf{z}, t)\nabla_{\mathbf{z}} b(\mathbf{z}, t)\right] \tag{5.2.28}$$

Since in most coding situations, the intensity $b(\mathbf{z}, t)$ is available only on a discrete sampling grid, $\nabla_{\mathbf{z}} b(\mathbf{z}, t)$ can be written approximately as:

$$\nabla_{\mathbf{z}} b(\mathbf{z}, t) = \begin{bmatrix} EDIF(\mathbf{z}) \\ LDIF(\mathbf{z}) \end{bmatrix} \tag{5.2.29}$$

where *EDIF* is a horizontal element difference given by

$$EDIF(\mathbf{z}) = \frac{1}{2}\left[b(\mathbf{z}+\Delta\mathbf{x}, t) - b(\mathbf{z}-\Delta\mathbf{x}, t)\right]$$

and *LDIF* is a vertical line difference given by

$$LDIF(\mathbf{z}) = \frac{1}{2}\left[b(\mathbf{z}+\Delta\mathbf{y}, t) - b(\mathbf{z}-\Delta\mathbf{y}, t)\right] \tag{5.2.30}$$

$\Delta\mathbf{x}$ and $\Delta\mathbf{y}$ are displacement vectors corresponding to one pel spacing horizontally and vertically within a field, respectively. Using these

$$\hat{\mathbf{D}} = -\begin{bmatrix} \sum EDIF^2(\mathbf{z}) & \sum EDIF(\mathbf{z})LDIF(\mathbf{z}) \\ \sum EDIF(\mathbf{z})LDIF(\mathbf{z}) & \sum LDIF^2(\mathbf{z}) \end{bmatrix}^{-1} \begin{bmatrix} \sum FDIF(\mathbf{z}, t)EDIF(\mathbf{z}) \\ \sum FDIF(\mathbf{z}, t)LDIF(\mathbf{z}) \end{bmatrix} \tag{5.2.31}$$

where all summations are over the moving area $\boldsymbol{R}$. The following assumption is made to convert the above matrix into a diagonal one.

$$\sum_R EDIF(\mathbf{z}) \cdot LDIF(\mathbf{z}) \approx 0 \tag{5.2.32}$$

Then

$$\hat{\mathbf{D}} \simeq - \begin{bmatrix} \sum FDIF(\mathbf{z}, t) EDIF(\mathbf{z}) / \sum EDIF^2(\mathbf{z}) \\ \sum FDIF(\mathbf{z}, t) LDIF(\mathbf{z}) / \sum LDIF^2(\mathbf{z}) \end{bmatrix} \tag{5.2.33}$$

This can be further approximated by avoiding the multiplications in the sum as:

$$\hat{\mathbf{D}} \simeq \begin{bmatrix} \dfrac{\sum FDIF(\mathbf{z}, t)\, sign(EDIF(\mathbf{z}))}{\sum |EDIF(\mathbf{z})|} \\ \\ \dfrac{\sum FDIF(\mathbf{z}, t)\, sign(LDIF(\mathbf{z}))}{\sum |LDIF(\mathbf{z})|} \end{bmatrix} \tag{5.2.34}$$

where

$$sign(u) = \begin{cases} \dfrac{u}{|u|} & , \text{ if } |u| \geq \text{Threshold} \\ \\ 0 & , \text{ otherwise} \end{cases} \tag{5.2.35}$$

The above operations are usually carried out on blocks of pels instead of the entire frame in order to estimate local displacement vectors. In practice, block sizes ranging from 4×8 to 16×32 pels have been used. Fig. 5.2.19 shows experimental results of speed measurements for three different scenes.[5.2.10] At higher values of displacement due to the linearization used, the estimates are less accurate and in general tend to be lower than the actual speed.

5.2.3h Recursive Methods

Recursive algorithms obtain an estimate of local displacement iteratively, i.e., based on previous estimates. Iterations may be performed at every pel or at each block of pels. Also, iterations may proceed along a scanning line, from line to line or from frame to frame. Recursive algorithms have the ability to overcome the problems of multiple moving objects as well as different parts of an object undergoing different displacements, provided the recursion has sufficiently rapid convergence. Also, the recursion can be arranged to minimize the pel prediction error instead of the displacement error, and thus it will be

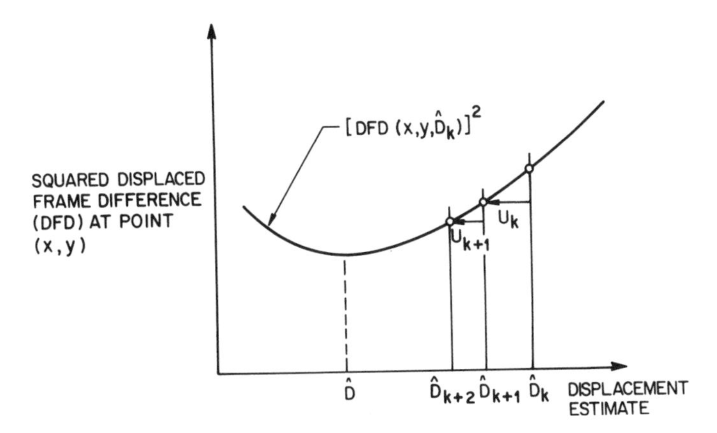

Fig. 5.2.20 Illustration of steepest descent technique used in recursive displacement estimation. The update term at each iteration is proportional to the gradient $\nabla_{\hat{D}_k}(DFD(x,y,\hat{D}_k))^2$.

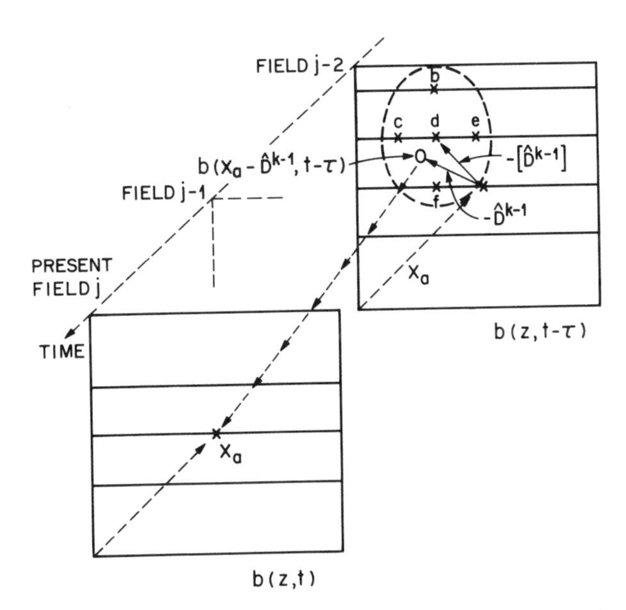

Fig. 5.2.21 Recursive motion estimation. Displacement estimate $\hat{D}_{k-1}$ is updated at pel a. Gradient of intensity $\nabla_z b(z-\hat{D}_k, t-\tau)$ is obtained by using intensities at pels b, c, d, e, f in the field $(j-2)$.

more useful for coding. Since the prediction error is calculated for transmission anyway, its use in the recursive estimation of displacement does not result in extra computations. Thus, if a pel at location $\mathbf{z}_n$ in the present frame is predicted with displacement $\hat{\mathbf{D}}_{k-1}$, i.e., intensity $b(\mathbf{z}_n - \hat{\mathbf{D}}_{k-1}, t-\tau)$, resulting in a certain amount of prediction error, the estimator should try to decrease the prediction error for $\mathbf{z}_{n+1}$ by proper choice of $\hat{\mathbf{D}}_k$. If the displacement is updated at each pel, then $k=n$. Otherwise k is related to n in some predetermined manner. It is assumed that the pel at location $\mathbf{z}_n$ is being predicted and displacement estimate $\hat{\mathbf{D}}_{k-1}$ is computed only with information transmitted prior to pel $\mathbf{z}_n$.

We define a quantity called *Displaced Frame Difference* (*DFD*) as follows

$$DFD(\mathbf{z}, \hat{\mathbf{D}}) = b(\mathbf{z}, t) - b(\mathbf{z}-\hat{\mathbf{D}}, t-\tau) \tag{5.2.36}$$

DFD is defined in terms of two variables: (i) the spatial location $\mathbf{z}$ at which *DFD* is evaluated and (ii) the estimated displacement $\hat{\mathbf{D}}$ used to compute an intensity in the previous frame. In the case of a two dimensional grid of discrete samples, an interpolation process must be used to evaluate $b(\mathbf{z}-\hat{\mathbf{D}}, t-\tau)$ for values of $\hat{\mathbf{D}}$ that are a noninteger number of pel spacings. In the ideal case, *DFD* converges to zero as $\hat{\mathbf{D}}$ converges to the actual displacement, $\mathbf{D}$.

Pel recursive displacement estimators attempt to minimize recursively $[DFD(\mathbf{z}, \hat{\mathbf{D}})]^2$ at each moving area pel using a steepest descent algorithm (see Fig. 5.2.20). Thus,

$$\hat{\mathbf{D}}_k = \hat{\mathbf{D}}_{k-1} - \frac{\varepsilon}{2} \nabla_{\hat{\mathbf{D}}_{k-1}} \left[DFD(\mathbf{z}_n, \hat{\mathbf{D}}_{k-1}) \right]^2 \tag{5.2.37}$$

where $\nabla_{\hat{\mathbf{D}}}$ is the two-dimensional gradient operator with respect to $\hat{\mathbf{D}}$. Carrying out the above operation using Eq. (5.2.36).

$$\hat{\mathbf{D}}_k = \hat{\mathbf{D}}_{k-1} - \varepsilon DFD(\mathbf{z}_n, \hat{\mathbf{D}}_{k-1}) \nabla_{\mathbf{z}} \left[b(\mathbf{z}_n - \hat{\mathbf{D}}_{k-1}, t-\tau) \right] \tag{5.2.38}$$

Thus, the new value for $\mathbf{D}$ is the old value plus an *update term*. It is interesting to observe that at every iteration we add to the old estimate a vector quantity that is either parallel to or opposite to the direction of the spatial gradient of image intensity depending on the sign of DFD, the motion compensated prediction error. If the actual displacement error $(\mathbf{D}-\hat{\mathbf{D}}_k)$ is orthogonal to the intensity gradient $\nabla_{\mathbf{z}} b$, the displaced frame difference is zero giving a zero update. This may happen even though the object has moved. Also if the gradient of intensity is zero, i.e. absence of edges, even if there is an error in the displacement estimation, the algorithm will do no updating. It is only through the occurrence of edges with differing orientations in real television scenes that convergence of

$\hat{\mathbf{D}}$ to the actual displacement $\mathbf{D}$ is possible. Fig. 5.2.21 shows the above recursion.

The above displacement estimator requires multiplication at each iteration, which is undesirable for hardware implementation. A simpler form is given by the following single pel steepest descent algorithm:

$$\hat{\mathbf{D}}_k = \hat{\mathbf{D}}_{k-1} - \varepsilon \text{ sign}\left[DFD(\mathbf{z}_n, \hat{\mathbf{D}}_{k-1})\right] \text{ sign}\left[\nabla_{\mathbf{z}}(b(\mathbf{z}_n - \hat{\mathbf{D}}_{k-1}, t-\tau))\right] \tag{5.2.39}$$

where the sign function is defined as in Eq. (5.2.35) for a threshold T and the components of the intensity gradient can be approximated as in Eqs. (5.2.29) and (5.2.30), but at a displaced location $(\mathbf{z}_n - [\hat{\mathbf{D}}_{k-1}])$ in the previous frame. $[\mathbf{D}]$ denotes $\mathbf{D}$ rounded to an integer number of pel spacings.

The above algorithm can be extended by computing the displaced frame differences at many picture elements in order to estimate $\mathbf{D}$. We give two examples of such algorithms. In the first, $\hat{\mathbf{D}}$ could be updated by a steepest-descent algorithm to minimize a weighted sum of the squared displaced frame differences at some previously transmitted neighboring pels. Thus,

$$\hat{\mathbf{D}}_k = \hat{\mathbf{D}}_{k-1} - \frac{\varepsilon}{2} \nabla_{\hat{D}_{k-1}} \left[\sum_{j=0}^{p} W_j (DFD(\mathbf{z}_{n-j}, \hat{\mathbf{D}}_{k-1}))^2\right], \tag{5.2.40}$$

where $W_j \geq 0$ and $\sum_{j=0}^{p} W_j = 1$. This can be expanded to give:

$$\hat{\mathbf{D}}_k = \hat{\mathbf{D}}_{k-1} - \varepsilon \left[\sum_{j=0}^{p} W_j DFD(\mathbf{z}_{n-j}, \hat{\mathbf{D}}_{k-1}) \nabla_{\mathbf{z}}\, b(\mathbf{z}_{n-j} - \hat{\mathbf{D}}_{k-1}, t-\tau)\right] \tag{5.2.41}$$

where $\nabla_{\mathbf{z}}(\cdot)$ can be approximated by finite differences as before.

This can be generalized further using a different error (metric) instead of the square function. The difference between Eqs. (5.2.41) and (5.2.38) is that, in order to estimate a new value $\hat{\mathbf{D}}_k$ from the old value $\hat{\mathbf{D}}_{k-1}$, the displaced frame difference is evaluated at several neighboring picture elements rather than just one picture element. This has the effect of smoothing the update term and making it less noise prone.

A second algorithm using multiple pels is a least-mean-square (LMS) estimator of displacement. Assuming Eq. (5.2.17), the displaced frame difference of Eq. (5.2.36) can be written as

$$DFD\,(\mathbf{z}_n, \hat{\mathbf{D}}_{k-1}) = b(\mathbf{z}_n - \mathbf{D}, t-\tau) - b(\mathbf{z}_n - \hat{\mathbf{D}}_{k-1}, t-\tau) \tag{5.2.42}$$

$$= -\,(\mathbf{D} - \hat{\mathbf{D}}_{k-1})' \nabla_{\mathbf{z}} b(\mathbf{z}_n - \mathbf{D}_{k-1}, t-\tau) + \text{other terms}.$$

The quantity $(\mathbf{D} - \hat{\mathbf{D}}_{k-1})$ can be estimated to a first order of approximation, using standard techniques of linear regression by evaluating Eq. (5.2.42) at several previously transmitted pels $\mathbf{z}_n$. This gives us the multiple LMS algorithm of the form

$$\hat{\mathbf{D}}_k = \hat{\mathbf{D}}_{k-1} - \varepsilon \left[\sum_{j=0}^{p} \nabla_{\mathbf{z}} b(\mathbf{z}_{n-j} - \hat{\mathbf{D}}_{k-1}, t-\tau) \nabla_{\mathbf{z}}' b(\mathbf{z}_{n-j} - \hat{\mathbf{D}}_{k-1}, t-\tau) \right]^{-1}$$

$$\times \left[\sum_{j=0}^{p} DFD(\mathbf{z}_{n-j}, \hat{\mathbf{D}}_{k-1}) \nabla_{\mathbf{z}} b(\mathbf{z}_{n-j} - \hat{\mathbf{D}}_{k-1}, t-\tau) \right] \tag{5.2.43}$$

The matrix inverse in the above equation can be approximated by

$$M_{k-1}^{-1} = \frac{1}{\Delta_{k-1}} \begin{bmatrix} \sum_{j=0}^{p} LDIF_j^2 & -\sum_{j=0}^{p} LDIF_j EDIF_j \\ -\sum_{j=0}^{p} LDIF_j EDIF_j & \sum_{j=0}^{p} EDIF_j^2 \end{bmatrix} \tag{5.2.44}$$

with

$$\Delta_{k-1} = \sum_{j=0}^{p} EDIF_j^2 \sum_{j=0}^{p} LDIF_j^2 - \sum_{j=0}^{p} EDIF_j\, LDIF_j \tag{5.2.45}$$

where

EDIF_j = Element difference at $(\mathbf{z}_{n-j} - \hat{\mathbf{D}}_{k-1})$ in the previous frame, approximating the horizontal component of $\nabla_{\mathbf{z}} b(\cdot)$.

$LDIF_j$ = Line difference at $(\mathbf{z}_{n-j} - \hat{\mathbf{D}}_{k-1})$ in the previous frame, approximating the vertical component of $\nabla_{\mathbf{z}} b(\cdot)$.

In evaluating the matrix inverse, one must be careful that the matrix is not singular. Singularity can be a result of not averaging enough samples. If, on the other hand, a large number of samples are averaged, then the local changes in

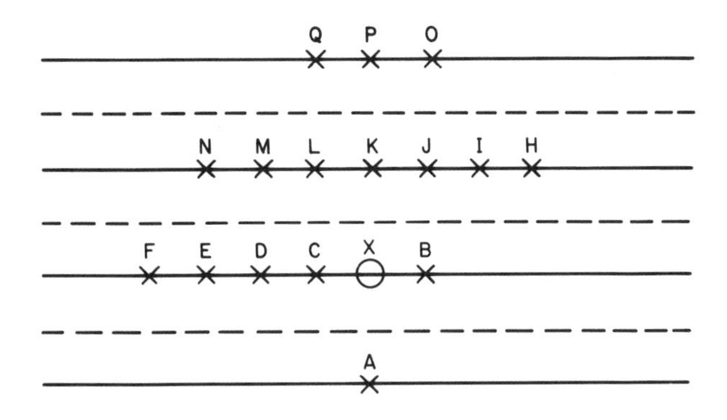

Fig. 5.2.22a Configuration of pels used for weighted error calculation. Pel X is being predicted. Dotted lines denote scan lines from previous field. Configuration 1 includes error term only at pel K; configurations 2, 3, 4 and 5 use displaced frame differences at pels $\{C,D,J,K,L\}$, $\{A,B,X,C,K\}$, $\{I,J,K,L,M,O,P,Q\}$ and $\{C,D,E,F,H,I,J,K,L,M,N,O,P,Q\}$, respectively. Note that configuration 3 uses pels not yet available to the receiver. (from Netravali and Robbins [5.2.12])

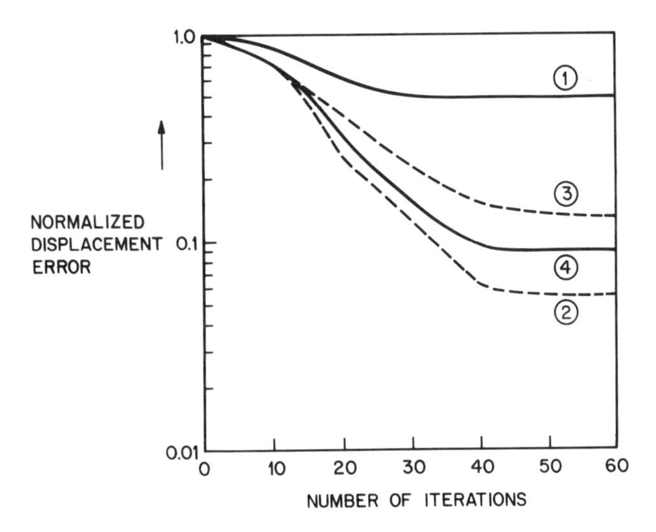

Fig. 5.2.22b Normalized displacement error ($||\mathbf{D}_k - \mathbf{D}|| / ||\mathbf{D}_0 - \mathbf{D}||$) is plotted against the iteration number for the algorithm of Eq. (5.2.41). Four configurations (1,2,3,4) as in Fig. 5.2.22a are shown. Configuration (1) corresponds to single pel recursion (same as Eq. (5.2.38). Other configurations include several surrounding pels. All iterations are started from the left-most pel in a scan line, $\mathbf{D}_0$ is taken to be zero, and $\mathbf{D}$ is 2 pels per frame in the horizontal direction. (from Netravali and Robbins [5.2.12])

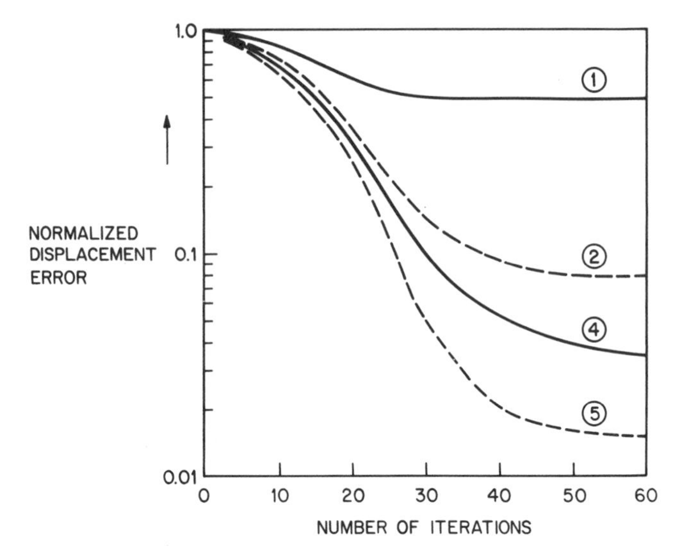

Fig. 5.2.22c Normalized displacement error against the iteration number for algorithm of Eq. (5.2.43). Four configurations are as in Fig. 5.2.22a. (from Netravali and Robbins [5.2.12])

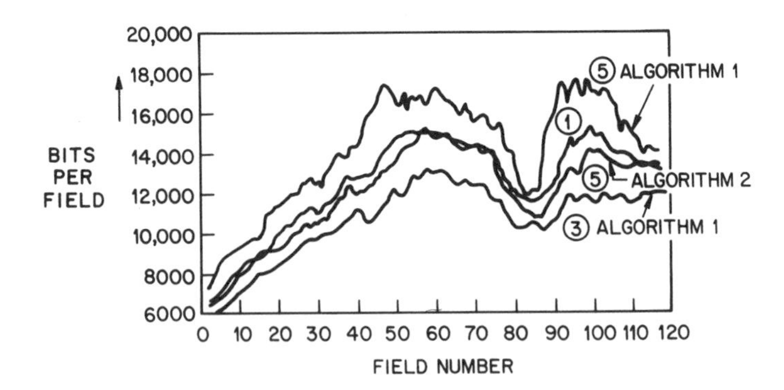

Fig. 5.2.23 Plots of bits per field for the scene "Judy". Configuration 1 corresponds to single-pel recursion of Eq. (5.2.38). Two algorithms with multiple-pel recursions, Eqs. (5.2.41) and (5.2.43), are also shown. Configuration 5 is shown for both the multiple pel algorithms. Configuration 3 uses some pels for displacement estimation that are not yet available to the receiver. It gives an indication of how much reduction in bit rate is possible if the displacement were known more accurately.

displacement will get averaged. To avoid this, whenever M_i is close to being singular (as indicated by its determinant), no update may be performed. The second factor of Eq. (5.2.43) is given by

$$\psi_{k-1} = \begin{bmatrix} \sum_{j=0}^{p} DFD(\mathbf{z}_{n-j} - \hat{\mathbf{D}}_{k-1}, t-\tau) EDIF_j \\ \\ \sum_{j=0}^{p} DFD(\mathbf{z}_{n-j} - \hat{\mathbf{D}}_{k-1}, t-\tau) LDIF_j \end{bmatrix} \tag{5.2.46}$$

Performance of both the single-pel and multiple-pel algorithms for estimation of displacement has been experimentally determined.[5.2.12] In these experiments, a moving pattern with a known displacement is generated by a computer. The displacement estimates are calculated using the above algorithms and compared to the true displacement. Fig. 5.2.22 shows the configuration of pels used and plots of normalized displacement error

$$||\hat{\mathbf{D}}_k - \mathbf{D}|| \;/\; ||\hat{\mathbf{D}}_0 - \mathbf{D}|| \tag{5.2.47}$$

with respect to the iteration number k. $\hat{\mathbf{D}}_0$ is the initial estimate assumed to be zero at the left most pel on the raster. Recursions are carried out pel to pel within a scan line.

Two things become clear from Fig. 5.2.22. The single-pel recursion exhibits slower convergence compared to the multiple-pel algorithms. Inclusion of more picture elements in the update term improves the convergence and the steady-state error. Also, among the two multiple-pel algorithms, the second algorithm shows better convergence and low steady-state error, but it is significantly more complex to implement. The significant difference between the *LMS* algorithm of Eq. (5.2.43) and other recursive algorithms is that the constant ε is replaced by $\varepsilon \cdot M_k^{-1}$ to achieve better adaptation to the local image statistics. There are other methods of incorporating local image statistics to improve convergence and reduce steady state error. Some of these will, no doubt, be tried and succeed in the future. Fig. 5.2.23 shows performance on a real scene from a TV camera. Having computed the displacement and the associated prediction, the prediction error is quantized and transmitted only if it is above a threshold of 3 (out of 255 corresponding to 8 bit original video signal). Entropy of the quantizer output of such transmitted pels is plotted in Fig. 5.2.23. Multiple-pel algorithms reduce the bit rate by about 10 percent compared to single-pel algorithms. Thus, although the rate of convergence of displacement iterations and the steady-state error are improved significantly for synthetic scenes by using multiple pels, the improvement in coder bit rate is not appreciable for real scenes. This is perhaps due to the existence of spatially nonuniform displacements in real scenes. The use of a larger number of picture elements in the update terms has the effect of averaging displacement, which may not be so useful if these displacements are spatially nonuniform.

5.2.3i Gain Compensation

Often the model of image intensity of Eq. (5.2.17) is not adequate. In some of these situations, a multiplicative factor (called *Gain*) may also affect the intensity changes. Thus, the model of Eq. (5.2.17) may be generalized to one of the following:

$$b(\mathbf{z}, t) = \rho_1 b(\mathbf{z}, t-\tau) \tag{5.2.48}$$

or

$$b(\mathbf{z}, t) = \rho_2 b(\mathbf{z}-\mathbf{D}, t-\tau) \tag{5.2.49}$$

depending on whether the frame-to-frame change is due to only a change in illumination or due to both movement and illumination changes. Eqs. (5.2.48) and (5.2.49) are alternative models of frame-to-frame intensity variation. In certain parts of the picture Eq. (5.2.48) is more appropriate than Eq. (5.2.49) and vice versa. As an example, Eq. (5.2.48) is more appropriate if there is no motion in the scene, but average illumination is changed; whereas Eq. (5.2.49) is useful if in addition to this there is also motion. One case that is easily handled by Eq. (5.2.49) is moving shadows caused by a moving object casting a shadow on some part of the scene.

The compensation for the above variation of intensity can be accomplished by estimating ρ_1, ρ_2 and $\mathbf{D}$ using gradient-type algorithms to minimize the square of the prediction error resulting from each model. The algorithm for compensation of Eq. (5.2.48) is:

$$\hat{\rho}_1^k = \hat{\rho}_1^{k-1} + \varepsilon_1 \left[b(\mathbf{z}_n, t) - \hat{\rho}_1^{k-1} b(\mathbf{z}_n, t-\tau) \right] b(\mathbf{z}_n, t-\tau) \ . \tag{5.2.50}$$

For compensation of Eq. (5.2.49), both ρ_2 and $\mathbf{D}$ need to be estimated. This is accomplished by

$$\hat{\rho}_2^k = \hat{\rho}_2^{k-1} + \varepsilon_1 DFD(\mathbf{z}_n, \hat{\rho}_2^{k-1}, \hat{\mathbf{D}}^{k-1}) b(\mathbf{z}_n - \hat{\mathbf{D}}^{k-1}, t-\tau) \tag{5.2.51}$$

$$\hat{\mathbf{D}}^k = \hat{\mathbf{D}}^{k-1} - \varepsilon_2 \rho_2^{k-1} DFD(\mathbf{z}_n, \hat{\rho}_2^{k-1}, \hat{\mathbf{D}}^{k-1}) \nabla_{\mathbf{z}} b(\mathbf{z}_n - \hat{\mathbf{D}}^{k-1}, t-\tau) \tag{5.2.52}$$

where $DFD(. , . , .)$ is defined by

$$DFD(\mathbf{z}, \hat{\rho}_2, \hat{\mathbf{D}}) = b(\mathbf{z}, t) - \hat{\rho}_2 b(\mathbf{z}-\hat{\mathbf{D}}, t-\tau) \tag{5.2.53}$$

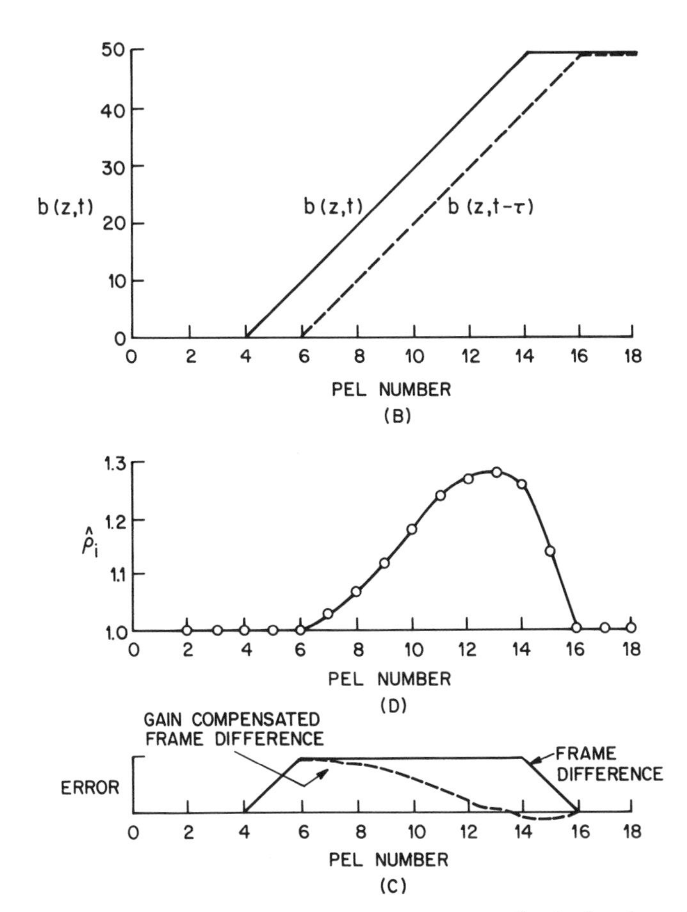

Fig. 5.2.24 A synthetic vertical edge whose intensity for a scan line is plotted as a function of horizontal position for two consecutive frames. The horizontal shift in the edge is 2 pels per frame. (b) Plot of the value of ρ_i generated recursively from Eq. (5.2.50), starting with a value of 1.0. Recursion is assumed to proceed pel by pel in the direction of scanning. (c) Plot of frame difference and the gain-compensated frame difference as a function of pel position.

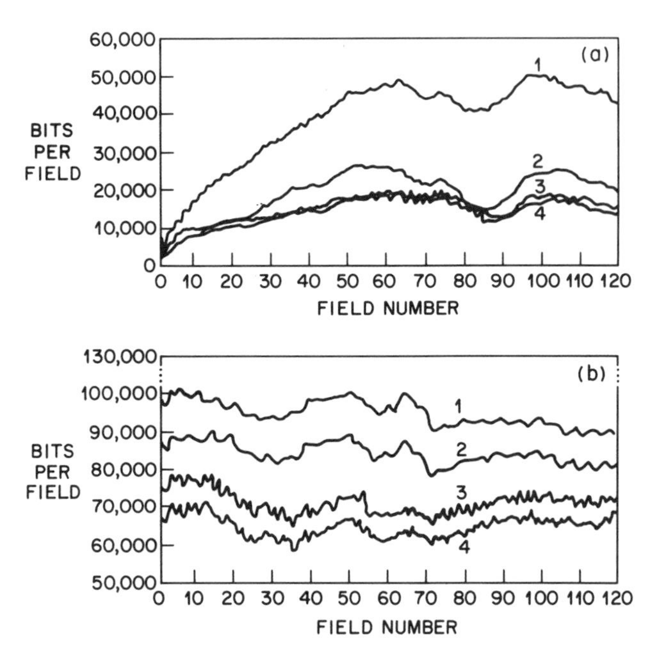

Fig. 5.2.25 Plots of bits/field versus the field number for the scenes (a) "Judy", and (b) "Mike and Nadine" using four prediction schemes: (1) frame difference; (2) gain compensation; (3) displacement compensation; (4) gain and displacement compensation. Scene "Judy" contains a head and shoulders view with moderate motion whereas the scene "Mike and Nadine" contains full view of two people in severe motion with spatially and temporarily changing illumination.

and ε_1 and ε_2 are positive scalar constants. The iterations may proceed from sample to sample along a scan line. However, ρ_1, ρ_2 and $\mathbf{D}$ may not be revised at every sample, and therefore, k may not be equal to n. Use of Eqs. (5.2.50) and (5.2.51) assumes that ρ_1, ρ_2, and $\mathbf{D}$ vary relatively slowly spatially, an assumption that appears to be often valid in practice.

Eq. (5.2.49) assumes that the intensity variation is due both to translational motion of objects and variation of illumination. For uniform illumination, ρ_2 will be unity and $\mathbf{D}$ will be equal to the displacement. However, in certain cases, Eq. (5.2.48) also provides an approximate description of object motion. This occurs if the parameter ρ_1 varies sufficiently slowly to be learned by Eq. (5.2.50). An example of this is shown in Fig. 5.2.24, which shows a vertical edge in intensity that is displaced from one frame to the next in the horizontal direction by two pels.[5.2.13] The recurrence relationship of Eq. (5.2.50) is carried out in Fig. 5.2.24, where a plot of $\hat{\rho}_1$ along the scan line is shown. Also shown are the frame difference and the gain-compensated frame difference GFDIF:

$$GFDIF(\mathbf{z}, \hat{\rho}_1, t) = b(\mathbf{z}, t) - \hat{\rho}_1 b(\mathbf{z}, t-\tau) . \tag{5.2.54}$$

It is seen then that the gain-compensated frame difference is able to provide a certain degree of displacement compensation and, therefore, results in prediction errors that are smaller than the frame differences.

Fig. 5.2.25 shows simulation results using gain and displacement compensation for two scenes. Curves 2, 3 and 4 of this figure are obtained by switching the prediction between several predictors. For gain compensation (2), switching is between frame difference and gain compensated difference; for displacement compensation (3), it is between frame difference and displacement compensated difference; and for gain and displacement compensation (4), switching is between frame, gain compensated, and displacement compensated difference. In each case, selection of the predictor is made by choosing the one that results in the least error for the adjacent previous elements. Thus there is no need to transmit additional predictor switching information to the receiver. For the (head and shoulders) scene "Judy", where the illumination is close to uniform, the gain-compensated coder (2) results in entropy that is about 50 percent below that of the frame difference conditional replenishment coder. The displacement-compensated coder (3), on the other hand, results in about a 61 percent decrease. Thus, gain compensation reduces the coder bit rate by a significant amount without the complexity associated with the displacement-compensated encoder. The gain- and displacement-compensated encoder (4) reduce the entropy by 63 percent for "Judy". This additional decrease due to gain compensation is small, perhaps, as a result of the relatively uniform illumination and the lack of shadows in the scene. For the scene "Mike and Nadine", again, most of the decrease in bit-rate is provided by displacement-

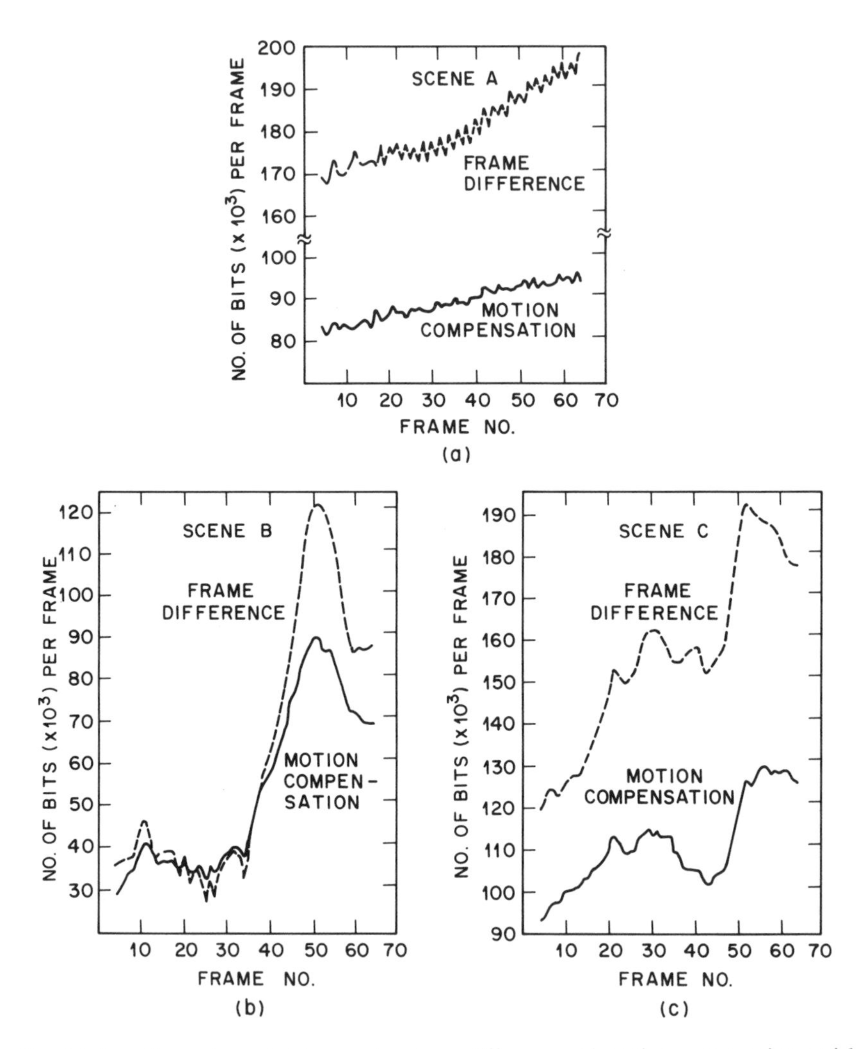

Fig. 5.2.26 Comparison of bit rates for frame difference and motion compensation-spatial prediction for color component signals from Scene A, Scene B and Scene C. The three scenes contain varying amount of motion and spatial detail.

compensation. Gain-compensation results in bit-rates that are between the frame difference conditional replenishment and displacement compensation. Also, combining gain and displacement compensation provides an additional decrease of only 7 percent.

5.2.3j Motion Compensation for Color Television

Motion compensation of color television signals can be performed either on component or composite signals. In the case of components, motion estimation and its use for coding could be done independently for each component. However, to keep the hardware simpler, it is desirable to estimate motion based only on the luminance component and then use such an estimate also for the chrominance components. Experimental evidence[5.2.14] indicates that correlation between the luminance and chrominance components may be exploited by:

a. assuming the chrominance displacements to be identical to those for the luminance

b. switching the predictor for the chrominance in exactly the same way as the luminance, i.e., switch between previous element, previous line, previous frame and displaced previous frame predictors for both luminance and chrominance

c. reducing the cost of addressing the unpredictable chrominance pels.

Fig. 5.2.26 shows simulation results for three scenes containing varying amount of motion and spatial detail. Scenes A and C contain simple uniform motion of large areas, and for this reason, motion compensation reduces the entropy by about 60 percent compared to frame difference prediction. In scene B, however, motion is restricted to small, but complex areas. This results in only a small difference between motion compensated and frame difference predictions.

For composite television signals, the problem of displacement estimation is complicated by the phase relationship of different samples. Two promising approaches are: (a) to treat samples of the composite signal of each phase separately and independently, (b) to derive an approximation of the luminance component of the signal by locally averaging the composite samples and using it for displacement estimation. Experimental evidence shows that the second approach is preferable over the first one.[5.2.15] However, due to the poor quality of displacement estimation and the resulting prediction, motion compensation of composite color signals is not as efficient as motion compensation of components. Simulations using composite signals indicate that a motion compensated predictive coder requires up to 25 percent fewer bits than a coder that uses the previous field prediction.

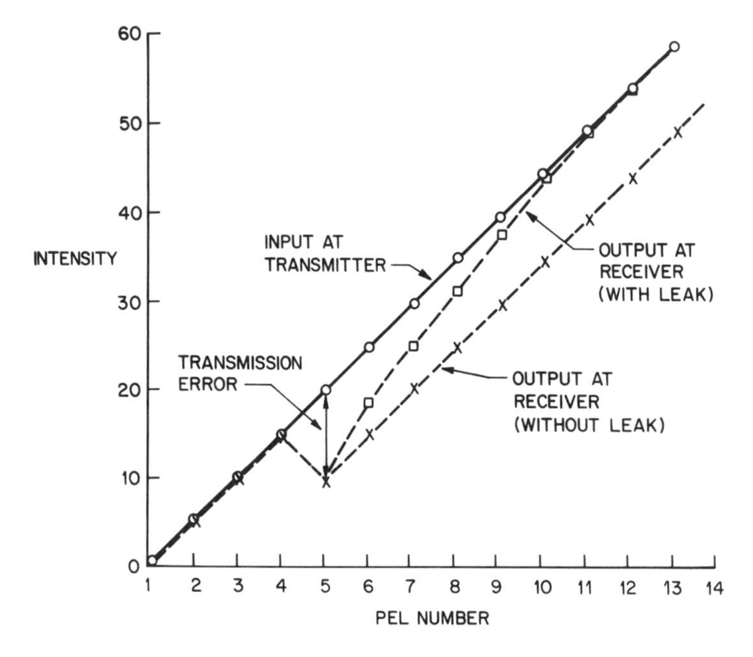

Fig. 5.2.27 Effect of transmission error on DPCM coding. An error of 10 introduced in transmitting pel number 5 propagates to all pels numbered five or more along the same scan line. A leak in the prediction limits the propagation of the transmission error.

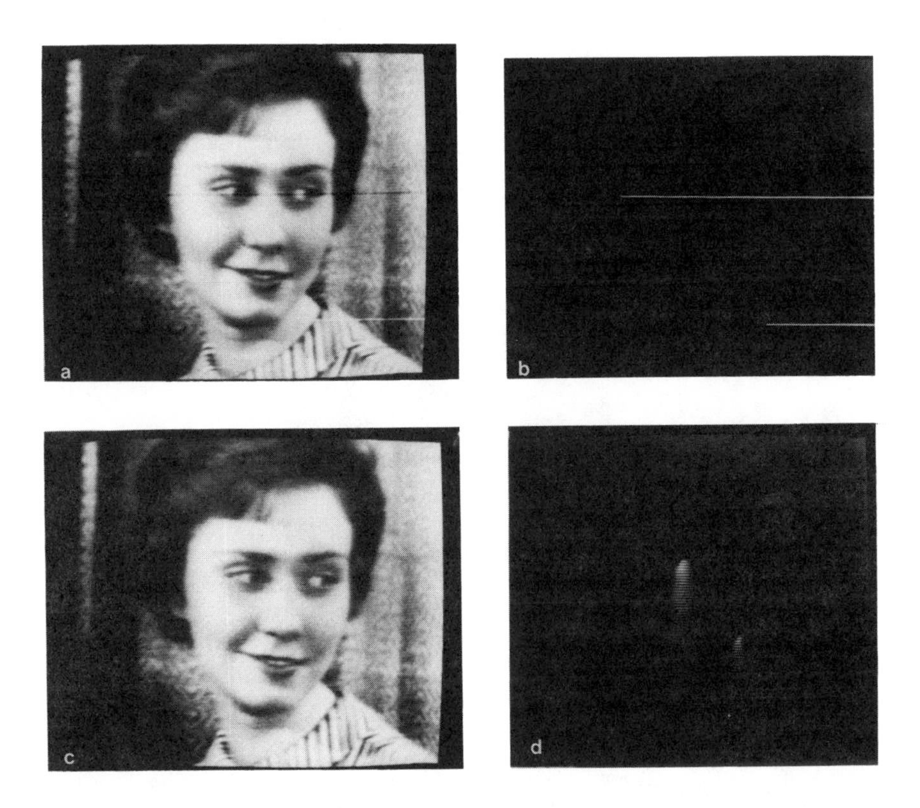

Fig. 5.2.28 Effect of transmission errors in DPCM coders (a) Previous element prediction. (b) Error patterns for previous element prediction. (c) Spatial average prediction. (d) Error patterns for spatial average prediction.

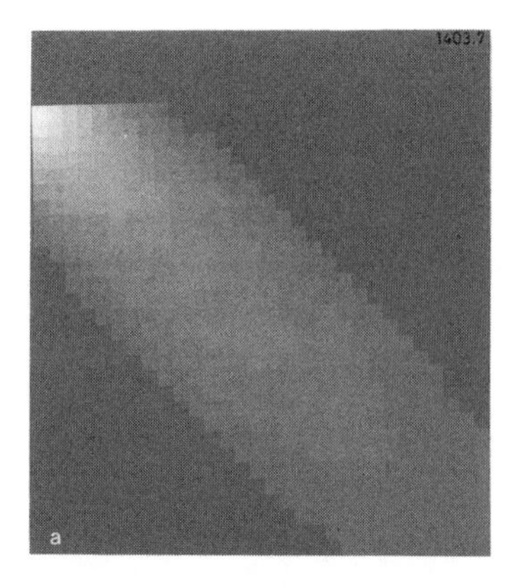

Fig. 5.2.29 (a),(b) Error patterns for the two-dimensional predictions of Eq. (5.2.55) and (5.2.55) in areas of uniform luminance; (c) and (d) are reconstructed pictures with channel error rate of 5×10^{-4}. (From Jung and Lippman [5.2.17])

5.2.3k Effects of Transmission Errors

One important aspect in the design of predictors is the effect of transmission bit errors in the reconstructed picture. In a DPCM coder, the prediction of each pel is made using certain previously transmitted pels. Thus, a transmission bit error in any pel could affect one or more future pels, which in turn could affect all of the subsequent pels. The amount and spread of this error propagation depends both upon the type and the functional definition of the predictor used. In the case of the optimum linear one-dimensional predictor for zero mean pels, it is known that for most types of correlations generally found in pictures, the predictor coefficients are such that the effect of transmission error decays as desired.

The previous element predictor with unity weighting coefficient does not limit the effects of transmission errors. This is clearly seen by an example. Let us assume that the quantizer used in the previous element DPCM coder maps input into output exactly. Then as shown in Fig. 5.2.27, a transmission error of 10 at pel number 5 gets propagated for all the subsequent pels along the scan line. In real coders, this appears as a horizontal streak in the reconstructed picture. However, the propagation errors can be limited by providing a *leak*, i.e., using a weighting coefficient less than one. In this case a fixed fraction of the previous element value is used for prediction. Fig. 5.2.27 also shows that with the leak, the transmission error has been reduced to zero after pel number 12. The rate of decay is determined by the predictor coefficients.

For two dimensional predictors the situation is more complex. The optimum* linear two dimensional predictors for most commonly used correlation functions may be unstable, i.e., transmission errors may propagate throughout the picture. Good criteria for stability of predictors are not yet known. However, there are some stable two-dimensional predictors for which the pattern of distortion produced by transmission errors is much less annoying than with one dimensional predictors. Fig. 5.2.28 shows error patterns produced by a previous element and a two dimensional predictor. Fig. 5.2.29 shows the propagation of transmission error for two different predictors. The predictors are: (using pel positions in Fig. 5.2.8)

$$\text{predictor I} \quad \triangleq 3/4\, A - 1/2\, B + 3/4\, C \qquad (5.2.55)$$

$$\text{predictor II} \quad \triangleq 3/4\, A - \frac{17}{32} B + 3/4\, C$$

* Optimum in the sense of minimum mean square prediction error.

Fig. 5.2.29(a) and (b) show response of the DPCM receiver to an impulse of transmission error in a pel at the upper left hand corner of the picture in an area of uniform luminance, for the two predictors. It is clear that rate of decay in the first predictor is slow since the sum of predictor coefficients is equal to one, whereas for the second predictor rate of decay is much faster, since the sum of predictor coefficients is less than one. Fig. 5.2.29(c) and (d) show the corresponding reconstructed pictures. It is not easy to perceive the differences between these two pictures due to loss of quality in the printing process, but on a real television monitor the differences are quite obvious.

In the case of adaptive predictors, a transmission error can have two deleterious effects - one due to incorrect values for the prediction pels, and the other due to selecting the wrong predictor at the receiver. Some improvement in adaptive predictor performance may be obtained by providing a leak in the rule of adaptation, i.e., by using a fraction of the adaptive prediction and a fixed predictor. However, in general, the criteria for stability of such predictors and the effects of transmission bit errors are much less understood. The only recourse may be the use of error detecting and correcting codes.

In interframe coding, the effect of transmission errors may last for several frames, and in most cases it is necessary to employ techniques of error correction and/or concealment. Several techniques have been investigated. These are in addition to the use of error detecting/correcting codes that may be used for very noisy channels. Two commonly used techniques are: (a) hybrid PCM/DPCM and (b) use of Remaining Redundancy. In hybrid PCM/DPCM, predictor coefficients are reset to zero periodically and during the resetting a PCM value is transmitted. A commonly used scheme is to transmit a PCM value for the first pel of every M-th line. This limits every transmission error impulse to M lines. The second technique utilizes the redundancy present in the picture signal to restore the error-corrupted DPCM-decoder output. Such techniques first attempt to detect transmission errors either by error detecting codes or by observing anomalies in the reconstructed picture at the receiver. If an error is detected, then some previously transmitted part of the picture (which is assumed to be error free) is repeated or averaged. One commonly employed scheme is to either repeat a previous scan line or a scan line from the previous frame at the same location. Better performance is obtained by replacing the error corrupted line by an average of some previous and future lines. Reliable detection of errors is important in these techniques, otherwise due to repeating or averaging the scan lines, there can be substantial loss of resolution.

5.2.4 Quantization

DPCM schemes achieve compression, to a large extent, by not quantizing the prediction error as finely as the original signal itself. Several methods of optimizing quantizers have been studied, but quantizer design still remains an art and somewhat ad hoc. Most of the work on systematic procedures for quantizer optimization has been for the previous-element DPCM coder, in which the approximate horizontal slope of the input signal is quantized. Although obvious extensions can be made to the case of two-dimensional and interframe predictors, they have not yet been thoroughly studied. For this reason, we shall discuss in detail quantizers for the previous-element DPCM and point out how extensions to the other cases can be made.

Three types of degradations may be seen as a result of DPCM quantization. These are referred to as *granular noise*, *slope overload* and *edge busyness* and are shown in Fig. 5.2.30. If the inner levels (corresponding to the small magnitudes of the differential signal) of the quantizer are too coarse, then the flat or low detail areas of the picture are coarsely quantized. If one of the output levels is zero (i.e., a mid-step quantizer), then such coarse quantization may show false contours in the reconstructed picture. Otherwise, it has the appearance of random or granular noise added to the picture. On the other hand, if the dynamic range (i.e., largest representative level) of the quantizer is too small, then for every high contrast edge, the decoder takes several samples for the output to catch up to the input, resulting in slope overload, which appears similar to low-pass filtering of the image. Finally, for edges whose contrast changes somewhat gradually, the quantizer output oscillates around the signal value and may change from line to line, or frame to frame, giving the appearance of a discontinuous jittery or "busy" edge.

Quantizers can be designed purely on a statistical basis or by using certain psychovisual measures. They may be adaptive or fixed. We shall discuss each of these cases in the following sections.

5.2.4a Nonadaptive Quantization

Optimum quantizers that are based on statistics have been derived using the minimum mean square error criterion.[5.2.18] Considering Fig. 5.2.31, if x with probability density $p(x)$, is the input to the DPCM quantizer, then quantizer parameters can be obtained to minimize the following measure of quantization error:

$$\mathbf{D} = \sum_{k=1}^{L} \int_{t_{k-1}}^{t_k} f(x - \ell_k)\cdot p(x)\,dx \qquad (5.2.56)$$

where $t_0 < t_1 < \ldots < t_L$ and $\ell_1 < \ell_2 < \ldots < \ell_L$ are decision and representative levels, respectively, and $f(\cdot)$ is a nonnegative error function. We assume, as shown in Fig. 5.2.31 that all inputs $t_{k-1} < x \leq t_k$ to the quantizer are represented as ℓ_k. If one of the representative levels is zero then the quantizer is termed *mid-tread,* and if one of the decision levels is zero then it is called *mid-riser*. Necessary conditions for the optimality with respect of t_k and ℓ_k for a fixed number of levels L are given by:

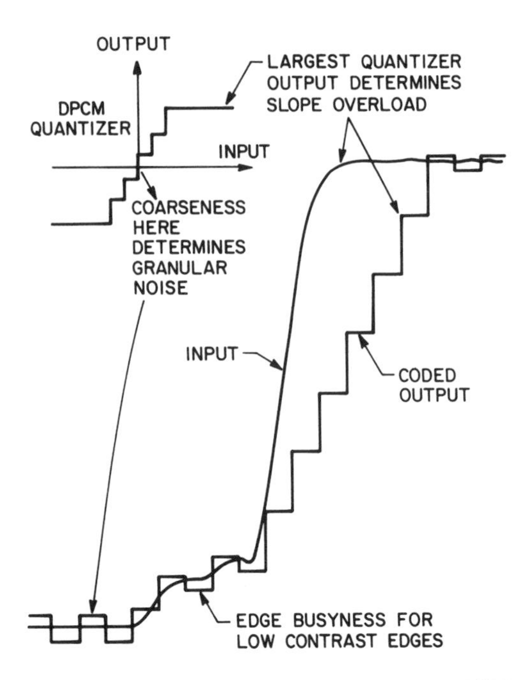

Fig. 5.2.30 An intuitive classification of quantizing distortion due to DPCM coding. Three classes of quantization noise are identified: granular noise, edge busyness and slope overload.

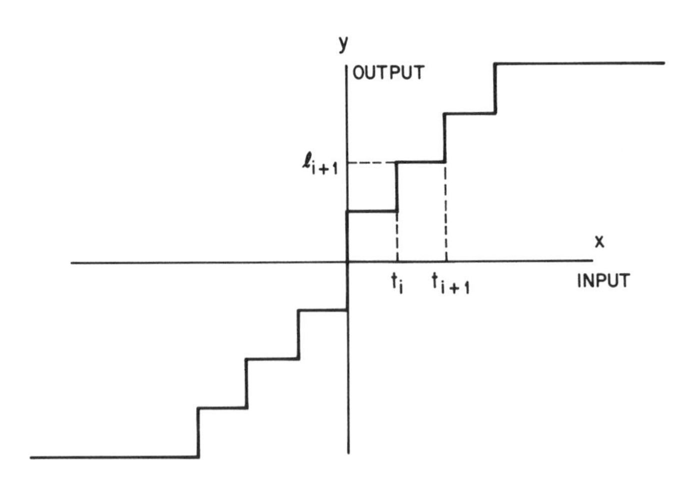

Fig. 5.2.31 Characteristics of a quantizer. x is the input and y is the output. $\{t_i\}$ and $\{\ell_i\}$ are decision and representative levels, respectively. Inputs between t_i and t_{i+1} are represented by ℓ_{i+1}.

$$f(t_{k-1} - \ell_{k-1}) = f(t_{k-1} - \ell_k), \quad k = 2 \ldots L \tag{5.2.57}$$

and

$$\int_{t_{k-1}}^{t_k} \frac{df(x - \ell_k)}{dx} \cdot p(x)\,dx = 0, \quad k = 1 \ldots L \tag{5.2.58}$$

assuming that $f(\cdot)$ is differentiable. In the case of the mean-square error criterion, $f(z) = z^2$, the above equations reduce to the Lloyd-Max quantizer[5.2.18] (see also Eq. (3.2.30))

$$t_k = 1/2\ (\ell_k + \ell_{k+1}) \tag{5.2.59}$$

and

$$\ell_k = \int_{t_{k-1}}^{t_k} x p(x)\ dx \Big/ \int_{t_{k-1}}^{t_k} p(x)\,dx. \tag{5.2.60}$$

Several algorithms and approximations exist for the solution of these equations. The probability density $p(x)$ for the case of previous pel and other differential coders can be approximated by a Laplacian density (see Chapter 3), and the optimum quantizer in this case is *companded*, i.e., the step size $(t_k - t_{k-1})$ increases with k. Since much of the time the picture signal has slope x close to zero, this results in a fine spacing of levels near $x = 0$ and a rather coarse spacing of levels as $|x|$ increases. For images, this generally results in overspecification of the low-detailed areas of the picture, and consequently only a small amount of granular noise, but relatively poor reproduction of detailed areas and edges. Due to the small dynamic range of such a quantizer, this is often visible as slope overload and edge busyness for high contrast, vertical edges. Thus for subjective reasons, the quantizers should not be designed based on the *MSE* criterion.

Minimization of a quantization error metric, i.e., distortion, for a fixed number of levels is relevant for DPCM systems that code and transmit the quantizer outputs using fixed-length binary words, since in that case the output bit-rate depends only on the logarithm of the number of levels of the quantizer. However, since the probability of occurrence of different quantizer levels is highly variable, there is a considerable advantage in using variable length words to represent quantizer outputs. In such a case, since the average bit-rate (in bits/pel) is lower bounded by the entropy of the quantizer output, it is more relevant to minimize the quantization distortion subject to a fixed entropy. It was stated in Chapter 3 that for Laplacian random variables and a mean square error distortion criterion, the optimum quantizers are uniform. Although such

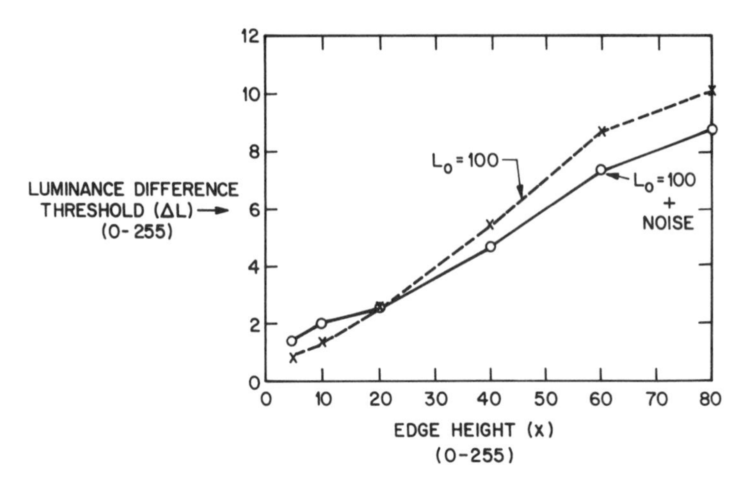

Fig. 5.2.32 Visibility threshold versus slope for a single luminance edge that rises to its peak value within one pel. Edge height and luminance difference threshold are plotted in units 1 to 255 (8-bit number) for two cases: 1) without any additive noise, and 2) with additive noise. (from Sharma and Netravali [5.2.19])

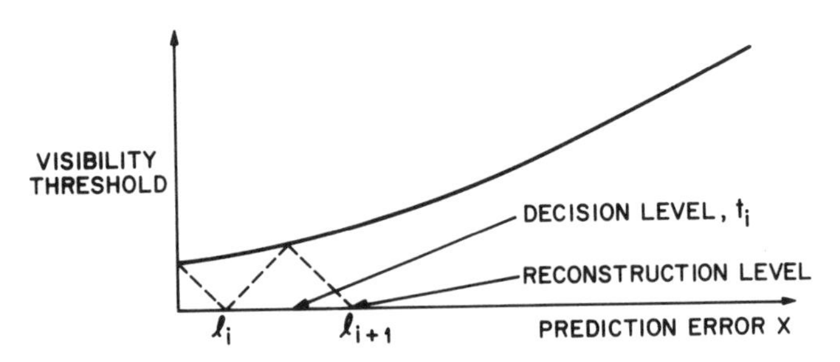

Fig. 5.2.33 Illustration of the quantizer design procedure for an even number L of reconstruction levels.

quantizers may provide adequate picture quality for sufficiently small step size, they may often be too complex to implement because of the large number of levels and the very long word-lengths required for the outer levels. Simplifications are necessary to keep the complexity small.

It has been realized for some time that for better picture quality, quantizers should be designed on the basis of psychovisual criteria. However, a debate continues on what criterion should be used, and expectedly so, considering the complexities of the human visual system. We will describe two approaches and point the reader to other works.

In the first method, a quantizer is designed such that the quantization error is at or below the threshold of visibility (using a given model of the error visibility) while minimizing either the number of quantizer levels or the entropy of the quantizer output. In Chapter 4 it was pointed out that the visibility of perturbations in image intensity is masked by spatial variations in intensity. The visibility threshold at an edge is the value by which the magnitude of the edge (i.e., the edge contrast) can be perturbed such that the perturbation is just visible. A typical relationship[5.2.19] is given in Fig. 5.2.32 for an edge width of one pel and a background luminance $L_o = 100$ (on a scale of 0-255) and variable height. Knowing this relationship between the edge visibility threshold and the slope of a single edge, the quantization error can be constrained to be below this threshold for all slopes of the signal. The algorithm for defining the characteristics of the quantizer satisfying this constraint is illustrated in Fig. 5.2.33, for an even L (number of quantizer output levels). In this case, $x = 0$ is a decision level, (i.e., mid-riser) and the geometric procedure starts at the corresponding point on the visibility threshold function with a dashed line of inclination -45°. The intersection of this line with the abscissa gives the first reconstruction level. From this point, a dashed line of inclination +45° is drawn. Its intersection with the visibility threshold function gives the next decision level, and so on. This procedure is continued until the last +45° line exceeds the amplitude range of x. The number of points at which this path meets the abscissa is the minimum number of levels of the desired quantizer. The resulting dashed zigzag line represents the magnitude of the quantization error as a function of the quantizer input signal x. If L is odd, (i.e., mid-tread) then the procedure starts with the reconstruction level at $x = 0$.

Experimental evidence indicates that quantizers satisfying the above threshold constraint turn out to be the same whether the number of levels is minimized or the entropy is minimized. For videotelephone type (i.e., 256×256 pels with 1 MHz bandwidth) monochrome signals with head and shoulders types of pictures, 27 levels are required for no visible error at normal (six times the picture height) viewing distance if previous element prediction is used. A larger number of levels are required for pictures containing a lot of sharp detail, e.g. the resolution charts. Using two dimensional prediction, this number may be reduced to 13 levels for natural (e.g. head and shoulders) pictures and 21 levels

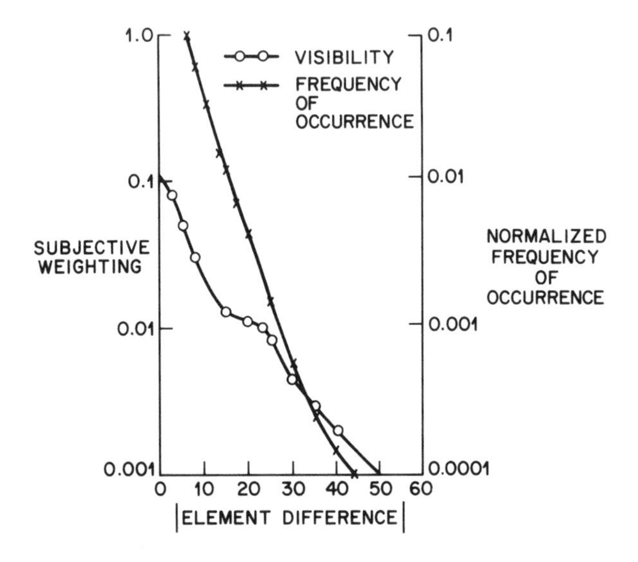

Fig. 5.2.34 Subjective weighting function for quantization noise visibility as related to the magnitude of the element difference for a typical head and shoulders view. Also shown is a histogram of the magnitude of the element differences. (from Netravali [5.2.20])

0	5	13	22	31	40	52
0 2	3 8	9 17	18 26	27 35	36 45	46 255

13 LEVEL MSE QUANTIZER

0	6	15	25	35	45	58
0 3	4 10	11 20	21 30	31 40	41 51	52 255

13 LEVEL MSSD QUANTIZER

0	4	9	14	19	24	31
0 2	3 6	7 11	12 16	17 21	22 27	28 34

36	45	53	62	71	80	91
35 41	42 48	49 57	58 66	67 75	76 84	85 255

MSSDE QUANTIZER WITH ENTROPY 2.6 BITS/PEL

(a)

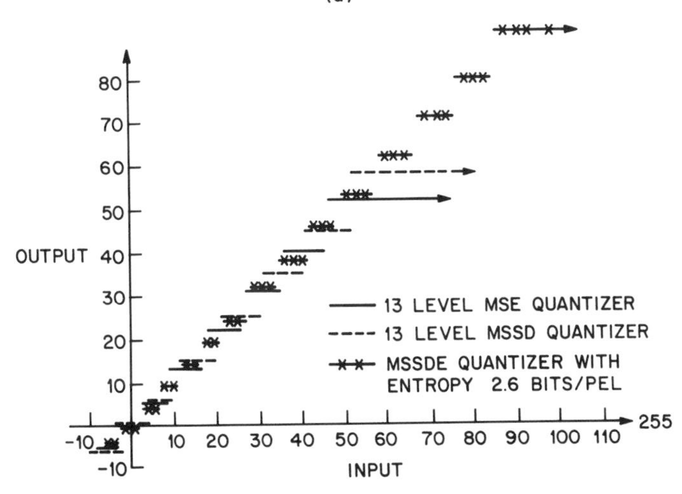

Fig. 5.2.35 Input and output characteristics of quantizers optimized under three criteria are shown: (1) 13 level minimum mean square error (MSE); (2) 13 level minimum subjective weighted mean square error (MSSD); (3) MSSD with a constraint on the entropy (MSSDE) of the quantizer output. For (2) and (3), weights are derived from subjective weighting functions such as in Fig. 5.2.34. (from Netravali [5.2.20])

for resolution charts. Chrominance components usually require fewer levels, e.g., 9 levels for $(R-Y)$ and 5 for $(B-Y)$.

A second method of designing psychovisual quantizers is to minimize a weighted square quantization error, where the weighting is derived from subjective tests. Such optimization would be similar to the MMSE quantization, where the probability density is replaced by a subjective weighting function obtained either from the threshold data (e.g., reciprocal of the threshold data, properly scaled) or from direct subjective tests using complex pictures as stimuli. A typical subjective weighting function for a head and shoulders type of picture is shown in Fig. 5.2.34 when a previous pel predictor is used.[5.2.20] Such weighting functions are picture dependent, but the variation for a class of pictures (e.g., head and shoulders views) is not significant. There has been and perhaps will continue to be considerable debate on the choice of the subjective weighting functions. Although the optimum choice of the weighting function is unknown, optimum quantizers designed on the basis of the above weighting functions have been quite successful. In general, quantizers that minimize the subjective weighted square error for a fixed number of levels (MSSD) or fixed entropy (MSSDE) are less companded than the minimum mean-square error (MMSE) quantizers, and reproduce edges more faithfully. Fig. 5.2.35 and Fig. 5.2.36 give characteristics of quantizers optimized under different criteria and the performance of these quantizers in terms of entropy of the quantizer output and the MSSD measure of picture quality.

Quantizer design for intraframe coders that use more sophisticated predictors has been either on the basis of MMSE or by trial and error. However, the methods of psychovisual quantizers discussed above can be extended easily for this case. For interframe coders, the situation is somewhat more complicated since the quantization error occurs predominantly in the moving areas of the picture, and its visibility depends upon the spatial as well as temporal variations in the scene. Because of the lack of suitable models for both the distribution of the prediction error and the visibility of quantizing error, the problems of designing both MMSE and psychovisual quantizers have not been studied to a great extent. Quantizers used in practice are based on trial and error.

In most applications, the picture signal is positive and has a limited range (say from 0 to 255). If in a DPCM system, the prediction of a pel is $\hat{b}$, then the range of possible inputs to the quantizer is $(-\hat{b},\ 255 - \hat{b})$. Thus, unless $\hat{b} = 128$, several levels of a symmetric quantizer may not get used. For example, if $\hat{b} = 253$, the only positive inputs to the quantizer will be 1 and 2, and therefore the quantizer levels corresponding to the prediction errors above 2 will be wasted. To overcome this, a "prediction-value-dependent" mapping of the quantizer characteristic can be designed. If the quantizer is implemented as a look-up-table (using random access memories), then several look-up-tables can be designed and the appropriate one can be selected depending upon the value of the prediction.

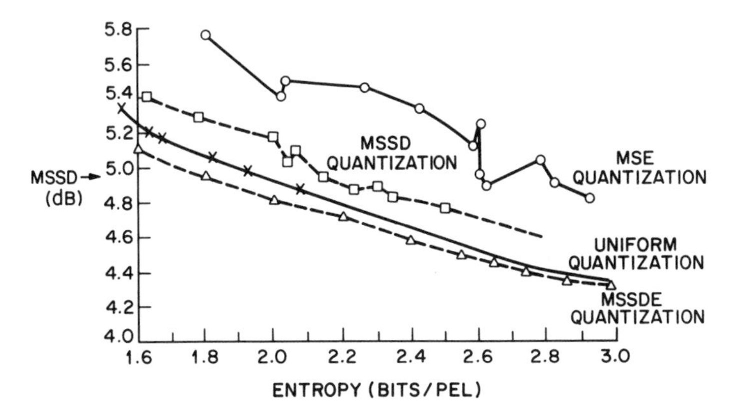

Fig. 5.2.36 Performance of quantizers optimized under different criteria for a typical head and shoulders picture. Mean-square subjective weighted quantization noise (MSSD) is plotted with respect to entropy. It is assumed that MSSD is a reasonable measure of picture quality. (from Netravali [5.2.20])

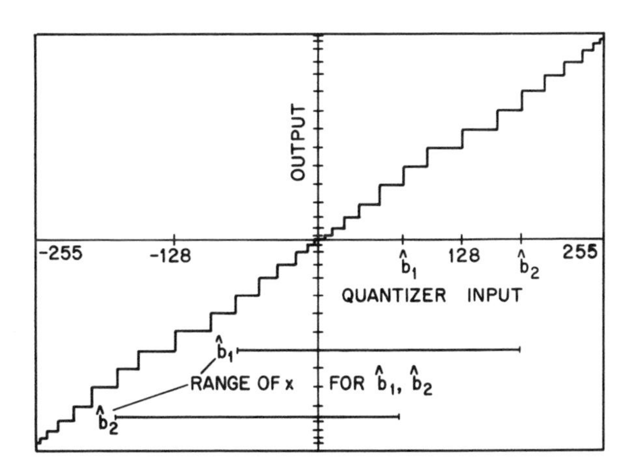

Fig. 5.2.37 Quantizing characteristics of a reflected quantizer.

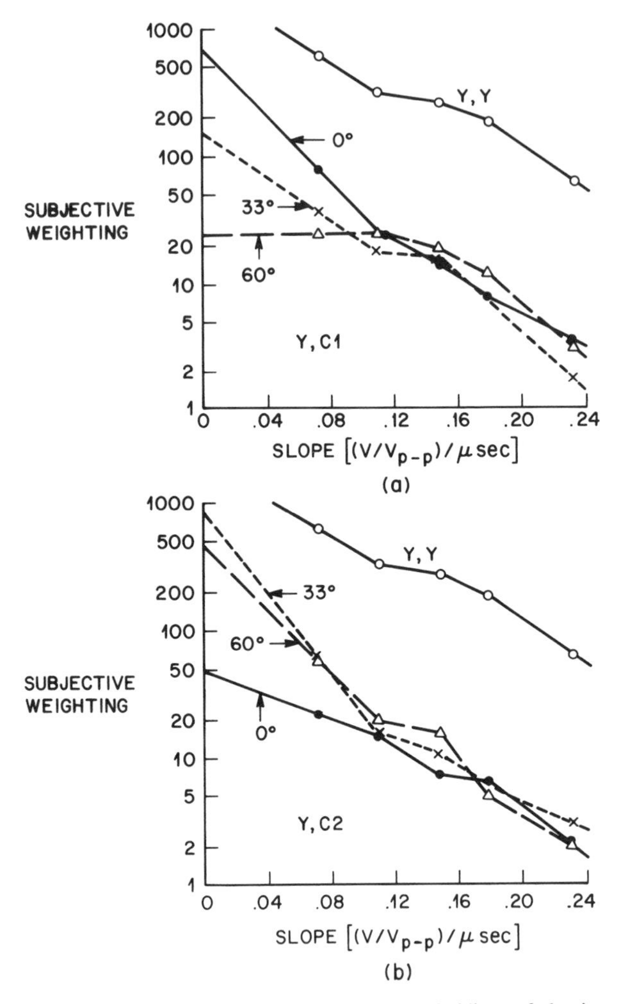

Fig. 5.2.38 Subjective weighting functions with thresholding of luminance slope. This corresponds to using previous element DPCM. Noise resembling quantization noise is added to Y (Y,Y curve), and to (a) $C1$ and (b) to $C2$ for different values of θ as shown. Abscissa is luminance slope measured as a ratio of voltage difference to peak voltage per unit time. Typical videotelephone standard was used, i.e., bandwidth of luminance and two chrominance components was 1 MHz and 250 kHz, respectively. (from Netravali and Rubinstein [5.2.21])

A *Reflected* quantizer[5.2.6] accomplishes this to some extent, first, by representing the prediction error in two's complement arithmetic so that same binary codeword represents values x and $256-x$. Next, a special quantizer characteristic shown in Fig. 5.2.37 is used. The range of possible prediction errors then covers a fixed number of levels with a unique set of code words for any given value of the prediction. Fig. 5.2.37 shows ranges of prediction error x for two values $\hat{b}_1$ and $\hat{b}_2$ of the prediction. In this example, 64 quantizing levels can be represented by only 5 bits, saving 1 bit per pel. Although there are other prediction-value-dependent mappings of the quantizer characteristics, the reflected quantizer is particularly simple to implement. Unfortunately, it has found only limited use since it forces the same step size for both small and large prediction errors, which is not consistent with desired companding characteristics based on psychophysics.

5.2.4b Nonadaptive Quantization for Color Signals

When the color signal is represented as luminance and chrominance components, each of them can be coded using a DPCM quantizer. In this case, procedures for either statistical optimization (MSE) or subjective optimization (MSSD) of the quantizer characteristics are the same as those outlined in the previous section for the luminance signal. However, the subjective data required for quantizer optimization is not yet available in many cases. Figs. 5.2.38 and 5.2.39 show subjective noise weighting functions for the color components when the previous pel predictor is used.[5.2.6] As in the case of monochrome pictures, such data are picture dependent, but they can be used for a class of pictures by minimizing the mean square subjective distortion (MSSD) to obtain the DPCM quantizers.

Weighting functions in these figures are given for three values of θ, where θ represents the rotation of the chrominance axes in the $(R-Y)/1.14$, $(B-Y)/2.03$ plane. $\theta = 33°$ corresponds to the NTSC I and Q signals. Fig. 5.2.38 gives the subjective weighting functions of noise added to the chrominance signal at those pels where the normalized horizontal slope of the luminance equals the abscissa. The top curve in each case shows the visibility of noise in the luminance as a function of the luminance slope. The weighting function for the luminance signal is much greater than that for the chrominance signal. This implies, as expected, that fewer levels are required for the chrominance quantization. The luminance weighting curve is used the same way as in the previous sections. Chrominance weighting plotted as a function of luminance slope can be used for adaptive quantization of chrominance and is discussed in Section 5.2.4c. Fig. 5.2.39 gives similar results with thresholding of the chrominance slope. Here chrominance noise weighting is plotted as a function of chrominance slope, and therefore these weighting functions can be used to derive MSSD quantizers as discussed in the previous section.

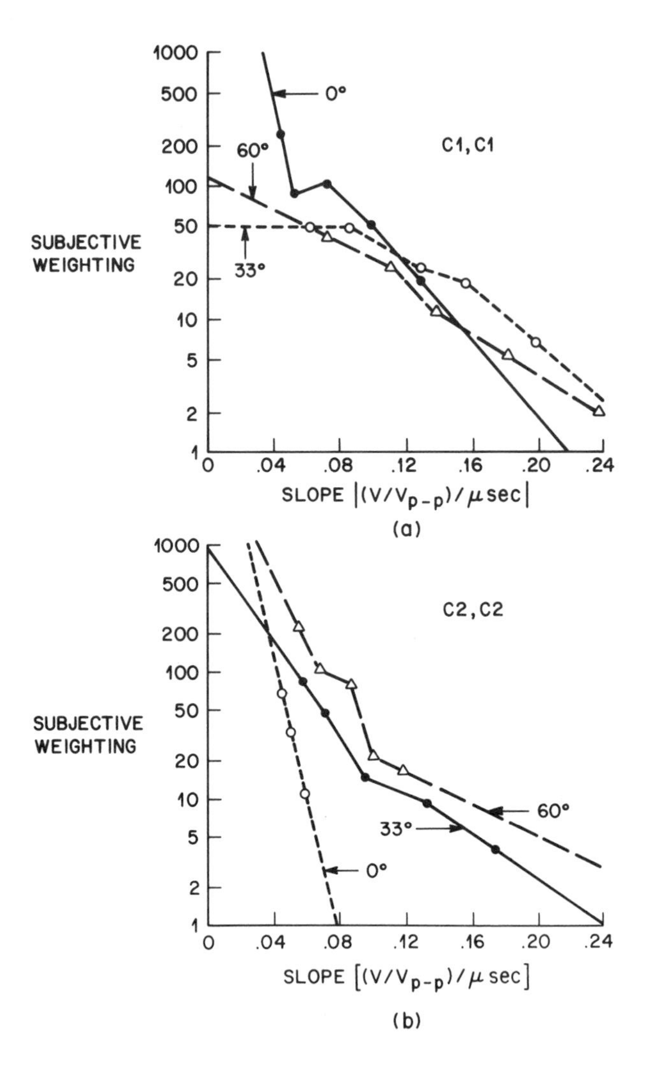

Fig. 5.2.39 Subjective weighting functions with chrominance slope thresholding. (a) Noise resembling quantization noise is added to $C1$ as a function of $C1$ slope and (b) to $C2$ as a function of $C2$ slope for different values of θ. (from Netravali and Rubinstein [5.2.21])

A comparison of Fig. 5.2.38 and Fig. 5.2.39 shows that in general the subjective weighting as a function of the luminance slope is lower than as a function of chrominance slope. This implies that a luminance slope provides a better masking of the noise in the chrominance signal than does the chrominance slope. This added masking due to the luminance is a result of limitations of visual perception, as well as the unequal filtering of the luminance and chrominance signals. It is used in adaptive quantization of the chrominance signal (see next section). It can also be observed from Fig. 5.2.39 that for most of the slope values, noise in $C1$ has higher visibility than noise in $C2$ for $\theta = 0°$ and $\theta = 33°$, with $\theta = 60°$ exhibiting somewhat equal visibility between the two chrominance components. The greater visibility for $C1$ noise implies that finer quantization of the $C1$ signal is necessary compared to the $C2$ signal.

These subjective weighting functions can be used to derive the MSSD quantizers with the help of Eqs. (5.2.57) and (5.2.58). Table 5.2.2 shows results of this type of quantizer design as well as the quantizers designed on the basis of minimum mean square error (MSE) for a fixed number of levels using the statistics of a head and shoulders type of image. As in the case of luminance signals (previous section) such subjective quantizers are much less companded than the MSE quantizers. MSE quantizers show over-design in the flat areas of the picture where the signal slope is small. They also have low dynamic range, which causes annoying slope overload and a consequent decrease in the resolution of the coded pictures. Also, because the inner levels of the *MSE* quantizers are more closely spaced than those of the subjective quantizer, the probability distribution of the quantized output of the *MSE* quantizer is more uniform. This results in the *MSE* quantizer output having higher entropy than the output of the subjective quantizer, for the same number of levels.

5.2.4c Adaptive Quantization

Due to the variation of the image statistics and the required fidelity of reproduction in different regions of the picture, adapting the DPCM quantizer is advantageous in terms of reducing the bit rate. In general, one would like to segment a picture into several subpictures such that within each subpicture both the quantization noise visibility and the statistical properties of the quantized differential signal are relatively stationary. However, this is an extremely difficult task since the perception of noise and the statistics may not be sufficiently related to each other. Several approximations of this idealized situation have been made, some purely statistical and some based on certain psychovisual criteria. For example, one can work in the spatial frequency domain and split the signal into two frequency bands in order to exploit the sensitivity of the eye to variations in image detail (see Chapter 4). In this case, the signal in the low-frequency band is spatially sampled at a low rate, but is quantized finely because of the high sensitivity of the eye to the noise in the low frequency region. The high frequency component of the video signal is spatially

Table 5.2.2 Quantizer Characteristics for DPCM Coding of Color Components. Picture is segmented in 2 or 4 segments and uniform quantizer of different step size is used for each segment.

		NO. OF LEVELS	ONE SIDED QUANTIZER CHARACTERISTIC (ON A SCALE OF 0-255)						
LUMINANCE CODING	VISIBILITY	13	0	3	8	15	22	31	44
			0 1	2 5	6 11	12 18	19 26	27 36	37 255
	MMSE	13	0	2	5	10	15	20	25
			0 1	2 3	4 7	8 12	13 17	18 22	23 255
CHROMINANCE CODING C1	VISIBILITY	9	0	4	10	18	37		
			0 2	3 7	8 14	15 27	28 255		
	MMSE	9	0	2	8	18	29		
			0 1	2 5	6 13	14 23	24 255		
CHROMINANCE CODING C2	VISIBILITY	7	0	3	6	12			
			0 1	2 4	5 9	10 255			
	MMSE	7	0	2	6	10			
			0 1	2 3	4 7	8 255			

* NOTATION z over x y INDICATES THAT SIGNAL BETWEEN x AND y (INCLUDING x, y) IS QUANTIZED AS z.

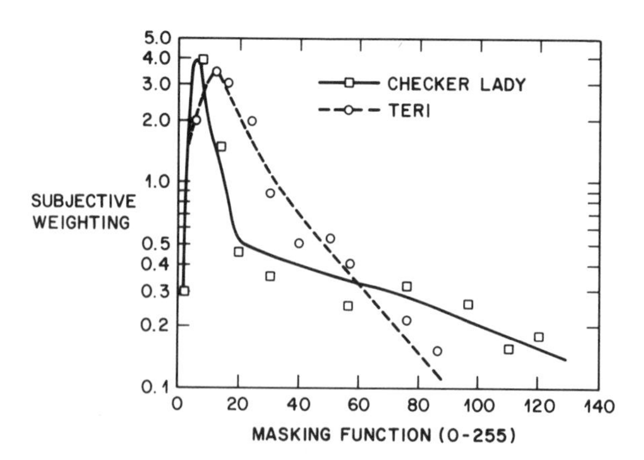

Fig. 5.2.40 Relationship between subjective visibility of noise and a measure of spatial detail (called the *Masking function*) that uses a combination of element and line differences in a 3×3 neighborhood. Dependence of the visibility on the picture content is shown for two head-and-shoulders views. (from Netravali and Prasada [5.2.22])

sampled at a higher rate, but is quantized coarsely due to reduced sensitivity to noise in highly detailed regions.

Pel domain approaches to adaptive quantization start out with a measure of spatial detail (called the *Masking Function**) and then obtain experimentally the relationship between the visibility of noise and the measure of spatial detail. In nonadaptive quantization of the previous section, the prediction error is used as the measure of spatial detail, but for adaptive quantization more complex measures are used. A typical relationship for head and shoulders pictures is shown in Fig. 5.2.40, where a weighted sum of slopes in a 3×3 neighborhood is used as a measure of local spatial detail.[5.2.22] If $\Delta_{i,j}^{H}$ and $\Delta_{i,j}^{V}$ are the horizontal and vertical slopes of the luminance at location (i, j) then the weighted sum of slopes is given by

$$M(i\ j) = \sum_{\ell, m=-1}^{+1} \alpha^{\sqrt{\ell^2+m^2}} \left[|\Delta_{i-\ell, j-m}^{H}| + |\Delta_{i-\ell, j-m}^{V}| \right] \tag{5.2.61}$$

where (i, j) are the coordinates relative to the pel being coded and $\alpha < 1$ is a factor based on psychovisual threshold data. In the experimental investigation of [5.2.22], α is taken to be 0.35, and therefore, exponentially decreasing weights are assigned to pels of greater distance from the current (to be coded) pel. Using the noise weighting function of Fig. 5.2.40, which relates the visibility of additive noise to the masking function, a picture is divided into (not necessarily contiguous) segments. Within each segment noise visibility is approximately constant, and an individual quantizer is designed for each one. The segmentation is controlled by the value of the masking function M, as indicated in Fig. 5.2.41. Fig. 5.2.42 shows a head and shoulders picture divided into four segments. Fig. 5.2.42a shows an original picture, and Figs. 5.2.42b–e show four subpictures (pels in these subpictures are shown white) which are in the order of increasing spatial detail, i.e., decreasing noise visibility. Thus, flat low detail areas of the picture constitute subpicture I (i.e., Fig. 5.2.42(b)) and sharp, high contrast edges are included in subpicture IV (i.e., Fig. 5.2.42(e)). Quantizers for each of the subpictures could be designed in a way similar to the earlier nonadaptive techniques.

One side benefit of such an adaptive quantization scheme is that the prediction error (i.e., the differential signal) statistics show a marked change from one segment to the other. In Fig. 5.2.43, prediction error histograms are shown for the four subpictures of Fig. 5.2.42. It is evident that since these histograms are quite different from one another, a different variable length code

---

* We will later refer to this as an *Activity Function*.

Table 5.2.3 Performance of Adaptive Chrominance Coding.

NO. OF SEGMENTS	ADAPTIVE QUANTIZER STEP SIZES CHROMINANCE 1	ADAPTIVE QUANTIZER STEP SIZES CHROMINANCE 2	EQUIVALENT PICTURE QUALITY	ADVANTAGE OF ADAPTION (BITS/CHROMINANCE PEL)
2	2,7	3,10	A	0.16
2	3,10	4,12	B	0.21
2	3,8	4,12	B	0.18
2	4,14	5,15	C	0.27
2	2,7	4,12	A	0.26
2	3,10	5,15	C	0.11
2	5,10	7,12	D	0.19
4	2,3,5,20	2,4,7,13	A	0.35
4	2,6,10,20	3,7,10,20	A	0.40
4	3,4,7,19	4,7,14,20	B	0.31
4	3,5,8,20	4,9,14,20	B	0.38
4	3,8,12,20	4,10,14,20	B	0.43
4	4,6,10,20	4,7,14,20	B	0.48
4	4,6,10,20	5,8,14,20	C	0.37

NOTES: EQUIVALENT PICTURE QUALITY JUDGED USING NONADAPTIVE QUANTIZER STEP SIZE OF 2,3(A); 3,4(B); 3,5(C); 4,6(D).

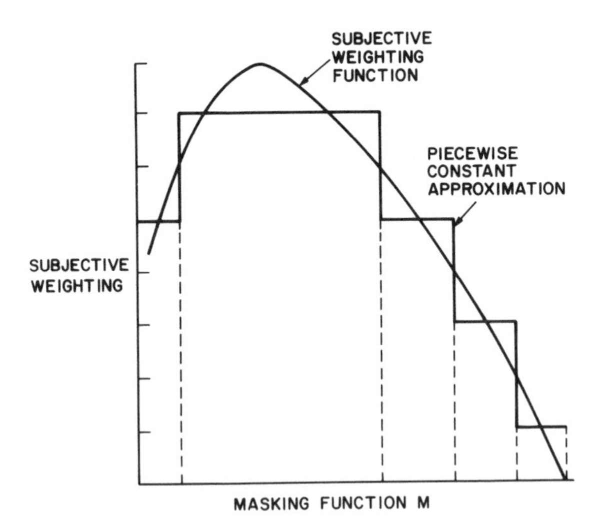

Fig. 5.2.41 Sketch of a typical subjective weighting function versus masking function M and its approximation by a four-level piecewise constant function. Using this approximation, images are segmented into four subpictures, I to IV, according to the value of the masking function M at each pel.

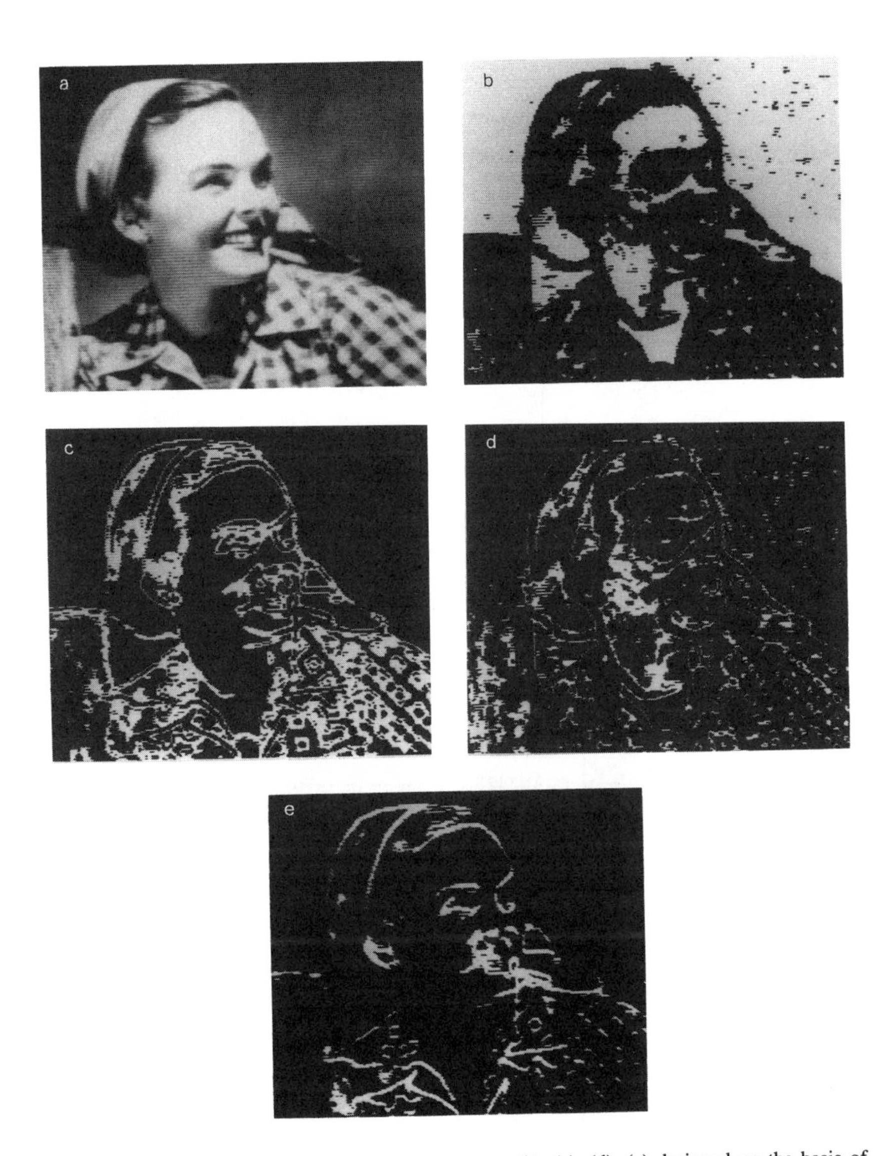

Fig. 5.2.42 Original picture (a) and its four segments (b), (c), (d), (e) designed on the basis of spatial detail (two-dimensional) and noise visibility. Segment (b) has the highest noise visibility, whereas segment (e) has the least noise visibility. (from Netravali and Prasada [5.2.22])

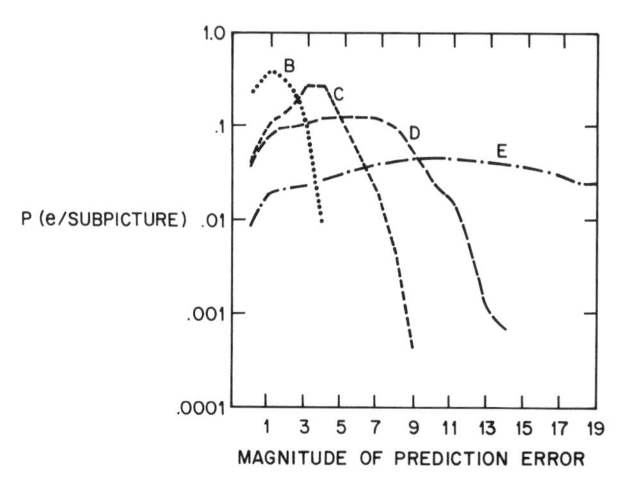

Fig. 5.2.43 Plot of conditional histograms of prediction errors conditioned on the four subpictures b, c, d, e as shown in Fig. 5.2.42. These subpictures are derived using a scheme similar to the one shown in Fig. 5.2.41. (from Netravali and Prasada [5.2.22])

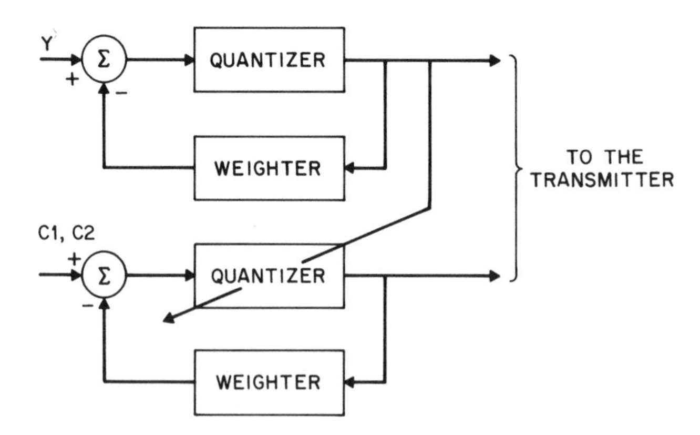

Fig. 5.2.44 Adaptive DPCM coding of the chrominance components. The chrominance quantizer is varied as a function of the luminance spatial activity as measured by the quantized luminance prediction error.

may be used to represent the quantizer outputs from each segment. This technique together with adaptive quantization is capable of reducing the bit-rate by about 30 percent compared with a nonadaptive previous pel DPCM coder. However, much of this reduction is a result of coarser quantization of different segments rather than the use of different variable word-length codes for different segments. It is also interesting to note that the advantage of adaptation increases with the number of segments used, up to about four, after which further improvement is small.

Another method of adapting the quantizer, called *delayed coding*, treats quantization as if following a path through an encoding tree determined by the quantizer steps. [See also Section 5.2.1 for a discussion on delayed coding in the context of delta modulation.] The path is chosen to minimize the weighted quantization error at several samples, including some subsequent samples that have yet to be transmitted. Since the representative levels are not varied from sample to sample, it is not necessary to send the adaptation information to the receiver. Thus the same input value to the quantizer may get mapped into a different representative level depending on which path on the tree will yield lower distortion. This is in contrast to the usual memoryless quantizer where every input value between the two decision levels is mapped into the representative level in between. As discussed in Section 5.2.1, such techniques have been used more successfully for delta modulation, since the growth of the tree is much smaller compared to DPCM with its many more quantizer levels.

In summary, for intraframe coding, although the optimum rules for adaptation are not known, rules based on intuition and trial-and-error have shown significant reduction in bit-rate compared with nonadaptive systems. In the case of interframe coders, systematic investigations of adaptive quantizers have yet to be made, primarily due to the lack of understanding of both statistics and the perceptual basis for adapting the quantizers. However, quantizers have been adapted in an ad hoc manner to facilitate matching of the variable bit-rate coder to the constant bit-rate channel. We shall discuss this in more detail in Section 6.6.

For coding of color components, since luminance edges provide better masking of the noise in the chrominance signal than do the chrominance edges (see Fig. 5.2.38), luminance edges should be used to compute the masking functions for adaptive coding of chrominance components. Fig. 5.2.44 shows a block diagram of a chrominance DPCM coder whose quantizer is varied as a function of the spatial activity of the luminance signal as measured by the quantized luminance prediction error. Experimental evidence indicates that by using the magnitude of the horizontal luminance slope as the measure of spatial activity, considerable advantage can be derived by adaptive coding. Table 5.2.3 summarizes the results. Here pictures coded with adaptive quantizers are subjectively compared to the nonadaptive quantizers. Differences in the entropies of the quantizer output for two (adaptive and nonadaptive) subjectively

LEVEL NO.	CODE WORD LENGTH	CODE
1	12	100101010101
2	10	1001010100
3	8	10010100
4	6	100100
5	4	1000
6	4	1111
7	3	110
8	2	01
9	2	00
10	3	101
11	4	1110
12	5	10011
13	7	1001011
14	9	100101011
15	11	10010101011
16	12	100101010100

Fig. 5.2.45 A typical variable-length code for a DPCM coded signal, with 16 quantizer levels. (from Pirsch [5.2.23])

equivalent pictures are tabulated. Clearly there is an advantage in chrominance coding with respect to luminance slopes. Using only two segments the advantage is of the order of 0.2 bits/chrominance pel, whereas using four segments the advantage is of the order of 0.35 bits/chrominance pel.

5.2.4d Code Assignment

We mentioned earlier that the probability distribution of the quantizer output levels is highly nonuniform for intraframe as well as interframe predictive coders. This naturally lends itself to representation using code words of variable lengths (e.g., Huffman codes). The average bit-rate for such a code is usually very close to and is lower bounded by the entropy of the quantizer output signal as we saw in Chapter 3. A typical variable length code [5.2.23] for a previous pel DPCM coder with 16 quantizer levels is shown in Fig. 5.2.45. Inner levels occur much more often and, therefore, are represented by a smaller length word.

A Huffman code (see Chapter 3) requires a fairly reliable knowledge of the probability distribution of the quantized output. However, there are other procedures that work with approximate probabilities and still remain reasonably efficient even when the probabilities depart somewhat from their assumed values. Also in practice, it is found that the Huffman code is not too sensitive to changes in the probability distribution that take place for different pictures, and therefore, Huffman codes based on an average of many pictures remain quite efficient compared with codes designed on the basis of knowledge of the statistics of an individual picture. For most pictures, the entropy of the previous pel DPCM quantizer output (and consequently the average bit-rate using Huffman coding) is about one bit/pel less than the corresponding bit-rate for a fixed length code. Stated differently, variable length coding results in an improvement of about 6 dB in signal-to-noise ratio (SNR), with no change in the average transmission bit-rate. A similar conclusion can be arrived at quantitatively for a Laplacian probability distribution, (a good approximation for the differential signal). In this case, variable length coding can improve the SNR ratio by about 5.6 dB if the number of quantizer levels is large.

One of the problems with the use of variable length codes is that the output bit-rate from the source coder changes with local picture content. In order to transmit a variable bit-rate digital signal over a constant bit-rate channel, the source coder output has to be held temporarily in a first-in-first-out *buffer* that can accept input at a nonuniform rate, but whose output is conveyed to the channel at a uniform rate. Since in a practical system a finite size buffer must be used, any coder design must take into account the possibility of buffer overflow and underflow, and this depends in a complicated way upon the size of the buffer, the variable length code used, the image statistics, and the channel bit-rate. Buffer underflow is usually not a problem since PCM values could be inserted to increase the coder output whenever required. By using a channel whose rate is somewhat higher than the entropy of the quantizer output, the probability of buffer overflow is markedly reduced. Also, by designing codes that minimize the probability of the length of the code words exceeding a certain number, the likelihood of buffer overflow can be reduced. However, since buffer overflow cannot always be prevented, strategies must be designed to gradually reduce the output bit-rate of the coder as the buffer begins to fill. This is particularly important in the case of interframe coders since the motion in a television scene is relatively bursty. Several techniques for reducing the bit-rate input to the buffer have been studied, e.g., reducing sampling rate, coarser quantization, etc. Such methods allow graceful degradation of picture quality when the source coder is overloaded, and will be discussed in more detail in Chapter 6.

5.3 Transform Coding of Blocks of Samples

With discrete transform coding, as we saw in Chapter 3, blocks of pels are transformed into another domain called the transform domain prior to coding and transmission. Transform coding has been found to have relatively good capability for bit-rate reduction, which comes about mainly from two mechanisms. First, not all of the transform domain coefficients need to be transmitted in order to maintain good image quality, and second, the coefficients that are coded need not be represented with full accuracy. Loosely speaking, transform coding is preferable to DPCM for compression to bit-rates below 2 or 3 bits per pel for single images. However, in those applications where complexity and cost are serious issues, low-pass filtering and reduction of pel sampling rate may be a preferable alternative.

5.3.1 Discrete Linear Orthonormal Transforms

With linear transforms each block of pels to be transformed is first arranged into a column vector $\boldsymbol{b}$ of length N, and with *orthonormal* (also known as unitary) linear transforms[†]

$$\boldsymbol{c} = \boldsymbol{T}\boldsymbol{b} \tag{5.3.1}$$

$$\boldsymbol{b} = \boldsymbol{T}'\boldsymbol{c} \tag{5.3.2}$$

where $\boldsymbol{T}$ is the $N \times N$ transform matrix, and $\boldsymbol{c}$ is the column vector of transform coefficients. If the mth column of matrix $\boldsymbol{T}'$ is denoted by $\boldsymbol{t}_m$ then[‡]

† Prime denotes conjugate transpose.

‡ $\delta_{mn} = 1$ if $m = n$, $\delta_{mn} = 0$ otherwise.

$$\boldsymbol{t}'_m \, \boldsymbol{t}_n = \delta_{mn} \tag{5.3.3}$$

which is where the term orthonormal arises. The vectors $\boldsymbol{t}_m$ are *orthonormal basis* vectors of the unitary transform $\boldsymbol{T}$, and using them Eq. (5.3.2) can be written as

$$\boldsymbol{b} = \sum_{m=1}^{N} c_m \boldsymbol{t}_m \tag{5.3.4}$$

where c_m is the mth element of $\boldsymbol{c}$, given by

$$c_m = \boldsymbol{t}'_m \, \boldsymbol{b} \tag{5.3.5}$$

The first basis vector $\boldsymbol{t}_1$ consists of constant values for practically all transforms of interest, i.e., it corresponds to zero spatial frequency and is given by

$$\boldsymbol{t}'_1 = \frac{1}{\sqrt{N}} \, (1,1,1...1) \tag{5.3.6}$$

For this reason c_1 is often called the DC coefficient. Thus, if $b_{\max}$ is the largest possible pel value, then $\sqrt{N} \; b_{\max}$ is the largest possible value* for c_1.

The vector $\boldsymbol{b}$ could be constructed from N successive pels of a raster scan line, in which case the transform coding exploits one-dimensional horizontal correlation between pels. Alternatively, $\boldsymbol{b}$ could be constructed from a two-dimensional $L \times L$ array of image pels by simply laying the L-pel columns (or rows) end to end to form a single column vector of length $N = L^2$, in which case the transform coding exploits both horizontal and vertical correlation in the image.

A third possibility is to denote the $L \times L$ array of image pels by the square matrix $\boldsymbol{B} = [b_{ij}]$ and *separate* the transform into two steps. The first step is to transform the *rows* of $\boldsymbol{B}$ with a length L transform in order to exploit horizontal correlation in the image. Next, the *columns* of $\boldsymbol{B}$ are transformed in order to exploit vertical image correlation. The combined operation can be written

* Some discussions of transform coding in the literature use $c_i / \sqrt{N}$ instead of the c_i that we have defined, in order that the DC coefficient have the same dynamic range as the original pels.

$$c_{mn} = \sum_{i=1}^{L} \sum_{j=1}^{L} b_{ij} t_{nj} t_{mi} \tag{5.3.7}$$

where the t's are the elements of the $L \times L$ transform matrix $\boldsymbol{T}$ and $\boldsymbol{C} = [c_{mn}]$ is the $L \times L$ matrix of transform coefficients. Assuming the one dimensional basis vectors are indexed in order of increasing spatial frequency, then subscript m indexes vertical spatial frequency, and n indexes horizontal spatial frequency. If f_m and f_n are the vertical and horizontal spatial frequencies of the respective basis vectors, then the spatial frequency corresponding to coefficient c_{mn} is given by

$$f_{mn} = \sqrt{f_m^2 + f_n^2} \ \frac{\text{cycles}}{\text{distance}} \tag{5.3.8}$$

For unitary transforms, the inverse operation is simply†

$$b_{ij} = \sum_{m=1}^{L} \sum_{n=1}^{L} c_{mn} t_{nj}^{*} t_{mi}^{*} \tag{5.3.9}$$

Matrices $\boldsymbol{C}$, $\boldsymbol{T}$ and $\boldsymbol{B}$ are related by

$$\boldsymbol{C} = \boldsymbol{TBT}^{t} \tag{5.3.10}$$

and for unitary $\boldsymbol{T}$ the inverse operation is simply

$$\boldsymbol{B} = \boldsymbol{T}'\boldsymbol{CT}^{*} \tag{5.3.11}$$

The separable transformation requires $2L^3$ multiplications and additions, whereas the column transformation of Eq. (5.3.1) requires $N^2 = L^4$ such operations. Thus, the separable transform would normally be preferred for the usual case of $L > 2$. Extension of separable transforms to dimension higher than two is straightforward.

If $\{\boldsymbol{t}_m\}$ are the basis vectors of the Lth order transform $\boldsymbol{T}$, then Eq. (5.3.11) can be written

$$\boldsymbol{B} = \sum_{m=1}^{L} \sum_{n=1}^{L} c_{mn} \boldsymbol{t}_m^{*} \boldsymbol{t}_n' \tag{5.3.12}$$

---

† Asterisk denotes complex conjugate (which is ignored for real transforms).

That is, $\boldsymbol{B}$ can be thought of as a linear combination of $L \times L$ *basis images*

$$\{ \boldsymbol{t}_m^* \boldsymbol{t}_n' \} \tag{5.3.13}$$

Eq. (5.3.12) is also known as an *outer product* expansion of $\boldsymbol{B}$.

We saw in Chapter 3 that linear transformation is an information preserving process. We also saw that a major objective of transform coding is to produce *statistically independent* (or at least uncorrelated) transform coefficients so that they can be coded independently of each other with good efficiency. Another objective is *energy compaction,* which means concentrating the energy in a few coefficients and thereby making as many coefficients as possible small enough that they need not be transmitted. Most of the unitary transforms discussed here achieve both objectives reasonably well.

We now define several unitary transforms of interest and discuss efficient methods for computing them.

5.3.1a Discrete Fourier Transform (DFT,RDFT)

The $N \times N$ unitary, symmetric Discrete Fourier Transform (DFT) matrix $\boldsymbol{T}$ has elements

$$t_{mi} = \frac{1}{\sqrt{N}} \exp \left[- \frac{2\pi}{N} \sqrt{-1}\ (i-1)(m-1) \right] \quad i,m = 1...N \tag{5.3.14}$$

For $N = 8$, matrix $\boldsymbol{T}$ is shown in Fig. 5.3.1. Since in our case $\boldsymbol{b}$ is real, the complex coefficients of $\boldsymbol{c}$ have conjugate symmetry, i.e.,

$$c_{2+p} = c_{N-p}' \qquad 0 \le p < N - 2 \tag{5.3.15}$$

Therefore, reconstruction of the N pels of $\boldsymbol{b}$ requires only the transmission of at most N real values.

The unitary *Real* Discrete Fourier Transform (RDFT), having N real coefficients, can be derived by normalizing the real and imaginary parts of the DFT basis vectors as follows:

1.000	1.000	1.000	1.000	1.000	1.000	1.000	1.000
1.000	.707	.000	-.707	-1.000	-.707	.000	.707
1.000	.000	-1.000	.000	1.000	.000	-1.000	.000
1.000	-.707	.000	.707	-1.000	.707	.000	-.707
1.000	-1.000	1.000	-1.000	1.000	-1.000	1.000	-1.000
1.000	-.707	.000	.707	-1.000	.707	.000	-.707
1.000	.000	-1.000	.000	1.000	.000	-1.000	.000
1.000	.707	.000	-.707	-1.000	-.707	.000	.707

.000	.000	.000	.000	.000	.000	.000	.000
.000	-.707	-1.000	-.707	.000	.707	1.000	.707
.000	-1.000	.000	1.000	.000	-1.000	.000	1.000
.000	-.707	1.000	-.707	.000	.707	-1.000	.707
.000	.000	.000	.000	.000	.000	.000	.000
.000	.707	-1.000	.707	.000	-.707	1.000	-.707
.000	1.000	.000	-1.000	.000	1.000	.000	-1.000
.000	.707	1.000	.707	.000	-.707	-1.000	-.707

Fig. 5.3.1 Transform matrix for the Discrete Fourier Transform (DFT) with $N=8$. The top is the real part; the bottom is the imaginary part.

$$t_1^R = t_1$$

$$t_2^R = \sqrt{2}\,\mathrm{Re}(t_2) \quad t_3^R = \sqrt{2}\,\mathrm{Im}(t_2)$$

$$t_4^R = \sqrt{2}\,\mathrm{Re}(t_3) \quad t_5^R = \sqrt{2}\,\mathrm{Im}(t_3) \tag{5.3.16}$$

$$\vdots$$

$$t_N^R = \begin{cases} \sqrt{2}\,\mathrm{Im}\left[t_{\frac{N+1}{2}}\right], & N \text{ odd} \\ t_{\frac{N+2}{2}}, & N \text{ even} \end{cases}$$

It is instructive to plot real basis vectors as waveforms, since amplitudes, phases and frequencies may then be readily compared. The RDFT basis vectors for $N = 16$ are shown in Fig. 5.3.2.

The DFT basis vectors t_m are complex sinusoids. Thus, the DFT is a spectral decomposition with the energy of frequency $(m-1)/N$ being given by $|c_m|^2$. We saw in Chapter 3 that for most images the spectral energy generally decreases as a function of frequency.† Thus, the DFT results in a compaction of energy into the lower index transform coefficients.

A major feature of the DFT is the existence of well known fast algorithms, called Fast Fourier Transforms or FFT's.[5.3.1−2] Whereas a direct matrix multiplication DFT requires approximately N^2 complex multiplications and additions, FFT's require only about $N\log_2 N$ such operations if N is a power of two. This efficiency is achieved by exploiting the periodicities that exist within the Fourier transform matrix $\boldsymbol{T}$. FFT algorithms have also been devised for the case where N is not a power of two. Other advantages of FFT's in comparison with the direct DFT include reduced storage requirements and reduced round-off error. Most reasonably sized computer facilities contain one or more FFT implementations.*

---

† For $m \le (N+2)/2$. Components above this frequency are post-aliasing terms due to the sampling process.

* Many FFT implementations do not include the factor $1/\sqrt{N}$ in Eq. (5.3.14). Instead they include a factor $1/N$ in the inverse FFT. Also, a few of them compute the transform coefficients in the order $m = 2,3,\ldots,N,N+1$.

Since $\boldsymbol{b}$ is real, the Discrete Fourier transformation can be made more efficient by taking advantage of the conjugate symmetry of the complex coefficients. This is illustrated (for N even)[5.3.1] by the following procedure:

1) Define the complex vector $\boldsymbol{w}$, having length $N/2$, to be

$$\mathrm{Re}\,(\boldsymbol{w}) = (b_1 b_3 b_5 \ldots b_{N-1})'$$

$$\mathrm{Im}\,(\boldsymbol{w}) = (b_2 b_4 b_6 \ldots b_N)' \tag{5.3.17}$$

2) Take the length $\frac{N}{2}$ DFT (or FFT‡) of $\boldsymbol{w}$ to obtain the complex coefficients $\{ W_m,\ m = 1 \ldots \frac{N}{2} \}$.

3) For $2 \le m < \frac{N+5}{4}$ form the intermediate complex values:

$$X_m = \frac{1}{\sqrt{8}} \left[W_m + W'_{\frac{N}{2}+2-m} \right] \tag{5.3.18}$$

$$Y_m = \frac{\sqrt{-1}}{\sqrt{8}} \left[W_m - W'_{\frac{N}{2}+2-m} \right] \exp\left[- \frac{2\pi\sqrt{-1}\,(m-1)}{N} \right]$$

4) Then the complex coefficients for the DFT of $\boldsymbol{b}$ are given by:

‡ Also, multiply by $\sqrt{\frac{2}{N}}$ if not included in the FFT.

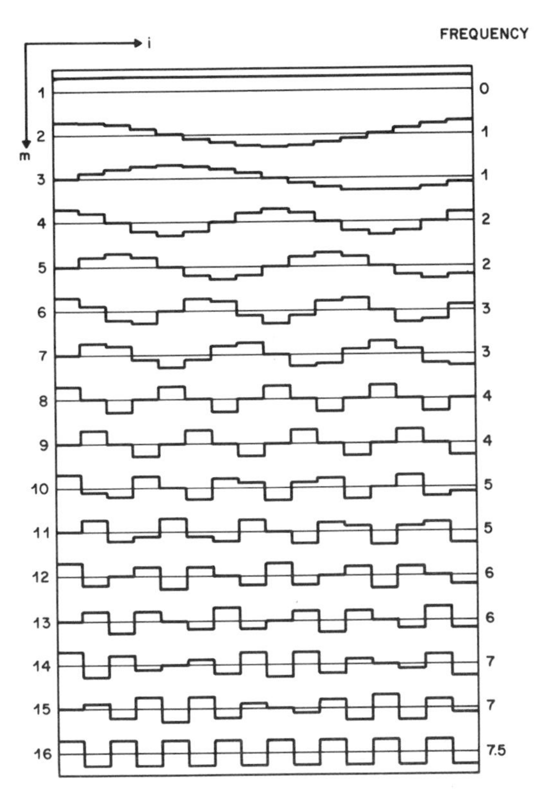

Fig. 5.3.2 Basis vectors for the Real Discrete Fourier Transform (RDFT) with $N=16$. For each m the elements of $\boldsymbol{t}_m^R$ are plotted.

$$\frac{1}{2}\begin{bmatrix} 1 & 1 & 1 & 1 \\ 1 & -1 & 1 & -1 \\ 1 & 1 & -1 & -1 \\ 1 & -1 & -1 & 1 \end{bmatrix}$$

$$\frac{1}{\sqrt{8}}\begin{bmatrix} 1 & 1 & 1 & 1 & 1 & 1 & 1 & 1 \\ 1 & -1 & 1 & -1 & 1 & -1 & 1 & -1 \\ 1 & 1 & -1 & -1 & 1 & 1 & -1 & -1 \\ 1 & -1 & -1 & 1 & 1 & -1 & -1 & 1 \\ 1 & 1 & 1 & 1 & -1 & -1 & -1 & -1 \\ 1 & -1 & 1 & -1 & -1 & 1 & -1 & 1 \\ 1 & 1 & -1 & -1 & -1 & -1 & 1 & 1 \\ 1 & -1 & -1 & 1 & -1 & 1 & 1 & -1 \end{bmatrix}$$

Fig. 5.3.3 Transform matrices of the Walsh Hadamard Transform (WHT) with $N=4$ and 8.

$$c_1 = \frac{1}{\sqrt{2}} [\text{Re}(W_1)+\text{Im}(W_1)]$$

$$\left\{ c_m = X_m - Y_m \ , \ 2 \le m < \frac{N+5}{4} \right\}$$

$$\left\{ c_{\frac{N}{2}+2-m} = X'_m + Y'_m \ , \ 2 \le m < \frac{N+5}{4} \right\} \qquad (5.3.19)$$

$$c_{\frac{N}{2}+1} = \frac{1}{\sqrt{2}} [\text{Re}(W_1)-\text{Im}(W_1)]$$

$$\left\{ c_{N-p} = c'_{2+p} \ , \ 0 \le p \le \frac{N}{2} - 2 \right\}$$

For coding purposes, only N real numbers need be computed. In particular, the coefficients of the unitary Real Discrete Fourier Transform (RDFT) are given (for N even) by:

$$c_1^R = c_1$$

$$c_2^R = \sqrt{2}\,\text{Re}(c_2) \quad c_3^R = \sqrt{2}\,\text{Im}(c_2)$$

$$c_4^R = \sqrt{2}\,\text{Re}(c_3) \quad c_5^R = \sqrt{2}\,\text{Im}(c_3) \qquad (5.3.20)$$

$$\vdots$$

$$c_N^R = c_{\frac{N}{2}+1}$$

In order to perform the inverse DFT (or inverse RDFT) we simply reverse the above procedure. That is, we solve for X_m and Y_m, then solve for W_m, then perform the inverse DFT (or inverse FFT) and finally reorder $\boldsymbol{w}$ to obtain $\boldsymbol{b}$.

5.3.1b Walsh-Hadamard Transform (WHT)

The WHT is most easily described recursively. Let $\boldsymbol{H}_1 = 1$, and for $N = 2^n$ define

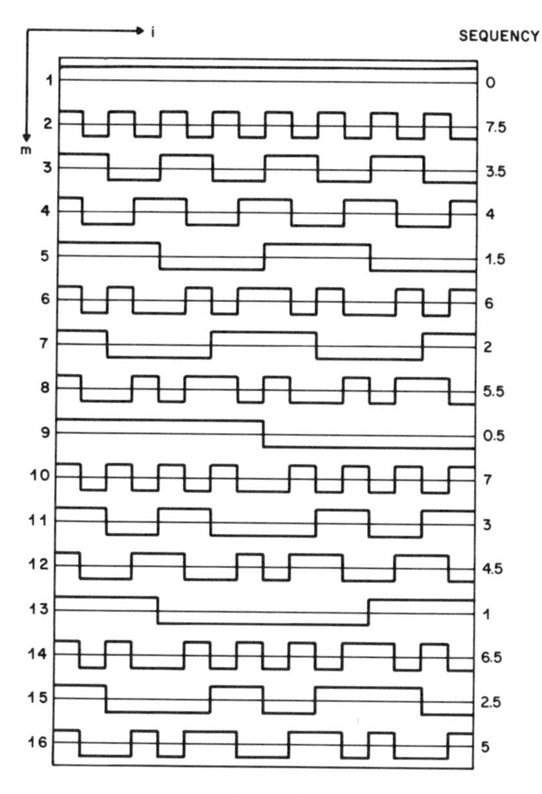

Fig. 5.3.4 Basis vectors for the WHT with $N = 16$. For each vector the sequency is the number of zero crossings divided by two.

$$\frac{1}{2}\begin{bmatrix} 1 & 1 & 1 & 1 \\ 1 & 1 & -1 & -1 \\ 1 & -1 & -1 & 1 \\ 1 & -1 & 1 & -1 \end{bmatrix}$$

$$\frac{1}{\sqrt{8}}\begin{bmatrix} 1 & 1 & 1 & 1 & 1 & 1 & 1 & 1 \\ 1 & 1 & 1 & 1 & -1 & -1 & -1 & -1 \\ 1 & 1 & -1 & -1 & -1 & -1 & 1 & 1 \\ 1 & 1 & -1 & -1 & 1 & 1 & -1 & -1 \\ 1 & -1 & -1 & 1 & 1 & -1 & -1 & 1 \\ 1 & -1 & -1 & 1 & -1 & 1 & 1 & -1 \\ 1 & -1 & 1 & -1 & -1 & 1 & -1 & 1 \\ 1 & -1 & 1 & -1 & 1 & -1 & 1 & -1 \end{bmatrix}$$

Fig. 5.3.5 Sequency ordered transform matrices for the WHT with $N = 4$ and 8.

$$\boldsymbol{H}_{2N} = \begin{bmatrix} \boldsymbol{H}_N & \boldsymbol{H}_N \\ \boldsymbol{H}_N & -\boldsymbol{H}_N \end{bmatrix} \tag{5.3.21}$$

Then the unitary, symmetric WH transform matrix is given by

$$\boldsymbol{T} = \frac{1}{\sqrt{N}} \boldsymbol{H}_N = \boldsymbol{T}' \tag{5.3.22}$$

For $N = 4$ and 8, matrices $\boldsymbol{T}$ are shown in Fig. 5.3.3 for the WHT. The main advantage of the WHT is that apart from the factor $1/\sqrt{N}$, the computation requires only addition and subtraction, as opposed to most other transforms that require multiplication as well. Moreover, a fast algorithm exists that requires only about $N \log_2 N$ operations.[5.3.4]

An alternative definition also exists that is often more useful for hardware implementation or computer calculation. For this representation we require the binary coefficients of integers less than N. We again assume $N = 2^n$, and for $k < N$ we write

$$\begin{aligned} k &= k_{n-1} 2^{n-1} + k_{n-2} 2^{n-2} + \ldots + k_1 2^1 + k_0 2^0 \\ &= \sum_{s=0}^{n-1} k_s 2^s \quad \text{where} \quad k_s = 0 \text{ or } 1 \ . \end{aligned} \tag{5.3.23}$$

Next, for integers $k, \ell < N$ we introduce the notation

$$\langle k, \ell \rangle \triangleq \sum_{s=0}^{n-1} k_s \ell_s \tag{5.3.24}$$

Then it can be shown that the WH transform matrix T has elements given by

$$t_{mi} = \frac{1}{\sqrt{N}} (-1)^{\langle m-1, i-1 \rangle} \quad m, i = 1 \ldots N \tag{5.3.25}$$

The concept of frequency can also be extended to the WHT, in which case it is called *sequency*. The sequency of a WHT basis vector $\boldsymbol{t}_m$ is defined as the number of sign changes within it divided by two. If the basis vectors $\boldsymbol{t}_m$ are plotted as 2-level waveforms, as in Fig. 5.3.4, then the sequency is half the number of zero crossings, which corresponds to the definition of frequency by sinusoids. As with the DFT, the energy in the WHT coefficients for images tends to decrease with sequency.

Unfortunately, the WHT definitions of Eqs. (5.3.22) and (5.3.25) do not have their basis vectors ordered according to sequency. To do so requires the following notation for $k,\ell < N$. Referring to Eq. (5.3.23)

$$\bar{k}_0 = k_{n-1}$$

$$\bar{k}_s = k_{n-s} + k_{n-s-1} \qquad s = 1 \ldots n-1 \tag{5.3.26}$$

$$\langle \bar{k},\ell \rangle = \sum_{s=0}^{n-1} \bar{k}_s \ell_s \tag{5.3.27}$$

Then it can be shown that the *sequency ordered* WH transform matrix $\boldsymbol{T}$ has elements given by

$$t_{mi} = \frac{1}{\sqrt{N}} (-1)^{\langle \overline{m-1}, i-1 \rangle} \tag{5.3.28}$$

$$m,i = 1 \ldots N$$

Sequency ordered WHT matrices are shown in Fig. 5.3.5 and basis vectors for $N = 16$ in Fig. 5.3.6. Note that T is still unitary and symmetric. A fast algorithm also exists for the sequency ordered WHT.[5.3.2]

WHT transform matrices when N is not a power of two also exist, although not for every value of N.[5.3.4] However, for image coding purposes $N = 2^n$ is almost universally used.

5.3.1c Karhunen-Loeve Transform (KLT)

The *KL* transform has coefficients that are uncorrelated. Its matrix $\boldsymbol{T}$ is derived from the $N \times N$ correlation matrix of the image pel blocks $\boldsymbol{b}$ that have been biased to have zero mean. That is,

$$\boldsymbol{R} \triangleq E[(\boldsymbol{b} - E\boldsymbol{b})(\boldsymbol{b} - E\boldsymbol{b})'] \tag{5.3.29}$$

or in terms of the elements of $\boldsymbol{R}$ and $\boldsymbol{b}$

$$r(i,j) = E[(b_i - Eb_i)(b_j - Eb_j)] \tag{5.3.30}$$

A slightly suboptimum procedure is to define the sample mean of each vector

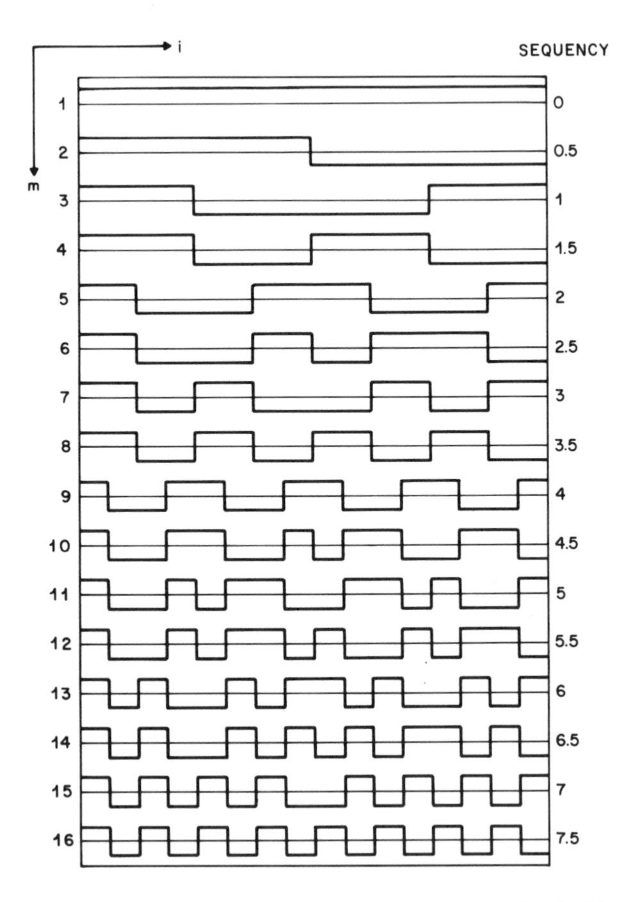

Fig. 5.3.6 Sequency ordered basis vectors for the WHT with $N = 16$.

Fig. 5.3.7 Images used for coding and statistics. (a) "Karen" has much more stationary statistics than (b) "Stripes."

$$\mu = \frac{1}{N} \sum_{i=1}^{N} Eb_i \tag{5.3.31}$$

and use the matrix[†]

$$r(i,j) = E[(b_i - \mu)(b_j - \mu)] \tag{5.3.32}$$

which is also positive definite and a legitimate correlation matrix. In fact, setting $\mu = 0$ in Eq. (5.3.32) does not usually cause significant penalty when coding images. In any event, the KLT basis vectors are the real, orthonormalized eigenvectors of $\boldsymbol{R}$, i.e.,

$$\begin{aligned} \boldsymbol{R}\boldsymbol{t}_m &= \lambda_m \boldsymbol{t}_m \ , \\ \boldsymbol{t}'_m \boldsymbol{t}_n &= \delta_{mn} \end{aligned} \tag{5.3.33}$$

where the eigenvalues $\{\lambda_m\}$ are nonnegative.

For example, Fig. 5.3.8 shows KLT eigenvectors for the picture "Karen" in Fig. 5.3.7a using one-dimensional blocks of pels with $N = 16$ and $\mu = 0$. The eigenvectors are arranged according to decreasing eigenvalue. Note that, for the most part, this also corresponds to ordering according to increasing frequency, i.e., number of zero crossings in the basis vector. Fig. 5.3.9 shows similar data for the first order Markov correlation model defined in Chapter 3. In this case, statistics are stationary with pel variance σ^2 and adjacent pel correlation $0 < \rho < 1$. The correlation function is given by

$$r(i,j) = \sigma^2 \rho^{|i-j|} \tag{5.3.34}$$

In the figure we assume a value of adjacent pel correlation $\rho = 0.95$. Note the similarity between the eigenvectors of Fig. 5.3.8 and Fig. 5.3.9. Such similarity is generally seen for images such as "Karen" that have "nearly stationary" statistics.

We now consider larger blocks of pels, both one dimensional and two dimensional. Fig. 5.3.10 shows the first few eigenvectors for the picture "Karen" using one-dimensional blocks of pels with $N = 64$. Fig. 5.3.11 shows results when the $N = 64$ blocks are made from $L \times L$ two-dimensional blocks of pels (here $L = 8$). Note that for the two-dimensional blocks many periodicities of length L are apparent. This reflects the vertical correlation between pels in successive scan lines, which occurs in most images.

---

† Note that in general $\mu \neq Eb_i \neq Eb_j$.

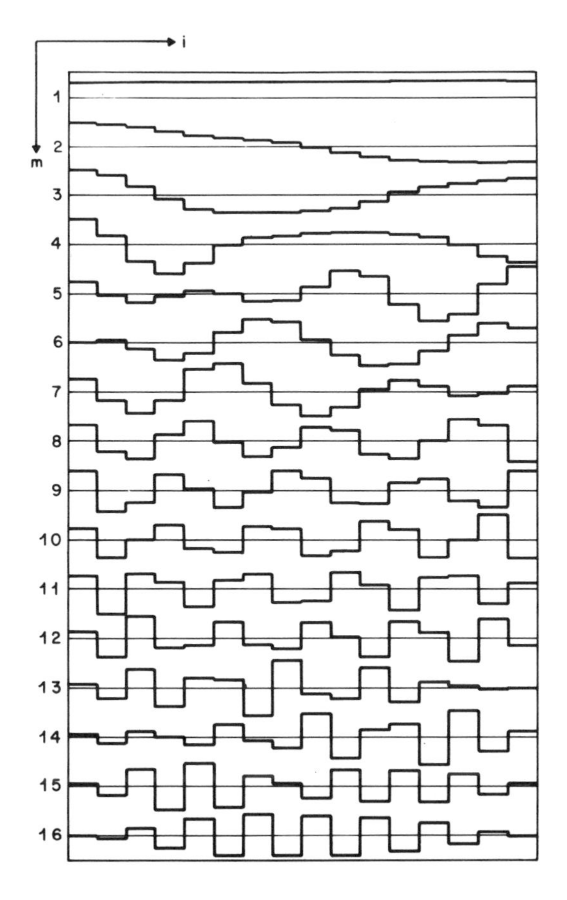

Fig. 5.3.8 KLT basis vectors for the image "Karen" using one-dimensional blocks with $N = 16$ and $\mu = 0$. For each m the elements of $\boldsymbol{t}_m$ are plotted.

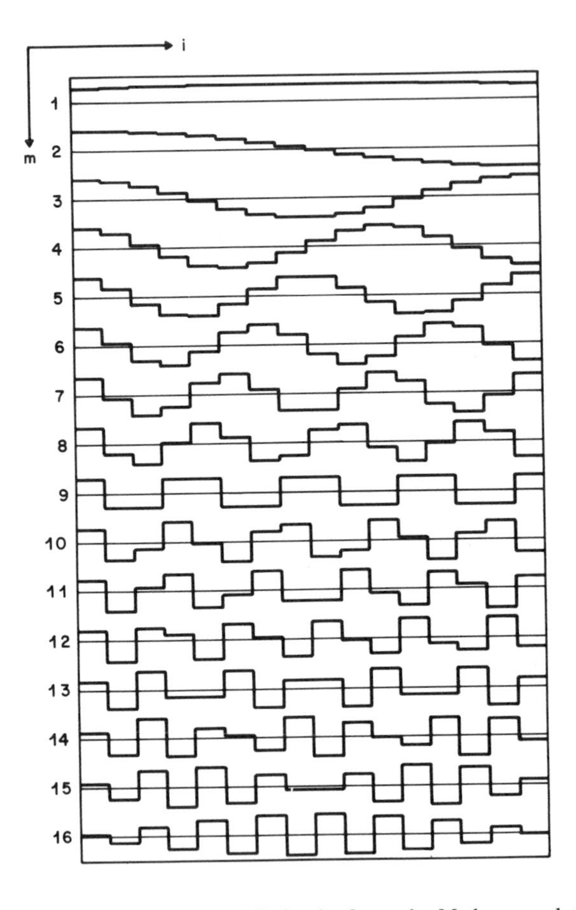

Fig. 5.3.9 KLT basis vectors ($1D$, $N=16$) for the first order Markov correlation model with adjacent pel correlation $\rho=0.95$.

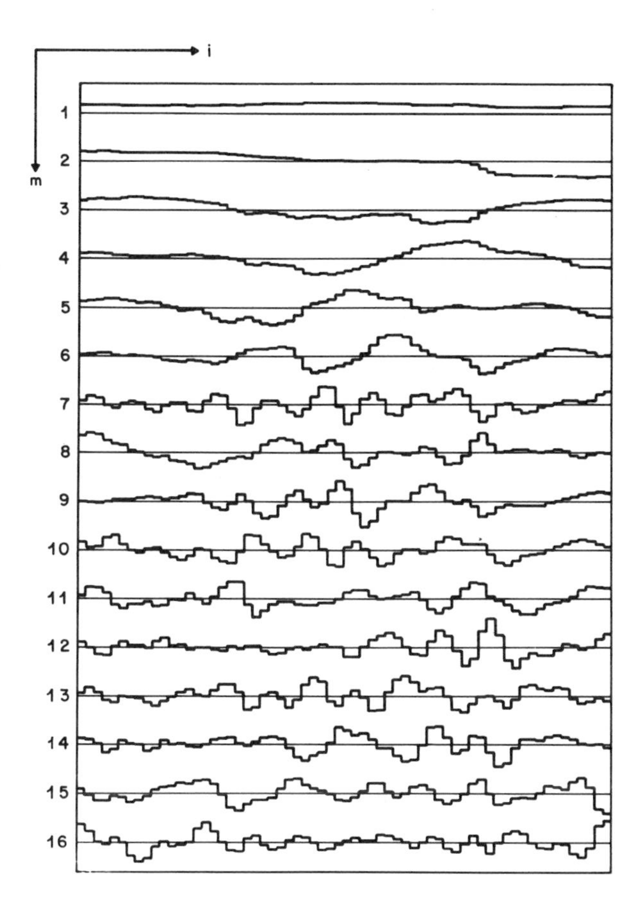

Fig. 5.3.10 First 16 KLT basis vectors for the image "Karen" using one-dimensional, $N=64$ blocks.

Fig. 5.3.12 shows one-dimensional, $N = 64$ eigenvectors for the picture "Stripes," which has highly nonstationary statistics. Note that these eigenvectors attempt to match the horizontal periodicities due to the stripes in the image, and are very different compared to the basis vectors of "Karen" shown in Fig. 5.3.10.

Once the KLT basis vectors have been determined, as we saw in Chapter 3, the transformation is then given by

$$\boldsymbol{c} = \boldsymbol{Tb} \tag{5.3.35}$$

and

$$E(c_m c_n) = \lambda_m \delta_{mn} \tag{5.3.36}$$

That is, the *KL* transform coefficients are uncorrelated, and their mean square values (MSV's) are equal to their respective eigenvalues. Thus, if the KLT basis vectors are ordered according to decreasing eigenvalues, i.e., λ_1 largest and λ_N smallest, then a compaction of energy into the lower index transform coefficients will result. It was shown in Chapter 3 that with the mean square error distortion criterion, the KLT achieves the best energy compaction of any linear transform.

A drawback of the KLT may occur with the coding of two dimensional $L \times L$ blocks of pels $\boldsymbol{B}$. If the L columns are laid end to end to make a length $N = L^2$ column, then a nonseparable KLT of order L^2 is required. If the separable two-dimensional transform on Eq. (5.3.10) is used, then a KLT only of order L is needed. However, if the image statistics are highly nonstationary then the performance of the separable transform may suffer.

5.3.1d Discrete Cosine Transform (DCT)

In Chapter 3 we defined the Discrete Cosine Transform or DCT.[5.3.6] In recent years, the DCT has become the most widely used of the unitary transforms, and under certain circumstances, as we shall see later, its performance can come close to that of the KLT. The elements of the DCT transform matrix $\boldsymbol{T}$ are given by $[\delta_0 = 1, \delta_p = 0, \text{for } p > 0]$

$$t_{mi} = \sqrt{\frac{2-\delta_{m-1}}{N}} \cos\left\{\frac{\pi}{N}\left[i - \frac{1}{2}\right](m-1)\right\} \tag{5.3.37}$$

$$i,m = 1 \ldots N$$

Thus, DCT basis vectors $\boldsymbol{t}_m$ are sinusoids with frequency indexed by m.

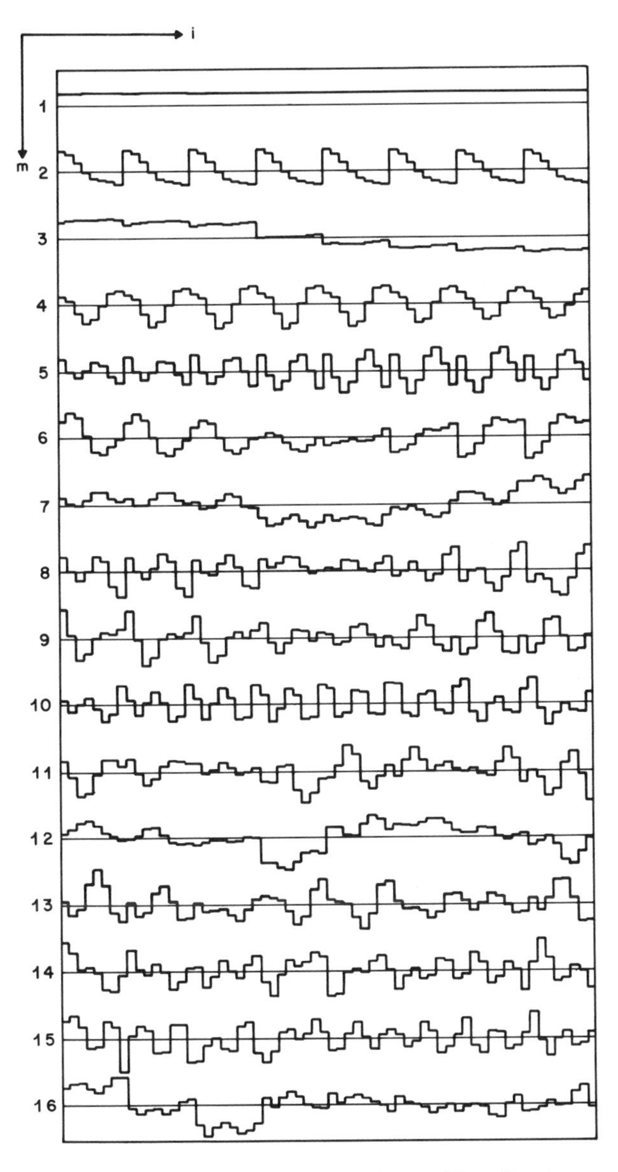

Fig. 5.3.11 First 16 KLT basis vectors for the image "Karen" using two-dimensional, nonseparable 8×8 blocks.

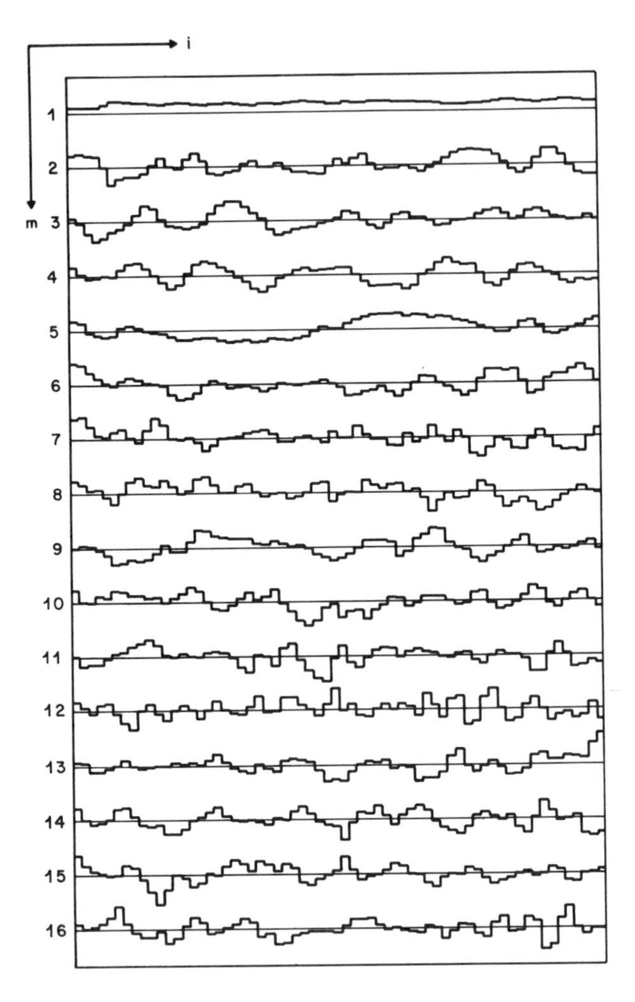

Fig. 5.3.12 First 16 KLT basis vectors for the image "Stripes" using one-dimensional $N=64$ blocks.

For $N = 16$ the DCT basis vectors are shown in Fig. 5.3.13. Note the similarity between the DCT basis vectors and the KLT basis vectors of Fig. 5.3.8 and Fig. 5.3.9. This experimentally observed property is, in part, responsible for the generally good performance of the DCT compared with other transforms, as we shall see later.

Computation of the DCT is greatly facilitated by observing that the elements of $\boldsymbol{T}$ in Eq. (5.3.37) can be written as the real part of

$$\frac{\sqrt{2(2-\delta_{m-1})}}{\sqrt{2N}} \quad \exp\left[\frac{2\pi\sqrt{-1}}{2N}\left(i-1+\frac{1}{2}\right)(m-1)\right]$$

$$= \sqrt{2(2-\delta_{m-1})}\ exp\left[\frac{\pi\sqrt{-1}}{2N}(m-1)\right]$$

$$\times \frac{1}{\sqrt{2N}}\ exp\left[\frac{2\pi\sqrt{-1}}{2N}(i-1)(m-1)\right] \qquad (5.3.38)$$

$$\triangleq F_m \times F_{mi}$$

The second factor F_{mi} is recognized as element (m,i) of the $2N$th order DFT matrix. Thus, the DCT can be efficiently computed as follows:[5.3.6]

1) Take the FFT* of the length $2N$ vector

$$\boldsymbol{b}^{+} \triangleq (b_1 b_2 ... b_N 000...0)' \qquad (5.3.39)$$

2) The first N of the resulting Fourier coefficients (indexed by m) are then multiplied by their respective F_m of Eq. (5.3.38).

3) Finally, the real part is taken to give the N DCT coefficients.

---

* Also, multiply by $\sqrt{\frac{2}{N}}$ if not included in the FFT.

The inverse DCT can be computed similarly since the elements t_{im} of the inversion matrix are also given by the real part of Eq. (5.3.38). The inverse procedure starts by constructing the length $2N$ vector of DCT coefficients

$$\mathbf{c}^{+} = (c_1 c_2 \ldots c_N 00 \ldots 0)' \tag{5.3.40}$$

Then each coefficient is multiplied by its respective F_m of Eq. (5.3.38) and the $2N$th order FFT taken. The real parts of the resulting first N values are the desired data.

An alternative and even faster DCT algorithm[5.3.7] is based on reordering the block of pels as follows:

$$\mathbf{v} \triangleq (b_1 b_3 b_5 b_7 \ldots b_8 b_6 b_4 b_2)' \tag{5.3.41}$$

that is odd pels first, followed by even pels in reverse order. Thus,

$$v_i = \begin{cases} b_{2i-1} & 1 \le i < \dfrac{N+1.5}{2} \\ \\ b_{2N-2i+2} & \dfrac{N+1.5}{2} < i \le N \end{cases} \tag{5.3.42}$$

The term $N+1.5$ allows one definition for both even and odd N. The DCT coefficients can then be written

$$\begin{aligned} c_m &= \sqrt{\frac{2-\delta_{m-1}}{N}} \; \mathrm{Re} \sum_{i=1}^{N} b_i \exp\left[-\frac{\pi}{N}\sqrt{-1}\,(i-\frac{1}{2})(m-1)\right] \\ &= \sqrt{\frac{2-\delta_{m-1}}{N}} \; \mathrm{Re}\left[\sum_{i \text{ odd}} \cdots + \sum_{i \text{ even}} \cdots\right] \end{aligned} \tag{5.3.43}$$

Replacing i by $2i - 1$ and $2N - 2i+2$ in the two summands, respectively, we get

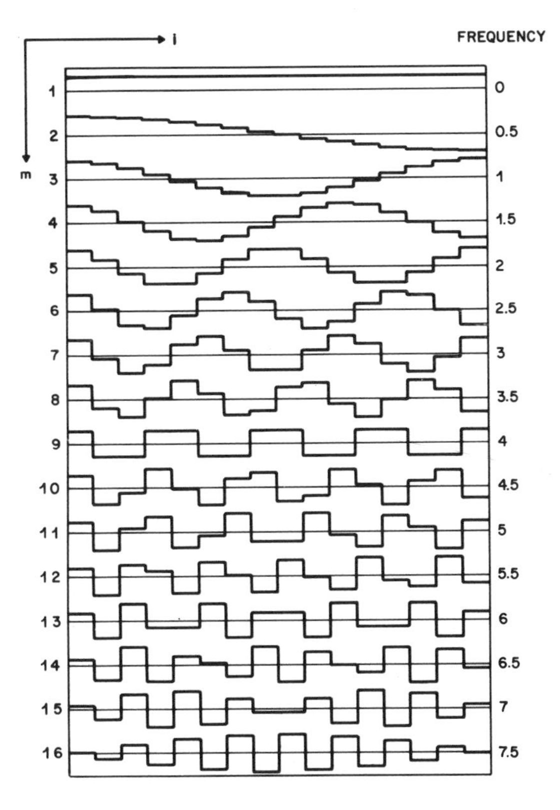

Fig. 5.3.13 Basis vectors for the DCT with $N = 16$.

N	Number of Computations For Matrix DCT	Fast DCT	Speedup Factor
4	16	8	2
8	64	24	2.7
16	256	64	4
32	1024	160	6.4
64	4096	384	10.7
138	16384	896	18.3
256	65536	2048	32

Fig. 5.3.14 Computational advantage of using the Fast DCT Algorithm compared with simple matrix multiplication.

$$c_m = \sqrt{\frac{2-\delta_{m-1}}{N}} \ \mathrm{Re} \sum_{i=1}^{N} v_i \exp\left[-\frac{2\pi}{N}\sqrt{-1}\left(i-\frac{3}{4}\right)(m-1)\right]$$

$$= \sqrt{2-\delta_{m-1}} \ \mathrm{Re}\,(G_m V_m) \tag{5.3.44}$$

where

$$G_m = \exp\left[-\frac{\pi}{2N}\sqrt{-1}\,(m-1)\right] \tag{5.3.45}$$

and $\{V_m\}$ is the Nth order DFT of the reordered block of pels $\mathbf{v}$. Also, since $V_{N-p} = V'_{2+p}$ and $G_{N-p} = -\sqrt{-1}\ G'_{2+p}$, we have

$$\mathrm{Re}(G_{N-p} V_{N-p}) = -\mathrm{Im}(G_{2+p} V_{2+p}) \tag{5.3.46}$$

and, as with the DFT (for N even), the DCT can be performed using only an $\frac{N}{2}$-point FFT. Thus we have (for N even):

A FAST DCT ALGORITHM

1) Reorder the pels to form the complex vector $\boldsymbol{w}$, having length $N/2$, as follows:

$$\mathrm{Re}(\boldsymbol{w}) = (b_1 b_5 b_9 ... b_{12} b_8 b_4)'$$
$$\mathrm{Im}(\boldsymbol{w}) = (b_3 b_7 b_{11} ... b_{10} b_6 b_2)' \tag{5.3.47}$$

2) Take the DFT (or FFT[†]) of $\boldsymbol{w}$ to obtain the complex coefficients $\{W_m,\ m = 1...N/2\}$, form the complex values X_m and Y_m from Eqs. (5.3.18), and compute the complex DFT coefficients $\{V_m,\ 1 \le m \le 1 + N/2\}$ using Eqs. (5.3.19).

---

† Also, multiply by $\sqrt{\frac{2}{N}}$ if not included in the FFT.

3) Form the complex products $\{U_m = G_m V_m,\ 2 \le m \le N/2\}$.

4) The DCT coefficients are then given by

$$
\begin{aligned}
c_1 &= V_1 \\
c_m &= \sqrt{2}\ \mathrm{Re}(U_m), \qquad 2 \le m \le \frac{N}{2} \\
c_{\frac{N}{2}+1} &= V_{\frac{N}{2}+1} \\
c_{N-p} &= -\sqrt{2}\ \mathrm{Im}(U_{2+p}), \qquad 0 \le p \le \frac{N}{2} - 2
\end{aligned}
\tag{5.3.48}
$$

The inverse DCT is the reverse of the above procedure. That is, we solve for U_m, then V_m, then X_m and Y_m, then W_m. An inverse $N/2$-point DFT (or inverse FFT) then gives $\boldsymbol{w}$, and a reordering produces $\boldsymbol{b}$. The computation time for the Fast DCT's increases as $N\log_2 N$, as opposed to the N^2 increase for direct matrix multiplication. For sizable N the computational advantage of using the Fast DCT Algorithm can be enormous, as shown in Fig. 5.3.14. For smaller N, other approaches may be more efficient. For example, the Appendix gives an implementation for block size 8 that uses *rotations*. See [5.3.7a] for details.

An indication of the compaction capabilities of the DCT can be obtained by examining the spectral energy distribution in the transform domain. Fig. 5.3.15 shows DCT coefficient mean square values (MSV) for the image "Karen", for various 1D and 2D block sizes, arranged according to increasing spatial frequency [see Eq. (5.3.8)] as well as arranged according to decreasing MSV. Note the roughly linear behavior of MSV's on a log-log scale for the separable 2D-DCT shown in Fig. 5.3.15c.

Results can also be derived using the mathematical models defined in Chapter 3. For example, with (real) separable 2D-transforms of $L \times L$ pel blocks $\boldsymbol{B} = [b_{ij}]$, we have from Eq. (5.3.7)

$$
c_{mn} = \sum_{i,j=1}^{L} b_{ij} t_{mi} t_{nj} \tag{5.3.49}
$$

From this, coefficient MSV's become

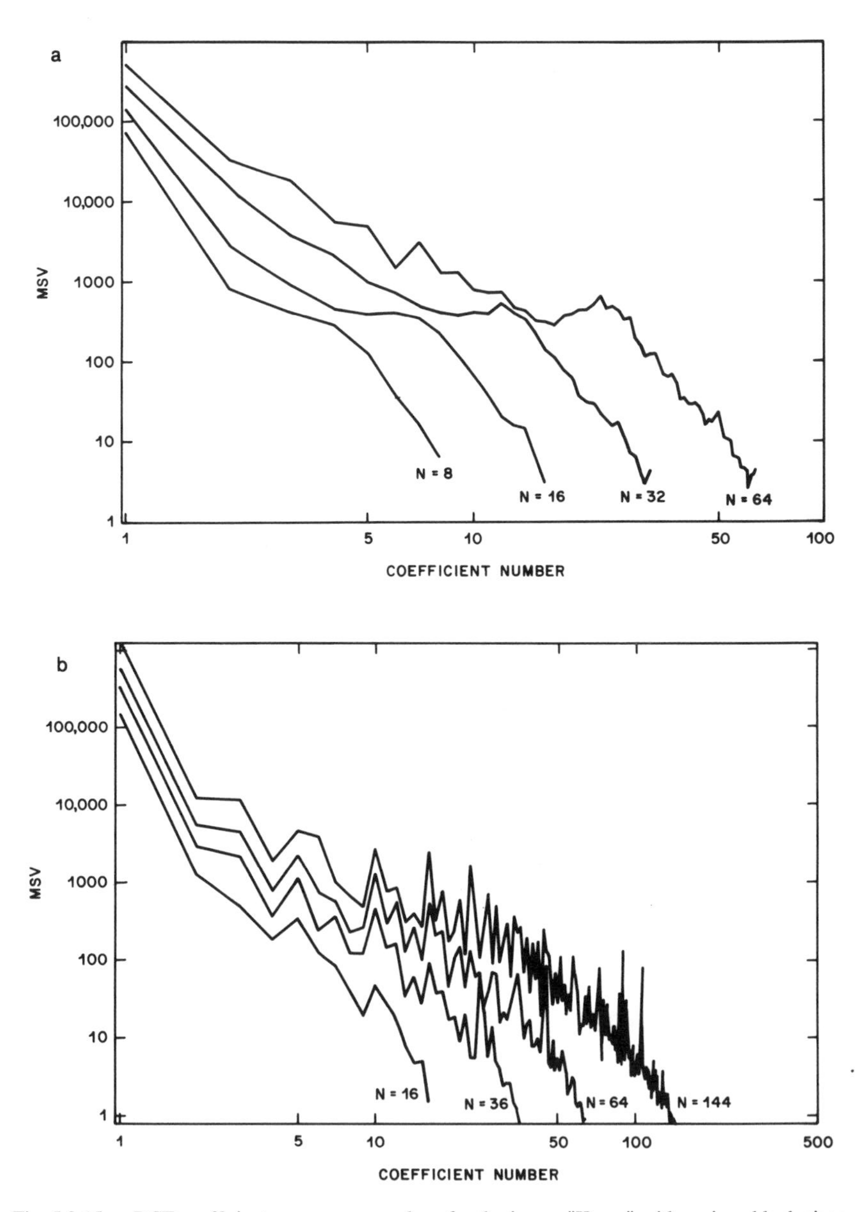

Fig. 5.3.15 DCT coefficient mean square values for the image "Karen" with various block sizes. (a) One-dimensional, arranged according to increasing frequency. (b) Separable two-dimensional, arranged according to increasing spatial frequency.

$$\sigma_{mn}^2 \triangleq E(c_{mn}^2) = E \sum_{i,j,k,\ell=1}^{L} b_{ij} b_{k\ell} t_{mi} t_{nj} t_{mk} t_{n\ell}$$

(5.3.50)

$$= \sum_{i,j,k,\ell=1}^{L} r_{ijk\ell} t_{mi} t_{nj} t_{mk} t_{n\ell}$$

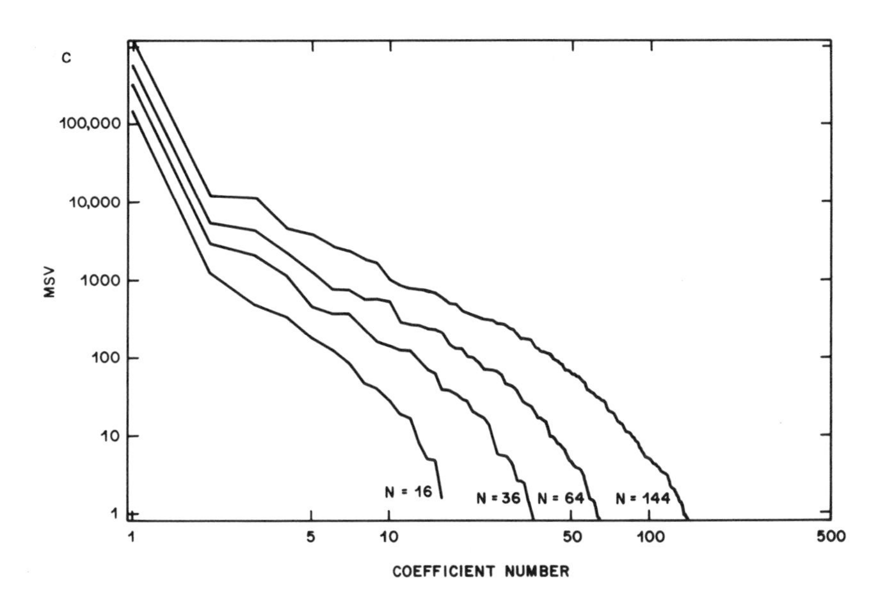

Fig. 5.3.15c Separable 2D arranged according to decreasing mean square value (MSV).

where the correlation $r_{i\,j\,k\ell} \triangleq E(b_{ij} b_{k\ell})$. For example, using an isotropic model with 8-bit PCM original data having range 0–255 and $MSV = 255^2/3$, the correlation function becomes

$$r_{ijk\ell} = \frac{255^2}{3} \rho^{\sqrt{(i-k)^2 + (j-\ell)^2}} \tag{5.3.51}$$

Fig. 5.3.16 shows DCT coefficient MSV's for the isotropic model with $\rho = 0.95$, for various 1D and 2D block sizes, again arranged according to increasing spatial frequency. Note again the linear behavior for the separable 2D-DCT.

5.3.1e Comparison of RDFT, WHT, KLT, DCT

So far, we have dwelt mainly on the computational aspects of discrete orthogonal transforms. We now compare the energy compaction performance of the four most often used transforms, namely the RDFT, WHT, KLT, and DCT. We have seen that unitary transformation preserves image energy or MSV, i.e.,

$$\sigma^2 = \frac{1}{N} \sum_{i=1}^{N} E(b_i^2) = \frac{1}{N} \sum_{m=1}^{N} E(c_m^2) \tag{5.3.52}$$

Also, if only a certain fixed fraction $p < 1$ of the transform coefficients are transmitted in each block and used to form the reconstructed pels $\tilde{\boldsymbol{b}}$ then the mean square error becomes

$$MSE = \frac{1}{N} \sum_{i=1}^{N} E\left[(b_i - \tilde{b}_i)^2\right]$$

$$= \frac{1}{N} \sum_{m=1}^{N} E\left[(c_m - \tilde{c}_m)^2\right] \tag{5.3.53}$$

Note that this does not include quantization error, which we will treat in the next section. If the coefficients are reindexed so that the fixed set of transmitted coefficients occur first, then

$$\tilde{c}_m = \begin{cases} c_m, & 1 \le m \le pN \\ 0, & m > pN \end{cases}$$

and (5.3.54)

$$MSE = \frac{1}{N} \sum_{m > pN}^{N} E\left[c_m^2\right]$$

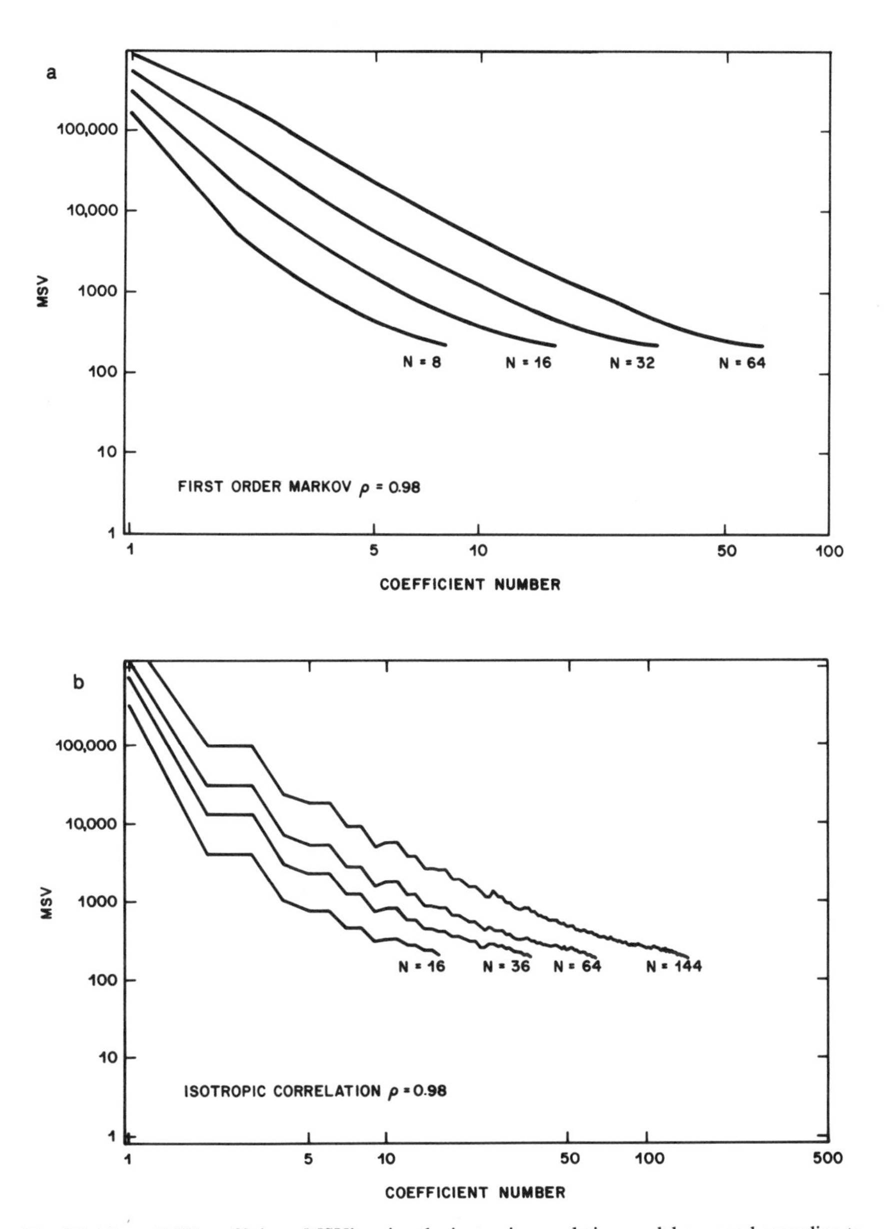

Fig. 5.3.16 DCT coefficients MSV's using the isotropic correlation model arranged according to increasing frequency ($\rho = 0.98$). (a) One-dimensional blocks. (b) Separable two-dimensional blocks.

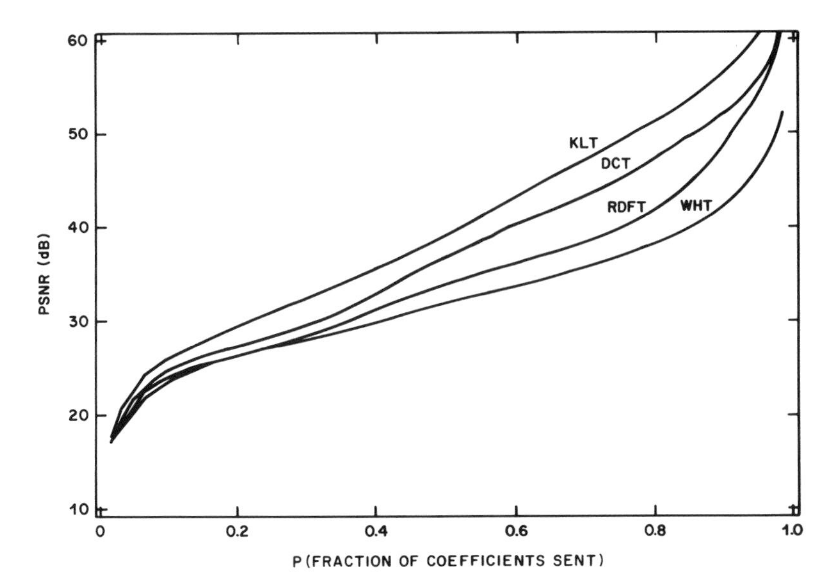

Fig. 5.3.17 Comparison of truncation errors for four different transforms. One-dimensional blocks, $N=64$, were used with the image "Karen". Peak signal to rms noise ratio (PSNR) is plotted versus the fraction of coefficients sent.

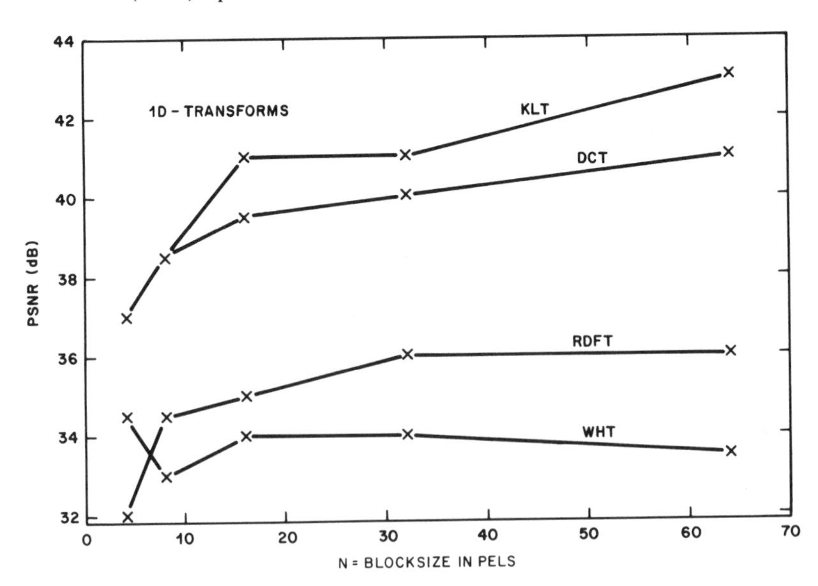

Fig. 5.3.18 Truncation Peak Signal to rms Noise ratio (PSNR) versus one-dimensional block size for the image "Karen" using four different transforms and keeping 60 percent of the coefficients ($p=0.6$).

For 8-bit PCM original data, the range of pel values is 0-255. The peak-signal to rms noise ratio then becomes (in dB)

$$PSNR = 10 \log \left[\frac{255^2}{MSE} \right] dB \tag{5.3.55}$$

Although unweighted PSNR (see Section 4.3.2) in this context has some value for comparison purposes, we should not pay too much attention to the actual PSNR values themselves. In the first place the impairment is not subjectively weighted. In addition, the distortion is in fact a low-pass filtering that appears to the viewer as blurring. Such impairment is subjectively not well parameterized by PSNR.

Fig. 5.3.17 shows PSNR results for the four transforms applied to "Karen" (1D-64 pel blocks) when only pN $(p<1)$ of the transform coefficients are transmitted. In each case the coefficients having the largest MSV's are transmitted, a technique called *zonal sampling*.* We see that the KLT is the best performing transform in terms of energy compaction, followed by the DCT, RDFT and WHT, respectively. For other block sizes Fig. 5.3.18 shows PSNR results for $p = 0.6$, i.e., 60 percent of the coefficients are transmitted. The energy compaction generally improves as block size increases, and the relative performance of the transforms generally remains the same. However, the improvement is small for block sizes larger than about 16 pels.

We now consider the case where $\boldsymbol{b}$ is constructed from a two-dimensional $L \times L$ block of pels by simply laying the L-pel rows end-to-end. The resulting vector of $N = L^2$ pels has not only high adjacent pel correlation, but also high correlation between pels with separation L. Thus for example, RDFT coefficients are relatively large not only at low frequencies, but also at frequencies $1/L$, $2/L$, etc. Fig. 5.3.19 shows PSNR results when two-dimensional blocks are transform coded in this manner and only pN of the coefficients having the largest MSV are transmitted. The curves are generally higher than those of Fig. 5.3.17 by several dB.

With two-dimensional $L \times L$ blocks of pels the most often used approach is to separate the transformation using two L-dimensional transforms as described earlier in this chapter. Recall that with separable transforms we use an $L \times L$ block of pels $\boldsymbol{B}$ and transform the rows and columns separately via

---

* Note that this is not equivalent to sending the pN coefficients of each block having the largest magnitude. More on this subject appears in Section 5.3.3.

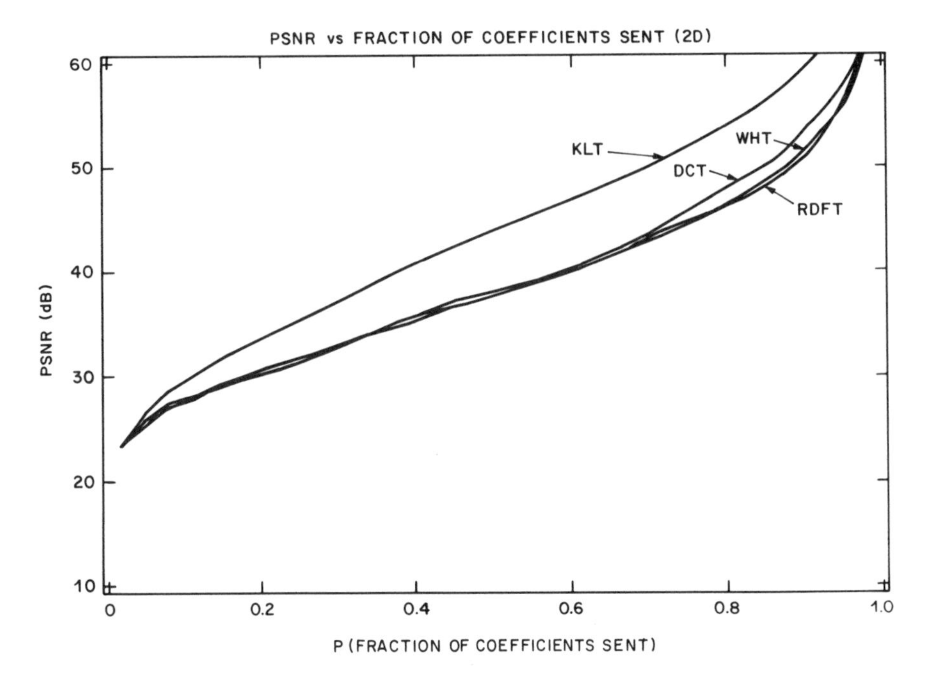

Fig. 5.3.19 Comparison of truncation errors using nonseparable, two-dimensional, 8×8 blocks with the image "Karen". The coefficients having the largest Mean Square Value (MSV) are transmitted.

6.0e+05	4.9e+04	2.3e+04	1.1e+04	4.5e+03	2.6e+03	1.0e+03	2.9e+02
5.9e+03	8.3e+02	4.4e+02	1.9e+02	1.0e+02	6.1e+01	1.9e+01	8.6e+00
5.8e+02	1.5e+02	8.4e+01	5.7e+01	3.1e+01	1.4e+01	7.6e+00	2.9e+00
4.1e+02	7.4e+01	5.0e+01	2.4e+01	1.2e+01	8.4e+00	3.6e+00	1.8e+00
1.7e+01	3.0e+01	2.0e+01	1.1e+01	6.6e+00	4.9e+00	2.4e+00	1.3e+00
1.3e+02	2.6e+01	1.7e+01	1.0e+01	5.6e+00	5.0e+00	2.6e+00	1.3e+00
4.5e+01	8.8e+00	9.0e+00	4.9e+00	3.2e+00	2.9e+00	1.5e+00	9.9e+01
1.4e+02	2.0e+01	1.3e+00	6.9e+00	3.4e+00	3.6e+00	1.5e+00	9.6e+01

Fig. 5.3.20 Coefficient MSV's using the separable 2D-DCT, 8×8 blocks with the image "Stripes", which has large correlation in the vertical direction.

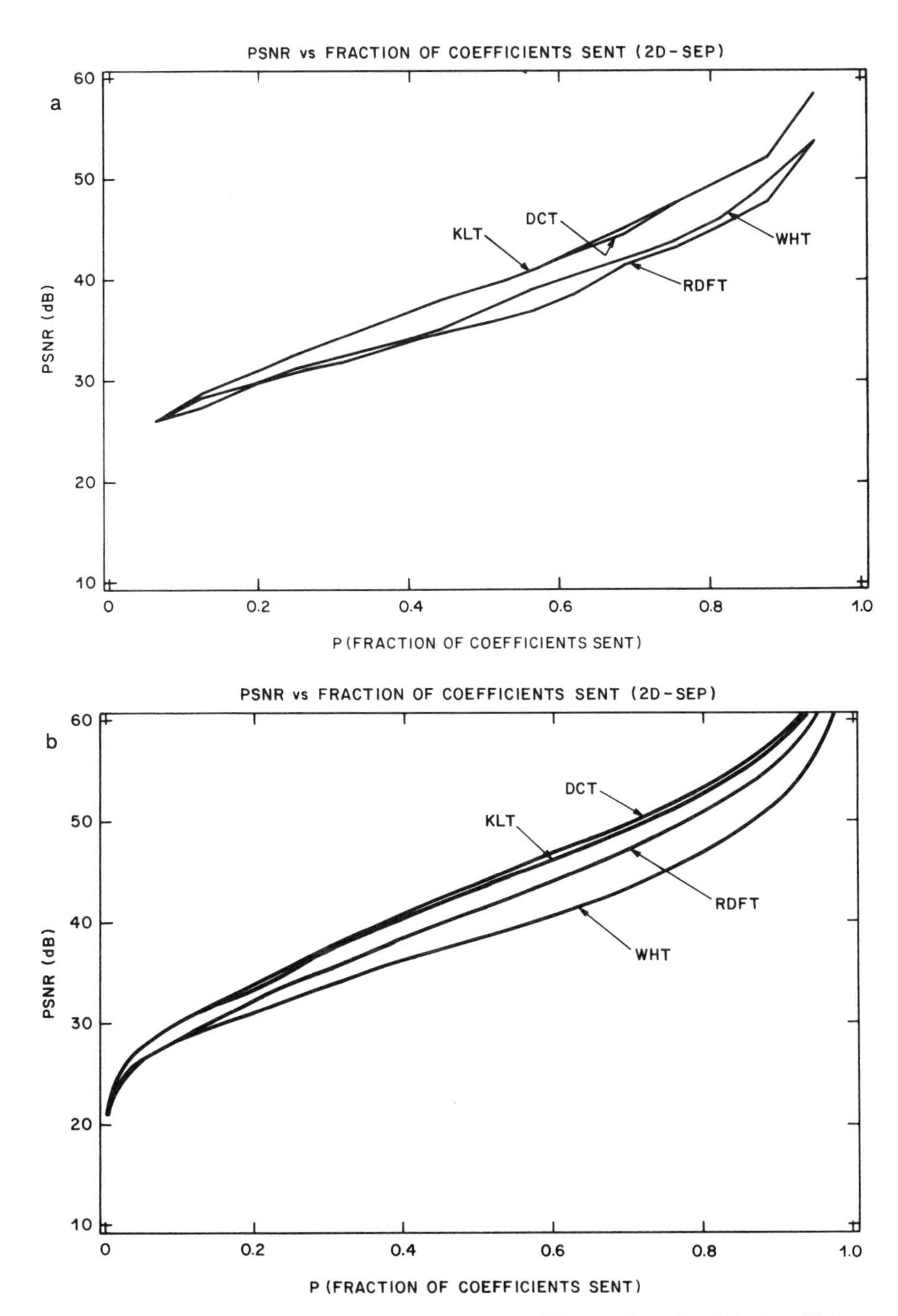

Fig. 5.3.21 Comparison of truncation errors using separable, two-dimensional blocks with the image "Karen". The coefficients having the largest MSV are transmitted. (a) 4×4 blocks, $N=16$. (b) 16×16 blocks, $N=256$.

$$C = TBT^t \tag{5.3.56}$$

to obtain an $L \times L$ matrix $\boldsymbol{C}$ containing the transform coefficients. For example, Fig. 5.3.20 shows the coefficient MSV's when the separable DCT (8x8 pel blocks) is applied to the 8-bit PCM data from "Stripes". Note the decreased energy in the vertical spatial frequencies compared with the horizontal frequencies. This reflects the high vertical correlation in this particular picture.

Fig. 5.3.21 shows for $L = 4$ and $L = 16$ the results of separable transformations of $L \times L$ blocks of pels on the picture "Karen" when only pN of the coefficients are transmitted. The KLT was derived from the correlation matrix of one-dimensional $1 \times L$ blocks of pels. We see that the separable DCT performs about as well as the separable KLT. This is probably due to the fact that the KLT cannot adapt very well to nonstationary image statistics in the relatively small block size of $1 \times L$ pels. Also, note that the performance of the separable DCT is only a few dB below that of the nonseparable KLT of Fig. 5.3.19. It is this characteristic behavior that makes the separable DCT the transform of choice in many coding applications.

For various block sizes Fig. 5.3.22 shows PSNR results for $p = 60\%$. As before the energy compaction generally improves as the block size increases. However, the improvement is relatively small for block sizes larger than about 8×8 pels.

The transform block size and shape are important parameters affecting the statistics of the transform coefficients although the picture itself is, of course, the overriding factor. Generally, the larger the block size, the higher the energy compaction achieved by the transform. Also two-dimensional blocks achieve more compaction than one-dimensional blocks. Experience has shown that over a wide range of pictures there is not much improvement in average energy compaction for two dimensional block sizes above 8×8 pels. However, individual pictures with higher nonstationary statistics can always be found for which this rule of thumb is violated (for example, compare the KLT curves of Fig. 5.3.17 and Fig.5.3.19). Also, considerable correlation may remain between blocks, even though the correlation between pels within a block is largely removed.[5.3.21] We shall return to this point in a later section.

5.3.1f Miscellaneous Transforms

Several other transforms have been studied. For example, the Haar Transform[5.3.8] can be computed from an orthogonal (but not orthonormal) matrix $\boldsymbol{T}$ that contains only +1's, −1's and zeros as shown in Fig. 5.3.23. This enables rather simple and speedy calculation, but at the expense of energy compaction performance.

The Slant Transform[5.3.9] has, in addition to the usual constant basis vector $\boldsymbol{t}_1$, a basis vector $\boldsymbol{t}_2$ given (for N even) by

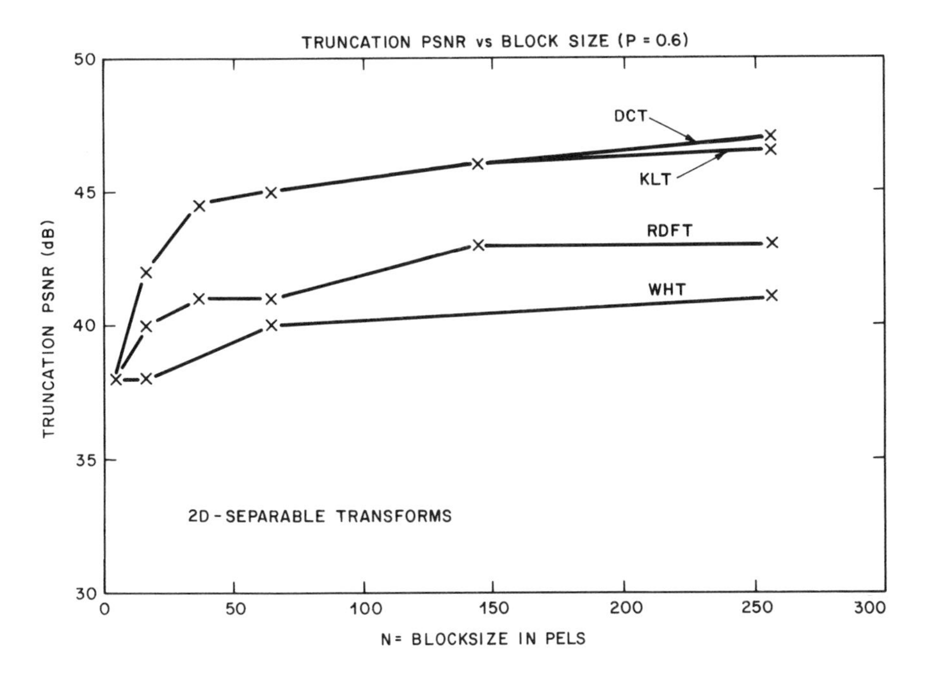

Fig. 5.3.22 Truncation PSNR versus block size for separable transforms with the image "Karen" when 60 percent of the coefficients are kept ($p = 0.6$).

$$\begin{bmatrix} 1 & 1 & 1 & 1 & 1 & 1 & 1 & 1 \\ 1 & 1 & 1 & 1 & -1 & -1 & -1 & -1 \\ 1 & 1 & -1 & -1 & 0 & 0 & 0 & 0 \\ 0 & 0 & 0 & 0 & 1 & 1 & -1 & -1 \\ 1 & -1 & 0 & 0 & 0 & 0 & 0 & 0 \\ 0 & 0 & 1 & -1 & 0 & 0 & 0 & 0 \\ 0 & 0 & 0 & 0 & 1 & -1 & 0 & 0 \\ 0 & 0 & 0 & 0 & 0 & 0 & 1 & -1 \end{bmatrix}$$

Fig. 5.3.23 Haar transform matrix. As with the WHT, transformation requires no multiplications.

$$\boldsymbol{t}_2 = \alpha(N-1,\ N-3,\ \ldots,\ 3,\ 1,\ -1,\ -3,\ldots,\ 1-N)' \tag{5.3.57}$$

where α is a normalization constant. This vector is thought to better approximate the local behavior of images and thus result in higher energy compaction. However, the $\boldsymbol{t}_2$ of the DCT is very similar, and so is the overall performance in most cases of interest. A fast algorithm has also been developed for the Slant Transform.

The Sine Transform[5.3.10] has matrix elements given by

$$t_{mi} = \sqrt{\frac{2}{N+1}} \sin \frac{mi\pi}{N+1} \tag{5.3.58}$$

$$m,i = 1 \ldots N$$

Its main utility arises when images are preprocessed and decomposed into a sum of two uncorrelated images, one of which has approximately stationary statistics with a KLT that is approximately the Sine Transform.

The Singular Value Decomposition (SVD)[5.3.11] begins by rewriting the separable inverse transform of Eq. (5.3.11) as

$$\boldsymbol{B} = \boldsymbol{U}'\boldsymbol{C}\boldsymbol{V} \tag{5.3.59}$$

where $\boldsymbol{U}$ and $\boldsymbol{V}$ are unitary Lth order transforms. If the basis vectors of $\boldsymbol{U}$ are chosen to be the eigenvectors of $\boldsymbol{B}\boldsymbol{B}'$, and the basis vectors of $\boldsymbol{V}$ the eigenvectors of $\boldsymbol{B}'\boldsymbol{B}$. then the matrix

$$\boldsymbol{C} = \boldsymbol{U}\boldsymbol{B}\boldsymbol{V}' \tag{5.3.60}$$

will be diagonal with $\{c_{mm}^2 = \lambda_m,\ 1 \le m \le L\}$ where $\lambda's$ are the L distinct eigenvalues resulting from the above operations. Eq. (5.3.12) then becomes

$$\boldsymbol{B} = \sum_{m=1}^{L} \lambda_m \boldsymbol{u}_m \boldsymbol{v}_m' \tag{5.3.61}$$

If the λ's are arranged in decreasing order and only the first pN of them are significant ($p < 1$), then specification of $\boldsymbol{B}$ requires the transmission of

$$\{\lambda_m,\ \boldsymbol{u}_m,\ \boldsymbol{v}_m,\ 1 \le m \le pN\} \tag{5.3.62}$$

So far, SVD has found more use in image restoration than in image coding.

5.3.2 Quantization of Transform Coefficients

We have seen that transform coding does a reasonably good job of exploiting *statistical* redundancy in images by decorrelating the pel data and by compacting information into the lower order coefficients. We now present a framework for the exploitation of *subjective* redundancy in pictures using transform coding.

Unlike DPCM, where a particular quantized sample affects relatively few pels in the reconstructed picture, quantization (or deletion) of a particular transform coefficient affects every pel in the transformed block. This *spreading* of the quantization error often makes the design of transform coefficient quantizers quite different than that of DPCM quantizers.

For example, consider the case where we employ a fairly sizable two-dimensional block as shown in Fig. 5.3.24. In many cases, part of the block will be low-detail flat area, and part will be high-detail busy area. Excessive quantization error in one coefficient will eventually cause a distortion pattern to become visible in the flat area, whose spatial frequency corresponds to the basis vector of the coefficient. Because of spatial masking, the distortion will probably not be visible in the busy area of the block.

If quantization error is to be kept below the visibility threshold and if the spatial frequencies of the basis vectors are far enough apart from each other that they are detected independently by the human visual system, then an appropriate quantization strategy might be

$$|c_m - \tilde{c}_m| < T_m \tag{5.3.63}$$

where T_m is the visibility threshold (see Fig. 4.3.11a) corresponding to basis vector $\boldsymbol{t}_m$.

However, for large block sizes the basis vector frequencies are quite close to each other, and they are not detected independently. In this case we might invoke the bandpass filter model of human vision (see Fig. 4.3.10) and partition the basis vectors into sets $\{S_i\}$ of similar spatial frequencies. For subthreshold distortion, the quantization strategy might be designed to satisfy*

$$\sum_{m \in S_i} (c_m - \tilde{c}_m)^2 < T_i \qquad \forall i \tag{5.3.64}$$

where T_i is the detection threshold corresponding to the spatial frequencies in S_i.

---

* Eq. (5.3.64) is only an approximate definition of nondetection by a human observer. Basis vectors having similar spatial frequencies combine in a complicated way, and an exact relationship for detectability is not yet known.

Eq. (5.3.64) applies to coefficients that have sizable MSV's, and in this case $E(c_m - \tilde{c}_m)^2$ is the mean square quantization error. However, we saw earlier that the MSV's of some coefficients may be so small that they need not be transmitted at all. In this case, the mean square error is $\sigma_m^2 = E(c_m^2)$, and is referred to as the *truncation error* due to that coefficient not being transmitted.

If a certain percentage of the low energy transform coefficients are set to zero and not transmitted in order to conserve bit-rate, then as the number of zeroed coefficients increases, distortion is first detected, not in the low-detail flat area, but in the high-detail busy area of the block or at the boundary between the two areas. The general effect is a blurring or loss of sharpness in regions of rapid luminance variation. For this reason, in Eq. (5.3.64), the threshold values corresponding to high spatial frequencies (low MSV) may have to be reduced somewhat below the values obtained from Fig. 4.3.11a, in order to preserve the desired degree of resolution in high-detail areas of the image.

Transform coefficient c_1 corresponds to a spatial frequency of zero, for which Fig. 4.3.11a does not apply. Quantization error in c_1 first appears at the block boundary and can cause a very visible discontinuity between adjacent blocks, especially in low detail areas of the image. The visibility of these discontinuities will determine the appropriate threshold value of T_1. Depending on the block size, quantization error in other low frequency coefficients may also contribute to these *block edge effects*, and this may require further adjustment of thresholds. The DCT and KLT generally have less block edge effects than other unitary transforms due to the fact that they have more slowly varying basis vectors.

In some cases, especially with small block sizes, a given block may contain little or no low-detail flat areas. Then, spatial masking by the high-detail luminance variations will allow for a larger quantization error than indicated in Eq. (5.3.64). We can take this into account for each block by defining a *Spatial Activity Function* A_F, as discussed in Section 4.3.1b, which measures the relative amount of detail in the block. A_F could be defined using gradient magnitudes as suggested in Chapter 4. However, in transform coding a more convenient definition often utilizes the non-DC transform coefficients. For example,

$$A_F = 1 + q \sum_{m=2}^{N} c_m^2 \tag{5.3.65}$$

where q is a suitable normalization constant. Alternatively, a subset of the non-DC coefficients might be used to simplify calculations somewhat, or instead of a direct summation of c_m^2, a weighted sum might be employed. In any event, using a spatial activity function the constraint on quantization error for subthreshold distortion becomes

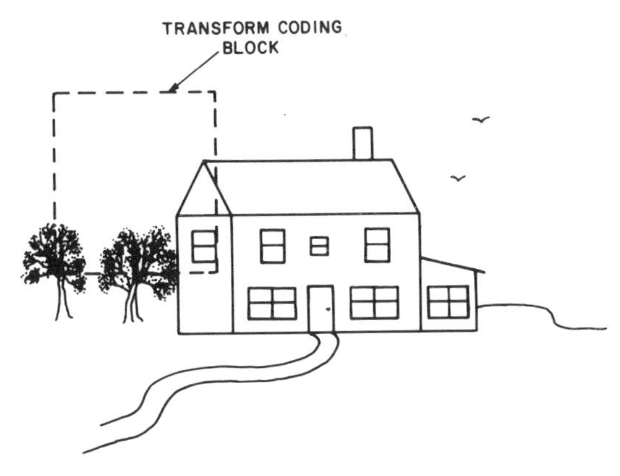

Fig. 5.3.24 In many cases, part of a transform coding block will be low-detail flat area, and part will be high-detail busy area.

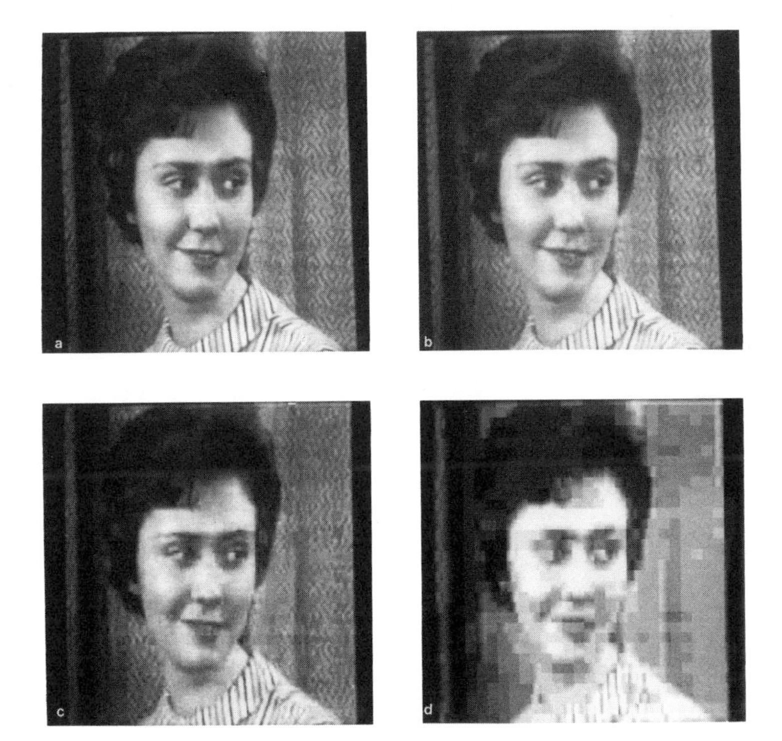

Fig. 5.3.25 Coded image "Karen" using the separable 2D-DCT with 8×8 blocks, uniform quantization and entropy coding. The coefficients having the largest MSV are sent. PSNR's and entropies are (left-right, top-bottom):

(a)PSNR = 38.0dB, 1.24 bits/pel (b)PSNR = 33.5dB, 0.67 bits/pel
(c)PSNR = 29.9dB, 0.36 bits/pel (d)PSNR = 25.7dB, 0.13 bits/pel

$$\sum_{m \in S_i} \frac{(c_m - \tilde{c}_m)^2}{A_F} < T_i \quad \forall i \tag{5.3.66}$$

or dividing through by T_i and defining $T_m = T_i$ for $m \in S_i$,

$$\sum_{m \in S_i} \frac{(c_m - \tilde{c}_m)^2}{A_F T_m} < 1 \qquad \forall i \tag{5.3.67}$$

If the bit-rate provided for transmission is not high enough to allow for sub-threshold distortion, then quantization must be made coarser and distortion will be above the threshold of visibility. In this case, we would like to spread the distortion over all of the spatial frequency bands in an effort to randomize it so that, in flat areas at least, it appears to a human observer to be like random noise. Thus, we relax Eq. (5.3.67) to get

$$\sum_{m \in S_i} \frac{(c_m - \tilde{c}_m)^2}{A_F T_m} < \left[1 + \varepsilon\right] \qquad \forall i \tag{5.3.68}$$

where ε is either zero or positive depending on whether the distortion is sub- or supra-threshold, respectively.

The quantization/coding task can then be stated as either:

1) For a fixed ε minimize the average bit-rate, or
2) for a fixed average bit-rate minimize ε.

The possible solutions to this coding problem have only been partially studied.[5.3.12] Here we simply state that they depend intimately on block-size, the particular transform used, quantization algorithm, buffer strategy, etc.

The quantization problem can be (and in practice usually is) simplified considerably by defining a block distortion

$$d = \frac{1}{N} \sum_{m=1}^{N} \frac{(c_m - \tilde{c}_m)^2}{A_F T_m} \tag{5.3.69}$$

and an average block distortion*

---

* As with DPCM, in some cases the statistical averaging operator E may be profitably replaced by a subjectively weighted average.

$$D = \frac{1}{N} \sum_{m=1}^{N} E\left[\frac{(c_m - \tilde{c}_m)^2}{A_F T_m}\right] \tag{5.3.70}$$

This is a noise visibility and spatial frequency weighted MSE distortion criterion, where the spatial activity function A_F is a measure of the masking and $1/T_m$ is a measure of the visibility of various spatial frequencies, e.g., see Fig. 4.3.11b in Chapter 4.

Consider, for now, only blocks having similar values for the spatial activity function A_F. These blocks would all be coded in the same way by adaptive systems. The distortion for this class of blocks is given by

$$D \approx \frac{1}{N} \sum_{m=1}^{N} \frac{\varepsilon_m}{A_F T_m} \tag{5.3.71}$$

where $\varepsilon_m \triangleq E[(c_m - \tilde{c}_m)^2]$ is the mean square quantization error (MSE) for coefficient c_m. Let r_m be the average number of bits used to code c_m. Then the average block bit-rate is

$$R = \frac{1}{N} \sum_{m=1}^{N} r_m \quad \text{bits/pel} \tag{5.3.72}$$

We saw in Chapter 3 that the functional relationship $\varepsilon_m(r_m)$ between coefficient MSE and bit-rate depends on the probability distribution (PD) of the coefficient c_m and on the quantization/coding strategy employed, e.g., uniform with fixed word-lengths, Lloyd-Max with fixed word-lengths, uniform with entropy coding, etc. Assuming $\varepsilon_m(\cdot)$ is a continuous function of its argument, we now show that for R fixed, a necessary condition to minimize D is (here prime denotes first derivative, not transpose):

$$\frac{\varepsilon_m'(r_m)}{A_F T_m} = \begin{cases} -\theta & r_m > 0 \\ s_m \geq -\theta & r_m = 0 \end{cases} \tag{5.3.73}$$

where θ is a nonnegative parameter. The proof is by contradiction. Suppose $r_m, r_n > 0$ and

$$\frac{\varepsilon_m'(r_m)}{A_F T_m} > \frac{\varepsilon_n'(r_n)}{A_F T_n} \tag{5.3.74}$$

Then reducing r_m by Δ and increasing r_n by Δ would leave R unchanged. However, D would undergo a net decrease, which contradicts our original assumption. Similarly, for $r_m > 0, r_n = 0$.

Eq. (5.3.73) is a fundamental result of transform coding theory for a frequency weighted MSE distortion measure. It tells us how to apportion the bits amongst the transform coefficients. If the MSE functions $\varepsilon_m(\cdot)$ are discontinuous, e.g., defined only by integer bit-rates, the problem is much more complicated,[5.3.13, 5.3.14] although approximate solutions can often be obtained by solving the appropriate continuous problem.

In Chapter 3 we saw that all transform coefficients except c_1 had approximately a Laplacian Probability Distribution (PD). Coefficient c_1 had a PD that could only be approximated by a uniform distribution, unless the block size was very large, in which case Gaussian was better. Also, quantization error in c_1 is not very well represented subjectively by MSE. For these reasons, c_1 is often quantized uniformly and coded using fixed word lengths.

Two quantization schemes are commonly used for the non-DC transform coefficients. They are 1) Lloyd-Max quantization with fixed-length code words and 2) Uniform quantization with variable-length code words, i.e., near optimum entropy coding. With Lloyd-Max quantization of zero mean coefficients having Laplacian PD's we saw in Chapter 3 that the MSE ε_m (which is in fact discontinuous) could be approximated using the piecewise linear relation

$$\ln \frac{\varepsilon_m(r_m)}{\sigma_m^2} \approx \begin{cases} -0.84 r_m & r_m < 2\text{bits} \\ 0.90 - 1.29 r_m & r_m \geq 2\text{bits} \end{cases} \tag{5.3.75}$$

where σ_m^2 is the MSV of c_m. This can be easily differentiated and substituted into Eq. (5.3.73) to give

$$\frac{\varepsilon_m}{A_F T_m} = \begin{cases} \dfrac{\theta}{1.29} & r_m \geq 2 \\ \dfrac{\theta}{0.84} & 0 < r_m < 2 \\ \dfrac{\sigma_m^2}{A_F T_m} & r_m = 0 \end{cases} \tag{5.3.76}$$

For a given value of the parameter θ we find the coefficient bit assignment by using Eq. (5.3.75) to solve the first of the above equations for r_m and seeing if it is 2 bits or more. If not, we solve the second and see if r_m is positive. If not, we set $r_m = 0$ and do not transmit that particular coefficient. This gives us the coefficient bit assignments for Lloyd-Max quantization with fixed word-length coding.

If the quantized transform coefficients are to be coded independently of each other then we must round each r_m to the nearest integer. However, in certain cases combined coding may be profitable. For example, if two coefficients are quantized with three and five levels, respectively, then they can be coded together (15 possible combinations) using only a 4-bit word.

With uniform quantization and entropy coding of Laplacian coefficients, we saw in Chapter 3 that for a given MSE ε_m could be approximated by

$$\ln \frac{\varepsilon_m(r_m)}{\sigma_m^2} \approx 0.19 - 1.39 r_m \tag{5.3.77}$$

where the entropy r_m is in bits/sample and the quantizer contains the zero† level. Differentiating and substituting into Eq. (5.3.73) we get

$$\frac{\varepsilon_m}{A_F T_m} = \begin{cases} \dfrac{\theta}{1.39} & r_m > 0 \\ \\ \dfrac{\sigma_m^2}{A_F T_m} \le \dfrac{\theta}{1.39} & r_m = 0 \end{cases} \tag{5.3.78}$$

This important result says that every coefficient that is transmitted should have the same weighted mean square error $\theta/1.39$, whereas untransmitted coefficients should each have a smaller weighted mean square error.

As before, for a given θ we use Eq. (5.3.77) to solve the first of the above equations for r_m, i.e.,

$$\begin{aligned} r_m &= \frac{1}{1.39}\left[0.19 + \ln \frac{1.39\sigma_m^2}{\theta A_F T_m}\right] \\ &\approx \frac{1}{1.39} \ln \frac{1.68\sigma_m^2}{\theta A_F T_m} \quad \text{bits} \end{aligned} \tag{5.3.79}$$

and see if it is positive.* If not, we set $r_m = 0$ and do not transmit the

---

† For a given interval size, mid-step quantization gives a smaller entropy than mid-riser quantization.

* In fact, $r_m \ge 1$ may be a more practical constraint since $r_m < 1$ requires block coding and may not be worth the added complexity.

coefficient. This gives us the coefficient bit assignments $\{r_m\}$ for uniform quantization with entropy coding.#

Suppose now that the coefficients are ordered according to decreasing σ_m^2/T_m and, utilizing Eq. (5.3.78), a certain fraction $p < 1$ of them are transmitted. (Note that a very good approximation to this ordering is to arrange the coefficients according to increasing spatial frequency of their basis vectors.) The relationship between θ and p can be estimated from Eq. (5.3.79) and the constraint

$$r_{pN} \geq 0 \geq r_{pN+1} \tag{5.3.80}$$

If the block size is not too small, we may assume $r_{pN} \gtrsim 0$ and from Eq. (5.3.79)

$$\theta \lesssim \frac{1.68\sigma_{pN}^2}{A_F T_{pN}}$$

and (5.3.81)

$$r_m \approx \frac{1}{1.39} \ln \frac{\sigma_m^2 T_{pN}}{\sigma_{pN}^2 T_m} \qquad m < pN$$

Thus,

$$R \approx \frac{1}{1.39N} \sum_{m=1}^{pN} \ln \frac{\sigma_m^2 \; T_{pN}}{\sigma_{pN}^2 \; T_m} \text{ bits/pel} \tag{5.3.82}$$

From Eq. (5.3.78) the weighted MSE distortion D then becomes

$$D \triangleq \frac{1}{N} \sum_{m=1}^{N} \frac{\varepsilon_m}{A_F T_m}$$

(5.3.83)

$$= \frac{p\theta}{1.39} + \frac{1}{N} \sum_{m>pN}^{N} \frac{\sigma_m^2}{A_F T_m}$$

Often, coefficient c_1 is positive and uniformly distributed, in which case Eq. (5.3.79) gives an r_1 that is too large by about 1.1 bits/pel. However, the discrepancy is small; and moreover it is on the conservative side.

That is, the distortion is the sum of two terms, the first being the *Quantization* error and the second being the *Truncation* error.

These two types of errors have a markedly different appearance in coded images having supra-threshold distortion. Quantization error is noiselike, whereas truncation error causes a loss of resolution. The tradeoff between these two types of distortion is not well understood, and depends to a large extent on the application, e.g., surveillance, television, etc. Thus, in practice, these equations must be used with care and the thresholds $\{T_m\}$ must be determined experimentally.

Note the similarity between Eqs. (5.3.78) and (5.3.68). In both cases, distortion is spread uniformly in all spatial frequency bands that are transmitted. Thus, even though the weighted MSE distortion measure of Eq. (5.3.69) does not take into account the way the error is distributed among the spatial frequencies, the subsequent optimization procedure nevertheless ends up distributing the error evenly over the transmitted frequencies.

An important corollary of Eq. (5.3.78) stems from the relationship derived in Chapter 3 between the MSE ε_m and the uniform quantizer step size Δ_m. We saw from Eq. (3.2.38) that for reasonable bit-rates, e.g., $r_m \geq 1$,

$$\Delta_m \approx \sqrt{12\varepsilon_m} \tag{5.3.84}$$

which is independent of the MSV of c_m. Thus, from Eq. (5.3.78), assuming $0 < r_m < 1$ is not allowed, we may write the uniform quantizer step-size directly as

$$\Delta_m = \sqrt{\frac{12A_F T_m \theta}{1.39}} \quad m < pN \tag{5.3.85}$$

Note that if T_m is assumed constant, then the same quantizer may be used for all transform coefficients. Alternatively, the coefficients c_m could be normalized by $\sqrt{T_m}$ prior to quantization.

However, the probability distribution of the quantizer outputs will vary considerably for coefficients having different MSV's. Thus, for optimum performance different variable-word-length codes should be used for coefficients having dissimilar variances.

At this point it is worthwhile to recapitulate these results for uniform quantization with entropy coding. If the coefficients are ordered according to decreasing σ_m^2/T_m, and a fraction $p < 1$ of them are sent, then the optimum quantization step size for c_m is given by Eq. (5.3.85), the distortion by Eq. (5.3.83), and the average bit-rate by Eq. (5.3.82).

Fig. 5.3.25 shows the picture "Karen" transform coded at several bit rates using the DCT with 8×8 blocks, uniform quantization and entropy coding. For simplicity, A_F and T_m were set to unity for all blocks, i.e., simple MSE was

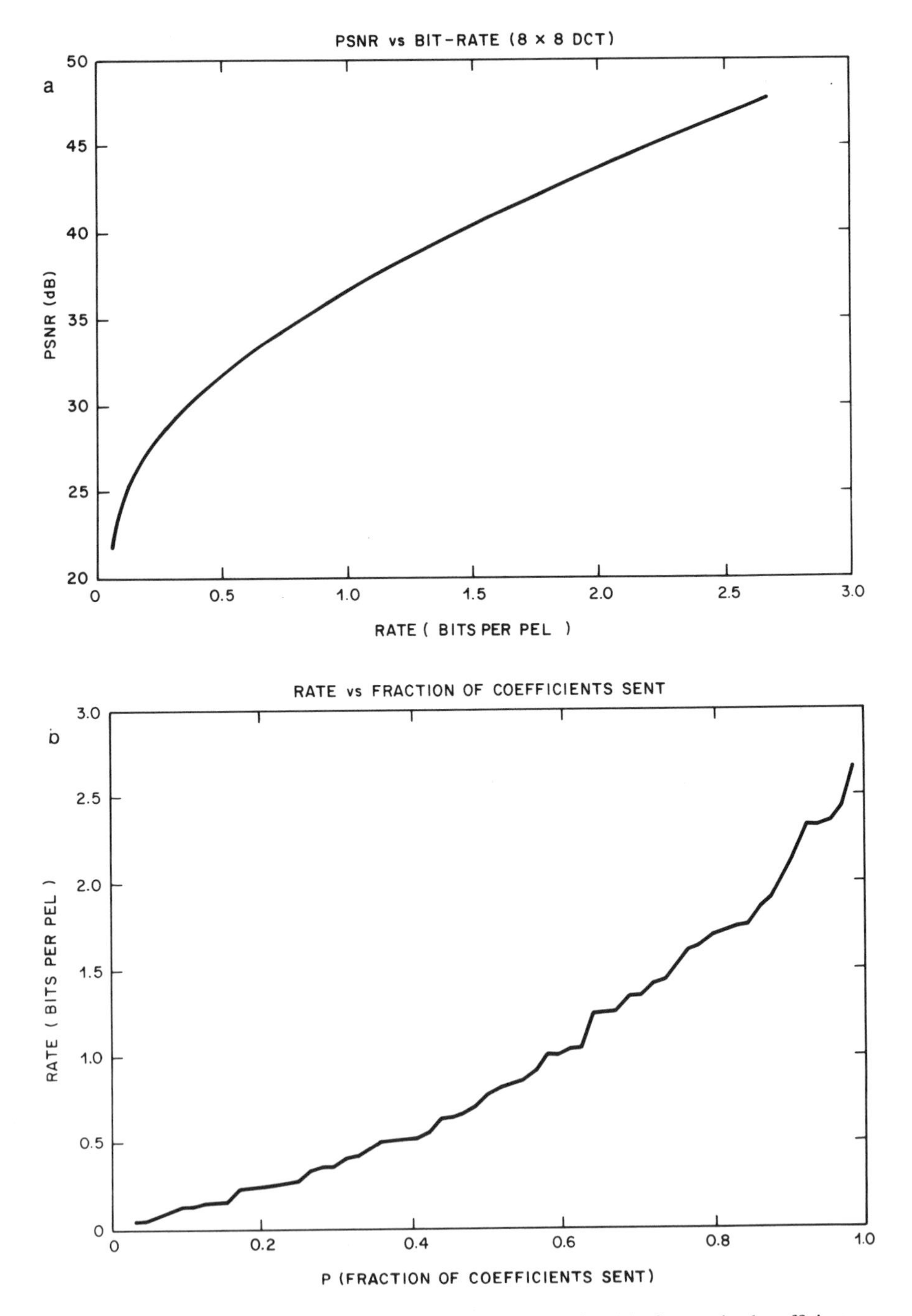

Fig. 5.3.26 Relationship between PSNR, bit-rate (R) and fraction (p) of transmitted coefficients of the images coded in Fig. 5.3.25 (a) PSNR vs. R, (b) R vs. p.

used as a distortion criterion. Also, $r_m < 1$ was not allowed.

Fig. 5.3.26 shows PSNR versus entropy R for the same picture. Also shown for each value of R is the fraction p of coefficients that are transmitted. Note the nearly linear relationship between R and p that exists over most of the range.

Similar results can be derived using the mathematical models defined in Chapter 3. For (real) separable 2D-transformation of $L \times L$ pel blocks $\boldsymbol{B} = [b_{ij}]$, the coefficient MSV's become [see Eq. (5.3.50)]

$$\sigma_{mn}^2 = \sum_{i\,j\,k\ell} r_{i\,j\,k\ell}\, t_{mi}\, t_{nj}\, t_{mk}\, t_{n\ell} \tag{5.3.86}$$

where the correlation $r_{i\,j\,k\ell} \triangleq E(b_{ij}, b_{k\ell})$. From these MSV's (arranged in decreasing order or increasing spatial frequency), the relation between rate R and distortion D can be found from Eqs. (5.3.78-83). For example, using an isotropic model with 8-bit PCM data

$$r_{i\,j\,k\ell} = \frac{255^2}{3}\, \rho^{\sqrt{(i-k)^2 + (j-\ell)^2}} \tag{5.3.87}$$

Fig. 5.3.27 shows 2D-DCT PSNR versus R for $\rho = 0.98$ and various values of L (again for $A_F T_m = 1$).

5.3.3 Adaptive Coding of Transform Coefficients

If the pictures to be coded have fairly similar statistics and the transform block-size is reasonably large, then the transform coefficient statistics will not change much from block to block. For example, aerial photos of urban regions might fall into this category. In this case, coefficient MSV's can be measured, picture quality determined, quantization parameters calculated according to the zonal sampling techniques of the previous Section, and the bit-rate can be estimated with high accuracy. This approach we term as *Nonadaptive* transform coding.

However, if image statistics vary widely and/or the block-size is relatively small so that some blocks are in low-detail areas of the picture while others are in high-detail areas, then nonadaptive techniques will often be unsatisfactory. For example, if the coefficients of each block are quantized using Lloyd-max quantization, and each coefficient is assigned a fixed number (possibly 0) of bits using Eq. (5.3.75) with $A_F = 1$ and some set of assumed coefficient MSV's, then the number of bits per block is constant, and very little data buffering is required. However, this simple zonal sampling approach is usually not much more efficient than DPCM in terms of bit-rate and picture quality. Moreover, it is far more complex and expensive to implement than DPCM.

The situation improves markedly if *Adaptive* transform coding techniques are employed. For example, using a technique called threshold sampling, only

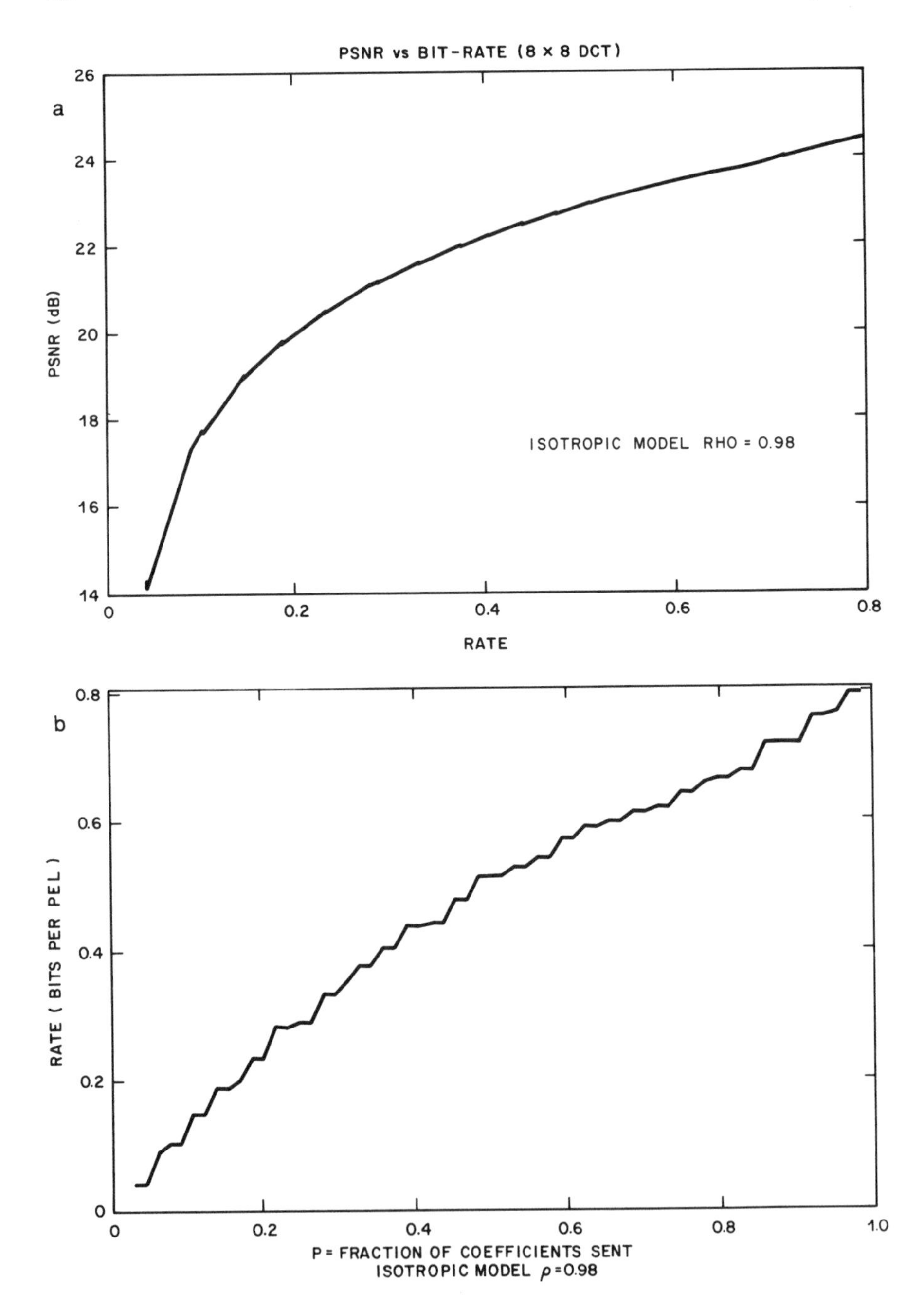

Fig. 5.3.27 Relationship between PSNR, R and p for the isotropic correlation model with adjacent pel correlation $\rho = 0.98$ and separable 2D-DCT coding of 8×8 blocks. (a) PSNR vs. R, (b) R vs. p.

those coefficients in each block whose magnitude exceeds a prescribed threshold are transmitted. With these methods low-detail blocks, having a large number of insignificant coefficients, are coded with far fewer bits than the higher-detail blocks, resulting in a significantly smaller bit-rate. Also, in order to exploit masking effects, high-detail blocks are coded with more quantization error than low detail blocks where masking does not occur and block boundary distortion is more visible. Indeed the performance of transform coding in general becomes attractive compared with DPCM only if adaptive techniques are used and the proportion of low-detail blocks is sizable, so that the overall bit-rate can be reduced.

A drawback of adaptive transform coding is an increased sensitivity to transmission bit errors. The effect of these can propagate over sizable areas of the received picture not only because each erroneous coefficient affects all pels of the block, but also because sending a variable number of coefficients and possibly using variable word-length codes makes it possible for the decoder to lose synchronization. Once this happens the receiver may decode too many or too few coefficients in several successive blocks. Also, blocks may be deleted or extra ones may be inserted into the received picture.

Adaptive coding, of sorts, occurs if we use uniform quantization plus entropy coding as defined in Eq. (5.3.78). Recall that here, we chose quantizer parameters according to an assumed A_F, some set of assumed coefficient MSV's and a value of θ that provides acceptable image quality. However, low-detail blocks will generally have coefficient MSV's that are much smaller than those assumed, and for optimum coding according to Eq. (5.3.79) the coefficients should be coded at lower bit-rates. Also, many more of them should be assigned $r_m = 0$ and not transmitted at all. However, entropy coding automatically reduces the bit-rate in this case since smaller coefficient values are assigned shorter code words, and if the MSV's decrease, so will the average code-word lengths. Moreover, if mid-step quantizers are used, then the smallest coefficients will be quantized to zero, and if these values are run-length coded their bit-rates are markedly reduced, although not to zero. Thus, with entropy coding, the bit-rate can indeed decline significantly for low-detail blocks having relatively small coefficient variances.

However, this scenario is not optimum. For one thing, a Huffman code designed for one probability distribution is not necessarily optimum for another. Also, in low-detail blocks the percentage p of transmitted (or nonzero) coefficients declines, and the quantizer step size Δ_m of Eq. (5.3.85) should be changed. However, offsetting this to some extent is the fact that low-detail blocks should also use a smaller value for the activity function A_F.

A more nearly optimum procedure is to *classify* each transform block according to its spatial activity into one of say L classes $\{C_\ell,\ \ell = 1...L\}$. Within each class C_ℓ the coefficient MSV's $\{{}_\ell\sigma_m^2,\ m = 1...N\}$ and the activity function ${}_\ell A_F$ should be about the same for each block. Thus, a nearly optimum code can be designed for each class such that the distortion is equalized over all classes. For example, with uniform quantization and entropy coding, Eqs. (5.3.78-85) may be used for each class to derive the percentage p_ℓ of transmitted coefficients and the step size $\{{}_\ell\Delta_m\}$ of the uniform quantizer, so that the average distortion

$$D = \frac{p_\ell \theta}{1.39} + \frac{1}{N} \sum_{m>Np_\ell}^{N} \frac{{}_\ell\sigma_m^2}{{}_\ell A_F T_m} \qquad (5.3.88)$$

is the same for all classes. For class C_ℓ the average bit-rate, from Eq. (5.3.79), then becomes

$$R_\ell = \frac{1}{1.39N} \sum_{m=1}^{Np_\ell} \ln \frac{1.68\,{}_\ell\sigma_m^2}{{}_\ell A_F T_m \theta} \qquad (5.3.89)$$

If a particular image has a fraction w_ℓ of its transform blocks in class C_ℓ, then the overall average bit-rate is

$$R = \sum_{\ell=1}^{L} w_\ell R_\ell \qquad (5.3.90)$$

Using the classification method of adaptive transform coding, the receiver either must be told or must be able to derive which class each block belongs to. If the block size is not too small and the number of classes not too large, then this *overhead* information can be transmitted directly without much penalty. Otherwise, alternative methods must be devised. For example, with small block sizes, spatial activity might be determined from the non-DC coefficients of adjacent, previously transmitted blocks.

So far in this Section, we have assumed that the allowable distortion D is fixed and that the average bit-rate R can vary in whatever way is necessary to achieve that distortion. This arrangement may be suitable for computer disk storage, facsimile transmission, store and forward services, etc. However, in many other applications, e.g., television, not only is the channel data-rate constant, but also the number of images (frames) per second that must be displayed is fixed. This means that if the advantages of adaptive coding are to be exploited, then data buffers must be provided at both the transmitter and receiver in order to smooth the data rate prior to transmission. The size, i.e., storage capacity, of these buffers is a major factor in determining the degree of adaptability and coding efficiency of the system. If the buffer size is too small, then the opportunity for bit-rate trade-off between low-detail and high-detail regions of the picture is severely restricted, and the required channel rate is determined essentially by the number of bits per block in high-detail regions. If the buffer is too large the transmission delay may be excessive for interactive communication applications such as videotelephone.

With large buffers and "normal" pictures having the assumed statistics, considerable improvement results from smoothing of the bit-rate between low- and high-detail regions of the picture. However, provision must also be made in any adaptive system for the occasional arrival of "abnormal" pictures that have large areas of high-detail and, with fixed distortion coding as described above, would produce too high a bit-rate to be transmitted without buffer overflow.

Abnormal pictures present two problems. First their presence must be detected, and second something must be done to reduce the bit-rate produced by the coder. Detection of too high a data rate is often accomplished by keeping track of how full the transmitter buffer is. When buffer overflow threatens, the coder bit-rate must be reduced, i.e., the distortion increased.

With transform coders the most effective way of decreasing the bit-rate is to reduce the number of transmitted coefficients. Indeed with the separable DCT, uniform quantization and entropy coding, we saw in Fig. 5.3.26 that the bit-rate R was approximately proportional to the fraction p of coefficients transmitted. This behavior is a direct consequence of the way the mean square coefficient values decline with spatial frequency. For example, suppose the percentage p of transmitted coefficients is increased by

$$\Delta p \triangleq \frac{1}{N} \tag{5.3.91}$$

and that θ is changed by $\Delta\theta$ in order to maintain optimal coding. Then from Eq. (5.3.79)

$$\Delta R \approx \frac{\Delta p}{1.39N} \ln \frac{1.68\ \sigma^2_{pN+1}}{\theta A_F\ T_{pN+1}} - \frac{p}{1.39} \frac{\Delta\theta}{\theta} \tag{5.3.92}$$

If we assume from Eq. (5.3.80) that the first term is small, then

$$\frac{\Delta R}{\Delta p} \approx - \frac{p}{1.39\theta} \frac{\Delta\theta}{\Delta p} \tag{5.3.94}$$

We saw in Fig. 5.3.15 and Fig. 5.3.16 that for the 2D-DCT, the coefficient MSV's exhibited an approximate linear behavior on a log-log plot over a wide range of spatial frequencies.

If this same behavior were to extend to the frequency weighted MSV's, i.e.,

$$\ln \frac{\sigma^2_{pN}}{A_F\ T_{pN}} \approx a_1 - a_2 \ln p \tag{5.3.95}$$

then from Eq. (5.3.81) we have

$$\ln \theta \approx \ln 1.68 + a_1 - a_2 \ln p$$

$$\triangleq \ln a_3 - a_2 \ln p \tag{5.3.96}$$

Differentiating we obtain

$$\frac{\Delta\theta}{\Delta p} \approx - a_2 \frac{\theta}{p} = -a_2 a_3 p^{-(1+a_2)} \tag{5.3.97}$$

Substituting in Eq. (5.3.94)

$$\frac{\Delta R}{\Delta p} \approx \frac{a_2}{1.39} \tag{5.3.98}$$

which is the observed linear relationship between R and p.

If we let $\Delta p = -1/N$ then the change in distortion is from Eq. (5.3.83)

$$\Delta D = \frac{\Delta p\theta + p\Delta\theta}{1.39} + \frac{1}{N} \frac{\sigma_{pN}^2}{A_F T_{pN}} \tag{5.3.99}$$

which from Eq. (5.3.81)

$$\approx \frac{\Delta p\theta + p\Delta\theta}{1.39} - \Delta p \frac{\theta}{1.68}$$

which from Eq. (5.3.97)

$$\approx \left[0.124 - \frac{a_2}{1.39}\right] \theta \, \Delta p$$

Thus, from Eq. (5.3.96)

$$\frac{\Delta D}{\Delta p} \approx - a_3 \left[\frac{a_2}{1.39} - 0.124\right] p^{-a_2} \tag{5.3.100}$$

The above equations provide a theoretical basis for a buffer control strategy when 2D-DCT coding is used with uniform quantization and entropy coding. If the transmitter buffer threatens to become too full or too empty, the percentage of transmitted coefficients is changed by Δp, and the quantizer step size is changed by [differentiate Eq. (5.3.85)]

$$\Delta(\Delta_m) \approx \frac{\Delta\theta}{2\Delta_m} \cdot \frac{12 A_F T_m}{1.39} \tag{5.3.101}$$

This results in a change in data rate of ΔR and a change in distortion of ΔD.

If adaptive transform coding is carried out by classification of blocks according to spatial activity, then buffer control is somewhat more complicated. In this case, each class C_ℓ has parameters p_ℓ and R_ℓ that must be optimized, and the simplest approach is usually to change θ in order to control rate and distortion. If buffer overflow or underflow threatens, then the p_ℓ are changed in such a way that the new distortion $D + \Delta D$ is the same in all classes. For example, with uniform quantization and entropy coding, for a given $\Delta\theta$ the new quantizer step sizes are found from Eq. (5.3.101). Then

$$\Delta\theta \approx -\, a_2 \frac{\theta_\ell}{p_\ell} \Delta p_\ell \tag{5.3.102}$$

determines Δp_ℓ for each class. Then*

$$\Delta D \approx -\, a_3 \left[\frac{a_2}{1.39} - 0.124 \right] p_\ell^{-a_2} \Delta p_\ell \tag{5.3.103}$$

The change in bit-rate for each class is [from Eq. (5.3.98)]

$$\Delta R_\ell \approx \frac{a_2}{1.39} \Delta p_\ell \tag{5.3.104}$$

and if a particular image has a fraction w_ℓ of its blocks in class C_ℓ, then the overall change in bit-rate is

$$\Delta R = \sum_{\ell=1}^{L} w_\ell \, \Delta R_\ell \tag{5.3.105}$$

The above techniques perform optimum coding under the assumed distortion criterion. However, there are countless suboptimal algorithms that have also been studied and whose performance may be adequate in some circumstances. For example, the quantization might be nonadaptive from block-to-block, whereas the percentage p of coefficients transmitted could vary according to spatial activity. Also, we would like to send only the coefficients

* Note that a_2 and a_3 may also be class dependent.

that are larger in magnitude than some predetermined threshold. This is called *Threshold Coding*. However, in practice the overhead information required to tell the receiver which coefficients were sent and which were not is sometimes excessive. Thus, a more efficient technique, called *Zonal Coding,* is simply to transmit the coefficients in order of increasing spatial frequency or some other predetermined order, and to send an end-of-block code-word when all the remaining coefficients are below threshold.

A simplification of adaptive coding with block classification is to make two passes over each image, and during the first pass measure the spatial activity of each block.[5.3.15] During the second pass each block is classified in such a way that the proportions $\{w_\ell\}$ of blocks in the classes $\{C_\ell\}$ equals some predetermined distribution $\{W_\ell\}$. If each class is then coded using, for example, Lloyd-Max fixed word-length codes, then the bit-rate R_ℓ for each class is constant, and so is the bit-rate R per image. More precisely,

$$R = \sum_{\ell=1}^{L} W_\ell R_\ell \text{ bits/pel} \tag{5.3.106}$$

The distortion is not constant, however, since an excessive number of high-detail blocks will result in some of them being classified into low-detail classes that are coded at a lower bit-rate and thus with higher distortion.

The preceding theoretical framework for transform coding should not be taken to be the final word on the subject. Indeed a great deal of experimental work remains undone, and much remains to be learned. For example, large block-sizes exploit statistical redundancy better than small ones. However, small blocks are better able to utilize subjective redundancy and to adapt to changing statistics within the image. Also, the mean square error distortion criterion treats quantization error and truncation error equally, whereas their subjective effects are quite different. Moreover, at low bit-rates, the edge of the blocks become visible, and this distortion tends to overshadow that of the interior pels. Receiver low-pass spatial filtering can reduce block-edge visibility. However, an approach that is often better in terms of picture quality is to simply filter at the transmitter and reduce the pel sampling rate prior to transform coding.

Intraframe transform coders generally aim for bit-rates below one bit/pel. In order to achieve this with adequate picture quality, the images must contain some proportion of low-detail, flat area blocks that can be coded at very low bit-rate. This then makes additional bits available for coding the high-detail blocks. Since for a given sampling rate DPCM generally lacks the ability to code low-detail areas at extremely low bit-rates, under these circumstances transform coding usually outperforms DPCM.

5.3.4 Transform Coding of Color Pictures

The most straightforward approach to transform coding of color pictures is to code each of the three color components* separately using the monochrome coding techniques described above. Typically, the first step is to convert the original *RGB* components to luminance and chrominance components such as *YIQ* that are reasonably uncorrelated with each other. We saw in Chapter 3 that for a given picture or class of pictures the Karhunen-Loeve techniques can be used to derive a linear transformation of *RGB* into three new components that are *completely* uncorrelated. However, we also saw that there are few benefits for doing so as compared with simply using *YIQ*. For multispectral images or pictures with unusual color distributions this may not be the case.

In this section we will be concerned mainly with coding the chrominance, since the previous sections have already discussed luminance coding. We would expect that the bit-rates required for the I and Q components, as compared with the luminance component Y, would be approximately proportional to their relative bandwidths. For example, corresponding to NTSC conventions the required I and Q signal bandwidths are approximately 0.36 and 0.12, respectively, of the Y signal bandwidth. Thus, a two-dimensional transform coder, which reduces bandwidth in both dimensions, should be able to code the I and Q components using about $(0.36)^2 + (0.12)^2 \approx 14\%$ of the bit-rate of the Y component. Moreover, since the chrominance signal power is significantly smaller than that of the luminance, we would expect the chrominance bit-rate to be even lower than this estimate.

Thus, as with PCM and DPCM color image coding, the first step in coding the I and Q signals should be some sort of bandlimiting operation. It could be done prior to transform coding by the use of digital and analog filters, and in this case the sampling rate could also be reduced. Alternatively, if the transform block size is large enough, bandlimiting may be effected by simply discarding high-frequency transform coefficients. However, if the block size is small, such action may lead to undersirable visibility of distortion at edges of the transform blocks.

After bandlimiting, the chrominance signals may be transform coded using the nonadaptive or adaptive techniques discussed earlier for monochrome signals. However, a complication arises in determining the relative proportion of bits to be allocated to color components. This problem has yet to be solved completely. However, a few approaches can be suggested as starting points. For example, given the luminance distortion D_Y, we may be able to uniquely

---

* Sometimes more than three parameters are used, e.g., multispectral satellite images (see Chapter 2).

specify the corresponding chrominance distortions. If I and Q have not yet been bandlimited and p_Y is the percentage of luminance 2D-transform coefficients sent, then we might choose

$$p_I = (0.36)^2 p_Y$$

(5.3.107)

$$p_Q = (0.12)^2 p_Y$$

which corresponds to the NTSC relative bandwidths. These would, in turn, determine the chrominance bit-rates and distortions, as for example in Eqs. (5.3.78-83). Alternatively, we may be able to define subjective chrominance spatial frequency weightings such that the overall image distortion is a sum of that in the components, i.e.,

$$D = D_Y + D_I + D_Q \tag{5.3.108}$$

The problem then reduces to one of simply coding $3N$ transform coefficients with minimum distortion or bit-rate, in which case the techniques for luminance transform coding described in previous sections may be used. For example, suppose the $3N$ transform coefficients are interleaved and reordered according to decreasing frequency weighted MSV, i.e., from Eq. (5.3.81)

$$\theta_m \triangleq \frac{1.68\sigma_m^2}{A_F\, T_m} \quad 1 \le m \le 3N \tag{5.3.109}$$

is monotonically nonincreasing. Then, for uniform quantization and entropy coding we have from Eqs. (5.3.78-83)

$$R = \frac{1}{1.39(3N)} \sum_{m=1}^{3pN} \ln \frac{1.68\, \sigma_m^2}{\theta_{3pN}\, A_F\, T_m} \text{ bits/pel}$$

(5.3.110)

$$D = \frac{p\theta_{3pN}}{1.39} + \frac{1}{N} \sum_{m>3pN}^{3N} \frac{\sigma_m^2}{A_F\, T_m}$$

where $p < 1$ is the fraction of the $3N$ coefficients that are transmitted.

In reality, as we have seen in Chapter 4, optimization of luminance/chrominance distortion is a very complex nonlinear problem that depends on many subjective relationships between luminance, chromaticity, spatial detail, etc. Many of the statistical and subjective phenomena described

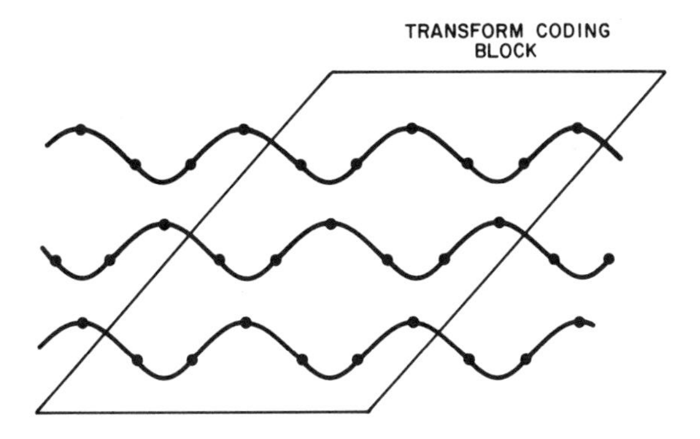

Fig. 5.3.28 Successive lines of an NTSC color waveform sampled at three times the color subcarrier frequency. If two-dimensional transform blocks are chosen appropriately, then line-to-line correlation between pels is high.

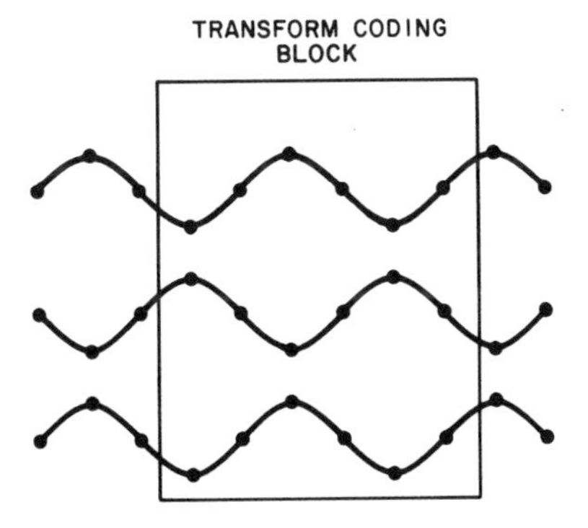

Fig. 5.3.29 Successive lines of an NTSC color waveform sampled at four times the color subcarrier frequency. Vertically adjacent pels are anticorrelated.

earlier for DPCM coding of color images may also be exploited for transform coding. For example, if a transform block has a high-detail luminance component, then it also usually has high-detail chrominance components. A high chrominance distortion may then be allowed not only because of self-masking by the chrominance, but also because of masking by the high-activity luminance component. Unlike DPCM however, the distortion is then spread throughout the block, which in the case of large block-sizes may limit the extent to which such masking may be exploited without undesirable degradation.

Transform coding has also been applied to composite color waveforms such as the NTSC and PAL color signals. In this application it is important that some of the basis vectors be chosen specifically to represent the amplitude and phase of the color subcarrier signal. This requires that the original sampling rate be synchronized in some way to the color subcarrier and that the block size be carefully chosen. It also requires both sine and cosine color subcarrier basis vectors, at least in the horizontal direction. These are present in the RDFT and KLT, but not in the DCT and WHT. In principle, a new set of orthonormal basis vectors could be constructed* from the two color subcarrier vectors plus most of the DCT vectors. However, this has not been studied in detail.

Line-to-line vertical correlation in the sampled composite color signal depends on the sampling rate, the block shape and the format of the composite waveform. For example, Fig. 5.3.28 shows an NTSC signal sampled at three times color subcarrier frequency. If the block boundary is chosen as shown, then the result is a high vertical correlation that can be exploited by a 2D transform. Fig. 5,3.29 shows an NTSC signal sampled at four times color subcarrier frequency. Because the scan line duration is 227.5 color subcarrier periods, vertically adjacent pels are anticorrelated. This behavior is exploited by the basis vector

$$\mathbf{t}' = (1 \ -1 \ \ 1 \ -1 \ \ 1 \ \ldots) \tag{5.3.111}$$

which is already present in most of the transforms discussed so far.

The PAL composite signal has a color subcarrier phase shift at the end of each scan line. This reduces the amount of vertical correlation when periodic sampling is employed. However, if the phase of the sampling is also shifted at the end of each scan line, high vertical correlation can be restored. The SECAM composite signal has different color subcarrier frequencies on adjacent scan lines, and therefore vertical correlation is lower than with NTSC or PAL.

---

* Using, for example, the Gram-Schmidt orthogonalization procedure.[5.3.16]

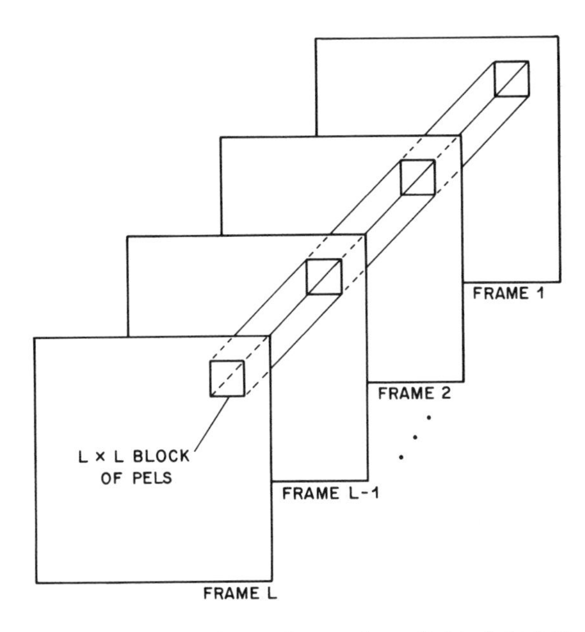

Fig. 5.3.30 Three-dimensional transform coding may be applied to $L \times L \times L$ blocks of pels taken from L successive frames of imagery.

Experimental results so far on intraframe transform coding of composite signals are mixed.[5.3.17] Picture quality versus bit-rate seems to be comparable to that which can be achieved with DPCM coding of composite waveforms. Sensitivity to transmission bit errors may be sometimes lower than DPCM. However, complexity and therefore cost are considerably higher.

As with DPCM, transform coding of the composite signal is somewhat simpler than component coding. However, the overall performance is not as good in terms of bit-rate and picture quality. Moreover, as with DPCM, component coding enjoys a considerable advantage in being relatively compatible with NTSC, PAL and SECAM color signals, once the composite-to-component conversion has been made. This aspect is very important in international broadcasting of color TV signals.

5.3.5 Interframe or Three-Dimensional Transform Coding

With multiple frame imagery, such as motion picture film or television, a logical extension of intraframe coding concepts is three-dimensional transform coding. With this approach the frames are coded L at a time, and the data is partitioned into $L{\times}L{\times}L$ blocks† of pels prior to transformation as shown in Fig. 5.3.30. Each block could then be arranged into a column vector of length $N = L^3$ and transformed via an Nth order transform. However, as before in the case of 2D-transforms, the usual approach is to use a *separable* 3D-transform, i.e., three successive applications of the L-th order transform to the horizontal, vertical and temporal axes of each pel block. Thus, if $\boldsymbol{B} = [b_{ijk}]$ is the three dimensional block of pels to be transformed, then the $L{\times}L{\times}L$ block of transform coefficients is $\boldsymbol{C} = [c_{\ell mn}]$, where

$$c_{\ell mn} = \sum_{i,j,k=1}^{L} b_{ijk}\, t_{\ell i}\, t_{mj}\, t_{nk} \tag{5.3.112}$$

and the t's are elements of the $L{\times}L$ transform matrix $\boldsymbol{T}$. If i, j and k are the vertical, horizontal and temporal coordinates, respectively, then ℓ and m are the vertical and horizontal spatial frequency coordinates, and n is the temporal frequency coordinate.

As with intraframe transforms, the interframe transform coefficients usually have Laplacian PD's, except for the DC coefficient, which is generally considered to be uniformly distributed.[5.3.18] However, interframe transform coding is affected not only by the differing amounts of spatial detail within a frame, but also by variations in the amount of movement and other temporal

† Generalization to non-cubic blocks is straightforward.

activity in different regions of the picture. Thus, there are two sources of statistical nonstationarity at work, and therefore it is usually *mandatory* that adaptive techniques be used in order to achieve good results. In fact, experimental evidence[5.3.19] indicates that nonadaptive interframe transform coding of $16\times16\times16$ pel blocks performs only slightly better than nonadaptive intraframe coding of 16×16 pel blocks, which is much less costly, by far, to implement.

If a region of the picture has low spatial detail then, as with intraframe transform coding, only the coefficients having low spatial frequency need be transmitted. Correspondingly, if a region has low or zero frame-to-frame changes then only the coefficients having low or zero temporal frequency, need be transmitted.

The problem of quantifying temporal activity for each three-dimensional block of pels and of adapting the coding of coefficients thereto is directly analogous to that of two-dimensional adaptive transform coding where spatial activity is the controlling parameter. However, due to the temporal averaging property of most TV cameras, it is very rare that a 3D-transform block will have simultaneously high spatial and high temporal activity. Thus, the non-DC coefficient energy is not a good activity measure for coding purposes. Instead, a selected set of spatial frequency coefficients [$n=0$ in Eq. (5.3.112)] should be used to indicate spatial activity, and a selected set of temporal frequency coefficients [$\ell=m=0$] should be used to indicate temporal activity.[5.3.20] Classification techniques may then be used that categorize each block according to the amount of spatial and/or temporal activity and, as in adaptive 2D-transform coding, code using a coefficient bit-assignment and quantization based on the subjective visibility of distortion in each class. Adaptive interframe coding experiments have yielded a halving of the bit-rate compared with nonadaptive coding.[5.3.19]

While, in principle, all of the above techniques will contribute to bit-rate reduction, in practice there are many pitfalls. For example, the frequency weighted mean square error distortion criterion is not well understood even for intraframe coding, and is even less understood as a measure of interframe coding artifacts and temporal resolution. In real-time, constant frame-rate applications the cost of implementing a three-dimensional transform coder can be prohibitively high compared with other coding techniques that may be equally or only slightly less efficient. For a block size of $L\times L\times L$ pels, L frames of memory are needed for the basic coding operation. In addition, some input buffer memory is required as well as the usual output data buffer associated with any adaptive-coding/constant-channel-rate system, and this causes additional transmission delay. With these and other impediments, it is not surprising that the technology of interframe block transform coding has been slow to develop.

5.4 Hybrid Transform Coding Techniques

We have seen from the previous sections that transform coding involves high complexity, both in terms of storage and computational requirements. For sources with stationary statistics, transform coding with large block size removes essentially all of the statistical redundancy present in the source signal. However, for real-world picture sources that do not have stationary statistics, block sizes should be smaller so that adaptive techniques can accommodate to the changes in local area statistics. Small block sizes should also be used if the adaptive coding is to take advantage of psychovisual properties of the human observer. However, with smaller block sizes, considerable block-to-block redundancy will exist even after transform coding. It is this dilemma that is partially alleviated by Hybrid Transform Coding. As we shall see, storage requirements and computational complexity are also reduced by hybrid transform coding.

5.4.1 Intraframe Hybrid Transform Coding

Basically, hybrid transform coding combines *transform* coding and *predictive* coding in order to reduce storage requirements and to increase adaptability and efficiency. Intraframe hybrid transform coding is typically implemented by first transform coding in one spatial dimension and then predictive (DPCM) coding in the other spatial dimension.[5.4.1]

For example, suppose each image to be coded is partitioned into vertical strips, L-pels wide as shown in Fig. 5.4.1. Let $\boldsymbol{B} = [b_{ij}]$ be such a strip. We then perform an Lth order 1D-transform on each row of $\boldsymbol{B}$ to obtain the coefficients

$$c_{im} = \sum_{j=1}^{L} b_{ij}\, t_{mj} \tag{5.4.1}$$

where the t's are the elements of the $L \times L$ transform matrix $\boldsymbol{T}$. This operation is a transformation in the horizontal (row) direction and serves to remove horizontal redundancy between pels that are in the same row of $\boldsymbol{B}$. We denote the vector of transform coefficients of row i, strip k by $\boldsymbol{c}_i^k$.

With intraframe hybrid transform coding we now remove vertical redundancy in $\boldsymbol{B}$ by employing DPCM in the vertical direction to code the transform coefficients. For example, with single element predictive coding, we would quantize and code the differential vector

$$\boldsymbol{e}_i^k \triangleq \boldsymbol{c}_i^k - \tilde{\boldsymbol{c}}_{i-1}^k \tag{5.4.2}$$

where $\tilde{\boldsymbol{c}}_{i-1}^k$ is the quantized version of the previous row transform coefficients in strip k. In the generalized case, all of the previously transmitted coefficients

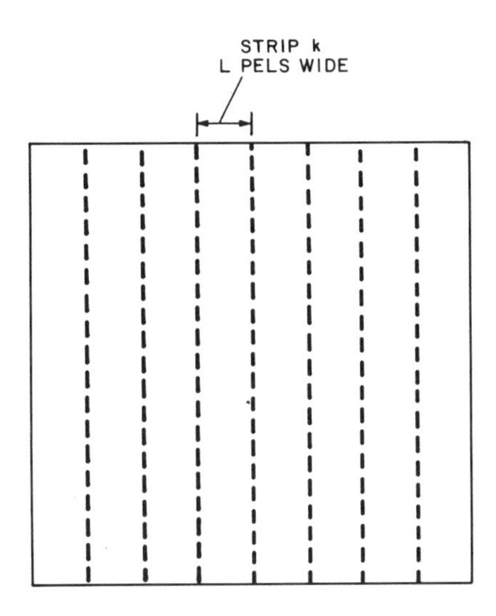

Fig. 5.4.1 With intraframe hybrid coding the image is first partitioned into vertical strips, each having a width of L pels.

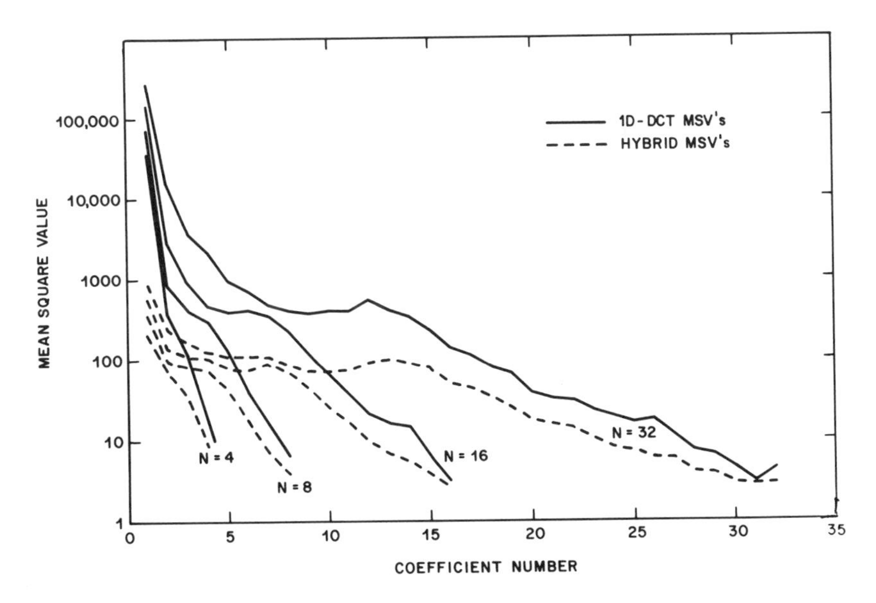

Fig. 5.4.2 Comparison of 1D-DCT coefficient MSV's with 2D hybrid MSV's for block lengths 4, 8, 16, 32. In all cases the hybrid MSV's are smaller. However, the difference is small for the higher order coefficients.

$$\tilde{c}_q^{\ell} = \begin{cases} q < i, \ \ell = \text{any} \\ \\ q = i, \ \ell < k \end{cases} \tag{5.4.3}$$

could be used to compute a prediction vector $\hat{c}_i^k$ for c_i^k, and the prediction error vector

$$e_i^k = c_i^k - \hat{c}_i^k \tag{5.4.4}$$

would then be quantized, coded and transmitted. We denote the quantized version by $\tilde{e}_i^k$.

In the absence of transmission errors, the receiver can compute $\hat{c}_i^k$ from the previously received information and reconstruct the block of pels as follows:

$$\begin{aligned} \tilde{c}_i^k &= \hat{c}_i^k + \tilde{e}_i^k \\ \\ \tilde{b}_i^k &= T^{-1} \, \tilde{c}_i^k \end{aligned} \tag{5.4.5}$$

assuming vectors are arranged in column matrices.

Hybrid intraframe transform coding with line-to-line DPCM requires only one line of memory instead of the L lines required of an $L \times L$, 2D transform coding. Moreover, with line-to-line processing the capability of adapting to local variations in the image is significantly enhanced compared with 2D transform coding.

Fig. 5.4.2 shows DCT hybrid transform statistics for the image "Karen" using various one-dimensional block sizes and simple line-to-line differences of transform coefficients. MSV's for both the differential signals and the respective one-dimensional transform coefficients are shown. The differential signal probability distributions all have the characteristic Laplacian shape, and for the lower order coefficients their MSV's are significantly smaller than the corresponding one-dimensional DCT coefficients. However, for the higher order coefficients the difference is not so great. In fact, for images containing high detail, the higher order coefficients have little or no correlation from line to line. In this case, the best predictor for them is zero, i.e., they should be coded independently of previously transmitted values.[5.4.3]

The theoretical framework for quantization and coding of hybrid transform values is very similar to that of transform coding itself. For example, consider a single strip $B = [b_{ij}]$ and simple line differential PCM, i.e.,

$$\hat{c}_i = \tilde{c}_{i-1} \tag{5.4.6}$$

Also, as in previous sections, suppose we utilize the frequency weighted mean square error distortion criterion*

$$D = \frac{1}{L} \sum_{m=1}^{L} \frac{E(c_{im} - \tilde{c}_{im})^2}{A_F T_m}$$

(5.4.7)

$$= \frac{1}{L} \sum_{m=1}^{L} \frac{E(e_{im} - \tilde{e}_{im})^2}{A_F T_m}$$

and that the e's are quantized and coded independently of each other. In Section 5.3.2 we saw that under these circumstances, the bit-rate depends only on the MSV's (arranged in decreasing order) of the quantities to be transmitted, i.e.,

$$\sigma_{im}^2 = E(e_{im}^2)$$

$$= E(c_{im} - c_{i-1,m})^2 \tag{5.4.8}$$

$$= E\left[\sum_{j=1}^{L} (b_{ij} - b_{i-1,j})\, t_{mj}\right]^2$$

where we assume that $\tilde{c}_{i-1,m} \approx c_{i-1,m}$. Thus,

$$\sigma_{im}^2 = E\left[\sum_{j=1}^{L} (b_{ij} - b_{i-1,j})\, t_{mj} \sum_{k=1}^{L} (b_{ik} - b_{i-1,k})\, t_{mk}\right]$$

$$= \sum_{j,k=1}^{L} \left\{ r_{ijik} - r_{i,j,i-1,k} - r_{i-1,j,i,k} + r_{i-1,j,i-1,k} \right\} t_{mj}\, t_{mk}$$

* As before, in some cases the statistical averaging operator E may be profitably replaced by a subjectively weighted averaging.

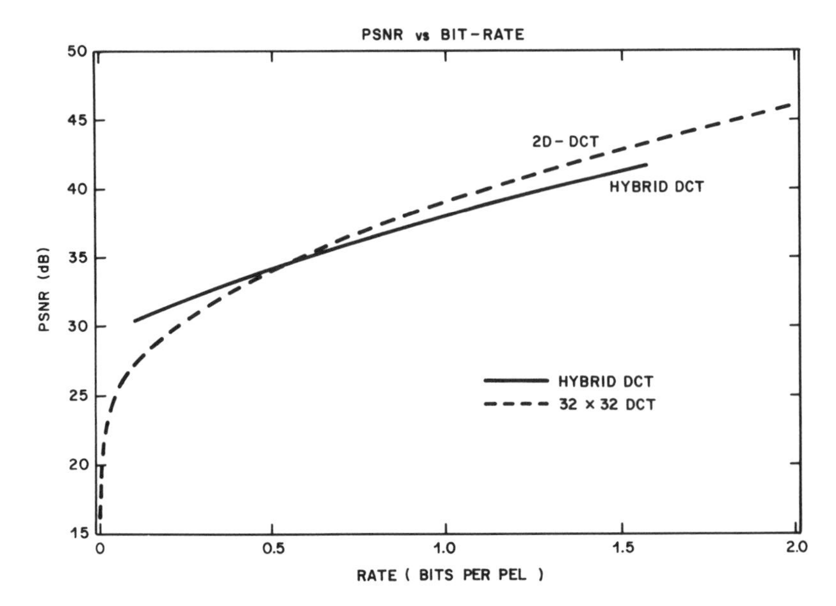

Fig. 5.4.3 Comparison of hybrid coding (DCT, L=32) with two-dimensional transform coding (separable DCT, 32×32 blocks) using the image "Karen". Both cases employed uniform quantization with entropy coding.

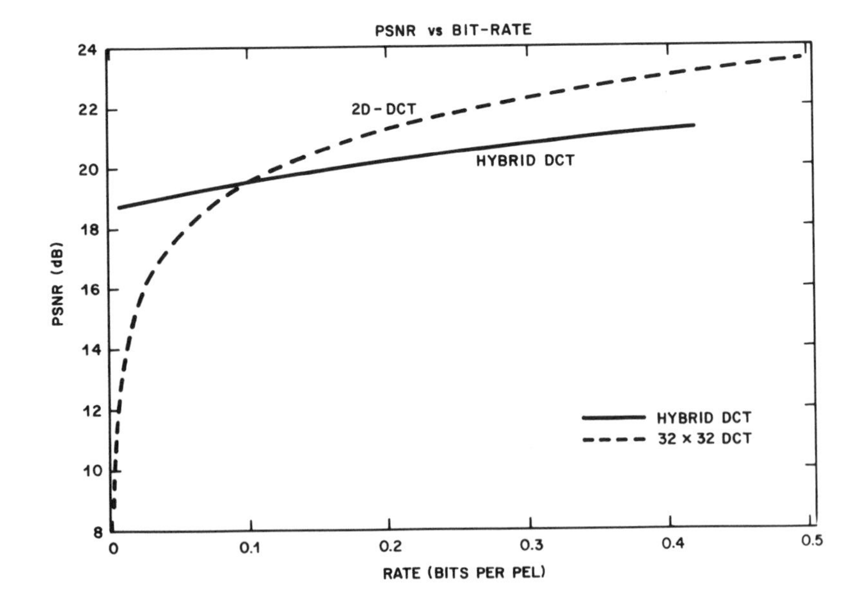

Fig. 5.4.4 Comparison of hybrid coding and transform coding using the isotropic correlation model with $\rho = 0.98$. The same coding as Fig. 5.4.3 was used.

where the correlations are given by

$$r_{ijk\ell} \triangleq E\,(b_{ij}\, b_{k\ell}) \tag{5.4.9}$$

For example, with uniform quantization and entropy coding the bit assignments and distortion are given by Eqs. (5.3.78-83), and the quantizer step size is given by Eq. (5.3.85). In particular, if the block-size is not too small and the fraction of transmitted coefficients is $p < 1$, then

$$\theta \approx \frac{1.68\sigma_{pL}^2}{A_F T_{pL}}$$

and

$$R \approx \frac{1}{1.39L} \sum_{m=1}^{pL} \ln \frac{\sigma_m^2\, T_{pL}}{T_m\, \sigma_{pL}^2} \quad \text{bits/pel}$$

(5.4.10)

$$D \approx \frac{1.21p\,\sigma_{pL}^2}{A_F\, T_{pL}} + \frac{1}{L} \sum_{m>pL}^{L} \frac{\sigma_m^2}{A_F\, T_m}$$

Fig. 5.4.3 shows *PSNR* vs *R* for "Karen" using $A_F\, T_m = 1$ and a block length $L = 32$ pels. Fig. 5.4.4 shows similar results for the isotropic correlation model

$$r_{ijk\ell} = \frac{255^2}{3}\, \rho^{\sqrt{(i-k)^2 + (j-\ell)^2}} \tag{5.4.11}$$

Also shown for reference in these figures are the results of the separable 2D-DCT coding using 32×32 pel blocks. We see that hybrid and 2D transform coding have comparable performance, at least for the MSE distortion criterion.

Fig. 5.4.5 shows a generalized block diagram for intraframe hybrid transform coding. The one-dimensional transform converts $1\times L$ blocks of pels $\boldsymbol{b}_i^k$ into blocks of coefficients $\boldsymbol{c}_i^k$ that are then DPCM coded. The predictor uses one or more previously transmitted coefficient blocks to compute the prediction vector $\hat{\boldsymbol{c}}_i^k$. In fact, previously transmitted coefficients in the *present* block could also be used to predict each coefficient. The differential vector $\boldsymbol{e}_i^k$ is then quantized, coded and transmitted to the receiver where DPCM decoding and inverse transformation take place in order to produce the pel vector $\tilde{\boldsymbol{b}}_i^k$.

The processor of Fig. 5.4.5 has all the elements necessary to carry out the adaptive coding techniques described in Section 5.3. In particular, the quantizer can be adapted to line-to-line variations in the picture, and coefficients can be discarded or quantized according to whatever quality criterion is used. If the

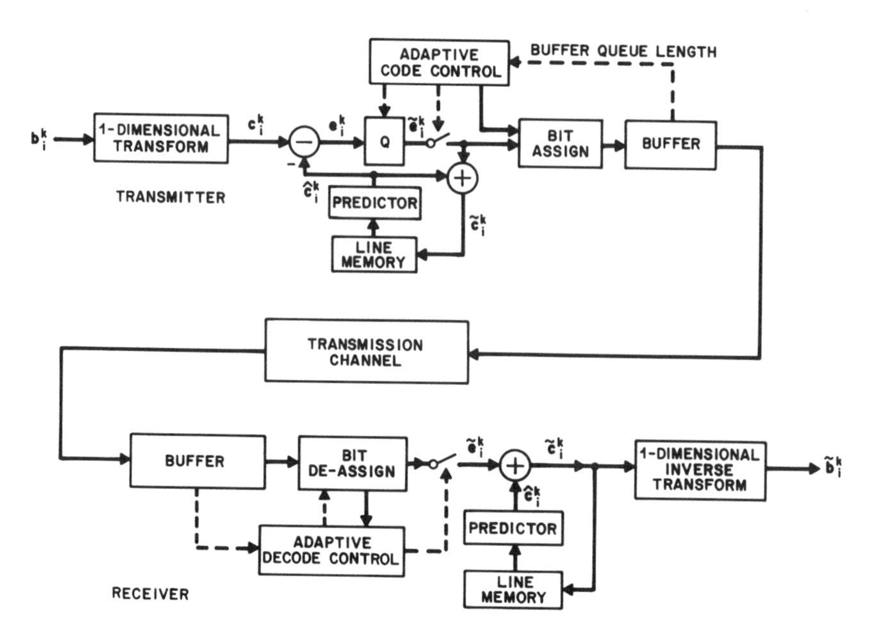

Fig. 5.4.5 Generalized intraframe hybrid coder. Line i of strip k is 1D-transformed into c_i^k, which is then DPCM coded into a quantized differential vector $\tilde{e}_i^k$. The inverse process takes place at the receiver.

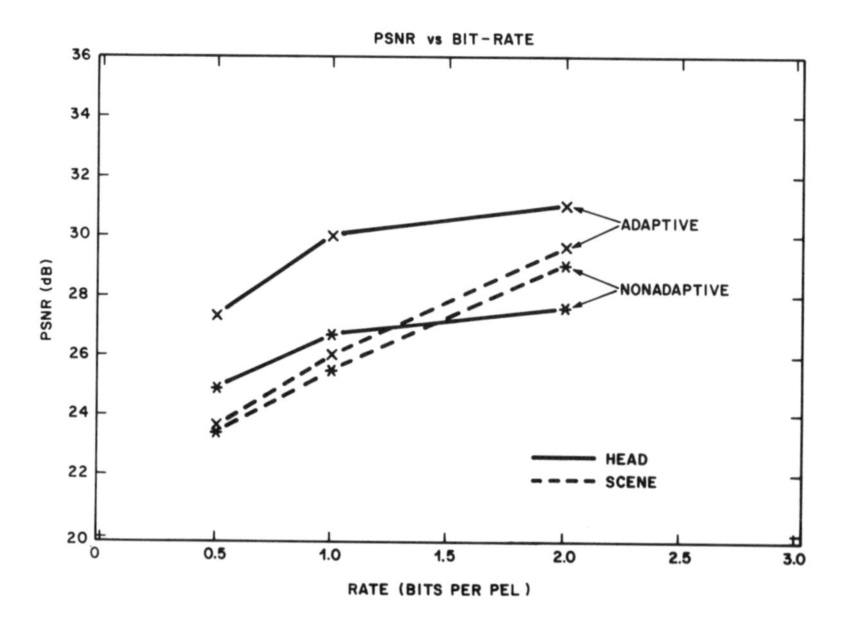

Fig. 5.4.6 Experimental results using intraframe hybrid coding on the "Head" and "Scene" images of Fig. 5.4.7. Data points * are for nonadaptive coding. Points × are for adaptive coding.

Fig. 5.4.7 Images used for hybrid coding studies. (top) Head. (bottom) Scene.

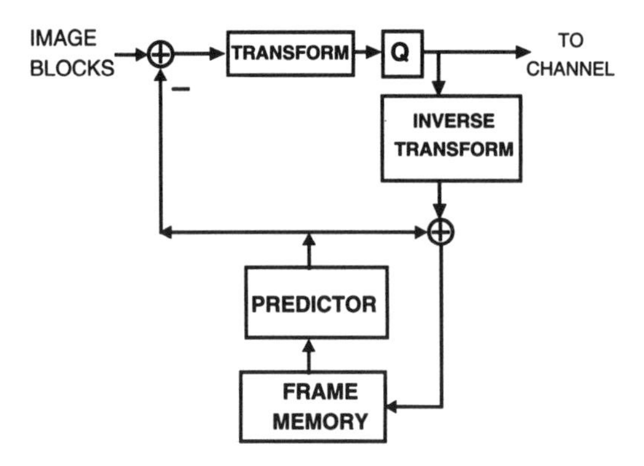

Fig. 5.4.8a Interframe hybrid transform coder for pel domain predictors. Here the prediction error is transformed instead the pels themselves.

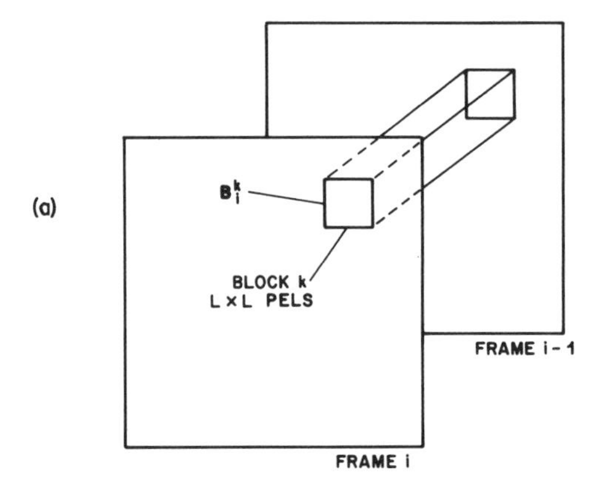

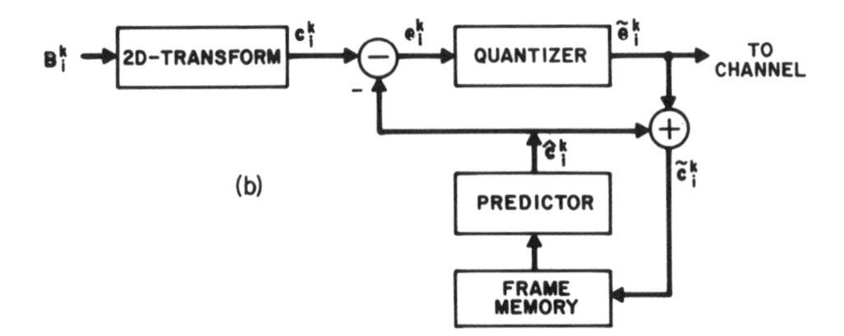

Fig. 5.4.8 Interframe hybrid coding. Blocks of $L \times L$ pels are first 2D-transformed. The resulting coefficients are then interframe DPCM coded and sent to the receiver where the inverse process takes place.

number of transmitted bits varies greatly from block-to-block and if the transmission channel operates at a constant bit-rate, then the buffers shown in Fig. 5.4.5 may have to be quite large in order to achieve a relatively low channel bit-rate. In this case the transmitter adaptive coding must, in addition to its other constraints, make sure that the transmitter buffer neither empties nor overflows. The receiver adaptive decoding must take into account the possibility of transmission bit errors, and in the event they occur the receiver must recover as fast as possible to avoid data being lost due to receiver buffer overflow.

The transmitter adaptive code controller will typically use some measure of spatial activity as well as the buffer queue length in order to determine the quantization accuracy and the number of coefficients that must be sent in a particular block. This controller may also modify the bit assignment algorithm, e.g., a Huffman or run-length code, in order to adapt to changing statistics of the transform coefficients and differential values. Such coding strategy overhead may have to be transmitted to the receiver, or in some cases it may be derived from the coded information itself.

Fig. 5.4.6 shows PSNR vs R results for adaptive intraframe hybrid transform coding of the two images shown in Fig. 5.4.7. One is fairly low-detail and the other is fairly high-detail.[5.4.2] In this coder the hybrid DCT was used with strip width $L = 32$ pels. In each strip the MSV's of the differential values were measured and used to design Lloyd-Max quantizers. Thus, adaptation to changing image statistics was only in the horizontal direction. Significant improvement over nonadaptive coding resulted only for the head and shoulders image, where detail was nonuniformly distributed in the picture.

5.4.2 Interframe Hybrid Transform Coding

Three dimensional $L \times L \times L$ block transform coding typically requires memory at the transmitter to store L frames of image data. A similar requirement exists at the receiver. This need can be reduced to approximately one frame of memory at the transmitter and one at the receiver by the use of interframe hybrid transform coding.

The first step is to partition each image frame into two-dimensional blocks of $L \times L$ pels. We represent the kth block of pels in the ith frame by $\boldsymbol{B}_i^k$. Fig. 5.4.8 shows a straightforward extension of intraframe hybrid coding in which each block is transformed by a 2D-transform producing a length L^2 vector of coefficients $\boldsymbol{c}_i^k$. As in the previous section, a predictor uses previously transmitted blocks of coefficients in order to compute a prediction vector $\hat{\boldsymbol{c}}_i^k$ that, in the absence of transmission errors, is available both at the transmitter and receiver. However, in this case the predictor has access to the previous frame blocks. For example, a single-block previous-frame prediction would produce

$$\boldsymbol{e}_i^k = \boldsymbol{c}_i^k - \tilde{\boldsymbol{c}}_{i-1}^k \tag{5.4.12}$$

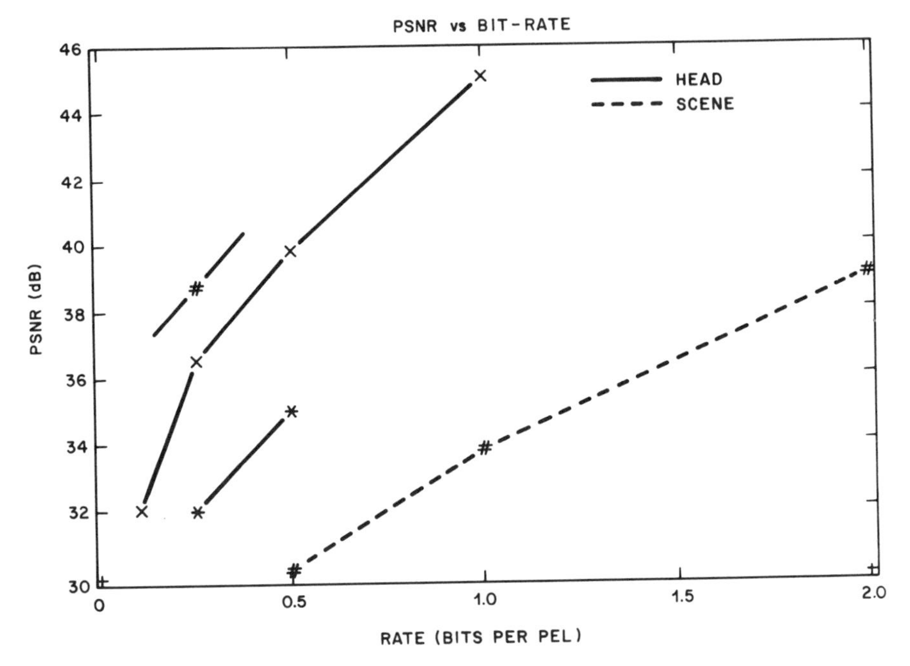

Fig. 5.4.9 Experimental results using interframe hybrid coding on the images of Fig. 5.4.7. Data points * are for nonadaptive coding, points × are for adaptive coding and points # are for adaptive coding with movement compensation.

Fig. 5.4.8a shows an equivalent interframe hybrid transform coding system, except that here the predictor operates in the pel domain instead of the transform domain. This arrangement is particularly useful when the prediction operation includes motion compensation, since motion estimation is difficult to perform in the transform domain.

Interframe hybrid transform coding differs significantly from its intraframe counterpart in that the non DC elements of the vector $\boldsymbol{c}_i^k$ are determined exclusively by the amount of *spatial* detail in the picture, whereas the DPCM differential transform vector $\boldsymbol{e}_i^k$ depends both on the amount of spatial detail and on the amount of frame-to-frame change, i.e., *temporal* activity in the scene. Thus, in regions of the picture containing low detail, many of the elements of $\boldsymbol{c}_i^k$ will be negligible, which in turn causes many of the elements of $\boldsymbol{e}_i^k$ to be negligible. In regions of the picture containing no movement, all elements of $\boldsymbol{e}_i^k$ should be negligible regardless of the amount of detail. However, camera noise and instabilities will often cause frame-to-frame differences even in the absence of movement. As with intraframe transform coding, adaptive coding is essential in order to achieve the greatest benefits of interframe hybrid transform coding. With interframe transform coding, nonadaptive operation is only slightly better than intraframe transform coding.[5.3.19]

With television, interlace is usually employed, with each frame consisting of two line-interleaved fields that are transmitted one after the other. Thus, if $\boldsymbol{B}_i^k$ represents the *k*th block of pels in the *i*th *field,* then for true frame differential operation we should define

$$\boldsymbol{e}_i^k = \boldsymbol{c}_i^k - \tilde{\boldsymbol{c}}_{i-2}^k \tag{5.4.13}$$

However, using the definition of Eq. (5.4.12) should not be too detrimental, except that highly detailed picture areas containing no movement will generate nonnegligible field differential vectors $\boldsymbol{e}_i^k$, whereas with the frame-differential of Eq. (5.4.13) this does not happen.

Adaptive interframe hybrid transform coding generally follows the same classification strategy described earlier for interframe transform coding. However, in this case temporal activity is measured by frame-to-frame coefficient or pel differences instead of transform coefficient MSV's. Experimental results of adaptive interframe hybrid transform coding are comparable to adaptive interframe nonhybrid transform coding,[5.3.19] which is much more complex and therefore more expensive to implement. Adaptation can reduce the required bit-rate by factors of 2 to 10, depending on the algorithm and the amount of spatial and temporal activity in the scene.

Fig. 5.4.9 shows *PSNR* vs *R* results for adaptive interframe hybrid transform coding of the images in Fig. 5.4.7. The separable 2D-DCT was used with block size 16×16. In each block the MSV's of the frame differential values were measured and used to design Lloyd-Max quantizers. As seen in the figure, adaptive interframe coding shows significant improvement over nonadaptive coding. Interframe hybrid coding will be discussed in more detail in Chapter 6.

With interframe hybrid transform coding, motion compensation can also be used to reduce the bit-rate significantly. The most straightforward approach is to first estimate the frame-to-frame translation for the block of pels to be coded using techniques for block matching motion estimation that were covered in Section 5.2.3. Then the block of pels to be used as a prediction in the previous frame is displaced by the measured amount, followed possibly by a recalculation of transform coefficients prior to subtraction from the incoming data. Alternatively, displaced frame differences could be computed prior to transformation, using techniques described earlier for DPCM. The advantage of sending transforms of differential signals instead of the differential signals themselves lies in the greater ability of transforms to send large areas of low spatial-frequency signals.

Experimental results[5.3.19] indicate that motion compensation can reduce the bit-rate by up to a factor of two compared with simple interframe hybrid transform coding (see Fig. 5.4.9). However, results are extremely dependent on the type of picture and the amount of motion. More will be said about motion compensated hybrid transform coding in later Chapters.

5.4.3 Interrelation Between Hybrid Transform Coding and Predictive Coding

Analytical results for hybrid transform coding, as with other areas of image analysis, rely heavily on the assumption of statistical stationarity and on the mean square error distortion criterion. While in many situations this does not reflect reality, the analysis may be useful in some aspects of adaptive coder design, particularly in the coding of classes having relatively constant spatial detail or temporal activity.

We begin by assuming that blocks of pels are rearranged into column vectors and that their statistics are wide-sense stationary, i.e., the correlation† matrix of the pel vectors

$$\boldsymbol{R}_b \triangleq E(\boldsymbol{bb}') \tag{5.4.14}$$

is constant and well defined. In this case the correlation matrix of the transform coefficients is easily shown to be

† Prime indicates conjugate transpose.

$$\boldsymbol{R}_c \triangleq E(\boldsymbol{c}\boldsymbol{c}') = \boldsymbol{T}\boldsymbol{R}_b\boldsymbol{T}' \tag{5.4.15}$$

The diagonal elements of $\boldsymbol{R}_c$ are then the variances of the transform coefficients.

We now denote by $\dot{\boldsymbol{c}} = \boldsymbol{T}\dot{\boldsymbol{b}}$ the already quantized and transmitted block of coefficients in the previous line, field or frame that is to be used in the prediction of $\boldsymbol{c}$. The cross correlation matrix between $\boldsymbol{c}$ and $\dot{\boldsymbol{c}}$ is then given by

$$\boldsymbol{R}_{c\dot{c}} \triangleq E(\boldsymbol{c}\dot{\boldsymbol{c}}') = \boldsymbol{T}\,\boldsymbol{R}_{b\dot{b}}\,\boldsymbol{T}' \tag{5.4.16}$$

where

$$\boldsymbol{R}_{b\dot{b}} \triangleq E(\boldsymbol{b}\dot{\boldsymbol{b}}') \tag{5.4.17}$$

is the corresponding cross correlation between $\boldsymbol{b}$ and the previously transmitted block $\dot{\boldsymbol{b}}$. If the effects of quantization are small, i.e., the received picture quality is high, then $\boldsymbol{R}_{b\dot{b}}$ can be found either by direct measurement or from the assumed statistical model of the original pels.

The minimum mean square error (MSE) linear prediction of $\boldsymbol{c}$ given $\dot{\boldsymbol{c}}$ is then found by elementary calculus (see also Sec. 3.1.5) to be

$$\hat{\boldsymbol{c}} = \boldsymbol{A}\dot{\boldsymbol{c}} \tag{5.4.18}$$

where

$$\begin{aligned} \boldsymbol{A} &= \boldsymbol{R}_{c\dot{c}}\,\boldsymbol{R}_c^{-1} \\ &= \boldsymbol{T}\,\boldsymbol{R}_{b\dot{b}}\,\boldsymbol{R}_b^{-1}\,\boldsymbol{T}' \end{aligned} \tag{5.4.19}$$

Similarly, we can show that

$$\hat{\boldsymbol{b}} \triangleq \boldsymbol{R}_{b\dot{b}}\,\boldsymbol{R}_b^{-1}\,\dot{\boldsymbol{b}} \tag{5.4.20}$$

is the minimum MSE linear predictor of $\boldsymbol{b}$ given $\dot{\boldsymbol{b}}$. Thus, Eq. (5.4.18) is equivalent to simply taking the transform of $\hat{\boldsymbol{b}}$. Indeed, more complicated predictors that use more than one previous block can be easily derived by simply taking the transform of the corresponding linear or other predictor of the pel vector $\boldsymbol{b}$. For example, suppose we use J previously transmitted blocks and derive a minimum MSE linear predictor for vector $\boldsymbol{b}$. Then we have

$$\hat{\boldsymbol{b}} = \sum_{j=1}^{J} \boldsymbol{A}_j \, \dot{\boldsymbol{b}}_j \tag{5.4.21}$$

where the $\boldsymbol{A}_j$ are found from the statistical model using the techniques of Section 3.1.6. The corresponding transform coefficient predictor is then simply

$$\hat{\boldsymbol{c}} = \sum_{j=1}^{J} \boldsymbol{T} \, \boldsymbol{A}_j \, \boldsymbol{T}' \, \dot{\boldsymbol{c}}_j \tag{5.4.22}$$

Note that if the transform produces truly uncorrelated coefficients then $\boldsymbol{R}_{c\dot{c}}$ and $\boldsymbol{R}_c$ are diagonal matrices, which in turn makes $\boldsymbol{A}$ in Eq. (5.4.19) a diagonal matrix. In this case the single-block minimum MSE linear predictor of the mth element of $\boldsymbol{c}$ is given by

$$\hat{c}_m = \frac{E(c_m \dot{c}_m')}{E(c_m c_m')} \, \dot{c}_m \tag{5.4.23}$$

In any event, for the single block predictor of Eq. (5.4.18) the element MSV's of the differential vector $\boldsymbol{e} = \boldsymbol{c} - \hat{\boldsymbol{c}}$ are the diagonal elements of the matrix

$$\begin{aligned} \boldsymbol{V} &= \boldsymbol{E}\{[\boldsymbol{c} - \boldsymbol{A}\dot{\boldsymbol{c}}] \, [\boldsymbol{c} - \boldsymbol{A}\dot{\boldsymbol{c}}]'\} \\ &= \boldsymbol{R}_c - R_{c\dot{c}} \, \boldsymbol{A}' - \boldsymbol{A}\boldsymbol{R}_{c\dot{c}}' + \boldsymbol{A}\boldsymbol{R}_c\boldsymbol{A}' \end{aligned} \tag{5.4.24}$$

If we choose $\boldsymbol{A}$ to be the identity matrix, i.e., use $\dot{\boldsymbol{c}}$ as the prediction, then

$$\boldsymbol{V} = \boldsymbol{T}\left[2\boldsymbol{R}_b - \boldsymbol{R}_{b\dot{b}} - \boldsymbol{R}_{b\dot{b}}'\right]\boldsymbol{T}' \tag{5.4.25}$$

whereas if we choose the optimum single block $\boldsymbol{A}$ of Eq. (5.4.19) then

$$\boldsymbol{V} = \boldsymbol{T}\left[\boldsymbol{R}_b - \boldsymbol{R}_{b\dot{b}} \, \boldsymbol{R}_b^{-1} \, \boldsymbol{R}_{b\dot{b}}'\right]\boldsymbol{T}' \tag{5.4.26}$$

We have seen that knowing the MSV's of the elements of $\boldsymbol{e} = \boldsymbol{c} - \hat{\boldsymbol{c}}$ determines how to quantize and code them. For example, if the mean square error distortion criterion is used then the overall pel MSE is equal to the sum of the mean square quantization errors of elements of $\boldsymbol{e}$. If (as is almost always the case) the elements are approximately Laplacian, then the results of Section 5.3.2 can be applied straightforwardly. However, if more realistic distortion criteria are used in order to exploit the properties of human vision, then other quantization strategies may have to be employed.

5.5 Other Block Coding

There are many other block coding methods besides transform coding and its variations. In this section we describe a few of them.

5.5.1 Vector Quantization and Variations

As we saw in Section 5.2.4, scalar quantization involves basically two operations, 1) *partitioning* the range of possible input values into a finite collection of subranges or subsets, and 2) for each subset choosing a *representative value* to be output when an input value is in that subrange, i.e., a member of that subset. With vector quantization,[5.5.1] the same two operations take place not in a one-dimensional scalar space, but in an N-dimensional vector space. Thus, if the vector space is partitioned into 2^{NR} subsets each having a corresponding representation value or *code vector,* then blocks $\boldsymbol{b}$ of N pels can be coded with RN bits per block or R bits per pel.

If we assume that filtering, gamma correction and other preliminary operations have already taken place, then it becomes plausible to define a difference distortion measure $d(\boldsymbol{b}-\tilde{\boldsymbol{b}})$ between an input vector $\boldsymbol{b}$ and a code vector $\tilde{\boldsymbol{b}}$. The problem of vector quantization is then to choose the partitioning and the code vectors in such a way as to minimize the overall distortion for the class of imagery to be handled.

Any of the difference distortion measures described previously can be used, at least in principle. These include mean square error, subjectively weighted error, frequency weighted error, etc. However, study of vector quantization of images has made headway only in recent years, and in practically all of these studies the mean square error (MSE) has been used exclusively, that is

$$
\begin{aligned}
d &= \frac{1}{N} (\boldsymbol{b}-\tilde{\boldsymbol{b}})' (\boldsymbol{b}-\tilde{\boldsymbol{b}}) \\
&= \frac{1}{N} \sum_{i=1}^{N} (b_i-\tilde{b}_i)^2
\end{aligned}
\tag{5.5.1}
$$

with the overall distortion being simply the expected value

$$D = E(d) \tag{5.5.2}$$

Utilization of non-difference distortions has not yet been attempted.

The design of image vector quantizers (VQ) generally requires a set of *training* pictures since accurate statistics are usually unavailable. The VQ code vectors and partitioning are then determined iteratively by repeated processing of the training set. Having designed the VQ, a block $\boldsymbol{b}$ of pels can be coded (for most distortion measures) by a simple nearest neighbor rule, that is, choose the code vector $\tilde{\boldsymbol{b}}$ that minimizes $d(\boldsymbol{b}-\tilde{\boldsymbol{b}})$ and transmit a code word index of NR bits to tell the receiver which code word to use. The decoder then simply displays the vector $\tilde{\boldsymbol{b}}$ as the coded representation.

The main advantage of VQ is the simple receiver structure, which consists only of a look up table or *codebook* containing the 2^{NR} code vectors. The disadvantages include coder complexity and the fact that images dissimilar to those in the training set may not be well represented by the code vectors in the codebook.

5.5.1a Design of VQ Codebook Using Training Images

One method proposed recently is the generalized Lloyd algorithm, also known as the LBG algorithm[5.5.2] or the *K-means* algorithm.[5.5.3] It requires an initial codebook and proceeds as follows.

1) Map the training blocks into the code words using the nearest neighbor rule. If the overall distortion of this mapping or quantization is low enough, quit. Otherwise,

2) For each code vector $\tilde{\boldsymbol{b}}$ determine the subset of training blocks that were mapped into it. Then replace $\tilde{\boldsymbol{b}}$ by another code word that reduces (ideally minimizes) the overall distortion for that subset of training blocks. Then go to step 1.

For the MSE distortion the best replacement code word in step 2 is simply the average of the corresponding subset of training blocks. For other distortions the best replacement code word may be much more difficult to compute. The algorithm terminates when the overall distortion stops decreasing significantly.

Note that for a given initial codebook the LBG algorithm only produces a *local* minimum distortion. Other initial codebooks may well converge to a better final codebook. In fact, choosing an initial codebook is a problem in itself. One plausible choice for an initial codebook is a subset of the training vectors. These should be chosen to be fairly dissimilar from each other, and obvious outliers should probably be avoided. Another initial codebook might be obtained simply by using a coarse scaler quantizer on the individual components of the training vectors and throwing out duplicates.

A third method for initial codebook selection called *splitting* starts with a single code word equal to the average of the training vectors. Another codeword is then generated by a small perturbation, and the LBG algorithm is invoked to optimize the two codewords (see Fig. 5.5.1). The process then continues in a similar manner, that is, given K codewords, generate $2K$ codewords by a small

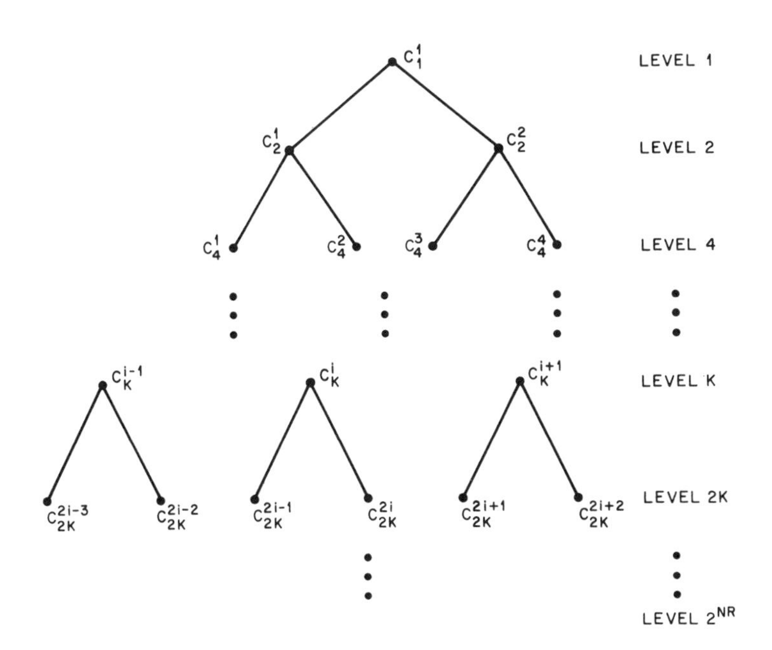

Fig. 5.5.1 The splitting method for generating a Vector Quantization codebook. At a given level K, each of the K code vectors is perturbed slightly to yield a set of $2K$ vectors, which is then optimized by the LBG algorithm to produce the codebook at level $2K$. The tree can also be used for coding. At each node C_K^i, the input vector is compared with code vectors C_{2K}^{2i-1} and C_{2K}^{2i}. The closer one is then chosen as the next node.

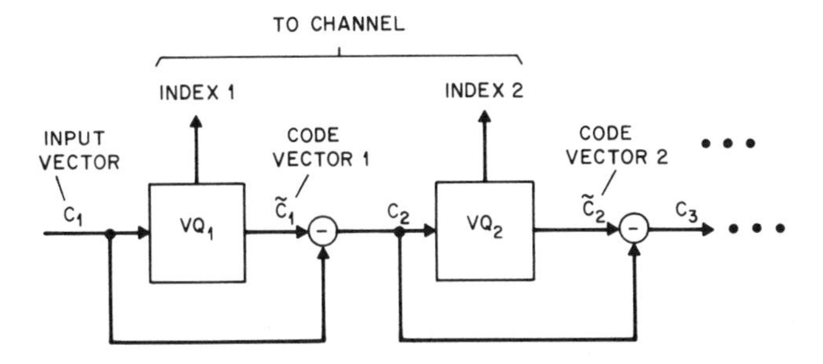

Fig. 5.5.2 Multistage Vector Quantization. At each state an error vector is computed, which is then used as the input to the next stage of VQ. The decoder merely computes a summation of the code vectors corresponding to the received indices.

perturbation of each, followed by the LBG algorithm to optimize them. Continue until 2^{NR} codewords are generated.

A problem with the LBG algorithm and its variants is its excessive convergence time in building the codebook. Techniques for reducing this time invariably tradeoff performance; however, since the LBG algorithm itself is only locally optimum this may not be much of a liability. One technique called *tree codebooks*[5.5.2] starts with one codeword and proceeds as in the splitting method above. However at each stage K of code book generation the set of training vectors is partitioned into subsets of training vectors that are mapped into the same code vector. Then at the next stage, as each code vector is perturbed slightly to produce another code vector, only those training vectors in the corresponding subset are used to optimize the resulting two vectors.

Another codebook generation technique developed by Equitz[5.5.3] is significantly faster with no apparent performance degradation. It is called the *Nearest Neighbor* or NN algorithm, and has computational complexity that grows only linearly with the size of the training set. The basic idea of this method is to start with the entire training set and, as a first step, merge the two vectors that are closest to each other into another vector equal to their mean. Successive application of this operation eventually reduces the set of vectors to the desired size, or alternatively raises the mean square quantization error to just short of an unacceptable level. The main computational efficiency of the NN algorithm is achieved by a preprocessing operation that partitions the training data into a *K-d Tree*. This enables not only much faster searching, but also multiple merges at each iteration.

The NN algorithm frequently produces a better codebook than the LBG algorithm, which shows the importance of initial codebook selection in the latter. On the other hand, an even better codebook is usually obtained by using the NN algorithm to derive an initial codebook for the LBG algorithm.

5.5.1b VQ Variants and Improvements

Several methods have been suggested for speeding up the coding operation, i.e., selecting the code vector that best approximates each input vector. Most of these include construction of a code tree of some type, which greatly speeds up the search, while sacrificing somewhat on performance. For example, the splitting technique tree codebook mentioned above for codebook generation can also be used for suboptimal coding. At each node a comparison is made between the input vector and the intermediate code vectors corresponding to that node in order to determine which branch to be traversed next (see Fig. 5.5.1). Similarly, the NN algorithm can be used to construct a code tree simply by letting the algorithm continue until only one code word is left. For a binary tree (two branches at each node) the number of comparisons is reduced from 2^{NR} to $2NR$, i.e., the computation time increases only linearly with the vector length. However, the storage requirement is essentially doubled compared with exhaustive search coding.

Multistage VQ (see Fig. 5.5.2) reduces both the search time and the storage requirement. At the first stage a low rate VQ is performed, and a code word index is transmitted. Then a first error vector is generated by subtracting the code vector from the original. This error vector is then coded by a different VQ, a code word index is transmitted and if necessary, a second error vector is generated. The receiver produces a decoded vector by simply summing up the code vectors corresponding to the received indices. In principle, coding error can be reduced in this way to any desired level. However, in practice the pels of the error vectors eventually become very uncorrelated and, thus, unsuited to efficient vector quantization.

Parameter extraction techniques, also called *Product Codes,* are capable of further reducing VQ coding complexity. For example, if the mean and variance of each input vector are computed and sent separately, then the mean can be subtracted and a gain factor applied to produce a zero mean, unity variance vector to be coded by VQ. The codebook for these vectors will presumably be much smaller than an equivalent quality simple VQ codebook for the original vectors. Also, the scalar parameters may be coded themselves, for example, by DPCM.

Block classification techniques may be used to good advantage in adaptive Vector Quantization. Here, blocks are classified into one of several classes according to spatial activity, and a different VQ codebook is used for each class. This approach not only produces codebooks that are better suited to the blocks they will be used to code, it also exhibits less sensitivity to changes in picture material. A drawback of this method is the overhead required to send the class of each block.

As with hybrid transform coding, prediction techniques can also be used with VQ. Simple predictive VQ would compute a prediction of the present block based on previously coded blocks just as in hybrid transform coding. Adjacent blocks either in the same frame or in the previous frame may be used in the prediction. The prediction error is then coded and sent by VQ in the same way as with nonpredictive VQ.

Vector quantization of color images raises the possibility of exploiting correlations between color components since, in principle, a block of $L \times L$ colored pels is simply an $L \times L \times 3$ array of numbers that can be rearranged as an $3L^2$ vector. VQ can then be carried out as before, except we have the ever present problem of finding a suitable distortion criterion. This approach can be quite efficient for block classifications techniques are used and each image to be coded contains relatively few colors. For then, the code books can be relatively small. Also, see Section 5.1.1a on color maps. However, if a wide variety of colors must be accommodated, then as we saw in Chapter 3 the *YIQ* color components are practically uncorrelated. In this case combined component coding is of little use, and separate component coding gives nearly optimum results.

5.5.1c Performance of VQ Codecs

The picture quality achievable by VQ coding depends on many factors, including the block size of the pel vectors, the amount of interpel correlation, the suitability of the codebook to the images to be coded, etc. VQ is useful mostly for low bit-rate coding of pictures, since high rates usually entail an impractical degree of complexity. For example, with blocks of size 4×4 pels, coding at a rate of one bit per pel implies a codebook containing 2^{16} code words. While this is not much of a problem for the decoder, the codebook search during coding can be prohibitively complex and lengthy, as can the process of constructing codebooks. Continuing the above example, with images of size 256 × 256 pels a training set of 16 pictures itself contains only 2^{16} blocks. This implies that a much larger training set is needed, thus further increasing the complexity of codebook generation.

Fig. 5.5.3 shows the results of VQ coding at 0.5 bits per pel using blocks of size 4×4 pels. The codebook was constructed using Equitz's NN algorithm and a training set of two images of size 512 × 512 pels. Note that the coding error of the training images is considerably smaller than for pictures outside the training set.

Optimizing the subjective image quality using vector quantization can only be described at present, as a worthy, but elusive goal. Mean square error is virtually the only criterion considered so far. However, it has been recognized that different types of picture material, e.g., low detail, texture, edges, should be coded and reproduced in different ways[5.5.4] and that vector quantization may be an ideal method for doing so.

5.5.2 Nonunitary Interpolative Transform Coding

Here we briefly discuss a somewhat different approach called nonunitary interpolative transform coding, which we illustrate by example. Suppose we partition the image into 4×4 blocks, with pels of each block labeled as follows:

$$\begin{matrix} A & B & C & D \\ E & F & G & H \\ I & J & K & L \\ M & N & O & P \end{matrix} \qquad (5.5.3)$$

For each block, we then code and transmit the first pel A unchanged. We

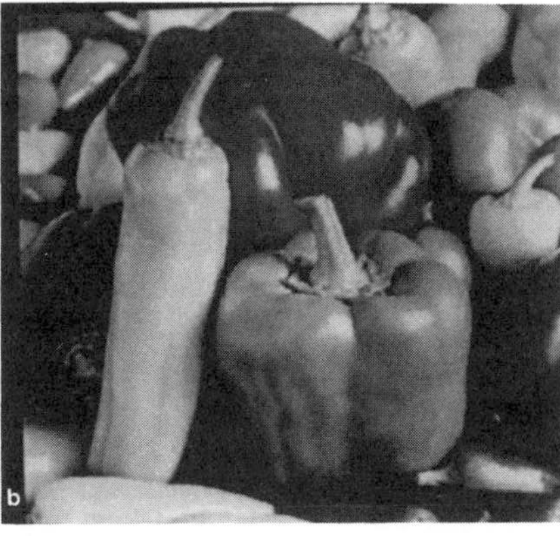

Fig. 5.5.3 Pictures of size 512×512 pels are coded at 0.5 bits per pel using Vector Quantization with block size 4×4 pels [from H-M. Hang]. The training set consisted of the blocks of images a and b. Image c is outside the training set. a) *PSNR* = 31.0 dB. b) *PSNR* = 30.7 dB. c) *PSNR* = 26.8 dB.

then code and transmit pel P via DPCM using[†] pel $\tilde{A}$ as a prediction and then send pels D and M via DPCM using $(\tilde{A}+\tilde{P})/2$ as a prediction. The remaining pels will then be coded and transmitted via DPCM using interpolations of previously sent nearest neighbors as predictions. Thus, for example, pel B would be coded using as a prediction the interpolated value $(2\tilde{A}+\tilde{D})/3$, and pel C would then use $(\tilde{B}+\tilde{D})/2$ as a prediction. In this way, all predictions use only pels that have been previously quantized and transmitted, thus avoiding a possible accumulation of unwanted quantization error. If we arrange the block of pels in Eq. (5.5.3) in a column vector

$$\boldsymbol{b} = [A\ B\ C\ D\ E\ F\ G\ H\ I\ J\ K\ L\ M\ N\ O\ P]' \tag{5.5.4}$$

then all of the aforementioned DPCM differential values can be computed by a 16×16 "transform" matrix. For example, the first six rows of the transform matrix

$$\begin{bmatrix} 1 & 0 & 0 & 0 & 0 & 0 & 0 & 0 & 0 & 0 & 0 & 0 & 0 & 0 & 0 & 0 \\ -1 & 0 & 0 & 0 & 0 & 0 & 0 & 0 & 0 & 0 & 0 & 0 & 0 & 0 & 0 & 1 \\ -\frac{1}{2} & 0 & 0 & 1 & 0 & 0 & 0 & 0 & 0 & 0 & 0 & 0 & 0 & 0 & 0 & -\frac{1}{2} \\ -\frac{1}{2} & 0 & 0 & 0 & 0 & 0 & 0 & 0 & 0 & 0 & 0 & 0 & 1 & 0 & 0 & -\frac{1}{2} \\ -\frac{2}{3} & 1 & 0 & -\frac{1}{3} & 0 & 0 & 0 & 0 & 0 & 0 & 0 & 0 & 0 & 0 & 0 & 0 \\ 0 & -\frac{1}{2} & 1 & -\frac{1}{2} & 0 & 0 & 0 & 0 & 0 & 0 & 0 & 0 & 0 & 0 & 0 & 0 \\ & & & & & & & \cdot & & & & & & & & \\ & & & & & & & \cdot & & & & & & & & \\ & & & & & & & \cdot & & & & & & & & \end{bmatrix} \tag{5.5.5}$$

correspond to interpolative DPCM coding of pels A, P, D, M, B and C.

The approach described so far differs somewhat from simple transform coding in that each transform coefficient is now associated with a particular pel of the block $\boldsymbol{b}$. Moreover, after each coefficient in quantized, the associated pel of $\boldsymbol{b}$ is replaced by its quantized (tilde) value so that all predictions may be accurately computed at the receiver. Since the transform matrix is very sparse, containing mostly zeros, the complexity can be considerably less than with

† Tildes denote quantized values.

orthogonal transforms. In most cases, complexity increases only linearly with block size.

A generalization of this approach is possible if, in formulating the predictions, we do not worry about whether or not neighboring pels have already been transmitted. For example, with the transform matrix whose first five rows are

$$\begin{bmatrix} 1 & 0 & 0 & 0 & 0 & 0 & 0 & 0 & 0 & 0 & \cdots \\ -1 & 0 & 0 & 1 & 0 & 0 & 0 & 0 & 0 & 0 & \cdots \\ -\frac{1}{2} & 1 & \frac{1}{2} & 0 & 0 & 0 & 0 & 0 & 0 & & \cdots \\ 0 & -\frac{1}{2} & 1 & \frac{1}{2} & 0 & 0 & 0 & 0 & 0 & & \cdots \\ -\frac{1}{3} & 0 & 0 & 0 & 1 & -\frac{1}{3} & 0 & 0 & -\frac{1}{3} & & \cdots \end{bmatrix} \tag{5.5.6}$$

the pels A and D are coded as before. However, the predictions for pels B, C and E use, as yet, untransmitted pels. Thus, as in the case with simple transform coding, more than one coefficient is required in order to decode some pels.

In general, the transform matrix can be made completely previous-pel predictive, i.e., each transmitted coefficient enables the decoding of one pel of the block, if through some reordering of rows and/or columns† the transform matrix can be made lower-diagonal. In this case the inverse transform matrix is also lower-diagonal, which also implies that the transform is non-unitary. If the transform matrix cannot be diagonalized in this fashion, the non-unitary transform operation

$$\boldsymbol{c} = \boldsymbol{T}\boldsymbol{b} \tag{5.5.7}$$

will be able to make only partial use of quantized (tilde) pel values in computing the transform coefficients.

A unitary transformation can be constructed by defining the columns of $\boldsymbol{V}'$ to be the normalized eigenvectors of $\boldsymbol{T}$. In this case Eq. (5.5.7) can be written

$$\boldsymbol{V}\boldsymbol{c} = \boldsymbol{V}\boldsymbol{T}\boldsymbol{V}'\ \boldsymbol{V}\boldsymbol{b} \tag{5.5.8}$$

where $\boldsymbol{V}\boldsymbol{T}\boldsymbol{V}'$ is a diagonal matrix of eigenvalues. The resulting unitary transform is

† Reordering columns implies a similar reordering of pels in $\boldsymbol{b}$.

$$c_u = VT^{-1}c = Vb \tag{5.5.9}$$

In most cases, little or no use of quantized (tilde) pel values will be possible with these unitary transforms. However, they still may be useful in certain adaptive coding situations where pel-to-pel correlations and dependencies change often, in a known way, and based upon this information it becomes possible to easily construct highly efficient interpolative transforms.

5.5.3 Block Truncation Coding (BTC)

The objective of this algorithm[5.5.5] is to quantize the pels of an $L \times L$ block in such a way as to preserve the mean and variance. In particular, a two-level quantizer is attractive because of its simplicity. The first step is to measure the mean and variance of the block

$$\bar{b} = \frac{1}{L^2} \sum_{i,j=1}^{L} b_{ij}$$

$$\sigma^2 = \frac{1}{L^2} \sum_{i,j=1}^{L} (b_{ij} - \bar{b})^2 \tag{5.5.10}$$

These are then quantized and transmitted. Next, the pels of the block are quantized to two levels using $\bar{b}$ as a threshold, i.e., a 1 or 0 is sent depending on whether b_{ij} is greater than or less than $\bar{b}$, respectively.

At the receiver the first step is to evaluate q, the number of 1's received for the block. The number of zeros is then $p = L^2 - q$. The next step is to evaluate the two quantizer levels that preserve mean and variance. These are easily shown to be

$$L_0 = \bar{b} - \sigma\sqrt{\frac{q}{p}}$$

$$L_1 = \bar{b} + \sigma\sqrt{\frac{p}{q}} \tag{5.5.11}$$

Finally, each pel is decoded to L_0 or L_1 depending on whether a 0 or 1 was received, respectively.

Using blocks of size 4×4 pels, and 8-bits each for $\bar{b}$ and σ, the coding bit-rate is 2-bits/pel. If $\bar{b}$ and σ are coded jointly using 10-bits, the data rate is 1.625 bits/pel.

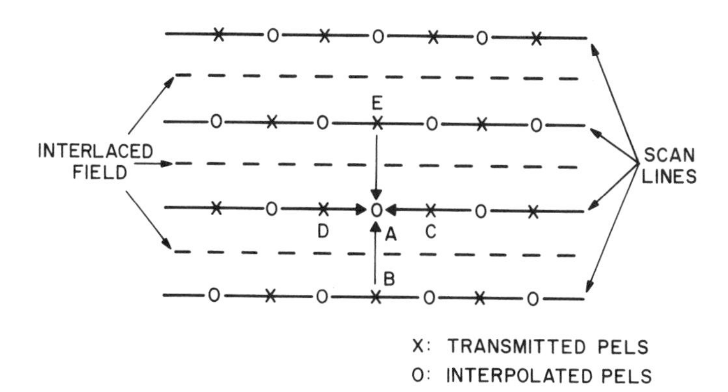

Fig. 5.6.1 An example of interpolative coding using 2:1 horizontal sub-sampling staggered from one line to the next.

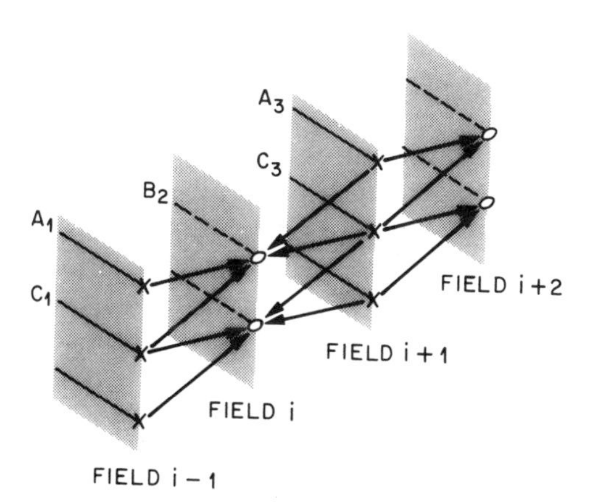

Fig. 5.6.2 Four-way averaging in which alternate fields (e.g. Field i, i+2) are not transmitted. At the receiver, the untransmitted fields are replaced by a four-way average of elements in the adjacent fields.

Decoded BTC images typically reproduce sharp edges of large objects fairly well. Also, random texture, although not reproduced with high accuracy, often looks good enough to be acceptable in many applications. However, in low detail areas of the picture where brightness and/or color vary slowly, there is often visible contouring and block-edge discontinuity due to the coarse quantization.

5.6 Miscellaneous Coding Methods

In this section we describe some other coding techniques that are currently under study. Some can be combined with methods already described, while others are stand-alone.

5.6.1 Interpolative and Extrapolative Coding

In interpolative and extrapolative coding, a subset of picture elements are first chosen for transmission. For this subset of pels any of the previously discussed coding schemes, such as PCM, DPCM or transform coding may be used. The pels that are not selected for transmission are reproduced by interpolation at the receiver using the information about the transmitted pels. In interpolative coding, some present and some future pels are transmitted and the rest are interpolated, whereas in extrapolative coding, pels in the immediate future are extrapolated from the past, one by one. The first pel for which the error of extrapolation goes above a previously set threshold causes the process of extrapolation to stop. That pel is then transmitted along with its address, and the following pels are extrapolated again as before.

Both interpolative and extrapolative methods can be divided into two classes: fixed and adaptive. In fixed interpolative methods, a fixed set of pels are selected for transmission and the rest are interpolated. Various types of samples can be chosen for transmission: For example, one may transmit every alternate sample, one sample out of four, every alternate scan line, every alternate field or frame, etc. The picture quality that results is a function of the number and type of samples that are dropped and the method of interpolation. Fig. 5.6.1 shows an example in which there is 2:1 subsampling along each scan line. The subsampling pattern is staggered from one line to the next and the dropped pels are interpolated by a four way average as shown by arrows, i.e., pel A is interpolated by averaging pels B, C, D, and E. More sophisticated interpolation can be used instead of linear weighted average. Interpolation based on higher degree polynomials or splines has been used, but experience shows that linear interpolation is quite effective and not much is gained by using polynomials of higher degree.

Adaptive interpolation can be more effective than fixed interpolation. As an example in Fig. 5.6.1, pel A may be reconstructed at the receiver by the following rule:

$$\hat{A} = \begin{cases} 0.5(C+D) \ , & \text{if } |C-D| \le |B-E| \\ \\ 0.5(B+E) \ , & \text{otherwise} \end{cases} \tag{5.6.1}$$

Such interpolations switch between horizontal, vertical, temporal or other directional averages depending upon the type of local correlation that exist in the data as measured by the transmitted pels.

An interpolative scheme that is commonly used in interframe coding is to drop alternate fields from transmission. This is shown in Fig. 5.6.2. Each pel in the dropped field is interpolated by taking a four way average of pels from adjacent fields that are spatially and temporally close. Such a four way average has the effect of reducing both the spatial as well as temporal resolution. It has been used effectively in interframe coding during buffer overloads in order to reduce the data rate.

Adaptive interpolative coding consists of three parts. a) choosing certain pels for transmission, b) constructing an interpolation (fixed or adaptive) of nontransmitted pels, and c) evaluating the interpolation error. If the error is below a certain preset threshold, then fewer pels are chosen for transmission; if it goes above the threshold, then a larger number of pels are chosen for transmission. The fewest number of pels that are needed to keep the interpolation error below the threshold is normally the desired characteristic. As an example, consider the video waveform along a scan line as shown in Fig. 5.6.3. Starting with sample A, one can go up to sample E and interpolate, using a straight line, all the in-between samples B, C, and D, keeping the interpolation error below the threshold. However, when samples A and F are chosen for transmission, then a simple straight line interpolation between pels A and F results in excessive interpolation error for pels B, C, D and E. Thus, pels A and E should be chosen for transmission, and pels B, C and D should be interpolated.

Several variations exist depending upon: a) the method used for transmission of pels such as A and E, b) interpolation and c) the method of judging the quality of interpolation. A more flexible variation is to treat this as a problem in waveform approximation in which values at certain pel positions (called knots) are transmitted and an approximation to the entire curve is made from the knot positions and the transmitted values (which may not coincide with the image intensities at the knot positions). The quality of interpolation is usually judged by the *MSE* between the interpolated signal and the true signal, since *MSE* is computationally straightforward. However, better results are obtained by using a quality criterion that is more closely related to human visual perception. As an example, better results are obtained by spatially filtering the error and then multiplying it by an appropriate masking function, before averaging the error over several pels. Unlike fixed interpolative coding, adaptive interpolative coding requires two types of information to be transmitted to the

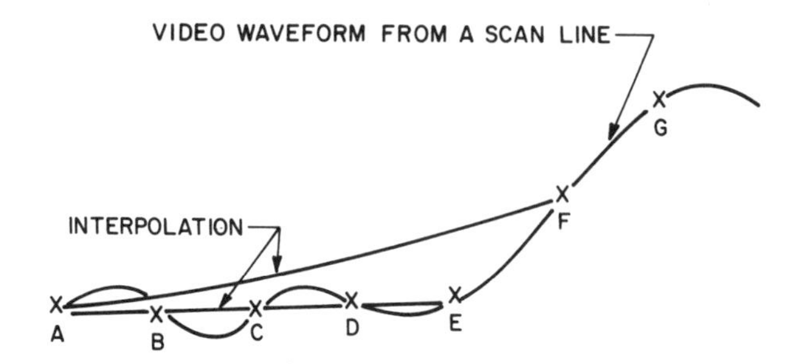

Fig. 5.6.3 An illustration of adaptive interpolation technique.

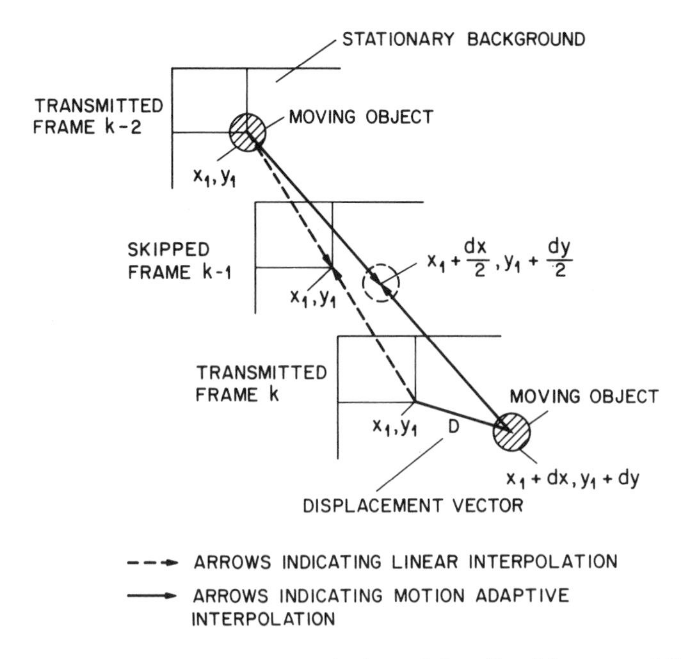

Fig. 5.6.4 Illustration of linear and motion interpolation. (from Musmann et al. [5.2.9])

receiver: a) addresses of pels chosen for transmission, and b) intensity of pels chosen for transmission.

The performance of the interpolative coders depends upon the technique used for transmitting samples and for interpolating. Interpolative coders show an improvement over nonadaptive DPCM and transform coding in terms of the number of bits required for a given picture quality. However, interpolative coders are inferior to both adaptive DPCM as well as adaptive transform techniques. Interpolative coders are useful for applications where a low bit rate is required, and where it might not be possible to encode each sample.

5.6.2 Motion Adaptive Interpolation

An adaptive interpolation that is very useful is called *motion adaptive* interpolation. Earlier in this section (Fig. 5.6.2) we described linear interpolation of dropped fields and remarked that four way linear average reduces spatial as well as temporal resolution. In some cases, it is possible to improve the quality of interpolation by using estimates of motion of objects in the scene. Fig. 5.6.4 shows an example where frame $(k-1)$ is skipped and is reconstructed at the receiver by using the frames transmitted at instants $(k-2)$ and (k). In this case, linear interpolation for a pel at location $(x_1 y_1)$ in frame $(k-1)$ would be a simple average of pels at the same location in frames $(k-2)$ and (k). This would blur the picture since, pels from background and moving object are averaged together. If however, the moving object was undergoing pure displacement as shown in the figure, then a pel at position $(x_1 + dx/2,$ $y_1 + dy/2)$ in the skipped frame should be interpolated using pels (x_1, y_1) and $(x_1 + dx, y_1 + dy)$ in frames $(k-2)$ and (k), respectively. This is indicated by arrows in the figure.

In order for such schemes to be successful, in addition to an accurate displacement estimator, it is necessary to segment the skipped frame into moving areas, uncovered area in the present frame relative to the previous frame and area to be uncovered in the next frame, so that proper interpolation can be applied. Obviously pels as well as coefficients used for interpolation of these different segments have to be different. This is a subject of current research. Preliminary simulations indicate that under moderate motion, a 3:1 frame dropping (i.e. skipping 2 out of 3 frames) and motion adaptive interpolation often gives reasonable motion rendition. In this as well as other interpolation schemes, since sometimes the interpolation may be inaccurate, techniques have been devised where the quality of interpolation is checked at the transmitter, and if the interpolation error is larger than a threshold, correction information is transmitted to the receiver. Due to unavoidable inaccuracies of the displacement estimator (e.g. complex translational and rotational motion) and the segmentation process, such side information is necessary to reduce artifacts that may otherwise be introduced due to faulty interpolation. Much more will be said about motion compensated interpolation in Chapter 9.

5.6.3 Pyramid Coding, Subband Coding

Another technique for image compression is called *pyramid* coding[5.6.1]. It represents the uncoded image as a series of band-pass images each sampled at successively lower rates. One implementation of this is to construct in the first stage a sequence of low-pass filtered images, $\{b_k(x_j,y_i)\}$, $k = 1 \ldots n$, such that (ignoring edge effects)

$$b_{k+1}(x_j,\, y_i) = \sum_{m=-p}^{+p} \sum_{n=-p}^{+p} h(m,n)\; b_k(x_{2j+n},\, y_{2i+m}) \qquad (5.6.2)$$

where weights $\{h(m,n)\}$ are chosen as an approximate Gaussian-looking function, and $b_0(x_j,y_i)$ is the original uncoded image. For simplicity, $h(m,n)$ is made separable

$$h(m,n) = h(m)\; h(n) \qquad (5.6.3)$$

For $p=2$, the weights are taken to be $h(0) = 0.4$, $h(1) = h(-1) = 0.25$ and $h(2) = h(-2) = 0.05$. In this case, $h(m,n)$ is a cartesian product of two identical symmetric triangular functions.* Image $b_{k+1}(\cdot)$ is a low-pass version of $b_k(\cdot)$ that has been subsampled at a lower rate (in this case by a factor of two).

The next stage of coding is to generate a sequence of error images by first expanding the low-pass, subsampled image $b_{k+1}(\cdot)$ by interpolation and then subtracting it from $b_k(\cdot)$. Thus,

$$L_k\;(x_j,y_i) = b_k(x_j,y_i) - b_{k+1,e}(x_j,y_i) \qquad (5.6.4)$$

where the interpolated pels of the expanded version $b_{k+1,e}(\cdot)$ are given by

$$b_{k+1,e}(x_j,y_i) = 4 \sum_{m=-2}^{+2} \sum_{n=-2}^{+2} h(m,n)\, b_{k+1}\left[x_{\frac{j-n}{2}},\; y_{\frac{i-m}{2}}\right] \qquad (5.6.5)$$

In the above equation, only integer subscripts of x and y are to be included in the summations.

Thus, we have created a sequence of band-pass filtered images, $L_k(\cdot)$. If we imagine these images as stacked one above the other, the result is a tapering pyramid type of data structure. The compression is then accomplished by quantizing $b_n(\cdot)$ followed by the $L_k(\cdot)$'s and then entropy coding the quantized outputs. Image $b_n(\cdot)$ may, in fact, contain only one pel. At the receiver, inverse operations can be performed easily. Simulations show that performance of about 1 bit/pel can be achieved for reasonable picture quality. In addition to the good compression ratio, the scheme also lends itself nicely to progressive

---

* See Fig. 1.3.6a.

transmission. In that case, the topmost level of the pyramid is sent first, and expanded by interpolation in the receiver to form an initial coarse image. This is followed by the transmission of the next lower level of the pyramid, which is added after expansion to the existing image in the receiver, thereby increasing the quality of the image at the receiver. The principal advantage is that the computations are simple, local, and may be performed in parallel. Moreover, the same computations are iterated to build the sequence of images constituting the pyramid.

In the procedure described above, the $L_k(\cdot)$ may be obtained not interactively, but instead as the output of n parallel bandpass filters. In this case the compression algorithm is called *subband coding*.[5.6.2] As above, the $L_k(\cdot)$ are quantized and coded separately, perhaps using run-length coding for strings of zero values. The decoded image is obtained by simply adding the received bandpass images.

5.7 Coding of Graphics

We have so far looked at coding of multilevel (gray level and color) pictures. While many of the methods of multilevel coding are applicable to two-level (black = 0 and white = 1) picture signals, other methods that are unique to two level signals tend to be more efficient. In this section we discuss such techniques. Coding methods of this section may be divided into two broad classes: (a) those that allow exact reproduction of the picture at the receiver assuming that there were no transmission errors, (sometimes called *information-lossless* or exact coding techniques) and (b) methods that make an approximation of the picture (referred to as *information-lossy* or approximate coding).

5.7.1 Exact Coding

We consider four methods of exact coding. These are: (a) run-length coding, (b) predictive coding, (c) line-to-line predictive differential coding, and (d) block coding. This classification is somewhat artificial since there are several coding schemes that combine the above methods into one. Run-length coding is primarily one-dimensional, whereas the other schemes are primarily two-dimensional.

5.7.1a Run-Length Coding

In run-length coding, a run of consecutive pels of the same color (i.e. black or white) is combined together and represented by a single code word for transmission. For example, in Fig. 5.7.1, the length of each run (black or white pels) is represented by binary words. Since runs of black and white pels alternate, except for the first run of each line, the color of the run need not be transmitted. Statistics of such run-lengths vary from document-to-document. In general, they are highly nonuniform (see Section 3.5). Variable length coding of

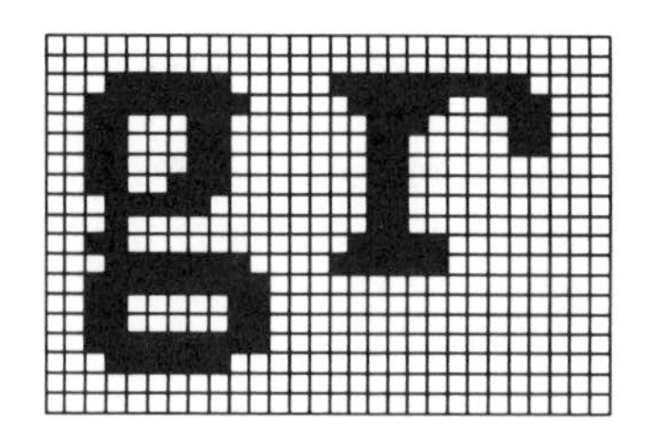

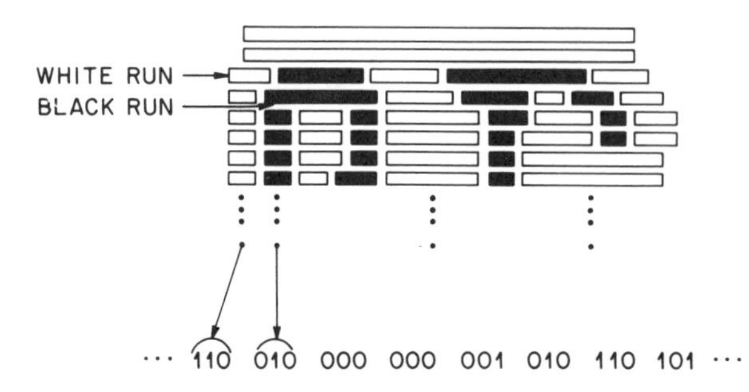

Fig. 5.7.1 A digitized black and white image on the pel grid and its representation in terms of horizontal runs.

DOC	$\bar{r}_w$	$\bar{r}_b$	H_w	H_b	Q_{max}
1	156.3	6.793	5.451	3.592	18.03
2	257.1	14.31	8.163	4.513	21.41
3	89.81	8.515	5.688	3.572	10.62
4	39.00	5.674	4.698	3.124	5.712
5	79.16	6.986	5.740	3.328	9.500
6	138.5	8.038	6.204	3.641	14.89
7	45.32	4.442	5.894	3.068	5.553
8	85.68	70.87	6.862	5.761	12.40

Fig. 5.7.2 Run-length distribution for the CCITT documents of Fig. 2.6.6. $\bar{r}_w$ and $\bar{r}_b$ are the average lengths of the white and black runs, respectively, and H_w and H_b are the corresponding entropies in bits/run. Q_{max} is the maximum compression ratio, i.e., it is the inverse of entropy, where entropy is expressed in bits/pel. (from Hunter and Robinson [5.7.1])

run-lengths takes advantage of the nonuniform probability distribution of run-lengths. Moreover, since the statistics of 0's are usually quite different from those of 1's, different code tables are often used, one for runs of 0's and another for 1's. Most of the variable length codes have been based on Huffman's procedure.

Since most documents have more than a thousand pels per line (the CCITT standard documents have 1728 as shown in Chapter 2), code tables could, in principle, contain a large number of code words. However, to avoid this complexity, a *modified* Huffman code is generally used in practical coders. In this modified Huffman code, every run-length greater than a given value (say l) is broken into two run-lengths: an initial run-length having a value Nl (where N is an integer) and a terminating run-length having a value between 0 and $(l-1)$. This reduces the number of code words in the table and the decoder implementation. Such a remapping of run-lengths modifies the run-length code statistics and results in a slight reduction of compression efficiency. Another advantage of the modified Huffman code is its ability to handle run-lengths of arbitrary size by proper extensions. In Chapter 6, we will give an example of the modified Huffman code that has been chosen as the CCITT standard.

Fig. 5.7.2 gives some relevant statistics that can be used to judge the efficiency of one-dimensional run-length codes for the CCITT standard images that are shown as Fig. 2.6.6. It is clear that for documents that have black information on a white background, the average white run-length ($\bar{r}_w$) is much larger than the average black run-length ($\bar{r}_b$). The compression ratio $(\text{entropy})^{-1}$ also varies widely depending upon the document. On the average the compression ratio is about 12.

5.7.1b Predictive Coding

This is quite similar to predictive coding for gray level pictures. There are, however, important differences in the way a predictor is constructed and how the prediction errors are encoded for transmission, e.g., linear predictive coding is not used. The first step in the prediction process is to choose several previously transmitted surrounding picture elements to be used for prediction of the present element. As an example, consider the seven previous pels A, B, C, D, E, F, G for predicting the present pel X as shown in Fig. 5.7.3. Since each of these pels can take on two values, they together define a possible set of $128 (= 2^7)$ patterns called *states*. The maximum likelihood prediction $\hat{X}(S_i)$ for a given state S_i is given by

$$\hat{X}(S_i) = \begin{cases} \text{black , if} \quad P(X = \text{black} \mid S = S_i) > 0.5 \\ \\ \text{white, otherwise} \end{cases} \tag{5.7.1}$$

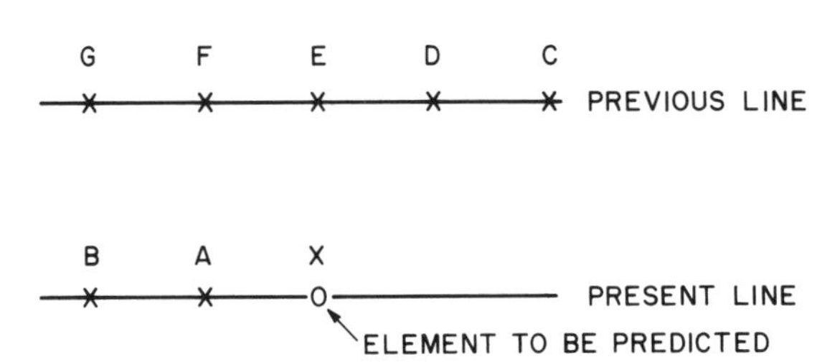

Fig. 5.7.3 The configuration of the seven elements constituting the state that is used to construct the predictor given in Fig. 5.7.4.

No.	State Configuration	FORWARD DIRECTION Prediction	Probability of Correct Prediction	Goodness (G or B)	State Configuration	REVERSE DIRECTION Prediction	Probability of Correct Prediction	Goodness (G or B)
0	00000 00X	0	0.999	G	00000 X00	0	0.999	G
1	00000 01X	1	0.834	B	00000 X10	1	0.833	B
2	00000 10X	0	0.982	G	00000 X01	0	0.983	G
3	00000 11X	1	0.767	B	00000 X11	1	0.770	B
4	00001 00X	0	0.965	G	10000 X00	0	0.963	G
5	00001 01X	1	0.931	G	10000 X10	1	0.917	G
6	00001 10X	0	0.918	G	10000 X01	0	0.928	G
7	00001 11X	1	0.924	G	10000 X11	1	0.900	G
8	00010 00X	0	0.776	B	01000 X00	0	0.745	B
9	00010 01X	1	0.951	G	01000 X10	1	0.939	G
10	00010 10X	0	0.673	B	01000 X01	0	0.724	B
11	00010 11X	1	0.960	G	01000 X11	1	0.959	G
12	00011 00X	0	0.833	B	11000 X00	0	0.846	B
13	00011 01X	1	0.985	G	11000 X10	1	0.983	G
14	00011 10X	0	0.787	B	11000 X01	0	0.763	B
15	00011 11X	1	0.973	G	11000 X11	1	0.955	G
16	00100 00X	0	0.602	B	00100 X00	0	0.607	B
17	00100 01X	1	0.948	G	00100 X10	1	0.979	G
18	00100 10X	0	0.617	B	00100 X01	0	0.540	B
19	00100 11X	1	0.936	G	00100 X11	1	0.945	G
20	00101 00X	0	0.629	B	10100 X00	0	0.640	B
21	00101 01X	1	0.793	B	10100 X10	1	0.878	B
22	00101 10X	1	0.571	B	10100 X01	1	0.545	B
23	00101 11X	1	0.972	G	10100 X11	1	0.886	B
24	00110 00X	1	0.819	B	01100 X00	1	0.784	B
25	00110 01X	1	0.992	G	01100 X10	1	0.994	G
26	00110 10X	1	0.623	B	01100 X01	1	0.656	B
27	00110 11X	1	0.967	G	01100 X11	1	0.976	G
28	00111 00X	1	0.649	B	11100 X00	1	0.639	B

Fig. 5.7.4a State-dependent prediction and its quality for 128 states based on seven surrounding pels. Forward and reverse directions correspond to scanning from left to right and vice versa. States 0–28.

No.	FORWARD DIRECTION State Configuration	Prediction	Probability of Correct Prediction	Goodness (G or B)	REVERSE DIRECTION State Configuration	Prediction	Probability of Correct Prediction	Goodness (G or B)
87	10101 11X	1	0.846	B	10101 X11	1	0.860	B
88	10110 00X	0	0.551	B	01101 X00	0	0.510	B
89	10110 01X	1	0.912	G	01101 X10	1	0.889	B
90	10110 10X	1	0.637	B	01101 X01	1	0.505	B
91	10110 11X	1	0.921	G	01101 X11	1	0.881	B
92	10111 00X	0	0.712	B	11101 X00	0	0.767	B
93	10111 01X	1	0.959	G	11101 X10	1	0.900	G
94	10111 10X	1	0.561	B	11101 X01	1	0.528	B
95	10111 11X	1	0.959	G	11101 X11	1	0.933	G
96	11000 00X	0	0.991	G	00011 X00	0	0.983	G
97	11000 01X	1	0.795	B	00011 X10	1	0.812	B
98	11000 10X	0	0.997	G	00011 X01	0	0.998	G
99	11000 11X	0	0.711	B	00011 X11	0	0.709	B
100	11001 00X	0	0.978	G	10011 X00	0	0.965	G
101	11001 01X	1	0.795	B	10011 X10	1	0.826	B
102	11001 10X	0	0.982	G	10011 X01	0	0.989	G
103	11001 11X	0	0.576	B	10011 X11	0	0.600	B
104	11010 00X	0	0.894	B	01011 X00	0	0.892	B
105	11010 01X	1	0.686	B	01011 X10	1	0.625	B
106	11010 10X	0	0.673	B	01011 X01	0	0.808	B
107	11010 11X	1	0.703	B	01011 X11	1	0.664	B
108	11011 00X	0	0.942	G	11011 X00	0	0.958	G
109	11011 01X	1	0.810	B	11011 X10	1	0.735	B
110	11011 10X	0	0.914	G	11011 X01	0	0.951	G
111	11011 11X	1	0.617	B	11011 X11	1	0.602	B
112	11100 00X	0	0.934	G	00111 X00	0	0.885	B
113	11100 01X	1	0.922	G	00111 X10	1	0.941	G
114	11100 10X	0	0.975	G	00111 X01	0	0.909	G
115	11100 11X	1	0.806	B	00111 X11	1	0.805	B

Fig. 5.7.4b States 29–57.

No.	State Configuration	FORWARD DIRECTION Prediction	Probability of Correct Prediction	Goodness (G or B)	REVERSE DIRECTION State Configuration	Prediction	Probability of Correct Prediction	Goodness (G or B)
58	01110 10X	0	0.594	B	01110 X01	1	0.565	B
59	01110 11X	1	0.950	G	01110 X11	1	0.976	G
60	01111 00X	1	0.530	B	11110 X00	1	0.502	B
61	01111 01X	1	0.996	G	11110 X10	1	0.996	G
62	01111 10X	0	0.768	B	11110 X01	0	0.615	B
63	01111 11X	1	0.961	G	11110 X11	1	0.975	G
64	10000 00X	0	0.996	G	00001 X00	0	0.996	G
65	10000 01X	1	0.768	B	00001 X10	1	0.637	B
66	10000 10X	0	0.997	G	00001 X01	0	0.999	G
67	10000 11X	0	0.648	B	00001 X11	0	0.665	B
68	10001 00X	0	0.982	G	10001 X00	0	0.986	G
69	10001 01X	1	0.757	B	10001 X10	1	0.689	B
70	10001 10X	0	0.988	G	10001 X01	0	0.991	G
71	10001 11X	1	0.614	B	10001 X11	1	0.607	B
72	10010 00X	0	0.902	G	01001 X00	0	0.890	B
73	10010 01X	1	0.625	B	01001 X10	1	0.682	B
74	10010 10X	0	0.717	B	01001 X01	0	0.843	B
75	10010 11X	1	0.880	B	01001 X11	1	0.828	B
76	10011 00X	0	0.924	G	11001 X00	0	0.937	G
77	10011 01X	1	0.748	B	11001 X10	1	0.767	B
78	10011 10X	0	0.906	G	11001 X01	0	0.927	G
79	10011 11X	1	0.826	B	11001 X11	1	0.789	B
80	10100 00X	0	0.879	B	00101 X00	0	0.875	B
81	10100 01X	1	0.660	B	00101 X10	1	0.719	B
82	10100 10X	0	0.531	B	00101 X01	1	0.595	B
83	10100 11X	1	0.882	B	00101 X11	1	0.930	G
84	10101 00X	0	0.897	B	10101 X00	0	0.768	B
85	10101 01X	1	0.739	B	10101 X10	1	0.700	B
86	10101 10X	1	0.623	B	10101 X01	1	0.611	B

Fig. 5.7.4c States 58–86.

No.	State Configuration	FORWARD DIRECTION Prediction	Probability of Correct Prediction	Goodness (G or B)	REVERSE DIRECTION State Configuration	Prediction	Probability of Correct Prediction	Goodness (G or B)
29	00111 01X	1	0.996	G	11100 X10	1	0.998	G
30	00111 10X	1	0.524	B	11100 X01	1	0.529	B
31	00111 11X	1	0.985	G	11100 X11	1	0.987	G
32	01000 00X	0	0.971	G	00010 X00	0	0.949	G
33	01000 01X	1	0.522	B	00010 X10	0	0.507	B
34	01000 10X	0	0.906	G	00010 X01	1	0.860	B
35	01000 11X	1	0.636	B	00010 X11	1	0.646	B
36	01001 00X	0	0.934	G	10010 X00	1	0.899	B
37	01001 01X	1	0.503	B	10010 X10	1	0.753	B
38	01001 10X	0	0.556	B	10010 X01	0	0.800	B
39	01001 11X	1	0.751	B	10010 X11	1	0.692	B
40	01010 00X	0	0.854	B	01010 X00	0	0.885	B
41	01010 01X	1	0.583	B	01010 X10	1	0.635	B
42	01010 10X	0	0.684	B	01010 X01	1	0.529	B
43	01010 11X	1	0.813	B	01010 X11	1	0.878	B
44	01011 00X	0	0.910	G	11010 X00	0	0.864	B
45	01011 01X	0	0.516	B	11010 X10	1	0.730	B
46	01011 10X	0	0.645	B	11010 X01	1	0.526	B
47	01011 11X	1	0.837	B	11010 X11	1	0.796	B
48	01100 00X	0	0.661	B	00110 X00	0	0.533	B
49	01100 01X	1	0.973	G	00110 X10	1	0.983	G
50	01100 10X	0	0.895	B	00110 X01	0	0.692	B
51	01100 11X	1	0.759	B	00110 X11	1	0.838	B
52	01101 00X	1	0.783	B	10110 X00	0	0.697	B
53	01101 01X	1	0.868	B	10110 X10	1	0.941	G
54	01101 10X	1	0.765	B	10110 X01	0	0.596	B
55	01101 11X	1	0.729	B	10110 X11	1	0.827	B
56	01110 00X	1	0.726	B	01110 X00	1	0.739	B
57	01110 01X	1	0.997	G	01110 X10	1	0.996	G

Fig. 5.7.4d States 87–115.

No.	State Configuration	FORWARD DIRECTION Prediction	Probability of Correct Prediction	Goodness (G or B)	REVERSE DIRECTION State Configuration	Prediction	Probability of Correct Prediction	Goodness (G or B)
116	11101 00X	0	0.951	G	10111 X00	0	0.900	G
117	11101 01X	1	0.885	B	10111 X10	1	0.910	G
118	11101 10X	0	0.920	G	10111 X01	0	0.908	G
119	11101 11X	1	0.748	B	10111 X11	1	0.753	B
120	11110 00X	0	0.756	B	01111 X00	0	0.724	B
121	11110 01X	1	0.975	G	01111 X10	1	0.972	G
122	11110 10X	0	0.849	B	01111 X01	0	0.813	B
123	11110 11X	1	0.929	G	01111 X11	1	0.926	G
124	11111 00X	0	0.830	B	11111 X00	0	0.835	B
125	11111 01X	0	0.957	G	11111 X10	0	0.953	G
126	11111 10X	0	0.848	B	11111 X01	0	0.847	B
127	11111 11X	1	0.989	G	11111 X11	1	0.989	G

Fig. 5.7.4e States 116–127.

PEL ROW	ℓ	$\ell+1$	$\ell+2$	$\ell+3$	
	8	136	40	168	LINE m
	200	72	232	104	LINE m + 1
	56	184	24	152	LINE m + 2
	248	120	216	88	LINE m + 3

Fig. 5.7.5 A 4×4 dither matrix for an image with intensity in the range 0–255 (8 bits). See Chapter 1.

where $P(\cdot|\cdot)$ is the experimentally determined conditional probability. In general, $P(\cdot|\cdot)$ is determined by computing histograms over a given set of documents. Using the 8 CCITT standard documents, the relevant statistics and a maximum likelihood predictor are given for each of the 128 states in Fig. 5.7.4. One obvious property of such a prediction scheme is that the probability of correct prediction is always higher than 0.5.

In the case of two-level pictures that are dithered to give the appearance of gray-level as described in Chapter 2, X can be predicted by including the value of the dither matrix in the definition of state. As an example, if we use a 4×4 dither matrix as shown in Fig. 5.7.5 there are 16 values of the dither matrix $\{D_{ij}\}$, $i, j = 1, \ldots, 4$. Using four surrounding elements A, D, E, and F as in Fig. 5.7.3 as well as the dither matrix value D_X, the state S can be defined by the five-tuple $S = (D_X, A, D, E, F)$. There are 256 possible states $\{S_i\}$. For a given pel, the maximum likelihood prediction is developed again as the most probable one given that a particular state has occurred. An example of such a predictor is given in Fig. 5.7.6, which shows only the first 54 states. The predictor, optimized for a particular picture, does show variation from picture-to-picture as a result of the nonstationarity of the image statistics.

For either two-level or dithered images, the next step in predictive coding is to compute the prediction error, which is obtained by a simple exclusive-or of the input sample and the prediction. The prediction error is then run-length coded, generally in the same direction as the scanning. Here again, variable length coding with two sets of code tables (one for 0's, and the other for 1's) is used.

Statistics of run-lengths of prediction errors conditioned on a particular state vary significantly from state to state. It is therefore more desirable to encode the conditional run-lengths, using a code table that is tailored to the statistics of that state. This is akin to state space coding as described in Chapter 3. Fig. 5.7.7 shows the conditional run-lengths for three states S_0, S_1 and S_2 for a scan line of hypothetical data. Here, instead of run-lengths of prediction error along the scan line, prediction errors of consecutive pels belonging to a particular state (or a group of states) define run-lengths. Complexity can be reduced to some extent by using only two groups of states and assigning a code table for each of these groups as explained below.

Another method of using the variation of conditional run-length statistics is *reordering*. In the reordering technique (see Chapter 3) the pels in a line are rearranged according to their state so as to improve the performance of the run-length coder. This may be done with only one code table or many code tables as in conditional run-length coding of the previous paragraph. Fig. 5.7.4 shows that the probability of making a correct prediction varies a great deal from state-to-state. For example, state 0 has a probability of correct prediction of 0.999, whereas, for state 45 it is 0.57. Thus, some states can be classified as *good* with respect to their ability to predict the next pel, and the others can be classified as

STATE NUMBER	STATE DEFINITION: Dither Matrix	F	E	D	A	PICTURE: KAREN Total Elements in State	Prediction	Probability of Prediction Error	PICTURE: ENGINEERING DRAWING Total Elements in State	Prediction	Probability of Prediction Error	PICTURE: HOUSE Total Elements in State	Prediction	Probability of Prediction Error
1	8	0	0	0	0	4026	1	0.018	3197	1	0.041	6225	1	0.000
2	8	0	0	0	1	12	1	0.000	197	1	0.000	0	0	0.000
3	8	0	0	1	0	77	1	0.000	760	1	0.011	32	1	0.000
4	8	0	0	1	1	2	1	0.000	97	1	0.000	0	0	0.000
5	8	0	1	0	0	0	0	0.000	0	0	0.000	0	0	0.000
6	8	0	1	0	1	0	0	0.000	0	0	0.000	0	0	0.000
7	8	0	1	1	0	0	0	0.000	0	0	0.000	0	0	0.000
8	8	0	1	1	1	0	0	0.000	0	0	0.000	0	0	0.000
9	8	1	0	0	0	2590	1	0.000	1489	1	0.003	5665	1	0.000
10	8	1	0	0	1	43	1	0.000	429	1	0.000	0	0	0.000
11	8	1	0	1	0	6121	1	0.000	2342	1	0.002	4204	1	0.000
12	8	1	0	1	1	3258	1	0.000	7618	1	0.000	3	1	0.000
13	8	1	1	0	0	0	0	0.000	0	0	0.000	0	0	0.000
14	8	1	1	0	1	0	0	0.000	0	0	0.000	0	0	0.000
15	8	1	1	1	0	0	0	0.000	0	0	0.000	0	0	0.000
16	8	1	1	1	1	0	0	0.000	0	0	0.000	0	0	0.000
17	24	0	0	0	0	2938	1	0.279	3259	1	0.417	4217	1	0.047
18	24	0	0	0	1	4	1	0.000	108	1	0.000	0	0	0.000
19	24	0	0	1	0	61	1	0.000	785	1	0.135	10	1	0.000
20	24	0	0	1	1	0	0	0.000	32	1	0.000	0	0	0.000
21	24	0	1	0	0	0	0	0.000	2	0	0.000	0	0	0.000
22	24	0	1	0	1	0	0	0.000	0	0	0.000	0	0	0.000
23	24	0	1	1	0	0	0	0.000	0	0	0.000	0	0	0.000
24	24	0	1	1	1	1	1	0.000	0	0	0.000	0	0	0.000
25	24	1	0	0	0	2302	1	0.003	1335	1	0.061	5193	1	0.000
26	24	1	0	0	1	40	1	0.025	267	1	0.011	0	0	0.000
27	24	1	0	1	0	0185	1	0.006	5154	1	0.009	6964	1	0.000

Fig. 5.7.6 Definition of state based on four surrounding pels and the value of the dither matrix. State-dependent prediction and its quality is given for three dithered pictures. Only the first 54 states are shown. (a) States 1–27.

STATE NUMBER	STATE DEFINITION					PICTURE: KAREN			PICTURE: ENGINEERING DRAWING			PICTURE: HOUSE		
	Dither Matrix	F	E	D	A	Total Elements in State	Prediction	Probability of Prediction Error	Total Elements in State	Prediction	Probability of Prediction Error	Total Elements in State	Prediction	Probability of Prediction Error
28	24	1	0	1	1	1181	1	0.007	5442	1	0.001	0	0	0.000
29	24	1	1	0	0	0	0	0.000	0	0	0.000	0	0	0.000
30	24	1	1	0	1	0	0	0.000	0	0	0.000	0	0	0.000
31	24	1	1	1	0	3	1	0.000	0	0	0.000	0	0	0.000
32	24	1	1	1	1	39	1	0.000	0	0	0.000	0	0	0.000
33	40	0	0	0	0	4038	1	0.403	3096	0	0.193	6317	1	0.113
34	40	0	0	0	1	24	1	0.000	396	1	0.063	9	1	0.000
35	40	0	0	1	0	2502	1	0.002	1327	1	0.251	5641	1	0.000
36	40	0	0	1	1	70	1	0.000	587	1	0.027	24	1	0.000
37	40	0	1	0	0	0	0	0.000	0	0	0.000	0	0	0.000
38	40	0	1	0	1	0	0	0.000	1	1	0.000	0	0	0.000
39	40	0	1	1	0	0	0	0.000	0	0	0.000	0	0	0.000
40	40	0	1	1	1	2	1	0.000	0	0	0.000	0	0	0.000
41	40	1	0	0	0	33	1	0.091	333	1	0.339	42	1	0.000
42	40	1	0	0	1	47	1	0.021	513	1	0.226	16	1	0.000
43	40	1	0	1	0	1996	1	0.001	1187	1	0.127	2319	1	0.000
44	40	1	0	1	1	7090	1	0.000	8774	1	0.002	1888	1	0.000
45	40	1	1	0	0	0	0	0.000	0	0	0.000	0	0	0.000
46	40	1	1	0	1	0	0	0.000	0	0	0.000	0	0	0.000
47	40	1	1	1	0	15	1	0.000	1	1	0.000	0	0	0.000
48	40	1	1	1	1	379	1	0.000	41	1	0.000	0	0	0.000
49	56	0	0	0	0	2943	0	0.271	3024	0	0.082	4136	0	0.487
50	56	0	0	0	1	9	1	0.000	229	1	0.066	0	0	0.000
51	56	0	0	1	0	2274	1	0.047	1349	1	0.375	5146	1	0.000
52	56	0	0	1	1	6	1	0.000	265	1	0.049	0	0	0.000
53	56	0	1	0	0	0	0	0.000	0	0	0.000	0	0	0.000
54	56	0	1	0	1	0	0	0.000	0	0	0.000	0	0	0.000

Fig. 5.7.6b States 28–54.

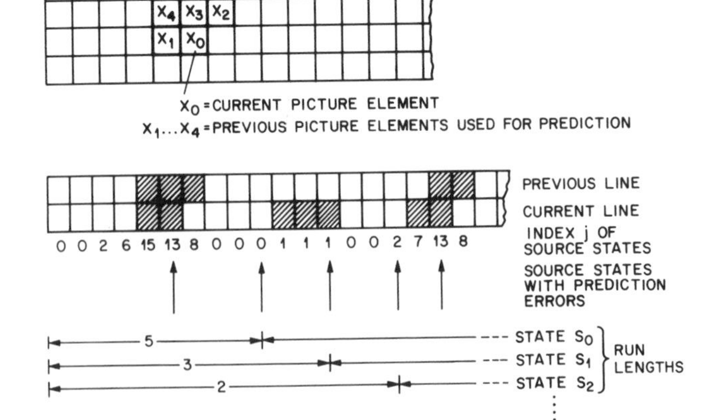

Fig. 5.7.7 Example of conditional run-length coding for three states S_0, S_1 and S_2. A conditional run is defined as a run of consecutive pels belonging to the same state. Conditional runs of length 5, 3 and 2 are shown for state S_0, S_1 and S_2.

bad. If the prediction errors for all of the good states are grouped together, then the run-lengths corresponding to prediction errors of these good states will be relatively long, and consequently a higher compression ratio can be achieved. To illustrate such a good/bad reordering process, consider a memory of 1728 cells, numbered from 1 to 1728 (equal to the number of pels per line). We classify good states as those for which the probability of correct prediction conditioned on that state is greater than or equal to a given threshold (called goodness threshold); all the other states are bad. In the process of reordering, if the first pel of the present line has a state that is classified as good, we put the prediction error corresponding to it in memory cell 1. If, on the other hand, the state is classified as bad, we put the prediction error corresponding to it in memory cell 1728. We continue in this manner: the prediction for the ith element of the present line is put in the unfilled memory cell with the smallest or the largest index, depending upon whether the state corresponding to the ith element is good or bad, respectively. When the memory is filled, the cells are read in numerical order, and the contents are run-length coded as usual. At the receiver the ordered data can be unscrambled uniquely (in the absence of transmission error), since the ordering sequence can be derived by the receiver.

Fig. 5.7.4 shows the classification of states into good and bad using a goodness threshold of 0.9. Fig. 5.7.8 shows one of the CCITT images, the corresponding prediction errors, and the good/bad reordered prediction errors, (a pel with prediction error is reproduced as black, and a pel without prediction error is reproduced as white). It is seen that the average run-length of the reordered data is much higher than that of the unordered data, resulting in better compression. Schemes have been developed in which only one code table is used for coding the run-length of the ordered data. Higher coding efficiency may be obtained by using separate code tables for runs in good and bad regions. The ordering process for dithered two-level images is identical to the one described above, with the proper definition of state, which includes the value of the dither matrix.

The performance of various predictive coding schemes is given in Fig. 5.7.9 for the eight CCITT standard images, where only entropies in bits/pel are given. The predictor used for these calculations consisted of four pels A, D, E, and F of Fig. 5.7.3 and the prediction rule was optimized for each picture. One dimensional run-length coding has a compression ratio between 5.7 and 22.4, whereas good/bad reordering the prediction errors, increases this ratio to 7.1 and 37.5. Using separate code tables for the good and bad region, the compression ratio can be increased by about 4–8 percent. Improvements of similar nature are obtained for dithered pictures.

5.7.1c Line-to-Line Run-Length Difference Coding

This scheme takes into account correlation between run-lengths in successive lines. Instead of transmitting run-lengths, differences between

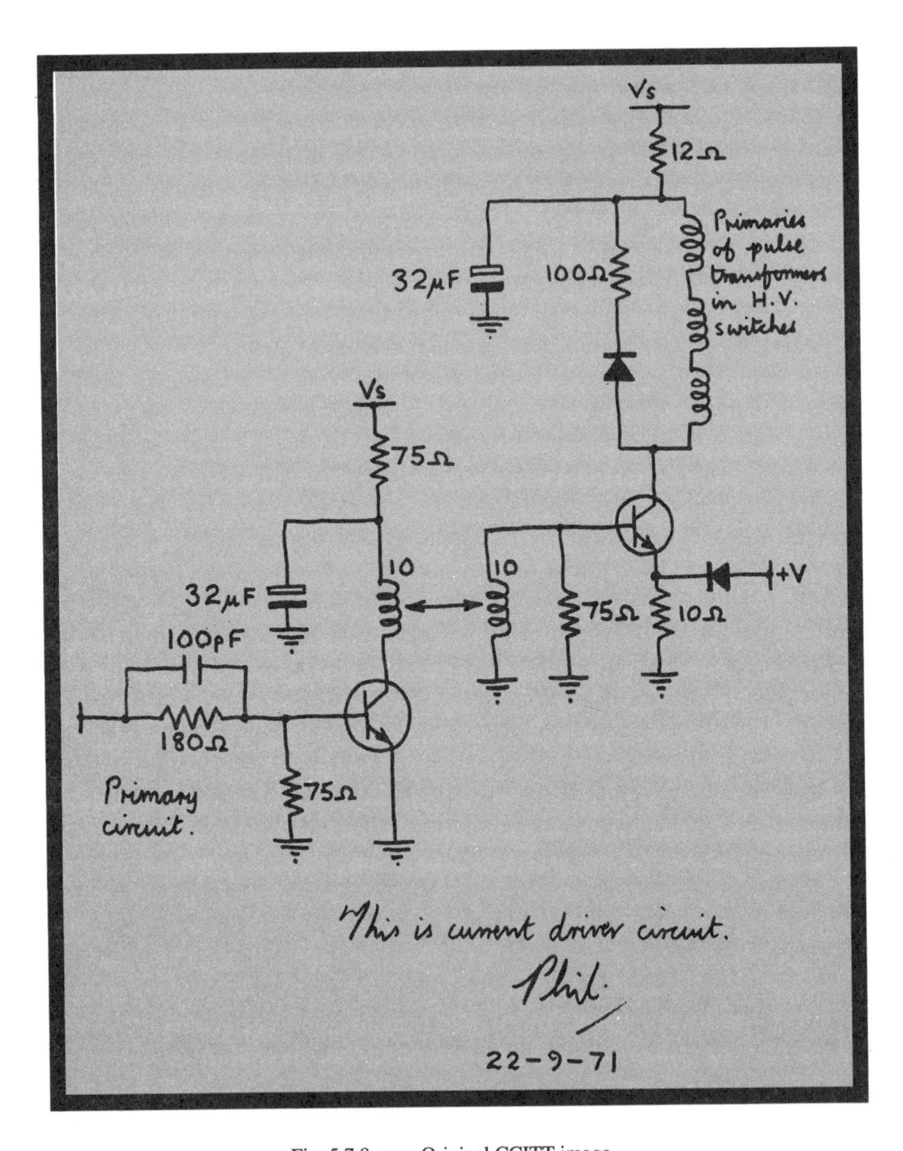

Fig. 5.7.8a Original CCITT image.

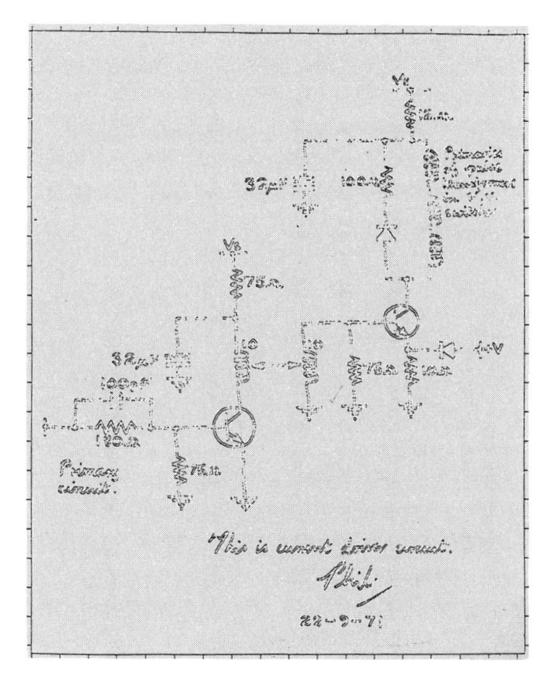

Fig. 5.7.8b Image of the prediction errors using the predictor of Fig. 5.7.4.

Fig. 5.7.8c The image of the reordered prediction errors. It is clear that the average run lengths here are larger than those in Fig. 5.7.8b.

CODING ALGORITHM	ENTROPY (BITS/PEL) CCITT IMAGE NUMBER							
	1	2	3	4	5	6	7	8
ONE-DIMENSIONAL RUN-LENGTH CODING	0.0505	0.0447	0.0914	0.1652	0.0988	0.0679	0.1791	0.087
RUN-LENGTH CODING OF PREDICTION ERRORS	0.0466	0.0373	0.0693	0.1640	0.0795	0.0482	0.1678	0.0671
RUN-LENGTH CODING OF ORDERED PREDICTION ERRORS; ONE SET OF CODES, GOODNESS THRESHOLD = 0.9	0.0390	0.0267	0.0571	0.1396	0.0652	0.0366	0.1400	0.0460
RUN-LENGTH CODING OF ORDERED PREDICTION ERRORS; TWO SETS OF CODES, GOODNESS THRESHOLD = 0.9;	0.0351	0.0233	0.0527	0.1282	0.0596	0 0355	0.1274	0.0419

Fig. 5.7.9 Comparative performance of Reordering Schemes for the eight CCITT Standard Images.

corresponding run-lengths of successive scan lines are transmitted. Fig. 5.7.10 shows the quantities Δ' and Δ'', and other messages such as new start, merge, etc., that are transmitted. Thus for each scan line, a transition element where the color changes from black to white (or white to black) is located and its location is transmitted with respect to a similar transition in the previous line. Several variations exist. For example, Fig. 5.7.10 shows transmission of white to black transitions (Δ') and black run-lengths (Δ''). Another variation of this principle is described in Chapter 6 as the CCITT standard two dimensional code. These schemes are essentially boundary coding schemes. The quantities Δ' and Δ'' indicate how the boundary propagates from one line to the next. We have seen in Chapter 3 that statistics of Δ' and Δ'' are highly nonuniform and, therefore, for higher efficiency, variable length codewords are used for representing Δ', Δ'', new start, and merge messages.

Schemes based on line-to-line run-length difference coding are quite efficient. Compression ratios of the same order as the best predictive scheme have been obtained. More details on the compression efficiency of a particular variation of this scheme (which has been chosen as a CCITT standard) are given in Chapter 6.

5.7.1d Block Coding

In block coding, a picture is divided into small rectangular blocks of $m \times n$ pels. Since each pel in this set of nm pels of the block can take on two values, there are 2^{nm} possible patterns of this block. However, not all these patterns are equally likely. Therefore, a variable length code is used to represent each of the patterns of the blocks. Optimum variable length codes based on Huffman construction become impractical for block sizes much larger than 3×3 (code table of size $2^9 = 512$). Therefore, much of the block coding literature is concerned with the assignment of suboptimum codes. One of these is indicated in Fig. 5.7.11. The code words for the most likely pattern (all zero or all white) is simply "0". The code words for the other patterns are obtained by the nm bits of the block preceded by the prefix "1". Thus, the code is a variable length code with only two possible lengths, namely 1 and $nm+1$. The compression ratio depends upon the size of the block and how many blocks consist of the all zero pattern. In general, block coding is less efficient than either predictive coding or line-to-line run-length differential coding.

5.7.1e Arithmetic Coding

We saw in Chapter 3 the basic principles of using arithmetic coding for bilevel images. First, we assumed the existence of a fairly good source model that could accurately estimate the conditional symbol probabilities given the recent past symbol values. Next, this source model was used to code the binary source symbols into a single real number x in the range 0 to 1. Fig. 3.1.10 showed an example of such coding for an initial four binary symbols.

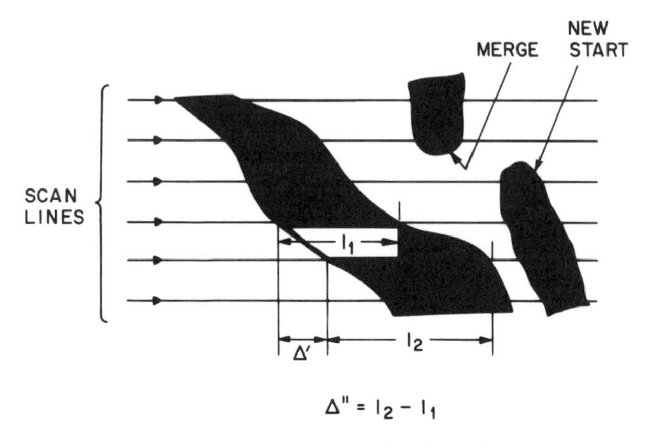

Fig. 5.7.10 Definition of transmitted quantities (e.g. Δ', Δ'') for line-to-line run-length difference coding.

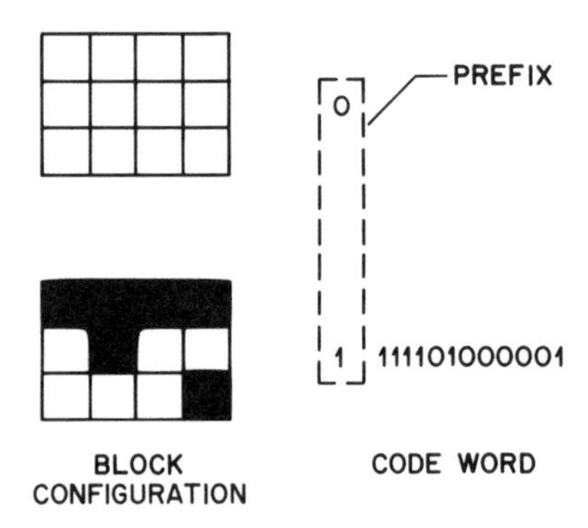

Fig. 5.7.11 Illustration of a suboptimum Block Code for a 4×3 block. The code for an all zero block is 0, since it occurs most often. There are 2^{12} codewords, only two of which are shown.

The portion of the unit interval on which x is known to lie after coding an initial sequence of symbols is known as the current coding interval. For each new binary input symbol the current coding interval is divided into two subintervals with sizes proportional to estimates of the conditional probabilities of symbol-value occurrences. The new current coding interval then becomes that portion of the old coding interval associated with the new symbol value actually occurring. Conceptually, this process continues until there are no more input symbols, after which a binary fraction representing some x within the last coding interval is transmitted.

Practically, of course, this process cannot be easily implemented as described, especially for long symbol strings. The accuracy required for the arithmetic simply become too large. However, certain approximations can be made to render the algorithm practical without seriously degrading performance. These will now be described.

By convention, the top subinterval of the partitioned current coding interval represents the Least Probable Symbol (LPS) probability, and the bottom subinterval the Most Probable Symbol (MPS) probability as shown in Fig. 5.7.12a. The size of the current coding interval is A. If p is the estimated probability of the LPS ($0<p<0.5$), then the size of the LPS subinterval is

$$LSZ = p \times A.$$

The current estimated value of x is C, which we take as the minimum value or *base* of the current coding interval. Before coding any symbols we set $C=0$, $A=1$.

Whenever an LPS is coded, the MPS subinterval $A-LSZ$ is added to C, and A is set to LSZ. Whenever an MPS is coded, the LPS subinterval LSZ is subtracted from A, and C is left unchanged. Thus, as coding proceeds, A always decreases while C either increases or stays the same.

In real implementations, of course, we can only store A and C values in registers having a finite number of bits. For A the problem is relatively easy. If after coding, A falls below 0.5 we successively *renormalize* by doubling both A and C until $A \geq 0.5$. Note that this may produce C values larger than 1. This is similar to floating point arithmetic.

The problem of finite accuracy storage of C values is not as easy to solve. What can be exploited, however, is the fact that as A decreases in value, there eventually becomes a time when some of the most significant bits of C stabilize to their final value and never change again. They can then be transmitted and no longer need to be stored within the encoder.

Thus, we partition the bit representation of C into four regions, as shown in Fig. 5.7.12b. The leftmost significant bits region is the Final (F) region for bits that have stabilized to their final value and are ready for transmission. The next region is the Tentative (T) region for bits that could still change, but will only do

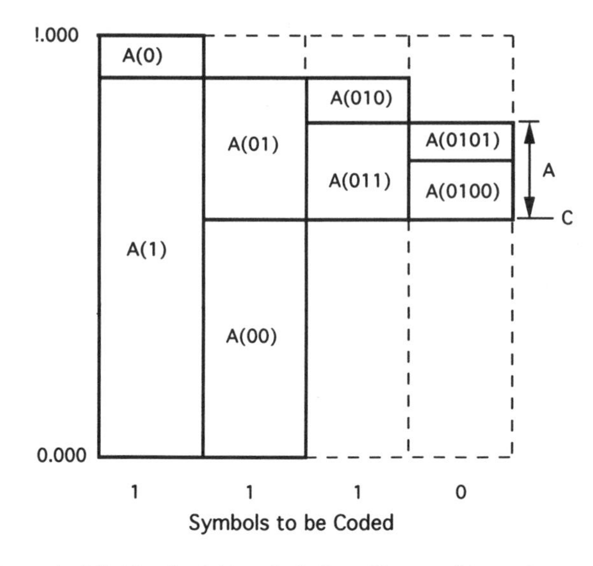

Fig. 5.7.12a Interval subdivision for Arithmetic Coding. The top subinterval represents the Least Probable Symbol (LPS), and the bottom subinterval represents the Most Probable Symbol (MPS).

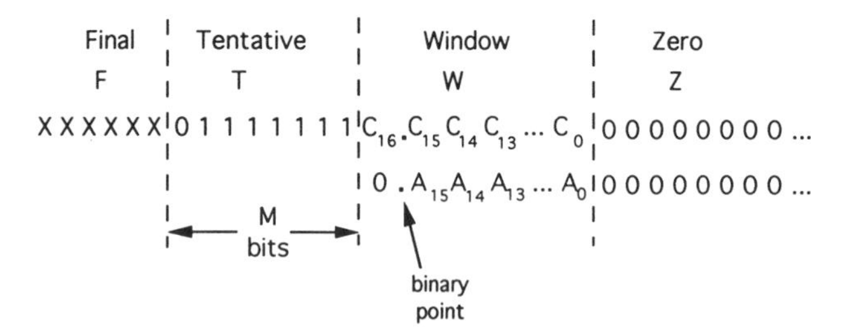

Fig. 5.7.12b Finite accuracy implementation of Arithmetic Coding. The current estimate C is partitioned into four regions - F, T, W and Z. Only the W region bits need be stored in the encoder.

so by a carry from the right. Next come the Window (W) region for bits that will change due to arithmetic operations, and finally the Zero (Z) region for bits that are always zero. The (floating) binary point is one bit from the left edge of the W region, as shown in the figure.

The W region is of fixed size, and for purposes of illustration we will assume 17 bits. The size of W also determines the minimum allowed value for *LSZ*, in our example 2^{-16}. Note that before coding any but the first symbol, $A_{16}=0$ and $A_{15}=1$.

A crucial property in this scenario is that before coding, the T region contains M bits, comprised of a zero on the left, followed by $M-1$ ones. Coding the next symbol then takes place as described above, with A decreasing and C either increasing (LPS) or staying the same (MPS). If coding causes a carry from W to T, then the T region changes to a one followed by $M-1$ zeros, and $C_{16}=0$. If this occurs, we then shift all the bits from the T region to the F region, and set $M=0$. Note this can only happen if an LPS is coded.

Following these coding operations, we then examine A_{15} to see if A has fallen below 0.5. If $A_{15}=0$ we renormalize. Note that coding an LPS always produces $A_{15}=0$.

Renormalization consists of two steps. First we examine C_{16}. If $C_{16}=0$ we shift all the bits, if any, from the T region to the F region, and set $M=0$. In the second step of renormalization, we shift the W region (and the floating binary point) one bit to the right and increment M by 1. The old C_{16} is thus shifted into the T region.

We continue renormalizing until $A_{15}=1$.

In the arrangement described above, the F region bits need to be stored within the encoder only until they are transmitted. Often one byte of storage is sufficient. All bits of the the W region need to be stored for the arithmetic operations. None of the bits for the Z region need to be stored.

The T region is a bit more complicated. Although none of the actual bits need to be stored, the value of M must be maintained. If a shift from T to F occurs due to coding, then a one followed by $M-1$ zeros must be shifted. If a shift occurs due to renormalization, then a zero followed by $M-1$ ones must be shifted.

After the last symbol is coded, all bits from the T and W regions can be shifted to the F region and transmitted.

Decoding is fairly straightforward. As soon as enough C bits arrive to indicate LSP or MPS, decode the symbol, recompute A and then renormalize if necessary. During renormalization, C bits to the left of the binary point can be discarded. Chapter 7 describes decoder implementation in more detail.

The compression performance of Arithmetic Coding is highly dependent on how good the probability estimates are. A probability model may exist from the start, or it may be developed on the fly as coding proceeds. In the latter case, typically, a running count is kept of symbol values conditioned on a predetermined number of previous symbols. Much more will be said about Arithmetic Coding and probability modelling in Chapter 7.

5.7.2 Approximate Coding

The encoding techniques of the previous section allow perfect reproduction of the digitized picture at the receiver (in the absence of transmission error). Of course, the irreversible transformation of the image intensity that takes place due to sampling and thresholding makes it impossible to reproduce the analog intensity exactly. However, once thresholded, the two-level image can be reproduced at the receiver with no change if there are no transmission errors. In this section we look at methods by which irreversible or information-lossy processing of two-level pictures can be made to improve the compression ratio without significant degradation in the picture quality.

In general, the irreversible processing is performed before the final coding stage. Since most of the coding schemes of the previous section depend upon the ability to predict the color of a pel or the progression of a contour from line-to-line, irreversible processing techniques generally attempt to reduce the prediction error by preserving the continuity of the contours from line-to-line. We discuss two techniques below: One that is useful with predictive coding and a second that is useful with block coding.

The color of a picture element can always be changed so that the number of pels having nonzero prediction error are reduced. However, doing so can degrade the picture quality. Fig. 5.7.13a shows certain patterns in which the pel marked X may be changed in color without serious degradation in the overall picture quality. In pattern 1, pel X is regarded as an isolated pel due to noise whereas in patterns 2,3 and 4, pel X is located on a slant edge, and a change of its color results in rectangular edges. Fig. 5.7.13b shows the effect of such processing on two-dimensional prediction. After applying the two-dimensional prediction, the number of 1's in the prediction error signal is reduced. However, the picture quality in such a case obviously depends upon the patterns used for changing the color of pel X. No systematic investigation has yet been made to relate picture quality to patterns used in the approximation process and the resulting savings in bit-rate.

Picture approximation can also be used with block coding. The compression efficiency of a block code of the type given in the previous section depends to a large extent on the probability of occurrence of the all zero block, since the length of the codeword assigned to it is small. The compression efficiency can therefore be improved by increasing the probability of occurrence of the all zero block. For example, this may be done by changing all blocks having not more than K 1's to the all zero block. The amount of distortion introduced by such an approximation obviously depends upon the value of K used. Fig. 5.7.14 shows portions of two original digitized pictures. Approximated pictures using 4×4 blocks with $K = 1$ and 2 are also shown in the

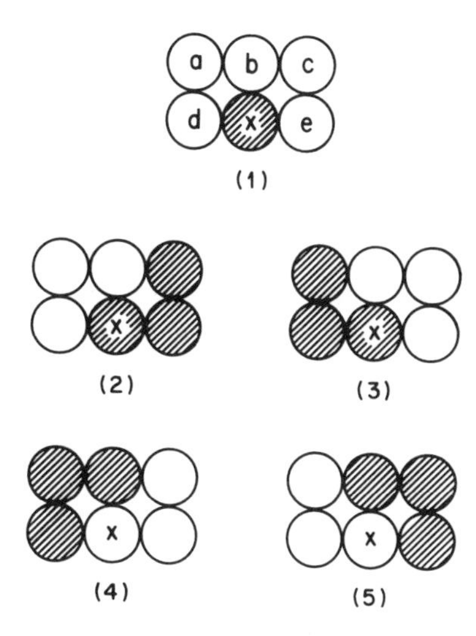

Fig. 5.7.13a Schemes for Modification of the pel *X* depending on the surrounding pattern.

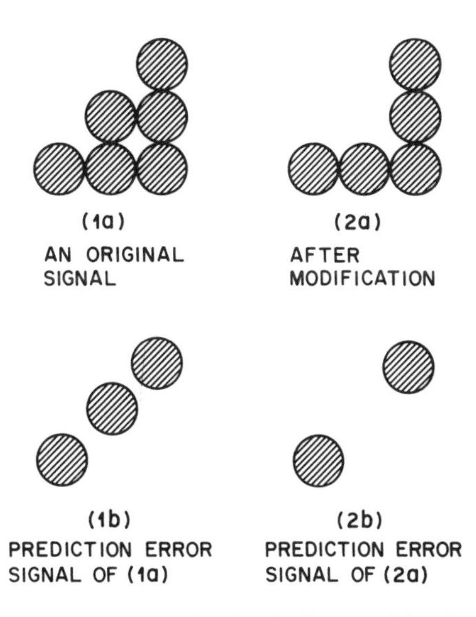

Fig. 5.7.13b An example showing the effectiveness of signal modification.

same figure. In general, the distortion introduced for $K=1$ is not noticeable, but the compression ratio is improved only by about 8 percent. Distortions are noticeable for $K=2$, and the compression ratio is improved by about 16 percent.

Another approximate block coding scheme is called *pattern matching*. This scheme is quite complex and still in its infancy. It is based on the observation that in documents containing text, several patterns (e.g., characters) occur repeatedly and, therefore, compression efficiency can be improved by identifying these patterns and transmitting to the receiver only their identification codes. First, a blocking operator is used to isolate patterns. Then, the first symbol encountered is placed in a library and its binary pattern is transmitted in a coded form. As each new pattern is detected, it is compared with entries of the library. If the comparison is within an error tolerance, the library identification code is transmitted along with the pattern location coordinates. Otherwise, the new pattern is placed in the library and its binary representation is transmitted in a coded form. Non-isolated patterns are left behind as residue, and are coded by conventional coding techniques. In one specific implementation based on this principle, a number of scan lines equal to about two to four times the average character height are stored in a scrolled buffer. This data is then examined line-by-line to determine if a black pel exists. If the entire line contains no black pels, this information is encoded by an end-of-line code. If a black pel exists, then a blocking operator is employed to isolate the pattern. For those isolated patterns, further processing is required to determine if a replica of the pattern under consideration already exists in the library. This process involves the extraction of a set of features, a screening operation to reject unpromising candidates, and finally a series of template matches. The first blocked pattern and its feature vector are put into the prototype library, and as each new blocked pattern is encountered, it is compared with each entry of the library that passes the screening test. If the comparison is successful, the library identification code is transmitted along with the location of the coordinates of the pattern. If the comparison is unsuccessful, the new pattern is both transmitted and placed in the library. Those areas of a document in which the blocker cannot isolate a valid pattern are assigned to a residue, and a conventional coding may be employed. The performance of the combined pattern matching scheme depends upon the matching algorithm and the tolerances used. In general, significant improvement is obtained for documents that are predominantly text (see Table 5.7.1). For other documents (e.g., containing drawings), the performance is about the same as predictive or line-to-line run-length difference coding. On the average, however, for the eight CCITT documents, the compression ratio of pattern matching[5.7.2] increases by almost a factor of two compared to the predictive or line-to-line run length difference coding.

Fig. 5.7.14 Approximate block coding. (a) original digitized image segments. Reconstructed pictures after block coding with $K = 1$ (b) and $K = 2$ (c).

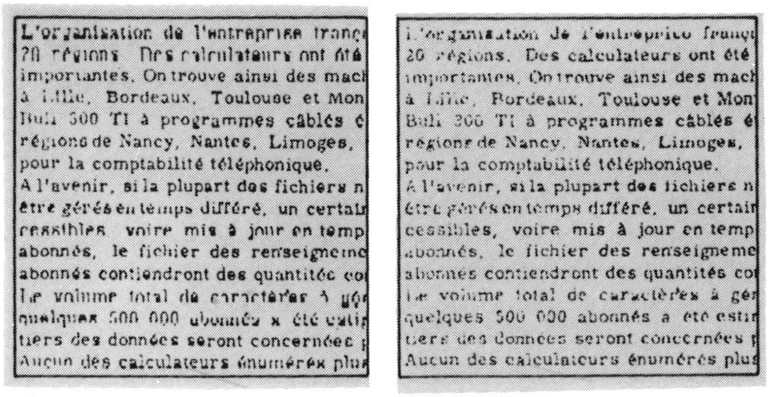

Fig. 5.7.15 Horizontal streaks caused by transmission errors in a picture coded with run-length codes. (a) original, (b) error image.

5.7.3 Transmission Errors

One penalty for efficient source coding is an increased sensitivity to errors in the transmission channel. This increased sensitivity is partially compensated for by the fact that after compression we have fewer bits per image and hence fewer chances for noise corruption. For example, at a resolution of about 8 pels/mm a standard CCITT image contains approximately 4 million bits. If we transmit such an image pel by pel through a channel with a bit error probability of 1 in 10^{-6}, we will get on the average 4 errors per image (assuming random errors rather than bursts). If an efficient coding scheme is used with a compression factor of 10, then we have 4 errors in every 10 images. The effect of the channel errors on the reproduced picture depends upon the encoding scheme and the technique used for combatting the channel errors.

In one-dimensional run-length coding algorithms a channel error will change a codeword so that the length of the corresponding run will be changed at the receiver. With variable length comma free codes, more information is lost because the decoder needs a certain interval to fall back into the correct codeword segmentation (i.e., regaining the synchronization). This causes a shift of the subsequent picture information. Fig. 5.7.15 shows an image that was coded by a one-dimensional run-length code and transmitted on a channel with average bit error rate of about 6×10^{-5}. The effect of a transmission error in this case is to cause horizontal streaks in the picture. These are generally limited to one scan line by introducing a synchronizing word at the end of each line. With two-dimensional codes, unless care is taken a single channel error can cause significant degradation in the rest of the picture.

In general, three methods have been considered to reduce the effects of transmission errors: a) Restriction of the damage caused by errors to as small an area as possible, b) Detection of errors and retransmission of corrupted blocks using an *Automatic Retransmission Request* (ARQ) System and c) Detection of errors and their correction at the receiver using a *Forward Acting Error Correcting* (FEC) code.

ARQ and FEC methods have not yet found wide acceptance since they are complex, add extra cost to the equipment and increase the transmission time. ARQ systems have the advantage of being very reliable and insensitive to changes in the channel error rate. Systems can be designed such that the probability of an undetected error in a block is less than 1 in 10^8. However, ARQ leads to reduction in the effective transmission rate. Also for channels that have inherently long transmission delay (e.g. satellite circuits) retransmission can become impractical. Forward error correcting codes, on the other hand, can have a higher effective transmission rate, but must be designed for the errors experienced on practical channels. Thus, they can be sensitive to changes in the error patterns.

In the case of one-dimensional run-length coding, several techniques exist to improve the quality of reconstructed pictures in the presence of transmission errors. The detection of a channel error can be easily performed by summing up the decoded run-lengths between two synchronizing words and comparing the result with the nominal length of a scan line. If an error is detected, one of the following error concealment techniques can be used to improve the reconstructed picture quality: a) replace the damaged line by an all white line, b) repeat the previous line, c) print the damaged line, d) use a line-to-line processing or correlation technique to reconstruct as much of the line as possible. Picture quality improvement can be rather high for (b). Correlation methods (d) take advantage of the fact that the error recovery period is often short, and thus, they attempt to retain as much as possible of the correctly decoded data on a damaged line. This is achieved by attempting to locate the damaged run-lengths. One method is to measure the correlation between group of pels on the damaged line with corresponding groups on the adjacent lines above and below. Where the correlation is good, generally at the beginning and end of the damaged scan line, the scan line data is used to reconstruct the line. The part of the scan line that is assumed to be damaged is then replaced by a corresponding part of the previous line. This method works well most of the time, but does result in a loss of vertical resolution.

In the case of two-dimensional coding, in order to prevent the vertical propagation of damage caused by transmission errors, no more than $K-1$ successive lines are two-dimensionally coded. The next line is one-dimensionally coded. Usually K is set equal to two at 4 pels/mm and set equal to four at 8 pels/mm resolution. Any transmission error thus will be localized to a group of K lines in the picture.

5.7.4 Soft Facsimile

Cathode ray tubes (CRT) and other erasable display media have proliferated in recent years. Also, the cost of refresh frame buffers has decreased dramatically in the past few years. This has given impetus to new techniques for transmission of still pictures. Traditionally, transmission of still pictures has used the regular scanning pattern, i.e., picture elements are transmitted sequentially along the scanning direction (from left to right and top to bottom). This scanning pattern cannot be easily changed when paper output is used. However, for low capacity channels, it is often more desirable to transmit and display the pictures in a progressive (or stage-by-stage) fashion. This is done by transmitting a crude representation of the image first, and then adding details later. This is possible since the contents of the refresh frame memory used at the receiver (to provide a flicker free display) can be changed in almost any random order, and not necessarily in a regular line-by-line scan. Progressive presentation is desirable where a viewer wishes to recognize image content as soon as possible, and perhaps halt the transmission of unnecessary detail, or initiate actions simultaneous to transmission of the remaining detail. Also, besides providing an aesthetically pleasing sequence that holds the viewer's attention, progressive presentation also allows more efficient browsing through a picture data base.

The usefulness of such progressive presentation methods is somewhat speculative at this time. Also, new techniques that are being proposed for hierarchical representation of pictures and subsequent coding are still in their infancy, and significant experience with them is lacking. For this reason, we give below a brief summary of some techniques that have been tried. The reader is referred to references for details. Fig. 5.7.16 shows results of computer simulation of one specific technique. Progressive transmission, as well as the top to bottom regular scanning method are shown. For each case, development of the pictures at various transmission times is shown. It is clear from the photographs that a quicker recognition of the picture may often be possible by progressive transmission.

Division of a picture into several subpictures (corresponding to the stages of progressive transmission) by various subsampling patterns can also be used for progressive transmission. In this case, the first stage may correspond to a known subsampling pattern, and the pels of this stage may be coded by one of the many standard techniques. At the receiver, the pels that are not transmitted in the first stage, are reproduced by interpolation. The quality of the pictures produced in different stages is a function of the subsampling pattern as well as the interpolation techniques. Transform coding methods can also be used for progressive transmission especially for multilevel pictures. This is easily done by transmitting sequentially one coefficient for all the blocks in an image and then transmitting the remaining coefficients in some sequence that is known to the receiver. The quality of the pictures will obviously depend upon the order of coefficient transmission. It is not clear whether lower order coefficients (which give low frequency representation) should be sent before the higher order coefficients (containing predominantly edge information). The problem is further complicated by the fact that different coefficients require different accuracy. In any case, transform coding methods apply more readily to progressive transmission of gray level images than to two-level images.

Another block-by-block method is to form a hierarchical structure of successively smaller sub divisions of an image, and to transmit the higher level data first. These can be applied to gray level as well as two-level images. Another method, primarily suitable for two-level signals, is based on the *growth geometry* technique. In this method, within one image, aggregates of pels are identified that are capable of being generated by applying specified growth processes to particular pels, called seed pels. The data transmitted is the location, type of growth, and the number of growth stages for each seed pel. At the receiver, pels in the neighborhood of each seed pel are attached according to the indicated growth process, and thus regenerate the image exactly. It is difficult to compare these techniques primarily because there is no good objective measurement of human-display interaction or the quality of browsing.

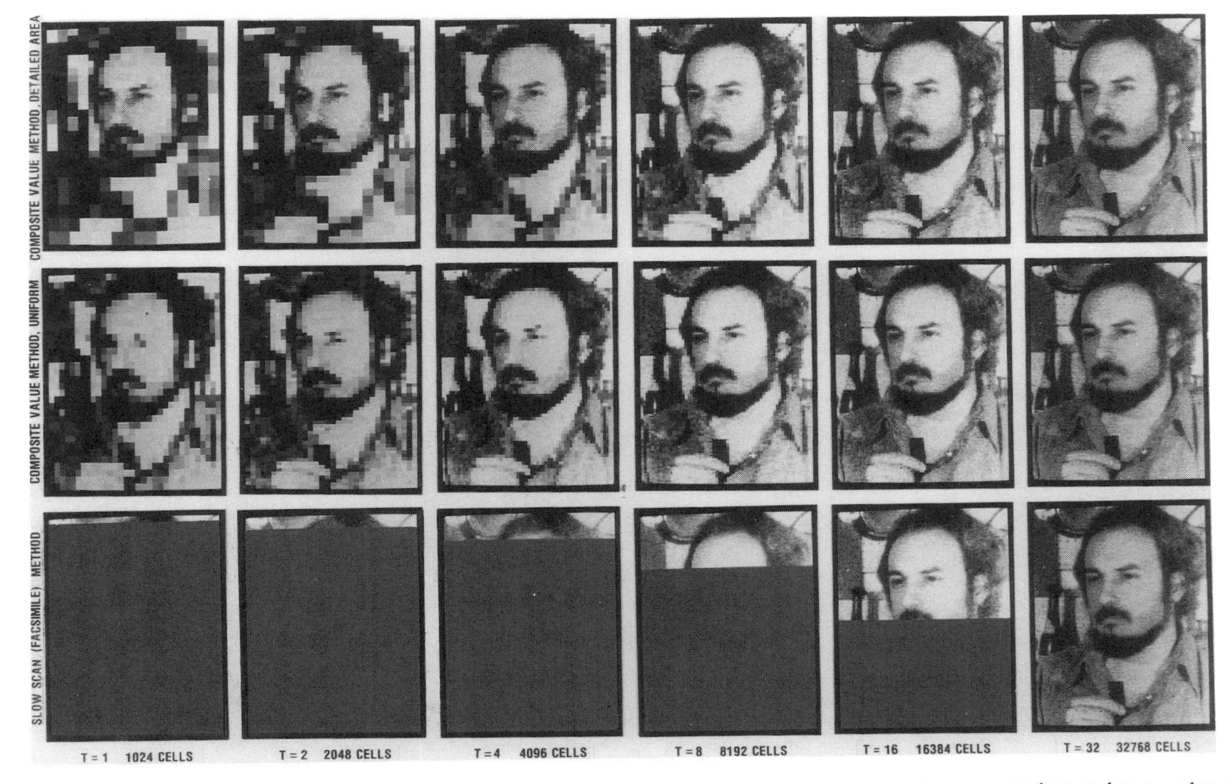

Fig. 5.7.16 Development of an image in a progressive transmission scheme and its comparison with a sequential top-to-bottom scheme (called slow-scan facsimile).

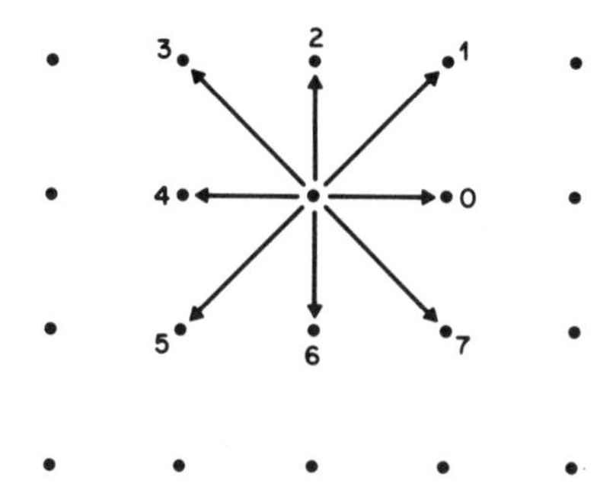

Fig. 5.7.17 Definition of Chain Code for line drawings. Eight directions shown by arrows are encoded by a three bit word. Dots represent the sampling grid.

Table 5.7.1 Comparison of compression ratios with pattern matching and the modified read code recommended by the CCITT (see Chapter 6 for more details).

TEST DOCUMENT	PATTERN MATCHING	MODIFIED READ
CCITT1	62.86	35.28
CCITT2	46.21	47.51
CCITT3	26.00	17.88
CCITT4	35.36	7.41
CCITT5	29.16	15.93
CCITT6	28.85	30.82
CCITT7	22.60	26.87
CCITT8	29.19	26.87
AVERAGE	29.19	15.54

Table 5.7.2 A Variable Length Code for Differential Chain of a Line Drawing.

DIRECTION OF THE DIFFERENTIAL CHAIN	CODEWORD
0	0
+1	01
−1	011
+2	0111
−2	01111
+3	011111
−3	0111111
+4	01111111
−4	011111111

5.7.5 Coding of Line Drawings

As we saw in Section 1.9, line drawings are a special subclass of two-level photographic images in which the picture consists essentially of interconnections of thin lines on a contrasting background. The thickness and color of the lines as well as the color and texture of the background are of little or, at most, symbolic significance. Such line drawings can be represented (see Fig. 1.9.5) on a rectangular grid by a sequence of pels with their corresponding (x,y) coordinates. However, specification of the picture by (x,y) coordinates of the pels is inefficient. A more efficient representation is by *chain codes* in which the vector joining the two successive pels is assigned a codeword. For a rectangular grid, since there are eight neighbors of any pel (see Fig. 5.7.17), only eight directions for this vector need to be specified, which requires 3 bits. More efficiency is obtained by *differential chain coding* in which each pel is represented by the difference between two successive absolute codes. Although, there are still eight codewords, in practice, they are not found to be equally likely, and therefore a variable length code of the form shown in Table 5.7.2 can be used. Simulation results show that such a code requires typically 1.8 to 1.9 bits/pel for alphanumeric characters and about 1.5 to 1.9 bits/pel for thinned edges of objects.

Appendix - Fast DCT

Here we provide fast 1D DCT and IDCT algorithms for blocksize 8. They are based on the theory of rotations[5.3.7a] and minimize the number of multiplications.

```
                /* 1D DCT block size 8*/
                /*Computed by rotations as in Reference [5.3.7a] */
#define PI      3.141592653589793238462643383279502884197169
#include <math.h>
dct(pels,coefs)
short   pels[8];            /* input values */
double  coefs[8];           /* output values */
{
        static int i;
        static double   c0,c1,c2,c3,c4,c5,c6,c7,c8,c9,c10,c11,c12;
        static double   alpha;
        static double   ts,m1,m2,m3;            /* temp variables */
        static double   s1[8], s2[8], s3[8], s4[8], s5[8];
        static int first=1;
        double cos(),sin(),sqrt();
```

```
if(first){
   first=0;
   alpha = 2*PI/(4*8)       ;
   c0  =  sin(4.0*alpha) - cos(4.0*alpha);
   c1  =  cos(4.0*alpha);
   c2  = - (sin(4.0*alpha) + cos(4.0*alpha));
   c3  =  sin(1.0*alpha) - cos(1.0*alpha);
   c4  =  cos(1.0*alpha);
   c5  = - (sin(1.0*alpha) + cos(1.0*alpha));
   c6  =  sin(2.0*alpha) - cos(2.0*alpha);
   c7  =  cos(2.0*alpha);
   c8  = - (sin(2.0*alpha) + cos(2.0*alpha));
   c9  =  sin(3.0*alpha) - cos(3.0*alpha);
   c10 =  cos(3.0*alpha);
   c11 = - (sin(3.0*alpha) + cos(3.0*alpha));
   c12 =  (double)1 / sqrt((double)2);
}
s1[0] = pels[0] + pels[7]; s1[1] = pels[0] - pels[7];
s1[2] = pels[2] + pels[1]; s1[3] = pels[2] - pels[1];
s1[4] = pels[4] + pels[3]; s1[5] = pels[4] - pels[3];
s1[6] = pels[5] + pels[6]; s1[7] = pels[5] - pels[6];

s2[0] = s1[1]; s2[1] = s1[5];
s2[2] = s1[0] + s1[4]; s2[3] = s1[0] - s1[4];
s2[4] = s1[2] + s1[6]; s2[5] = s1[2] - s1[6];
s2[6] = s1[3] + s1[7]; s2[7] = s1[3] - s1[7];

s3[0] = s2[0]; s3[1] = s2[1];
s3[2] = s2[2]; s3[3] = s2[5]*c12;
s3[4] = s2[3]; s3[5] = s2[4];
s3[6] = s2[6]; s3[7] = s2[7]*c12;

/*The following can be skipped and incorporated
  into the quantization process by doubling
  the quantizer step size                  */
for (i=0; i<8; ++i) s3[i] = s3[i]/2.0;

s4[0] = s3[2]; s4[1] = s3[4];
s4[2] = s3[5]; s4[3] = s3[6];
s4[4] = s3[0] + s3[3]; s4[5] = s3[0] - s3[3];
s4[6] = s3[7] + s3[1]; s4[7] = s3[7] - s3[1];
```

```
        ts = s4[0] + s4[2]; m1 = s4[0] * c0; m2 = ts * c1;
        m3 = s4[2] * c2; s5[0] = m2 + m3; s5[1] = m1 + m2;

        ts = s4[4] + s4[6]; m1 = s4[4] * c3; m2 = ts * c4;
        m3 = s4[6] * c5; s5[2] = m2 + m3; s5[3] = m1 + m2;

        ts = s4[1] + s4[3]; m1 = s4[1] * c6; m2 = ts * c7;
        m3 = s4[3] * c8; s5[4] = m2 + m3; s5[5] = m1 + m2;

        ts = s4[5] + s4[7]; m1 = s4[5] * c9; m2 = ts * c10;
        m3 = s4[7] * c11; s5[6] = m2 + m3; s5[7] = m1 + m2;

        coefs[0] = s5[1]; coefs[1] = s5[2]; coefs[2] = s5[4];
        coefs[3] = s5[6]; coefs[4] = s5[0]; coefs[5] = s5[7];
        coefs[6] = s5[5]; coefs[7] = s5[3];
return(0);
}
                                /*1d idct blocksize 8*/
#define PI      3.141592653589793238462643383279502884197169
#include <math.h>

idct(coefs,pels)
double coefs[8];
short pels[8];
{
        static double dout[8];
        static double   c1,c2,c3,c4,c5,c6,c7,c8,c9;
        static double   s1[8],s2[8],s3[8],s4[8],s5[8];
        static double   nmax2, alpha;
        static int i;
        static int first=1;

        if(first){
          first=0;
          nmax2 = 4;
          alpha = 2*PI/(4*8)      ;

          c1 = cos(nmax2*alpha); c2 = sin(nmax2*alpha);
          c3 = cos(alpha); c4 = sin(alpha);
          c5 = cos(2*alpha); c6 = sin(2*alpha);
          c7 = cos(3*alpha); c8 = sin(3*alpha);
          c9 = sqrt((double)2);
        }
```

```
        s1[0] = coefs[4]; s1[1] = coefs[0]; s1[2] = coefs[1];
        s1[3] = coefs[7]; s1[4] = coefs[2]; s1[5] = coefs[6];
        s1[6] = coefs[3]; s1[7] = coefs[5];

        s2[0] = c1*2*s1[0] + c2*2*s1[1]; s2[1] = c5*2*s1[4] + c6*2*s1[5];
        s2[2] = c1*2*s1[1] - c2*2*s1[0]; s2[3] = c5*2*s1[5] - c6*2*s1[4];
        s2[4] = c3*2*s1[2] + c4*2*s1[3]; s2[5] = c7*2*s1[6] + c8*2*s1[7];
        s2[6] = c3*2*s1[3] - c4*2*s1[2]; s2[7] = c7*2*s1[7] - c8*2*s1[6];

        s3[0] = 0.5*s2[4] + 0.5*s2[5]; s3[1] = 0.5*s2[6] - 0.5*s2[7];
        s3[2] = s2[0]; s3[3] = 0.5*s2[4] - 0.5*s2[5];
        s3[4] = s2[1]; s3[5] = s2[2]; s3[6] = s2[3];
        s3[7] = 0.5*s2[6] + 0.5*s2[7];

        s4[0] = s3[0]; s4[1] = s3[1]; s4[2] = s3[2];
        s4[3] = s3[4]; s4[4] = s3[5]; s4[5] = s3[3]*c9;
        s4[6] = s3[6]; s4[7] = s3[7]*c9;

        s5[0] = 0.5*s4[2] + 0.5*s4[3]; s5[1] = s4[0];
        s5[2] = 0.5*s4[4] + 0.5*s4[5]; s5[3] = 0.5*s4[6] + 0.5*s4[7];
        s5[4] = 0.5*s4[2] - 0.5*s4[3]; s5[5] = s4[1];
        s5[6] = 0.5*s4[4] - 0.5*s4[5]; s5[7] = 0.5*s4[6] - 0.5*s4[7];

        dout[0] = 0.5*s5[0] + 0.5*s5[1]; dout[1] = 0.5*s5[2] - 0.5*s5[3];
        dout[2] = 0.5*s5[2] + 0.5*s5[3]; dout[3] = 0.5*s5[4] - 0.5*s5[5];
        dout[4] = 0.5*s5[4] + 0.5*s5[5]; dout[5] = 0.5*s5[6] + 0.5*s5[7];
        dout[6] = 0.5*s5[6] - 0.5*s5[7]; dout[7] = 0.5*s5[0] - 0.5*s5[1];

        for(i=0; i<=7; i++){
                if( dout[i]<0.0 )pels[i] = dout[i] - 0.5;
                else             pels[i] = dout[i] + 0.5;
        }
        return(0);
}
```

References

5.1.1 B. M. Oliver, J. R. Pierce and C. E. Shannon, "The Philosophy of PCM", *Proceedings of IRE,* Vol. 36, Oct. 1948, pp. 1324-1331.

5.1.2 W. M. Goodall, "Television by Pulse Code Modulation", *Bell System Technical Journal,* Vol. 30, Jan. 1951, pp. 33-49.

5.1.3 L. G. Roberts, "Picture Coding Using Pseudo-Random Noise", *IRE Trans. on Information Theory*, Vol. IT-8, February 1962, pp. 145-154.

5.1.4 A. K. Bhushan, "Efficient Transmission and Coding of Color Components", M.S. Thesis, Massachusetts Institute of Technology, Cambridge, MA, June 1977.

5.1.5 W. Frei, P. A. Jaeger and P. A. Probst, "Quantization of Pictorial Color Information", *Nachrichtentech Z*, Vol 61, 1972, pp. 401-404.

5.1.6 B. Marti, "Preliminary Processing of Color Images", CCETT ATA/T/3/73, September 5, 1973.

5.1.7 L. Stenger, "Quantization of TV Chrominance Signals Considering the Visibility of Small Color Differences", *IEEE Trans. on Communications*, Vol. COM-25, No. 11, November 1977.

5.2.1 C. C. Cutler, "Differential Quantization of Communication Signals", U.S. Patent 2 605 361, July 1952.

5.2.2 F. deJager, "Delta Modulation, A Method of PCM Transmission using a 7-Unit Code", Philips Research Reports, Dec. 1952, pp. 442-466.

5.2.3 C. W. Harrison, "Experiments with Linear Prediction in Television", *Bell System Technical Journal*, Vol. 31, July 1952, pp. 764-783.

5.2.4 T. Kailath, *Linear Systems*, Prentice-Hall, Englewood Cliffs, NJ, 1980.

5.2.5 A. Habibi, "Comparison of Nth order DPCM encoder with Linear Transformations and Block Quantization Techniques", *IEEE Trans. on Communication Technology*, Vol. COM-19, Dec. 1971, pp. 948-956.

5.2.6 H. G. Musmann, "Predictive Coding" Chapter in the book *Image Transmission Techniques*, edited by W. K. Pratt, Academic Press 1979.

5.2.7 R. E. Graham, "Predictive Quantizing of Television Signals", *IRE Wescon Convention Record*, Vol. 2, pt 4, 1958, pp. 147-157.

5.2.8 B. G. Haskell, "Entropy Measurements for Nonadaptive and Adaptive, Frame-to-Frame, Linear Predictive Coding of Video Telephone Signals", *Bell System Technical Journal*, Vol. 54, No. 6, August 1975, pp. 1155-1174.

5.2.9 H. G. Musmann, P. Pirsch, and H.-J. Grallert, "Advances in Picture Coding", *Proceedings of IEEE*, April 1985.

5.2.10 J. O. Limb, and H. A. Murphy, "Measuring the Speed of Moving Objects from Television Signals", *IEEE Trans. on Comm.,* April 1975.

5.2.11 A. N. Netravali and J. D. Robbins, "Motion-Compensated Television Coding: Part I", *Bell System Technical Journal*, Vol. 58, No. 33, March 1979, pp. 631-669.

5.2.12 A. N. Netravali and J. D. Robbins, "Motion Compensated Coding: Some New Results", *Bell System Technical Journal*, Nov. 1980.

5.2.13 J. A. Stuller, A. N. Netravali and J. D. Robbins, "Interframe Television Coding Using Gain and Displacement Compensation", *Bell System Technical Journal*, Sept. 1980.

5.2.14 K. A. Prabhu and A. N. Netravali, "Motion Compensated Components Color Coding," *IEEE Trans. Comm.*, Dec. 1982.

5.2.15 K. A. Prabhu and A. N. Netravali, "Motion Compensated Composite Color Coding", *IEEE Trans. Comm.*, Feb. 1983.

5.2.16 D. J. Connor, R. C. Brainard and J. O. Limb, "Intraframe Coding for Picture Transmission", *IEEE Proceedings*, July 1972.

5.2.17 R. Jung and R. Lippman, "Error Response of DPCM Decoders", *Sonderdruck aus Nachrichtentechnische Zeitschrift*, Bd. 28, 1974, pp. 431-436.

5.2.18 J. Max, "Quantizing for Minimum Distortion", *IEEE Trans. on Information Theory*, Vol. IT-6, March 1960, pp. 7-12. Also, S. P. Lloyd," Least Squares Quantization in PCM", *IEEE Trans on Information Theory*, March 1982, pp. 129-136.

5.2.19 D. K. Sharma and A. N. Netravali, "Design of Quantizers for DPCM Coding of Picture Signals", *IEEE Trans. on Communication*, Vol. COM-25, Nov. 1977, pp. 1267-1274.

5.2.20 A. N. Netravali, "On Quantizers for DPCM Coding of Picture Signals", *IEEE Trans. on Information Theory*, Vol. IT-23, No. 3, May 1977, pp. 360-370.

5.2.21 A. N. Netravali and C. B. Rubinstein, "Quantization of Color Signals", *Proceedings of IEEE*, Vol. 65, No. 3, August 1977, pp. 1177-1187.

5.2.22 A. N. Netravali and B. Prasada, "Adaptive Quantization of Picture Signals Using Spatial Masking", *Proceedings of IEEE*, Vol. 65, April 1977, pp. 536-548.

5.2.23 P. Pirsch, "Block Coding of Color Video Signals", in Proceedings of National Telecommunications Conference, 1977, pp. 10.5.1-10.5.5.

5.3.1 E. O. Brigham, *The Fast Fourier Transform*, Prentice-Hall, Englewood Cliffs, N.J., 1984.

5.3.2 *Programs for Digital Signal Processing*, Edited by IEEE Acoustics, Speech and Signal Processing Society, IEEE Press, New York, 1979.

5.3.3 H. F. Silverman, "An Introduction to Programming the Winograd Fourier Transform Algorithm (WFTA)," *IEEE Trans. on Acoustics, Speech and Signal Processing*, v. ASSP-25, No. 2, April 1977, pp. 152-165.

5.3.4 W. K. Pratt, J. Kane and H. C. Andrews, "Hadmard Transform Image Coding," *Proc. IEEE*, v. 57, No. 1, Jan. 1969, pp. 58-68.

5.3.5 R. D. Brown, "A Recursive Algorithm for Sequency Ordered Fast Walsh Transforms," *IEEE Trans. Computers*, v. C-26, No. 8, Aug. 1977, pp. 819-822.

5.3.6 N. Ahmed and K. R. Rao, *Orthogonal Transform for Digital Signal Processing*, Springer-Verlag, New York, 1975.

5.3.7 J. Makhoul, "A Fast Cosine Transform in One and Two Dimensions," *IEEE Trans. Acoustics, Speech and Signal Processing*, v. ASSP-28, No. 1, Feb. 1980, pp. 27-34.

5.3.7a C. Loeffler, A. Ligtenberg and G. S. Moschytz, "Practical fast 1D DCT algorithm with 11 multiplications," *Proceedings IEEE ICASSP-89*, v.2, pp. 988-991, Feb. 1989.

5.3.8 B. Fino, "An Experimental Study of Image Coding Utilizing the Haar Transform and the Hadamard Complex Transform," *Ann. Telecommunications*, v. 27 (5-6) pp. 185-208, 1972 (in French).

5.3.9 W. K. Pratt, W. Chen and L. R. Welch, "Slant Transform Image Coding," *IEEE Trans. Communications,* V. COM-22, pp. 1075-1093, Aug. 1974.

5.3.10 A. K. Jain, "Image Data Compression: A Review," *Proc. IEEE,* v. 69, No. 3, pp. 349-389, March 1981.

5.3.11 H. C. Andrews and C. L. Patterson, "Outer Product Expansions and Their Uses in Digital Image Processing," *IEEE Trans. Computers,* v. C-25, No. 2 pp. 140-148, Feb. 1976.

5.3.12 F. W. Mounts, A. N. Netravali and B. Prasada, "Design of Quantizers for Real-Time Hadamard Transform Coding of Pictures," *Bell Sys. Tech. J.,* v. 56, No. 1, January 1977, pp. 21-48.

5.3.13 A. K. Jain and S. H. Wang, "Stochastic Image Models and Hybrid Coding," Final Report Contract No. N00953-77-C-003 MJE, Dept. of Elec. Engin, State U. of New York at Buffalo, October 1977.

5.3.14 J. Huang and P. Schultheiss, "Block Quantization of Correlated Gaussian Random Variables," IEEE Trans. Communications, COM-11, 1963, pp. 286-296.

5.3.15 W. H. Chen and C. N. Smith, "Adaptive Coding of Monochrome and Color Images," *IEEE Trans. Communications,* v. COM-25, No. 11, November 1977, pp. 1285-1292.

5.3.16 J. M. Wozencraft and I. M. Jacobs, *Principles of Communication Engineering,* John Wiley & Sons, Inc., New York, 1965.

5.3.17 J. O. Limb, C. B. Rubinstein and J. E. Thompson, "Digital Coding of Color Video Signals - A Review," *IEEE Trans. Communications,* v COM-25, No. 11, November 1977, pp. 1349-1385.

5.3.18 T. R. Natarajan and N. Ahmed, "On Interframe Transform Coding," *IEEE Trans. Communications,* v. COM-25, No. 11, November 1977, pp. 1323-1329.

5.3.19 J. R. Jain and A. K. Jain, "Interframe Adaptive Data Compression Techniques for Images," Signal and Image Processing Laboratory—Dept. of Elec. and Computer Engin., University of Calif., Davis, CA., AD-A078841.

5.3.20 S. C. Knauer, "Real-Time Video Compression Algorithm for Hadamard Transform Processing," *IEEE Trans. Electromagnetic Compatibility,* v. EMC-18, No. 1, February 1976, pp. 28-36.

5.3.21 A. G. Tescher, "Transform Image Coding," Chapter 4 of *Image Transmission Techniques,* W. K. Pratt ed., Academic Press, New York, 1979.

5.4.1 A. Habibi, "Hybrid Coding of Pictorial Data," *IEEE Trans. Communications,* v. COM-22, No. 5, May 1974, pp. 614-624.

5.4.2 J. A. Roese, "Hybrid Transform/Predictive Image Coding," Chapter 5 of *Image Transmission Techniques,* W. K. Pratt ed., Academic Press, New York, 1979.

5.4.3 R. J. Clarke, "Hybrid Intraframe Transform Coding of Image Data," *IEE Proc.,* v. 131 part F, No. 1, Feb. 1984, pp. 2-6.

5.5.1 A. Gersho and R. Gray, *Vector Quantization and Signal Compression,* Kluwer Academic Publishers, 1992.

5.5.2 Y. Linde, A. Buzo and R. Gray, "An Algorithm for Vector Quantizer Designs," *IEEE Trans Commun.,* COM-28, Jan. 1980, pp. 84-95.

5.5.3 W. H. Equitz, "Fast Algorithms for Vector Quantization Picture Coding," M. S. Thesis, MIT, June 1984. Also see "Some Methods for Classification and Analysis of Multivariate Observations", MacQueen, Proc. 5th Berkeley Symp. on Math, Statistics and Probability, v. 1, pp. 281-296, 1967.

5.5.4 A. Gersho and B. Ramamurthi, "Image Coding using Vector Quantization," Proc.of Int. Conf. ASSP, Paris, 1982, pp. 428-431.

5.5.5 D. J. Healy and D. R. Mitchell, "Digital Video Bandwidth Compression Using Block Truncation Coding", *IEEE Trans. Commun.*, COM-29, Dec. 1981, pp. 1809-1823.

5.6.1 P. J. Burt and E. H. Adelson, "The Laplacian Pyramid as a Compact Image Code," *IEEE Trans. on Communications,* COM-31, April 1983, pp. 532-540.

5.6.2 J. W. Woods and Sean D. O'Neil, "Subband Coding of Images," *IEEE Trans. on Acoustics, Speech and Signal Proc.*, ASSP-34, Oct. 1986, pp. 1278-1288.

5.7.1 R. Hunter and A. H. Robinson, "International Digital Facsimile Coding Standards," *Proceedings of IEEE,* July 1980.

5.7.2 O. Johnsen, J. Segen and G. L. Cash, "Coding of Two-Level Pictures of Pattern Matching and Substitution," *Bell System Technical J.*, Vol. 62, Oct. 1983, pp. 2513-2545.

Questions for Understanding

5.1 What is the quantization noise PSNR in dB for 7 bit, 8 bit and 9 bit PCM? What are the impairments for each on the CCIR 5-point scale?

5.2 Explain the difference between 1-bit PCM and DM. Which is better and why?

5.3 Explain granular noise and slope overload. How does compand DM improve these impairments?

5.4 What are instantaneous and syllabic DM companding?

5.5 What are the advantages and disadvantages of minimum MSE linear predictor in image coding?

5.6 What factors should be considered in choosing linear predictor weighting coefficients?

5.7 Explain the operation of adaptive intrafield predictive coders.

5.8 What are the advantages and disadvantages of interframe predictive coding?

5.9 What are the underlying assumptions of block matching and pel recursive motion estimation?

5.10 When would gain compensation be useful?

5.11 What methods are used in predictive coding to deal with transmission bit errors?

5.12 What causes edge busyness?

5.13 What are mid-step and mid-riser quantizers? When would each be used?

5.14 How are visibility thresholds used to design quantizers?

5.15 What are the relative advantages and disadvantages of uniform and companded quantization in predictive coding? Also, variable length coding and fixed length coding?

5.16 What are the differences between MMSE and MSSDE quantizers?

5.17 Explain reflected quantizers.

5.18 What are masking functions and how are they used in adaptive quantizers for predictive coding? What are the advantages and disadvantages of this approach?

5.19 What are the advantages of separable transform coding?

5.20 Explain the relative merits of RDFT, WHT, DCT and KLT.

5.21 How is quantization different for DPCM and transform coding?

5.22 Explain quantization error and truncation error in optimum transform coding.

5.23 Derive Eq. (5.3.86).

5.24 Explain adaptive transform coding with classes.

5.25 What are threshold coding and zonal coding?

5.26 What are the advantages and disadvantages of 3D transform coding?

5.27 Explain the merits of intra and inter frame hybrid transform coding.

5.28 Explain motion compensated hybrid transform coding.

5.29 How is vector quantization different than transform coding?

5.30 Explain the LBG algorithm for VQ. What are the difficulties of using it?

5.31 What is multistage VQ?

5.32 Explain BTC.

5.33 What is interpolative coding? How is motion adaptation used?

5.34 Explain pyramid coding. Also, subband coding.

5.35 What is run length coding?

5.36 Explain reordering, as applied to graphics coding.

5.37 Explain line-to-line run-length difference coding.

5.38 What are the merits and demerits of approximate graphics coding?

5.39 What is pattern matching coding?

5.40 What are ARQ and FEC?

5.41 Explain the chain code algorithm.

6

Examples of Codec Designs

In this chapter, we describe several image codec designs in some detail. Our purpose is not to present the current state of the art, for that changes almost on a daily basis. Rather, we wish to show how the basic principles discussed in previous chapters might be applied in an interrelated way to practical systems. Some applications require a simple codec without a high degree of data compression. In other cases we need a low bit-rate because of expensive transmission channels, and codec cost is less of a consideration. In still other situations we want very high picture quality, with less concern about low complexity and compression capabilities. The techniques described in this chapter cover a range of complexity, picture quality and data compression in order to assist in the making of sound engineering decisions when designing image communication systems. However, some of the implementations also contain a certain amount of ad hoc procedures that can only be optimized by experimentation.

We start (see Fig. 6.0.1) by describing a fairly simple intrafield DPCM codec for composite NTSC color television. This design has been implemented in VLSI and provides a good quality videoconferencing picture. Next, we discuss an interfield adaptive DPCM coding technique applied to broadcast quality, NTSC color television. The methodology of this codec is also applicable to component color and monochrome pictures. However, in these cases the bit-rates, predictors and quantizers will be somewhat different. Next, we describe an interfield hybrid transform coder for component color TV. This approach, in a modified form, is also applicable to single color images. Section 6.5 describes a scene decomposition algorithm for coding single images. The next section describes interframe coders suitable for videoconferencing, and the final section discusses two-level graphics coding.

Section	Image Source	Principal Technique	Uncoded Bit-Rate	Coded Bit-Rate	Complexity	Image Quality
6.1	NTSC Color Video	Intrafield DPCM	86 Mbits/s	43 Mbits/s	Low	4
6.2	NTSC Color Video	Adaptive Interfield DPCM	129 Mbits/s	42 Mbits/s	Moderate	5
6.3	Component Color Video	Interfield Hybrid Transform	64 Mbits/s	16-24 Mbits/s	High	3-4
6.4	Component Color Video	Motion Compensation Hybrid	210 Mbits/s	34 or 45 Mbits/s	High	4.5-5 Depending on Motion
6.5	Monochrome 256×256	Scene Decompose	8 bits per pel	≈0.8 bits per pel	High	3-4
6.6	Component Color Video	Conditional Replenishment DPCM	17.6 Mbits/s	2.0 or 1.5 Mbits/s	Moderate	3-5 Depending on Motion
6.7.1	Black & White Document	CCITT 1 Dimen. Coding	1-bit per pel	Compress. Ratio 5-16	Low	Same as Original
6.7.2	Black & White Document	CCITT 2 Dimen. Coding	1-bit per pel	Compress. Ratio 6-32	Moderate	Same as Original

Fig. 6.0.1 Summary of characteristics and capabilities of the systems described in this chapter. Image quality is according to the five point CCIR scale of Table 4.1.

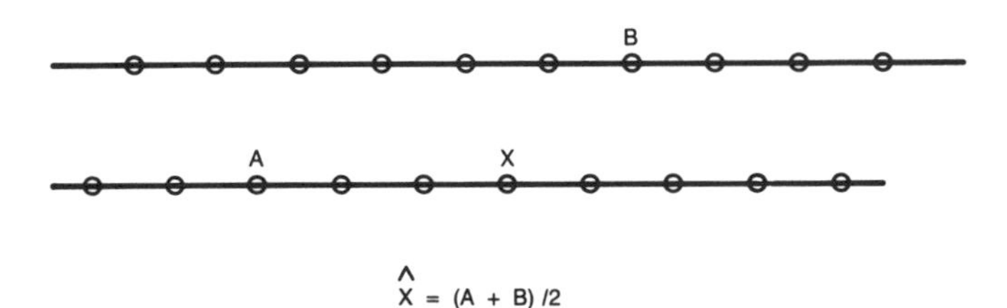

Fig. 6.1.1 Two successive lines of NTSC video signal sampled at three times the color subcarrier frequency of 3579545 Hz. Pel X occurs 3 pels after A and 681 pels after B, and therefore all three pels have the same color subcarrier phase. DPCM uses an average of A and B as $\hat{X}$ the prediction of X.

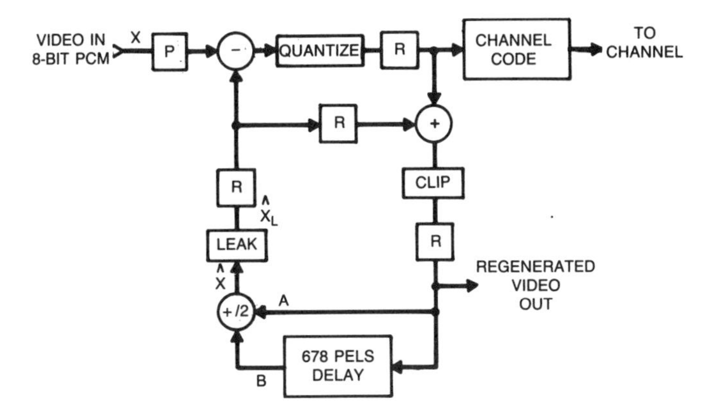

Fig. 6.1.2 Circuit configuration for two-dimensional DPCM Coding. *R* denotes a register. Quantization and Channel coding can be realized via simple Read Only Memories (ROM's). The leak factor is 31/32.

Range of Differential Magnitudes $\lvert X - \hat{X}_L \rvert$	Quantized Values
0-2	±1
3-5	±3
6-10	±7
11-18	±14
19-28	±23
29-40	±34
41-55	±47
56-255	±63

Fig. 6.1.3 Quantization table for prediction errors.

6.1 45 Mbs Intrafield DPCM Codec for NTSC Color Video

The system described here[6.1,6.2] is a fairly simple color TV codec for situations that require a subjectively pleasing picture at the receiver, but which do not need the so called "transparent" coding quality typical of TV studio or network transmission applications. The output bit-rate for video and auxiliary data closely matches the DS3 rate (44.736 Megabits/sec) used in North America. It may also be suitable for some local-area fiberguide or cable TV networks.

The first step is to PCM code the NTSC composite video signal at a sampling rate of three times the color subcarrier frequency using 8-bits per pel. The pel configuration of two successive scan lines is shown in Fig. 6.1.1, and pels A, B and X all have the same color subcarrier phase, i.e., they are equal in regions of constant color. Thus, for DPCM a reasonably good prediction of pel X is the average of pels A and B.

Fig. 6.1.2 shows a block diagram of the DPCM coder. The ROM quantizer outputs one of a limited set of values, 16 in this case, as shown in Fig. 6.1.3. The ROM channel coder then outputs 4-bits per pel to give a video bit-rate of about 43 Megabits/sec, which allows a considerable margin for audio, additional data and stuffing bits to be added prior to the DS3 channel. Fig. 6.1.4 shows the corresponding decoder, for which only a channel decoder ROM is needed.

As was described in Sec. 5.2.3, resilience to transmission bit errors is improved considerably by the use of *leak*, i.e., attenuating the prediction $\hat{X}$ by a small amount before computing the prediction error. However, this can have a detrimental effect on the quality of prediction if the leak is too large (see Sec. 5.2.3k). As a compromise, the leak factor in this system was chosen as 31/32, and was implemented to cause the prediction to decay toward the center of the dynamic range, i.e.,

$$\hat{X}_L = 128 + \frac{31}{32}(\hat{X} - 128) \tag{6.1.1}$$

Note that this is equivalent to subtracting 128 from the input pels and using a standard leak configuration. This arrangement gives virtually no perceptible image degradation for single-bit errors up to an error rate of 10^{-6}.

A clipping circuit at both the coder and decoder prevents overflow and underflow that can be caused by the quantized differential values. For overflow the output is 255; for underflow it is 0. The clipping function also performs another very important service at the decoder during startup and following catastrophic transmission errors. Because of the clipping, discrepancies between the coder and decoder prediction values tend to disappear quite rapidly (within a few line periods), assuming the video signal has a reasonable magnitude.

This system has been implemented in VLSI. The quantizer, channel coder and channel decoder functions were obtained using standard ROM. The remaining operations were designed into a single custom CMOS VLSI using 2.5 micron design rules.

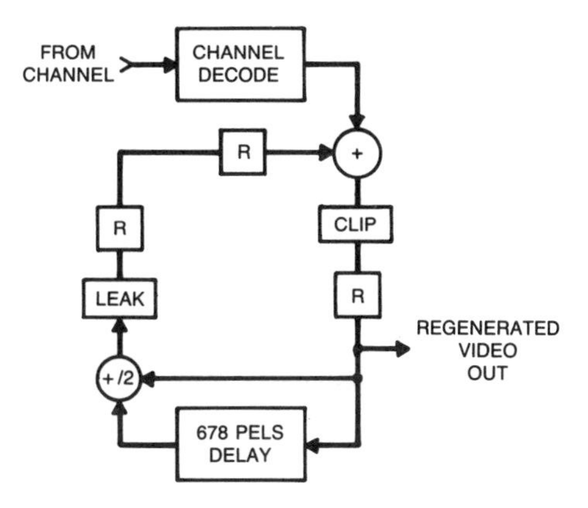

Fig. 6.1.4 Two dimensional DPCM decoder.

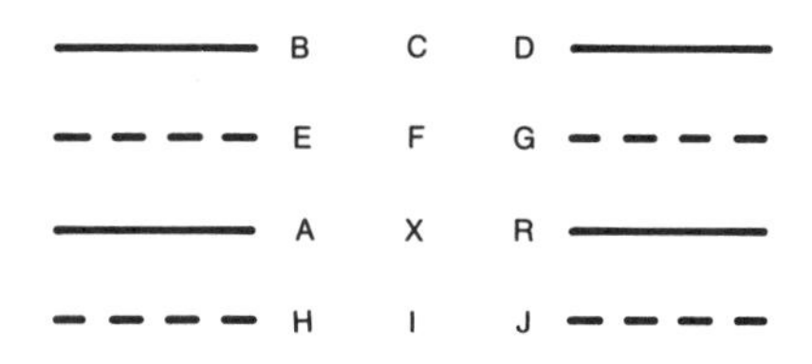

$$\hat{X}_1 = (A + D)/2$$

$$\hat{X}_2 = (F + I)/2$$

Fig. 6.2.1 Pel configuration for adaptive DPCM coding. $\tilde{X}_1$ is an intrafield predictor suitable for moving areas, whereas $\tilde{X}_2$ is an interfield predictor suitable for non-moving areas of the picture.

6.2 Adaptive Predictive Interfield Coder for NTSC Color TV

Conceptually, the idea of adaptive prediction is simple, as we saw in Chapter 5. Within each small region of the image the coder must keep track of the directions of maximum correlation between the pels, and then adapt the predictor accordingly. However, for sequentially scanned single images, simple adaptive prediction does not give a marked reduction in bit-rate. The problem is that in low-detail areas of the picture the correlation is fairly high in all directions, and adaptation is of little use, whereas in high-detail areas of the picture the correlation directions change so fast that either the adaptation algorithm cannot keep up or excessive overhead information must be transmitted. The net result is that there has been very little improvement over Graham's original suggestion[5.2.7] of keeping a running track of which is larger, horizontal correlation or vertical correlation, and accordingly using the pel to the left or the pel directly above as a prediction (see Sec. 5.2.3d).

With a 2:1 interlaced television signal however, adaptive predictive coding has a much more marked effect than with sequentially scanned pictures. For example, Fig. 6.2.1 shows a pel configuration along with an intrafield predictor $\hat{X}_1 = (A+D)/2$ and an interfield predictor $\hat{X}_2 = (F+I)/2$. For areas of the picture in which there is no movement $\hat{X}_2$ is the better predictor, whereas in areas of the picture containing significant movement $\hat{X}_1$ is better. This is due largely to the fact that pel X is spatially closer to pels F and I, but temporally closer to pels A and D. An efficient adaptation algorithm simply examines which predictor gave a better prediction of (quantized) pel A and then use that predictor for pel X. This works well because moving areas are usually contiguous in an image.

Such a coder[6.3] is shown in Fig. 6.2.2a. From $\tilde{A}$, the quantized version of pel A, and the two predictions $\hat{A}_1$ and $\hat{A}_2$, which were computed for pel A, a decision is made as to whether $\hat{X}_1$ or $\hat{X}_2$ is the best prediction for pel X. The identical decision can also be made at the receiver with no additional transmitted information as shown in Fig. 6.2.2b. The prediction $\hat{X}$ is then used in the normal manner for DPCM coders.

The use of predictive coding with composite color television signals must take into account the presence of the sinusoidal color subcarrier in the video signal as we saw in Chapter 5. Moreover, the relationship between the pel X (which is to be predicted) and the surrounding previously transmitted pels depends on the sampling pattern of the initial digitization of the composite signal. For example, if the initial sampling rate is exactly four times the NTSC color subcarrier frequency, i.e., $\approx$14.3 MHz, then in Fig. 6.2.3 pels K, L, M and I are approximately equal to pel X in non-moving regions of constant color. However, this does not preclude the use of other pels in a linear prediction as we saw in Section 5.2.3c.

The simplest intrafield *in-phase* predictor is given by

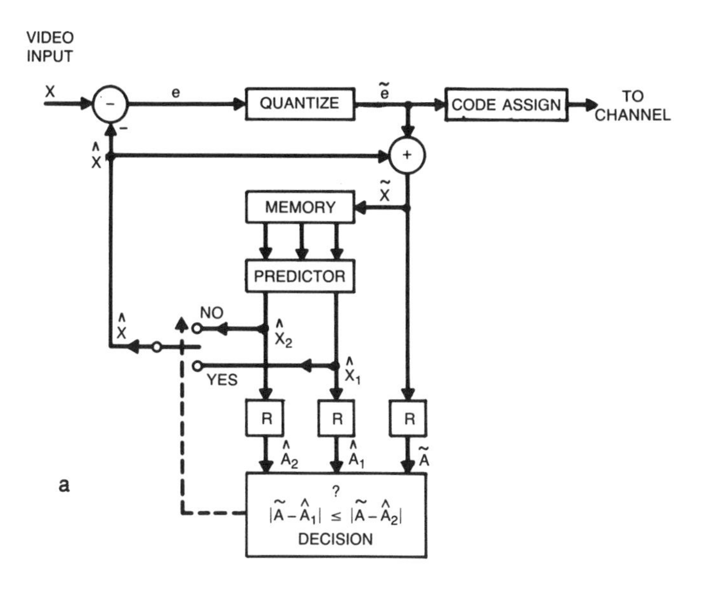

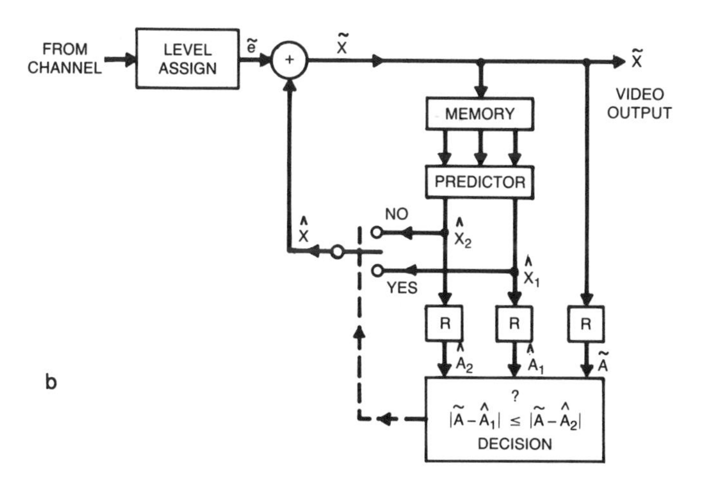

Fig. 6.2.2 Adaptive predictive codec. (a) Coder: Registers *R* are 1 pel delays. (b) Decoder.

$$\hat{X}_1 = M \tag{6.2.1}$$

where pel M of Fig. 6.2.3 is the fourth previous pel along the scan line. This prediction should work well in moving areas of the picture where there is a reduction in spatial resolution.

The suitable interfield in-phase predictor is

$$\hat{X}_2 = \frac{1}{2} I + \frac{1}{4} (K+L) \tag{6.2.2}$$

where pels K, L, and I are in the adjacent lines of the previous field as shown in Fig. 6.2.3. This predictor should work well in stationary areas of the image.

Choosing the best predictor for pel X could be done as shown in Fig. 6.2.2, i.e., choose whichever predictor was best for pel A. However, it has been found that sometimes pel A by itself is not sufficient to enable a good choice of predictors and that additional information should also be taken into account such as which prediction was best for pel C of Fig. 6.2.3. For example, if C_1 was a better prediction of pel C and $\hat{A}_1 \approx \hat{A}_2$, then $\hat{X}_1$ would be chosen as the best predictor for pel X.

In coding a composite color signal with broadcast quality, the usual distortion criterion is that there be no visible quantization noise. Thus, in this system the first step is to PCM code the composite signal at a sampling rate of four times color subcarrier frequency using 9-bits per pel. In addition, the DPCM quantization error must at all times be below the threshold of visibility (see Fig. 5.2.33). Two quantizers were designed for the 10-bit prediction error according to this experimental procedure, and their characteristics are given in Fig. 6.2.4. One is slightly coarser than the other and is used during periods of moderate motion as described below. Both produce subjectively invisible quantization noise. However, measured signal-to-noise is smaller with the coarse quantizer.

If the codec of Fig. 6.2.2 is implemented with variable word-length coding and buffers at the coder and decoder, a buffer overflow strategy must be devised to deal with scenes containing rapid motion. A very straightforward technique is subsampling, e.g., sending only 75 percent or 50 percent of the samples, in order to reduce the data rate until buffer overflow no longer threatens. However, with a composite signal, interpolation is not as simple as with color components or monochrome signals. For example, with 4:3 subsampling, pel P in Fig. 6.2.3 could be interpolated from its (quantized) neighbors by

$$\begin{aligned}\tilde{P} &= \frac{1}{2}(\tilde{N}+\tilde{A}) + \frac{1}{2}(\tilde{N}-\tilde{M}) + \frac{1}{2}(\tilde{A}-\tilde{X}) \\ &= \tilde{N} + \tilde{A} - \frac{1}{2}\tilde{M} - \frac{1}{2}\tilde{X}\end{aligned} \tag{6.2.3}$$

With 2:1 subsampling, better picture quality results if the sampling pattern is shifted between fields, and therefore a suitable interpolation for pel P would be

B C D

K E F G L

M N P A X R

U V W H I J

Fig. 6.2.3 Pel configuration for NTSC composite color TV signal sampled at four times color subcarrier frequency. Pels K, L, M, I and U have the same color subcarrier phase as pel X.

Fine Quantizer		Coarse Quantizer	
Input Magnitude	Quantized Values	Input Magnitude	Quantized Values
0	0	0-1	0
1-2	±1	2-4	±3
3-5	±4	5-7	±6
6-9	±7	8-11	±9
10-14	±12	12-16	±14
15-19	±17	17-21	±19
20-25	±22	22-27	±24
26-32	±29	28-34	±31
33-39	±36	35-42	±38
40-49	±45	43-51	±47
50-59	±54	52-61	±56
60-71	±65	62-73	±67
72-85	±78	74-87	±80
86-101	±93	88-103	±95
102-119	±110	104-121	±112
120-139	±129	122-141	±131
140-162	±150	142-164	±152
163-188	±175	165-194	±177
189-217	±202	195-222	±206
218-251	±233	223-257	±239
252-290	±270	258-297	±276
291-511	±311	298-511	±319

Fig. 6.2.4 Quantization (symmetric) characteristic for 10-bit prediction error input signal. The coarse quantizer is used during periods of moderate motion.

$$\tilde{P} = \frac{1}{2}(\tilde{N}+\tilde{A}) + \frac{1}{4}(\tilde{W}-\tilde{U}) + \frac{1}{4}(\tilde{W}-\tilde{I}) \tag{6.2.4}$$

$$= \frac{1}{2}(\tilde{N}+\tilde{A}) + \frac{1}{4}(2\tilde{W}-\tilde{U}-\tilde{I})$$

The first term is a luminance estimate obtained from the two pels that are spatially and temporally closest. The second term is a chrominance estimate obtained from the line below in the previous field.

The effects of 4:3 subsampling on picture quality are only visible in areas of high detail where large pel-to-pel luminance transitions occur. Such conditions are usually present only in electronically generated graphics of test waveforms and are extremely rare in moving areas of the picture, which is where subsampling would normally be employed in order to reduce buffer filling.

The effects of 2:1 subsampling are visible at large pel-to-pel luminance transitions and at large line-to-line chrominance transitions. With moderate to rapid motion, which is where 2:1 subsampling would normally be employed, such conditions are extremely rare. The overall codec is shown in Fig. 6.2.5.

Five modes of operation, shown in Fig. 6.2.6, are utilized by the codec in order to keep the buffer from emptying or overflowing. Mode 0 is used exclusively to prevent buffer underflow. With some modification it could also help to alleviate the effects of transmission bit errors. Modes 1 to 3 utilize variable word-length codes to lower the data rate for the quantized prediction errors. For best performance, a slightly different code should be used for the two quantizers so that the data rate is significantly reduced when the coarse quantizer is utilized. Mode 4 has such a low bit-rate that variable word-length coding is not needed.

The buffer size in this system should be large enough to smooth the data over at least one field interval, i.e., 17 msec. At a bit-rate of 42 Mbits/sec, this implies a buffer size greater than 714 kbits. At the receiver, a larger size is required in order to deal with transmission bit errors.

Fig. 6.2.7 shows a mode control strategy based on the fullness of an 800 kbit buffer. Switching to a higher mode occurs when the buffer fullness exceeds the prescribed level, but switching to a lower mode does not take place until the buffer fullness drops below some lower level, as shown. This prevents oscillation between modes, which is detrimental to image quality in some cases.

Operating at a channel bit-rate of about 43 Megabits/sec (3 bits/pel), very high quality images can be transmitted that are free of any visible coding distortion. For normal entertainment television pictures, the codec spends a large majority of its time (more than 80 percent) in mode 1 and most of the remaining time (about 12 percent) in mode 2. Mode 0 is used primarily for low detail pictures, and mode 3 is used only for rapidly moving regions. At a bit rate of 43 Mbits/sec, mode 4 is needed only for anomalous images such as rapidly changing electronically generated signals, and in this case the slightly larger

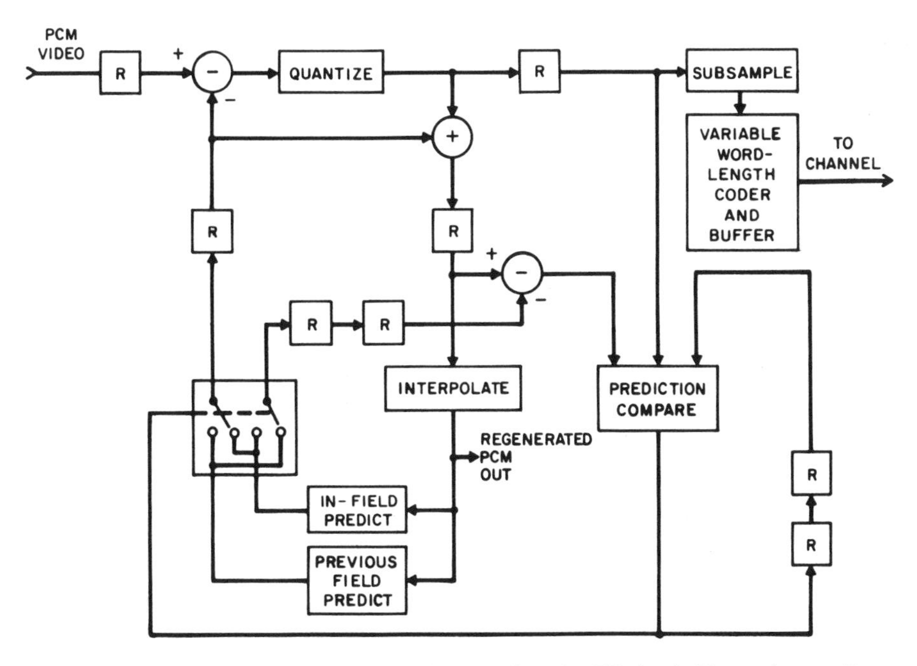

Fig. 6.2.5 Adaptive DPCM coder for NTSC composite color TV signal. The mode controller varies the coarseness of the quantizer and invokes subsampling in order to prevent buffer overflow.

Mode	Code Word Length	Subsample	Quantizer
0	6-bits	none	fine
1	variable	none	fine
2	variable	none	coarse
3	variable	4:3	fine
4	6-bits	2:1	fine

Fig. 6.2.6 Modes used for the codec operating at 42 Mbits/sec.

coding distortion is completely masked by the picture and is invisible to the viewer.

Some additional measures must be taken if there are transmission bit errors. Forward error correction (FEC) channel codecs are the most straightforward solution, and these help to eliminate audio as well as video distortions. However, even with the FEC, occasional catastrophic errors will occur. Leak may be less effective with adaptive coding. In this case, error detection at the receiver and transmission of some PCM data by the transmitter would normally be required in a practical system. More will be said later in this chapter about error control.

Further improvement of adaptive predictive coding for broadcast quality video is certainly possible with more elaborate processing. For example, motion compensation methods are highly desirable for the slow panning and zooming found so often in entertainment television. However, applying motion compensation to a composite color video signal has some fundamental problems, the most important of which is that corresponding translated pels in successive frames do not usually have the same color subcarrier phase (see Sec. 5.2.3j). The usual cure is to either implicitly or explicitly demodulate the composite signal into its color components, and then apply motion compensation to the components[6.4] before reconstructing the composite signal. Unfortunately, this extra composite demodulation almost always introduces some visible artifacts into the picture. Thus, further significant bit-rate reduction for broadcast quality images may have to wait for the conversion of studio equipment to handle digital component signals. Then, codecs will not have to process composite waveforms at all.

6.3 Interfield Hybrid Transform Coder for Component Color TV

In this section, we discuss an interfield coder design for television (similar to [6.5]) that utilizes hybrid transform coding techniques. Resulting bit-rates are in the range of 10 to 20 Mbits/sec with image quality that is very good, but not transparent as required for network broadcasts. This bit-rate is somewhat below that which can normally be achieved with simple DPCM.

Here we assume that the luminance signal Y as well as two chrominance signals, e.g., I and Q, are available to the coder. This would be the case, for example, in a videoconferencing situation where the camera and coder were in close proximity, or alternatively in a networking application where the studio produced digital component signals. Thus, artifacts due to modulation and demodulation of a color composite signal need not be taken into account in the coding algorithm.

However, interlace does affect both the performance and the design of the coder. We could, in principle, convert each TV frame from a 2:1 interlaced format to a sequentially scanned, non-interlaced format, and then employ intraframe transform or intraframe hybrid transform coding. However, such an

Switch Mode From	To	Buffer Content (kbits)
0	1	>102
1	2	>614
2	3	>691
3	4	>768
4	3	<717
3	2	<640
2	1	<563
1	0	<77

Fig. 6.2.7 Mode control parameters for a buffer size of 800 kbits.

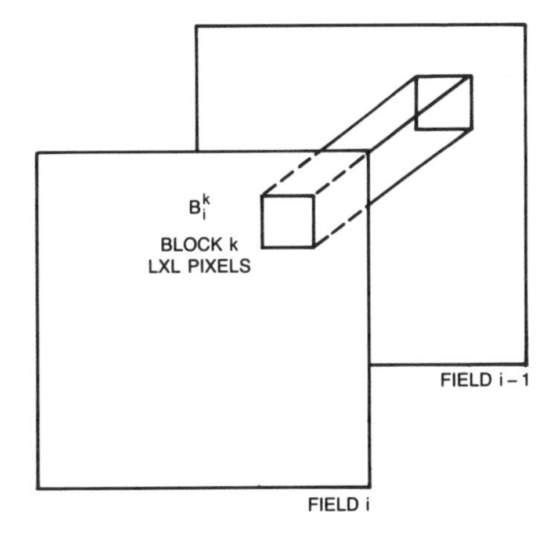

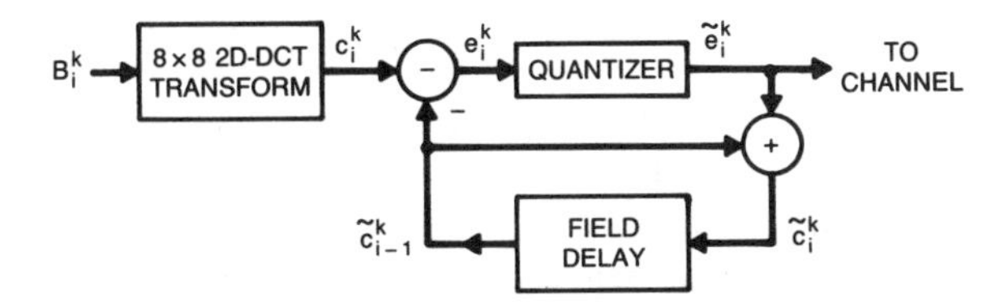

Fig. 6.3.1 Field differential hybrid transform coder for component color TV. Blocks of 8×8 pels are transformed via the 2D-DCT. The coefficients are then coded using field differential PCM.

approach would add another level of complexity without much promise of improved performance, especially in light of the fact that movement in the scene would markedly reduce line-to-line correlation upon which intraframe coders depend so heavily.

Thus, the basic approach of this coder (shown in Fig. 6.3.1) is to partition each field into two-dimensional blocks, transform each block using the separable discrete-cosine-transform (DCT) and then code each transform coefficient using simple field-differential DPCM, i.e., predictive coding using the corresponding coefficient of the previous field as a prediction. In the following, we will dwell mostly on the luminance, since most of the coding bits will be devoted to that component. Chrominance coding will be treated briefly thereafter.

The choice of block size is affected by several factors. A large size leads to a smaller bit-rate in low-detail areas of the picture. However, it also means higher complexity and less ability to adapt to changing picture conditions (see Sec. 5.3.1d). A block size of 8×8 pels has been found to be a reasonable compromise, and this is shown in Fig. 6.3.1.

With interlace, horizontal correlation should exceed vertical correlation within a transform block. However, line-to-line spacing within a *frame* is half the line spacing within a field. Thus, correlation between blocks in successive fields will exceed the pel-to-pel correlation within a block, unless there is rapid horizontal movement in the scene, in which case intra-block pel-to-pel correlation will be increased significantly because of camera integration, which causes blurring of moving areas.

The NTSC luminance signal,* sampled at ≈8.4 MHz produces about 534 pels per scan line. However, nearly 18 percent of these are in the horizontal blanking period, and therefore only about 440 pels per scan line need be coded and transmitted. Similarly, of the 262.5 lines per field, 5 to 8 percent of them are in the blanking period, and therefore only 248 lines/field need be sent. Thus, each field consists of a 31×55 array of blocks.

Improvement in performance with this coder would require at least adaptive quantization, variable word-length (entropy) coding and very large buffer memories at the transmitter and receiver, all of which lead to considerable expense and system complexity. Some compromises are clearly in order.

We saw in Sections 5.3-4 that if the mean square values $\{ \sigma_m^2 \}$ are known for the N quantities $\{ c_m \}$ that are to be coded and transmitted, then the mean square error could be minimized by sending some fraction $p < 1$ of the coefficients and coding each of them at an average rate of r_m bits per coefficient. In particular, using uniform quantization and entropy coding (MSV's arranged

---

* Line scan rate ≈15734 Hz, bandwidth ≈4.2 MHz, 525 lines per frame, image aspect ratio 4:3.

Luminance DCT Coefficient MSV's (16 pix average)

1.9e+05	1.1e+04	2.5e+03	1.3e+03	5.4e+02	2.2e+02	1.3e+02	3.9e+01
1.2e+04	1.3e+03	3.7e+02	1.4e+02	5.8e+01	2.3e+01	9.0e+00	4.9e+00
4.6e+03	5.5e+02	2.2e+02	9.3e+01	4.6e+01	2.4e+01	9.2e+00	1.3e+01
1.5e+03	2.6e+02	1.2e+02	5.8e+01	3.0e+01	1.5e+01	6.4e+00	6.0e+00
5.4e+02	1.4e+02	7.7e+01	4.3e+01	2.5e+01	1.5e+01	6.2e+00	8.0e+00
5.1e+02	8.2e+01	5.3e+01	3.1e+01	2.1e+01	1.4e+01	5.9e+00	9.8e+00
4.9e+02	5.1e+01	3.1e+01	2.0e+01	1.2e+01	7.2e+00	3.5e+00	3.4e+00
2.5e+02	3.6e+01	2.9e+01	2.0e+01	1.6e+01	1.3e+01	5.5e+00	1.2e+01

Corresponding Field Differential MSV's

6.7e+02	5.8e+01	2.7e+01	1.6e+01	1.2e+01	1.2e+01	5.1e+00	1.2e+01
8.3e+02	8.5e+01	3.6e+01	1.7e+01	9.3e+00	5.4e+00	3.1e+00	2.7e+00
4.3e+02	9.9e+01	5.4e+01	3.2e+01	2.3e+01	2.1e+01	8.2e+00	2.1e+01
3.1e+02	9.2e+01	5.0e+01	2.9e+01	1.8e+01	1.2e+01	5.8e+00	9.3e+00
5.5e+02	8.8e+01	5.6e+01	3.4e+01	2.4e+01	1.7e+01	7.4e+00	1.4e+01
8.1e+02	8.6e+01	6.1e+01	3.7e+01	2.8e+01	2.2e+01	9.2e+00	1.8e+01
6.8e+02	6.6e+01	4.5e+01	2.9e+01	1.9e+01	1.2e+01	6.0e+00	6.2e+00
3.6e+02	6.2e+01	5.4e+01	3.7e+01	2.9e+01	2.6e+01	1.0e+01	2.5e+01

Fig. 6.3.2 Mean Square Values (MSV's) of luminance DCT coefficients and their corresponding field differentials averaged over 16 images containing no movement.

Y	I	Q
5 4 3 2 1 0 1 1	4 2 1 0 0 0 0 0	3 1 0 0 0 0 0 0
4 3 3 2 1 0 1 1	3 1 0 0 0 0 0 0	2 1 0 0 0 0 0 0
4 3 2 1 1 0 1 1	2 1 0 0 0 0 0 0	2 0 0 0 0 0 0 0
4 3 2 1 1 0 1 1	2 1 0 0 0 0 0 0	1 0 0 0 0 0 0 0
4 2 2 1 1 0 0 1	2 0 0 0 0 0 0 0	1 0 0 0 0 0 0 0
4 2 2 1 0 0 0 0	1 0 0 0 0 0 0 0	1 0 0 0 0 0 0 0
3 2 2 1 0 0 0 0	0 0 0 0 0 0 0 0	0 0 0 0 0 0 0 0
3 2 2 1 0 0 1 1	0 0 0 0 0 0 0 0	0 0 0 0 0 0 0 0

Fig. 6.3.3 Example of nonadaptive hybrid field differential quantizer bit allocation for the Y, I, and Q color components. Average bit-rate is 2 bits/pel.

BITS/PEL	PSNR $= 20\log\left(\frac{255}{\sqrt{MSE}}\right)$
0.5	29 dB
1.0	32 dB
2.0	38.5 dB
3.0	42.6 dB

Fig. 6.3.4 Performance of nonadaptive hybrid field differential coding. Peak signal to rms noise ratio (unweighted PSNR) is shown versus bit-rate for a scene containing fairly rapid motion.

Luminance DCT Coefficient MSV's (16 pix's containing movement)

7.3e+04	5.0e+03	1.8e+03	8.6e+02	6.0e+02	3.4e+02	5.1e+02	1.2e+02
1.1e+04	8.9e+02	3.3e+02	1.9e+02	1.2e+02	6.3e+01	4.9e+01	2.9e+01
4.0e+03	3.7e+02	1.8e+02	9.4e+01	5.3e+01	3.3e+01	2.4e+01	1.5e+01
1.8e+03	3.0e+02	1.6e+02	7.7e+01	4.6e+01	3.0e+01	2.4e+01	1.6e+01
1.3e+03	1.9e+02	1.1e+02	5.8e+01	4.0e+01	2.7e+01	2.0e+01	1.4e+01
5.2e+02	1.1e+02	6.3e+01	4.5e+01	3.3e+01	2.0e+01	1.5e+01	9.5e+00
1.0e+03	1.3e+02	7.3e+01	4.9e+01	3.5e+01	2.1e+01	1.8e+01	1.3e+01
8.3e+02	1.1e+02	6.0e+01	3.7e+01	2.6e+01	1.8e+01	1.6e+01	1.4e+01

Corresponding Field Differential MSV's

2.0e+03	6.8e+02	3.7e+02	1.8e+02	1.1e+02	6.8e+01	6.3e+01	4.6e+01
1.1e+03	3.6e+02	2.2e+02	1.2e+02	6.4e+01	3.8e+01	3.4e+01	2.7e+01
6.6e+02	2.5e+02	1.5e+02	8.9e+01	4.8e+01	3.0e+01	3.0e+01	2.3e+01
1.6e+03	3.1e+02	1.9e+02	9.9e+01	6.0e+01	4.2e+01	3.7e+01	2.9e+01
1.1e+03	2.2e+02	1.4e+02	8.0e+01	5.4e+01	3.8e+01	3.4e+01	2.2e+01
4.0e+02	1.4e+02	9.5e+01	7.0e+01	5.3e+01	3.7e+01	2.9e+01	1.7e+01
1.4e+03	2.3e+02	1.3e+02	8.1e+01	6.0e+01	4.1e+01	3.4e+01	2.5e+01
1.2e+03	2.1e+02	1.1e+02	7.2e+01	5.1e+01	3.4e+01	3.0e+01	2.5e+01

Fig. 6.3.5 Mean Square Values (MSV's) of luminance DCT coefficients and field differentials averaged over 16 frames of a scene containing motion.

Luminance DCT Coefficient MSV's (16 pix's moving area only)

6.6e+04	6.2e+03	1.7e+03	6.4e+02	3.5e+02	2.9e+02	2.6e+02	5.7e+01
2.1e+04	1.6e+03	5.4e+02	2.4e+02	1.4e+02	7.9e+01	6.4e+01	3.1e+01
1.1e+04	7.5e+02	2.5e+02	1.1e+02	5.1e+01	3.6e+01	2.9e+01	1.6e+01
5.4e+03	5.6e+02	2.6e+02	1.3e+02	5.2e+01	5.5e+01	4.0e+01	2.4e+01
3.2e+03	3.7e+02	1.5e+02	8.0e+01	3.5e+01	2.9e+01	2.3e+01	1.5e+01
1.4e+03	1.5e+02	8.6e+01	3.7e+01	1.6e+01	1.2e+01	1.0e+01	6.0e+00
4.4e+03	2.7e+02	1.0e+02	7.5e+01	3.7e+01	3.8e+01	2.9e+01	1.8e+01
3.2e+03	2.6e+02	1.0e+02	5.7e+01	3.0e+01	3.1e+01	2.8e+01	1.5e+01

Corresponding Field Differential MSV's

1.0e+04	2.2e+03	1.1e+03	4.5e+02	2.2e+02	1.1e+02	1.5e+02	1.1e+02
4.0e+03	9.9e+02	6.3e+02	2.6e+02	1.1e+02	5.1e+01	5.0e+01	4.1e+01
2.0e+03	6.9e+02	3.9e+02	1.8e+02	7.1e+01	5.2e+01	3.9e+01	3.5e+01
7.0e+03	1.0e+03	6.0e+02	2.4e+02	1.2e+02	9.2e+01	6.7e+01	6.1e+01
4.5e+03	6.1e+02	3.6e+02	1.5e+02	7.4e+01	4.6e+01	4.1e+01	3.6e+01
8.8e+02	2.1e+02	1.3e+02	8.7e+01	3.7e+01	3.0e+01	1.9e+01	1.7e+01
6.3e+03	7.1e+02	3.7e+02	1.7e+02	1.0e+02	6.6e+01	5.3e+01	5.2e+01
5.8e+03	6.7e+02	3.4e+02	1.5e+02	8.9e+01	6.2e+01	5.4e+01	4.5e+01

Fig. 6.3.6 MSV's of luminance DCT coefficients and field differentials averaged over the moving areas of 16 frames containing movement.

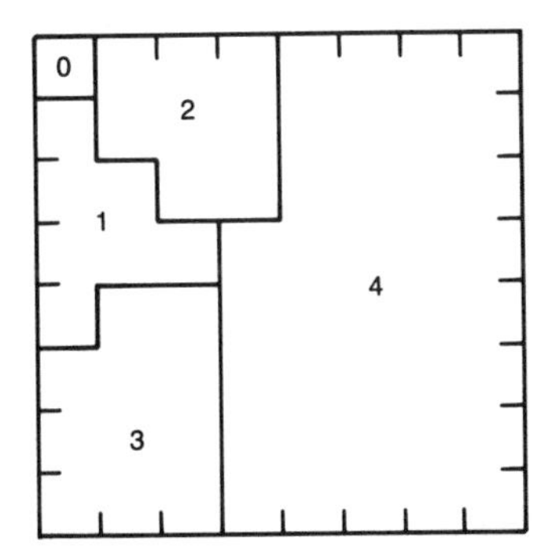

Fig. 6.3.7 Each block of 8×8 DCT coefficients is partitioned into five sub-blocks, and the field differential MSV's are measured within each sub-block to enable adaptive quantization depending on temporal and spatial activity. Sub-blocks are not used for chrominance.

in decreasing order)

$$r_m \approx \frac{1}{1.39} \ln \frac{\sigma_m^2}{\sigma_{pN}^2} \text{ bits.} \tag{6.3.1}$$

The quantizer step size for all coefficients is given by

$$\Delta \approx 3.81 \ \sigma_{pN} \tag{6.3.2}$$

Similar results were obtained for the Lloyd-Max quantizer.

As an example, consider Fig. 6.3.2, which shows DCT luminance coefficient MSV's plus the corresponding hybrid field differential MSV's averaged over 16 monochrome pictures containing no movement. As expected, the vertical spatial frequencies show higher energies than the horizontal because of the larger spacing between lines of a field. Also as expected, the total energy in the hybrid field differentials is significantly less than that of the DCT coefficients themselves. However, we also see that many of the hybrid field differential MSV's (particularly at the higher vertical frequencies) are larger than their DCT counterparts, which indicates that there is not much field-to-field correlation for those particular coefficients.

If Lloyd-Max quantization is used, the number of bits per coefficient is constant. Fig. 6.3.3 shows an example of hybrid field-differential quantizer bit allocation for the *Y*, *I*, and *Q* signals. In this case the average bit-rate is two bits/pel. Fig. 6.3.4 shows the luminance PSNR (unweighted) at various bit-rates using Lloyd-Max quantizers to code a scene containing fairly rapid motion.[6.5] We see that for reasonably good picture quality (35-40 dB PSNR) close to two bits/pel or 13 Mbs are required. For very good picture quality (40-45 dB PSNR) about 3 bits/pel or 20 Mbs are needed for luminance.

Fig. 6.3.5 shows DCT luminance coefficient MSV's along with hybrid field differential MSV's averaged over 16 frames containing two people in motion. Although the DCT coefficient MSV's are comparable to those of Fig. 6.3.2, we see that the hybrid field differential MSV's are somewhat larger. In fact, if we measure MSV's only over the moving areas of the pictures, as shown in Fig. 6.3.6, we see that the moving-area hybrid field differentials are, for the most part, considerably larger than stationary-area field differentials of Fig. 6.3.2. Moreover, most of the moving-area field differentials are larger than their corresponding DCT coefficients, although the total moving-area field differential energy is still less than the coefficient energy.

Therefore, not only do we have to code the transmitted values differently from each other because of dissimilar MSV's, we should also code the moving-areas of the picture differently than the stationary areas. Moreover, some of the blocks are "low-detail" with low MSV's, while other blocks are "high-detail" having larger MSV's.

One way of adapting to these changing conditions is to partition the block of hybrid differentials to be coded into sub-blocks, and measure the sum of the

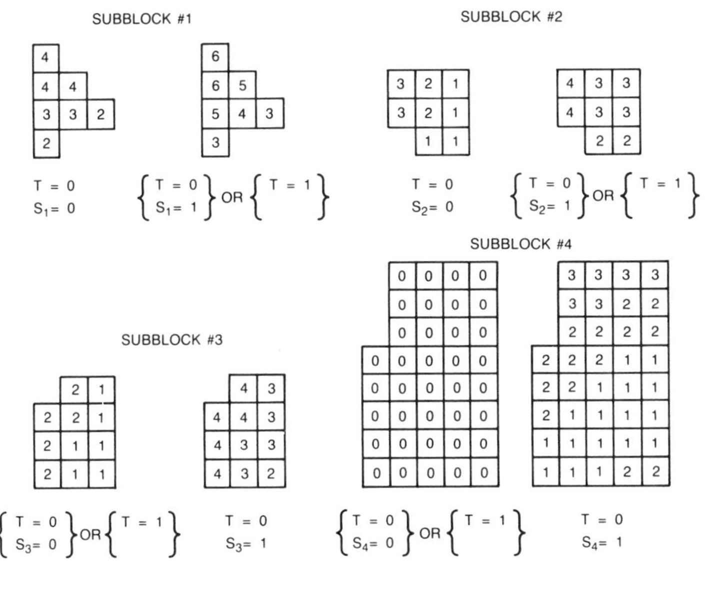

I, T = 0

3	1	0	0	0	0	0	0
2	0	0	0	0	0	0	0
2	0	0	0	0	0	0	0
1	0	0	0	0	0	0	0
1	0	0	0	0	0	0	0
0	0	0	0	0	0	0	0
0	0	0	0	0	0	0	0
0	0	0	0	0	0	0	0

I, T = 1

4	2	1	0	0	0	0	0
3	2	1	0	0	0	0	0
3	2	0	0	0	0	0	0
2	1	0	0	0	0	0	0
2	1	0	0	0	0	0	0
1	0	0	0	0	0	0	0
1	0	0	0	0	0	0	0
0	0	0	0	0	0	0	0

Q, T = 0

2	0	0	0	0	0	0	0
1	0	0	0	0	0	0	0
1	0	0	0	0	0	0	0
1	0	0	0	0	0	0	0
1	0	0	0	0	0	0	0
0	0	0	0	0	0	0	0
0	0	0	0	0	0	0	0
0	0	0	0	0	0	0	0

Q, T = 1

3	1	0	0	0	0	0	0
2	1	0	0	0	0	0	0
2	1	0	0	0	0	0	0
2	0	0	0	0	0	0	0
1	0	0	0	0	0	0	0
1	0	0	0	0	0	0	0
0	0	0	0	0	0	0	0
0	0	0	0	0	0	0	0

CHROMINANCE BIT ASSIGNMENTS

Fig. 6.3.8 Quantizer bit assignments for the sub-blocks of Fig. 6.4.7. For temporally active blocks, sub-blocks 1 and 2 are coded with more bits. Otherwise, coding is according to spatial activity. Sub-blocks are not used for chrominance.

BITS/PEL	PSNR
0.89	33.7 dB
1.37	36.4 dB
1.87	39.5 dB

Fig. 6.3.9 Performance of adaptive hybrid field differential coding. Unweighted PSNR is shown versus bit-rate for the same scene used in Fig. 6.4.4.

squared differential values within each sub-block in order to determine how the coefficients should be coded. For example, Fig. 6.3.7 shows partitioning of an 8×8 block into five sub-blocks.[6.5] If the DC squared field differential value of sub-block 0 exceeds a predefined threshold, then the entire block is considered temporally active, and parameter T is set to 1. Otherwise, $T = 0$. If the sum of the squared field differential values of sub-block i exceeds a predefined threshold, then that sub-block is considered active and $S_i = 1$. Otherwise, $S_i = 0$. The sub-blocks then have their field differential values quantized and coded according to their activity classification.

Fig. 6.3.8 shows an adaptive Lloyd-Max quantizer bit assignment designed for an overall average bit-rate of 2.0 bits/pel. The DC field differential value (sub-block 0) is always coded with 8-bits. For temporally active blocks ($T=1$), sub-blocks 1 and 2 are coded using more bits while sub-blocks 3 and 4 are coded using fewer bits. For temporally inactive blocks ($T=0$), the sub-blocks are coded individually using variable bits depending on whether they are active or non-active, respectively. The chrominance differential values are coded with variable bits depending only on whether the block is temporally active or temporally non-active, respectively.

Fig. 6.3.9 shows luminance PSNR (unweighted) for this adaptive coding algorithm[6.5] operating on the same color TV sequence used in Fig. 6.3.4. Comparison shows an improvement with adaptive coding of only 1 to 2 dB. However, the adaptive algorithm is less sensitive to image statistics and handles scene changes better than the nonadaptive algorithm.

6.4 Adaptive Interframe Transform Coder for Contribution-Quality Applications - CCIR 723

The CCIR* has standardized a component video coder, designated as CCIR 723, for video applications requiring contribution quality, i.e., near transparent reproduction. The source video is CCIR 601 Y Cb Cr components, with pel values biased downward to the range -128 to 127 prior to coding. CCIR 723 is commonly used at bit rates in excess of 30 Mbs for 50 fields/sec and 40 Mbs for 60 fields/sec.

The coding algorithm is a straightforward extension of the previous Section, with several additional features for improving coding efficiency. Most significantly, the prediction is carried out in the pel domain. This requires an additional 8×8 inverse discrete cosine transform (IDCT), as shown in Fig. 6.4.1. However, the ability to easily perform pel-domain motion-compensated prediction more than offsets the additional complexity of the IDCT. CCIR 723

---

* Now known as International Telecommunication Union - Radio Sector, abbreviated ITU-R.

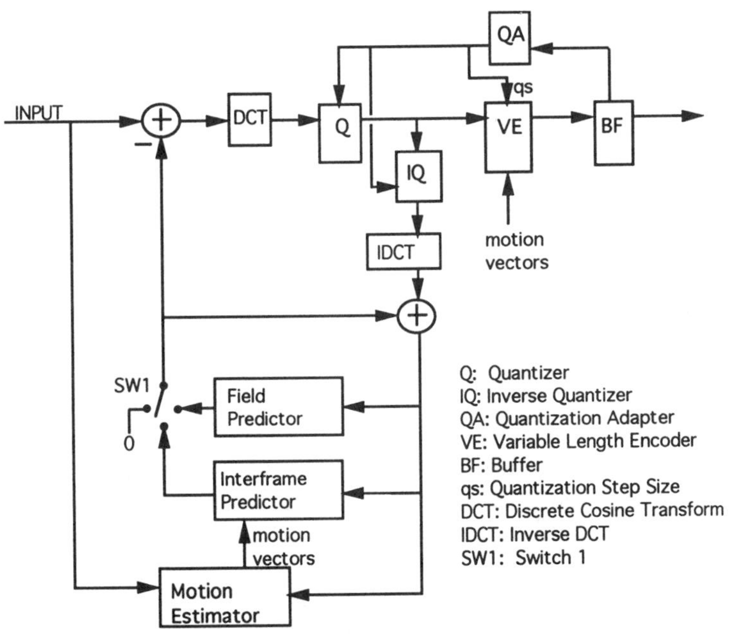

Fig. 6.4.1 Adaptative interframe transform coder CCIR 723. For each Macroblock three coding modes are available - intrafield, interfield and motion compensation using decoded pels from two fields back.

defines a Block as 8×8 samples. It also defines a Macroblock as 16 pels by 8 lines, including luminance and chrominance. Thus, a Macroblock consists of two horizontally adjacent Y-Blocks, plus one Cb-Block and one Cr-Block, i.e., 256 samples in all. A Stripe is a complete horizontal row of 45 Macroblocks. Each field of video has either 288 coded lines (50 Hz) or 248 coded lines (60 Hz). For the latter, the first five lines are not active and are discarded by the decoder.

For each Macroblock, three coding modes are available. The first is simple Intrafield transform coding, where Switch 1 of Fig. 6.4.1 is set to the zero position for the duration of the Macroblock. The second is Interfield prediction, where each pel of the Macroblock is predicted using an average of the pel above and the pel below in the previous decoded field.

The third coding mode incorporates Motion Compensated Interframe prediction using the decoded pels from two fields back in time. One Motion Vector per Macroblock is transmitted and used to predict all pels in the Macroblock. The Motion Vector range is ± 15.5 Y pels horizontally, and ± 7.5 lines vertically, with half pel accuracy. The predictions for noninteger Motion Vectors are computed using bilinear interpolation, i.e., linear interpolation horizontally followed by linear interpolation vertically. Note that for color pels, horizontal interpolation with quarter pel accuracy is needed.

The method used by an encoder for choosing the Macroblock coding mode is not specified by the standard. One possibility is to encode each Macroblock by all three modes and select the one producing the fewest bits. A somewhat simpler method is to compute the mean square (or absolute) prediction error for each mode and select the mode producing the smallest value.

Motion Vectors are DPCM coded (no quantization, however) using the Motion Vector to the left as a prediction. Zero prediction is used for the leftmost Macroblock. The horizontal and vertical components of the differential Motion Vectors are then coded separately using Huffman coding. Codewords are needed for all values between ± 31.0. Codewords range from 2 to 14 bits in length.

Adaptive quantization for CCIR 723 utilizes many of the principles described in previous chapters. First, frequency dependent log Visibility Threshold matrices are defined for luminance and chrominance as shown in Fig. 6.4.2. The large asymmetry of the Y matrix is due to the Y pels being much closer horizontally than vertically. Thus, horizontal luminance DCT spatial frequencies can withstand more quantization noise than equivalent vertical spatial frequencies.

Next, a four level Activity Function is defined that is used to augment the Visibility Thresholds for each Macroblock according to some measure of local masking. The masking measure is outside the standard. In principle, spatial and/or temporal masking could be used to determine the Activity Function.

m \ n	0	1	2	3	4	5	6	7
0	00	00	02	08	12	18	22	28
1	00	06	06	10	16	18	22	34
2	00	06	10	14	18	20	24	38
3	02	06	12	16	18	20	26	40
4	06	12	14	16	20	22	28	42
5	10	14	14	18	22	24	30	42
6	14	16	16	18	22	24	34	44
7	14	18	18	20	24	30	38	44

Initial Luminance Thresholds

m \ n	0	1	2	3	4	5	6	7
0	00	00	03	04	06	08	08	11
1	00	01	02	03	06	08	09	13
2	02	02	03	04	07	09	10	16
3	03	04	05	05	08	10	12	16
4	05	06	06	07	09	11	13	17
5	08	07	09	09	11	14	16	21
6	10	11	11	11	14	16	19	24
7	12	12	12	12	17	18	20	26

Initial Chrominance Thresholds

Fig. 6.4.2 Frequency dependent log Visibility Thresholds for luminance and chrominance DCT coefficients. A higher Threshold implies a lower visibility of quantization noise. These values are augmented according to a local Activity Function to obtain T_{mn}.

Quantizer Index	Quantized Value	Quantizer Index	Quantized Value
0	0	384	513
1	1	385	517
2	2	386	521
.	.	.	.
.	.	.	.
.	.	.	.
255	255	511	1021
256	256	512	1027
257	258	513	1035
258	260	514	1043
.	.	.	.
.	.	.	.
.	.	.	.
383	510	639	2043

Fig. 6.4.3 Companded quantizer characteristic. Uniformly quantized integer values in the range -2048 to 2032 are rounded to the nearest allowed representation level. Only positive levels are shown. The 1279 levels are numbered by a quantization index I_{mn} in the range -639 to +639.

m \ n	0	1	2	3	4	5	6	7
0	00	02	06	12	20	28	36	44
1	01	05	11	19	27	35	43	51
2	03	07	13	21	29	37	45	52
3	04	10	18	26	34	42	50	57
4	08	14	22	30	38	46	53	58
5	09	17	25	33	41	49	56	61
6	15	23	31	39	47	54	59	62
7	16	24	32	40	48	55	60	63

Scanning Order for Luminance

m \ n	0	1	2	3	4	5	6	7
0	00	02	03	09	10	20	21	35
1	01	04	08	11	19	22	34	36
2	05	07	12	18	23	33	37	48
3	06	13	17	24	32	38	47	49
4	14	16	25	31	39	46	50	57
5	15	26	30	40	45	51	56	58
6	27	29	41	44	52	55	59	62
7	28	42	43	53	54	60	61	63

Scanning Order for Chrominance

Fig. 6.4.4 Scanning path for quantized DCT coefficients.

For the lowest Activity level, Y thresholds are clipped to a maximum of 24, and Cb/Cr are clipped to a maximum of 9. For the next Activity level, Y thresholds are clipped to a maximum of 34 and Cb/Cr are clipped to a maximum of 16. For the next Activity level all thresholds are incremented by 2, and for the highest Activity level all thresholds are incremented by 8. The value of the Activity Function is transmitted for each Macroblock so that the decoder is also able to derive the augmented log Visibility Thresholds T_{mn}.

The next control of quantization is effected by a buffer fullness factor F that is transmitted once per stripe. The range of F is 0 to 175.

Given the buffer fullness factor F and the augmented log Visibilities T_{mn}, we first compute an interim log Visibility q_{mn} using an empirically derived relation

$$q_{mn} = Min[2T_{mn} - 48, F] + F \tag{6.4.1}$$

Following this, q_{00} is clipped to be in the range 0 to 48, q_{mn} is clipped to be in the range 0 to 175, and the actual quantization step size for each DCT coefficient is computed by

$$\Delta_{mn} = 2^{[q_{mn}/16 - 1]} \tag{6.4.2}$$

The range of Δ_{mn} is 0.5 to 980. It is computed to an accuracy of 12-bits.

A first mid-step uniform quantization is carried out by simply dividing each DCT coefficient by Δ_{mn} and rounding to the nearest integer. A final companded quantization is then effected by rounding the magnitude to the nearest representation level in Fig. 6.4.3. At the decoder, the quantized DCT coefficients $\tilde{c}_{mn}$ are simply the product of the (signed) representation levels and Δ_{mn}.

Assuming prediction errors are clipped to the range -128 to 127, DCT coefficients have the range -1024 to 1016 before quantization, -2048 to 2032 after uniform quantization, and an allowable range of -2043 to +2043 after companded quantization.

Each of the 64 coefficients in a block is thus represented by a quantizer index I_{mn} in the range -639 to +639. The quantizer indices for each block are transmitted in the order shown in Fig. 6.4.4, which approximates the order of decreasing coefficient variance for luminance and chrominance blocks.

Each nonzero index is assigned its own variable length codeword. However, zero indices are runlength coded with variable length codes assigned to each possible runlength between 1 and 63. The last runlength, if any, of each block is not transmitted. Instead an End-of-Block (EOB) code tells the decoder that all remaining coefficients of the block are zero. An EOB is sent for each block even if there is no last runlength.

If two runlengths are separated by one or more indices of value +1, one of the indices is not transmitted. At the decoder an extra index of value +1 is inserted.

CODEWORD (base 4)	Indices and Runlengths (r) Y	Y&C	C
1		1	
30		2	
31		r1	
320		r3	
321	3		r5
330	4		r7
331		EOB1	
3220	r5		3
3221	r7		4
3230		r9	
3231		r11	
3320		5	
3321	6		r13
3330	7		r15
3331	8		r17
32220	r13		6
32221	r15		7
32230	r17		8
32231		r19	
32320		r21	
32321		r23	
32330		r25	
32331		r27	
33220		9	
33221		10	
33230		11	
33231		12	
33320		13	
33321		14	
33330		15	
33331		16	
322220		r29	
322221		r31	
322230		r33	
322231		r35	
322320		r37	
322321		r39	
322330		r41	
322331		r42	
323220		r45	
323221		r47	
323230		r49	
323231		r51	
323320		r53	
323321		r55	
323330		r57	
323331		r59	
332220		r61	
332221		r63	

CODEWORD (base 4)	Indices and Runlengths (r) Y	Y&C	C
0		-1	
21		-2	
20		r2	
231		r4	
230	-3		r6
221	-4		r8
220		EOB0	
2331	r6		-3
2330	r8		-4
2321		r10	
2320		r12	
2231		-5	
2230	-6		r14
2221	-7		r16
2220	-8		r18
23331	r14		-6
23330	r16		-7
23321	r18		-8
23320		r20	
23231		r22	
23230		r24	
23221		r26	
23220		r28	
22331		-9	
22330		-10	
22321		-11	
22320		-12	
22231		-13	
22230		-14	
22221		-15	
22220		-16	
233330		r30	
233331		r32	
233320		r34	
233321		r36	
233230		r38	
233231		r40	
233220		r42	
233221		r44	
232330		r46	
232331		r48	
232320		r50	
232321		r52	
232230		r54	
232231		r56	
232220		r58	
232221		r60	
223330		r62	
223331		null	

Fig. 6.4.5 A portion of the Variable Length Code for Quantizer Indices and Runlengths. Only the first 98 codewords are shown. Codewords are shown in base 4, i.e., two bits per quarternary symbol.

For the shorter codewords, different variable length codes are used for luminance and chrominance coefficients. Fig. 6.4.5 shows some of these. For index values larger in magnitude than 16, an algorithmic construction of codewords is provided to alleviate the need for large lookup tables.

Two EOB codewords are provided for synchronization purposes. For each block one or the other is chosen based upon a predetermined selection rule known to both encoder and decoder. If the wrong EOB is received, the decoder knows an error occurred.

6.5 Decomposition of Single Images into Edges and Texture

We have seen that under virtually no circumstances can images (of any interest) be modeled as two-dimensional, stationary, Gaussian random fields. Instead they tend to consist of distinct objects, such as persons, lungs or grassy fields, which have fairly well defined and abrupt boundaries and whose interiors are more or less random texture. Unfortunately, we have also seen that rate-distortion theory is most easily applied to stationary, Gaussian random processes. Moreover, transform coding is also most efficient on images that do not contain many large luminance transitions such as occur at object boundaries. One approach to resolving this dilemma is to decompose the visual information into two components, i.e.,

$$B(x,y) = d(x,y) + r(x,y) \tag{6.5.1}$$

where $d(x,y)$ is an image containing basically objects with their interior texture removed, and $r(x,y)$ is a remainder image containing texture, surface roughness and other irregularities of the object interiors.[6.6] Thus, in the $d(x,y)$ image, relatively large luminance transitions occur at the object boundaries, whereas the interiors of objects are relatively smooth and uniformly colored. In the $r(x,y)$ image, the luminance transitions are much smaller and noise-like.

Hopefully, $d(x,y)$ can then be coded efficiently using algorithms such as interpolative coding, whereas $r(x,y)$ will approximate a Gaussian random field that can be efficiently represented using transform coding techniques. We also desire that mean square error (or at least frequency-weighted MSE) be a meaningful distortion criterion for the transform coding. This is facilitated if a compression-type nonlinearity, e.g., log or cube-root, is applied either to $B(x,y)$ or $r(x,y)$ prior to coding so that at least Weber's Law effects are compensated.

Construction of the object boundary function $d(x,y)$ is not straightforward for several reasons. For example, at some points on the object boundary, the object intensity may be nearly equal to that of the background scene behind the object, which can lead to an incorrect object boundary definition. Also, distinguishing between a distinct small object and a piece of texture within a large object, is a design decision that depends on the final desired picture quality as well as the relative proportion of bits devoted to $d(x,y)$ and $r(x,y)$. If the smallest allowable object size is too small, then $d(x,y)$ will require too many bits, whereas if only large objects are allowed, then $r(x,y)$ will require too many bits.

Thus, the edge detection algorithm implicit in the construction of $d(x,y)$ must include a smoothing operation as well as spatial differentiation. Many edge detection schemes are contained in the pattern recognition literature. An example is shown in Fig. 6.5.1. First, a spatial differentiation is carried out on the image.* In this case, each pel is replaced by the difference between itself and an average of its four nearest neighbors. The magnitudes of the differential values are then thresholded with the result shown in Fig. 6.5.1b, i.e., pels for which the absolute value of the differential exceeds some predetermined threshold are shown in black. A smoothing operation is then carried out in order to eliminate small objects and unclosed contours. The result defines the object boundaries, and is shown in Fig. 6.5.1c.

The next step in constructing $d(x,y)$ is to approximate the image intensity on even numbered scan lines (exemplified in Fig. 6.5.2) by a piecewise linear function whose breakpoints are at or near the boundary pels defined previously. Note that in this case the *boundary* may consist of several pels, and that a linear approximation of the boundary pels is calculated as well.

For each odd numbered scan line, an interpolated piecewise linear approximation is formed from those of the adjacent even lines above or below. A check is then carried out to see how close this interpolated approximation is to the original intensity. If it is not good enough (according to some predefined criterion), then a separate piecewise linear approximation is formed for that odd line in the same way as was done for even lines. The resulting piecewise linear approximations on each scan line then define the *object* image intensity $d(x,y)$.

The breakpoints of $d(x,y)$ may be coded using any of the two-level graphics coding methods described in earlier chapters (see Sec. 5.7). For example, in [6.6] a two-dimensional, run-length, Huffman coding scheme is utilized. The amplitudes of $d(x,y)$, which are usually quantized more coarsely than the original image, must be sent separately.

Coding the remainder function $r(x,y)$ can be accomplished with good efficiency by the use of transform coding. For a given picture quality, the techniques described in Chapter 5 can be used straightforwardly. For example, in [6.6] $r(x,y)$ is clipped so that its absolute value does not exceed 40 out of a possible 255. This clipping causes errors only near sharp edges where it is much less visible. Next, the entire $r(x,y)$ is transform coded using the RDFT (see Sec. 5.3.1) with uniform quantization and a frequency weighted MSE distortion criterion. A single variable word-length code is then used for coding each of the

* Gamma corrected and digitized by non-interlaced raster scanning and 8-bit PCM quantization.

Fig. 6.5.1 (a) Original Image. (b) Example of contours of thresholded spatial derivative. (c) Remaining objects after elimination of unclosed contours and small objects (from Kocher and Kunt [6.7]).

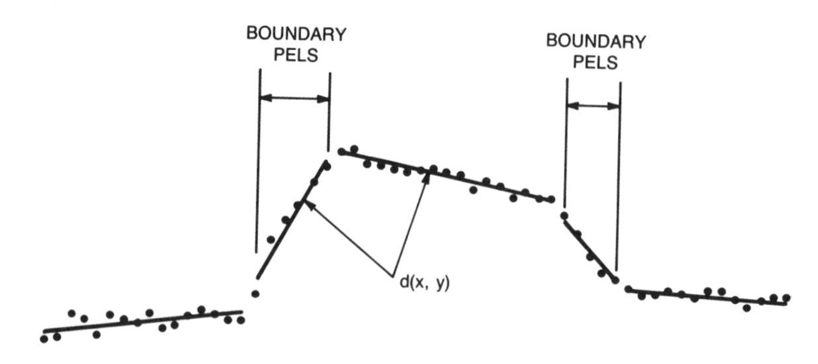

Fig. 6.5.2 Piecewise linear approximation to pel values along a scan line. The resulting discontinuous function is used to form $d(x,y)$.

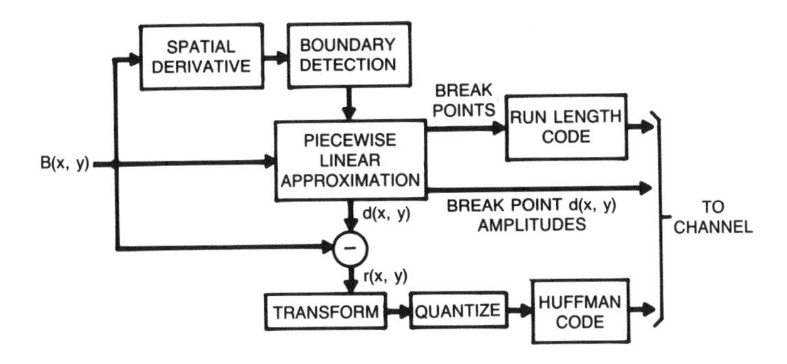

Fig. 6.5.3 System for coding images by decomposing them into edges and texture. For simplicity, only the luminance coding is shown.

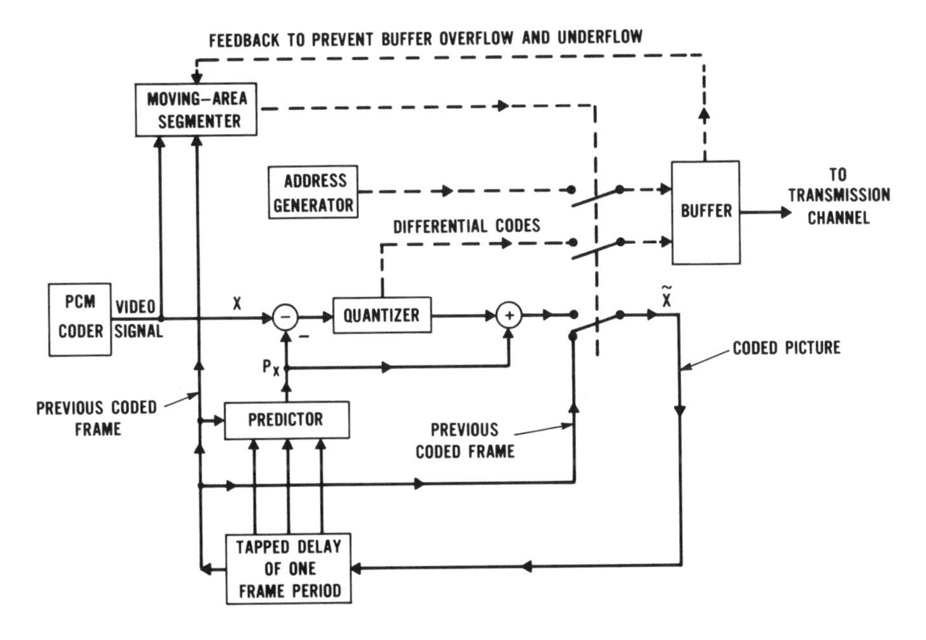

Fig. 6.6.1 Conditional replenishment coder that sends moving-area picture elements (pels) via predictive coding. For each pel X, the predictor uses previously coded pels to compute a prediction P_x. The differential signal $X-P_x$ is quantized and coded. If X is a moving-area pel as determined by the segmenter, the differential code plus addressing information is fed to the buffer to await transmission, and a new pel value $\tilde{X}$ is passed to the frame memory to be used later by the coder.

transmitted coefficients. Using this algorithm, which is shown diagrammatically in Fig. 6.5.3, images have been coded with moderate quality at rates around 0.8 bits/pel.

For a fixed picture quality, image decomposition achieves a lower bit-rate than using transform coding by itself. However, the resulting bit-rates will vary markedly from picture to picture. If a fixed bit-rate is desired for both $r(x,y)$ and $d(x,y)$, then the complexity of image decomposition rises significantly since the edge detection algorithm used to define $d(x,y)$ must then adapt itself to produce an approximately constant number of breakpoints. An alternative would be to use a nonadaptive edge detection algorithm, accepting whatever bit-rate results for $d(x,y)$ and then simply assigning whatever bits are left to the coding of $r(x,y)$. However, achievable picture quality would not be as high.

6.6 Conditional Replenishment DPCM Codecs

These systems typically have the generic form shown in Fig. 6.6.1. Moving-area pels (as defined by the segmenter) are transmitted via predictive coding, whereas stationary-area pels are not transmitted at all. Instead their previous frame values are repeated both at the transmitter and receiver. The quantizer might be a scalar quantizer operating on individual pels, or it might be a block quantizer operating on two-dimensional blocks of pels either in the pel domain or in some transform domain. The predictor might be a simple previous frame predictor, or it might be scene adaptive incorporating such algorithms as motion compensation. With highly efficient predictors, a further segmentation may be utilized that classifies moving-area pels as either *predictable*, in which case the prediction error is nearly zero and need not be sent, or *unpredictable*, in which case prediction errors must be transmitted.

The data rate produced by the coder is directly related to the amount of movement in the scene. As movement increases, the buffer would normally tend to fill unless steps are taken to reduce the data rate. These include spatial subsampling, i.e., only sending a subset of the moving-area pels, coarser quantization in order to reduce the number of bits per transmitted pel, and field/frame repeating, i.e., sending nothing for an entire field/frame period.

In many countries of the world, digital transmission facilities abound at certain *primary* bit-rates. In North America and Japan, one of the most plentiful is the so called DS1 rate of 1.544 Mbits/sec. In Europe, the corresponding rate is 2.044 Mbits/sec. Thus, there is great incentive to use these fairly ubiquitous bit-rates for videoconferencing or videotelephone services, and indeed standards have already been set by the CCITT[6.8] for these rates.

In videoconferencing applications the television camera is fixed. Picture data is then produced only if there are moving objects in the scene. In this case the amount of moving-area in the picture is generally much smaller than the amount of stationary area, and significant bit-rate reduction is possible using conditional frame replenishment. In those rare instances where there is a lot of movement, the data rate generated by the coder can be lowered and buffer overflow prevented by reducing the moving-area resolution (either spatial, temporal, or amplitude) and keeping the stationary background area at full resolution.

Some videoconferencing systems have several cameras, and at any given time the one pointing toward the current or most recent speaker is switched into the transmission. In this case, data is produced also by scene changes when the camera is switched. However, we have seen in Chapter 4 that a new full-resolution picture need not be sent in one frame period. Viewers cannot perceive full resolution for several frame periods after a scene change. Thus, a frame replenishment coder need only transmit a very low-resolution picture immediately following a scene change. If full resolution can be built up within a reasonable time (8-10 frame periods), picture degradation will not be objectionable.

For purposes of illustration in this section, we will excerpt design details from one of the CCITT algorithms[6.8], in particular, the "Inter-Regional Part 2" codec, which can input either 525-line or 625-line video and operate at either 1.544 Mbits/sec or 2.044 Mbits/sec. This algorithm is a compromise design that accomplishes compatibility between the industrial countries of the world and can be implemented with relative simplicity. For this reason, its picture resolution is somewhat less than might be accomplished if compatibility and simplicity were sacrificed

The input interlaced video consists of luminance and chrominance $Y\ Cb\ Cr$ components as defined in Section 2.2.5. These are sampled with 8-bits per pel at a sampling rate (blanking included) of 320 pels per line for luminance and 64 pels per line for chrominance. Pel amplitudes are restricted to the range 16–239. Next, a spatial averaging process is carried out to reduce the number of active lines per field to 143. The chrominance is then line interleaved as in Section 2.2.5, so that only one chrominance component per line need be transmitted. This results in 256 active luminance pels and 52 active chrominance pels per line, with an uncompressed bit-rate of 17.6 Mbits/second. Interpolation at the receiver restores the correct number of lines for display.

The codec is capable of operating at frame rates of 25 Hz or 30 Hz. If the input and output frame rates are different, it transmits 25 frames per second. With a 30 Hz input frame rate, this is accomplished by not coding one out of six frames. With a 30 Hz output, a separate frame memory is used to convert from 25 Hz by means of temporal interpolation.

We saw from Fig. 3.2.12b that among the simple linear interframe predictors, the element difference of field difference had the lowest entropy, with element difference of frame difference and line difference of frame difference coming second. At higher speeds of movement intrafield predictors such as the simple element difference or the intrafield $(A+D)/2$ predictor of Fig. 6.3.1 also perform well.

However, with most interframe codecs horizontal and vertical spatial subsampling of the moving area is an important technique for data rate reduction when buffer overflow threatens (with interlace vertical subsampling is equivalent to sending moving-area pels only in alternate fields). Moreover, with subsampling, the moving-area pels that are not transmitted are interpolated from their neighbors, and in this case, the predictor should only use non-subsampled pels that are available with fairly good accuracy. Otherwise, the prediction values may be corrupted by interpolation error.

During intervals of low movement when little data is being generated, predictor efficiency is of less concern than during periods of high movement. Since intrafield and interfield predictors perform about the same during high movement, the inter-regional CCITT codec uses a relatively simple intrafield prediction to code moving-area pels. For luminance pels the $(A+D)/2$ predictor is used, whereas for chrominance, pel A is used as a prediction. During periods of horizontal subsampling, if either pel A or pel D were obtained by interpolation, then the non-interpolated pel to the left is used instead in the prediction. In normal operation, prediction errors are quantized to 16 levels and coded with variable word-lengths as shown in Fig. 6.6.2.

The inter-regional codec uses adaptive horizontal subsampling to reduce the data rate when necessary. Normally, even numbered pels are sent during even lines, and odd pels during odd lines. Transmitted prediction errors are quantized and coded with 8 levels, as shown in Fig. 6.6.3. However, some of the intervening pels may also be sent in order to reduce the interpolation error. Differential values for these pels are coded using the "extra" variable word-length codes given in Fig. 6.6.3. Pels that are horizontally subsampled are obtained by interpolating the two horizontally adjacent pels.

As mentioned above, vertical subsampling results in not sending anything for an entire field period. In principle, all pels of the field could be replaced by an average of pels in the adjacent fields. For example, in Fig. 6.3.1 each pel X could be replaced by a four-way average of pels F and I in the two adjacent fields. However, this would cause a noticeable resolution reduction in non-moving areas of the picture. Thus, the inter-regional codec only performs the interpolation of pel X if all four temporally nearby pels (F and I in the adjacent fields) were nonmoving. Otherwise, pel X is repeated from two fields back, thus maintaining full spatial resolution in the stationary background.

If spatio-temporal subsampling is to be used as a data reduction mechanism then, in principle, we should low-pass filter the signal in order to avoid aliasing due to too low a sampling rate. Spatial filtering is very difficult to implement only in the moving area. However, temporal filtering is not, and moreover it has almost the same effect. Fig. 6.6.4(a) shows a one-stage recursive temporal filter where the value of α determines the amount of filtering, and the frame memory is the same one used by the coder (see Fig. 6.6.1). Fig. 6.6.4(b) shows a temporal filter containing its own frame memory. In this case, coding error does

Input Levels	Output Levels	Variable-length Code	Code No.
−255 to −125	−141	1 0 0 0 0 0 0 0 0 1	17
−124 to −95	−108	1 0 0 0 0 0 0 0 1	16
−94 to −70	−81	1 0 0 0 0 0 0 1	15
−69 to −49	−58	1 0 0 0 0 0 1	14
−48 to −32	−39	1 0 0 0 0 1	13
−31 to −19	−24	1 0 0 0 1	12
−18 to −9	−13	1 0 1	10
−8 to −1	−4	1 1	9
0 to 7	+3	0 1	1
8 to 17	+12	0 0 1	2
18 to 30	+23	0 0 0 1	3
31 to 47	+38	0 0 0 0 1	4
48 to 68	+57	0 0 0 0 0 1	5
69 to 93	+80	0 0 0 0 0 0 1	6
94 to 123	+107	0 0 0 0 0 0 0 1	7
124 to 255	+140	0 0 0 0 0 0 0 0 1	8

Fig. 6.6.2 Quantization parameters and code words for prediction errors during full horizontal sampling. The end-of-cluster (EOC) code number 11 is 1001.

Quantization		Variable-Length Codes			
Input Range	Output Levels	Normal Elements	Code No.	Extra Elements	Code No.
−255 to −41	−50	10000001	15	1000000001	17
−40 to −24	−31	100001	13	100000001	16
−23 to −11	−16	101	10	1000001	14
−10 to −1	−5	11	9	10001	12
0 to +9	+4	01	1	0001	3
10 to 22	+15	001	2	000001	5
23 to 39	+30	00001	4	00000001	7
40 to 255	+49	0000001	6	000000001	8

Fig. 6.6.3 Quantization parameters and code words during horizontal subsampling. Intervening pels can be sent, if desired, using the *extra* code words.

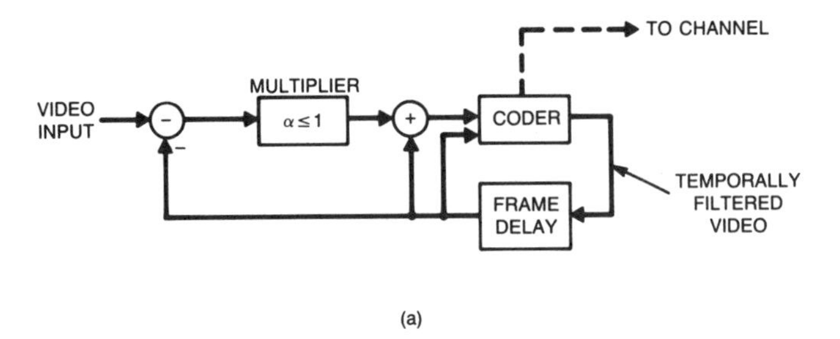

(a)

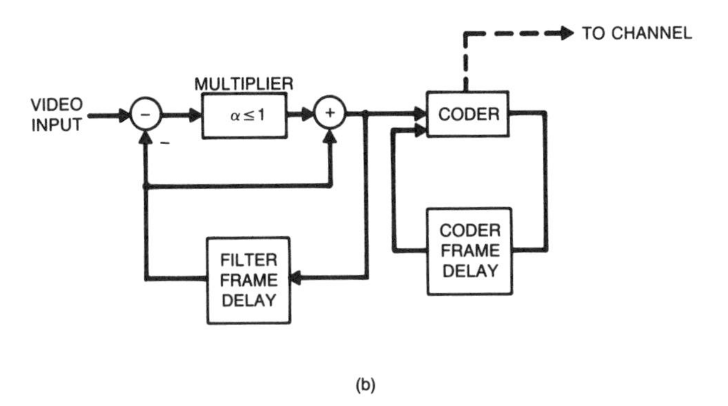

(b)

Fig. 6.6.4 Interframe coder preceded by a single-pole temporal filter. The lower the value of α, the more the temporal filtering. In (a) the filter and coder share the same frame memory. In (b) the temporal filter has its own memory.

not propagate through the temporal filter. Additional memories, i.e., a higher order filter, give improved performance, but at a higher cost. Also, replacing the constant gain multiplier by a nonlinear characteristic having lower gain for small input and larger gain for large input can still reduce the input noise, yet not cause as much blurring of the moving areas as occurs with a constant gain. Temporal filtering not only make subsampling effects less visible, it also reduces the level of frame-to-frame noise in stationary-areas thus making moving-area segmentation much easier. In addition, because temporal filtering causes blurring in the moving-area, it reduces the entropy of the differential signal — especially for intrafield predictors such as the ones used by the inter-regional codec.

An additional problem for vertical and horizontal subsampling is that the high data rates requiring subsampling may be due to either relatively small areas moving rapidly or large areas moving slowly. In the latter case, subsampling degradation is easily visible since temporal filtering (either camera integration or external) does not reduce the spatial resolution sufficiently to mask the subsampling effects.

The codec has the capability of sending entire lines as 8-bit PCM instead of as DPCM. This is called "forced updating" and is necessary to accommodate transmission bit errors. It is also convenient in preventing buffer underflow.

Transmission of each field starts with a 28-bit field start code indicating whether it is an even field or an odd field. Vertical subsampling is signaled simply by a missing field start code. Transmission of each line starts with a 20-bit line-start code that also indicates whether or not horizontal subsampling is to be used. Clusters of moving-area luminance pels, as defined by the segmenter, are then sent, followed by clusters of moving-area chrominance pels.

Each cluster begins with the 8-bit PCM value of the first pel of the cluster followed by its 8-bit address along the line. Then comes the coded DPCM data of the remaining pels of the cluster, followed finally by the end of cluster (EOC) code. Just before the first chrominance cluster of the line, an 8-bit color escape word 00001001 is sent, which is outside the range of allowable PCM values and therefore distinguishable from them. Signaling of an entire line of PCM values is accomplished with 11111111, another escape code.

The coder buffer size is chosen as 96 kbits, which is small enough to assure a one-way delay of less than 200 msec.

Adaptive moving-area resolution control is used to keep the generated data rate more constant than it would be otherwise, and therefore avoid requiring extremely large buffer memories (and therefore large delay) or overflowing buffers of more practical size. The most convenient measure of whether or not moving-area resolution should be altered is the likelihood of near term buffer overflow or underflow. Typically, if the buffer is filling rapidly, resolution is decreased, and vice versa if the buffer is nearly empty. This measure is not an instantaneous indication of movement. However, if the buffer is of moderate

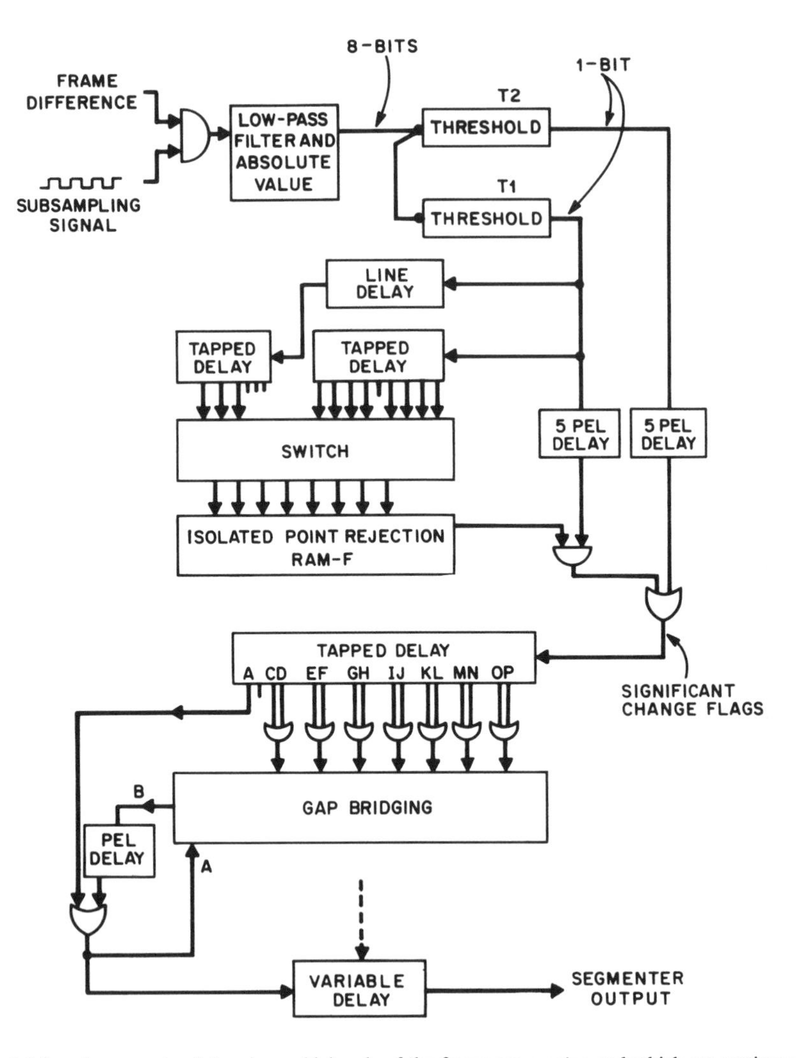

Fig. 6.6.5 A segmenter determines which pels of the frame are *moving* and which are *stationary background*. Frame difference magnitudes larger than $T2$ indicate *moving* pels. Frame difference magnitudes between $T1$ and $T2$ that are isolated are assumed to be due to noise, and do not indicate moving pels.

size, e.g., less than a few frames of transmitted data, the time lag is not a serious handicap. A possible serious drawback of using this measure is that it is difficult to distinguish between small areas moving fast and large areas moving slowly.

It is not known how to optimally control moving-area resolution for a given channel bit-rate and buffer size. Most resolution control algorithms have been designed on an ad hoc basis by trial and error. Reducing resolution in the moving area as buffer fullness exceeds certain thresholds gives reasonably good results as long as some hysteresis is built in to avoid oscillations between different resolutions. Without hysteresis it is possible for the top and bottom of a picture to be displayed with different resolutions, which may be objectionable if the difference is visible. A simple technique for avoiding oscillations is to reduce the moving area resolution when the buffer queue length exceeds a certain threshold. However, only when the queue length falls below a much lower threshold is the moving-area resolution increased again. Alternatively, mode-switching could be done only in the vertical blanking interval.

What follows is an example of how to combine these techniques in a multimode conditional replenishment codec[6.9,6.10] for coding a 525 line, component color picture at 1.5 Mb/s (0.19 bits/pel at 8 MHz sampling rate).

Segmenting in such a system is fairly straightforward as shown in Fig. 6.6.5. The luminance frame difference is low-pass filtered and threshold detected with two alterable thresholds, $T2 \geq T1$. Differences larger than $T2$ are immediately classified as significant, whereas differences exceeding $T1$ must undergo isolated point rejection before being classified. Following this, small gaps between clusters of changes are redefined as *changed* to increase addressing efficiency. A variable amount of temporal filtering is also included to reduce entropies of differential signals, noise on the input signal and quantization noise.

Nine modes (0-8) are used in coding both the luminance and chrominance as shown in Fig. 6.6.6. In general, switching to the next higher (movement) mode takes place as soon as the buffer queue length exceeds an upper limit specified for that mode. However, switching to a lower (movement) mode only occurs at the end of a coded field if at that time the buffer queue length is below a lower limit specified for that mode. This avoids oscillations between modes, which otherwise could cause unnecessary impairment and needless filling of the buffer.

In the lower modes, where relatively high resolution is maintained, 16-level quantization of the differential signal is used. However, for the higher modes 8-level quantization is used. Variable word-length coding is used for both quantizers, with one 4-bit word reserved for end-of-cluster signaling. Varying the quantization and temporal filtering requires that changes be made in the segmenting, as well. Thus, different modes have different frame difference low-pass filters and different thresholds. Subsampling is utilized straightforwardly under buffer control in the following order: vertical 2:1 (field

Mode	Buffer Threshold (bits) Lower	Buffer Threshold (bits) Upper	Number of Quantizer Levels	Sub Sampling 2:1 Vert	Sub Sampling 2:1 Horiz	Comments
0	0	6400	256	No	No	8-bit PCM
1	1600	38400	16	No	No	Low Threshold T1, T2
2	1600	38400	16	No	No	Higher T1, T2
3	3200	38400	16	Yes	No	Alternate Fields Dropped
4	3200	51200	16	Yes	No	More Temporal Filtering
5	3200	54400	8	Yes	Yes	Adaptive Horiz. Subsamp.
6	10000	70400	8	Yes	Yes	Horiz. Subsamp. All Pels
7	32000	83200	8	Yes	Yes	More Temporal Filtering
8	32000	96000	0			Drop Frames

Fig. 6.6.6 Modes used by the coder are controlled by the fullness of the buffer.

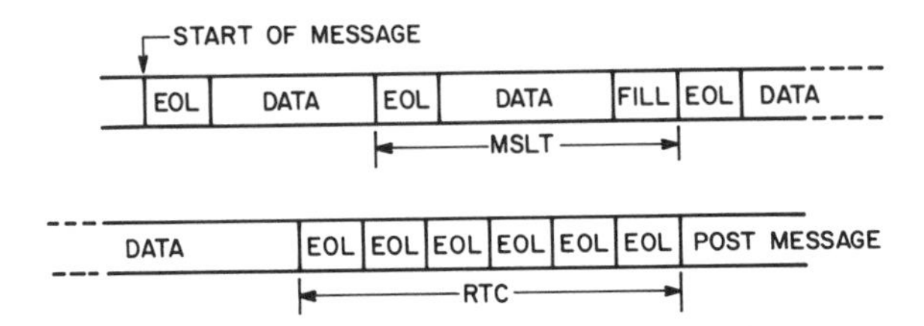

Fig. 6.7.1 Format of transmitted messages for several lines. Minimum scan line time (MSLT) is obtained by using fill bits as shown on the top line. End of the document transmission is signalled by 6 consecutive end of line (EOL) codewords, which indicates return to control (RTC).

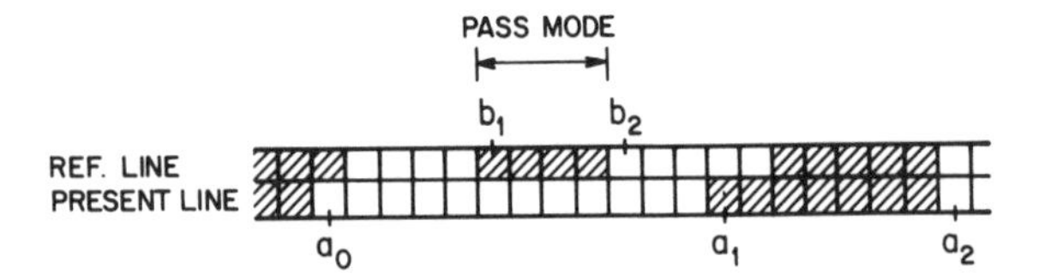

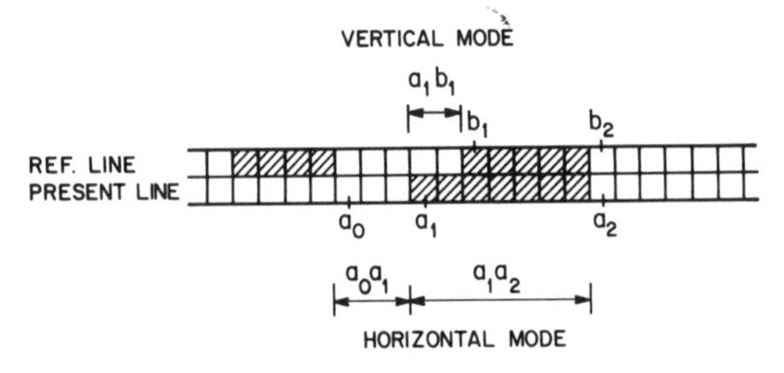

Fig. 6.7.2 Examples of coding modes for the two dimensional coding scheme. Top configuration shows the pass mode and the bottom one shows the horizontal and vertical modes. (from Hunter and Robinson [2.6.2])

dropping), horizontal 2:1, frame repeating. The last is used only during camera movement or scene changes.

6.6.1 Error Control

With intraframe coding where each frame is transmitted separately, the effect of a digital transmission error lasts for at most a frame period. With interframe coding such effects can last much longer. In general, the more one removes redundancy from a television signal through the use of sophisticated coding techniques, the more important each bit of transmitted information becomes, and the more noticeable are the effects of digital transmission errors. For example, with simple frame differential transmission of every pel, an error appears in only one pel and lasts indefinitely unless some error correction strategy is employed, e.g., incorporating leak into the feedback loop as described in Sec. 5.2.3.[6.11,6.12] With frame replenishment, leak is not very effective since not all pels are sent in every frame. Furthermore, if higher order DPCM and/or variable word-length coding are used, defects due to transmission errors can spread beyond one pel, possibly into adjacent lines.

Forward acting error control has been incorporated into frame replenishment coders in which effects due to an error are confined for the most part to one line. Channel error correcting codes might be used such as BCH or convolutional codes.[6.13] Special protection of sync and cluster demarcation words is also worthwhile in order to minimize the area affected by a transmission error.[6.14] In spite of these precautions, however, errors are bound to appear in the pictures sooner or later, and further steps must be taken to deal with them.

Typically, one or more lines or blocks per frame are sent via PCM or some other intraframe coding method that does not rely on pels from surrounding lines or blocks. This is called *forced updating*.[6.13] In this way, in a few seconds, every pel in the frame can be updated with correct values, thus wiping out any residual effects due to errors that survive the channel error correcting codes.

Several seconds is a long time for a highly visible error defect to remain at the display. Through the use of highly reliable error detection codes, it is possible for the receiver to tell which lines contain errors. If these lines are then replaced, for example, with an average of adjacent lines, then the effects of errors (if their rate is reasonably low) can be made much less visible.[6.12] This technique usually requires that once an erroneous line is replaced by an average, replacement of that line must continue during succeeding frames until PCM values arrive for that line.

Error detection followed by a request for retransmission is often not feasible for several reasons. First, for the high data rates usually entailed in video transmission, very large buffers would be required. Second, for the long distances often involved, too much delay would be introduced for two-way video communication. However, if substitution of an erroneous line, as described above, were followed by a request for PCM transmission of that line during some later frame, then the duration of error effects could be markedly reduced compared with strictly forward acting techniques.

If the effects of a digital transmission error propagate to adjacent lines, as they would for example with DPCM that uses pels from previous lines, then line substitution does not work. Substitution from the previous frame would not be very disturbing if movement were slow. However, in rapidly moving areas, picture breakup would occur, and substitution of the entire field would probably be preferable. Subdivision of the picture into smaller blocks that are coded independently of one another might alleviate the situation somewhat. In any event, substitution could not be carried out for very long — one or two frames at most. A request for retransmission would have to be issued to minimize the visibility of effects due to transmission errors.

6.7 Graphics Coders

In this section we give details of two coding schemes standardized by the CCITT* along with their performance on the CCITT pictures. These pictures have already been described in Chapter 2 and are shown in Fig. 2.6.6. They have played a key role in optimizing the parameters of the adopted standard coding schemes. The first coding scheme uses one-dimensional run-length coding, whereas the second exploits two-dimensional correlation in the data. These schemes, called Group 3[6.15] (or G3) schemes, are used primarily for transmitting a A4 size (210 mm $\times$ 298 mm) document over a public switched telephone network. Since the telephone network is prone to transmission errors, two forms of redundancy are provided to avoid degradation of the image when errors occur. These include a special code at the end of each line and transmission of one-dimensional compressed code every two or four lines.

The CCITT has also developed recommendations called Group 4[6.16] (or G4), applicable for more general pictures and communications media. As examples of this, for digital communication networks containing error control, G4 schemes are obtained from G3 schemes by removing the above two forms of redundancy (i.e., end of line code and periodic transmission of one-dimensional compressed data) since the network provides error free transmission. Also, G4 schemes allow more general pictures by incorporating four possible resolutions: 200, 240, 300 and 400 pels per inch. A mixed mode operation containing symbols and graphics is also allowed. Those parts of the image containing alphanumerics are transmitted as characters using formats such as ASCII, whereas other parts of the image containing non-character information such as

---

* Now known as ITU-T.

Table 6.7.1 Modified Huffman code for the run-lengths in the one-dimensional coding scheme. Both terminating codes (TC) and make-up codewords (MUC) are shown.

Terminating Codewords (TC)

Run-length	White Runs	Black Runs
0	00110101	0000110111
1	000111	010
2	0111	11
3	1000	10
4	1011	011
5	1100	0011
6	1110	0010
7	1111	00011
8	10011	000101
9	10100	000100
10	00111	0000100
11	01000	0000101
12	001000	0000111
13	000011	00000100
14	110100	00000111
15	110101	000011000
16	101010	0000010111
17	101011	0000011000
18	0100111	0000001000
19	0001100	00001100111
20	0001000	00001101000
21	0010111	00001101100
22	0000011	00000110111
23	0000100	00000101000
24	0101000	00000010111
25	0101011	00000011000
26	0010011	000011001010
27	0100100	000011001011
28	0011000	000011001100
29	00000010	000011001101
30	00000011	000001101000
31	00011010	000001101001
32	00011011	000001101010
33	00010010	000001101011
34	00010011	000011010010
35	00010100	000011010011
36	00010101	000011010100
37	00010110	000011010101
38	00010111	000011010110
39	00101000	000011010111
40	00101001	000001101100
41	00101010	000001101101
42	00101011	000011011010
43	00101100	000011011011
44	00101101	000001010100
45	00000100	000001010101

Terminating Codewords (TC) Con't.

Run-length	White Runs	Black Runs
46	00000101	000001010110
47	00001010	000001010111
48	00001011	000001100100
49	01010010	000001100101
50	01010011	000001010010
51	01010100	000001010011
52	01010101	000000100100
53	00100100	000000110111
54	00100101	000000111000
55	01011000	000000100111
56	01011001	000000101000
57	01011010	000001011000
58	01011011	000001011001
59	01001010	000000101011
60	01001011	000000101100
61	00110010	000001011010
62	00110011	000001100110
63	00110100	000001100111
	Make-up Codewords (MUC)	
64	11011	0000001111
128	10010	000011001000
192	010111	000011001001
256	0110111	000001011011
320	00110110	000000110011
384	00110111	000000110100
448	01100100	000000110101
512	01100101	0000001101100
576	01101000	0000001101101
640	01100111	0000001001010
704	011001100	0000001001011
768	011001101	0000001001100
832	011010010	0000001001101
896	011010011	0000001110010
960	011010100	0000001110011
1024	011010101	0000001110100
1088	011010110	0000001110101
1152	011010111	0000001110110
1216	011011000	0000001110111
1280	011011001	0000001010010
1344	011011010	0000001010011
1408	011011011	0000001010100
1472	010011000	0000001010101
1536	010011001	0000001011010
1600	010011010	0000001011011
1664	011000	0000001100100
1728	010011011	0000001100101
EOL	000000000001	000000000001

drawings and handwriting are coded by graphics coding schemes. In any case, the graphics coding schemes used for G4 are minor variations of the basic G3 schemes that are described in detail below.

Another scheme, described in Chapter 7, was developed by the ISO Joint Bilevel Image Group (JBIG). It is more efficient than G3/G4 and also has the capability of progressive coding.

6.7.1 CCITT G3/G4 — One-Dimensional Coding Scheme

Each scan line is coded by assigning words to the runs of "white" and "black" elements that alternate along a scan line. The lines are assumed to begin with a white run; if the first actual run on a line is black, then a white run of zero length is assumed and the corresponding code word is transmitted at the beginning of the line. Separate code tables based on a modified Huffman procedure using the statistics of the eight CCITT pictures are used to represent the black and white runs. The code table is given in Table 6.7.1. The code table contains codes for run-lengths up to 1728 pels (length of a scan line). The code words are of two types: terminating codes (TC) and make-up codes (MUC). Run-lengths between 0 and 63 are transmitted using a single terminating codeword. Run-lengths between 64 and 1728 are transmitted by a MUC followed by a TC. The MUC represents a run-length value of $64\times N$ (where N is an integer between 1 and 27), which is equal to, or shorter than, the value of the run to be transmitted. The difference between the MUC and the actual value of the run-length is specified by the trailing TC. Each coded line is followed by the end-of-line (EOL) codeword, which is a unique sequence that cannot occur within a valid line of coded data. The codeword chosen for EOL is 000000000001, eleven 0's followed by a 1. If the number of coded bits in a line is fewer than a certain minimum value, then "fill" bits consisting of strings of 0's of varying length are inserted between the line of coded data and the EOL codeword. This ensures that each coded scan line will require more than a specified amount of transmission time so that transmitters and receivers can keep in step, and mechanical limitations with respect to scanning or printing can be overcome. The recommended standard minimum is 96 bits per line (at 4800 bits/sec transmission rate) with options for 48, 24 and 0 bits. The fill bits are easily recognized and discarded by the receiver. Fig. 6.7.1 shows the format of the data for several coded lines. The end of document transmission is indicated by six consecutive EOL codewords that form the return control signal. The compression performance for this run-length coding scheme is given in Table 6.7.2 for the eight CCITT documents. The average compression factor is 9.7.

6.7.2 CCITT — Two-Dimensional Coding Scheme

This scheme is known as the "modified READ code" (Relative Element Address Designate), and has evolved through the various proposals submitted to

Table 6.7.2 Performance and the relevant statistics of the one dimensional coding scheme for the eight CCITT documents. Maximum compression ratio is computed by Eq. (3.4.9) using the measured entropies of black and white runs for each document. Compression ratio is computed using the actual modified Huffman code, but without end of line and fill bits. (from Hunter and Robinson [2.6.2])

CCITT DOCUMENT NO.	AVERAGE WHITE RUN-LENGTH	AVERAGE BLACK RUN-LENGTH	ENTROPY OF WHITE RUNS	ENTROPY OF BLACK RUNS	MAXIMUM COMPRESSION RATIO	COMPRESSION RATIO
1	134.6	6.790	5.230	3.592	16.02	15.16
2	167.9	14.02	5.989	4.457	17.41	16.67
3	71.50	8.468	5.189	3.587	9.112	8.350
4	36.38	5.673	4.574	3.126	5.461	4.911
5	66.41	6.966	5.280	3.339	8.513	7.927
6	90.65	8.001	5.063	3.651	11.32	10.78
7	39.07	4.442	5.320	3.068	5.188	4.990
8	64.30	60.56	4.427	5.310	11.52	8.665

Table 6.7.3 Code table for the two-dimensional coding scheme. $M(a_0a_1)$ and $M(a_1a_2)$ are codewords for the appropriate run-lengths taken from Table 6.7.1. (from Hunter and Robinson [2.6.2])

MODE	CHANGING ELEMENTS TO BE CODED		NOTATION	CODEWORD
PASS	b_1, b_2		P	0001
HORIZONTAL	a_0a_1, a_1a_2		H	$001+M(a_0a_1)+M(a_1a_2)$
	a_1 JUST UNDER b_1	$a_1b_1=0$	V(0)	1
VERTICAL	a_1 TO THE RIGHT OF b_1	$a_1b_1=2$	$V_R(2)$	000011
		$a_1b_1=1$	$V_R(3)$	0000011
	a_1 TO THE LEFT OF b_1	$a_1b_1=1$	$V_L(1)$	010
		$a_1b_1=2$	$V_L(2)$	000010
		$a_1b_1=3$	$V_L(3)$	000010
END-OF-LINE CODEWORD (EOL)				000000000001
1-D CODING OF NEXT LINE 2-D CODING OF NEXT LINE				EOL + '1' EOL + 'O'

the CCITT until 1979. It is a line-by-line scheme in which the position of each *changing* element on the *present line* is coded with respect to either the position of a corresponding changing element on the *reference line*, which lies immediately above the present line, or with respect to the preceding changing element on the present line. After the present line has been coded, it becomes the reference line for the next line.

The changing element is an element of different color from that of the previous element along the same line. The coding procedure uses five changing elements defined as in Fig. 6.7.2.

1. a_0 : the first changing element on the present line.
2. a_1 : the next changing element on the present line; by our definition, it has opposite color to a_0 and gets coded next.
3. a_2 : the changing element following a_1 on the present line.
4. b_1 : the changing element on the reference line to the right of a_0 with the same color as a_1.
5. b_2 : the changing element following b_1 on the reference line.

Depending on the relative position of the changing element that is being coded, the coder operates in three modes.

(i) Pass Mode
In this mode, the element of b_2 lies horizontally to the left of a_1. Using our definitions above, it occurs whenever the white or black runs on the reference line are not adjacent to corresponding white or black runs on the present line. The pass mode is represented by a single codeword.

(ii) Vertical Mode
In this mode, the element a_1 is sufficiently close to b_1 and is therefore coded relative to the position of b_1. It is used only if a_1 is to the left or right of b_1 by at most 3 pels and hence, the relative distance $a_1 b_1$ can take on any of seven values $V(0)$, $V_R(1)$, $V_R(2)$, $V_R(3)$, $V_L(1)$, $V_L(2)$, $V_L(3)$. The subscript R is used if a_1 is to the right of b_1. Subscript L is used otherwise. The number in parentheses indicates the distance $a_1 b_1$ in pels.

(iii) Horizontal Mode
If a_1 is not sufficiently close to b_1, then its position must be coded by horizontal mode. Thus, the run-lengths $a_0 a_1$ and $a_1 a_2$ are coded using the concatenation of three codewords H, $M(a_0 a_1)$ and $M(a_1 a_2)$. Codeword H, taken to be 001, serves as a prefix or flag, and $M(a_0 a_1)$ and $M(a_1 a_2)$ are taken from the code tables to represent the colors and values of the run-lengths $a_0 a_1$ and $a_1 a_2$. The Table is based on the modified Huffman procedure as in CCITT scheme 1.

Table 6.7.4 Performance of the two-dimensional coding schemes for low and high resolutions. Actual coded bits are shown for each of the CCITT documents for $K = 2, 4$ and ∞.

CCITT DOCUMENT NUMBER	LOW RESOLUTION (100 PELS/INCH)	HIGH RESOLUTION (200 PELS/INCH)		
	K = 2	K = 4	K = ∞	COMPRESSION RATIO
1	130684	207660	175704	21.6
2	106851	175163	117304	32.4
3	207584	326297	260527	14.6
4	408261	654436	585074	6.5
5	226285	353172	288655	13.2
6	150572	225879	164085	23.2
7	402333	651643	585135	6.5
8	184369	264029	183674	20.7
AVERAGE	227117	355034	295019	12.9

In practice, the coder selects first the coding mode and then the appropriate codeword from Table 6.7.3. The two steps are:

Step 1

If b_2 is detected before a_1 then the pass mode is selected and the codeword 0001 is transmitted. The reference pel a_0 is set on the pel below b_2 in preparation for the next coding. If a pass mode is not detected, step 2 is followed.

Step 2

The horizontal distance between a_1 and b_1 is determined. If the distance is at most three, then vertical mode is used, and a_0 is set on the position of a_1 in preparation for the coding of the next changing element. Otherwise, the horizontal mode is used.

The performance of the two-dimensional code is given in terms of the number of coded bits for each CCITT document in Table 6.7.4. Use of vertical correlation makes it possible for the effects of any transmission error to propagate vertically down. To combat this, every Kth line is coded using the one-dimensional code. At the resolution of 100 pels/inch, with $K=2$, at a transmission rate of 4800 bits/sec, the average transmission time is 47.3 sec, whereas at high resolution with $K=4$ the average transmission time is 74 sec. At $K=\infty$ and high resolution, the average transmission time is 61.5 sec. Thus, there is a significant penalty for using the one-dimensional code every fourth line. However, in general, for high resolution there is a considerable decrease in transmission time by using the two-dimensional code as compared to the one-dimensional code.

References

6.1 R. C. Brainard and J. H. Othmer, "VLSI Implementation of a DPCM Compression Algorithm for Digital TV," *IEEE Trans. Communications,* August 1987, pp. 854-856.

6.2 N. S. Jayant and P. Noll, *Digital Coding of Waveforms*, Chapter 6, Prentice Hall, New York, 1984.

6.3 R. C. Brainard and A. N. Netravali, "Digital Broadcast TV at 45 Mb/s," GLOBECOM '82 Proceedings, Miami 1982, pp. B6.1.1-4. Also, J. SMPTE.

6.4 H. Murakami, et al., "A 15 Mbit/s Universal Codec for TV signals Using a Median Adaptive Prediction Coding Method," Sixth Int. Conf. Digital Satellite Communications, September 19-23, 1983, Phoenix, Arizona, pp. VII-A-17 to VII-A-24.

6.5 F. A. Kamangar and K. R. Rao, "Interfield Hybrid Coding of Component Television Signals," *IEEE Trans. Communications,* COM-29, December 1981, pp. 1740-1753.

6.6 J. K. Yan and D. J. Sakrison, "Encoding of Images Based on a Two-Component Source Model," *IEEE Trans. Commun.,* COM-25, November, 1977, pp. 1315-1322.

6.7 M. Kocher and M. Kunt, "Image Data Compression by Contour Texture Modelling", *Proceedings of SPIE,* v. 397, April 19-22, 1983, Geneva, pp. 132-139.

6.8 International Telegraph and Telephone Consultative Committee (CCITT) Recommendations H.110, H.120, and H.130; Parts 1 to 3.

6.9 B. G. Haskell, et al., "Interframe Coding of 525-Line Monochrome Television at 1.5 Mbits/s," *IEEE Trans. Commun.,* COM-25, November, 1977, pp. 1339-1348.

6.10 T. S. Duffy and R. C. Nicol, "A Codec for International Visual Teleconferencing," Globecom '82 Conference Record, v. 3, December, 1982, Miami, pp. E2.5.1-.5.

6.11 D. J. Connor, R. C. Brainard, and J. O. Limb, "Intraframe Coding for Picture Transmission," *Proc. IEEE,* 60, No. 7, July, 1972, pp. 780-791.

6.12 D. J. Connor, "Techniques for Reducing the Visibility of Transmission Errors in Digitally Encoded Video Signals," *IEEE Transactions on Communications,* COM-21, No. 3, June, 1973, pp. 695-706.

6.13 T. Ishiguro, K. Iinuma, Y. Iijima, T. Koga and H. Kaneko, "NETEC System: Interframe Encoder for NTSC Color Television Signals," *Third International Conference on Digital Satellite Communications Record,* November, 1975.

6.14 H. Yasuda, F. Kanaya and H. Kawanishi, "1.544 Mbit/s Transmission of TV Signals by Interframe Coding System," *IEEE Transmissions on Communications,* COM-24, No. 10, October, 1976, pp. 1175-1180.

6.15 CCITT Recommendation T.4, *Standardization of Group 3 facsimile apparatus for document transmission.*

6.16 CCITT Recommendation T.6, *Facsimile coding schemes and coding control functions for Group 4 facsimile apparatus.*

Questions for Understanding

6.1 In Fig. 6.1.1 why is pel A so far from pel X? Would a closer pel ever be a better predictor? When?

6.2 Which statistical and perceptual redundancies are exploited by the Adaptive Predictive Interfield Coder of Sec. 6.2?

6.3 Which statistical and perceptual redundancies are exploited by the Interfield Hybrid Transform Coder of Sec. 6.3?

6.4 Which statistical and perceptual redundancies are exploited by the three coding modes of the CCIR 723 coding algorithm?

6.5 In the CCIR 723 coder, suppose a Macroblock has the lowest Activity level and a Buffer Fullness Factor 50. Also, the first four transmitted DCT coefficients have unquantized values -1000, 150, -875 and 690. What are the first four transmitted quantizer indices?

6.6 Which statistical and perceptual redundancies are exploited by the Conditional Replenishment DPCM Coders of Sec. 6.6?

6.7 Explain *Hysteresis* in buffer control of interframe coders. What does it accomplish?

6.8 Explain *Forced Updating* in interframe coders. What does it accomplish?

6.9 Which statistical redundancies are exploited by the bilevel image coders of Sec. 6.7?

7

Still Image Coding Standards - ISO JBIG and JPEG

7.1 JBIG Coding

The bi-level coding algorithm most recently standardized by ISO is the JBIG (Joint Bi-level Imaging Group) algorithm[7.1]. JBIG coding is more complex than G3/G4 coding, but offers two compensating advantages.

One is superior compression. On text or line-art images, JBIG achieves about 20 percent greater compression than G4, the most efficient of the G3/G4 algorithms. The JBIG compression advantage is much greater, generally ranging between a factor of two and ten, on halftone bi-level images rendering grayscale. The JBIG algorithm is adaptive and indirectly recognizes and adjusts to input images rendering grayscale. In addition, special features are built in to recognize and exploit any periodicities present in grayscale renderings.

The second advantage of JBIG coding is that, if desired, it can be parameterized for progressive coding.

7.1.1 Progressive Coding

Progressive, or synonymously, *multiresolution*, coding captures images as a compression of a low resolution rendition plus a sequence of *delta* files that each provide another level of resolution enhancement. In the case of JBIG, each delta file contains the information needed to double both the vertical and horizontal resolution.

Progressive coding is attractive in database applications where output devices of varying resolution capability (say PC screens, workstation screens,

---

This chapter contains excerpts taken with permission from AT&T Technical Journal[7.17].

and laser printers) are all to be served. Only the bottom-layer coding and as many delta files as are of use are transmitted and processsed. In contrast, if images are stored non-progressively, it is necessary to either store compressions at all resolutions or to store only compressions at the highest resolution but require output devices to first decode to high resolution and then map down to display resolution. The first alternative wastes storage capacity. The second wastes transmission capacity and processing power.

Another application for progressive coding is in image browsing over medium rate communication links. A low-resolution rendition of an image can be made available quickly and then followed by as much resolution enhancement as is desired. A user on seeing at low resolution that the image being developed is not that desired, can quickly interrupt its transmission and move on to the next. This advantage for progressive coding only accrues on medium rate links, roughly those with speeds between 9.6 kb/s and 64 kb/s when bi-level images are being retrieved. On slower links, no user would have the patience required for image browsing no matter what the form of presentation. On higher speed links, the image comes so fast relative to human reaction times that how it develops is immaterial.

A third potential application for progressive coding is on packet networks on which packets can or must be priority classified as droppable or non-droppable. The packets carrying the information for the final resolution doubling would be sent at low priority and if the network were to drop them, the only penalty would be a slightly less sharp image. No entire regions would be lost or destroyed.

The JBIG specification uses the symbol D to denote the number of resolution doublings (delta files) that are to be available. The parameter D is unrestricted and may be chosen as 0 when progression is of no benefit. Making this choice generally improves compression by about 5 percent, but surprisingly, can degrade compression by as much as 50% on grayscale renderings created by periodicity generating algorithms.

Common choices for D when progression is wanted are 4, 5, and 6. If the original is a 400 dpi scanning of an 8.5 by 11 inch page and D is chosen as 5, the bottom-layer image will be 107 by 138 pels in size. Normally such a small image would be blown up by pel replication (or perhaps something more sophisticated) before display. If instead, however, it is displayed dot for dot on something like a 1000 by 1000 pel display, it can serve nicely as an icon.

7.1.2 Functional Blocks

The JBIG algorithm is complex and describing it completely here is not possible. Such detail is best obtained from the specification itself[7.1]. As in the discussion of the G3/G4 algorithms, the goal here is only to convey the basics of the algorithm.

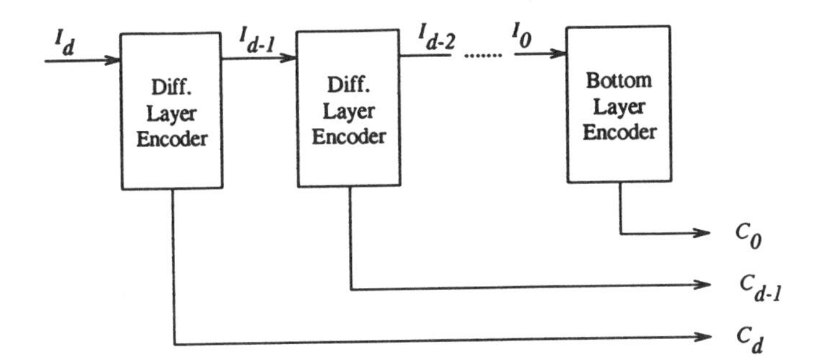

Fig. 7.1.1 Decomposition of a JBIG encoder.

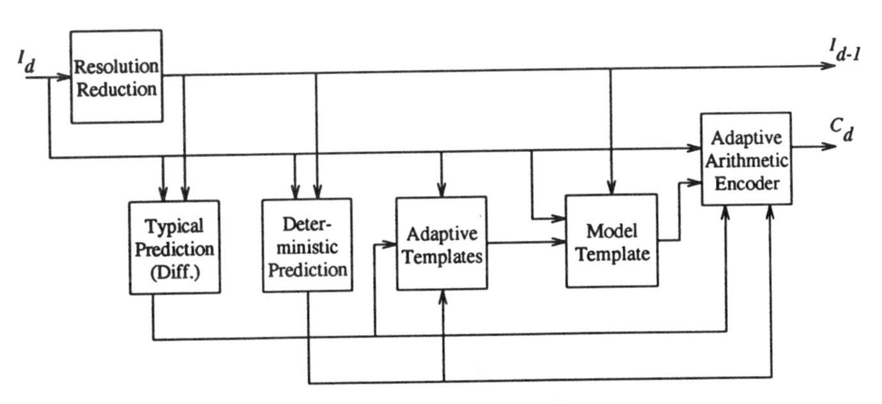

Fig. 7.1.2 Functional blocks in a JBIG differential-layer encoder.

Conceptually a JBIG encoder can be decomposed (see Fig. 7.1.1) into a chain of D differential layer encoders followed by a bottom-layer encoder. In Fig. 7.1.1, I_d denotes the image at layer d and C_d denotes its coding. A hardware implementation would in all likelihood time-share one physical differential-layer encoder, but for heuristic purposes the decomposition of Fig. 7.1.1 is helpful. Each differential layer encoder can be decomposed into the functional blocks shown in Fig. 7.1.2. The bottom-layer encoder has the somewhat simpler decomposition shown in Fig. 7.1.3.

7.1.3 Resolution Reduction

The resolution reduction block in a differential-layer encoder accepts the high-resolution image I_d and creates the low-resolution image I_{d-1} with, as nearly as possible, half as many rows and half as many columns.* A simple way of performing this resolution reduction would be by subsampling, that is, discarding every other row and every other column. Subsampling is simple, but the low-resolution images it creates are poorer in subjective quality than is necessitated by their diminished resolution alone. On images with line art, whole lines can be lost if they happen to lie on rows or columns being discarded. Grayscale renderings with periodicities frequently display aliasing artifacts caused by beats between the superpel frequency and the subsampling by two.

JBIG's resolution reduction algorithm works remarkably well for all image types — text, line art, and grayscale renderings. It is a table based algorithm. The low-resolution image is created pel by pel in the usual raster scan row order. The color of any given low-resolution pel is then dictated by the colors of nine particular high-resolution neighbors and three causally positioned low-resolution neighbors.

7.1.4 Arithmetic Coding

The remaining functional blocks in Fig. 7.1.2 and 7.1.3 implement the compression algorithm. The heart of this functionality in both the bottom-layer coder and the differential-layer coder is an adaptive arithmetic coder[7.2]. As we saw in Chapters 3 and 5, arithmetic coders are distinguished from other entropy coders such as Huffman coders and Ziv-Lempel coders in that, conceptually at least, they map a string of binary symbols to be coded into a real number x on the unit interval [0.0,1.0). In the JBIG application, the string of symbols to be coded is the pels of the image I_d presented in raster scan order. What is transmitted or stored instead of the image I_d is a binary representation of x.

---

* In general, there need not be even numbers of rows and columns in I_d.

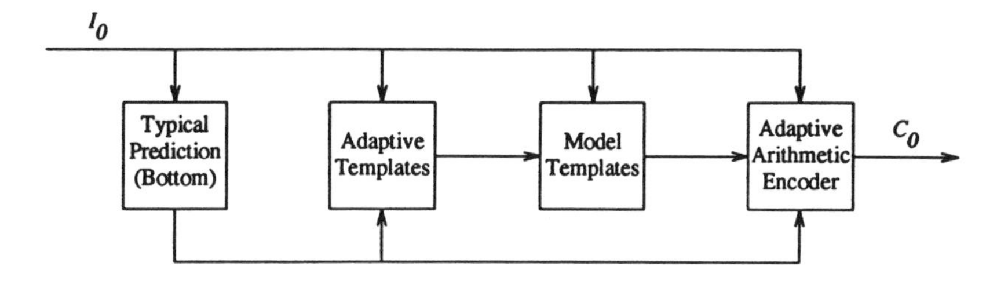

Fig. 7.1.3 Functional blocks in a JBIG bottom-layer encoder.

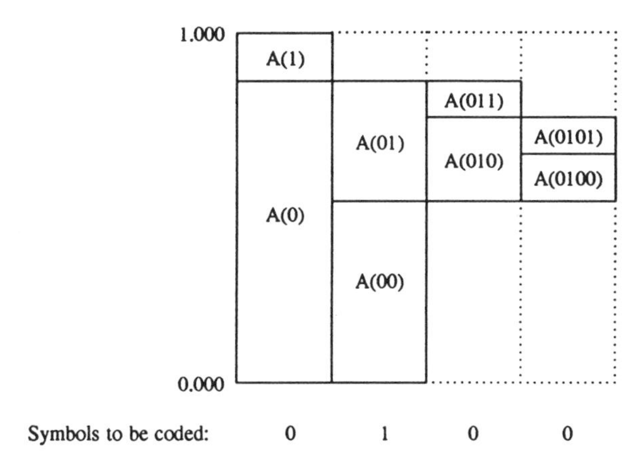

Fig. 7.1.4 Example of interval subdivision.

The particular number x to which the input sequence maps is determined by the recursive probability interval subdivision of the Elias coder[7.3] as described in Sec. 3.1.7 and Sec. 5.7.1e. Fig. 7.1.4 shows an example of such interval division through an initial sequence 0,1,0,0 to be coded.

The portion of the unit interval on which x is known to lie after coding each symbol is known as the *current coding interval*, and we designate its size by A. We define the current estimate of x to be the base of the current interval and denote this by C. For each new binary input the current coding interval is divided into two sub-intervals with sizes pA and $(1-p)A$, where p is the conditional probability of the Least Probable Symbol (LPS). The new current coding interval becomes that portion of the old coding interval associated with the new symbol value actually occurring. The fact that, most frequently, the symbol to be coded will be the Most Probable Symbol (MPS) and that by design this symbol codes into the bottom of the two sub-intervals means that most of the time the coding interval A is not reduced in size by as much as 1/2, C does not change and an input symbol is coded without generating any bits in the binary expansion of x. It is true that whenever an LPS does happen to occur, more than one output bit is generated. However, a central result of information theory is that on the average there is still less than one bit generated for each input symbol.

The Elias coder was conceived soon after Shannon's seminal paper gave birth to Information Theory in 1948[7.4]. It was of little practical use, however, until ways were found in the mid 1970's to perform all the necessary arithmetic with finite precision arithmetic and in a pipelined fashion so that a decoder could start its processing without having to wait for an encoder to finish its processing[7.5].

Sec. 5.7.1e described a finite precision arithmetic coding technique that implements the above interval subdivision algorithm. However, in JBIG the multiplication needed to compute pA is deemed too costly for inexpensive implementation. Instead, JBIG assumes that after normalization, A is usually not too far from some value $\bar{A}$, measured by averaging A over a wide variety of test documents. JBIG assumes a value of approximately 0.72 for $\bar{A}$.

With this assumption, the conditional probability tables that contain p can be replaced with values $p \times \bar{A}$, which we denote by $\bar{p}$. Thus, in the interval subdivision, LPS has size $\bar{p}$, and MPS has size $A-\bar{p}$, which avoids any multiplication in the algorithm.

One complication of this approach is that for some values of A we can have $\bar{p} > A-\bar{p}$, i.e., the nominal LPS size can exceed the nominal MPS size. When this happens, JBIG performs what is called a *conditional exchange*, which basically reverses the assignment of symbols to intervals so that the symbol with higher probability is always assigned to the largest subinterval. The complete operation of coding a symbol is as follows:

```
if (p̄ > A/2) exchange = true      //need conditional exchange
   else exchange = false
if( (!exchange && MPS) || (exchange && LPS) ){
   A = A − p̄                      //code bottom subinterval
}
else{                               //code top subinterval
   A = p̄
   C = C + A − p̄
}
```

Following the symbol coding, A and C are renormalized as described previously.

7.1.5 Model Templates

For each high-resolution pel to be coded, the model-templates block provides the arithmetic coder with an integer S called the context. For differential-layer coding S is determined by the colors of six adjacent previously transmitted pels in the causal high-resolution image, also by the colors of four particular pels in the already available low-resolution image, and also by the *spatial phase* of the pel being coded. The term spatial phase denotes which of the four possible positions the high-resolution pel has with respect to its corresponding low-resolution pel. Thus, there are $4 \times 2^6 \times 2^4 = 4096$ possible contexts for differential-layer coding. The six particular high-resolution pels and four particular low-resolution pels whose colors (along with spatial phase) define the context are known as the coding template or model template.

For bottom-layer coding, the coding template only includes 10 high-resolution pels for a total of 1024 contexts. There are no low-resolution pels to incorporate nor is there an analog of the spatial-phase concept.

The arithmetic coder generates for each context an estimate of the conditional probability $\bar{p}$ of the LPS given that context. Doing this well requires carefully balancing speed of adaptation against quality of estimation[7.6]. The greatest coding gain is achieved when this probability estimate is both accurate and close to 0. Thus, good templates must have good predictive value so that when the values of the pels in the template are known, the value of the pel to be coded is highly predictable.

The simplest implementation of Arithmetic Coding would utilize a fixed lookup table containing one $\bar{p}$ value for each context. However, JBIG allows for adaptation of the context probability values in order to better track the actual statistics of the image being coded. Basically, this adaptation works by decrementing $\bar{p}$ slightly whenever an MPS renormalization occurs, and incrementing $\bar{p}$ slightly whenever an LPS renormalization occurs. If any $\bar{p}$ value becomes too large, indicating an erroneous LPS symbol definition, then the LPS and MPS symbol definitions are interchanged. The adaptation process is completely defined by the Probability Estimation State Machine shown in Fig. 7.1.5.

JBIG only allows for 112 possible $\bar{p}$ values. These are shown in the second column of Fig. 7.1.5 in hexadecimal, where the binary point is on the left. Thus, context statistics can be stored in two arrays Index[ ] and MPS[ ] of size 4096 containing, respectively, the indices of the $\bar{p}$ values and the most probable symbol values (0 or 1). At the start of coding, both arrays are initialized to zero. During coding of a symbol having context S, the LPS probability and most probable symbol value are $\bar{p}$[Index[S]] and MPS[S], respectively.

If coding causes a renormalization, then context statistics are adjusted according to the Fig. 7.1.5 state machine via the following algorithm:

```
                        //after renormalization...
I = Index[S]            //save current p̄ index
if( symbol==MPS[S]){    //an MPS was coded
Index[S] = NextMPS[I]   //update p̄
}
else{                   //an LPS was coded
Index[S] = NextLPS[I]   //update p̄
if( Switch[I]==1 )      //an LPS was coded and p̄ is large
MPS[S] = ~MPS[S]        //interchange LPS and MPS definition
}
```

Note that the smallest $\bar{p}$ is 2^{-16}. Also, indices 0 to 8 are normally used only during startup of the statistical adaptation.

7.1.6 Adaptive Templates

The comparatively good compressions obtained by JBIG on dithered grayscale renderings is partially attributable to the adaptation within the arithmetic coder. The adaptive-templates (AT) block adds a factor of approximately two improvement on dithered grayscale renderings with periodicities. AT looks for periodicities in the image and, upon finding a strong indication for one, changes the template so that, as nearly as possible, a pel offset by this periodicity is incorporated into the new template. Such a pel of course has excellent predictive value.

Any changes of this sort to the template must only be made infrequently since a period of readaptation in the probability estimator ensues, and compression degrades while it occurs. An algorithm for determining the if, when, and how of any template rearrangement normally has substantial hysteresis so that template changes only occur when there is an extremely strong indication that a new template offers substantial improvement over the current one.

Index	$\overline{p}$	NextLPS	NextMPS	Switch
0	.5a1d	1	1	1
1	.2586	14	2	0
2	.1114	16	3	0
3	.080b	18	4	0
4	.03d8	20	5	0
5	.01da	23	6	0
6	.00e5	25	7	0
7	.006f	28	8	0
8	.0036	30	9	0
9	.001a	33	10	0
10	.000d	35	11	0
11	.0006	9	12	0
12	.0003	10	13	0
13	.0001	12	13	0
14	.5a7f	15	15	1
15	.3f25	36	16	0
16	.2cf2	38	17	0
17	.207c	39	18	0
18	.17b9	40	19	0
19	.1182	42	20	0
20	.0cef	43	21	0
21	.09a1	45	22	0
22	.072f	46	23	0
23	.055c	48	24	0
24	.0406	49	25	0
25	.0303	51	26	0
26	.0240	52	27	0
27	.01b1	54	28	0
28	.0144	56	29	0
29	.00f5	57	30	0
30	.00b7	59	31	0
31	.008a	60	32	0
32	.0068	62	33	0
33	.004e	63	34	0
34	.003b	32	35	0
35	.002c	33	9	0
36	.5ae1	37	37	1
37	.484c	64	38	0
38	.3a0d	65	39	0
39	.2ef1	67	40	0
40	.261f	68	41	0
41	.1f33	69	42	0
42	.19a8	70	43	0
43	.1518	72	44	0
44	.1177	73	45	0
45	.0e74	74	46	0
46	.0bfb	75	47	0
47	.09f8	77	48	0
48	.0861	78	49	0
49	.0706	79	50	0
50	.05cd	48	51	0
51	.04de	50	52	0
52	.040f	50	53	0
53	.0363	51	54	0
54	.02d4	52	55	0
55	.025c	53	56	0
56	.01f8	54	57	0
57	.01a4	55	58	0
58	.0160	56	59	0
59	.0125	57	60	0
60	.00f6	58	61	0
61	.00cb	59	62	0
62	.00ab	61	63	0
63	.008f	61	32	0
64	.5b12	65	65	1
65	.4d04	80	66	0
66	.412c	81	67	0
67	.37d8	82	68	0
68	.2fe8	83	69	0
69	.293c	84	70	0
70	.2379	86	71	0
71	.1edf	87	72	0
72	.1aa9	87	73	0
73	.174e	72	74	0
74	.1424	72	75	0
75	.119c	74	76	0
76	.0f6b	74	77	0
77	.0d51	75	78	0
78	.0bb6	77	79	0
79	.0a40	77	48	0
80	.5832	80	81	1
81	.4d1c	88	82	0
82	.438e	89	83	0
83	.3bdd	90	84	0
84	.34ee	91	85	0
85	.2eae	92	86	0
86	.299a	93	87	0
87	.2516	86	71	0
88	.5570	88	89	1
89	.4ca9	95	90	0
90	.44d9	96	91	0
91	.3e22	97	92	0
92	.3824	99	93	0
93	.32b4	99	94	0
94	.2e17	93	86	0
95	.56a8	95	96	1
96	.4f46	101	97	0
97	.47e5	102	98	0
98	.41cf	103	99	0
99	.3c3d	104	100	0
100	.375e	99	93	0
101	.5231	105	102	0
102	.4c0f	106	103	0
103	.4639	107	104	0
104	.415e	103	99	0
105	.5627	105	106	1
106	.50e7	108	107	0
107	.4b85	109	103	0
108	.5597	110	109	0
109	.504f	111	107	0
110	.5a10	110	111	1
111	.5522	112	109	0
112	.59eb	112	111	1

Fig. 7.1.5 JBIG Probability Estimation State Machine. Probability values are hexadecimal with the binary point at the left.

7.1.7 Differential-Layer Typical Prediction

The differential-layer typical prediction (TP) block provides some coding gain, but its primary purpose is to speed implementations. Differential-layer TP looks for regions of solid color and when it finds that a given current high-resolution pel is in such a region, none of the processing normally done in the deterministic prediction, adaptive templates, model templates, or arithmetic coding blocks is needed. On text and line-art images differential-layer TP usually makes it possible to avoid coding about 95 percent of the pels. Grayscale renderings do not generally have the sought-after large regions of continuous color and do not allow processing savings like this.

The key idea behind differential-layer TP is that if a low resolution pel and all the pels in an eight-pel neighborhood of it are the same color, then it is extremely likely that all four high-resolution pels to be associated with it are that same color also. Unfortunately, it is not certain that this is so, but exceptions occur infrequently enough that it is efficient and reasonable to flag them. In particular, an encoder notes at the beginning of each high-resolution line pair whether or not a decoder would ever go wrong on that line pair if it always were to "typically expand" any low-resolution pels it found to be within a common-color eight-neighborhood into four high-resolution pels of that color. The failure or success of this strategy over the line pair is coded and sent to the decoder. Note that failure here is not that some low-resolution pel in the line pair is not within a common-color eight-neighborhood, but rather that some low-resolution pel to be associated with the line pair is within a common-color eight-neighborhood but will not have associated high-resolution pels of that color. Failure in this sense is extremely rare and on many images never occurs. When success over the line pair is coded, both the encoder and decoder skip over high-resolution pels associated with any low-resolution pels found to be within a common-color eight-neighborhood. If failure must be coded, the only penalty is that no skipping can be done for the line pair and everything must be coded.

7.1.8 Bottom-Layer Typical Prediction

Bottom-layer typical prediction, like differential-layer typical prediction, tries to exploit solid color regions of the image to save processing effort. However, the algorithms are quite different. The bottom-layer algorithm is a line-skipping algorithm. A given line is said to be "typical" and all its pels are declared *typical*, if it is identical to the line above it. Which lines are typical is again transmitted to the decoder. Both the encoder and decoder skip the coding of all pels in typical lines and generate them instead by line duplication.

It is not possible in bottom-layer coding to skip as large a percentage of pels as it is in differential-layer coding. On text and line-art images savings are generally about 40 percent.

7.1.9 Deterministic Prediction

When images are reduced in resolution by the JBIG resolution-reduction algorithm, it sometimes happens that the value of the particular high-resolution pel being coded is inferable from the pels already known to both the encoder and decoder, i.e., all the pels in the low-resolution image and those in the high-resolution image that are causally related in a raster sense to the current pel. When this occurs, the current pel is said to be deterministically predictable. The deterministic prediction (DP) block flags any such pels and inhibits their coding by the arithmetic coder. DP is a table driven algorithm. The values of particular surrounding pels in the low-resolution image and causal high-resolution image are used to index into a table to check for determinicity and, when it is present, obtain the deterministic prediction. DP provides about a 7 percent coding gain.

7.1.10 Compression Comparison

Fig. 7.1.6 shows compression performance on the eight standard CCITT test images and one additional image. The one additional image is a binary image rendering grayscale via halftoning. It is a picture of a Japanese woman holding flowers. The eight CCITT images are all sampled at 200 dpi (dot-per-inch) and contain 1728 by 2376 pels. The halftoned image contains 2304 by 2896 pels. Compressed-file byte counts are provided for coding with one-dimensional G3, two-dimensional G3 (k factor of 4), G4, non-progressive JBIG, and progressive JBIG with 4 delta layers.

Over the eight CCITT images, non-progressive JBIG coding has about a 22% coding advantage over G4, the most efficient of the G3/G4 algorithms. The progressive JBIG algorithm provides progressivity while at the same time still showing an average 15% coding gain over G4.

It's widely recognized that the G3/G4 algorithms are not suitable for coding bi-level images rendering grayscale via halftoning. This is readily apparent in the last row of Fig. 7.1.6 where the JBIG compression advantage is about a factor of five.

7.2 JPEG Still-Color-Image Coding

The need for an international standard for continuous-tone still image compression resulted, in 1986, in the formation of JPEG, which stands for Joint Photographic Experts Group. This group was chartered by ISO and the CCITT to develop a general-purpose standard suitable for as many applications as possible. After thorough evaluation and subjective testing of a number of proposed image-compression algorithms, the group agreed, in 1988, on a DCT-based technique. From 1988 to 1990, the JPEG committee refined several methods incorporating the DCT for lossy compression. A lossless method was also defined. The committee's work has been published in two parts: "Part 1: Requirements and guidelines"[7.7] describes the JPEG compression and decompression method. "Part 2: Compliance Testing"[7.8] describes tests to

Image	Bytes					
	Raw	G3D1	G3D2	G4	JBIGD0	JBIGD4
CCITT #1	513216	37423	25967	18103	14715	16771
CCITT #2	513216	34367	19656	10803	8545	8933
CCITT #3	513216	65034	40797	28706	21988	23710
CCITT #4	513216	108075	81815	69275	54356	58656
CCITT #5	513216	68317	44157	32222	25877	28086
CCITT #6	513216	51171	28245	16651	12589	13455
CCITT #7	513216	106420	81465	69282	56253	60770
CCITT #8	513216	62806	33025	19114	14278	15227
Halftone	834048	483265	572259	591628	131479	103267

Fig. 7.1.6. Compressed file sizes in bytes for various coding algorithms.

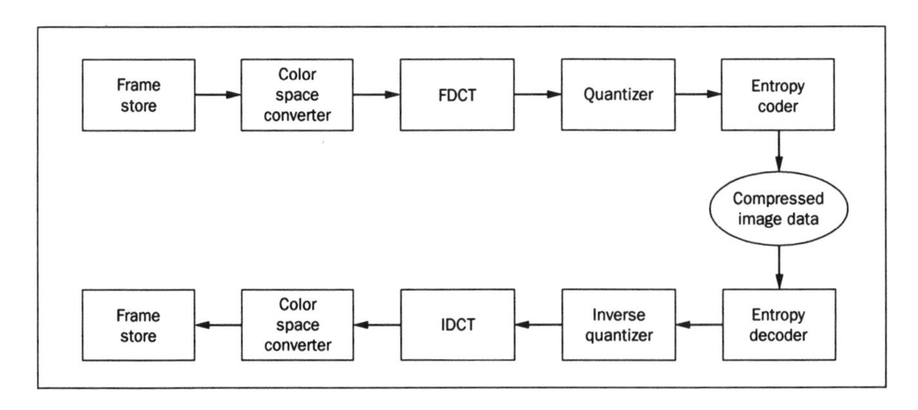

Fig. 7.2.1. Sequential DCT codec.

verify whether an encoder-decoder (codec) has implemented the JPEG algorithms correctly. Reference[7.9] describes the JPEG algorithms in detail.

To appreciate the need for a standard image compression algorithm, consider the storage/transmission requirements of an uncompressed image. A typical VGA digital color image has 640 pels × 480 lines. At three bytes per pel (one each for the red, green and blue components), such an image requires 921,600 bytes of storage space. To transmit the uncompressed image over a 64-kb/s ISDN channel takes about 1.9 minutes. The JPEG algorithms offer ''excellent'' quality for most images compressed to about 1.0 bit/pel. This 24:1 compression ratio reduces the required storage of the 640 × 480 color image to 38,400 bytes, and its transmission time to about 4.8 seconds. Applications for image compression may be found in desktop publishing, education, real estate, and security, to name a few.

In the next section, we give an overview of the JPEG algorithms. In subsequent sections, we present some operating parameters and definitions, and describe each of the JPEG operating modes in more detail.

7.2.1 Overview of the JPEG Algorithms

The JPEG committee could not satisfy the requirements of every still-image compression application with one algorithm. As a result, the committee defined four different modes of operation:

- *Sequential DCT-based* — Fig. 7.2.1 presents a simplified diagram of a sequential DCT codec. In this mode, 8 × 8 blocks of the input image are formatted for compression by scanning the image left to right and top to bottom. A block consists of 64 samples of one color component that make up the image. Each block of 64 samples is transformed to a block of 64 coefficients by the forward discrete cosine transform (DCT). The DCT concentrates most of the energy of the component samples' block into a few coefficients, usually in the top-left corner of the DCT block. The coefficient in the immediate top-left corner is called the DC coefficient because it is proportional to the average intensity of the block of spatial domain samples. AC coefficients corresponding to increasingly higher frequencies of the sample block progress away from the DC coefficient. The coefficients are then quantized and entropy-coded. A subset of the Sequential mode is the *Baseline Mode*, which is required to be present in all JPEG implementations.

- *Progressive DCT-based* — This mode offers a means of producing a quick ''rough'' decoded image when the medium separating the encoder and decoder has a low bandwidth. The method is similar to the sequential DCT-based algorithm, but the quantized coefficients are partially encoded in multiple scans.

- *Lossless* — In this mode, the decoder receives an exact reproduction of the digital input image. The differences between input samples and predicted values, where the predicted values are combinations of one to three neighboring samples, are entropy-coded.
- *Hierarchical* — This mode, also known as Pyramid coding, is used to encode an input image as a sequence of increasingly higher-resolution frames in exactly the same way as in JBIG. The first frame is a reduced resolution version of the original. Subsequent frames are coded as higher-resolution differential frames.

The color space conversion process in Fig. 7.2.1 is not a part of the standard. In fact, JPEG is color-space-independent. As a first step in the compression process, many image-compression schemes take advantage of the human visual system's low sensitivity to high-frequency chrominance information by reducing the chrominance resolution. Many images (usually RGB) are typically converted to a luminance-chrominance representation, e.g., YCbCr, before this processing takes place. Input pel values having precision n bits are unsigned integers in the range 0 to $2^n - 1$.

Either Huffman or arithmetic coding techniques can be used for entropy coding in any of the JPEG modes of operation (except the *baseline*, where Huffman coding is mandatory). There are no standard JPEG Huffman codes. The encoder must transmit its Huffman code tables (up to 4 AC, and 4 DC) according to a prescribed format. As described in Chapter 3, a Huffman coder compresses a series of input symbols by assigning short code words to frequently occurring symbols and long code words to improbable symbols.[7.10,7.11] The output of an arithmetic coder, on the other hand, is a single real number for each color component image. Unlike a Huffman coder, an arithmetic coder does not require an integral number of bits to represent an input symbol. As a result, arithmetic coders are usually more efficient than Huffman coders.[7.12,7.13] For the JPEG test images, Huffman coding (using fixed tables) resulted in compressed data requiring, on average, 13.2 percent more storage than arithmetic coding.

7.2.2 JPEG Operating Parameters and Definitions

A number of parameters related to the source image and the coding process may be customized to meet the user's needs. In this section, we discuss some of the important variable parameters and their allowable ranges. Also, as an aid to the algorithm descriptions in the following sections, we define some JPEG terms and present the hierarchical structure of the compressed data.

Parameters

An image to be encoded using any JPEG mode may have from 1 to 65,535 lines and from 1 to 65,535 pels per line. Each pel may have from 1 to 255 color components (at most 4 components are allowed for progressive mode). The operating mode determines the allowable pel precision of the color component. For the DCT modes, either 8 or 12 bits of (unsigned) pel precision are supported (only 8 bit precision is allowed for *baseline*). Prior to coding, unsigned pel values, having precision n bits, are shifted downward by subtracting 2^{n-1} to give a signed range of -2^{n-1} to $2^{n-1}-1$. The lossless mode pel precision may range from 2 to 16 bits. If a DCT operating mode has been selected, the coefficient quantizer precision must also be defined. For 8-bit component precision, the quantizer precision is fixed at 8 bits. Twelve-bit components require either 8- or 16-bit quantizer precision.

Data interleaving

To reduce the processing delay and/or buffer requirements, JPEG can interleave up to four color components in a single scan (for progressive mode, only the DC scan may have interleaved components). A data structure called the *minimum-coded unit* (MCU) has been defined to support this interleaving. An MCU consists of one or more data units, where a data unit is a component sample for the lossless mode, and an 8×8 block of component samples for the DCT modes. If a scan contains only one component, then its MCU is equal to one data unit. For multiple component scans, the MCU for the scan consists of interleaved data units. The maximum number of data units per MCU is 10. As an interleaving example, consider a CCIR-601 digital image in which the chrominance components are subsampled 2:1 horizontally. For a DCT coder, a CCIR-601 MCU could consist of two Y blocks, followed by a Cb block and a Cr block.

Marker codes

JPEG has defined a number of two-byte marker codes to delineate the various sections of a compressed data stream. All marker codes begin with a byte-aligned hexadecimal ''FF'' byte, making it easy to scan and extract parts of the compressed data without actually decoding it. It is also possible to create by chance a byte-aligned hexadecimal ''FF'' byte within the normal entropy-coded data. Thus, the encoder must detect this situation and follow a byte aligned ''FF'' byte with a zero byte in order to avoid ambiguity. When the decoder encounters the byte aligned hexadecimal ''FF00'' combination, the zero byte must be removed.

Compressed-image data structure

At the top level of the compressed data hierarchy is the *image* (see Fig. 7.2.2). A nonhierarchical mode image consists of a *frame* surrounded by ''Start'' and ''End of Image'' marker codes. There will be multiple *frames* in a hierarchical mode image. Within a frame, a start of frame (SOF) marker identifies the coding mode to be used. The SOF marker is followed by a number of parameters (see Reference 7.7), and then by one or more *scans*. Each scan begins with a header identifying the components to be contained within the scan, frequency coefficients contained in the scan for progressive mode, and more parameters. The scan header is followed by an entropy-coded segment (ECS). An option exists to break the ECS into chunks of MCUs called *restart intervals*. The restart interval structure is useful for identifying select portions of a scan, and for recovery from limited corruption of the entropy-coded data. Quantization and entropy-coding tables may either be included with the compressed image data or communicated separately.

7.2.3 Baseline Sequential DCT

The sequential DCT mode offers very good compression ratios, while maintaining excellent image quality. A subset of the sequential DCT capabilities has been identified by JPEG as a *Baseline System* (8-bit pels, Huffman coding, 8-bit quantizer precision, up to 2 AC and 2 DC Huffman Tables). All DCT-based JPEG implementations are required to include baseline capability. This requirement should help to ensure interoperability between codecs from different vendors.

The following subsections describe the processing steps for a baseline coder. A decoder is formed by reversing the coder steps.

DCT and quantization

All JPEG DCT-based coders begin the coding process by partitioning the input image into non-overlapping 8×8 blocks of component samples. After level-shifting the 8-bit samples so that they range from -128 to +127, the blocks are transformed to the frequency domain using the DCT[7.14,7.15] as defined in Chapter 5. The equations for the forward and inverse orthonormal discrete cosine transforms* are given by:

$$DCT: \quad c_{uv} = \frac{1}{4} K_u K_v \sum_{x=0}^{7} \sum_{y=0}^{7} b_{xy} \cos\frac{\pi u(2x+1)}{16} \cos\frac{\pi v(2y+1)}{16} \tag{7.1}$$

* These equations are exactly equivalent to the matrix representations defined in Chapters 3 and 5.

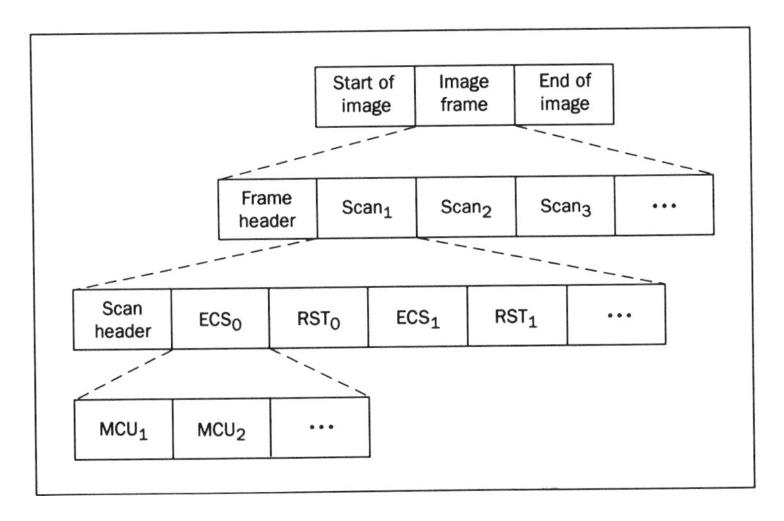

Fig. 7.2.2. Compressed-image data structure.

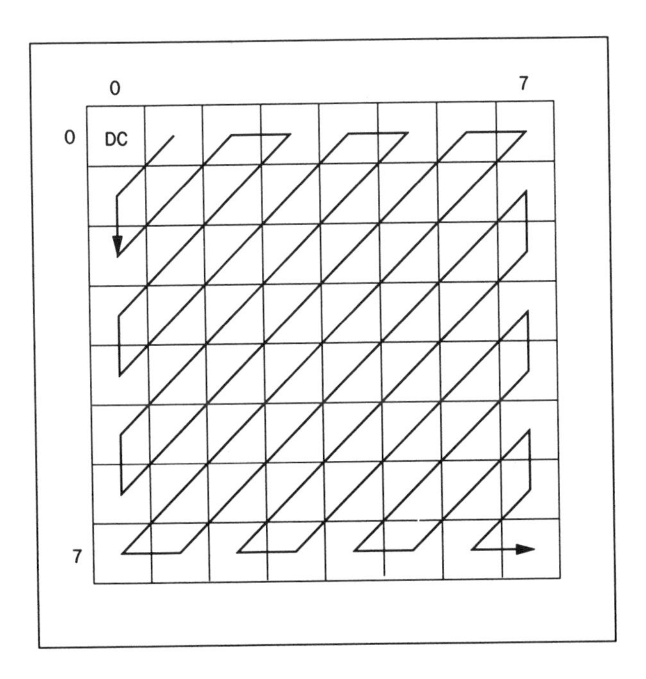

Fig. 7.2.3. Zigzag scan.

$$IDCT: \quad b_{xy} = \frac{1}{4} \sum_{u=0}^{7} \sum_{v=0}^{7} K_u K_v c_{uv}$$

$$\cos\frac{\pi u(2x+1)}{16} \cos\frac{\pi v(2y+1)}{16} \tag{7.2}$$

where

$$K_w = \frac{1}{\sqrt{2}} \text{ for } w = 0;\ K_w = 1 \textit{ otherwise.}$$

The next step in the process, quantization, is the key to most of the JPEG compression. For each picture, a 64-element quantization matrix $Q[u,v]$ is defined, where each element corresponds to a coefficient in the DCT block. $Q[u,v]$ is used to reduce the amplitude of the coefficients, and thus to increase the number of zero-value coefficients after quantization. Although there are no standard JPEG quantization matrices, JPEG includes a set that gives good results for CCIR-601 type images. The encoder must transmit its quantization matrices (up to 4) according to a prescribed format. The quantization and dequantization is performed according to Eq. (7.3) and Eq. (7.4), respectively.

$$FQ_{uv} = round\left(\frac{c_{uv}}{Q[u,v]}\right) \tag{7.3}$$

$$\tilde{c}_{uv} = FQ_{uv} \times Q[u,v] \tag{7.4}$$

A carefully designed quantization matrix will produce high compression ratios while introducing negligible "visible" distortion.[7.16] Many JPEG implementations control the compression ratio (and output image quality) by using a *q-factor*, which is usually just a scale factor applied to the quantization matrices.

DC coefficient entropy coding

Greater compression efficiency can be obtained if a simple predictive method is used to entropy-code the DC coefficient separately from the AC coefficients. Recall that the DC coefficient corresponds to the average intensity of the component block. Since adjacent blocks will probably have similar average intensities., it is advantageous to encode the *differences* between the DC coefficients of adjacent blocks rather than their values. Each differential DC value is coded using a variable-length code (VLC) followed by a variable-length integer (VLI). The VLC, which is Huffman coded, indicates the size in bits of the VLI. The VLI simply gives the differential DC value.

Zigzag scan and AC coefficient entropy coding

After quantization, the coefficient blocks usually contain many zero-value AC coefficients. If the coefficients are reordered, using the zig-zag scan illustrated in Fig. 7.2.3, there will be a tendency to have long runs of zeroes. Only the nonzero AC coefficients are entropy-coded. As in the DC coefficient coding, a VLC-VLI pair results from the coding of an AC coefficient. However, in this case the AC VLC corresponds to two pieces of information: the number of zeroes (run) since the last nonzero coefficient, and the size of the VLI following the VLC. Two additional symbols are also assigned Huffman codes. The End-of-Block (EOB) symbol indicates that all remaining coefficients in the zigzag scan are zero. If the last coefficient of the zigzag scan is nonzero, the EOB is not sent. The zero-run-length (ZRL) symbol indicate a run of 16 zeros, which is needed since VLC can only code zero runs up to length 15.

7.2.4 Sequential DCT with Arithmetic Coding

As stated previously, entropy coding of DCT coefficients can use either Huffman or Arithmetic coding, with Arithmetic Coding giving somewhat better coding efficiency. As with Huffman Coding, Arithmetic Coding codes the DC difference first, followed by the AC coefficients in the zigzag order of Fig. 7.2.3. Coding typically ends after the last nonzero coefficient of the block.

The first step is to convert the DC difference (*DIFF*) into a string of bits amenable to Arithmetic Coding. If $DIFF = 0$, the first and only bit (except for EOB) is zero. Otherwise, the first bit is one, and coding continues by defining the decremented absolute value $Sz = |DIFF| - 1$. The second bit is the sign bit. The next set of bits corresponds to the Huffman Coding VLC bits, and is related to the size in bits of Sz. The VLC bit pattern consists of $0 \le P \le 15$ ones followed by a zero, and indicates that $2^{P-1} < |DIFF| \le 2^P$. If $P \ge 2$, the VLC bits are followed by the VLI, which consists of the $P-1$ least significant bits of Sz. Finally, comes the EOB bit, which is one if there are no more AC coefficients to code, and is zero otherwise.

This DC bit string is then Arithmetic Coded using a 50 entry context table along with the JBIG probability table of Fig. 7.1.5. The first bit is coded using one of five contexts depending on an integer parameter $0 \le D_a \le 4$, which in turn depends on the magnitude and sign of the DC difference in the previous block. Similarly, for the sign bit. The context of the first VLC bit depends not only on D_a, but also on the sign bit, for a total of 10 possible contexts. The remaining VLC bits each have their own context, independent of anything else. The VLI bits are all coded with the same context that depends only on P. The DC EOB bit has its own context.

Each AC coefficient value (*ACVAL*) is then coded in a very similar manner. First it is converted to a string of bits in the same way as the DC difference, with two important exceptions. If $ACVAL = 0$, no EOB bit is produced since the decoder knows by convention that another AC coefficient will be sent. Also, if all 64 coefficients of the block are sent, no EOB bit is produced after the last coefficient.

This AC bit string is then Arithmetic Coded using a 244 entry context table along with the JBIG probability table of Fig. 7.1.5. The first bit is coded using one of 63 contexts depending on the position $1 \leq K \leq 63$ of the coefficient in the zigzag sequence. The sign bit is almost completely random. Therefore, it is coded without context using a fixed probability ($MPS=0$, $\bar{p}=$ .5A1D hexidecimal). The first two VLC bits use the same context, which again depends only on K, thus requiring another 63 contexts. The remaining VLC bits (up to 14) each have two contexts that depend on whether K exceeds a predefined threshold K_x, thus requiring another 28 contexts. The VLI bits (up to 14) are all coded with the same context that depends only on P and whether $K > K_x$. This requires another 28 contexts. Finally, the EOB bit context depends only on K, thus requiring another 62 contexts ($K=63$ has no EOB bit).

7.2.5 Progressive DCT

A progressive DCT mode has been defined by JPEG to satisfy the need for a fast decoded picture when a low-bandwidth medium separates an encoder and decoder. By partially encoding the quantized DCT coefficients in multiple scans, the decoded image quality builds progressively from a coarse level to the quality attainable with the quantization matrices. Either spectral selection, successive approximation, or a combination of the two is used to code the quantized coefficients.

Spectral selection: In this method, the quantized DCT coefficients of a block are first partitioned into non-overlapping bands along the zigzag block scan. The bands are then coded in separate component scans. Before an AC coefficient band of a component may be coded, its DC coefficient must be coded. DC coefficients from as many as four components may be interleaved in a single scan. Interleaving is not permitted for AC bands.

Successive approximation: With this method, the precision of the coefficients is successively increased during multiple scans. Following a scan for a specified number of most significant bits of the quantized coefficients, subsequent scans increase the precision in increments of one bit until the least significant bits have been coded.

7.2.6 Lossless Mode

The lossless mode was defined for applications in which output pels from a decoder must be identical to the input pels to the encoder. The compression ratios achievable with the lossless mode, typically around 2:1, are much smaller than those afforded by the lossy modes. It does not use the DCT. Instead it uses DPCM in a way similar to that used to code the DC coefficients in the DCT-based modes, except that the predictor is selectable from one of seven choices, as shown in Fig. 7.2.4. Samples a, b, and c in the table correspond to neighbors

Selection value	Prediction
0	No prediction
1	a
2	b
3	c
4	a + b - c
5	a + ((b - c)/2)
6	b + ((a - c)/2)
7	(a + b)/2

Fig. 7.2.4. Lossless Mode Predictors.

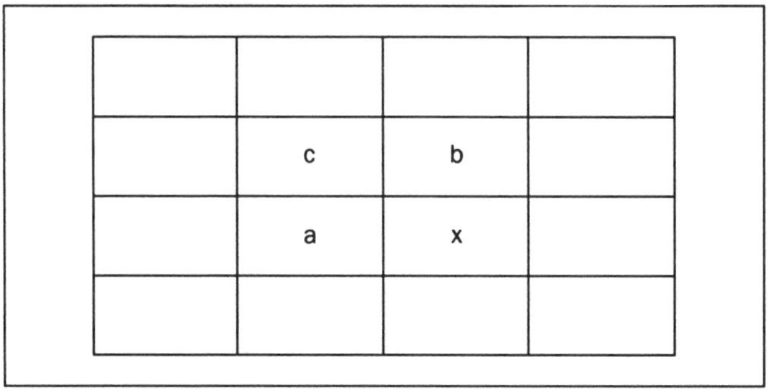

	c	b	
	a	x	

Fig. 7.2.5. Prediction neighborhood.

of the sample x to be predicted. Fig. 7.2.5 illustrates the prediction neighborhood. Entries 1 to 3 in Fig. 7.2.4 are used for one-dimensional predictive coding, and 4 through 7 form two-dimensional predictors. Entry 0 identifies differential coding for the hierarchical mode. As in the DC coefficient entropy coding described earlier, differences between the actual and predicted values are entropy-coded.

7.2.7 Hierarchical Mode

In the hierarchical mode, an image is coded as a succession of increasingly higher-resolution frames. Also known as *pyramidal* coding, this approach offers a higher quality alternative to the previously described methods for achieving progression. However, it is more expensive to implement. It also allows decoders with different resolution capabilities to use the same compressed data stream.

The first coded frame is created by reducing the resolution of the input image by a power of two in one or both dimensions, and then processing the lower resolution image using one of the lossy or lossless techniques of the other operating modes. Subsequent frames are formed by upsampling the decoded image by a factor of two in the dimension(s) having reduced resolution, subtracting the upsampled image from the input image at the same resolution, and coding the difference. This process continues until the decoded image has the same resolution as the full-resolution input image. After that, one or more full-resolution difference images may be coded. A hierarchical decoder may abort the decoding process after it has decoded a frame that provides the desired resolution.

Any coding methods described in the other three modes of operation may be used to code the hierarchical mode frames, with the following restrictions:

- If a lossy method is chosen, all but the last frame must be coded using that method. A lossless method optionally may be used to code the last frame.
- If a lossless method is chosen, all frames must be coded with that method.
- The same entropy-coding technique (Huffman or arithmetic) must be used for all frames.

The hierarchical coding/decoding process is not symmetrical. Indeed, a hierarchical coder must also include the greater part of a decoder. However, a hierarchical decoder is more complex than a nonhierarchical decoder because it must provide a way to upsample and add. This increased complexity may be justified, given the flexibility afforded in matching the decoder to the application. This type of codec is well suited for *one-to-many* applications, as in a number of decoders (possibly having different resolution capabilities) accessing a database of images precoded by a hierarchical coder.

References

7.1 ISO Draft International Standard 11544, *Coded representation of picture and audio information — Progressive bi-level image compression*, 1992.

7.2 T. C. Bell, J. G. Cleary, and I. H. Witten, *Text Compression*, Prentice Hall, 1990.

7.3 N. Abramson, *Information Theory and Coding*, pp. 61-62, McGraw-Hill, 1963.

7.4 C. E. Shannon, "A mathematical theory of communication," *Bell System Technical Journal*, vol. 27, pp. 379-423 and 623-656, 1948.

7.5 F. Rubin, "Arithmetic stream coding using fixed precision registers," *IEEE Trans. Inf. Theory*, vol. IT-25, no. 6, pp. 672-675, November 1979.

7.6 C. Chamzas and D. Duttweiler, "Probability estimation in arithmetic and adaptive-Huffman entropy coders," submitted to *IEEE Transactions on Acoustics, Speech, and Signal Processing*.

7.7 *Digital Compression and Coding of Continuous-Tone Still Images, Part 1: Requirements and Guidelines*, ISO/IEC IS 10918-1, 1991.

7.8 *Digital Compression and Coding of Continuous-Tone Still Images, Part 2: Compliance Testing*, ISO/IEC IS 10918-2, 1991.

7.9 W. B. Pennebaker and J. L. Mitchell, *JPEG - Still Image Data Compression Standard*, Van Nostrand Reinhold, new York, 1993.

7.10 D. A. Huffman, ''A Method for the Construction of Minimum-Redundancy Codes,'' *Proc. IRE*, No. 40, September 1952, pp. 1098-1101.

7.11 J. Amsterdam, ''Data Compression with Huffman Coding,'' *BYTE*, Vol. 11, No. 5, May 1986, pp. 99-108.

7.12 G. G. Langdon, Jr., ''An Introduction to Arithmetic Coding,'' *IBM J. Res. Develop.*, Vol. 28, No. 2, March 1984, pp . 135-149.

7.13 I. H. Witten, R. M. Neal, and J. G. Cleary, ''Arithmetic Coding for Data Compression,'' *Commun. ACM*, Vol. 30, No. 6, June 1987, pp. 520-540.

7.14 N. Ahmed, T. Natarajan, and K. R. Rao, ''Discrete Cosine Transform,'' *IEEE Transactions on Computers*, Vol. C-23, No. 1, January 1974, pp. 90-93.

7.15 R. J. Clarke, *Transform Coding of Images*, Academic Press, Orlando, Florida, 1985.

7.16 H. Lohscheller, ''A subjectively adapted image communication system,'' *IEEE Transactions on Communications*, Vol. COM-32, December 1984, pp. 1316-1322.

7.17 R. Aravind, *et al*, ''Image and Video Coding Standards,'' *AT&T Technical J.*, Vol. 72, No. 1, January 1993, pp. 67-89.

Questions for Understanding

7.1 Why is the G3/G4 coding algorithm inefficient on dithered halftone bilevel images?

7.2 What advantages does progressive coding offer?

7.3 What is $\overline{p}$ in JBIG coding? Why does JBIG use $\overline{p}$ instead of p?

7.4 Why is *conditional exchange* needed in JBIG?

7.5 What is *context* in JBIG coding? How is it computed? How many possible context values are there for layered and non layered coding?

7.6 For a context value of 1346, what is the starting value of $\overline{p}$?

7.7 Suppose in the example above the next three pels have context values 1580, 1346, 1346 and all are LPS. What $\overline{p}$ values are used in coding these three pels?

7.8 Explain how Adaptive Templates work in coding dithered gray scale images.

7.9 Explain how Typical Prediction works in JBIG.

7.10 What are the four operating modes of JPEG coding?

7.11 How are Huffman codes determined in JPEG?

7.12 Of what use is data interleaving in JPEG? What is an MCU?

7.13 Why are Marker Codes used? How are they inserted into the compressed bit stream?

7.14 Briefly, describe the JPEG Baseline coding algorithm.

7.15 Write the bit strings used by JPEG Arithmetic Coding to represent DCDIFF or AC values between -20 and 20.

7.16 Write the bit string input to a JPEG Arithmetic Coder when the DCDIFF and first five AC values are -10, 8, 0, 0, -5, 20.

7.17 Explain the Hierarchical JPEG mode.

8

CCITT H.261 (P*64) Videoconferencing Coding Standards

8.1 Video Coding at Sub-Primary Rates

In most countries there is only limited availability of bit-rates below the primary rates, although the situation is changing rapidly. Two bit-rates that have been established for Integrated Services Digital Networks (ISDN) are of interest for image transmission. They are the so called B-channel of 64 kbits/s and the H0-channel of 384 kbits/s.

Many countries are offering a so called *Basic Rate* ISDN, which consists of two B-channels plus a signalling channel (2B+D). Basic Rate ISDN enables video coding at 112 kbits/s and speech coding at 16 kbits/s. H0 ISDN allows video coding at 320 kbits/s and audio at 64 kbits/s.

Standard international video coding algorithms have been established for ISDN bit-rates. In this section, we will outline some of the methodology that is used for video codecs operating at sub-primary bit-rates.

Coding of motion video at these rates usually requires not only conditional replenishment, but also fairly high data compression down to rates well below 1 bit per moving-area pel. The most straightforward way of achieving this is to code two-dimensional blocks of data using a transform such as the Discrete Cosine Transform (DCT) followed by Variable Length Codes (VLCs) such as Huffman Codes. Additional efficiency can be achieved by using motion compensation to subtract predictions from the moving-area pel values prior to transformation, as well as adaptive quantization and coding to maximize the received image quality for the channel capacity at hand.

This chapter contains excerpts taken with permission from AT&T Technical Journal[7.17].

8.2 From JPEG to H.261 (P*64)

From an algorithmic point of view, the extension from JPEG intraframe DCT coding to H.261[8.1] motion-compensated DCT video coding is a rather natural one. However, historically H.261 was developed long before JPEG. In December 1984, CCITT* Study Group XV (Transmission Systems and Equipment) established a "Specialists Group on Coding for Visual Telephony." The development of this video transmission standard for low-bit-rate ISDN services has gone through several stages. At the beginning, the design target was for $m \times 384$ kbits channels, where m was between 1 and 5. Later, $n \times 64$ kbits transmission rates (n from 1 to 5) were considered. However, by late 1989, the final CCITT recommendations were made for a $p \times 64$ kbits/s video codec, where p is between 1 and 30, whence came the name P*64.

In fact, P*64 (or H series) is a group of audiovisual teleservices standards (or recommendations) consisting of H.221 frame structure; H.230 frame synchronous control; H.242 communication between audiovisual terminals; H.320 systems and terminal equipment; and H.261 video codec. Audio codecs at several bit rates have also been specified by other ITU-T recommendations, such as G.711, G.722, G.728. In this chapter, we concentrate on the H.261 video codec system.

Both JPEG baseline and H.261 codecs use DCT and VLC techniques. The major difference between JPEG and H.261 is the way each handles motion information. In JPEG, incoming picture frames are processed independently using intraframe DCT, while H.261 uses a block-based, motion-compensated scheme as described in Section 5.2.3g. That is, the picture data in the previous frame can be used to predict the image blocks in the current frame, as shown in Fig. 8.1. As a result, only differences, typically of small magnitude, between the displaced previous block and the current block have to be transmitted.

There are several interesting characteristics or design considerations in H.261.

- First, it defines essentially only the *decoder*. However, the *encoder*, which is not completely and explicitly specified by the standard, is expected to be compatible with the well-defined decoder.
- Second, because H.261 is designed for real-time communications, it uses only the closest previous frame for motion picture sequence coding to reduce the encoding delay.
- Third, it tries to balance the hardware complexities between the encoder and the decoder, because they are both needed for real-time videophone

* The CCITT has since renamed itself the ITU-T.

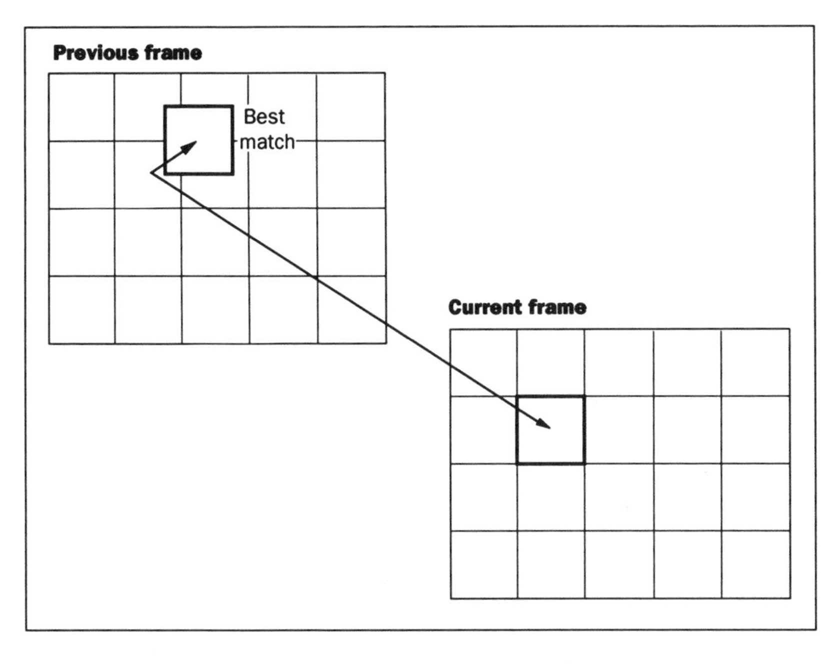

Fig. 8.1. Block motion compensation.

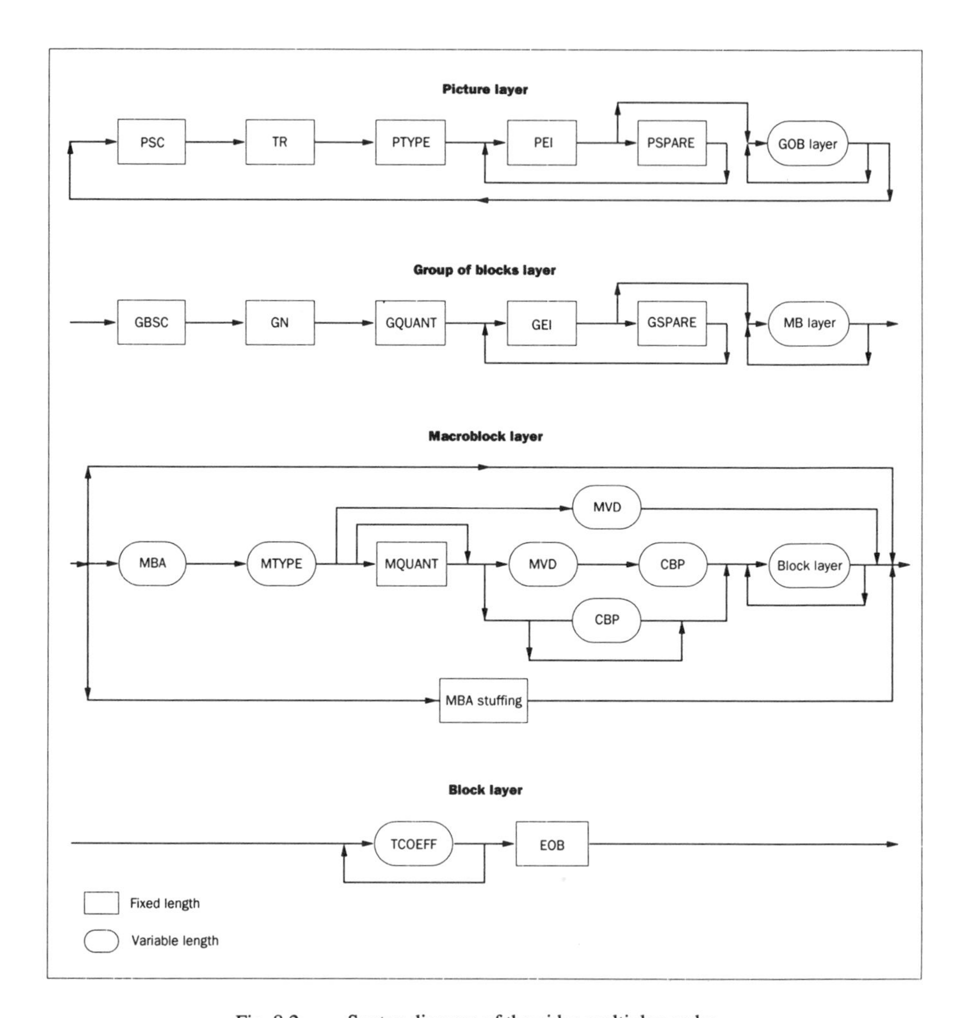

Fig. 8.2. Syntax diagram of the video multiplex coder.

applications. Other coding schemes, such as vector quantization (VQ), may have a rather simple decoder, but a very complex encoder.

- Fourth, H.261 is a compromise between coding performance, real-time requirements, implementation complexity, and system robustness. Motion-compensated DCT coding is a mature algorithm, and after years of study, quite general and robust in that it can handle various types of pictures.
- Fifth, the final coding structures and parameters are tuned more toward low-bit-rate applications. This choice is logical, because selection of the coding structure and coding parameters is more critical to codec performance at very low bit rates. At higher bit rates, less-than-optimal parameters values do not affect codec performance very much.

8.3 Decoder Structures and Components

H.261 specifies a set of protocols that every compressed video bit stream has to follow, and a set of operations that every standard compatible decoder must be able to perform. The actual hardware codec implementation and the encoder structure can vary drastically from one design to another. We will first explain briefly the data structure in an H.261 video bit stream and then the functional elements in an H.261 decoder.

The compressed H.261 video bit stream[8.1] contains several layers (see Fig. 8.2). They are *picture* layer, group of blocks (*GOB*) layer, Macroblock (*MB*) layer, and *block* layer. The higher layer consists of its own header followed by a number of lower layers.

Only two picture formats — common intermediate format (CIF) and quarter-CIF (QCIF) — are allowed. CIF pictures are made of three components: luminance Y and color differences C_B and C_R, as defined in CCIR Recommendation 601. The CIF picture size for Y is 352 pels per line by 288 lines per frame. The two-color difference signals are subsampled to 176 pels per line and 144 lines per frame. Fig. 8.3 shows the sampling pattern of Y, C_B, and C_R. The image aspect ratio is 4(horizontal):3(vertical), and the picture rate is 29.97 non-interlaced frames per second. Fig. 8.4 shows specifications for CIF and QCIF. All standard codecs must be able to operate with QCIF; CIF is optional.

A picture frame is partitioned into 8 line by 8 pel image blocks. A so-called Macroblock (MB) is made of 4 Y blocks, one C_B block, and one C_R block at the same location, as shown in Fig. 8.5a. Fig. 8.5b contains 33 MBs grouped into a GOB. Therefore, one CIF frame contains 12 GOBs and one QCIF frame 3 GOBs, as shown in Fig. 8.5c.

Picture Layer

In a compressed video bit stream, we start with the picture layer. Its header contains:

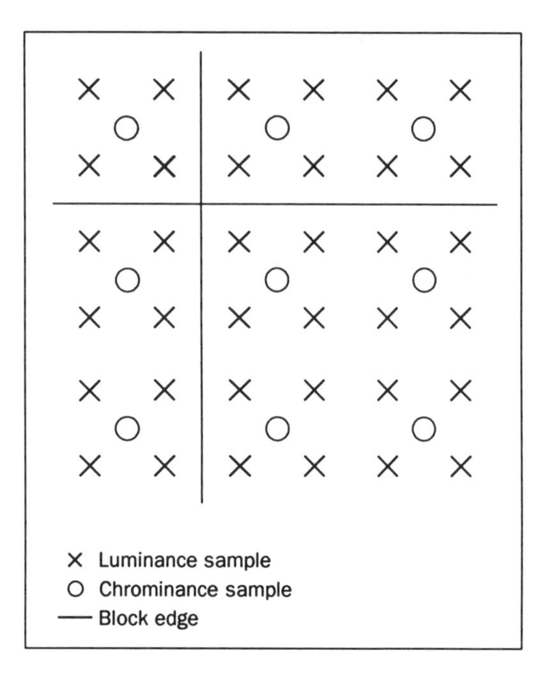

Fig. 8.3. Positioning of luminance and chrominance samples.

	CIF	QCIF
Luminance Pels per Line	352	176
Chrominance Pels per Line	176	88
Luminance Lines per Field	288	144
Chrominance Lines per Field	144	72
Fields per Second	29.97	29.97
Interlace	1:1	1:1
Color Components	$Y\ C_B C_R$	$Y\ C_B C_R$
Luminance Range	16 - 235	16 - 235
Chrominance Range	16 - 240	16 - 240
Zero Color Difference Level	128	128
Bits per Pel	8	8

Fig. 8.4. Internationally compatible videoconferencing video formats.

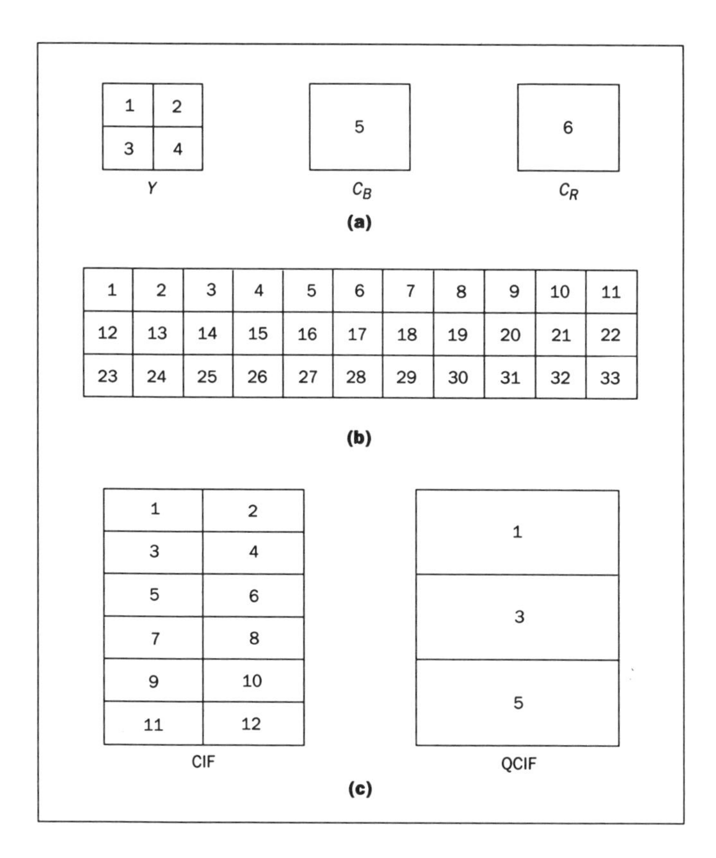

Fig. 8.5. Successive arrangement of (a) blocks in a Macroblock, (b) Macroblocks in a GOB, and (c) GOBs in a picture.

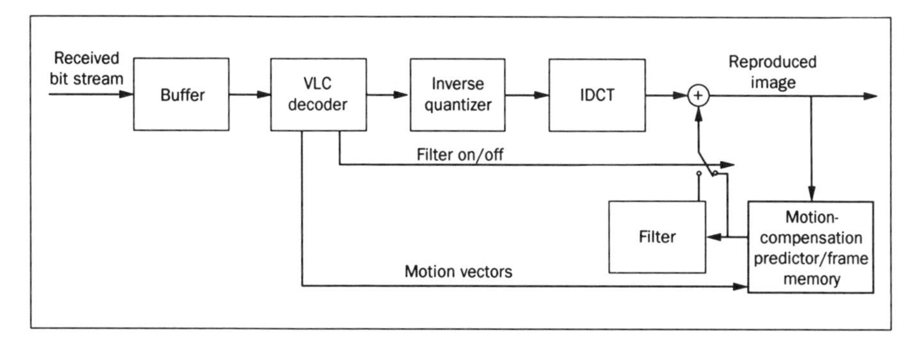

Fig. 8.6. A typical H.261 decoder.

- Picture start code (PSC) — a 20-bit pattern.
- Temporal reference (TR) — a 5-bit input frame number.
- Type information (PTYPE) — such as CIF/QCIF selection.
- Spare bits to be defined in later versions.

GOB Layer

At the GOB layer, a GOB header contains:

- Group of blocks start code (GBSC) — a 16-bit pattern.
- Group number (GN) — a 4-bit GOB address.
- Quantizer information (GQUANT) — Initial quantizer step size normalized to the range 1 to 31. At the start of a GOB, we set QUANT = GQUANT.
- Spare bits to be defined in later versions of the standard.

Next, come a number of MB layers. An 11-bit stuffing pattern can be inserted repetitively right after a GOB header or after a transmitted Macroblock.

Macroblock (MB) Layer

At the MB layer, the header contains:

- Macroblock address (MBA) — location of this MB relative to the previously coded MB inside the GOB. MBA equals one plus the number of skipped MBs preceding the current MB in the GOB.
- Type information (MTYPE) — 10 types in total.
- Quantizer (MQUANT) — normalized quantizer step size to be used until the next MQUANT or GQUANT. If MQUANT is received we set QUANT = MQUANT. Range is 1 to 31.
- Motion vector data (MVD) — differential displacement vector.
- Coded block pattern (CBP) — Indicates which blocks in the MB are coded. Blocks not coded are assumed to contain all zero coefficients.

Block Layer

The lowest layer is the block layer, consisting of quantized transform coefficients (TCOEFF), followed by the end of block (EOB) symbol. All coded blocks have the EOB symbol.

Not all header information need be present. For example, at the MB layer, if an MB is not Inter motion-compensated (as indicated by MTYPE), MVD does not exist. Also, MQUANT is optional. Most of the header information is coded using Variable Length Codewords.

Differential Motion Vectors	Code Words
-16 & 16	0000 0011 001
-15 & 17	0000 0011 011
-14 & 18	0000 0011 101
-13 & 19	0000 0011 111
-12 & 20	0000 0100 001
-11 & 21	0000 0100 011
-10 & 22	0000 0100 11
-9 & 23	0000 0100 01
-8 & 24	0000 0101 11
-7 & 25	0000 0101
-6 & 26	0000 1001
-5 & 27	0000 1011
-4 & 28	0000 111
-3 & 29	0001
-2 & 30	0011
-1	011
0	1
1	10
2 & -30	0010
3 & -29	0001 0
4 & -28	0000 110
5 & -27	0000 1010
6 & -26	0000 1000
7 & -25	0000 0110
8 & -24	0000 0101 10
9 & -23	0000 0101 00
10 & -22	0000 0100 10
11 & -21	0000 0100 010
12 & -20	0000 0100 000
13 & -19	0000 0011 110
14 & -18	0000 0011 100
15 & -17	0000 0011 010

Fig. 8.7. Variable Length Codes for differential motion vectors. Since the motion vector range is limited to ±15 pels, only one of the differential vector values can be legal.

Increasing cycles-per-picture width →

Increasing cycles-per-picture height ↓

1	2	6	7	15	16	28	29
3	5	8	14	17	27	30	43
4	9	13	18	26	31	42	44
10	12	19	25	32	41	45	54
11	20	24	33	40	46	53	55
21	23	34	39	47	52	56	61
22	35	38	48	51	57	60	62
36	37	49	50	58	59	63	64

Fig. 8.8. Zigzag transmission order for transform coefficients.

Fig. 8.6 is a functional diagram of a typical H.261 decoder. The received video bit stream is first kept in the receiver buffer. The VLD decodes the compressed bits and distributes the decoded information to the elements that need that information. Further processing proceeds as described below.

There are essentially four types of coded MBs as indicated by MTYPE:

- Intra — original pels are transform-coded.
- Inter — frame difference pels (with zero-motion vectors) are coded. Skipped MBs are considered inter by default.
- Inter_MC — displaced (nonzero-motion vectors) frame differences are coded.
- Inter_MC_with_filter — the displaced blocks are filtered by a predefined loop filter, which may help reduce visible coding artifacts at very low bit rates.

Certain MB types in this list allow the optional transmission of MQUANT and TCOEFF information. The received MTYPE information controls various switches at the decoder to produce the correct combination.

A single-motion vector (horizontal and vertical displacement) is transmitted for one Inter_MC MB. That is, the four Y blocks, one C_B, and one C_R block all share the same motion vector. The range of motion vectors is ± 15 Y pels with integer values. For color blocks, the motion vector is obtained by halving the transmitted vector and truncating the magnitude to an integer value.

Motion vectors are differentially coded using, in most cases, the motion vector of the MB to the left as a prediction. Zero is used as a prediction for the leftmost MBs of the GOB, and also if the MB to the left has no motion vector. Differential motion vectors are coded using the VLC of Fig. 8.7. Thus, using both MVD and MTYPE information, the predictor is able to choose the right pels for prediction.

The transform coefficients of either the original (Intra) or the differential (Inter) pels are ordered according to the zigzag scanning pattern in Fig. 8.8. These transform coefficients are selected and quantized at the encoder, and then coded using variable-length codewords (VLCs) and/or fixed-length codewords (FLC), depending on the values. Just as with JPEG, successive zeros between two nonzero coefficients are counted and called a *RUN*. The value of a transmitted nonzero quantized coefficient is called a *LEVEL*. The most likely occurring combinations of (RUN, LEVEL) are encoded with the VLC table shown in Fig. 8.9, with the sign bit terminating the RUN-LEVEL VLC codeword. The other combinations are coded with a 20-bit FLC consisting of a 6-bit ESCAPE code, 6 bits RUN, and 8 bits LEVEL. However, Intra DC coefficients use a different FLC. EOB is always appended to the last codeword, indicating the end of a block.

Run	Level	Code Word
0	1	1s (if first inter coefficient)
0	1	11s (otherwise)
0	2	0100 s
0	3	0010 1s
0	4	0000 110s
0	5	0010 0110s
0	6	0010 0001s
0	7	0000 0010 10s
0	8	0000 0001 1101s
0	9	0000 0001 1000s
0	10	0000 0001 0011s
0	11	0000 0001 0000s
0	12	0000 0000 1101s
0	13	0000 0000 1100s
0	14	0000 0000 1100s
0	15	0000 0000 1011s
1	1	011s
1	2	0001 10s
1	3	0010 0101s
1	4	0000 0011 00s
1	5	0000 0001 1011s
1	6	0000 0000 1011 0s
1	7	0000 0000 1010 1s
2	1	0101 s
2	2	0000 100s
2	3	0000 0010 11s
2	4	0000 0001 0100s
2	5	0000 0000 1010 0s
3	1	0011 1s
3	2	0010 0100s
3	3	0000 0001 1100s
3	4	0000 0000 1001 1s

Run	Level	Code Word
4	1	0011 0s
4	2	0000 0011 11s
4	3	0000 0001 0010s
5	1	0001 11s
5	2	0000 0010 01s
5	3	0000 0000 1001 0s
6	1	0001 01s
6	2	0000 0001 1110s
7	1	0001 00s
7	2	0000 0001 0101s
8	1	0000 111s
8	2	0000 0001 0001s
9	1	0000 101s
9	2	0000 0000 1000 1s
10	1	0010 0111 s
10	2	0000 0000 1000 0s
11	1	0010 0011 s
12	1	0010 0010 s
13	1	0010 0000 s
14	1	0000 0011 10s
15	1	0000 0011 01s
16	1	0000 0010 00s
17	1	0000 0000 1111s
18	1	0000 0001 1010s
19	1	0000 0001 1001s
20	1	0000 0001 0111s
21	1	0000 0001 0110s
22	1	0000 0000 1111 1s
23	1	0000 0000 1111 0s
24	1	0000 0000 1110 1s
25	1	0000 0000 1110 0s
26	1	0000 0000 1101 1s

EOB 10
Escape 0000 01

Fig. 8.9. Variable Length Codes (VLCs) for zero-run lengths (RUN) and quantized coefficient values (LEVEL). The sign bit s terminates the VLC word. Other combinations are coded with a 20-bit fixed length code (FLC), consisting of the ESCAPE code, followed by a 6-bit RUN and an 8-bit LEVEL. The End of Block (EOB) symbol is inserted after the last codeword of every block.

The inverse quantizer or the reconstruction process for all the coefficients other than the intra DC is defined by the following formula:

$$REC = QUANT \times (2 \times LEVEL + Sign), \text{ for } odd\ QUANT \tag{8.1}$$

$$REC = QUANT \times (2 \times LEVEL + Sign) - Sign, \text{ for } even\ QUANT$$

where *Sign* = +1, -1 or 0 depending on whether *LEVEL* is positive, negative or zero, respectively.

QUANT is the half step size determined by GQUANT or MQUANT, and REC is the reconstructed value of a quantized coefficient clipped to the range -2048 to 2047. Almost all the reconstruction levels are odd numbers to reduce problems of mismatch between encoders and decoders from different manufacturers. The intra-DC coefficient is uniformly quantized with a fixed step size of 8, and coded with 8 bits.

The standard requires a compatible IDCT (inverse DCT) to be close to the ideal 64-bit floating point IDCT. H.261 specifies a measuring process for checking a valid IDCT. The error in pel values between the ideal IDCT and the IDCT under test must be less than certain allowable limits given in the standard, e.g., peak error ≤ 1, mean error ≤ 0.0015, and mean square error ≤ 0.02.

A few other items are also required by the standard. One of them is the image-block updating rate. To prevent mismatched IDCT error as well as channel error propagation, every MB should be intra-coded at least once in every 132 transmitted picture frames.

The contents of the transmitted video bit stream must also meet the requirements of the *hypothetical reference decoder* (HRD). For CIF pictures, every coded frame is limited to fewer than 256 Kbits; for QCIF, the limit is 64 Kbits, where K = 1024. The HRD receiving buffer size is $B + 256$ Kbits, where $B = 4 \times R_{max} / 29.97$ and R_{max} is the maximum connection (channel) rate. At every picture interval (1/29.97 sec), the HRD buffer is examined. If at least one complete coded picture is in the buffer, then the earliest picture bits are removed from the buffer and decoded. The buffer occupancy, right after the above bits have been removed, must be less than B.

Note that this definition of HRD allows for varying the delay during the video coding. If the encoder assigns a relatively large number of bits to a particular frame, the decoder must repeat a few frames waiting for the arrival of those bits, thus increasing the delay. Conversely, if the encoder assigns relatively fewer bits to a frame, the decoder can decode early, thus reducing the delay.

8.4 Multipoint Features of P*64

Several signalling features have been provided in P*64 to facilitate video conferencing between several separated points. For example, the *Freeze Picture* signal requests the decoder to stop decoding and to continually repeat the most recently received frame in preparation for a disruption in the video bit stream. A

Freeze Release signal tells the decoder that decoding can resume with the reception of the current picture. A *Fast Update* signal tells the encoder to encode the entire next frame using only Intra MBs.

These signals are used typically by a centrally located Multipoint Control Unit (MCU) that interconnects the participants in the videoconference. For example, if it is desired that all participants see the person who is talking, then the MCU must detect whenever there is a new speaker. It then issues a Freeze Picture request to all decoders, routes the speaker's encoder output bit stream to all decoders, issues a Fast Update signal to the speaker's encoder which issues a Freeze Release signal to all decoders.

8.5 Encoder Constraints and Options

Fig. 8.10 shows a typical encoder structure. The input standard color components, e.g., YC_BC_R of Sec. 2.2.5, are spatially and temporally filtered so that they can be resampled in an internationally compatible common intermediate format. The spatio-temporal filter may also adaptively effect additional filtering, as indicated by the data-rate control, in order to allow for reduced resolution or a reduced frame rate for actual coding. Such control information must also be passed to the multiplexer and sent to the decoder to enable proper interpretation of the received bits.

The uncoded pels from the current frame and the coded pels from the frame memory both pass to the motion estimator, which then calculates one displacement vector per luminance motion estimation block, i.e., 16×16 Y pels. The motion estimator would probably use one of the block matching methods of Section 5.2.3 in its calculation and produce a motion vector with single pel displacement accuracy and a maximum displacement range of ±15 vertically and horizontally. Chrominance blocks use half the displacement of the corresponding luminance blocks, thus eliminating the need to compute and send chrominance displacements.

The motion estimation then passes to the predictor, which applies the indicated displacement to the contents of the frame memory and passes the result to the subtractor. The output of the subtractor is the motion compensated frame difference, also known as the Displaced Frame Difference (DFD), which is then partitioned into two-dimensional 8×8 blocks for transformation.

The data-rate control unit must compare the uncoded pels with the DFD pels in order to decide whether the MB is to be coded Intra or Inter. Whichever minimizes a simple sum of magnitudes is often a sufficiently good decision.

If the decision is Inter and the motion vector in nonzero, the data-rate control unit must decide whether to invoke the loop filter. A simple threshold on the sum of DFD magnitudes is often sufficient.

The transform coefficients are then adaptively quantized, variable word-length coded and passed to the multiplexer for transmission. All coefficients (except the Intra DC coefficient) are quantized with the same quantizer shown in

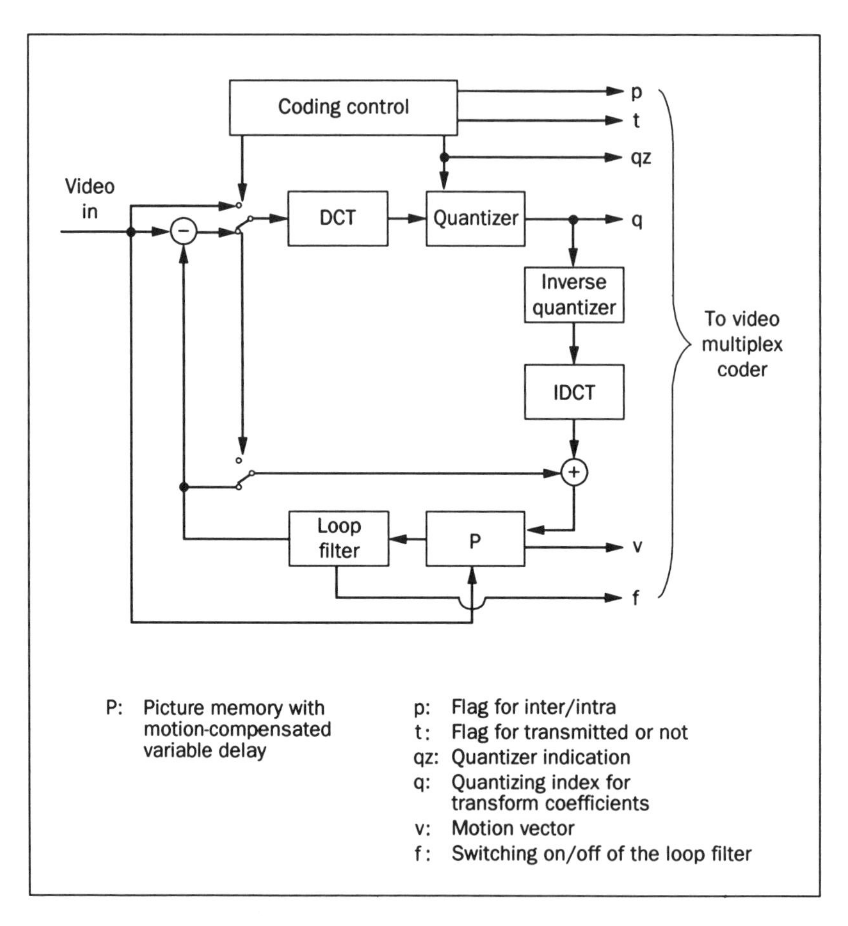

Fig. 8.10. Generic form of sub-primary rate coder. Input is assumed to be color components of the local standard. The spatio-temporal filter enables resampling to a common intermediate format. A subsampler (not shown) adaptively adjusts the frame rate, as instructed by the data-rate control. Motion compensated frame differences are transformed in 8×8 blocks, adaptively quantized under control of the buffer fullness, variable word-length coded, multiplexed with the side information and fed to a buffer (not shown) to await transmission. The coded image is reconstructed, fed to a frame memory and thence to a predictor and motion estimator to be used in coding the next frame. SI is side information that is sent to enable correct decoding at the receiver.

Fig. 8.11, whose step size $L = 2 \times QUANT$ is variable from MB to MB, as set by the data-rate control. The quantized coefficients are then inverse transformed and added to the prediction (zero for Intra) to form a reconstructed picture that is fed to the frame memory to be used in coding the next frame.

The output of the multiplexer is fed to a buffer that smooths the data rate prior to sending to the channel coder. The channel coder computes parity bits for a (511,493) BCH code, adds a framing bit and sends the result to the audio/video/signalling multiplexer and hence to the ISDN channel.

For the purpose of this discussion, the elements inside a standard compatible encoder can be classified, based on their functionalities, into two categories:

- The *basic coding operation* units, such as motion estimator, quantizer, transform, and variable-word-length encoder (VLE).
- The *coding parameter decision* units, such as the coding control in Fig. 8.10. These units select the parameter values of the basic operation units, including motion vectors, quantization step size, and picture frame rate.

Although H.261 does not explicitly specify a standard encoder, most basic operation elements are strongly constrained by the standard. However, other crucial elements, such as the parameter decision unit, are still open to the design engineers. We briefly outline some observations below.

The VLE is pretty much fixed because the VLC tables are given. The forward DCT is not specified, but it is expected that the IDCT inside the encoder matches fairly closely the decoder IDCT. Moreover, the encoder forward DCT should match its own IDCT.

The control examines the quantized coefficients to see if they are all or nearly all zero. If so, that block may not be coded. If all six blocks of an Inter (no MC) MB have insignificant coefficients, the entire MB may be skipped.

Also, the control can force the predictor to output zeros, thus causing the transform to take place on the pels themselves (Intra) rather than on the displaced frame differences (Inter). This intraframe coding mode might be used following a scene change or for forced updating to accommodate for transmission bit errors.

The control decides which spatial and temporal resolution reductions are appropriate, and signals the spatio-temporal filter accordingly. With lower spatial resolution, fewer transform coefficients need be sent per block. In fact, some implementations effect the spatial filtering by simply reducing the number of transmitted coefficients. However, with 8×8 blocks, this can cause deleterious effects due to the visibility of block boundaries as the bit-rate declines.

Because the inverse quantizer (IQ) is defined at the decoder, variations of the encoder quantizer are quite limited. From a theoretical viewpoint, however, it is not necessary for the decision levels of the encoder quantizer to be in the

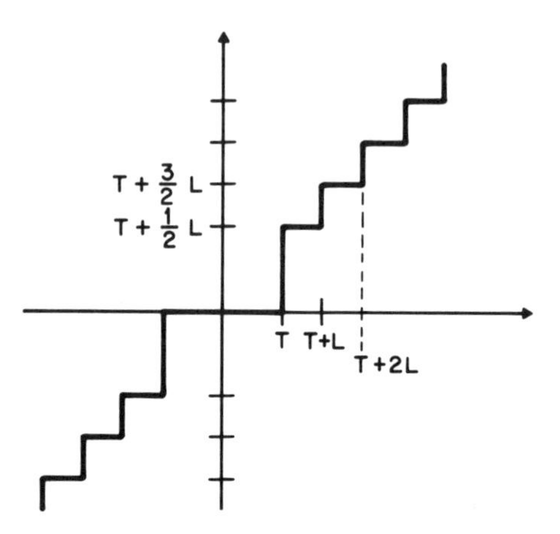

Fig. 8.11. Quantizer used for all transform coefficients except Intra DC. The characteristic is uniform except for the dead zone around zero. (See Eq. 8.1) This dead zone helps to decrease sensitivity to noise in the input. The step size $L = 2 \times QUANT$ varies from 2 to 62 in intervals of 2. To alleviate mismatch, representation levels are made odd by decrementing toward zero, if necessary.

middle of two reconstruction levels. Also, encoder designers determine the criterion (a fixed or an adaptive threshold, for example) for selecting transform coefficients to be coded.

If motion compensation is selected, the motion estimator must be able to produce one motion vector for the entire MB. However, the displacement-estimating algorithm is completely open to the designer. Block matching is a popular scheme, but many other block-motion estimation algorithms exist, such as gradient-based methods. Even within the category of block-matching algorithms, there are several variations, such as hierarchical-motion estimation.

Because of the HRD model required by the standard, the encoder output bits of every coded frame must be regulated carefully. For example, many successive low bit frames may violate the HRD requirement.

For video teleconferencing the size of the buffer must be limited to avoid excessive transmission delay. For a channel rate R and Buffer Size BS the one-way delay due to buffering is BS/R. Thus, for a one-way delay of 250 msec the buffer size should not exceed 28 kbits for Basic Rate ISDN (112 kbits/s video) and approximately 80 kbits for the H0 channel (320 kbits/s video). Although the decoder buffer size specified by H.261 is much larger than these numbers, it is the encoder that decides how much of the buffer to use and, therefore, how much delay is incurred by coding.

Although individual basic coding elements may affect the overall coding performance, the most critical and global influence on the encoder performance comes from the parameter decision units. The questions they have to answer are:

- How many frames should be transmitted, or conversely, how many should be skipped?
- What MTYPE should each Macroblock use?
- What is the proper quantization step size?
- How do we control the buffer fullness so that it does not produce long delay and does not violate the HRD requirements?

Researchers have studied many of the above questions, and various implementations are in the market. It is important to keep the hardware relatively simple for practical applications. Thus, different products will make different trade-offs of performance and cost. Since the H.261 standard does not specify the encoder, a wide latitude in capabilities is possible. Moreover, as time goes on and circuit costs decline, the performance of codecs will improve even though the standard changes little.

References

8.1 ITU-T (CCITT), *Recommendation H.261---Video Codec for Audiovisual Services at p*64 kbit/s*, Geneva, August 1990.

Questions for Understanding

8.1 What is the luminance pel sampling rate for CIF? QCIF?

8.2 What is the purpose of stuffing information?

8.3 If the first two MBs in a GOB have MVs (2,-2) and (-3,4), what code words are used to code the MVs?

8.4 What are the MBAs for the above example?

8.5 If the first five quantized DCT coefficients of an Inter block have levels 20,0,0,-3,22 what code words are used to code them?

8.6 If QUANT = 4, what are the reconstructed coefficient values for the above example?

8.7 If QUANT = 5, what are the reconstructed coefficient values for the above example?

8.8 If QUANT=5 and the first five unquantized DCT coefficients of an Inter block have values -27, 33, 138, -100 and 86, what are the quantized levels? What are the reconstructed values? What are the code words?

8.9 Suppose an encoder codes the first four frames using 4000, 3000, 2000 and 1000 bits, respectively. At a bit rate of 64 kbits/sec, plot the fullness of the HRD buffer.

8.10 In the above example, plot the fullness of the encoder buffer assuming the encoder produces a constant number of bits per pel.

8.11 In the above example, plot the fullness of the encoder buffer assuming the encoder produces all the pel bits at the top of the picture.

8.12 In the above example, plot the fullness of the encoder buffer assuming the encoder produces all the pel bits at the bottom of the picture.

8.13 How does the HRD buffer fullness change in the above three examples?

9

Coding for Entertainment Video - ISO MPEG

9.1 Moving Picture Experts Group (MPEG)

This group was chartered by the International Standards Organization (ISO) to standardize a coded representation of video and associated audio suitable for digital storage and transmission media. Digital storage media include magnetic computer disks, optical compact disk read-only-memory (CD-ROM), digital audio tape (DAT), etc. Transmission media include telecommunications networks, home coaxial cable TV (CATV), and over-the-air digital video. The group's goal has been to develop a *generic* coding standard that can be used in many digital video implementations. Thus, particular applications will typically require some further specification and refinement to be useful. As of this writing, MPEG has produced two standards, known colloquially as MPEG1 and MPEG2. Work also continues on methods and applications for future coding standards.

9.2 Video Coding at 1.5 Mbits/s (MPEG1)

MPEG1[9.1-9.3] is an international standard for the coded representation of digital video and associated audio at bit-rates up to about 1.5 Mbits/s. The colloquial name of the standard comes from the experts group that developed the standard, MPEG. The official name is ISO/IEC 11172.

If video is coded at about 1.1 Mbits/s and stereo audio is coded at 128 kbits/s per channel, then the total audio/video digital signal will fit onto the CD-ROM bit-rate of approximately 1.4 Mbits/s as well as the North American ISDN

This chapter contains excerpts taken with permission from AT&T Technical Journal[7.17].

Primary Rate (23 B-channels) of 1.47 Mbits/s. The specified bit-rate of 1.5 Mbits/s is not a hard upper limit. In fact, MPEG1 allows rates as high as 100 Mbits/s. However, during the course of MPEG1 algorithm development, coded image quality was optimized at a rate of 1.1 Mbits/s using progressive* scanned color images. Two *Source Input Formats* (SIF) were used for optimization. One corresponding to NTSC was 352 pels, 240 lines, 29.97 frames/sec. The other corresponding to PAL, was 352 pels, 288 lines, 25 frames/sec. SIF uses 2:1 color subsampling, both horizontally and vertically, in the same 4:2:0 format as H.261.

The MPEG1 standard currently has four parts (others will follow):

- Part 1 "Systems" concerns the synchronization and multiplexing of video, audio and ancillary data.
- Part 2 "Video" describes the coded video bitstream syntax and semantics.
- Part 3 "Audio" describes the coded audio bitstream syntax and semantics.
- Part 4 "Compliance Testing" describes tests for verifying if bitstreams and/or decoders meet the requirements specified in the first three parts.

An overview of the video part of the MPEG1 standard follows.

9.2.1 Requirements of the MPEG1 Video Standard

Uncompressed digital video requires an extremely high transmission bandwidth. Digitized NTSC resolution video, for example, has a bit-rate of approximately 100 Mbits/s. With digital video, compression is necessary to reduce the bit-rate to suit most applications. The required degree of compression is achieved by exploiting the spatial and temporal redundancy present in a video signal. However, the compression process is inherently lossy, and the signal reconstructed from the compressed bit stream is not identical to the input video signal. Compression typically introduces some artifacts into the decoded signal.

The primary requirement of the MPEG1 video standard is that it should achieve the highest possible quality of the decoded motion video at a given bit-rate. In addition to picture quality under normal play conditions, different applications have additional requirements. For instance, multimedia applications may require the ability to randomly access and decode any single video picture† in the bitstream. Also, the ability to perform fast search directly on the bit stream, both forward and backward, is extremely desirable if the storage medium has ''seek'' capabilities. It is also useful to be able to edit

---

* The term *NonInterlaced* is also used for progressively scanned pictures.

† Frames and pictures are synonymous in MPEG1.

compressed bit streams directly while maintaining decodability. And finally, a variety of video formats should be supported.

9.2.2 Compression Algorithm Overview

Both spatial and temporal redundancy reduction are needed for the high compression requirements of MPEG1. Most techniques used by MPEG1 have been described in previous chapters.

9.2.2a Exploiting spatial redundancy

The compression approach of MPEG1 video uses a combination of the ISO JPEG (still image) and CCITT H.261 (videoconferencing) standards. Because video is a sequence of still images, it is possible to achieve some compression using techniques similar to JPEG. Such methods of compression are called intraframe coding techniques, where each picture of video is individually and independently compressed or encoded. Intraframe coding exploits the spatial redundancy that exists between adjacent pels of a picture. Pictures coded using only intraframe coding are called *I-pictures*.

As in JPEG and H.261, the MPEG1 video-coding algorithm employs a Block-based two-dimensional DCT. A picture is first divided into 8×8 Blocks of pels, and the two-dimensional DCT is then applied independently on each Block. This operation results in an 8×8 Block of DCT coefficients in which most of the energy in the original (pel) Block is typically concentrated in a few low-frequency coefficients. The coefficients are scanned and transmitted in the same zigzag order as JPEG and H.261.

A quantizer is applied to the DCT coefficients, which sets many of them to zero. This quantization is responsible for the lossy nature of the compression algorithms in JPEG, H.261 and MPEG1 video. Compression is achieved by transmitting only the coefficients that survive the quantization operation and by entropy-coding their locations and amplitudes.

9.2.2b Exploiting temporal redundancy

Many of the interactive requirements can be satisfied by intraframe coding. However, as in H.261, the quality achieved by intraframe coding alone is not sufficient for typical video signals at bit-rates around 1.1 Mbits/s.

Temporal redundancy results from a high degree of correlation between adjacent pictures. The MPEG1 algorithm exploits this redundancy by computing an interframe difference signal called the *prediction error*. In computing the prediction error, the technique of motion compensation is employed to correct for motion. A Macroblock (MB) approach is adopted for motion compensation.

In unidirectional motion estimation, called *Forward Prediction*, a *Target* Macroblock in the picture to be encoded is matched with a set of blocks of the same size in a past picture called the *Reference* picture. As in H.261 the

Macroblock in the Reference picture that "best matches" the Target Macroblock is used as the *Prediction* Macroblock. The prediction error is then computed as the difference between the Target Macroblock and the Prediction Macroblock.* The position of this best-matching Prediction Macroblock is indicated by a motion vector that describes the displacement between it and the Target Macroblock. The motion vector information is also encoded and transmitted along with the prediction error. Pictures coded using Forward Prediction are called *P-pictures.*

The prediction error itself is transmitted using the DCT-based intraframe encoding technique summarized above. In MPEG1 video (as in H.261), the Macroblock size for motion compensation is chosen to be 16×16, representing a reasonable trade-off between the compression provided by motion compensation and the cost associated with transmitting the motion vectors.

9.2.2c Bidirectional temporal prediction

Bidirectional temporal prediction, also called Motion-Compensated Interpolation, is a key feature of MPEG1 video. Pictures coded with Bidirectional prediction use two Reference pictures, one in the past and one in the future. A Target Macroblock in bidirectionally coded pictures can be predicted by a Macroblock from the past Reference picture (*Forward Prediction*), or one from the future Reference picture (*Backward Prediction*), or by an average of two Macroblocks, one from each Reference picture (*Interpolation*). In every case, a Prediction Macroblock from a Reference picture is associated with a motion vector, so that up to two motion vectors per MB may be used with Bidirectional prediction. Motion-Compensated Interpolation for a Macroblock in a Bidirectionally predicted picture is illustrated in Fig. 9.1.

Pictures coded using Bidirectional Prediction are called *B-pictures.* Pictures that are Bidirectionally predicted are never themselves used as Reference pictures, i.e., Reference pictures for B-pictures must be either P-pictures or I-pictures. Similarly, Reference pictures for P-pictures must also be either P-pictures or I-pictures.

Bidirectional prediction provides a number of advantages. The primary one is that the compression obtained is typically higher than can be obtained from Forward (unidirectional) prediction alone. To obtain the same picture quality, Bidirectionally predicted pictures can be encoded with fewer bits than pictures using only Forward prediction.

---

* Prediction MBs do not, in general, align with coded MB boundaries in the Reference frame.

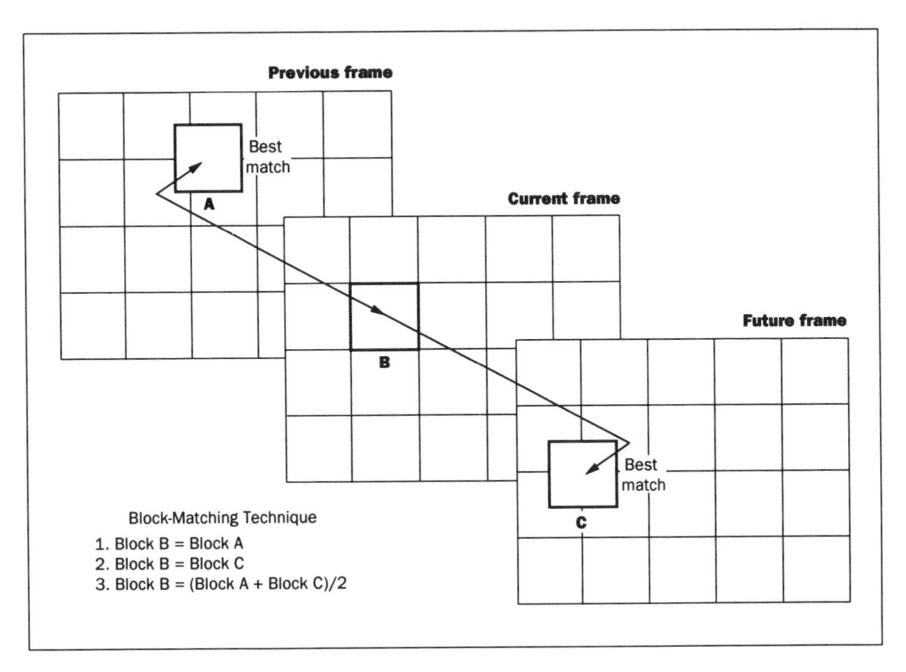

Fig. 9.1 Motion-Compensated Interpolation using Bi-directional prediction.

Table 9.1 Six Headers of MPEG1 Video Bit-Stream Syntax

Syntax Header	Functionality
Sequence	Context unit
Group of pictures	Random access unit
Picture	Primary coding unit
Slice	Resynchronization unit
Macroblock	Motion compensation unit
Block	DCT unit

However, Bidirectional prediction does introduce extra delay in the encoding process, because pictures must be encoded out of sequence. Further, it entails extra encoding complexity because Macroblock matching (the most computationally intensive encoding procedure) has to be performed twice for each Target Macroblock, once with the past Reference picture and once with the future Reference picture.

9.2.3 Features of the MPEG1 Bit-Stream Syntax

The MPEG1 video standard specifies the *syntax* and *semantics* of the compressed bit stream produced by the video encoder. The standard also specifies how this bit stream is to be parsed and decoded to produce a decompressed video signal.

Many encoder procedures are not specified, i.e., different algorithms can be employed at the encoder as long as the resulting bit stream is consistent with the specified syntax. For example, the details of the motion estimation matching procedure are not part of the standard. However, as with H.261 there is a strong limitation on the variation in bits/picture in the case of constant bit-rate operation. This is enforced through a *Video Buffer Verifier* (VBV), which corresponds to the Hypothetical Reference Decoder of H.261. Any MPEG1 bit stream is prohibited from overflowing or underflowing the buffer of this VBV. Thus, unlike H.261, there is no picture skipping allowed in MPEG1.

The bit-stream syntax is flexible in order to support the variety of applications envisaged for the MPEG1 video standard. To this end, the overall syntax is constructed in a hierarchy* of several Headers, each performing a different logical function. The different Headers in the syntax and their use are illustrated in Table 9.1.[9.3]

Video Sequence Header

The outermost Header is called the *Video Sequence* Header, which contains basic parameters such as the size of the video pictures, Pel Aspect Ratio (PAR), picture rate, bit-rate, assumed VBV buffer size and certain other global parameters. This Header also allows for the optional transmission of JPEG style Quantizer Matrices, one for Intra coded pictures and one for Non-Intra coded pictures. Unlike JPEG, if one or both quantizer matrices are not sent, default values are defined. These are shown in Fig. 9.2. Private *user data* can also be sent in the Sequence Header as long as it does not contain a *Start Code Header* , which MPEG1 defines as a string of 23 or more zeros.

---

* As in H.261, MPEG1 uses the term *Layers* for this hierarchy. However, Layer has another meaning in MPEG2. Thus, to avoid confusion we will not use Layers in this section.

8	16	19	22	26	27	29	34
16	16	22	24	47	49	34	37
19	22	26	27	29	34	34	38
22	22	26	27	29	34	37	40
22	26	27	29	32	35	40	48
26	27	29	32	35	40	48	58
26	27	29	34	38	46	56	69
27	29	35	38	46	56	69	83

Intra

16	16	16	16	16	16	16	16
16	16	16	16	16	16	16	16
16	16	16	16	16	16	16	16
16	16	16	16	16	16	16	16
16	16	16	16	16	16	16	16
16	16	16	16	16	16	16	16
16	16	16	16	16	16	16	16
16	16	16	16	16	16	16	16

Inter

Fig. 9.2 Default Quantizer Matrices for Intra and NonIntra Pictures.

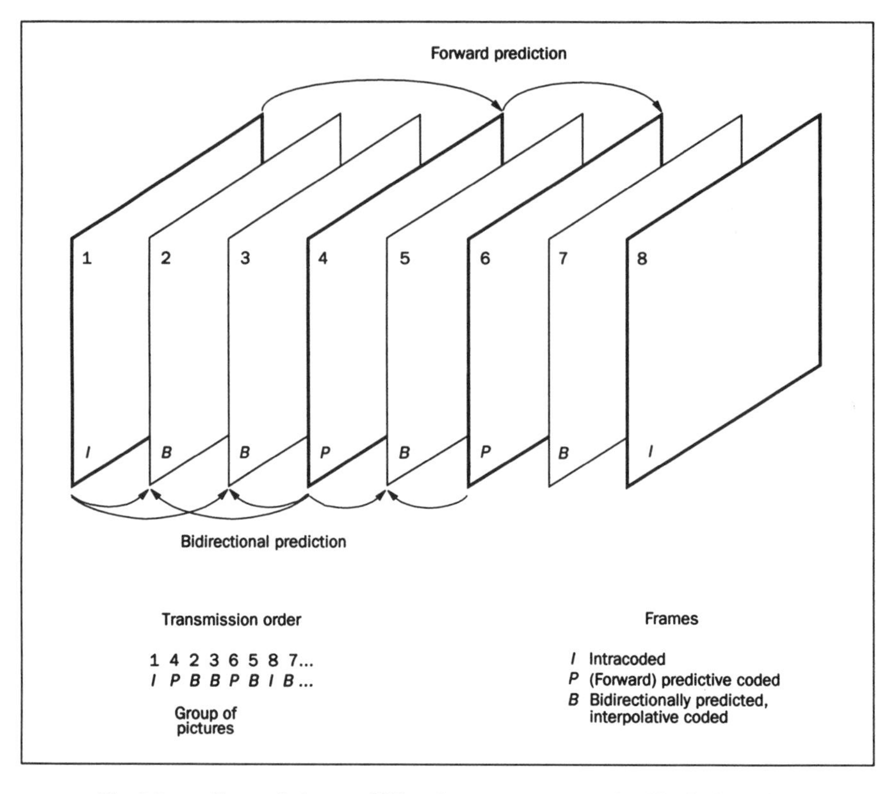

Fig. 9.3 Group of pictures. Video pictures are not transmitted in display order.

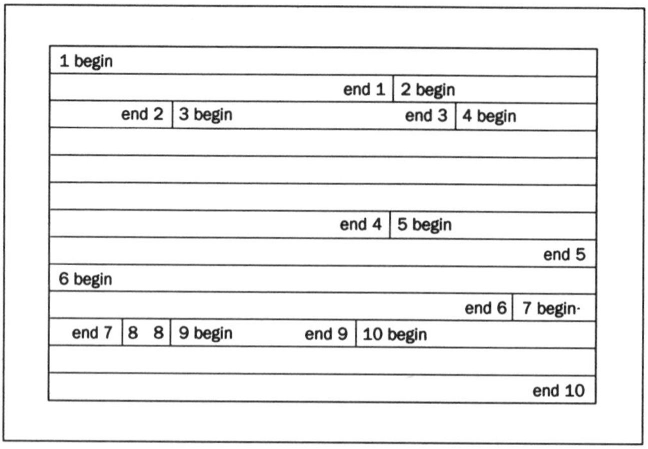

1 begin
end 1 | 2 begin
end 2 | 3 begin end 3 | 4 begin
end 4 | 5 begin
end 5
6 begin
end 6 | 7 begin
end 7 | 8 8 | 9 begin end 9 | 10 begin
end 10

Fig. 9.4 Possible arrangement of Slices in a 256 × 192 picture.

Group of Pictures (GOP) Header

Below the Video Sequence Header is the *Group of Pictures* (GOP) Header, which provides support for random access, fast search, and editing. A sequence of transmitted video pictures is divided into a series of GOPs, where each GOP contains an intra-coded picture (I-picture) followed by an arrangement of Forward predictive-coded pictures (P-pictures) and Bidirectionally predicted pictures (B-pictures). Fig. 9.3 shows a GOP example with six pictures, 1 to 6. This GOP contains I-picture 1, P-pictures 4 and 6, and B-pictures 2, 3 and 5. The encoding/transmission order of the pictures in this GOP is shown at the bottom of Fig. 9.3. B-pictures 2 and 3 are encoded after P-picture 4, using P-picture 4 and I-picture 1 as reference. Note that B-picture 7 in Fig. 9.3 is part of the next GOP because it is encoded after I-picture 8.

Random access and fast search are enabled by the availability of the I-pictures, which can be decoded independently and serve as starting points for further decoding. The MPEG1 video standard allows GOPs to be of arbitrary structure and length. The GOP Header is the basic unit for editing an MPEG1 video bit stream.

Picture Header

Below the GOP is the *Picture* Header, which contains the type of picture that is present, e.g., I, P or B, as well as a *Temporal Reference* indicating the position of the picture in display order within the GOP. It also contains a parameter called *vbv_delay* that indicates how long to wait after a random access before starting to decode. Without this information, a decoder buffer could underflow or overflow following a random access.

Slice Header

A *Slice* is a string of consecutive Macroblocks of arbitrary length running from left to right and top to bottom across the picture. The Slice Header is intended to be used for re-synchronization in the event of transmission bit errors. Prediction registers used in the differential encoding of motion vectors and DC Intra coefficients are reset at the start of a Slice. It is again the responsibility of the encoder to choose the length of each Slice depending on the expected bit error conditions. The first and last MBs of a Slice cannot be skipped MBs, and gaps are not allowed between Slices. The Slice Header contains the vertical position of the Slice within the picture, as well as a *quantizer_scale* parameter (corresponding to GQUANT in H.261). Fig. 9.4 shows an example in which Slice lengths vary throughout the picture.

Macroblock Header

The Macroblock (MB) is the 16×16 motion compensation unit. In the Macroblock Header, the horizontal position (in MBs) of the first MB of each Slice is coded with the MB Address VLC. The positions of additional transmitted MBs are coded differentially with respect to the most recently transmitted MB, also using the MB Address VLC. Skipped MBs are not allowed in I-pictures. In P-pictures, skipped MBs are assumed NonIntra with zero coefficients and zero motion vectors. In B-pictures, skipped MBs are assumed NonIntra with zero coefficients and motion vectors the same as the previous MB. Also included in the Macroblock Header are MB Type (Intra, NonIntra, etc.), *quantizer_scale* (corresponding to MQUANT in H.261), motion vectors and coded block pattern. As with other Headers, these parameters may or may not be present, depending on MB Type.

Block

A *Block* consists of the data for the quantized DCT coefficients of an 8×8 Block in the Macroblock. It is VLC coded as described in the next sections. For noncoded Blocks, the DCT coefficients are assumed to be zero.

9.2.4 Quantization of DCT Coefficients

As with JPEG, MPEG1 Intra coded blocks have their DC coefficients coded differentially with respect to the previous block of the same YCbCr type, unless the previous block is nonIntra or belongs to another Slice, in which case the DC coefficient is coded differentially with respect to the value 8*128=1024. The range of unquantized Intra DC coefficients is 0 to 8*255, which means the range of differential values is -2040 to 2040. Intra DC Differentials are quantized with a uniform mid-step quantizer with fixed step size 8. Thus, quantization is normally implemented simply by dividing by 8 and rounding to the nearest integer. Quantized Levels have a range of -255 to 255.

Intra AC coefficients are quantized with a nearly uniform mid-step quantizer having variable step size. The slight nonuniformity is because representation Levels are constrained to odd integer values by decrementing toward zero by one, if necessary. This constraint alleviates mismatch. For coefficient $c[m][n]$ the step size is given by

$$(2 * quantizer\_scale * IntraQmatrix[m][n])/16, \tag{9.1}$$

where *quantizer_scale* is set by a Slice or MB Header and $IntraQmatrix[m][n]$ is the Intra quantizer matrix value for DCT frequency $[m][n]$. Again, quantization is usually performed by division and rounding.

All NonIntra coefficients are quantized in the same way as in H.261 (See Eq. 8.1) using the stepsize

$$2 * QUANT = (2 * quantizer\_scale * NonIntraQmatrix[m][n])/16, \tag{9.2}$$

where $NonIntraQmatrix[m][n]$ is the NonIntra quantizer matrix value for DCT frequency $[m][n]$. In this case, the quantizer is nonuniform, with a *dead zone* of

Differential DC	SIZE	VLC Luminance	VLC Chrominance	VLI
-255 to -128	8	1111 110	1111 1110	00000000 to 01111111
-127 to -64	7	1111 10	1111 110	0000000 to 0111111
-63 to -32	6	1111 0	1111 10	000000 to 011111
-31 to -16	5	1110	1111 0	00000 to 01111
-15 to -8	4	110	1110	0000 to 0111
-7 to -4	3	101	110	000 to 011
-3 to -2	2	01	10	00 to 01
-1	1	00	01	0
0	0	100	00	
1	1	00	01	1
2 to 3	2	01	10	10 to 11
4 to 7	3	101	110	100 to 111
8 to 15	4	110	1110	1000 to 1111
16 to 31	5	1110	1111 0	10000 to 11111
32 to 63	6	1111 0	1111 10	100000 to 111111
64 to 127	7	1111 10	1111 110	1000000 to 1111111
128 to 255	8	1111 110	1111 1110	10000000 to 11111111

Fig. 9.5 Variable Length Codes (VLC) and Variable Length Integers (VLI) for quantized Intra DC differential values, which have a range of -255 to 255.

Table 9.2 Bounds for MPEG1 Constrained Parameter Bit-Streams (K=1024)

Parameter	Upper Bound
horizontal size	768 pels
vertical size	576 lines
number of MBs	396
MB rate	396*25 MB/s
frame rate	30 Hz
VBV Buffer	40 Kbytes
bit-rate	1.856 Mbits/s

half a step size around zero, as shown in Fig. 8.11. Thus, coefficient values should be shifted toward zero by at least *QUANT* prior to dividing by the step size 2**QUANT* and rounding.

Quantized Levels have a range of -255 to 255. In principle, this could be enforced by clipping after division by the step size. However, clipping may cause unacceptable picture distortion. A better solution is usually to increase the step size.

9.2.5 Variable Length Coding of Quantized DCT Coefficients

The Intra DC is coded similar to JPEG. Differential DC levels are coded as a VLC size followed by a VLI DIFF, where the VLC indicates the size in bits of the VLI. The VLCs and VLIs are given in Fig. 9.5 for quantized Intra DC DIFFs, which have a range of -255 to 255.

The remaining coefficients, including nonIntra DC, are coded in zigzag order using a combined runlength+level VLC, as in H.261. In fact, the first portion of the MPEG1 runlength-level VLC is exactly the same as in H.261. The remainder is shown in Fig. 9.6. Runlength-Level combinations that are not in the VLC are coded using a 6-bit Escape, 6-bit FLC for Runlength, an 8-bit FLC for Levels in the range -127 to 127 and a 16-bit FLC for remaining Levels. The 6-bit and 8-bit FLCs, as well as the Escape are the same as H.261. Quantized Levels have a range of -255 to 255. Every transmitted block ends with EOB.

9.2.6 Reconstruction of Quantized DCT Coefficients

Intra DC Differential values are obtained by simple multiplication of received Levels by 8. Intra AC coefficient reconstruction is complicated by the mismatch requirement of only odd integer values. The complete reconstruction of Intra AC coefficients is

$$\begin{aligned}&\tilde{c}[m][n]=(2*Level[m][n]*quantizer\_scale*IntraQmatrix[m][n])/16\\&\text{if}(\tilde{c}[m][n]\,\&1\ ==\ 0)\,\tilde{c}[m][n]-=Sign[m][n]\end{aligned}\tag{9.3}$$

where *Sign* = +1, -1 or 0 depending on whether *Level* is positive, negative or zero, respectively. All NonIntra coefficients are reconstructed by

$$\begin{aligned}&\tilde{c}[m][n]=((2*Level[m][n]\\&\quad+Sign[m][n])*quantizer\_scale*NonIntraQmatrix[m][n])/16\\&\text{if}(\tilde{c}[m][n]\,\&1\ ==\ 0)\ \tilde{c}[m][n]-=Sign[m][n]\end{aligned}\tag{9.4}$$

Values are then clipped to the range -2048 to 2047.

Run	Level	Code Word	Run	Level	Code Word
0	16	8*0 0111 11s	1	8	8*0 0011 111s
0	17	8*0 0111 10s	1	9	8*0 0011 110s
0	18	8*0 0111 01s	1	10	8*0 0011 101s
0	19	8*0 0111 00s	1	11	8*0 0011 100s
0	20	8*0 0110 11s	1	12	8*0 0011 011s
0	21	8*0 0110 10s	1	13	8*0 0011 010s
0	22	8*0 0110 01s	1	14	8*0 0011 001s
0	23	8*0 0110 00s	1	15	8*0 0001 0011s
0	24	8*0 0101 11s	1	16	8*0 0001 0010s
0	25	8*0 0101 10s	1	17	8*0 0001 0001s
0	26	8*0 0101 01s	1	18	8*0 0001 0000s
0	27	8*0 0101 00s	6	3	8*0 0001 0100s
0	28	8*0 0100 11s	11	2	8*0 0001 1010s
0	29	8*0 0100 10s	12	2	8*0 0001 1001s
0	30	8*0 0100 01s	13	2	8*0 0001 1000s
0	31	8*0 0100 00s	14	2	8*0 0001 0111s
0	32	8*0 0011 000s	15	2	8*0 0001 0110s
0	33	8*0 0010 111s	16	2	8*0 0001 0101s
0	34	8*0 0010 110s	27	1	8*0 0001 1111s
0	35	8*0 0010 101s	28	1	8*0 0001 1110s
0	36	8*0 0010 100s	29	1	8*0 0001 1101s
0	37	8*0 0010 011s	30	1	8*0 0001 1100s
0	38	8*0 0010 010s	31	1	8*0 0001 1011s
0	39	8*0 0010 001s			
0	40	8*0 0010 000s			

8*0 = 0000 0000

Fig. 9.6 Second half of MPEG1 VLC for RUNS and LEVELS. The first half is the same as for H.261, as shown in Fig. 8.9.

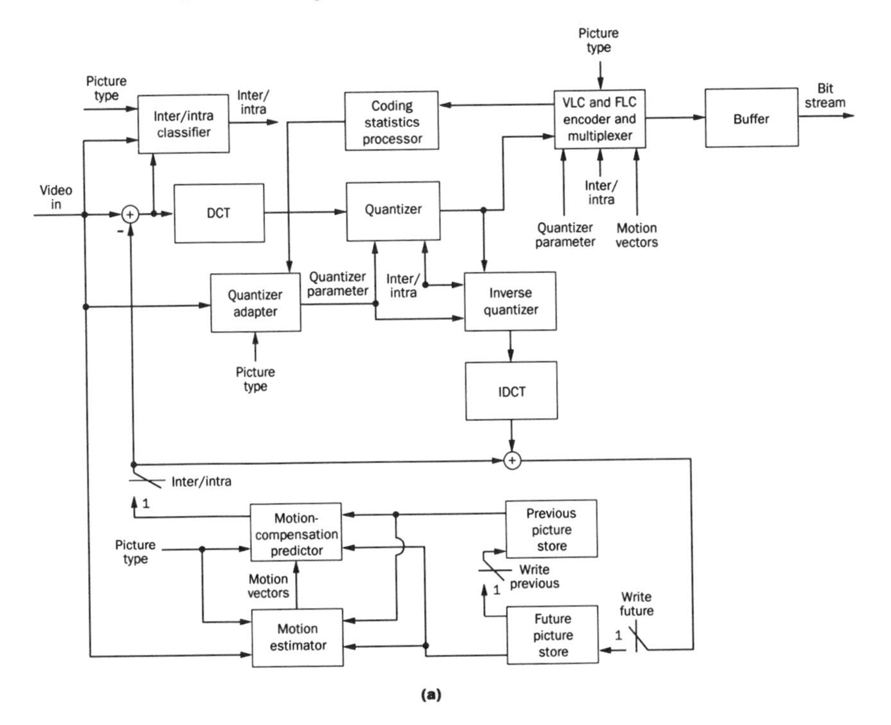

Fig. 9.7a A typical MPEG-1 encoder.

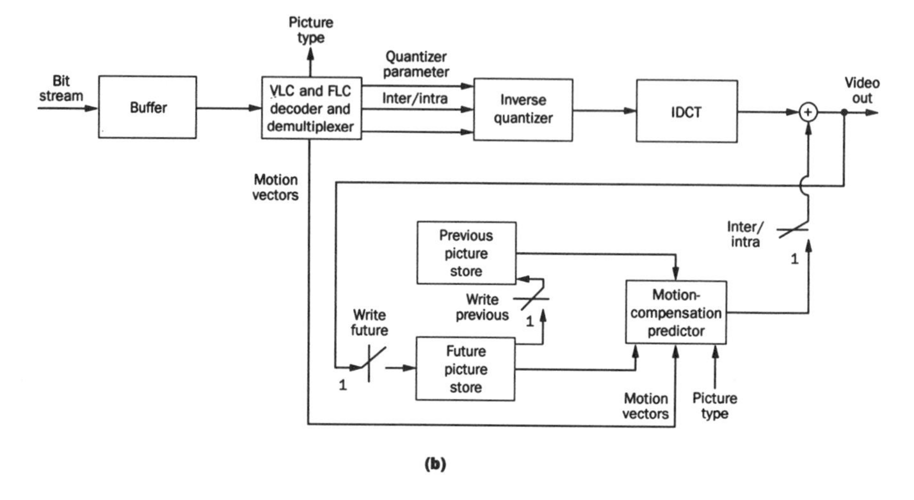

Fig. 9.7b A typical MPEG-1 decoder.

Table 9.3 Parameter Bounds for MPEG2 MP@ML Bit-Streams (K=1024)

Parameter	Upper Bound
horizontal size	720 pels
vertical size	576 lines
frame rate	30 Hz
pel rate	10.368 MHz
VBV Buffer	224 Kbytes
bit-rate	15.0 Mbits/s
color subsampling	MPEG2 4:2:0

9.2.7 Constrained Parameter Bit Streams

In order to allow manufacturers to build standard decoder chips for the most popular applications, MPEG1 has defined a set of parameter bounds for bitstreams, as well as a constrained parameters flag in the Sequence Header indicating that the bitstream conforms to those parameters. These parameter bounds are shown in Table 9.2.

9.2.8 Typical Encoder Architecture

Fig. 9.7a shows a typical MPEG1 video encoder. It is assumed that frame reordering takes place before coding, i.e., I- or P-pictures used for B-picture prediction must be coded and transmitted before any of the corresponding B-pictures, as shown in Fig. 9.3.

Input video is fed to a Motion Compensation Estimator/Predictor that feeds a prediction to the minus input of the Subtractor. For each MB, the Inter/Intra Classifier then compares the input pels with the prediction error output of the Subtractor. Typically, if the mean square prediction error exceeds the mean square pel value, an Intra MB is decided. More complicated comparisons involving DCT of both the pels and the prediction error yield somewhat better performance, but are not usually deemed worth the cost.

For Intra MBs the prediction is set to zero. Otherwise, it comes from the Predictor, as described above. The prediction error is then passed through the DCT and Quantizer before being coded, multiplexed and sent to the Buffer.

Quantized Levels are converted to reconstructed DCT coefficients by the Inverse Quantizer and then inverse transformed by the IDCT to produce a coded prediction error. The Adder adds the prediction to the prediction error and clips the result to the range 0 to 255 to produce coded pel values.

For B-pictures the Motion Compensation Estimator/Predictor uses both the Previous Picture and the Future Picture. These are kept in picture stores and remain unchanged during B-picture coding. Thus, in fact, the Inverse Quantizer, IDCT and Adder may be disabled during B-picture coding.

For I and P-pictures the coded pels output by the Adder are written to the Future Picture Store, while at the same time the old pels are copied from the Future Picture Store to the Previous Picture Store. In practice this is usually accomplished by a simple change of memory addresses.

The Coding Statistics Processor in conjunction with the Quantizer Adapter control the output bit-rate in order to conform to the Video Buffer Verifier (VBV) and to optimize the picture quality as much as possible. A simple control that works reasonably well is to define a target buffer fullness for each picture in the GOP. For each picture the *quantizer_scale* value is then adjusted periodically to try to make the actual buffer fullness meet the assigned value. More complicated controls would, in addition, exploit spatio-temporal masking in choosing the *quantizer_scale* parameter for each MB.

9.2.9 Typical Decoder Architecture

Fig. 9.7b shows a typical MPEG1 video decoder. It is basically identical to the pel reconstruction portion of the encoder. It is assumed that frame reordering takes place after decoding and video output. However, extra memory for reordering can often be avoided if during a write of the Previous Picture Store, the pels are routed also to the display.

The decoder cannot tell the size of the GOP from the bitstream parameters. Indeed it does not know until the Picture Header whether the picture is I, P or B Type. This could present problems in synchronizing audio and video were it not for the Systems part of MPEG1, which provides Time Stamps for audio, video and ancillary data. By presenting decoded information at the proper time as indicated by the Time Stamps, synchronization is assured.

Decoders often must accommodate occasional transmission bit errors. Typically, when errors are detected either through external means or internally through the arrival of illegal data, the decoder replaces the data with skipped MBs until the next Slice is detected. The visibility of errors is then rather low unless they occur in an I-picture, in which case they may be very visible throughout the GOB. A cure for this problem was developed for MPEG2 and will be described later.

9.2.10 Performance of MPEG1

In demonstrations of MPEG1 video at a bit-rate of 1.1 Mbits/s, SIF resolution pictures have been used. This resolution is roughly equivalent to one field of an interlaced NTSC or PAL picture. The quality achieved by the MPEG1 video encoder at this bit-rate has often been compared to that of VHS* videotape playback. Although the MPEG1 video standard was originally intended for operation in the neighborhood of the above bit-rate, a much wider range of resolution and bit-rates is supported by the syntax. The MPEG1 video standard thus provides a generic bit-stream syntax that can be used for a variety of applications. The MPEG1 Video Part 2 of ISO/IEC 11172 provides all the details of the syntax, complete with informative sections on encoder procedures that are outside the scope of the standard.

9.3 MPEG Second Work Item (MPEG2)

As of this writing, a second MPEG work item is nearly finished, which is colloquially called MPEG2. It is aimed at video bit-rates above 2 Mbits/s. As with MPEG1, the MPEG2 standard is written in four parts, Systems, Video,

---

* VHS, an abbreviation for Video Home System, is a registered trademark of the Victor Company of Japan.

Audio and Compliance Testing. More parts will follow. Originally, a third work item MPEG3 was planned for High Definition TV. However, the HDTV work was folded into MPEG2 at an early stage. The official name of MPEG2 is ISO/IEC 13818.

The original objective of MPEG2 was to code interlaced CCIR 601 video at a bit-rate that would serve a large number of consumer applications, and in fact one of the main differences between MPEG1 and MPEG2 is that MPEG2 handles interlace efficiently. Since the picture resolution of CCIR 601 is about four times that of the SIF of MPEG1, the bit-rate chosen for MPEG2 optimization was 4 Mbits/s. However, MPEG2 allows rates as high as 429 Gbits/s.

A bit-rate of 4 Mbits/s was deemed too low to enable high quality transmission of every CCIR 601 color sample. Thus, an MPEG2 4:2:0 format was defined to allow for 2:1 vertical subsampling of the color, in addition to the normal 2:1 horizontal color subsampling of CCIR 601. Pel positions are shown in Fig. 9.8, and as we can see, it is only slightly different than the CIF of H.261 and the SIF of MPEG1.

For interlace, the temporal integrity of the chrominance samples must be maintained. Thus, MPEG2 normally defines the first, third, etc. rows of 4:2:0 chrominance CbCr samples to be temporally the same as the first, third, etc. rows of luminance Y samples. The second, fourth, etc. rows of chrominance CbCr samples are temporally the same as the second, fourth, etc. rows of luminance Y samples. However, an override capability is available to indicate that the 4:2:0 chrominance samples are all temporally the same as the temporally first field of the frame.

At higher bit-rates the full 4:2:2 color format of CCIR 601 may be used, in which the first luminance and chrominance samples of each line are co-sited. MPEG2 also allows for a 4:4:4 color format.

9.3.1 Requirements of the MPEG2 Video Standard

The primary requirement of the MPEG2 video standard is that it should achieve the highest possible quality of the decoded video during normal play at a given bit-rate. In addition to picture quality, different applications have additional requirements even beyond those provided by MPEG1. For instance, multipoint network communications may require the ability to communicate simultaneously with SIF and CCIR 601 decoders. Communication over packet networks may require prioritization so that the network can drop low priority packets in case of congestion. Broadcasters may wish to send the same program to CCIR 601 decoders as well as to progressive scanned HDTV decoders. In order to satisfy all these requirements MPEG2 has defined a large number of capabilities.

However, not all applications will require all the features of MPEG2. Thus, to promote interoperability amongst applications, MPEG2 has designated several sets of constrained parameters using a two-dimensional rank ordering. One of the dimensions, called *Profile* specifies the coding features supported. The other dimension, called *Level*, specifies the picture resolutions, bit-rates, etc. that can be handled. A number of Profile-Level combinations have been defined, the most important of which is called *Main Profile at Main Level*, or MP@ML for short. Parameter constraints for MP@ML are shown in Table 9.3.

9.3.2 Main Profile at Main Level - Algorithm Overview

As with MPEG1, both spatial and temporal redundancy reduction are needed for the high compression requirements of MPEG2. For progressive scanned video there is very little difference between MPEG1 and MPEG2 compression capabilities. However, interlace presents complications in removing both types of redundancy, and many features have been added to deal specifically with it. MP@ML only allows 4:2:0 color sampling.

For interlace, MPEG2 specifies a choice of two picture structures. *Field Pictures* consist of fields that are coded independently. With *Frame Pictures*, on the other hand, field pairs are combined into frames before coding. MPEG2 requires interlace to be decoded as alternate top and bottom fields.* However, either field can be temporally first within a frame.

MPEG2 allows for progressive coded pictures, but interlaced display. In fact, coding of 24 frame/s film source with 3:2 pulldown display (see Sec. 2.2.4) for 525/60 video is also supported.

9.3.2a Features of the MPEG2 Bit-Stream Syntax

The MPEG2 video standard specifies the *syntax* and *semantics* of the compressed bit stream produced by the video encoder. The standard also specifies how this bit stream is to be parsed and decoded to produce a decompressed video signal. Most of MPEG2 consists of additions to MPEG1. The Header structure of MPEG1 was maintained with many additions within each Header. For constant bit-rates, the Video Buffer Verifier (VBV) was also kept. However, unlike MPEG1, picture skipping as in H.261 is allowed.

Video Sequence Header

The *Video Sequence* Header contains basic parameters such as the size of the coded video pictures, size of the displayed video pictures, Pel Aspect Ratio (PAR), picture rate, bit-rate, VBV buffer size, Intra and NonIntra Quantizer Matrices (defaults are the same as MPEG1), Profile/Level Identification, Interlace/Progressive Display Indication, Private user data, and certain other global parameters.

---

* The top field contains the top line of the frame. The bottom field contains the second (and bottom) line of the frame.

Group of Pictures (GOP) Header

Below the Video Sequence Header is the *Group of Pictures* (GOP) Header, which provides support for random access, fast search, and editing. The GOP Header contains a time code (hours, minutes, seconds, frames) used by certain recording devices. It also contains editing flags to indicate whether the B-pictures following the first I-picture can be decoded following a random access.

Picture Header

Below the GOP is the *Picture* Header, which contains the type of picture that is present, e.g., I, P or B, as well as a *Temporal Reference* indicating the position of the picture in display order within the GOP. It also contains the parameter *vbv_delay* that indicates how long to wait after a random access before starting to decode. Without this information, a decoder buffer could underflow or overflow following a random access.

Within the Picture Header several picture coding extensions are allowed. For example, the Intra DC coefficient precision may be increased from the 8 bits of MPEG1 to as much as 10 bits. A 3:2 pulldown flag indicates, for frame pictures, that the first field of the picture should be displayed one more time following the display of the second field. An alternative scan to the DCT zigzag scan may be specified. And the presence of error concealment information may be indicated. Other information includes Picture Structure (field or frame), field temporal order (for frame pictures), progressive frame indicator, and information for reconstruction of a composite NTSC or PAL analog waveform.

Within the Picture Header a picture display extension allows for the position of a *display rectangle* to be moved on a picture by picture basis. This feature would be useful, for example, when coded pictures having IAR 16:9 are to be also received by conventional TV having IAR 4:3. This capability is also known as *Pan and Scan.*

MPEG2 Slice Header

The *Slice* Header is intended to be used for re-synchronization in the event of transmission bit errors. It is the responsibility of the encoder to choose the length of each Slice depending on the expected bit error conditions. Prediction registers used in differential encoding are reset at the start of a Slice. Each horizontal row of MBs must start with a new Slice, and the first and last MBs of a Slice cannot be skipped. In MP@ML, gaps are not allowed between Slices. The Slice Header contains the vertical position of the Slice within the picture, as well as a *quantizer_scale* parameter (corresponding to GQUANT in H.261). The Slice Header may also contain an indicator for Slices that contain only Intra MBs. These may be used in certain Fast Forward and Fast Reverse display applications.

Macroblock Header

In the *Macroblock* Header, the horizontal position (in MBs) of the first MB of each Slice is coded with the MB Address VLC. The positions of additional transmitted MBs are coded differentially with respect to the most recently transmitted MB also using the MB Address VLC. Skipped MBs are not allowed in I-pictures. In P-pictures, skipped MBs are assumed NonIntra with zero coefficients and zero motion vectors. In B-pictures, skipped MBs are assumed NonIntra with zero coefficients and motion vectors the same as the previous MB. Also included in the Macroblock Header are MB Type (Intra, NonIntra, etc.), Motion Vector Type, DCT Type, *quantizer_scale* (corresponding to MQUANT in H.261), motion vectors and coded block pattern. As with other Headers, many parameters may or may not be present, depending on MB Type.

MPEG2 has many more MB Types than MPEG1, due to the additional features provided as well as to the complexities of coding interlaced video. Some of these will be discussed below.

Block

A *Block* consists of the data for the quantized DCT coefficients of an 8×8 Block in the Macroblock. It is VLC coded as described in the next sections. MP@ML has six blocks per MB. For noncoded Blocks, the DCT coefficients are assumed to be zero.

9.3.2b Exploiting spatial redundancy

As in MPEG1, the MPEG2 video-coding algorithm employs an 8×8 Block-based two-dimensional DCT. A quantizer is applied to each DCT coefficient, which sets many of them to zero. Compression is then achieved by transmitting only the nonzero coefficients and by entropy-coding their locations and amplitudes.

The main effect of interlace in frame pictures is that since alternate scan lines come from different fields, vertical correlation is reduced when there is motion in the scene. MPEG2 provides two features for dealing with this.

First, with reduced vertical correlation, the zigzag scanning order for DCT coefficients shown in Fig. 8.8 is no longer optimum. Thus, MPEG2 has an *Alternate Scan,* shown in Fig. 9.9, that may be specified by the encoder on a picture by picture basis. A threshold on the sum of absolute vertical line differences is often sufficient for deciding when to use the alternate scan. Alternatively, DCT coefficient amplitudes after transformation and quantization could be examined.

Second, a capability for field coding within a frame picture MB is provided. That is, just prior to performing the DCT, the encoder may reorder the luminance lines within a MB so that the first 8 lines come from the top field, and the last 8 lines come from the bottom field. This reordering is undone after the IDCT in the encoder and the decoder. Chrominance blocks are not reordered in

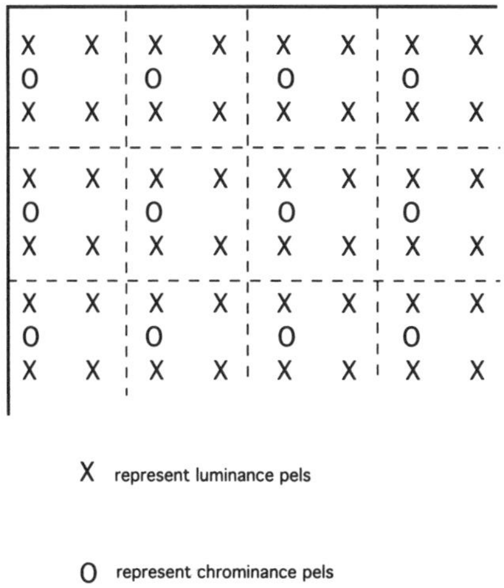

```
X   X | X   X | X   X | X   X
O     | O     | O     | O
X   X | X   X | X   X | X   X
------+-------+-------+------
X   X | X   X | X   X | X   X
O     | O     | O     | O
X   X | X   X | X   X | X   X
------+-------+-------+------
X   X | X   X | X   X | X   X
O     | O     | O     | O
X   X | X   X | X   X | X   X
```

X represent luminance pels

O represent chrominance pels

Fig. 9.8 The position of luminance and chrominance samples in MPEG2 4:2:0 format. For interlace, alternate rows of chrominance come from alternate fields.

m \ n	0	1	2	3	4	5	6	7
0	00	04	06	20	22	36	38	52
1	01	05	07	21	23	37	39	53
2	02	08	19	24	34	40	50	54
3	03	09	18	25	35	41	51	55
4	10	17	26	30	42	46	56	60
5	11	16	27	31	43	47	57	61
6	12	15	28	32	44	48	58	62
7	13	14	29	33	45	49	59	63

Fig. 9.9 MPEG2 Alternate Scan. For interlaced frame structure pictures containing motion, the alternate scan may be used instead of the zigzag scan of Fig. 8.8.

MP@ML. The effect of this reordering is to increase the vertical correlation within the luminance blocks and thus decrease the DCT coefficient energy. Again, a threshold on the sum of absolute vertical line differences is often sufficient for deciding when to use field DCT coding in a MB.

9.3.2c Exploiting temporal redundancy

As in MPEG1, the quality achieved by intraframe coding alone is not sufficient for typical MPEG2 video signals at bit-rates around 4 Mbits/s. Thus, MPEG2 uses all the temporal redundancy reduction techniques of MPEG1, plus other methods to deal with interlace.

Frame Prediction for Frame Pictures

MPEG1 exploits temporal redundancy by means of motion compensated MBs and the use of P-pictures and B-pictures. MPEG2 refers to these methods as *Frame Prediction for Frame Pictures*, and, as in MPEG1, assigns up to one motion vector to Forward Predicted Target MBs and up to two motion vectors to Bidirectionally Predicted Target MBs. Prediction MBs are taken from previously coded Reference frames, which must be either P-pictures or I-pictures. Frame Prediction works well for slow to moderate motion, as well as panning over a detailed background. Frame Prediction cannot be used in Field Pictures.

Field Prediction for Field Pictures

A second mode of prediction provided by MPEG2 for interlaced video is called *Field Prediction for Field Pictures*. This mode is conceptually similar to Frame Prediction, except that Target MB pels all come from the same field. Prediction MB pels also come from one field, which may or may not be of the same polarity (top or bottom) as the Target MB field. For P-pictures, the Prediction MBs may come from either of the two most recently coded I- or P-fields. For B-pictures, the Prediction MBs are taken from the two most recently coded I- or P-frames. Up to one motion vector is assigned to Forward Predicted Target MBs and up to two motion vectors to Bidirectionally Predicted Target MBs. This mode is not used for Frame Pictures and is useful mainly for its simplicity.

Field Prediction for Frame Pictures

A third mode of prediction for interlaced video is called *Field Prediction for Frame Pictures*. With this mode, the Target MB is first split into top-field pels and bottom-field pels. Field prediction, as defined above, is then carried out independently on each of the two parts of the Target MB. Thus, up to two motion vectors are assigned to Forward Predicted Target MBs and up to four motion vectors to Bidirectionally Predicted Target MBs. This mode is not used for Field Pictures and is useful mainly for rapid motion.

Dual Prime for P-pictures

A fourth mode of prediction, called *Dual Prime for P-pictures*, transmits one motion vector per MB. From this motion vector two preliminary predictions are computed, which are then averaged together to form the final prediction. The previous picture cannot be a B-picture for this mode. The first preliminary prediction is identical to Field Prediction, except that the Prediction pels all come from the previously coded fields having the same polarity (top or bottom) as the Target pels. Prediction pels, which are obtained using the transmitted MB motion vector, come from one field for Field Pictures and from two fields for Frame Pictures.

The second preliminary prediction is derived using computed motion vectors plus a small motion vector correction. For this prediction, Prediction pels come from the opposite polarity field as the first preliminary prediction. The computed motion vectors are obtained by a temporal scaling of the transmitted motion vector to match the field in which the Prediction pels lie. The small motion vector correction of up to one-half pel is transmitted in the MB Header.

The performance of Dual Prime prediction can rival that of B-picture prediction under some circumstances and has the advantage of lower encoding delay. However, memory bandwidth requirements are usually higher for Dual Prime.

16×8 MC for Field Pictures

A fifth mode of prediction is called 16×8 *MC for Field Pictures*. Basically, this mode splits the Field Picture MB into an upper half and lower half, and performs a separate Field Prediction for each half. Up to two motion vectors are transmitted per P-picture MB, and up to four motion vectors per B-picture MB.

It is the responsibility of the Encoder to decide which prediction mode is best. Needless to say, this can be an extremely complex process if done with full optimality. Thus, in practice numerous shortcuts are usually employed for economical implementation.

9.3.2d Quantization of DCT Coefficients

As with MPEG1, Intra coded blocks have their DC coefficients coded differentially with respect to the previous block of the same YCbCr type, unless the previous block is nonIntra or belongs to another Slice. The range of unquantized Intra DC coefficients is 0 to 8*255, which means the range of differential values is -2040 to 2040. Intra DC Differentials are quantized with a uniform mid-step quantizer with step size 8, 4 or 2, as specified in the Picture Header. Quantized Intra Differential DC Levels for MP@ML have a range of -1020 to 1020.

Intra AC coefficients are quantized with a uniform mid-step quantizer having variable step size. Unlike MPEG1, MPEG2 does not require odd representative levels. For coefficient $c[m][n]$ the IntraAC step size is given by

$$(2*quantizer\_scale*IntraQmatrix[m][n])/32, \quad (9.5)$$

where $quantizer\_scale$ is set by a Slice or MB Header and $IntraQmatrix[m][n]$ is the Intra quantizer matrix value for DCT frequency $[m][n]$. Again, quantization is usually performed by division and rounding.

All NonIntra coefficients are quantized in a way similar to H.261 and MPEG1, except that MPEG2 does not require odd representative levels. The NonIntra step size is given by

$$2*QUANT=(2*quantizer\_scale*NonIntraQmatrix[m][n])/32, \quad (9.6)$$

where $NonIntraQmatrix[m][n]$ is the NonIntra quantizer matrix value for DCT frequency $[m][n]$. The quantizer is nonuniform with a *dead zone* around zero, as shown in Fig. 8.11. Thus, coefficient values should be shifted toward zero by at least $QUANT$ prior to dividing by $2*QUANT$ when computing quantized Levels.

Quantized Levels have a range of -2047 to 2047. In principle, this could be enforced by clipping after division by the step size. However, clipping may cause unacceptable picture distortion. A better solution is usually to increase the step size.

9.3.2e Variable Length Coding of Quantized DCT Coefficients

The Intra DC VLC is similar to MPEG1, i.e., Levels are coded as a VLC size followed by a VLI Level, where the VLC indicates the size in bits of the VLI. Quantized Intra DC Levels have a range of -1020 to 1020.

The remaining coefficients are coded either in zigzag order or alternate scan order using one of two available VLCs. Runlength-Level combinations that are not in the VLC are coded using a a 6-bit Escape, 6-bit FLC for Runlength, and a 12-bit FLC for Levels. Quantized Levels have a range of -2047 to 2047.

9.3.2f Reconstruction of Quantized DCT Coefficients

Intra DC Differential values are obtained by simple multiplication of received Levels by the specified step size. Intra AC coefficient reconstruction is simply

$$\tilde{c}[m][n]=(2*Level[m][n]*quantizer\_scale*IntraQmatrix[m][n])/32 \quad (9.7)$$

All NonIntra coefficients are reconstructed by

$$\tilde{c}[m][n]=((2*Level[m][n] +Sign[m][n])*quantizer\_scale*NonIntraQmatrix[m][n])/32 \quad (9.8)$$

All reconstructed coefficients are then clipped to the range -2048 to 2047, followed by mismatch control. Mismatch control is very different than MPEG1 or H.261. MPEG2 basically adds all the reconstructed coefficients of each block, and if the sum is even, the least significant bit of the highest frequency coefficient is changed.

9.3.2g MPEG2 Error Concealment

MPEG2 has added an important feature for concealing the effects of transmission bit-errors in MP@ML. Intra pictures may optionally contain coded motion vectors, which are used only for error concealment. Typically, a Slice that is known to be erroneous is replaced by motion compensated pels from the previous I or P-picture. Motion vectors from the slice above could be used for this process.

9.3.2h Typical Encoder/Decoder Architecture

An MPEG2 codec architecture is nearly identical in form to the MPEG1 codec of Fig. 9.7. The main difference is in the complexity of the motion estimation and the Coding Statistics Processor, which for full optimality must decide amongst many many alternatives. Most of these decisions can utilize the mean absolute prediction error of the various modes.

9.3.2i Performance of MPEG2 MP@ML

In demonstrations of MPEG2 video at a bit-rate of 4.0 Mbits/s, interlaced CCIR 601 pictures have been used, with 2:1 color subsampling both horizontally and vertically. The quality achieved by the MPEG2 video encoder is very good to excellent, depending on scene content. Error resilience has been found to be very good at bit error rates up to 10^{-4} and usable up to 10^{-3}.

9.3.3 MPEG2 Scalability Extensions

Several additions to the basic Main Profile have been defined for those applications that require increased functionality. The *Scalability* extensions basically provide for two or more separate bitstreams, or *Layers*, that can be combined to provide a single high quality video signal.

The Lower Layer or *Base Layer* bitstream can, by definition, be decoded all by itself to provide a lower quality video signal. With the incorporation of *Enhancement Layer* bitstreams, the Base Layer image quality can be increased significantly, depending on the relative bit-rates assigned to the various Layers.

Scalability is especially useful in packet transmission systems, such as Asynchronous Transfer Mode (ATM) networks, that allow for dual priority packetization. The Base Layer would be sent at high priority and the Enhancement Layer at low priority. Then in case of low priority packet loss, a reasonably good video signal could still be recovered.

9.3.3a SNR Scalability

With SNR Scalability, the Enhancement Layer provides a direct improvement in DCT coefficient accuracy by adding correction values just prior to the Base Layer decoder IDCT. Thus, the extra processing needed by the Enhancement Layer consists of variable length decoding, inverse scan, and finally, reconstruction of quantized differential coefficient values. This extension also provides for simulcast 4:2:2 color information to be coded in the Enhancement Layer, with 4:2:0 color in the Base Layer.

SNR Scalability might be used in video distribution networks, where the full enhanced video would be required for editing and overlays prior to broadcast, whereas the Base Layer by itself would be sufficient for broadcast of nonedited video.

9.3.3b Data Partitioning

Data Partitioning is basically a simplified version of SNR Scalability. With this extension, up to a certain number (specified in the Slice Header) of low frequency run-level codewords in a block are sent in the Base Layer. The remaining high frequency codewords, if any, are sent in the Enhancement Layer.

The main attraction of Data Partitioning is its simplicity and reduced overhead compared with other scalability extensions. The only extra processing needed is a codeword counter to control switching from one layer to the other. Because of its reduced overhead, the enhanced image quality of Data Partitioning is typically very close to that of single layer coding.

9.3.3c Spatial Scalability

With Spatial Scalability, the Base Layer video has a lower spatial resolution than the original video. The Enhancement Layer is then combined with the Base Layer to produce a decoded image with the original resolution. This feature can be used for interworking between video standards, e.g., SIF Base Layer and CCIR 601 original, or CCIR 601 Base Layer and HDTV original.

Simple Pyramid Coding can be used to implement Spatial Scalability, and MPEG2 allows for this. First, the resolution of the decoded Base Layer is increased to that of the Enhancement Layer by interpolation and filtering. Then the difference between the original video and the upsampled Base Layer is coded in the Enhancement Layer.

However, MPEG2 also provides for a much more powerful spatio-temporal prediction process than simple pyramid coding. In coding the Enhancement Layer MBs, the predictor has three choices. It can use the pyramid coding spatial prediction described above, or it can use the normal MPEG2 motion compensated temporal prediction, or it can use a weighted average of the two. The result is a considerable improvement in coding efficiency.

9.3.3d Temporal Scalability

With Temporal Scalability, the Base Layer video has a lower temporal resolution, i.e., frame rate, than the original video. The Enhancement Layer is then combined with the Base Layer to produce a decoded image with the original frame rate. For example this feature can be used for interworking between an interlaced CCIR 601 Base Layer and progressive scanned original video at 50 or 60 Hz frame rate.

Temporal Scalability also uses a powerful temporal prediction process for encoding the Enhancement Layer. The motion compensated prediction can be taken from adjacent Base Layer pictures as well as adjacent Enhancement Layer pictures.

9.3.4 Other Profiles and Levels

In addition to the Main Profile, several other MPEG2 Profiles have been defined as of this writing. The *Simple* Profile is basically the Main Profile without B-pictures. The *SNR Scalable* Profile adds SNR Scalability to the main Profile. The *Spatially Scalable* Profile adds Spatial Scalability to the SNR Scalable Profile, and the *High* Profile adds 4:2:2 color to the Spatially Scalable Profile. All scalable Profiles are limited to at most three layers.

As stated previously, the Main Level is defined basically for CCIR 601 video. In addition, the *Simple* Level is defined for SIF video. Two higher levels for HDTV are the *High-1440* Level, with a maximum of 1440 pels/line, and the *High* Level, with a maximum of 1920 pels/line.

Not all Profile/Level combinations have been standardized. In the future, additional Profiles and Levels will be defined for various applications, as demand dictates.

References

9.1 ISO/IEC 11172, Information Technology-Coding of moving pictures and associated audio for digital storage media up to about 1.5 Mbit/s

9.2 D. J. LeGall, "MPEG: A Video Compression Standard for Multimedia Applications," *Communications of the ACM*, Vol. 34, No.4, April 1991, pp. 47-58.

9.3 "Digital Video," *IEEE Spectrum*, Vol. 29, No.3, March 1992.

9.4 ISO/IEC 13818, Information Technology - Generic coding of moving pictures and associated audio information.

Questions for Understanding

9.1 Describe the two SIF video formats? How do they differ from CIF?

9.2 How does JPEG coding differ from that of MPEG1 I-pictures?

9.3 How does P*64 coding differ from that of MPEG1 P-pictures?

9.4 What are advantages and disadvantages of MPEG1 B-Pictures?

9.5 How does MPEG support random access in stored bit streams.

9.6 How are MPEG1 Slices different than P*64 GOBs?

9.7 If the first five quantized DCT coefficients of an MPEG1 Inter block have levels 20, 0, 18, -3 and 22, what code words are used to code them? What are the code words if the block is the first Intra block of the Slice? Assume no quantizer matrices were transmitted.

9.8 If QUANT = 4, what are the reconstructed coefficient values for the above example?

9.9 If QUANT = 5, what are the reconstructed coefficient values for the above example?

9.10 Repeat the above three examples for MPEG2.

9.11 What are the MPEG2 picture structures?

9.12 How is the MPEG2 3:2 pulldown flag used?

9.13 Describe the five MPEG2 temporal prediction modes.

9.14 Describe the four MPEG2 Scalability extensions.

10

High Definition Television

10.1 Introduction

Television was initially intended for sports, news and entertainment applications and was almost entirely distributed by terrestrial broadcasts. In North America, since the adoption of the NTSC standard in 1941, only two significant modifications have been made to the television signals, namely the addition of color in 1953 and stereo sound in 1984. Both of these modifications were made in an backward compatible manner, i.e., by modifying the transmitted signal in such a way that existing receivers could still decode the new signal, while newer receivers have enhanced functionality due to the signal modifications. Television is now finding applications in telecommunications (e.g. videoconferencing), computing (e.g. desktop video), medicine (e.g. telemedicine) and education (e.g. distance learning). At the same time, television is being distributed by other means, such as cable (fiber or coaxial), video tapes or disks (analog or digital) and satellites (e.g. direct broadcast). Moreover, the television signal packaged for terrestrial broadcasting is very wasteful of the broadcast spectrum and can benefit from modern signal processing. In addition, the average size of television sets has grown steadily, and the average viewer is getting closer to video displays, particularly in applications other than transmitted entertainment television. These newer applications, delivery media and viewing habits are exposing various limitations of the television standard established over five decades ago. High definition television systems now being designed are expected to be more suitable for these newer applications and to interoperate easily with the evolving telecommunications and computing infra-structure. HDTV will be an entirely new high-resolution and wide-screen television system producing much better quality pictures and sound, thereby creating a higher sensation of reality that far exceeds that of the current television systems. In this chapter, we describe the main characteristics of HDTV, the current state in Europe, Japan, and the United States, and the expected future evolution.

10.2 What is HDTV?

There have been a number of proposals for HDTV, and there is only partial agreement on the definition of HDTV at this time. However, there are a number of commonly agreed aspects that will be described in this section. HDTV is characterized by increased spatial and temporal resolution; increased image aspect ratio (i.e., wider image); multichannel CD-quality surround sound; reduced artifacts compared with current composite analog television; bandwidth compression and channel encoding to make better utilization of the terrestrial spectrum; and finally, better interoperability with the evolving telecommunications and computing infrastructure. We will look at each of these aspects in this chapter.

First and foremost, HDTV, as produced in a studio and received in consumer homes, must create a much better picture. This is done by increasing the spatial resolution by approximately a factor of two in both the horizontal and vertical dimensions. Thus, we get a picture that has about 1000 scan lines and more than 1000 pels per scan line. In addition, it is widely accepted (but not fully agreed to) that progressive (noninterlace) scanning at about 60 frames/sec will render better fast-action sports, while also eliminating the artifacts associated with interlace.* The result of this is that the number of active pels in an HDTV signal increases by about a factor of five, with a corresponding increase in the analog bandwidth. From consumers' experience in cinema, a wider image aspect ratio is also preferred. Most HDTV systems specify the image aspect ratio to be 16:9.

Improvement in audio is also crucial to HDTV. This means multi (4 to 6) channel surround sound, in which each channel is independently transmitted. In digital HDTV, each audio channel is independently compressed, and these compressed bits are multiplexed with compressed bits from video, plus bits for closed captioning, teletext, encryption, addressing, program identification and other data services in a layered fashion. Audio fidelity is nearly as good as that obtainable from the current generation of CDs.

---

* Proponents of interlace argue that interlace represents a better trade-off between spatial and temporal resolution (i.e., for a manageably low overall pel-rate, interlace allows 1000 lines), interlace-equipment exists (e.g. 1125/60 based on the SMPTE 240 system) and is cheaper. Therefore, they argue, one should start with an interlaced signal and migrate to progressive scan at a later time.

10.3 History of HDTV

In this section we give a brief history of the development of HDTV in Japan, Europe and the United States.

HDTV in Japan

In the 1970's, the Japan Broadcasting Corporation[10.1] (NHK) began to design the next generation television system. The main systems work was done by NHK, but the substantial infra-structure equipment required for a complete system (e.g. cameras, recorders and displays) was developed by other Japanese companies. The NHK developed system had 1125, 2:1 interlaced scan lines and frequency multiplexing of color (but not in the same bandwidth as the luminance). The initial system was intended for a wideband (100 MHz) satellite channel to supplement (not replace) NTSC. Further developments led to the MUSE* compression system that would fit into a normal 27 MHz satellite transponder.[10.2] In the mid 80's, it was realized that the production system could be distinct from the transmission system. Since then, the NHK 1125/60 system has been proposed as a production and international exchange standard, on the basis that much of the equipment has already been developed. However, there is still considerable disagreement in selection of the scanning parameters of the production standard. Meanwhile, satellite HDTV has been in service in Japan for several years, and the receiver costs have come down steadily (but still remain very high). MUSE was further modified to fit into the 6MHz terrestrial channel (called Narrow-MUSE) in order to compete in the FCC-sponsored competition in the United States. Even though the NHK standard itself remains controversial, it gave rise to substantial developments in much of the equipment required for both professional and consumer HDTV. More recently, Japanese efforts are being revived to make HDTV digital using both compression and digital modulation for channel coding.

European HDTV

Although the European Broadcasting Union (EBU), had given some indication of support[10.3] for the NHK-1125/60 system for professional studio use, in 1986 the broadcasting and manufacturing interests in Europe agreed on their own high definition standards, called HD-MAC.† HD-MAC was expected to be delivered exclusively by direct broadcast satellite (DBS), and it was configured to be backward compatible with the newer versions of PAL and SECAM (i.e., D-MAC). Thus, HD-MAC would supplement PAL and SECAM

---

* MUSE = Multiple Sub-Nyquist sampling encoding.

† HD-MAC = High Definition – Multiplexed Analogue Components.

and, therefore, would never result in retiring them. Significant expenditures were made by government and industry to develop HD-MAC under project Eureka-95. Nevertheless, the future of HD-MAC appears bleak. Although it has not been officially withdrawn, sentiment in Europe is now leaning toward a system using digital compression as well as digital modulation for transmission. There are many opportunities to select a standard, or major elements of a standard, such that interoperability within Europe as well as with Japan and the U.S. could be facilitated.

HDTV in North America

Since television is delivered over many media, the service providers for each medium were concerned that HDTV might succeed over an alternative medium, thus taking their audience away. Broadcasters, under pressure from the alternative media (since HDTV may not increase an already saturated audience's total viewing hours), petitioned the FCC (Federal Communications Commission of the United States) in 1987 to assess the effect of HDTV on the existing broadcasting structure. Broadcasters were also feeling the pressure from the mobile telecommunications service providers who wanted the unused (taboo channel) spectrum for wireless services. Before the petition, in spite of pressure from several industry groups such as SMPTE 240M (See Chapter 2), a slightly modified version of the NHK 1125/60 system was rejected as the American National Standard (ANSI) for production. In September 1987, the FCC established the Advisory Committee on Advanced Television Service (ACATS) to make a recommendation on advanced television (including HDTV). ACATS proposed hardware testing the best of the 23 proposed systems then vying to be the standard and set up a process to conduct such tests. As the testing drew closer, only six systems remained. Compared to the previously proposed systems, four out of these six systems emphasized simulcasting in the taboo channels, digital compression and transmission. When the testing concluded in late 1992, all four digital systems remained in the running, but a clear winner could not be chosen. In May 1993, the remaining competitors agreed to pool their knowledge to create a combined "best of the best" system called the Grand Alliance HDTV (GA-HDTV) system.

10.4 Problems of Current Analog TV

Current composite analog television has many defects and artifacts. Some are inherent and some are due to the addition of color on a subcarrier without increasing the overall bandwidth. HDTV attempts to alleviate many of these artifacts. For example, **inter-line flicker, line crawl and vertical aliasing** are largely a result of interlaced scanning. These effects are more pronounced in pictures containing slowly moving high-resolution graphics and text with sharp edges. The computer industry has largely abandoned interlace scanning in order to avoid these problems. Progressive scanning and transmission of HDTV will eliminate these effects almost entirely.

Large area flicker as seen from a close distance or on large screens is due to our ability to resolve temporal brightness variations at frequencies even beyond 60Hz when the peripheral areas of vision are engaged. The cost of overcoming this problem in terms of bandwidth (i.e., by increasing the frame rate) is too large and is, therefore, not addressed in any current HDTV proposal. Computer displays employ much higher frame rates (up to 72 Hz) to overcome large area flicker.

Static raster visibility is more apparent on larger displays since viewers can make out individual scan lines. Increased spatial resolution in HDTV will reduce this effect substantially.

Finally, **cross-color and cross-luminance** are a result of color modulation used to create the composite signal. HDTV will use component signals. Moreover, in digital HDTV, each color component will be compressed separately, with the compressed bits multiplexed in time. Thus, these two effects will be eliminated entirely.

With the present analog television, a number of other artifacts also arise, due to transmission. For example, **ghosts**, which are due to multiple receptions of the same signal, cannot be entirely removed from the analog signal. In digital HDTV, ghost cancellation can be done more successfully, generally leaving no impact on the transmitted bit stream.

Another transmission problem is interference from other TV signals. As an example, the NTSC signal, containing high energy and sync pulses with sharp transitions, has poor spectrum utilization due to a high degree of **co-channel*** and **adjacent channel** interference. Co-channel interference constrains the minimum geographic distance between transmitters on the same frequency band. Adjacent channel interference precludes using spectrally adjacent channels to serve the same area unless the transmitters are located together, thus assuring near-equivalent, non-interfering signal strengths at all receiving locations. Both of these interferences are reduced by keeping adjacent channels in the spectrum vacant (i.e. taboo) at any one place, while using these vacant channels at other surrounding places. As a result, in the United States only about 20 channels are usable in each location out of a total of 68. In the United Kingdom, only about 4 out of 44 channels are usable.

10.5 Advantages of Digital HDTV

Unfortunately, increasing the spatial and temporal resolution of the HDTV signal also increases its analog bandwidth. In the currently allocated terrestrial

---

* Two signals from different places, but in the same frequency band.

broadcast spectrum, such an analog signal could not be accommodated in any location. Therefore, all the newer HDTV proposals employ digital bandwidth compression. Digital HDTV reduces the bit-rate from approximately 1 Gbits/s to about 20Mbits/s. This digital signal is incompatible with the current television system and, therefore, cannot be decoded by today's television sets.

Such a compression, by itself, however, is not enough. In order to improve spectrum utilization, it is also necessary to reduce interference to and from other broadcast signals, including the analog TV channels.

Digital HDTV modulates the compressed audio and video bits into a signal that has much lower power, while requiring the same analog bandwidth as the current analog television. In addition, most HDTV proposals simulcast the HDTV signal along with the current analog television signal having the same content. Simulcasting is shown in Figs. 10.1 and 10.2. A low power, modulated digital HDTV signal will be transmitted in a taboo channel that is currently unused because of co-channel and adjacent channel interference considerations. Simultaneously, in an existing channel, the same content (down converted to current television resolution) will be transmitted and received by today's analog television sets. By proper shaping of the modulated HDTV signal, it is possible to avoid its interference into distant co-channel as well as nearby adjacent channel analog television signals.

At the same time, distant co-channel and adjacent channel analog television signals must not degrade the HDTV pictures significantly. This allows previously unused terrestrial spectrum to be used for HDTV as well as digital conventional TV. Consumers with current television sets will receive television as they do now. However, consumers willing to spend an additional amount will be able to receive HDTV in currently unused channels. As the HDTV service takes hold in the market place, spectrally inefficient analog television can be retired to release additional spectrum for additional HDTV channels or other services.*

Another desirable feature for HDTV systems is easy interoperability with the evolving telecommunications and computing infra-structure. Such interoperability will allow easy introduction of different services and applications, and at the same time reduce costs due to commonality of components across different applications. There are several characteristics that improve interoperability. Progressive scanning, square pels and digital representation improve interoperability with computers. In addition, they facilitate signal processing required to convert HDTV signals to other formats either for editing, storage, special effects, or down-conversions to current

* In the United States, the FCC plans to retire NTSC after 15 years from the start of the HDTV broadcasts.

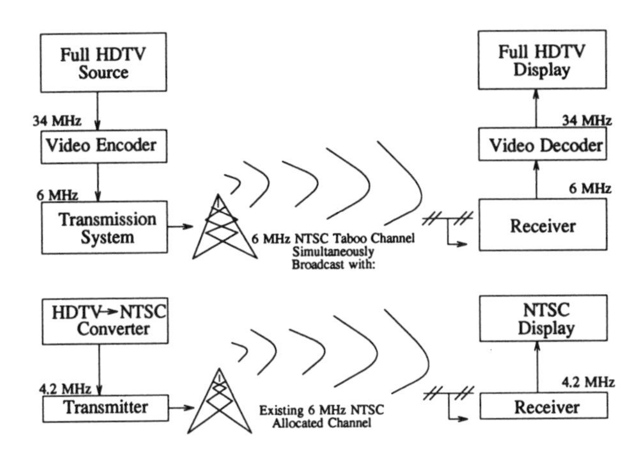

Fig. 10.1 Simulcasting of Digital HDTV with NTSC. Under current channel assignments, many channels remain vacant (i.e. taboo) to avoid co-channel and adjacent channel interference. Since digital signals have lower transmission power, they can use the taboo channels in the new channel assignment.

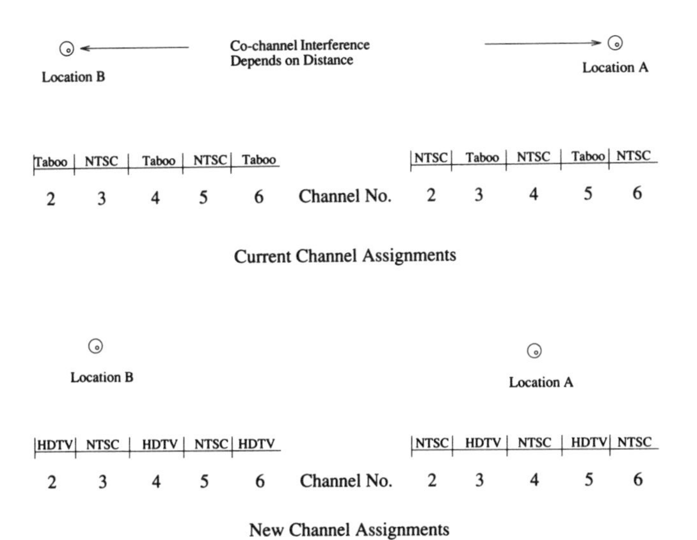

Fig. 10.2 HDTV Channel Assignments. This picture is somewhat simplified. In each geographical area, one needs to examine the channel assignments based on co-channel and adjacent channel interference between current TV and HDTV.

television. The use of a standard compression algorithm (e.g. MPEG) transforms raw, high-definition audio and video into a coded bit stream, which is nothing more than a set of computer instructions and data that can be executed by the receiver to create sound and pictures.

Further improvement in interoperability at the transport layer is achieved by encapsulating audio and video bit-streams into fixed-size packets. This facilitates the use of forward error correction (necessary to overcome transmission errors) and provides synchronization and extensibility by the use of packet identifiers, headers and descriptors. In addition, it allows easy conversion of HDTV packets into other constant size packets adopted by the telecommunications industry, e.g., the Asynchronous Transfer Mode (ATM) standard for data transport.

Fitting the HDTV signal into an existing television channel (e.g. 6MHz for NTSC) improves interoperability with existing terrestial broadcasting, with cable TV, and with certain kinds of video-tape recorder technology in which the modulated digital HDTV signal is stored as an analog signal.*

10.6 GA-HDTV System[10.4]

The Grand Alliance HDTV system is a proposed transmission standard considered by the FCC for North America. It is not a production standard, display standard or a consumer interface standard. It is expected that a production standard that is optimized for operation within a studio, but at the same time interoperable with the transmission standard, will emerge once the transmission standard is approved. Efforts to this end are underway within the Advanced Television Systems Committee (ATSC). Similarly using the transmitted signal, different manufacturers of consumer television receivers will create optimum displays consistent with picture quality obtainable for a given cost.

The GA-HDTV supports signals in multiple formats as follows:

> 1280 pels × 720 lines at frame rates of 24, 30, 60 Hz progressive

and

> 1920 pels × 1080 lines at frame rates of 24Hz, 30Hz progressive and 60Hz interlaced.†

Frame rates may also be divided by 1.001 in systems that require compatibility with NTSC. With a 16:9 image aspect ratio these formats create square pels for

---

* Current VCR's might be modified without changing the basic analog technology, but existing recorders would not be directly interoperable

† Also under consideration is 1440 pels, 60 Hz interlaced, for bit rates that cannot support 1920 pels.

both 720 and 1080 line rasters. Such flexibility in scanning allows different industries, program producers, application developers and users to optimize among resolution, frame rate, compression and interlace artifacts.

The Grand Alliance would have wished to include a 1920 × 1080 progressive scan format at 60Hz. However, the 6MHz terrestrial broadcast channel does not allow for the required bit rate. As the NTSC audience declines and the interference into NTSC becomes less of a concern, there will be an opportunity to increase the data rate to support, in a compatible manner, 1920 × 1080 progressive scan at a 60Hz frame rate. The GA-HDTV system may allow for temporal scalability, so that an enhanced bit stream to handle 1920 × 1080 progressive scan at a 60Hz frame rate can be created in the future.

The GA-HDTV system incorporates a subset of the MPEG-2 compression standard described in Chapter 9. In addition, it includes several enhancements in the encoder built for initial FCC testing, e.g., the GA-HDTV system can process film originated material at 24 or 30 frames/s by adaptively adjusting the compression algorithm at the encoder. Also, compression efficiency is improved by appropriate prefiltering/subsampling to handle signals of different formats and of varying quality (e.g. noisy signals). Among the compression techniques used are: DCT of motion compensated signals with field/frame prediction and transform, flexible refreshing (both I-frame as well as pseudo-random intra-blocks), bidirectional prediction (B-frames) and forward analysis/perceptual selection for rate control and best picture quality.

The GA-HDTV system uses five-channel audio with surround sound, which compresses into a 384 kbits/s bit stream.* Audio, video and auxiliary data are packaged into fixed-length packets in conformity with the MPEG2 Systems transport layer. MPEG2 transport packets are 188 bytes long, consisting of up to 184 bytes of payload and 4 bytes of header. These packets easily interoperate with ATM since they have approximately a 4:1 ratio with ATM packets, which are 53 bytes long with 5 bytes of header.

The modulation scheme chosen for over-the-air is 8-VSB, which gives 32.28 Mbits/s of raw data in a 6 MHz bandwidth. After error correction, 19.3 Mbits/s are available. Coaxial cable can carry more than this in a 6 Mhz channel, since there is no co-channel and very little adjacent channel interference. Cable standards are under discussion.

Thus, the GA-HDTV system attempts to make the best compromise to suit different applications and provide for maximum interoperability.

---

* This is composed of five full bandwidth and one lower bandwidth audio channels.

10.7 Future of HDTV

HDTV faces significant challenges to its widespread deployment. Many people believe that the current resolution of television is adequate for most applications and that television should evolve by being more interactive. Some believe that better use of bandwidth could be made by filling it with a larger number of lower resolution programs and then interactively choosing between them. For example, using 20 Mbits/s, which is possible in the 6MHz bandwidth, one can accommodate 4 to 10 compressed digital 525-line NTSC programs. Viewers could then select between them with ease, provided there is a good user interface. Many people also believe that the precious spectrum should not be used to transmit television to fixed locations. Larger and larger fractions of modern industrialized countries are installing fiber-optic cables having enormous traffic capability. Television could reach many people through such fixed wired connections. Some argue that terrestrial spectrum should be reserved for mobile services, which are in great demand as a result of cost reductions due to low power, compact electronics and better batteries.

The high cost of HDTV sets may be another inhibiting factor, although with growing demand, both the electronics and displays should become cheaper over time. ACATS and the Grand Alliance companies in the United States have done much to quantify the costs to broadcasters of implementing HDTV. While the costs look more manageable than ever before, broadcasters are still unenthusiastic about the business value of HDTV. However, they are concerned that other media (tape, DBS, cable) might deliver HDTV to the public before them or that the FCC might give the taboo channel spectrum to other services if the broadcasters do not use it.

HDTV may also grow by other means. In the United States, the national information infrastructure (NII) is being created. Digital HDTV, with its flexible data transport structure, is an immediate opportunity to deploy an important part of the interconnected web of information super highways that make up the NII. GA-HDTV's interoperability with computing and telecommunication platforms may allow many uses of HDTV for medicine, information access, and education. Thus, it is clear that many of the technical impediments to HDTV are being removed. Customers and the market place will decide which way it will evolve.

References

10.1 T. Fujio, "High-Definition Television System", Proceedings of IEEE vol. 73, No. 4, April 1985, pp. 647-655.

10.2 Y. Ninoniya, et. al., "HDTV Broadcasting and Transmission Systems - MUSE", Proc. of 3rd Int'l. Colloq. on Advanced TV Systems, Ottawa, Canada, Oct 1987, pp. 4.1.1 - 4.1.31.

10.3 D. I. Crawford, "High Definition Television - Parameters and Transmission", British Telecom Technology Journal, Vol. 5, No. 4, Oct. 1987, pp. 76-83.

10.4 Grand Alliance High Definition Television System, GA-HDTV Document, Nov. 1993.

Questions for Understanding

10.1 What are the pros and cons of interlaced and progressive video formats?

10.2 What are the pros and cons of digital transmission of HDTV?

10.3 What are taboo channels? Why are they needed?

10.4 How is interoperability enhanced between HDTV, telecommunications and computers?

10.5 Which video formats are defined by the GA-HDTV system?

11

Postscript

In order to transmit video information at the minimum bit rate for a given quality of reproduction it is necessary to exploit our understanding of many branches of science. Ideally one should have an appreciation of vision, signal theory, display devices, and so on. As engineers we are concerned with complex stimuli, their perception, as well as the final utilization of the perceived information. Knowledge of these is often unavailable or sketchy, forcing us to design encoders based on a relatively primitive understanding of the problem. The ultimate limits of bit-rate compression will only be approached, we believe, as our knowledge of stimuli, perception and utilization increases.

Thus, in a discussion on the directions and limitations of video bit-rate compression we are very aware of *our* limitations. In this book our modest objective of defining the state-of-the-art is, we are well aware, open to the criticisms of over-simplification, serious omissions and factual disagreement. As for where the subject is heading and its inherent limitations, we confess myopia and will not be surprised by a discovery that could not have been extrapolated from existing thinking and known ignorances.

The conventional representation of a digital communication link for the transmission of pictorial information is shown in Fig. 11.1. The function of the source encoder is to operate on an analog signal $x(t)$, and to convert it into a stream of binary digits $s(t)$. The source decoder at the receiver accepts a binary signal $S(t)$ and produces a continuous signal $X(t)$. It may not be necessary to ensure $X(t) \simeq x(t)$, but what does matter is that after transduction, $X(t)$ should be perceived as $x(t)$ subject to an acceptable quality criterion. Although $x(t)$ does not always have to be identical to $X(t)$, system engineers do prefer that $s(t) = S(t)$, i.e., the channel appears ideal. Most practical channels contain imperfections that are largely overcome by pre-processing and post-processing of the binary signals $s(t)$ and $S(t)$ by the channel codec and terminal equipment.

The purpose of this chapter is to discuss limitations of source encoding. However, Fig. 11.1 demonstrates that $X(t)$ is also dependent on the channel terminal equipment, the channel codec and, of course, the channel. Thus, encoding picture signals is not merely a source encoding problem, but engulfs the complete communication system. For example, if the channel is known to result in a high bit-error-rate (ber) then the effect on the recovered signal $X(t)$ may be mitigated by altering the modulation and regeneration strategies, increasing the length of the check bits in the channel coding words, altering the source encoding algorithm, or combinations of all of these. The conventional arrangement of source and channel codecs may be altered, even merged. Post processing of $X(t)$ can also be successfully employed.

However, in this short discourse we confine ourselves to the limitations of the source codec, assuming the channel codec has been designed to ensure that $s(t) = S(t)$. We thereby ignore the limitations imposed by the channel.

11.1 Signal Sources

Video processing or transmission systems typically start with a two-dimensional distribution of light intensity. Thus, three dimensional scenes must first be projected onto a two dimensional plane by an optical imaging system. Color pictures can usually be represented by such light intensity distributions in three primary bands of wavelengths. If moving objects are to be accommodated, all three light intensities change with time.

Television raster scanning produces 25 to 30 frames per second with 2-to-1 interlacing. Each frame is composed of many lines, e.g. 525 (Japan, North America), 625 (elsewhere), $\approx$280 for videotelephone. Over 2,000 lines are often used for documents and photographs. Television is bandlimited between dc and 4 or 6 MHz, with each scan line having frequency components up to about 250 cycles. Photographs and documents typically require 500 to 2,000 cycles per scan line, the width of the item being the determining factor. In color pictures, chrominance bandwidth is typically about one-fourth the luminance bandwidth.

Video is characterized by high signal energy in low to mid-band frequencies, and low energies at high frequencies. In raster scanned video, energy peaks occur at multiples of the line scan rate, with significant gaps between the peaks. Each of the line-rate harmonic peaks is made up of many sub-peaks separated by the field rate. Color multiplexing techniques often use these gaps to convey color information in the composite waveform.

Video statistics are non-stationary even over the short term. Long-term autocorrelation functions for raster scanned pictures have periodic components that depend on the scan and field rates. Two dimensional picture autocorrelation functions typically decrease monotonically with respect to spatial distance. The probability density function (PDF) of color component signals is too non-stationary to be quantified, although component differential signals tend to be Laplacian. PDF's are often used in quantizer design to minimize the mean

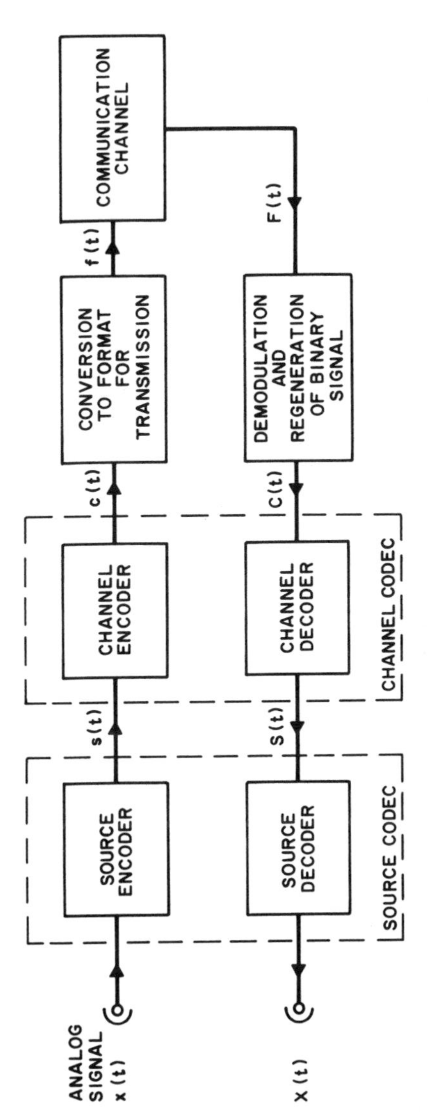

Fig. 11.1 Basic transmission system.

square error. However, many workers in the field (including the authors) reject this criterion, preferring instead criteria based on human perception. Activity factors may be useful in video, e.g., low-detail/high-detail areas, stationary/moving areas. The frequency spectra are related to both the activity factors and autocorrelation functions.

Pictures tend to have many local areas of similar color, and boundaries of objects are, more or less, continuous. In television, luminance levels in most areas do not change by large amounts between frames. Black/white graphics have large areas of constant luminance.

Such predictable structure in the video waveforms is termed *statistical redundancy.* Given the short term past, it enables the prediction, with reasonable accuracy, of the short term future. Removal of this redundancy can greatly reduce the bit-rate required for representing the video information.

11.2 The Human Receiver

The eye is the organ of sight, having at its rear an inner nervous coating known as the retina. Rays of light pass through the cornea, aqueous humour, lens and vitreous body to form an image on the retina. The central area of the retina, known as the fovea, provides high resolution and good color vision in about 1° of solid angle. The images on the two retinas are sent along the two optic nerves, one for each eye, until they meet at the optic chiasma, where half the fibers of each nerve diverge to opposite sides of the brain. This enables stereo vision and perception.

The eye behaves as a two-dimensional low-pass filter for spatial patterns, with a high-frequency cut-off of about 60 cycles per degree of foveal vision and significant low-frequency attenuation below about 0.5 cycles. Thus, high spatial frequencies in the image are not seen and need not be transmitted. The eye also acts as a temporal band-pass filter having a high-frequency cut-off between 50 and 70 Hz depending on viewing conditions. Flicker is more disturbing at high display luminances and low spatial frequencies.

Noise and amplitude distortion are generally less visible at high luminance levels than at mid and low luminance values, again depending on viewing conditions such as overall scene brightness and ambient room lighting. High and low frequency noise is less visible than mid-frequency noise. Amplitude distortions are also less visible near color transitions, such as occur at boundaries of objects in a scene. This is termed *spatial masking*, since the transitions *mask* the distortions. Although amplitude distortions are less visible near color transitions, small spatial shifts in the transitions themselves are easily visible and annoying.

Temporal masking also occurs. For example, shortly after a television scene change, the viewer is relatively insensitive to distortion and loss of resolution. This is also true of objects in a scene that are moving in an erratic and unpredictable fashion. However, if a viewer is able to track a moving object, then resolution and distortion requirements are the same as for stationary areas of a picture.

As the variety of encoding algorithms increases so do the types of resulting degradations. If perception of these degradations were thoroughly understood, the quality of reproduction of a particular image encoding strategy could be ascertained by objective measurements of signal parameters. The current situation is one of ad-hoc objective measurements,[11.1] each trying to relate subjective observations with each new encoding algorithm or each type of degradation. Old methods of signal-to-noise ratio, spectral distance measures, pulse shapes etc., are frequently inadequate. To postulate a new objective measure, subjective testing must be done. Here, tests are made on a small sample of the population, and by statistical methods the effect on the entire population is estimated, i.e. a *statistical eye* is determined. Subjective testing is controversial. Should simple grading, bad to excellent in 5 steps, or multi-dimensional analysis be used? What form should the test take, e.g., type of picture detail, amount of motion, etc.? However, what is even more in dispute is relating subjective testing results to objective measurements. Our inability to do this, implicit in our ignorance of perception, is a serious impediment both to communication between research scientists and to source encoding itself. Only when perception is properly understood will we have accurate objective measures. However the day when we can, with confidence, objectively evaluate a *new* impairment without recourse to subjective testing seems very remote.

11.3 Waveform Encoding

In waveform coding[11.2,11.3], a continuous *analog* signal $x(t)$ is encoded into a stream of bits by the source encoder, and, from these bits, a decoder recovers a signal $X(t)$. The design objective in waveform encoding is that, for a given bit-rate, $X(t)$ should be as close a replica of $x(t)$ as possible. Since many $x(t)$'s can produce the same $X(t)$ the difference $n(t) = x(t) - X(t)$ cannot, in general, be zero all the time. This *quantization noise* is a fundamental limitation of finite bit-rate coding. For example, in pulse code modulation (PCM), $x(t)$ is sampled at the Nyquist rate, and each sample is represented by a binary number, i.e., quantized. The decoder converts the binary numbers back to analog samples and low-pass filters them to obtain $X(t)$. The bit-rate is the product of the sampling rate and the binary word length of each sample, the latter fixing the accuracy of conversion.

The signals from most TV cameras are already companded, i.e., a compressed nonlinear function of scene luminance*. Eight-bit uniform quantization of this companded signal gives imperceptible quantization noise in most cases. Single pictures, e.g., photographs, typically require one bit less quantization accuracy than television, where the quantization noise is time

* Henceforth, we will consider mainly luminance signal encoding since it consumes most of the channel capacity. The smaller bandwidth chrominance signal can be encoded similarly.

varying and, therefore, much more visible. For black/white images only one-bit quantization is required.

If perceptible quantization noise can be tolerated, then coarser quantization can be used and the bit rate reduced. The addition of random, or pseudo random noise prior to quantization, called dithering, changes the quantization error from being conditional on the input signal to approximately white noise and gives improved subjective results.

Successive pels of video are often highly correlated. In addition, periodicities exist in the waveform and lead to high correlations between pels that are separated, in some cases, by many sampling epochs, e.g., line, field or frame period. Predictive coding[11.2] (also called differential PCM or DPCM) exploits these correlations by using previously transmitted quantized pels to form a prediction of the current pel to be encoded. The difference between the actual pel value and its prediction is quantized, binary encoded and transmitted. The decoder is able to form the same prediction as the encoder (in the absence of transmission errors) because it has access to the same quantized pels. By adding the received quantized differential signal to the prediction the decoded quantized pel is obtained, and the signal is recovered by low pass filtering.

The predictor may be a linear, nonlinear or an adaptive function of previously encoded samples. Similarly the quantizer may be uniform, non-uniform or adaptive. Using the previously quantized pel value as a prediction, companded 5-bit quantization achieves imperceptible distortions in monochrome television.

The predictor enables the variance and correlation of the differential signal to be significantly less than that of the original signal, enabling the quantization noise to be reduced for a given number of quantization levels. Further gains can be made by *entropy coding* the quantized differential signal, i.e., assigning short code words to small, but frequently occurring, values and longer code words to the seldomly occurring large values. However, with entropy coding, the bit-rate depends on the input signal, and unless protective measures are taken there is a chance of some signals producing a bit-rate that exceeds the channel capacity causing severe distortion in the recovered signal.

In some cases it pays to represent *groups* of pels with a single code word. For example, with black/white graphics and text, long strings of identical pels occur both with and without predictive encoding. Considerable savings occur from coding such strings with a single binary word, called *run-length-coding*. Entropy coding of runs yields further gains. In television, conditional frame replenishment coder-decoders use previous frame prediction in non-moving areas of the picture. These areas are efficiently encoded as groups of pels.

DPCM performance can be improved significantly by using adaptive predictors, adaptive quantizers or both. Adaptive predictors attempt to optimize the prediction depending on the local waveform properties. For example, interframe coders typically use previous frame prediction in non-moving areas, but in moving areas use linear combinations of pels in both the previous and present frame as a prediction. By adapting the moving-area predictor to the speed and direction of motion, further improvement is achieved. Adaptive DPCM (ADPCM) systems are ultimately limited by the predictor making predictions from pels corrupted by quantization noise. Thus, the design of the predictor should take into account the characteristics of the quantizer and vice versa.

With adaptive quantization, the effective number of quantization levels for a given bit-rate and quality of encoding is greatly increased. Adaptive quantizers used in picture encoding usually span the range of the input signal, and adaptively discard some of their levels as a function of the video signal. Sharp transitions in the waveforms need not be represented as accurately as when variations are relatively slow, i.e., visibility of quantization noise is markedly less at edges of objects than in flat, low detail areas. Such subjective phenomena enable adaptive quantization to save a bit or more per Nyquist sample. With a two-dimensional predictor, adaptive quantization and entropy coding, a good quality picture can be obtained with 1.5–2 bits/Nyquist sample. For conference scenes containing low motion, interframe motion compensated prediction is used with adaptive quantization and entropy coding. Only 0.1–0.2 bits/Nyquist sample are sufficient for good quality picture.

Correlations and periodicities can also be exploited by transform coding. With this approach the pels to be coded are first partitioned into blocks. Pels within a block need not be contiguous in time, i.e., groups of pels may be chosen from adjacent lines and adjacent frames in order to make up a block. Each block is then linearly transformed into another domain, e.g., frequency, having the desirable property that signal energy is concentrated in relatively few transform coefficients compared with the number of pels in the original block. Furthermore, all of the coefficients need not be quantized to the same accuracy to achieve a given quality of reproduction. By encoding only the significant coefficients with an accuracy dependent on human perception considerable bit-rate reductions are possible.

With adaptive transform coding (ATC), the coefficients selected for transmission change from block to block as does the quantization strategy for each coefficient. For single pictures, many blocks contain little or no picture detail, i.e., they contain energy only at low frequencies and require only a few bits for encoding. Other blocks contain high frequency components and produce more coding bits. Pictures containing an average amount of detail can be encoded using intraframe ATC at around 2 bits per pel with imperceptible distortion, i.e., *excellent* quality, and 1.5 bits per Nyquist sample with perceptible but not annoying distortion, i.e., *good* quality. In television three dimensional ATC operates on blocks from several frames yielding large reductions in bit-rate, particularly in non-moving areas of the picture.

Hybrid encoding involves transform coding of blocks of samples followed by block-to-block DPCM encoding of the resulting coefficients. Used in picture encoding, its performance is similar to transform coding with larger blocks; however, implementation is simpler. Interframe adaptive hybrid coding of pictures with low movement has yielded bit rates of 1 bit per Nyquist sample with *excellent* quality and 0.5 bits per sample with *good* quality[11.4].

ATC, ADPCM using entropy coding, and interframe coders all have the property that for a fixed quality, the bit-rate depends very much on the input waveform. This is undesirable for communication channels with a fixed channel capacity. Buffers can be used to accommodate the variable bit-rate generation to the constant bit-rate transmission. However, they introduce delay that may be intolerable in certain two-way communications. This may be reduced by sacrificing quality during periods of excessive bit-rate generation. Interframe coders take this approach by reducing moving-area picture quality during periods of rapid movement. In ATC it is customary to adaptively assign the bits to the coefficients in order to ensure constant bit-rate, i.e., allow the quality to vary from block to block. It must be emphasized that all constant bit-rate waveform codecs produce a variable quality either on a block basis, as with ATC, or on a per pel basis, as with DPCM. Another solution is for many encoders to share the same data channel, thus taking advantage of the low likelihood of several encoders producing high bit-rates at the same time.

Finally we note that subjective testing of DPCM speech and picture signals reveal some intriguing comparisons[11.5,11.6]. The effect of slope overload is to cause muffling in speech and blurring of the edges in pictures. With no slope overload, the quantization noise is often referred to as granular noise because it is perceived visually to have a granular texture. For a given mean noise power, slope overload noise is generally less annoying than granular noise for both pictures and speech. Also, moderate signal echo, i.e. reverberation in speech and ghosts in pictures, is usually more acceptable for a given distortion power than quantization noise. Interestingly, both listeners and viewers tend to have similar high order processing behavior. They perform a judgement as to whether the speech or picture is clear or unclear, and in the latter case whether the distortion is associated with the signal (slope overload, echo, bandlimiting) or the background (*snow* in pictures, hiss in audio).

11.4 Model-based Coding

In model-based coding of speech (known as vocoding), the signal is analyzed in terms of a model of the vocal mechanism, and the parameters of the model transmitted. The receiver then uses the parameters to synthesize a speech signal that is perceptually similar to the original speech. Unlike waveform encoding, no attempt is made to preserve the integrity of the individual samples of the waveform. There is no equivalent of vocoding in picture encoding because of the difference in the nature of the sources. In several constrained situations, it is possible to construct models of the scene. For example, in videophone applications, most scenes contain a head-and-shoulders view of a person covering a static background. Head-and-shoulders views of people have been modeled as wire-frames with some success. This type of work is still in its infancy and faces formidable challenges in building models, finding applications, and dealing with the complexity of the encoder/decoder. Nonetheless, promising research is continuing.

11.4.1 Model-based Fax Coding

For black/white text consisting of typed characters, optical character recognition (OCR) algorithms identify each character and its position as two parameters that are then encoded and transmitted. This gives large compression gains compared to waveform encoding. Typically a standard 8.5×11 inch typewritten page (80×66 characters) can be transmitted with 8-bits per character, i.e., about 0.01 bits/Nyquist sample*. Black/white graphics can also be handled using OCR, but the character alphabet must be constructed adaptively for each document or class of documents to be coded. This usually necessitates the transmission of side information. Test documents have been encoded at ≈ 0.025 bits/Nyquist sample[11.4].

11.4.2 Model-based Coding of Video

As mentioned earlier, advances in computer vision and graphics are allowing us to progress from pel oriented coding (or waveform coding) to coding of objects in the scene. Techniques of computer vision can be used to identify a variety of objects from a complex scene. Each of these objects may contain hundreds of pels and can be described by a modeling process available from computer graphics.

In order to achieve a high degree of compression, proper models must be used. Currently, models that can handle arbitrary scenes do not exist. Therefore, two approaches are being pursued. In the first approach, a very limited class of scenes is considered in which the dominant object (e.g. human face) can be modeled easily. In the second approach, instead of constructing specific models, an image is segmented into regions using techniques from computer vision, and each region is coded separately. Regions are not necessarily rectangular blocks of pels as in block-transform coding. A brief description of each of these approaches is given below.

---

* We assume a sampling rate of 60 per mm^2. This is one of several rates in use and corresponds to *high resolution*. Also, "Nyquist sample" is not strictly correct since bandlimiting is not usually done.

One of the objects that is dominant in videophone applications is a human face. Modeling of human faces has been studied for some time[11.7–11.9]. Recent methods require only a few parameters in the model and yet create good quality images in most situations, resulting in good compression. Fig. 11.2a shows a block diagram of a model-based encoder. Each frame of the incoming video is analyzed with the assumption that the principal object in it is a human face. Computer vision techniques are used to extract and parametrize properties such as the shape of the head, geometric properties, color shading and texture. These parameters are used by both the transmitter and the receiver to create a three-dimensional model of the human head. As the head moves from frame to frame, these parameters are updated. The 3-D motion of the head is also estimated by techniques similar to the ones defined in Section 5.2. Based on the motion and the model parameters, both the transmitter and receiver reconstruct each frame. At the transmitter this frame is subtracted from the input frame, and the residue is coded by standard waveform coding techniques. At the receiver, the reconstructed image is added to the decoded residue image to create the image to be displayed.

The principal difficulty in this approach is to construct a three-dimensional model of the human face using the least number of parameters. Further, the model should be easily manipulated and easily synthesized by the receiver into an image that is a close replica of the input image. Wireframe models with triangular patches[11.8,11.9] are widely used for these reasons. See Fig. 11.2b.

Having found the model parameters, facial expressions need to be characterized. The Facial Action Coding System (FACS) developed by Ekman and Friesen[11.7] has a set of about 60 parameters that describe facial expressions. Wireframe models and FACS have been integrated by a group at Linkoping University led by R. Forcheimer[11.10]. The transmitter must estimate the specific values of each of these facial expression parameters for each frame and include them in the model parameters that are transmitted. The receiver then uses them to synthesize the image. Thus, the overall head motion is estimated by the 3-D motion estimation techniques similar to those of Section 5.2, whereas animation of the face is handled by FACS. As long as the unknown objects, background, etc., that are coded by the residue coder require a small fraction of the coded bits, the model based coder shows improvements in performance.

The second approach does not require any a priori knowledge or assumption about specific objects in the scene, and therefore can be applied to general classes of scenes. While there are many variations in this fast evolving area[11.11], the general characteristics are similar, and the differences are in the details. Basically, each frame of video is described in terms of moving objects and coded by three sets of parameters consisting of color (shading), contour (boundaries of objects) and motion.

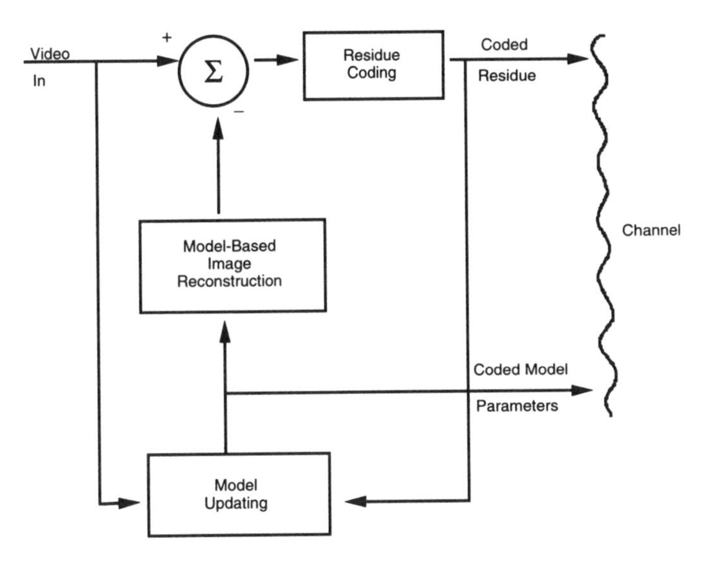

Fig. 11.2a Model Based Encoder.

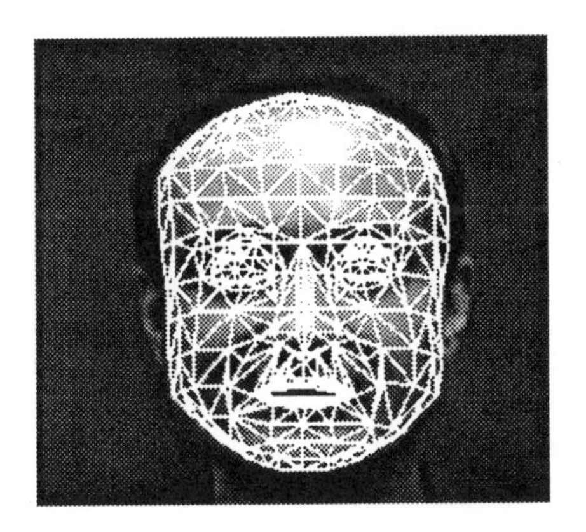

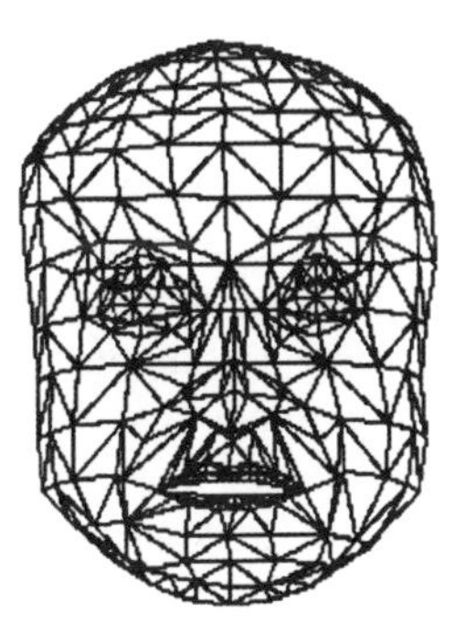

Fig. 11.2b Wire Frame Model.

A region-based object coder is shown in Fig. 11.3. Techniques from computer vision are used to segment each new frame into regions containing similar color or texture pels. Spatial and temporal consistency as well as edge continuity are used to generate well-defined contours. Both the new uncoded frame and the decoded frame (also available at the receiver) are necessary to maintain temporal consistency. Each region is then coded by standard waveform coding methods (e.g. DCT of motion compensated prediction error) specialized for non-rectangular regions.

The motion estimation, transform, quantization and rate-control all must be adapted for pel-regions of arbitrary shape. Many schemes have been proposed for this. In addition, unlike in the case of rectangular blocks, region boundaries must be coded. Here again many variations exist (for example, see [11.12], [11.13]).

The principal difficulty in region-based schemes described above is the ability of the algorithm to segment the image into areas of importance consistent with the regions perceived by the human eye. Additionally, for efficiency of compression, fewer regions are desirable. Extensive research is being carried out currently with the hope that impressive gains in performance can be made.

11.5 Bit Rates Versus Complexity

Fig. 11.4 is our attempt to summarize where we are today in picture coding. To accommodate sources of different bandwidths we use bits per Nyquist sample (η), and plot this as a function of codec complexity for different quality criteria. To convert η to bit-rate the following multiplication factors (sampling rates) apply: 288 line videotelephone –3 MHz, 525 line TV – 8.4 MHz, 625 line TV – 10 to 12 MHz. TV resolution facsimile is about 512×512 samples per picture; photographic resolution and black/white graphics use about 60 samples per mm^2.

The complexity allocated to each codec should not be taken too literally; rather it is a crude estimate relative to the complexity of an adaptive delta modulation (ADM) codec. The relation of cost to complexity is governed by an evolving technology, and codecs with high complexity that are prohibitively costly today may be inexpensive in the future. The different qualities of reproduction are more meaningful than the level of complexity. However, qualities are subjective measures and open to dispute, particularly for the newer codecs where rigorous subjective testing and comparisons are still to be made. The top curve corresponds to broadcast television quality, the middle curve to cassette videotape quality and the bottom curve to a quality that is usable, but with visible distortions. The dashed line corresponds to black/white text and graphics.

For single monochrome pictures (gray scale facsimile) the bit-rates can be reduced by roughly 20% to account for the fact that the "frozen noise" that appears in a coded single picture is much less visible than the time varying noise

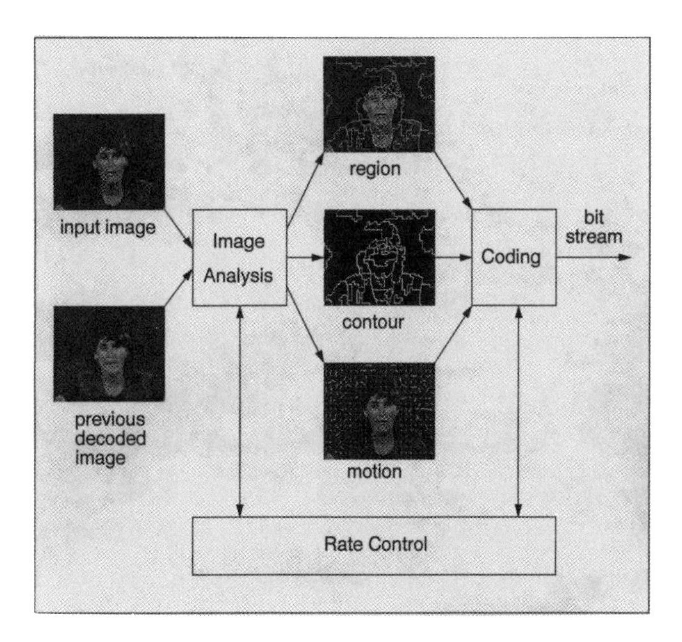

Fig. 11.3 Region Based Encoder.

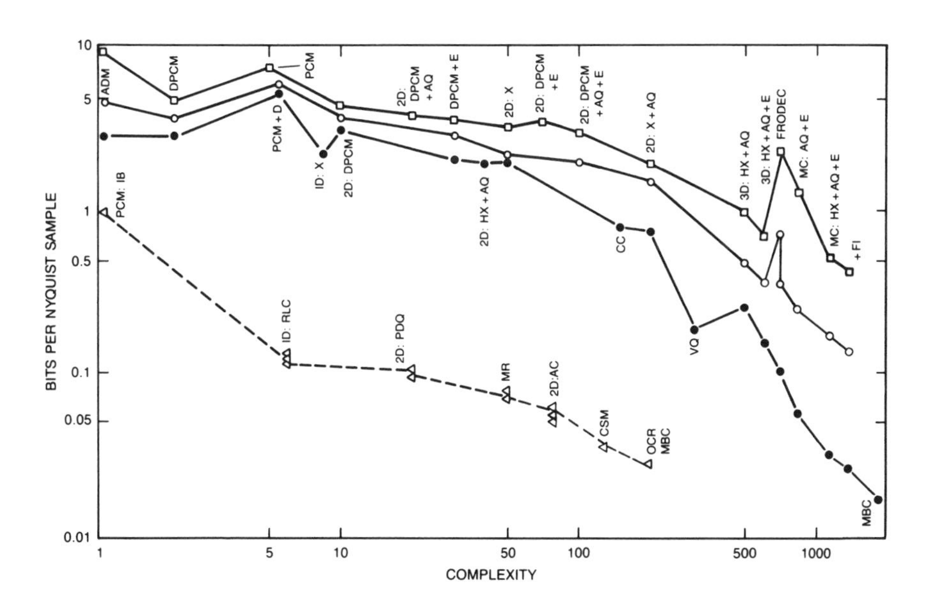

Fig. 11.4a Video bits per Nyquist sample η vs Complexity for monochrome television and black/white graphics. The three picture qualities correspond roughly to peak-signal-to-rms-unweighted-noise ratios (PSNR) of 45 dB (excellent), 40 dB (good), 35 dB (fair).

ADM	Adaptive Delta Modulation
DPCM	Differential pulse code modulation (previous element)
PCM	Pulse code modulation
PCM+D	PCM plus dither
1D:X	One dimensional transform coding
2D:DPCM	Two dimensional DPCM
2D:DPCM+AQ	2D:DPCM plus adaptive quantizing
DPCM+E	DPCM plus entropy coding
2D:HX+AQ	Two dimensional hybrid transform coding plus adaptive quantization.
2D:X	Two dimensional transform coding
2D:DPCM+E	2D:DPCM plus entropy coding
2D:DPCM+AQ+E	... plus adaptive quantizing and entropy coding.
CC	Contour coding (two component model [11.6])
2D:X+AQ	2D:X plus adaptive quantizing (JPEG Standard)
VQ	Vector Quantization
MC:AQ+E	Motion compensation, adaptive quantization and entropy coding
3D:HX+AQ	Interframe hybrid transform coding plus adaptive quantizing.
3D:HX+AQ+E	... plus adaptive quantizing and entropy coding
MC: HX+AQ+E	... with motion compensation (P*64 Standard)
+FI	... with spatial and temporal subsampling and interpolation
FRODEC	Conditional Frame replenishment coder-decoder
MBC	Model Based Coding
PCM:1B	PCM with one-bit (two level) quantizing
1D:RLC	One dimensional run length coding. (1D - CCITT standard)
2D:PDQ	Two dimensional predictive differential quantizing
MR	Modified READ code (2D - CCITT standard)
2D:AC	Two-dimensional Arithmetic Coding (JBIG Standard)
CSM	Combined Symbol Matching
OCR	Optical character recognition
MBC	Model-Based Coding

Fig. 11.4b Codec abbreviations in Fig. 11.4a.

or distortion in coded television. For color pictures the bit-rates should be increased by 15 to 25%.

Points toward the left in Fig. 11.4 are generally reliable. The algorithms have been studied in great detail, with a large variety of pictures and, in most cases, subjected to rigorous subjective testing. This tends to be less so for the more recently studied codecs of higher complexity, and many workers in the field (including the authors) would disagree with some of these published results. The non-monotonic nature of the curves in this region is probably attributed to the lack of refinement of the most complex systems and to the lack of detailed subjective evaluation.

11.6 Future Directions and Ultimate Limits

The usual approach to predicting the future is to extrapolate from the immediate past. The pitfalls of this are many. Constraints that we understand poorly or are completely unaware of will come into play eventually in an unpredictable manner. Unforeseen developments or inventions may occur that invalidate many, if not all, of the old assumptions. Also, the temptation to extrapolate techniques that, even today, are not well understood is usually too much to resist. In spite of all these pitfalls, we present the following predictions.

For some of the codecs discussed here we could have offered as performance limits information theory results[11.14–11.16] of SNR versus bit rate for input signals having stationary statistical properties. Unfortunately, and in spite of its wide use, SNR (with its numerous definitions) is a coarse yard-stick. A system with a high SNR will in general give good performance, but different codecs with their individual signal impairments, having similarly high SNR's will often be perceived quite differently. Low SNR (< 10 dB) values are meaningless in terms of codec performance. Fig. 11.4, therefore, relates to perceptual observations, not objective measures, and it is in perceptual terms that we must think of limits in the transmission bit-rate.

11.6.1 Waveform Encoding

In picture waveform encoding[11.3], intraframe ADPCM will be refined to the point where edges of objects are tracked by the prediction algorithm and quantizers adapt to the local picture characteristics in order to optimize the quantization noise according to subjective criteria. However, in a quasi-optimal system using entropy coding, edges contribute to overall bit-rate in varying amounts depending on the picture. With intraframe transform encoding the same may be said. Since most systems must accommodate both high and low detail pictures, bit-rates should not fall drastically.

Interframe ADPCM has advanced to the point where object motion is tracked by means of prediction. Adaptive filters and quantizers will optimize the displayed resolution (temporal and spatial) and quantization noise to the subjective requirements of the viewer. With these techniques, camera motion (zooming and panning) will have little effect on overall bit-rate, with the result that the middle and upper curves of Fig. 11.4 will be essentially parallel to the lower curve.

In an interframe encoder with entropy coding, the long-term average and the short-term peak bit-rates differ considerably. For purposes of digital recording this is of little consequence. However, for present day real-time communication, where data peaks cannot be "buffered out" via the use of large memories and long delays, either excess channel capacity has to be provided or picture quality has to be compromised. In the future, there will be considerably more video traffic making it feasible for many video sources to share the same communication channel. Advantage can then be taken of the fact that simultaneous data peaks in several sources rarely occur, and the allocated *per source* channel capacity can be made much closer to the long term average data rate without introducing long delays due to buffering.

Ultimately, such channel sharing arrangements (equivalent to packet switching with variable length packets) appear to be the only way that real-time video can take advantage of highly adaptive waveform coders that produce low bit-rates for low detail or low movement pictures, but require higher rates otherwise. While it may be true that the "average" picture has average detail and average movement, few systems will be successful unless they can accommodate the full range of pictures that the average *viewer* finds interesting.

11.6.2 Model-Based Coding

Model-based coding has the prospect of further reductions in bit-rate compared to waveform coding. We consider model-based coding to include recognition of one or more global attributes of the objects in the image that enable more efficient coding while still achieving the quality objective.

For black/white text, the ultimate in parameter coding is character recognition. The visual information consists of symbols from a finite alphabet (which may be large as in the case of Chinese characters) and once identified they are easily coded in an efficient manner using entropy coding. For example, English text has an entropy between two and four bits/symbol.[11.17] Thus, an 8.5×11 inch typewritten sheet, double spaced, with one inch margins, containing about 1660 symbols requires about 5000 bits for coding or $\eta = 0.0014$ bits per pel at normal sampling rates. Multiple fonts and character sizes do not increase the bit-rate significantly.

In the future, the recognition aspect of text encoding will disappear. Instead, coded characters will pass directly from source terminal to local storage module or centralized data store, and will be displayed in soft or hard copy form only as needed.

For black/white non-text graphics, pattern recognition is also the key to ultra-efficient coding. However, this approach rapidly merges with the science of computer generated graphics and the study of specialized graphics languages.

Generally, the more specialized the language, the greater the efficiency of representation. For example, the mathematical expression

$$\left[\mathbf{A}(i,j,k)\right]_{a=0}^{a=5} = \int \int_{x+3}^{\frac{y}{2}} \left\{ \frac{g(x,y)+cos(2\pi ft+\phi)}{\tan^{-1}[h(v,t)]} \right\} dt\ d\theta$$

requires 1386 bits for specification under NROFF, the UNIX® word processing language. This is about $\eta = 0.0037$ bits/sample at 60 samples per mm^2.

Many other specialized graphics languages currently exist. For example, integrated circuit layouts never reside in computer memory in pel-by-pel form. Instead, they are built up from basic blocks according to instructions written in graphics language. Implementation of a generalized graphics language that will handle any and all forms of input is a long way away. However, we believe it is only through this approach that the ultimate limits of graphics encoding will be achieved.

The graphics language approach will ultimately benefit the encoding of gray scale pictures as well. Complete modeling of video scenes will probably not be possible, except for specialized situations like cartoons, human faces, etc. However, important features such as boundaries, locations and motion of objects are essentially graphical information, and is possible to represent them as such with very efficient codes. Motion of objects as well as shape changes can be representable by fairly efficient parameter codes.

Replication of the detail, shading, etc., within boundaries of objects requires additional *residue* data to be transmitted. The *amount* of residue needed depends on the application. In some cases, e.g. surveillance, very little is necessary; however, in broadcast television full cosmetic restoration must be maintained. The *fidelity* of the residue depends on context. For example, the texture in a grassy field may well be replaceable with random patterns having similar statistics, whereas detail in a human face may have to be replicated exactly.

Trying to estimate the ultimate bit-rate achievable by model-based coding of pictures suffers the same problem as with waveform coding, namely different pictures produce different bit-rates, and different applications require different fidelities of reproduction. In addition, how closely and what fraction of the image can be approximated by good models becomes additional factors. However, from the dashed curve of Fig. 11.4, the graphical parameters should require on average about 0.08 bits per pel for a single picture and perhaps a quarter of that (or less) for moving video depending on the amount of motion rendition required. The bit-rate needed for residue is much more elusive due to the large variation in pictures and fidelity requirements. We guess it should range downward from 1 bit per pel for excellent quality single pictures and a quarter of that (or less) for moving video where interframe redundancy can be exploited.

11.6.3 Ultimate Display

The ultimate display would be a projection onto a large enclosed spherical surface with the viewer positioned at the center. The picture would fill the entire field of vision and, with appropriate control of temperature, odors, audio, gravity, etc., viewers would be unable to tell that they were not witnessing the original scene. Assuming human spatial vision cutoff at 60 cycles per degree, Nyquist sampling at 120 per degree yields about 594 million samples for the entire sphere. Further, assuming a picture presentation rate of 60 Hz, PCM luminance quantization of 10-bits, 30% additional bit-rate for chrominance and 75% additional bit-rate for stereo 3-D, the ultimate display represents a bit-rate of about 810 Gigabits per second.

11.7 Concluding Remarks

To extrapolate the future of picture compression encoding we have reviewed the current situation by classifying the performance of codecs in terms of bit-rate per Nyquist sample, complexity and perceptual quality. Both audio and video stimuli are electrical signals, and many encoding techniques, particularly those involving waveform encoding, are similar for both types of signals. This may seem somewhat surprising considering the radically different nature of the sources and the perceptions of audio and visual phenomena. The more efficient the compression algorithm, the more it is likely to be specific and unique to either speech or pictures. The ease at which encoding engineers can currently move between encoding audio and video signals, making comparatively minor concessions to source statistics and perceptual observations, demonstrates that substantial improvements are to be made. Picture compression is the interface between us and electronic systems, and the limits of encoding performance are ultimately set by how well we understand human perception, that is, ourselves.

References

[11.1] J. S. Goodman and D. E. Pearson, "Multidimensional Scaling of Multiply-Impaired Television Pictures", IEEE Trans. on Systems, Man, and Cybernetics, Vol. SMC-9, No. 6, June 1979, pp. 353-356.

[11.2] N. S. Jayant, Ed., "Waveform Quantization and Coding", New York: IEEE Press Selected Reprint Series, 1976.

[11.3] A. N. Netravali and J. O. Limb, "Picture Coding: A Review", Proc. IEEE, Vol. 68, No. 3, March 1980, pp. 366-406.

[11.4] W. K. Pratt, Ed., "Image Transmission Techniques", Academic Press, New York, 1979.

[11.5] B. J. McDermott, C. Scagliola and D. J. Goodman, "Perceptual and Objective Evaluation of Speech Processed by Adaptive Differential PCM", Bell Syst. Tech. J., Vol. 57, 1978, pp. 1597-1618.

[11.6] P. Ekman and W. V. Friesen, "Facial Action Coding System", Consulting Psychologist Press, 1977.

[11.7] S. M. Park and N. I. Badler, "Animating Facial Expressions", IEEE Computer Graphics, Vol. 13, August 1981, pp. 245-252.

[11.8] F. I. Parke, "Parameterized Models For Facial Animation", IEEE Computer Graphics, Vol. 14, November 1992.

[11.9] R. Forcheimer and T. Kvonander, "Image Coding - Frame Waveforms to Animation", Vol. 37, No. 12, December 1989, pp. 2008-2023.

[11.10] H. G. Musmann, H. Hoffer, and J. Osterman, "Object-Oriented Analysis-Synthesis Coding of Moving Images", Signal Processing: Image Communication, Vol. 1, 1989, pp. 117-138.

[11.11] H. H. Chen, C. Horne, and B. G. Haskell, "A Region-Based Approach to Very Low Bit-rate Video Coding", (to be published), December 1993.

[11.12] M. Eden and M. Kocher, "On th Performance of a Contour Coding Algorithms in the Context of Image Coding - Part 1: Contour Segment Coding", Signal Processing, Vol. 8, 1985, pp. 381-386.

[11.13] J. K. Yan and D. J. Sakrison, "Encoding of Images Based on a Two-Component Source Model", IEEE Transactions on Communications, Vol. COM-25, No. 11, November 1977, pp. 1315-1322.

[11.14] H. G. Musmann and J. Klie, "TV Transmission Using a 64 kbit/s Transmission Rate", International Conference on Communications, 1979, pp. 23.3.1-23.3.5.

[11.15] A. D. Wyner, "Information Theory", Proc. IEEE, February 1981.

[11.16] C. E. Shannon, "Prediction and Entropy Printed English", BSTJ, Vol. 30, 1, January 1951; pp. 50-64.

Index